C0-CCE-065

Cartesian Coordinate System:

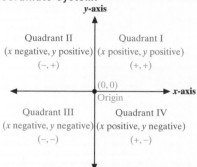

Summary of Formulas and Properties of Straight Lines:

1. $Ax + By = C$ where A and B do not both equal 0.   Standard form

2. $m = \dfrac{y_2 - y_1}{x_2 - x_1}$ where $x_1 \neq x_2$.   Slope of a line

3. $y = mx + b$ Slope-intercept form
(with slope m, and y-intercept $(0, b)$)

4. $y - y_1 = m(x - x_1)$ Point-slope form

5. $y = b$ Horizontal line, slope 0

6. $x = a$ Vertical line, undefined slope

7. Parallel lines have the same slope.

8. Perpendicular lines have slopes that are negative reciprocals of each other.

Relation, Domain, and Range:

Relation: A **relation** is a set of ordered pairs of real numbers.

Domain: The **domain**, D, of a relation is the set of all first coordinates in the relation.

Range: The **range**, R, of a relation is the set of all second coordinates in the relation.

Function:

A **function** is a relation in which each domain element has exactly one corresponding range element.

Vertical Line Test:

If **any** vertical line intersects the graph of a relation at more than one point, then the relation graphed is **not** a function.

Linear Inequality Terminology:

Half-plane: A straight line separates a plane into two **half-planes**.

Boundary line: The line itself is called the **boundary line**.

Closed half-plane: If the boundary line is included, then the half-plane is said to be **closed**.

Open half-plane: If the boundary line is not included, then the half-plane is said to be **open**.

Systems of Linear Equations (Two Variables):

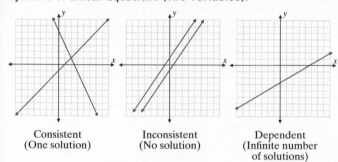

Consistent
(One solution)

Inconsistent
(No solution)

Dependent
(Infinite number
of solutions)

Matrices:

System of Linear Equations	Coefficient Matrix	Augmented Matrix

$$\begin{cases} a_1x + b_1y = c_1 \\ a_2x + b_2y = c_2 \end{cases} \qquad \begin{bmatrix} a_1 & b_1 \\ a_2 & b_2 \end{bmatrix} \qquad \begin{bmatrix} a_1 & b_1 & \vdots & c_1 \\ a_2 & b_2 & \vdots & c_2 \end{bmatrix}$$

Elementary Row Operations:

1. Interchange two rows.

2. Multiply a row by a nonzero constant.

3. Add a multiple of a row to another row.

Upper Triangular Form and Row Echelon Form:

$$\begin{bmatrix} a_{11} & a_{12} & a_{13} \\ 0 & a_{22} & a_{23} \\ 0 & 0 & a_{33} \end{bmatrix}$$

A matrix is in **upper triangular form** if its entries in the lower left triangular region are all 0's. If a_{11}, a_{22}, and a_{33} (the entries along the main diagonal) all equal 1 when the matrix is in upper triangular form, then the matrix is also in **row echelon form** (or **ref**).

Gaussian Elimination:

1. Write the augmented matrix for the system.

2. Use elementary row operations to transform the matrix into row echelon form.

3. Solve the corresponding system of equations by using back substitution.

Determinants:

Value of a 2 × 2 Determinant:

For the square matrix, $A = \begin{bmatrix} a_{11} & a_{12} \\ a_{21} & a_{22} \end{bmatrix}$, $\det(A) = \begin{vmatrix} a_{11} & a_{12} \\ a_{21} & a_{22} \end{vmatrix} = a_{11}a_{22} - a_{21}a_{12}$.

Value of a 3 × 3 Determinant:

For the square matrix, $A = \begin{bmatrix} a_{11} & a_{12} & a_{13} \\ a_{21} & a_{22} & a_{23} \\ a_{31} & a_{32} & a_{33} \end{bmatrix}$, $\det(A) = \begin{vmatrix} a_{11} & a_{12} & a_{13} \\ a_{21} & a_{22} & a_{23} \\ a_{31} & a_{32} & a_{33} \end{vmatrix}$

$= a_{11}(\text{minor of } a_{11}) - a_{12}(\text{minor of } a_{12}) + a_{13}(\text{minor of } a_{13})$

$= a_{11}\begin{vmatrix} a_{22} & a_{23} \\ a_{32} & a_{33} \end{vmatrix} - a_{12}\begin{vmatrix} a_{21} & a_{23} \\ a_{31} & a_{33} \end{vmatrix} + a_{13}\begin{vmatrix} a_{21} & a_{22} \\ a_{31} & a_{32} \end{vmatrix}$.

Cramer's Rule:

For the system $\begin{cases} a_{11}x + a_{12}y = k_1 \\ a_{21}x + a_{22}y = k_2 \end{cases}$

where $D = \begin{vmatrix} a_{11} & a_{12} \\ a_{21} & a_{22} \end{vmatrix}$, $D_x = \begin{vmatrix} k_1 & a_{12} \\ k_2 & a_{22} \end{vmatrix}$, and $D_y = \begin{vmatrix} a_{11} & k_1 \\ a_{21} & k_2 \end{vmatrix}$,

if $D \neq 0$, then $x = \dfrac{D_x}{D}$ and $y = \dfrac{D_y}{D}$ is the unique solution to the system.

CHAPTER 4 — Exponents and Polynomials

Properties of Exponents:
For nonzero real numbers a and b and integers m and n,

The Exponent 1: $a = a^1$ (a is any real number.)

The Exponent 0: $a^0 = 1$ $(a \neq 0)$

Product Rule: $a^m \cdot a^n = a^{m+n}$

Quotient Rule: $\dfrac{a^m}{a^n} = a^{m-n}$

Negative Exponents: $a^{-n} = \dfrac{1}{a^n}$

Power Rule: $\left(a^m\right)^n = a^{mn}$

Power Rule for Products: $(ab)^n = a^n b^n$

Power Rule for Fractions: $\left(\dfrac{a}{b}\right)^n = \dfrac{a^n}{b^n}$

Scientific Notation:
$N = a \times 10^n$ where N is a decimal number, $1 \le a < 10$, and n is an integer.

Classification of Polynomials:
Monomial: polynomial with one term

Binomial: polynomial with two terms

Trinomial: polynomial with three terms

Degree: The **degree of a term** is the sum of the exponents of its variables. The **degree of a polynomial** is the largest of the degrees of its terms.

Leading Coefficient: The coefficient of the term with the largest degree.

FOIL Method:

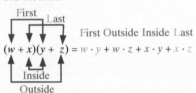

First Outside Inside Last

$(w + x)(y + z) = w \cdot y + w \cdot z + x \cdot y + x \cdot z$

Division Algorithm:
For polynomials P and D, $\dfrac{P}{D} = Q + \dfrac{R}{D}$, $(D \neq 0)$ where Q and R are polynomials and the degree of $R <$ the degree of D.

Factoring out the GCF:
1. Find the variable(s) of highest degree and the largest integer coefficient that is a factor of each term of the polynomial. (This is one factor.)
2. Divide this monomial factor into each term of the polynomial resulting in another polynomial factor.

Special Products of Polynomials:
1. $(x+a)(x-a) = x^2 - a^2$: Difference of two squares
2. $(x+a)^2 = x^2 + 2ax + a^2$: Square of a binomial sum
3. $(x-a)^2 = x^2 - 2ax + a^2$: Square of a binomial difference
4. $(x-a)\left(x^2 + ax + a^2\right) = x^3 - a^3$: Difference of two cubes
5. $(x+a)\left(x^2 - ax + a^2\right) = x^3 + a^3$: Sum of two cubes

Quadratic Equation:
An equation that can be written in the form $ax^2 + bx + c = 0$ where a, b, and c are constants and $a \neq 0$.

Zero-Factor Property:
If a and b are real numbers, and $a \cdot b = 0$, then $a = 0$ or $b = 0$ or both.

Factor Theorem:
If $x = c$ is a root of a polynomial equation in the form $P(x) = 0$, then $x - c$ is a factor of the polynomial $P(x)$.

Consecutive Integers:
$n, n+1, n+2, \ldots$

The Pythagorean Theorem:
In a right triangle, the square of the hypotenuse is equal to the sum of the squares of the legs.

$$c^2 = a^2 + b^2$$

CHAPTER 5 — Rational Expressions and Rational Equations

Rational Expression:
A **rational expression** is an expression of the form $\dfrac{P}{Q}$ where P and Q are polynomials and $Q \neq 0$.

Fundamental Principle of Rational Expressions:
If $\dfrac{P}{Q}$ is a rational expression where $Q \neq 0$ and K is a polynomial where $K \neq 0$, then $\dfrac{P}{Q} = \dfrac{P \cdot K}{Q \cdot K}$.

Opposites in Rational Expressions:
For a polynomial P, $\dfrac{-P}{P} = -1$ where $P \neq 0$.

In particular, $\dfrac{a-x}{x-a} = \dfrac{-(x-a)}{x-a} = -1$ where $x \neq a$.

Multiplication with Rational Expressions:
If P, Q, R, and S are polynomials and $Q, S \neq 0$, then $\dfrac{P}{Q} \cdot \dfrac{R}{S} = \dfrac{P \cdot R}{Q \cdot S}$.

Division with Rational Expressions:
If P, Q, R, and S are polynomials and $Q, R, S \neq 0$, then $\dfrac{P}{Q} \div \dfrac{R}{S} = \dfrac{P}{Q} \cdot \dfrac{S}{R}$.

Addition and Subtraction with Rational Expressions:
$\dfrac{P}{Q} + \dfrac{R}{Q} = \dfrac{P+R}{Q}$ and $\dfrac{P}{Q} - \dfrac{R}{Q} = \dfrac{P-R}{Q}$ where $Q \neq 0$.

Negative Signs in Rational Expressions:
$-\dfrac{P}{Q} = \dfrac{P}{-Q} = \dfrac{-P}{Q}$

Work Problems:
To solve this type of problem, represent what part of the work is done in one unit of time.

Distance-Rate-Time Problems:
Use the formula $d = rt$, where $d =$ the distance traveled, $r =$ the rate, and $t =$ the time taken, to solve this type of problem.

Variation:
Direct Variation: A variable quantity y varies directly as a variable x if there is a constant k such that $\dfrac{y}{x} = k$ or $y = kx$.

Inverse Variation: A variable quantity y varies inversely as a variable x if there is a constant k such that $x \cdot y = k$ or $y = \dfrac{k}{x}$.

INTERMEDIATE
ALGEBRA

Sixth Edition

D. FRANKLIN WRIGHT
CERRITOS COLLEGE

HAWKES
LEARNING
SYSTEMS

Editor: Nina Waldron
Vice President, Development: Marcel Prevuznak
Production Editor: Kara Roché
Associate Editor: Barry Wright, III
Editorial Assistant: Liz Allen
Copy Editors: Jessica Ballance, Bethany Bates, Phillip Bushkar, Kimberly Cumbie, Taylor Hamrick,
          Susan Niese, Sundar Parthasarathy, Claudia Vance, Colin Williams
Answer Key Editors: Ashley Godbold, Bill Epperson, Kirk Boyer, Michael Lane
Contributor: Elizabeth Thomas
Layout: E. Jeevan Kumar, D. Kanthi, U. Nagesh, B. Syamprasad
Art: Ayvin Samonte
Cover Art and Design: Johnson Design

HAWKES
LEARNING
SYSTEMS

A division of Quant Systems, Inc.
1023 Wappoo Road, A6, Charleston, SC 29407

Library of Congress Control Number: 2009921407

Printed in the United States of America

ISBN:
Student Textbook: 978-1-932628-43-2
Student Textbook and Software Bundle: 978-1-932628-48-7

TABLE OF CONTENTS

Chapter 1

Real Numbers and Solving Equations   1

Chapter 2

Linear Equations and Functions   103

Chapter 3

Systems of Linear Equations   191

Chapter 7

Quadratic Equations and Quadratic Functions   547

Chapter 8

Exponential and Logarithmic Functions   637

Chapter 9

Conic Sections   745

PREFACE

Purpose and Style

Intermediate Algebra (sixth edition) provides a solid base for further studies in mathematics. In particular, business and social science majors who will continue their studies in statistics and calculus will be well prepared for success in those courses. With feedback from users, insightful comments from reviewers, and skillful editing and design by the editorial staff at Hawkes Learning Systems, we have confidence that students and instructors alike will find that this text is indeed a superior teaching and learning tool. The text may be used independently or in conjunction with the software package *Hawkes Learning Systems: Intermediate Algebra* developed by Quant Systems.

We have provided very little overlap with material covered in a beginning algebra course. While Chapter 1 provides a review of topics from beginning algebra, students will find that the review is comprehensive and that the pace of coverage is somewhat faster and in more depth than they have seen in previous courses. As with any text in mathematics, students should read the text carefully and thoroughly.

The style of the text is informal and nontechnical while maintaining mathematical accuracy. Each topic is developed in a straightforward step-by-step manner. Each section contains many carefully developed and worked out examples to lead the students successfully through the exercises and prepare them for examinations. Whenever appropriate, information is presented in list form for organized learning and easy reference. Common errors are highlighted and explained so that students can avoid such pitfalls and better understand the correct corresponding techniques. Practice problems with answers are provided in nearly every section to allow the students to "warm up" and to provide the instructor with immediate classroom feedback.

A special feature in many sections is "Writing and Thinking About Mathematics." The related questions are placed at the end of the Exercises for the section and ask the students to delve deeper into mathematical concepts and to become accustomed to organizing their thoughts and writing about mathematics in their own words.

The NCTM and AMATYC curriculum standards have been taken into consideration in the development of the topics throughout the text. In particular:

- there is an emphasis on reading and writing skills as they relate to mathematics,
- techniques for using a graphing calculator are discussed early,
- a special effort has been made to make the exercises motivating and interesting,
- geometric concepts are integrated throughout, and
- statistical concepts, such as interpreting bar graphs and calculating elementary statistics, are included where appropriate.

Real Numbers and Solving Equations

C H A P T E R

1

Did You Know?

In Chapter 1, you will find a great many symbols defined, as well as rules of manipulation for these symbolic expressions. Most people think that algebra has always existed, complete with all the common symbols in use today. That is not the case, since modern symbols did not appear consistently until the beginning of the sixteenth century. Prior to that, algebra was rhetorical. That is, all problems were written out in words using either Latin, Arabic, or Greek, and some nonstandard abbreviations. Numbers were written out. The common use of Hindu-Arabic numerals did not begin until the sixteenth century, although these numerals had been introduced into Europe in the twelfth century.

The sign for addition, +, was a contraction of the Latin *et*, which means "and." Gradually, the *e* was contracted and the crossed *t* became the plus sign. The minus sign or bar, –, is thought to be derived from the habit of early scribes of using a bar to represent the letter *m*. Thus the word *summa* was often written *suma* and the bar came to represent the missing *m*, the first letter of the word *minus*. The radical symbol, $\sqrt{}$, is derived from a small printed *r*, which stood for the Latin word *radix*, or root. The symbol for times, a cross, ×, was developed from cross multiplication or for the purpose of indicating products in proportions. Thus

$$\frac{2}{3} \times \frac{6}{9} \text{ stood for } \frac{2}{3} = \frac{6}{9}.$$

The cross is not well suited for algebra, since it resembles the symbol *x*, which is used for variables. Therefore, a dot is usually used to indicate multiplication in algebra. The dot seems to have developed from an Italian practice of separating columns in multiplication tables with a dot. Exponents were used as early as the fourteenth century by the mathematician Nicole Oresme (1320? – 1382), who gave the first known use of the rules for fractional exponents in a textbook he wrote. The equal sign is attributed to Robert Recorde (1510? – 1558), who wrote, "I will sette as I doe often in woorke use, a paire of parallels, or Gemowe [twin] lines of one lengthe, thus: = , bicause noe .2. thynges, can be moare equalle." As you can tell, the development of algebraic symbols occurred over a long period of time, and symbols became standardized through usage and convenience.

1.1   Prime Numbers, Exponents, and LCM

1.2   Introduction to Real Numbers

1.3   Operations with Real Numbers

1.4   Linear Equations in One Variable: $ax + b = c$

1.5

1.6

1.7

"The Ma
flood of
purely fo
of endles
the physi

Karl Pear

Introduction

Presented before the first section of every chapter, this feature prefaces the subject of the chapter and its purpose.

Did You Know?

A feature at the beginning of every chapter that presents some interesting math history related to the chapter at hand.

Welcome to your second course in algebra. You will find that many of the topics presented, such as solving equations, factoring, working with fractional expressions, and solving word problems, are familiar because they were discussed in beginning algebra. However, you will also find that the coverage of each topic, including many new topics, is in greater depth and that the pace is somewhat faster. The intent is to prepare you for success with the algebraic content in your future studies in mathematics, science, business, and economics. You should be prepared to spend considerable time and effort on this course throughout the semester.

We begin in Chapter 1 with a review of prime numbers and least common multiple (LCM) and continue with a review of the properties of and operations with real numbers (positive and negative numbers included). These ideas form the foundation for the study of algebra and need to be understood thoroughly. Appendix A.1 gives a complete review of operations with fractions (positive numbers only). As an example of a deeper analysis of familiar topics, we will discuss solving first-degree equations, then expand the techniques to include solving absolute value equations and solving formulas for specified variables. Also, we will review the topic of intervals of real numbers. Intervals and interval notation are related to graphs and are used frequently in higher level mathematics courses.

There is much to look forward to in this course. With hard work and perseverance, you should have a very rewarding experience.

1.1   **Prime Numbers, Exponents, and LCM**

- *Understand the terms* factor, *prime factorization*, multiple, *and* exponent.
- *Recognize prime numbers and composite numbers.*
- *Use exponents to write the prime factorization of a composite number.*
- *Find the least common multiple (LCM) of a set of counting numbers.*

Objectives

The objectives provide the students with a clear and concise list of skills presented in each section.

Prime Numbers

The **whole numbers** consist of the number 0 and the *natural numbers* (also called the *counting numbers*). We use **N** to represent the set of natural numbers and **W** to represent the set of whole numbers.

$$\textbf{Natural Numbers} = \textbf{N} = \{1, 2, 3, 4, 5, 6, 7, 8, 9, 10, 11, ...\}$$

$$\textbf{Whole Numbers} = \textbf{W} = \{0, 1, 2, 3, 4, 5, 6, 7, 8, 9, 10, 11, ...\}$$

The three dots ... are called an ellipsis and are used to indicate that the pattern continues without end.

Examples

Examples are denoted with titled headers indicating the problem solving skill being presented. Each section contains many carefully explained examples with appropriate tables, diagrams, and graphs. Examples are presented in an easy to understand, step-by-step fashion and annotated with notes for additional clarification.

Example 3: Addition with Unlike Signs

a. $(-10) + (+3) = -(|-10| - |+3|) = -(10 - 3) = -7$

b. $(+10) + (-3) = +(|+10| - |-3|) = +(10 - 3) = 7$

c. $-\dfrac{7}{11} + \dfrac{5}{11} = -\left(\left|-\dfrac{7}{11}\right| - \left|\dfrac{5}{11}\right|\right) = -\left(\dfrac{7}{11} - \dfrac{5}{11}\right) = -\dfrac{2}{11}$

 NOTES If a negative number occurs after an addition symbol then we must place the number in parentheses so that the two operation symbols are not next to each other.

Notes

Notes highlight common mistakes and give additional clarification to more subtle details.

He wants to buy $3\dfrac{1}{2}$ lbs ground beef, 3 lbs pork chops, . How much meat does Jeremy intend to buy?

ing fractions, see Appendix A.1 for a review.]

$+ 3 + 2 + \dfrac{13}{4}$

$\dfrac{12}{4} + \dfrac{8}{4} + \dfrac{13}{4}$

$= \dfrac{47}{4}$

$= 11\dfrac{3}{4}$

Jeremy will need to buy $11\dfrac{3}{4}$ lbs of meat.

Note: The improper fraction $\dfrac{47}{4}$ is a correct solution and

algebraic situations. However, in practical applications, the is more appropriate and understandable to most people.

Numbers that can be written as fractions and whose numerators are integers and denominators are nonzero integers have the technical name **rational numbers**. Integers and terminating and infinite, repeating decimal numbers can also be classified as rational numbers. For example, the following numbers are all rational numbers:

$$1.3 = \frac{13}{10}, \quad 5 = \frac{5}{1}, \quad -4 = \frac{-4}{1}, \quad \frac{3}{8}, \quad \text{and} \quad 2.8333\ldots = \frac{17}{6}.$$

Variables are needed to be able to state rules and definitions in general forms. The definition is restated here for easy reference and for emphasis.

Variable

*A **variable** is a symbol (generally a letter of the alphabet) that is used to represent an unknown number or any one of several numbers.*

The variables a and b are used to represent integers in the following definition of a **rational number**.

Rational Number

*A **rational number** is any number that can be written in the form $\dfrac{a}{b}$ where a and b are integers and $b \neq 0$. (The letter Q represents the set of all rational numbers.)*

Example 1: Rational Numbers

a. $\dfrac{2}{3}, \dfrac{7}{1}, \dfrac{-5}{3}, \dfrac{27}{10},$ and $\dfrac{3}{-10}$ are all rational numbers. Each is in the form $\dfrac{a}{b}$ where a and b are integers and $b \neq 0$.

b. $1\dfrac{3}{4}$ and 2.33 are also rational numbers. They are not in the form $\dfrac{a}{b}$, but they **can be written** in that form:

$$1\dfrac{3}{4} = \dfrac{7}{4} \quad \text{and} \quad 2.33 = 2\dfrac{33}{100} = \dfrac{233}{100}.$$

c. $\dfrac{\pi}{6}$ and $\dfrac{\sqrt{2}}{3}$ are in the form $\dfrac{a}{b}$, but the numerator in each case is not an integer and cannot be written as an integer. Thus, the fractions are **not** rational numbers. Instead, as we will see, they are called **irrational numbers**.

Definition Boxes

Definitions are presented in highly visible boxes for easy reference.

Step 3: Go to the \ and hit **ENTER** three times so that the display appears as follows:

Step 4: Press **GRAPH** and (using the standard WINDOW settings) the following graph should appear on the display.

b. Graph the linear inequality $-5x + 4y > -8$.

Solution:

Step 1: Solving the inequality for y gives: $y > \dfrac{5}{4}x - 2$.

Step 2: Press the key ▬▬ and enter the function: $\backslash Y_1 = (5/4)X - 2$.

Step 3: Go to the \ and hit **ENTER** two times so that the display appears as follows:

Calculator Instruction

Step-by-step instructions are presented to introduce students to basic graphing skills with a TI-84 Plus calculator along with actual screen shots of a TI-84 Plus for visual reference.

Practice Problems

Practice Problems with answers are presented at the end of almost every section giving the students an opportunity to practice their newly acquired skills.

Practice Problems

1. Combine like terms: $3x^2 - \left[x^2 + \left(3x - 2x^2\right)\right]$.

Solve the following equations.

2. $3x - 4 = 2x + 6 - x$

3. $6(x - 4) + x = 4(1 - x) + 4x$

4. $\dfrac{x+4}{2} + \dfrac{1}{4} = \dfrac{x+52}{12}$

5. $|2x - 1| = 8.2$

1.4 Exercises

In each of the expressions in Exercises 1 – 16, simplify by combining like terms.

1. $-2x + 5y + 6x - 2y$

2. $4x + 2x - 3y - x$

3. $4x - 3y + 2(x + 2y)$

4. $5(x - y) + 2x - 3y$

5. $(3x^2 + x) - (7x^2 - 2x)$

6. $-(4x^2 - 2x) - (5x^2 + 2x)$

7. $4x - [5x + 3 - (7x - 4)]$

8. $3x - [2y - (3x + 4y)]$

9. $\dfrac{-4x - 2x}{3} + 7x$

10. $\dfrac{2(4x - x)}{3} - \dfrac{3(6x - x)}{3}$

11. $\dfrac{8(5x + 2x)}{7} - \dfrac{2(3x + x)}{4}$

12. $\dfrac{6(5x - x)}{8} + \dfrac{5(x + 5x)}{3}$

13. $2x + [9x - 4(3x + 2) - 7]$

14. $7x - 3[4 - (6x - 1) + x]$

15. $6x^2 - [9 - 2(3x^2 - 1) + 7x^2]$

16. $2x^2 + [4x^2 - (8x^2 - 3x) + (2x^2 + 7x)]$

Exercises

Each section includes a variety of paired and graded exercises to give the students much needed practice applying and reinforcing the skills learned in the section. More than 4600 carefully selected exercises are provided in the sections. The exercises progress from relatively easy problems to more difficult problems. Each chapter has a new set of Review Exercises labeled to match the corresponding section in the chapter.

Solve each of the equations in Exercises 17 – 60.

17. $7x - 4 = 17$

18. $9x + 6 = -21$

19. $4 - 3x = 19$

20. $18 + 11x = 23$

21. $6x - 2.5 = 1.1$

22. $5x + 4.06 = 2.31$

23. $7x + 3.4 = -1.5$

24. $8.2 = 2.6 + 8x$

25. $7x - 6 = 2x + 9$

26. $x - 7 = 4x + 11$

27. $\dfrac{x}{5} - 1 = -6$

28. $\dfrac{3x}{4} + 11 = 20$

to Practice Problems: 1. $4x^2 - 3x$ 2. $x = 5$ 3. $x = 4$ 4. $x = 5$ 5. $x = 4.6$ or $x = -3.6$

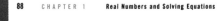

88 CHAPTER 1 Real Numbers and Solving Equations

Chapter 1 Index of Key Ideas and Terms

Section 1.1 Prime Numbers, Exponents, and LCM

Natural Numbers (Counting Numbers) page 2
$N = \{1,2,3,4,5,6,7,8,9,10,11,...\}$

Whole Numbers page 2
$W = \{0,1,2,3,4,5,6,7,8,9,10,11,...\}$

Factor page 3
A **factor** (or **divisor**) divides into a product with a remainder of 0.

Prime Numbers page 3
A **prime number** has exactly two different factors, 1 and the number itself. The prime numbers less than 50 are: 2, 3, 5, 7, 11, 13, 17, 19, 23, 29, 31, 37, 41, 43, 47.

Composite Numbers page 3
A **composite number** has more than two different factors.

To Find the Prime Factorization of a Composite Number page 4
1. Factor the composite number into any two factors.
2. Factor each factor that is not prime into two more factors.
3. Continue this process until all factors are prime.
The **prime factorization** is the product of all the prime factors.

Exponents
An **exponent** is a number written to the right and slightly above another number (called the **base**) that indicates repeated multiplication of the base by itself.

Multiples
The **multiples** of a number are the products of that number with the counting numbers.

Index of Key Ideas and Terms

Each chapter contains an index highlighting the main concepts and skills presented in the chapter, along with complete definitions and page numbers for easy reference.

---

24 CHAPTER 1 Real Numbers and Solving Equations

Complete the expressions in Exercises 69 – 80 by using the given property.

69. $x + 7 =$ _____ Commutative property of addition

70. $x \cdot 3 =$ _____ Commutative property of multiplication

71. $x \cdot (6 + y) =$ _____ Distributive property

72. $x + (3 + y) =$ _____ Associative property of addition

73. $3 \cdot (x \cdot z) =$ _____ Associative property of multiplication

74. Either $x < y, x > y,$ or _____. Trichotomy property

75. $2 \cdot (y + 3) =$ _____ Distributive property

76. If $x < a$ and $a < 10$, then _____. Transitive property

77. The multiplicative inverse of 6 is _____ because _____. Inverse of multiplication

78. The reciprocal of -4 is _____ because _____. Inverse of multiplication

79. The additive inverse of -7 is _____ because _____. Inverse of addition

80. The additive inverse of 15 is _____ because _____. Inverse of addition

Writing and Thinking About Mathematics

81. Explain the difference between graphing a set of integers and graphing a set of real numbers.

82. The inequality $x > 5$ is used to indicate all real numbers greater than 5. Is there a "first" real number greater than 5? Or, in other words, is there a real number greater than 5 that is "closest" to 5? Write, in your own words, a paragraph or two with examples to indicate your understanding of this question. Show your analysis to a friend to see if he or she understands and agrees or disagrees with your thinking. A related question would be "What is the real number closest to 0?".

HAWKES LEARNING SYSTEMS: INTERMEDIATE ALGEBRA SOFTWARE

- Name that Real Number
- Properties of Real Numbers

Writing and Thinking About Mathematics

These exercises provide the student an opportunity to independently explore and expand on concepts presented in the chapter.

Additional Features

Calculator Problems: Each problem is designed to highlight the usefulness of a calculator in solving certain complex problems, but maintain the necessity of understanding the concepts behind the problem.

Chapter Review: Provides extra problems organized by section. These problems give the student an opportunity to review concepts presented throughout the chapter and to identify strengths and potential weaknesses before taking an exam.

Chapter Test: Provides an opportunity for the students to practice the skills presented in the chapter in a test format.

Cumulative Review: As new concepts build on previous concepts, the cumulative review provides the student with an opportunity to continually reinforce existing skills while practicing newer skills.

Answers: Answers are provided for odd numbered section exercises and for all even and odd numbered exercises in the chapter reviews, chapter tests, and cumulative reviews.

Instructor's Annotated Edition:

Answers: Answers to all the exercises are conveniently located in the margins next to the problems.

Teaching Notes: Suggestions for more in depth classroom discussions and alternate methods and techniques are located in the margins.

Changes included in the new edition:
- New, reader-friendly, layout and improved use of color
- Rearrangement of chapters for better flow, continuity, and progression
- Progressive Chapter Review problems at the end of each chapter
- Calculator Instructions (new emphasis on the TI-84 Plus graphing calculator)

Content

Calculators: The TI-84 Plus graphing calculator has been made an integral part of many of the presentations in this textbook. To get maximum benefits from the use of this text, the student must have one of these calculators (or a calculator with similar features). Directions are given for using the related calculator commands as they are needed throughout. Generally, calculator discussions and exercises are placed at the end of a section to give the instructor flexibility in what specific calculator usage to include in the course.

Chapter 1, Real Numbers and Solving Equations, is a review of topics from beginning algebra. Prime numbers, exponents, LCM (least common multiple), real numbers and their properties, sets and set-builder notation, order of operations, solving equations, and solving for terms in formulas are all reviewed. Other topics include a variety of word problems, interval notation (possibly new for some students), solving linear and absolute value inequalities, and graphing the solution sets of inequalities.

Chapter 2, Linear Equations and Functions, includes complete discussions on the three basic forms for equations of straight lines in a plane: the standard form, the slope-intercept form, and the point-slope form. Slope is discussed for parallel and perpendicular lines and treated as a rate of change. Functions and function notation are introduced and the vertical line test is used to tell whether or not a graph represents a function. Use of a TI-84 Plus graphing calculator is an integral part of this introduction to functions, as well as part of graphing linear inequalities in the last section.

Chapter 3, Systems of Linear Equations, covers solving systems of two equations in two variables and systems of three equations in three variables. The basic methods of graphing, substitution, and addition are included along with matrices and Gaussian elimination, determinants, and Cramer's Rule. Double subscript notation is now used with matrices for an easy transition to the use of matrices in solving systems of equations with the TI-84 Plus calculator. Applications involve mixture, interest, work, algebra, and geometry. The last section discusses half-planes and graphing systems of linear inequalities, again including the use of a graphing calculator.

Chapter 4, Exponents and Polynomials, discusses exponents and scientific notation and the operations of addition, subtraction, multiplication, and division with polynomials. Factoring of polynomials includes factoring the greatest common factor, the trial-and-error method and recognizing special forms such as perfect square trinomials and the sum and difference of cubes. Equations are solved by factoring and applications involve topics such as consecutive integers and the Pythagorean Theorem. The last section discusses using a graphing calculator to solve equations and inequalities.

Chapter 5, Rational Expressions and Rational Equations, applies the factoring skills from Chapter 4 in operations with rational expressions. Included are the basic operations of multiplication, division, addition, and subtraction with rational expressions, simplifying complex fractions and solving equations containing rational expressions. Applications are related to work, distance-rate-time, and variation.

Chapter 6, Roots, Radicals, and Complex Numbers, introduces roots and fractional exponents and the use of a calculator to find estimated values. Arithmetic with radicals includes simplifying radical expressions, addition, subtraction, and rationalizing denominators. A section on functions with radicals shows how to analyze the domain and range of radical functions and how to graph these functions by using a graphing calculator. Complex numbers are introduced along with the basic operations of addition, subtraction, multiplication, and division. These are skills needed for the work with quadratic equations and quadratic functions in Chapters 7.

Chapter 7, Quadratic Equations and Quadratic Functions, reviews solving quadratic equations by factoring and introduces the methods of using the square root property and completing the square. The quadratic formula is developed by completing the square and students are encouraged to use the most efficient method for solving any particular quadratic equation. Applications are related to the Pythagorean Theorem, projectiles, geometry, and cost. The fourth section covers solving equations in quadratic form and the last two sections deal with quadratic functions (graphing parabolas) and solving quadratic inequalities.

Chapter 8, Exponential and Logarithmic Functions, begins with a section on the algebra of functions and leads to the development of the composition of functions and methods for finding the inverses of one-to-one functions. This introduction lays the groundwork for understanding the relationship between exponential functions and logarithmic functions. While the properties of real exponents and logarithms are presented completely, most numerical calculations are performed with the aid of a calculator. Special emphasis is placed on the number e and applications with natural logarithms. Students will find the applications with exponential and logarithmic functions among the most interesting and useful in their mathematical studies. Those students who plan to take a course in calculus should be aware that many of the applications found in calculus involve exponential and logarithmic expressions in some form.

Chapter 9, Conic Sections, provides a basic understanding of conic sections (parabolas, circles, ellipses, and hyperbolas) and their graphs. The first section gives detailed analyses of translations involving horizontal and vertical shifting. Function notation is used in discussing reflections and translations of a variety of types of functions. Vertical and horizontal parabolas are developed as conic sections. The distance formula is developed and used to find equations of circles. The thorough development of ellipses and hyperbolas includes graphs with centers not at the origin. Solving systems with nonlinear equations is the final topic of the chapter.

Chapter 10, Sequences, Series, and the Binomial Theorem, provides flexibility for the instructor and reference material for the students. The topics presented here are likely to appear in courses in probability and statistics, finite mathematics, and higher level courses in mathematics. Any of these topics covered at this time will give students additional mathematical experience and insight for future studies.

I recommend that the topics be covered in the order presented because most sections assume knowledge of the material in previous sections. This is particularly true of the cumulative review sections at the end of each chapter. Of course, time and other circumstances may dictate another sequence of topics. For example, in some programs, Chapters 1 and 2 might be considered review or Chapters 9 and 10 might be considered part of a college algebra course. In case of any changes, the instructor should be sure that the students are somewhat familiar with a graphing calculator.

Acknowledgements

I would like to thank Editor Nina Waldron, Production Editor Kara Roché and Vice President of Development, Marcel Prevuznak for their hard work and invaluable assistance in the development and production of this text.

Many thanks go to the following manuscript reviewers who offered their constructive and critical comments:

Donna Ahlrich, *Holmes Community College*
Diane Allgood, *Holmes Community College*
Lisa Anglin, *Holmes Community College*
Margaret Bailey, *Holmes Community College*
Dana Bingham, *Arkansas State University*
Stephanie Blue, *Holmes Community College*
Stephanie Burton, *Holmes Community College*
Cynthia Carter, *Lindsey Wilson College*
James Cochran, *Kirkwood Community College*
Scott Dillery, *Lindsey Wilson College*
Dr. Hamidullah Farhat, *Hampton University*
Theresa Hert, *Mount San Jacinto College*
Marlene Kustesky, *Virginia Commonwealth University*
Denise Manning, *Arkansas State University*
Adrienne Palmer, *Green River Community College*
Jennie Pegg, *Holmes Community College*
Kendra Schroeder, *Morehead State University*
Beth St. Jean, *McMurry University*
Gail Stringer, *Somerset Community College*
Patricia Treloar, *University of Mississippi*
Luther Yost, *Idaho State University*

Finally, special thanks go to James Hawkes for his faith in this sixth edition and his willingness to commit so many resources to guarantee a top-quality product for students and teachers.

D. Franklin Wright

TO THE STUDENT

The goal of this text and of your instructor is for you to succeed in intermediate algebra. Certainly, you should make this your goal as well. What follows is a brief discussion about developing good work habits and using the features of this text to your best advantage. For you to achieve the greatest return on your investment of time and energy you should practice the following three rules of learning:

1. Reserve a block of time for study every day.
2. Study what you don't know.
3. Don't be afraid to make mistakes.

How to use this book

The following seven-step guide will not only make using this book a more worthwhile and efficient task, but it will also help you benefit more from classroom lectures or the assistance that you receive in a math lab.

1. Try to look over the assigned section(s) before attending class or lab. In this way, you will be more comfortable with new ideas presented in class or lab. This will also help you anticipate where you need to ask questions about material that is difficult for you to understand.

2. Read examples carefully. They have been chosen and written to show all of the problem-solving steps that you need to be familiar with. You might even try to solve example problems independently before studying the solutions that are given.

3. Work the section exercises as they are assigned. Problem-solving practice is the single most important element in achieving success in any math class, and there is no good substitute for actually doing this work yourself. Demonstrating that you can think independently through each step of each type of problem will also build your confidence in your ability to answer questions on quizzes and exams. Check the Answer Key periodically while working section exercises to be sure that you have the right ideas and are proceeding in the right manner. Identify and correct your mistakes as you work.

4. Use the Writing and Thinking About Mathematics questions as an opportunity to explore the way that you think about math. A big part of learning and understanding mathematics is being able to communicate mathematical ideas and the thinking that occurs to you as you approach new concepts and problems. These questions can help you analyze your own approach to mathematics and, in class or group discussions, learn from ideas expressed by your fellow students.

5. Use the Chapter Index of Key Ideas and Terms as a recap when you begin to prepare for a Chapter Test. It will reference all the major ideas that you should be familiar with from that chapter and indicate where you can turn if review is needed. You can also use the Chapter Index as a final checklist once you feel you have completed your review and are prepared for the Chapter Test.

6. Chapter Tests are provided so that you can practice for the tests that are actually given in class or lab. To simulate a test situation, block out a one-hour, uninterrupted period in a quiet place where your only focus is on accurately completing the Chapter Test. Use the Answer Key at the back of the book as a self-check only after you have completed all of the questions on the test.

7. Chapter Reviews and Cumulative Reviews will help you retain the skills that you acquired in studying earlier material. Approach them in much the same manner as you would the Chapter Tests in order to keep all of your skills sharp throughout the entire course.

How to Prepare for an Exam

Gaining Skill and Confidence

The stress that many students feel while trying to succeed in mathematics is what you have probably heard called "math anxiety." It is a real-life phenomenon, and many students experience such a high level of anxiety during mathematics exams in particular that they simply cannot perform to the best of their abilities. It is possible to overcome this stress simply by building your confidence in your ability to do mathematics and by minimizing your fears of making mistakes.

No matter how much it may seem that in mathematics you must either be right or wrong, with no middle ground, you should realize that you can be learning just as much from the times that you make mistakes as you can from the times that your work is correct. Success will come. Don't think that making mistakes at first means that you'll never be any good at mathematics. Learning mathematics requires lots of practice. Most importantly, it requires a true confidence in yourself and in the fact that, with practice and persistence, the mistakes will become fewer, the successes will become greater, and you will be able to say, "I can do this."

Showing What You Know

If you have attended class or lab regularly, taken good notes, read your textbook, kept up with homework exercises, and asked for help when it was needed, then you have already made significant progress in preparing for an exam and conquering any anxiety. Here are a few other suggestions to maximize your preparedness and minimize your stress.

1. Give yourself enough time to review. You will generally have several days notice before an exam. Set aside a block of time each day with the goal of reviewing a manageable portion of the material that the test will cover. Don't cram!

2. Work lots of problems to refresh your memory and sharpen you skills. Go back to rework selected exercises from all of your homework assignments.

3. Reread your text and your notes, and use the Chapter Index of Key Ideas and Terms, the Chapter Review, and the Chapter Test to recap major ideas and do a self-evaluated test simulation.

4. Study with a friend or classmate. Peer tutoring almost always helps in gaining better understanding of concepts and developing better problem solving skills.

5. Be sure that you are well-rested so that you can be alert and focused during the exam.

6. Don't study up to the last minute. Give yourself some time to wind down before the exam. This will help you to organize your thoughts and feel more calm as the test begins.

7. As you take the test, realize that its purpose is not to trick you, but to give you and your instructor an accurate idea of what you have learned. Good study habits, a positive attitude, and confidence in your own ability will be reflected in your performance on any exam.

8. Finally, you should realize that your responsibility does not end with taking the exam. When your instructor returns your corrected exam, you should review your instructor's comments and any mistakes that you might have made. Take the opportunity to learn from this important feedback about what you have accomplished, where you could work harder, and how you can best prepare for future exams.

HAWKES LEARNING SYSTEMS: INTERMEDIATE ALGEBRA

Overview

This multimedia courseware allows students to become better problem-solvers by creating a mastery level of learning in the classroom. The software includes a(n) "Instruct," "Practice," "Tutor," and "Certify" mode in each lesson, allowing students to learn through step-by-step interactions with the software. The automated homework system's tutorial and assessment modes extend instructional influence beyond the classroom. Intelligence is what makes the tutorials so unique. By offering intelligent tutoring and mastery level testing to measure what has been learned, the software extends the instructor's ability to influence students to solve problems. This courseware can be ordered either seperately or bundled together with this text.

Minimum Requirements

In order to run *HLS: Intermediate Algebra*, you will need:

1 GHz or faster processor
Windows® XP or later
256 MB RAM
500 MB hard drive space
800x600 resolution (1024x768 recommended)
Internet Explorer 6.0 or later
CD-ROM drive

Getting Started

Before you can run *HLS: Intermediate Algebra*, you will need an access code. This 30 character code is your personal access code. To obtain an access code, go to **http://www.hawkeslearning.com** and follow the links to the access code request page (unless directed otherwise by your instructor).

Installation

Insert the *HLS: Intermediate Algebra* installation CD-ROM into the CD-ROM drive. Select the Start>Run command, type in the CD-ROM drive letter followed by :\setup. exe. (For example, d:\setup.exe where d is the CD-ROM drive letter.)

The complete installation may use over 200 MB of hard drive space and will install the entire product, except the multimedia files, on your hard drive.

After selecting the desired installation option, follow the on-screen instructions to complete your installation of *HLS: Intermediate Algebra.*

Starting the Courseware

After you install *HLS: Intermediate Algebra* on your computer, to run the courseware select Start>Programs>Hawkes Learning Systems>Intermediate Algebra.

You will be prompted to enter your access code with a message box similar to the following:

Type your entire access code in the box. When you are finished, press OK.

If you typed in your access code correctly, you will be prompted to save the code to disk. If you choose to save your code to disk, typing in the access code each time you run *HLS: Intermediate Algebra* will not be necessary. Instead, select the Load from File button when prompted to enter your access code and choose the path to your saved access code.

Now that you have entered your access code and saved it to disk, you are ready to run a lesson. From the table of contents screen, choose the appropriate chapter and then choose the lesson you wish to run.

Features

Each lesson in **HLS: Intermediate Algebra** has four modes: Instruct, Practice, Tutor, and Certify.

Instruct: Instruct provides an exposition on the material covered in the lesson in a multimedia environment. This same instruct mode can be accessed via the tutor mode.

Practice: Practice allows you to hone your problem-solving skills. It provides an unlimited number of randomly generated problems. Practice also provides access to the Tutor mode by selecting the Tutor button located next to the Submit button.

Tutor: Tutor mode is broken up into several parts: Instruct, Explain Error, Step by Step, and Solution.

1. **Instruct**, which can also be selected directly from Practice mode, contains a multimedia lecture of the material covered in a lesson.

2. **Explain Error** is active whenever a problem is incorrectly answered. It will attempt to explain the error that caused you to incorrectly answer the problem.

3. **Step by Step** is an interactive "step through" of the problem. It breaks each problem into several steps, explains to you each step in solving the problem, and asks you a question about the step. After you answer the last step correctly, you have solved the problem.

4. **Solution** will provide you with a detailed "worked-out" solution to the problem.

Throughout the Tutor, you will see words or phrases colored green with a dashed underline. These are called Hot Words. Clicking on a Hot Word will provide you with more information on these words or phrases.

Certify: Certify is the testing mode. You are given a finite number of problems and a certain number of strikes (problems you can get wrong). If you answer the required number of questions, you will receive a certification code and a certificate. Write down your certification code and/or print out your certificate. The certification code will be used by your instructor to update your records. Note that the Tutor is not available in Certify.

Integration of Courseware and Textbook

Throughout the text you will find references that will help you integrate the Intermediate Algebra textbook and the *HLS: Intermediate Algebra* courseware. At the end of each section and again at the end of each chapter you will find a list of which *HLS: Intermediate Algebra* lessons you should run in order to test yourself on the subject material and to review the contents of a chapter.

Support

If you have questions about *HLS: Intermediate Algebra* or are having technical difficulties, we can be contacted as follows:

Phone: (843) 571-2825
Email: support@hawkeslearning.com
Web: hawkeslearning.com

Our support hours are 8:30 a.m. to 5:30 p.m., EST, Monday through Friday.

Real Numbers and Solving Equations

Did You Know?

In Chapter 1, you will find a great many symbols defined, as well as rules of manipulation for these symbolic expressions. Most people think that algebra has always existed, complete with all the common symbols in use today. That is not the case, since modern symbols did not appear consistently until the beginning of the sixteenth century. Prior to that, algebra was rhetorical. That is, all problems were written out in words using either Latin, Arabic, or Greek, and some nonstandard abbreviations. Numbers were written out. The common use of Hindu-Arabic numerals did not begin until the sixteenth century, although these numerals had been introduced into Europe in the twelfth century.

The sign for addition, +, was a contraction of the Latin *et*, which means "and." Gradually, the *e* was contracted and the crossed *t* became the plus sign. The minus sign or bar, −, is thought to be derived from the habit of early scribes of using a bar to represent the letter *m*. Thus the word *summa* was often written *suma* and the bar came to represent the missing *m*, the first letter of the word *minus*. The radical symbol, $\sqrt{}$, is derived from a small printed *r*, which stood for the Latin word *radix*, or root. The symbol for times, a cross, ×, was developed from cross multiplication or for the purpose of indicating products in proportions. Thus

$$\frac{2}{3}\times\frac{6}{9} \text{ stood for } \frac{2}{3}=\frac{6}{9}.$$

The cross is not well suited for algebra, since it resembles the symbol *x*, which is used for variables. Therefore, a dot is usually used to indicate multiplication in algebra. The dot seems to have developed from an Italian practice of separating columns in multiplication tables with a dot. Exponents were used as early as the fourteenth century by the mathematician Nicole Oresme (1320? – 1382), who gave the first known use of the rules for fractional exponents in a textbook he wrote. The equal sign is attributed to Robert Recorde (1510? – 1558), who wrote, "I will sette as I doe often in woorke use, a paire of paralleles, or Gemowe [twin] lines of one lengthe, thus: = , bicause noe .2. thynges, can be moare equalle." As you can tell, the development of algebraic symbols occurred over a long period of time, and symbols became standardized through usage and convenience. If you are interested in the history of numerical symbolism, you will find more information in D. E. Smith's *History of Mathematics*, Volume II.

Recorde

"The Mathematician, carried along on his flood of symbols, dealing apparently with purely formal truths, may still reach results of endless importance for our description of the physical universe."

Karl Pearson (1857 – 1936)

Welcome to your second course in algebra. You will find that many of the topics presented, such as solving equations, factoring, working with fractional expressions, and solving word problems, are familiar because they were discussed in beginning algebra. However, you will also find that the coverage of each topic, including many new topics, is in greater depth and that the pace is somewhat faster. The intent is to prepare you for success with the algebraic content in your future studies in mathematics, science, business, and economics. You should be prepared to spend considerable time and effort on this course throughout the semester.

We begin in Chapter 1 with a review of prime numbers and least common multiple (LCM) and continue with a review of the properties of and operations with real numbers (positive and negative numbers included). These ideas form the foundation for the study of algebra and need to be understood thoroughly. Appendix A.1 gives a complete review of operations with fractions (positive numbers only). As an example of a deeper analysis of familiar topics, we will discuss solving first-degree equations, then expand the techniques to include solving absolute value equations and solving formulas for specified variables. Also, we will review the topic of intervals of real numbers. Intervals and interval notation are related to graphs and are used frequently in higher level mathematics courses.

There is much to look forward to in this course. With hard work and perseverance, you should have a very rewarding experience.

1.1   Prime Numbers, Exponents, and LCM

- *Understand the terms **factor**, **prime factorization**, **multiple**, and **exponent**.*
- *Recognize prime numbers and composite numbers.*
- *Use exponents to write the prime factorization of a composite number.*
- *Find the least common multiple (LCM) of a set of counting numbers.*

Prime Numbers

The **whole numbers** consist of the number 0 and the *natural numbers* (also called the *counting numbers*). We use $\mathbb{N}$ to represent the set of natural numbers and $\mathbb{W}$ to represent the set of whole numbers.

$$\textbf{Natural Numbers} = \mathbb{N} = \left\{1, 2, 3, 4, 5, 6, 7, 8, 9, 10, 11, ...\right\}$$

$$\textbf{Whole Numbers} = \mathbb{W} = \left\{0, 1, 2, 3, 4, 5, 6, 7, 8, 9, 10, 11, ...\right\}$$

The three dots ... are called an ellipsis and are used to indicate that the pattern continues without end.

The operation of multiplication with whole numbers can be indicated by writing the numbers horizontally separated by a raised dot ($\cdot$), separated by a times sign ($\times$), or by writing a number next to parentheses containing another number. The numbers being multiplied are called **factors** and the result is called the **product**. For example,

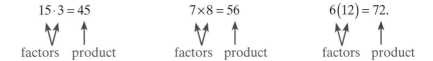

$$15 \cdot 3 = 45 \qquad\qquad 7 \times 8 = 56 \qquad\qquad 6(12) = 72.$$

factors product   factors product   factors product

You may remember the following about even and odd numbers:
1. If 2 is a factor (the units digit is 0, 2, 4, 6, or 8), the number is called an **even number**.
2. If 2 is not a factor, the number is called an **odd number**.

Because of the close relationship between multiplication and division, each factor of a number is also called a **divisor** of the number. That is, each factor will divide into the product with a remainder of 0.

Every counting number, except the number 1, has **at least two factors** (or **divisors**), as illustrated by the following list.

Examples of Counting Numbers	Factors
44 ------------------------------	$1, 2, 4, 11, 22, 44$
7 ------------------------------	$1, 7$
6 ------------------------------	$1, 2, 3, 6$
19 ------------------------------	$1, 19$
86 ------------------------------	$1, 2, 43, 86$
41 ------------------------------	$1, 41$

In the list above 7, 19, and 41 have **exactly two different factors**, 1 and the number itself. Such numbers are called **prime numbers**. The other numbers in the list (44, 6, and 86) are called **composite numbers**. (Note that 0 and 1 are neither prime nor composite.)

The prime numbers less than 50 are

$$2, 3, 5, 7, 11, 13, 17, 19, 23, 29, 31, 37, 41, 43, 47.$$

For convenience and ease in working with fractions, you should memorize, or at least recognize, these primes.

Using basic knowledge of multiplication and division, we can find factors of relatively small counting numbers. For example, basic multiplication facts give

$$99 = 9 \cdot 11.$$

However, for use in dealing with fractions, we need to find a **factorization** of 99 in which all of the factors are prime numbers. In this example 9 is not a prime number. By factoring 9, we can write

$$99 = 3 \cdot 3 \cdot 11.$$

This last product $(3 \cdot 3 \cdot 11)$ contains only prime factors and is called the **prime factorization** of 99. Regardless of the method used to find a prime factorization, **there is only one prime factorization for any composite number**. (Note that a different ordering of the same factors represents the same prime factorization.)

To Find the Prime Factorization of a Composite Number

1. *Factor the composite number into any two factors.*
2. *Factor each factor that is not prime into two more factors.*
3. *Continue this process until all factors are prime.*

The **prime factorization** *is the product of all the prime factors.*

Example 1: Prime Factorization

Find the prime factorization of each number.

a. 75   **b.** 294

Solutions:

a. $75 = 3 \cdot 25$ 3 is prime but 25 is not. Continue to factor 25.

         $= 3 \cdot 5 \cdot 5$ Now all of the factors are prime. This is the prime factorization.

b. $294 = 2 \cdot 147$ 2 is prime but 147 is not. Continue to factor 147.

          $= 2 \cdot 3 \cdot 49$ 3 is a prime factor of 147. Continue to factor 49.

          $= 2 \cdot 3 \cdot 7 \cdot 7$ Now all of the factors are prime. This is the prime factorization.

Exponents

A whole number **exponent** is a number used to indicate repeated multiplication of a number with itself. An exponent is written to the right and slightly above the number. The number being multiplied by itself is called the **base** of the exponent. [**Note:** Properties of exponents will be discussed in detail in Section 4.1.]

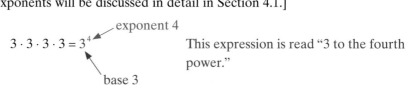

$$3 \cdot 3 \cdot 3 \cdot 3 = 3^4$$

This expression is read "3 to the fourth power."

A **variable** is a symbol (generally a letter of the alphabet) that is used to represent an unknown number or any one of several numbers. The letters n and a are used as variables in the following discussion of exponents.

Exponent

In general, for any counting number n and any real number a,

$$\underbrace{a \cdot a \cdot a \cdot \ldots \cdot a}_{n \text{ factors}} = a^n.$$

exponent

base

If no exponent is written, the exponent is understood to be 1. That is $a = a^1$.

The base is said to be "squared" if the exponent is 2 and "cubed" if the exponent is 3. Thus

$$5^2 = 25 \quad \text{is read "5 squared is equal to 25"}$$

and

$$2^3 = 8 \quad \text{is read "2 cubed is equal to 8."}$$

Example 2: Exponents

a. $2^5 = 2 \cdot 2 \cdot 2 \cdot 2 \cdot 2 = 32$ 2 is the base of the exponent 5.

b. $6^3 = 6 \cdot 6 \cdot 6 = 216$ 6 is the base of the exponent 3.

Example 3: Exponents

Use exponents in writing the prime factorization of each number.

a. $375 = 25 \cdot 15 = 5 \cdot 5 \cdot 3 \cdot 5 = 3 \cdot 5^3$

b. $2700 = 27 \cdot 100 = 3 \cdot 9 \cdot 10 \cdot 10 = 3 \cdot 3 \cdot 3 \cdot 2 \cdot 5 \cdot 2 \cdot 5 = 2^2 \cdot 3^3 \cdot 5^2$

Least Common Multiple (LCM)

The **multiples** of a number are the products of that number with the counting numbers. Thus the first multiple of any number is the number itself. All other multiples are larger than that number. For example, the multiples of 8 are:

$$8, 16, 24, 32, 40, 48, 56, 64, 72, 80, 88, 96, \ldots$$

We are interested in finding common multiples and more particularly the **least common multiple (LCM)** for a set of counting numbers. Listing all the multiples for two or more numbers and then choosing the least common multiple (LCM) is not very efficient. The following technique involving prime factorizations is generally much easier to use. Study it carefully. It is particularly useful throughout all mathematical discussions of fractions.

To Find the LCM of a Set of Counting Numbers

1. *Find the prime factorization of each number.*

2. *List the prime factors that appear in any one of the prime factorizations.*

3. *Find the product of these primes using each prime the greatest number of times it appears in any one of the prime factorizations.*

Example 4: Least Common Multiple (LCM)

Find the least common multiple (LCM) of 27, 30, and 50.

Solution:

Step 1: Prime factorizations:

$$27 = 3 \cdot 3 \cdot 3 \qquad \text{three 3's}$$
$$30 = 2 \cdot 3 \cdot 5 \qquad \text{one 2, one 3, one 5}$$
$$50 = 2 \cdot 5 \cdot 5 \qquad \text{one 2, two 5's}$$

Step 2: Prime factors present are 2, 3, and 5.

Step 3: The most number of times each factor is used in any one factorization:

One 2   (in 30 and 50)
Three 3's   (in 27)
Two 5's   (in 50)

Find the product of these primes.

$$\text{LCM} = 2 \cdot 3 \cdot 3 \cdot 3 \cdot 5 \cdot 5 = 2 \cdot 3^3 \cdot 5^2 = 1350$$

1350 is the smallest number divisible by all three of the numbers 27, 30, and 50.

1.1 Exercises

1. List the prime numbers less than 50.

2. Describe, in your own words, the meaning of the word *factor*.

3. Define *composite number*.

4. Describe, in your own words, how to find the LCM of a set of counting numbers.

Find the prime factorization of each of the composite numbers in Exercises 5 – 20. Use exponents to represent repeated factors.

5. 36   **6.** 45   **7.** 48   **8.** 60

9. 66   **10.** 72   **11.** 144   **12.** 155

13. 270   **14.** 315   **15.** 336   **16.** 460

17. 550   **18.** 624   **19.** 675   **20.** 1692

Find the least common multiple (LCM) of each set of whole numbers in Exercises 21 – 40.

21. {5, 25}   **22.** {7, 42}   **23.** {45, 75}   **24.** {14, 49}

25. {50, 65}   **26.** {30, 48}   **27.** {4, 7, 14}   **28.** {9, 12, 18}

29. {3, 5, 11}   **30.** {2, 7, 13}   **31.** {5, 14, 35}   **32.** {6, 10, 20}

33. {10, 12, 25}   **34.** {14, 49, 56}   **35.** {18, 24, 64}   **36.** {27, 34, 51}

37. {55, 121, 110}   **38.** {26, 39, 91}   **39.** {24, 60, 72, 96}   **40.** {25, 27, 45, 50}

 HAWKES LEARNING SYSTEMS: INTERMEDIATE ALGEBRA SOFTWARE

▪ Prime Numbers, Exponents, and LCM

<table>
</table>

1.2 Introduction to Real Numbers

- *Identify given numbers as members of one or more of the following sets: natural numbers, whole numbers, integers, rational numbers, irrational numbers, and real numbers.*
- *Write rational numbers as infinite repeating decimals.*
- *Graph sets of numbers on real number lines.*
- *Describe sets of numbers using set-builder notation given their graphs.*
- *Name the properties of real numbers that justify given statements.*
- *Complete statements using the real number properties.*

We begin with a development of the terminology and properties of numbers that form the foundation for the study of algebra. The following kinds of numbers are studied in some detail in beginning algebra courses.

Types of Numbers

Natural Numbers (or **Counting Numbers**)

$$\mathbb{N} = \{\, 1, 2, 3, 4, 5, 6, \ldots \,\}$$

Whole Numbers (The number 0 is added to the set of natural numbers.)

$$\mathbb{W} = \{\, 0, 1, 2, 3, 4, 5, 6, \ldots \,\}$$

Integers

$$\mathbb{Z} = \{\, \ldots,\ -4, -3, -2, -1, 0, 1, 2, 3, 4, \ldots \,\}$$

The integers are one of the important stepping stones from arithmetic to algebra since the concept of positive and negative numbers is basic to algebra. Integers can be represented on a number line by marking 0 at some point and then marking the **positive integers** to the right of 0 and their **opposites** or **negative integers** to the left of 0 (Figure 1.1).

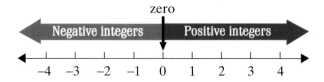

Figure 1.1

Numbers that can be written as fractions and whose numerators are integers and denominators are nonzero integers have the technical name **rational numbers**. Integers and terminating and infinite, repeating decimal numbers can also be classified as rational numbers. For example, the following numbers are all rational numbers:

$$1.3 = \frac{13}{10}, \quad 5 = \frac{5}{1}, \quad -4 = \frac{-4}{1}, \quad \frac{3}{8}, \quad \text{and} \quad 2.8333... = \frac{17}{6}.$$

Variables are needed to be able to state rules and definitions in general forms. The definition is restated here for easy reference and for emphasis.

Variable

*A **variable** is a symbol (generally a letter of the alphabet) that is used to represent an unknown number or any one of several numbers.*

The variables a and b are used to represent integers in the following definition of a **rational number**.

Rational Number

*A **rational number** is any number that can be written in the form $\frac{a}{b}$ where a and b are integers and $b \neq 0$. (The letter $\mathbb{Q}$ represents the set of all rational numbers.)*

Example 1: Rational Numbers

a. $\frac{2}{3}, \frac{7}{1}, \frac{-5}{3}, \frac{27}{10}$, and $\frac{3}{-10}$ are all rational numbers. Each is in the form $\frac{a}{b}$ where a and b are integers and $b \neq 0$.

b. $1\frac{3}{4}$ and 2.33 are also rational numbers. They are not in the form $\frac{a}{b}$, but they **can be written** in that form:

$$1\frac{3}{4} = \frac{7}{4} \quad \text{and} \quad 2.33 = 2\frac{33}{100} = \frac{233}{100}.$$

c. $\frac{\pi}{6}$ and $\frac{\sqrt{2}}{3}$ are in the form $\frac{a}{b}$, but the numerator in each case is not an integer and cannot be written as an integer. Thus, the fractions are **not** rational numbers. Instead, as we will see, they are called **irrational numbers**.

Rational numbers have been defined as numbers that can be written in the form $\dfrac{a}{b}$ where a and b are integers and $b \neq 0$. With this definition, we can show that in decimal form a rational number is either

> **1.** a terminating decimal, or
> **2.** an infinite repeating decimal.

Examples of terminating decimals are:

$$\frac{1}{4} = 0.25, \quad \frac{3}{8} = 0.375, \quad 1\frac{4}{5} = \frac{9}{5} = 1.8, \quad \text{and} \quad \frac{6}{1} = 6.$$

Examples of infinite repeating decimals are: $\dfrac{2}{3}, \dfrac{1}{7},$ and $\dfrac{4}{11}.$ Long division shows the repeating decimal pattern for each.

$$
\begin{array}{r}
0.6666... \\
3\overline{)2.0000...} \\
\underline{18} \\
20 \\
\underline{18} \\
20 \\
\underline{18} \\
20 \\
\underline{18} \\
2
\end{array}
\qquad
\begin{array}{r}
0.14285714... \\
7\overline{)1.00000000...} \\
\underline{7} \\
30 \\
\underline{28} \\
20 \\
\underline{14} \\
60 \\
\underline{56} \\
40 \\
\underline{35} \\
50 \\
\underline{49} \\
10 \\
\underline{7} \\
30 \\
\underline{28} \\
2
\end{array}
\qquad
\begin{array}{r}
0.3636... \\
11\overline{)4.0000...} \\
\underline{3\,3} \\
70 \\
\underline{66} \\
40 \\
\underline{33} \\
70 \\
\underline{66} \\
4
\end{array}
$$

Thus

$$\frac{2}{3} = 0.6666..., \quad \frac{1}{7} = 0.14285714..., \quad \text{and} \quad \frac{4}{11} = 0.3636....$$

Or we can write a bar over the repeating pattern of digits as:

$$\frac{2}{3} = 0.\overline{6}, \quad \frac{1}{7} = 0.\overline{142857}, \quad \text{and} \quad \frac{4}{11} = 0.\overline{36}.$$

As the following definition indicates, **irrational numbers** are infinite decimal numbers with no repeating pattern to their digits. Irrational numbers cannot be written in fraction form, $\dfrac{a}{b}$ where a and b are integers.

Irrational Number

*An **irrational number** is any number that can be written as an infinite, nonrepeating decimal.*

Example 2: Irrational Numbers

The following are irrational numbers. Note that there is no repeating pattern in their decimal representation. The three dots indicate that the digits continue indefinitely.

a. $\pi = 3.14159265358979\ldots$ π has no repeating pattern in its decimal form.

b. $\sqrt{2} = 1.414213562\ldots$ The square root of 2 has no repeating pattern in its decimal form.

c. $e = 2.718281828459045\ldots$ e is a number used in higher mathematics and engineering courses.

d. $0.01001000100001\ldots$ Even though there is a pattern to the digits, the pattern is not repeating.

The notation for π and e is attributed to Swiss mathematician Leonhard Euler (1707 – 1783).

NOTES   The number π is particularly fascinating to mathematicians and has recently been represented to 1.24 trillion decimal places. A discussion of π and a representation to 3742 decimal places is in Appendix 3 at the back of this text.

Together, the rational numbers and irrational numbers form the set of **real numbers** ($\mathbb{R}$). The relationships between the various types of real numbers can be seen in the diagram in Figure 1.2.

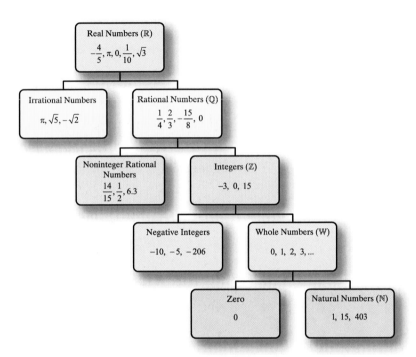

Figure 1.2

Understanding these relationships is critical for success in algebra. For example, if a problem calls for integer solutions and your solution is $\frac{3}{5}$, then you need to understand that you have not found an integer solution.

Summary of Relationships Among Various Types of Numbers

1. *Every natural number is also a whole number, an integer, a rational number, and a real number.*

2. *Every whole number is also an integer, a rational number, and a real number.*

3. *Every integer is also a rational number and a real number.*

4. *Every rational number is also a real number.*

5. *Every irrational number is also a real number.*

Example 3: Classifying Numbers

Given the set $\left\{-3, \frac{1}{2}, 4, \sqrt{10}\right\}$ tell what classification each number fits.

Solution: -3 : integer, rational number, real number

$\frac{1}{2}$: rational number, real number

4 : natural number, whole number, integer, rational number, real number

$\sqrt{10}$: irrational number, real number

Example 4: Identifying Types of Numbers

Given the set of real numbers $\left\{-2, -\sqrt{3}, -1.1, -\frac{1}{2}, 0, \frac{5}{8}, \sqrt{1.7}, 1.7\right\}$:

a. Tell which numbers are integers.

   Solution: -2 and 0 are integers.

b. Tell which numbers are rational numbers.

   Solution: $-2, -1.1, -\frac{1}{2}, 0, \frac{5}{8}$, and 1.7 are rational numbers.

c. Tell which numbers are irrational numbers.

   Solution: $-\sqrt{3}$ and $\sqrt{1.7}$ are irrational numbers. As you can find with a calculator, $-\sqrt{3}$ is approximately -1.732 and $\sqrt{1.7}$ is approximately 1.304.

Practice Problems

Given the set of real numbers $A = \left\{-3, -2.5, -2, -\sqrt{2}, -1, 0, \frac{3}{4}, \sqrt{6}, \pi, 4.3, 5\right\}$:

1. *List the numbers in A that are integers.*

2. *List the numbers in A that are rational numbers.*

3. *List the numbers in A that are irrational numbers.*

Real Number Lines

There is a one-to-one correspondence between the real numbers and the points on a line. Thus number lines are also called **real number lines** (Figure 1.3), and every rational and irrational number has a corresponding point on a real number line.

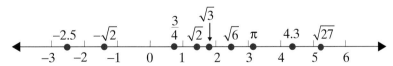

Figure 1.3

Irrational numbers in the form of various roots, such as $\sqrt{3}$, will be discussed thoroughly in Chapter 6. For now, to estimate the placement of numbers such as $\sqrt{3}$, $\sqrt{6}$, or $\sqrt{27}$ on a real number line, you can note their relationships to the square roots of perfect

Answers to Practice Problems: **1.** $\{-3, -2, -1, 0, 5\}$ **2.** $\left\{-3, -2.5, -2, -1, 0, \frac{3}{4}, 4.3, 5\right\}$ **3.** $\left\{-\sqrt{2}, \sqrt{6}, \pi\right\}$

square integers such as 1, 4, 9, 16, 25 and so on, or use a calculator to find a decimal estimation. Thus

$$\sqrt{3} \text{ is slightly less than } \sqrt{4} = 2. \quad \text{(With a calculator: } \sqrt{3} = 1.732050808...)$$
$$\sqrt{6} \text{ is slightly more than } \sqrt{4} = 2. \quad \text{(With a calculator: } \sqrt{6} = 2.449489743...)$$
$$\sqrt{27} \text{ is slightly more than } \sqrt{25} = 5. \quad \text{(With a calculator: } \sqrt{27} = 5.196152423...)$$

To understand how an infinite, nonrepeating decimal corresponds to a single point on a line, we will illustrate how $\pi = 3.14159265\ldots$ can be marked. If a circle has a diameter of 1 unit, then its circumference is π units because the formula for circumference ($C = \pi d$) gives $C = \pi \cdot 1 = \pi$. By rolling such a circle along a line, the number π can be located (Figure 1.4).

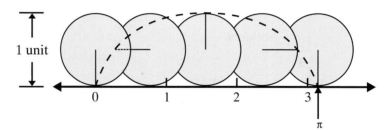

Figure 1.4

Example 5: Graphing Real Numbers on a Real Number Line

Given the set of real numbers $\left\{-2, -\sqrt{3}, -1.1, -\dfrac{1}{2}, 0, \dfrac{5}{8}, \sqrt{1.7}, 1.7\right\}$, graph the numbers on a number line.

Solution:

Sets and Set-Builder Notation

A **set** is a collection of objects or numbers. The items in the set are called **elements**, and sets are indicated with braces, { }, and named with capital letters. If the elements are listed within the braces, as we have done earlier with the set of natural numbers, $\mathbb{N}$, the set of whole numbers, $\mathbb{W}$, and the set of integers, $\mathbb{Z}$, the set is said to be in **roster form**. The symbol $\in$ is read "is an element of" and is used to indicate that a particular number belongs to a set. For example, $0 \in \mathbb{W}$ and $-3 \in \mathbb{Z}$.

If the elements in a set can be counted, the set is said to be **finite**. If the elements cannot be counted, as for $\mathbb{N}$, $\mathbb{W}$, and $\mathbb{Z}$, the set is said to be **infinite**. If a set has absolutely no elements, it is called the **empty set** or **null set** and is written in the form { } or with the special symbol $\varnothing$. For example, the set of all people over 12 feet tall is the empty set, $\varnothing$.

The notation $\{x|\quad\}$ is read "the set of all x such that …" and is called **set-builder notation**. The vertical bar (|) is read "such that." A statement following the bar gives a condition (or restriction) for the variable x. For example,

> $\{x|x$ is an even integer$\}$ is read "the set of all x such that x is an even integer."
>
> $\{x|x \in \mathbb{Z}\}$ is read "the set of all x such that x is an element of the set of integers."

The following inequality symbols can be used to indicate relationships between numbers and, along with set notation, to indicate entire sets of real numbers.

Inequality Symbols			
$<$	*"less than"*	$\leq$	*"less than or equal to"*
$>$	*"greater than"*	$\geq$	*"greater than or equal to"*

Table 1.1

Using Set Notation and Graphs to Indicate Sets of Real Numbers

Set Notation	Meaning	Graph
$\{x\|x \leq a\}$	"the set of all x such that x is less than or equal to a"	
$\{x\|x \geq b\}$	"the set of all x such that x is greater than or equal to b"	
$\{x\|x < a \text{ or } x > b\}$ (This is also known as the **union** of two sets of numbers.)	"the set of all x such that x is less than a **or** x is greater than b"	
$\{x\|x > a \text{ and } x < b\}$ or $\{x\|a < x < b\}$ (This is also known as the **intersection** of two sets of numbers.)	"the set of all x such that x is greater than a **and** x is less than b"	

Table 1.2

NOTES

Comment on Graphing Techniques

Note that we have shown two types of symbols for indicating whether a number is or is not included in a graph. Parentheses such as) and (are used to indicate that a number **is not** included in a graph. Brackets such as] and [are used to indicate that a number **is** included in a graph. Similarly, open circles ○ can be used in place of parentheses to indicate a point **is not** included, and closed circles ● can be used in place of brackets to indicate a point **is** included. In this text we will use the parentheses and brackets as illustrated in Example 6.

NOTES

Comment about Union and Intersection

The concepts of union and intersection are part of set theory which is very useful in a variety of courses including abstract algebra, probability, and statistics. These concepts are also used in analyzing inequalities and analyzing relationships among sets in general. The **union** (symbolized ∪, as in A ∪ B) of two (or more) sets is the set of all elements that belong to either one set or the other set or to both sets. The **intersection** (symbolized ∩, as in A ∩ B) of two (or more) sets is the set of all elements that belong to both sets. The word **or** is used to indicate union and the word **and** is used to indicate intersection. For example, if A = {1, 2, 3} and B = {2, 3, 4}, then the numbers that belong to A **or** B is the set A ∪ B = {1, 2, 3, 4}. The set of numbers that belong to A **and** B is the set A ∩ B = {2, 3}. These relationships can be illustrated using the following Venn diagram.

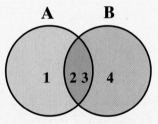

Similarly, union and intersection notation can be used for sets with inequalities.

For example, $\{x \mid x < a \text{ or } x > b\}$ can be written in the form

$$\{x \mid x < a\} \cup \{x \mid x > b\}.$$

Also, $\{x \mid x > a \text{ and } x < b\}$ can be written in the form

$$\{x \mid x > a\} \cap \{x \mid x < b\} \text{ or } \{x \mid a < x < b\}.$$

The following examples show how various sets of real numbers can be graphed on a real number line. **Note the use of parentheses to indicate that a point is not included in the set and the use of brackets to indicate that a point is included in the set.**

Example 6: Graphs Indicated by Inequalities

a. Graph the set of real numbers $\{x \mid -1 \leq x < 2\}$.

Solution:

This graph shows all points between −1 and 2, including −1 but not 2.

b. Graph the set $\{x \mid x > 3\}$.

Solution:

This graph shows all points greater than 3, but not including 3.

In Example 6c note carefully the use of the word **and** to indicate an intersection ∩, and in Example 6d the use of **or** to indicate a union ∪.

c. Graph the set $\{x \mid x \leq 2 \textbf{ and } x \geq 0\}$. The word **and** implies those values of x that satisfy **both** inequalities. The solution graph shows the intersection ∩ of the first two graphs.

Solution: $x \leq 2$

$x \geq 0$

$x \leq 2$ **and** $x \geq 0$

Note that the third graph shows the points in common between the first two graphs in this example.

This set can also be indicated as $\{x \mid 0 \leq x \leq 2\}$.

d. Graph the set $\{x \mid x > 5 \textbf{ or } x \leq 4\}$. The word **or** implies those values of x that satisfy **at least one** of the inequalities. The solution graph shows the union ∪ of the first two graphs.

Solution: $x > 5$

$x \leq 4$

$x > 5$ **or** $x \leq 4$

Properties of Addition and Multiplication with Real Numbers

The various properties of the operations of addition and multiplication with real numbers are summarized here. These properties are used throughout algebra and mathematics in developing formulas and general concepts.

Properties of Addition and Multiplication

For real numbers a, b, and c:

For Addition	*Name of Property*	*For Multiplication*
$a + b = b + a$ $4 + 2 = 2 + 4$	*Commutative Property*	$a \cdot b = b \cdot a$ $6 \cdot 3 = 3 \cdot 6$
$(a + b) + c = a + (b + c)$ $(6 + 3) + 5 = 6 + (3 + 5)$	*Associative Property*	$a \cdot (b \cdot c) = (a \cdot b) \cdot c$ $3 \cdot (4 \cdot 5) = (3 \cdot 4) \cdot 5$
$a + 0 = 0 + a = a$ $10 + 0 = 0 + 10 = 10$	*Identity*	$a \cdot 1 = 1 \cdot a = a$ $-5 \cdot 1 = 1 \cdot (-5) = -5$
$a + (-a) = 0$ $15 + (-15) = 0$	*Inverse*	$a \cdot \dfrac{1}{a} = 1 \; \left(for\ a \neq 0 \right)$ $4 \cdot \dfrac{1}{4} = 1$

Zero Factor Law

$$a \cdot 0 = 0 \cdot a = 0 \qquad\qquad -2 \cdot 0 = 0 \cdot (-2) = 0$$

Distributive Property of Multiplication over Addition

$$a(b + c) = ab + ac \qquad\qquad 2(x + 4) = 2 \cdot x + 2 \cdot 4$$

NOTES

The number **0** is called the **additive identity** because when 0 is added to a number the result is the same number. Likewise, the number **1** is called the **multiplicative identity** because when 1 is multiplied by a number the result is the same number. Also, the **additive inverse** of a number *a* is its **opposite**, −*a*.

The **multiplicative inverse** of a nonzero number *a* is its **reciprocal**, $\dfrac{1}{a}$.

Example 7: Identify the Property

Tell which property justifies each statement.

a. $-3 \cdot 1 = -3$ Identity property of multiplication

b. $\dfrac{1}{2} + 0 = \dfrac{1}{2}$ Identity property of addition

c. $7 + 1.6 = 1.6 + 7$ Commutative property of addition

d. $(2 \cdot 3) \cdot 8 = 2 \cdot (3 \cdot 8)$ Associative property of multiplication

e. $6 + (-6) = 0$ Inverse property of addition

f. $9 \cdot \dfrac{1}{9} = 1$ Inverse property of multiplication

g. $4 \cdot (2 + 3) = 4 \cdot 2 + 4 \cdot 3$ Distributive property

h. $5 \cdot (x + 3) = 5 \cdot x + 5 \cdot 3$ Distributive property

As the following examples illustrate, subtraction and division are neither commutative nor associative.

Subtraction is not commutative: $a - b \neq b - a$

     Example: $6 - 2 \neq 2 - 6$

                  because $6 - 2 = 4$

                  and $2 - 6 = -4$

                  and $4 \neq -4$.

Subtraction is not associative: $a - (b - c) \neq (a - b) - c$

     Example: $10 - (5 - 3) \neq (10 - 5) - 3$

                  because $10 - (5 - 3) = 10 - (2) = 8$

                  and $(10 - 5) - 3 = 5 - 3 = 2$

                  and $8 \neq 2$.

Division is not commutative: $a \div b \neq b \div a$

     Example: $6 \div 2 \neq 2 \div 6$

                  because $6 \div 2 = 3$

                  and $2 \div 6 = \dfrac{1}{3}$

                  and $3 \neq \dfrac{1}{3}$.

Division is not associative: $a \div (b \div c) \neq (a \div b) \div c$

Example: $24 \div (4 \div 2) \neq (24 \div 4) \div 2$

because $24 \div (4 \div 2) = 24 \div (2) = 12$

and $(24 \div 4) \div 2 = (6) \div 2 = 3$

and $12 \neq 3$.

The number 0 is the only real number that does not have a reciprocal. We say that

$$\frac{1}{0} \text{ is undefined}.$$

[To help in understanding this idea, suppose that $\frac{1}{0} = x$. In this case we would need to have $1 = 0 \cdot x$. But this is not possible because by the zero factor law $0 \cdot x = 0$ for all real values of x. Therefore, $\frac{1}{0}$ is undefined.]

There are two basic properties related to inequalities (or order) with real numbers. You may have seen these properties before.

Properties of Inequality (Order)

For real numbers a, b, and c:

Trichotomy Property: *Exactly one of the following is true: $a < b$, $a = b$, or $a > b$.*

Transitive Property: *If $a < b$ and $b < c$, then $a < c$.*

Note: *The transitive property applies to $>$, $\geq$, and $\leq$ as well.*

Example 8: Identify the Property

State which property of inequality or order is illustrated.

a.   If $x < 3$ and $3 < y$, then $x < y$.   Transitive property (of order)

b.   If a is a real number, then either   Trichotomy property (of order)
$a = 5$ or $a > 5$ or $a < 5$.

c.   If x and y are real numbers, then either   Trichotomy property (of order)
$x = y$ or $x > y$ or $x < y$.

d.   If $t > x$ and $x > -7$, then $t > -7$.   Transitive property (of order)
Note that this property applies to
$>$, $\geq$, $<$, and $\leq$.

Practice Problems

1. List the integers.
2. What type of number is π?
3. Graph the set $\{x | 0 < x \le 1\}$.
4. $a + 3 = 3 + a$ illustrates which property of addition?
5. $2(x + y) = 2x + 2y$ illustrates which property?

1.2 Exercises

For Exercises 1 – 6, list the numbers in the set $A = \left\{ -8, -\sqrt{5}, -\sqrt{4}, -\dfrac{4}{3}, -1.2, -\dfrac{\sqrt{3}}{2}, 0, \dfrac{4}{5}, \sqrt{3}, \right.$ $\left. \sqrt{11}, \sqrt{16}, 4.2, 6 \right\}$ that are described in each exercise.

1. $\{x | x$ is a whole number$\}$
2. $\{x | x$ is a natural number$\}$

3. $\{x | x$ is an integer$\}$
4. $\{x | x$ is an irrational number$\}$

5. $\{x | x$ is a rational number$\}$
6. $\{x | x$ is a real number$\}$

In Exercises 7 – 12, choose the word that correctly completes each statement.

7. If x is a rational number, then x is (always, sometimes, never) a real number.

8. If x is a rational number, then x is (always, sometimes, never) an irrational number.

9. If x is an integer, then x is (always, sometimes, never) a whole number.

10. If x is a real number, then x is (always, sometimes, never) a rational number.

11. If x is a rational number, then x is (always, sometimes, never) an integer.

12. If x is a natural number, then x is (always, sometimes, never) a whole number.

Answers to Practice Problems: 1. $\{ \dots, -4, -3, -2, -1, 0, 1, 2, \dots \}$ 2. Irrational number, real number

3. $\begin{array}{ccccc} & & & & \\ -1 & 0 & 1 & 2 \end{array}$ 4. Commutative property of addition

5. Distributive property

Write each of the rational numbers in Exercises 13 – 18 as a terminating decimal or an infinite repeating decimal.

13. $\dfrac{5}{8}$

14. $\dfrac{9}{16}$

15. $-\dfrac{7}{3}$

16. $-\dfrac{8}{9}$

17. $\dfrac{71}{20}$

18. $\dfrac{5}{7}$

Graph each of the sets of numbers in Exercises 19 – 26 on a number line. Note that in each exercise a condition is placed on the variable.

19. $\left\{ x \mid x < 7,\ x \text{ is a whole number} \right\}$

20. $\left\{ x \mid x < -12,\ x \text{ is an integer} \right\}$

21. $\left\{ x \mid x \geq -4,\ x \text{ is an integer} \right\}$

22. $\left\{ x \mid -9 < x \leq 2,\ x \text{ is an integer} \right\}$

23. $\left\{ x \mid -5 < x < 6,\ x \text{ is a natural number} \right\}$

24. $\left\{ x \mid -8 < x < 0,\ x \text{ is a whole number} \right\}$ 25. $\left\{ x \mid x < 12 \text{ and } x > 0,\ x \text{ is an integer} \right\}$

26. $\left\{ x \mid x > 0 \text{ and } x \leq 8,\ x \text{ is an integer} \right\}$

For Exercises 27 – 30, use set-builder notation to indicate each set of numbers as described.

27. the set of all real numbers between 3 and 5, including 3

28. the set of all real numbers between –4 and 4

29. the set of all real numbers greater than or equal to –2.5

30. the set of all real numbers between –1.8 and 5, including both of these numbers

Graph each of the sets of real numbers in Exercises 31 – 38 on a number line. Note that since no restriction is placed on the variable, it is understood to represent real numbers.

31. $\left\{ x \mid x < 2 \text{ or } x > 8 \right\}$

32. $\left\{ x \mid x \leq -5 \text{ or } x \geq \dfrac{9}{5} \right\}$

33. $\left\{ x \mid -\sqrt{2} < x < 0 \right\}$

34. $\left\{ x \mid -4 \leq x < -\sqrt{5} \right\}$

35. $\left\{ x \mid x \geq -1 \text{ and } -3 \leq x \leq 0 \right\}$

36. $\left\{ x \mid -\dfrac{7}{4} < x \leq 2 \text{ and } x < 1 \right\}$

37. $\left\{ x \mid -1.6 < x < 0 \text{ or } 2 \leq x \leq 3.7 \right\}$

38. $\left\{ x \mid -\dfrac{3}{5} < x < 0 \text{ or } 0 \leq x < \pi \right\}$

In Exercises 39 – 46, the graphs of sets of real numbers are given. Use set-builder notation to indicate the set of numbers shown in each graph.

39.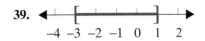
$$-4 \ -3 \ -2 \ -1 \ \ 0 \ \ 1 \ \ 2$$

40.
$$-4 \quad -2 \quad 0 \quad 2 \quad 4$$

41.
$$-8 \quad -6 \quad -4 \quad -2 \quad 0$$

42.
$$-4 \ -2 \ \ 0 \ \ 2 \ \ 4 \ \ 6$$

43.
$$-2 \ -1 \ \ 0 \ \ 1 \ \ 2$$

44.
$$-2 \quad 0 \quad 2 \quad 4 \quad 6$$

45.
$$-4 \ -2 \ \ 0 \ \ 2 \ \ 4$$

46.
$$-9 \quad -6 \quad -3 \quad 0 \quad 3$$

State the property of real numbers that is illustrated in Exercises 47 – 68. All variables represent real numbers and no denominator is 0.

47. $9 + (-9) = 0$

48. Either $x < y, x = y$, or $x > y$.

49. $4 + 0 = 4$

50. $5 + (a + b) = (5 + a) + b$

51. $9 \cdot (x + 5) = 9 \cdot x + 45$

52. $3 \cdot y = y \cdot 3$

53. $7 \cdot \dfrac{1}{7} = 1$

54. Exactly one of the following is true:
$$x < 5, x = 5, \text{ or } x > 5.$$

55. If $x < 11$ and $11 < y$, then $x < y$.

56. $\left(\sqrt{2} \cdot x\right) \cdot y = \sqrt{2} \cdot (x \cdot y)$

57. $6 + y = y + 6$

58. Either $s = t, s > t$, or $s < t$.

59. $x \cdot (y + 5) = x \cdot y + x \cdot 5$

60. $\sqrt{7} + \left(-\sqrt{7}\right) = 0$

61. $(y + z)(1) = y + z$

62. $x + (y + 7) = (x + y) + 7$

63. If $a < -2$ and $-2 < b$, then $a < b$.

64. $8 \cdot x + 3 \cdot x = (8 + 3) \cdot x$

65. $11 + (y + 4) = (11 + y) + 4$

66. $(x + y) \cdot \dfrac{1}{x + y} = 1$

67. $s + (-s) = 0$

68. If $a < b$ and $b < (x - 2)$, then $a < (x - 2)$.

Complete the expressions in Exercises 69 – 80 by using the given property.

69. $x + 7 =$ _____   Commutative property of addition

70. $x \cdot 3 =$ _____   Commutative property of multiplication

71. $x \cdot (6 + y) =$ _____   Distributive property

72. $x + (3 + y) =$ _____   Associative property of addition

73. $3 \cdot (x \cdot z) =$ _____   Associative property of multiplication

74. Either $x < y$, $x > y$, or _____.   Trichotomy property

75. $2 \cdot (y + 3) =$ _____   Distributive property

76. If $x < a$ and $a < 10$, then _____. Transitive property

77. The multiplicative inverse of 6   Inverse of multiplication
is _____ because _____.

78. The reciprocal of –4 is _____   Inverse of multiplication
because _____.

79. The additive inverse of –7 is _____   Inverse of addition
because _____.

80. The additive inverse of 15 is _____   Inverse of addition
because _____.

Writing and Thinking About Mathematics

81. Explain the difference between graphing a set of integers and graphing a set of real numbers.

82. The inequality $x > 5$ is used to indicate all real numbers greater than 5. Is there a "first" real number greater than 5? Or, in other words, is there a real number greater than 5 that is "closest to" 5? Write, in your own words, a paragraph or two with examples to indicate your understanding of this question. Show your analysis to a friend to see if he or she understands and agrees or disagrees with your thinking. A related question would be "What is the real number closest to 0?".

 HAWKES LEARNING SYSTEMS: INTERMEDIATE ALGEBRA SOFTWARE

- Name that Real Number
- Properties of Real Numbers

Operations with Real Numbers

1.3

- *Evaluate absolute value expressions.*
- *Determine the values, if any, that satisfy absolute value equations.*
- *Add, subtract, multiply, and divide real numbers.*
- *Evaluate expressions by using the rules for order of operations.*

Absolute Value

On the real number line, any number and its **opposite** lie the same number of units from 0 on the number line. The **distance a number is from 0 on a number line** is called its **absolute value** and is symbolized by two vertical bars, $|\ \ |$. Thus, because both 3 and –3 are 3 units from 0, we have $|3| = 3$ and $|-3| = 3$. (See Figure 1.5.)

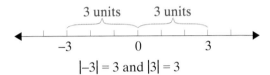

Figure 1.5 $|-3| = 3$ and $|3| = 3$

NOTES

For a variable a, the symbol $-a$ should be thought of as "the **opposite** of a." Thus, if a represents a **positive** number, then $-a$ represents a **negative** number. However, if a represents a **negative** number, then $-a$ represents a **positive** number. For example,

if $a = 3$, then $-a = -3$.

But, if $a = -5$, then $-a = -(-5) = 5$.

Because distance (similar to length) is never negative, the **absolute value of a number is never negative**. Or, the absolute value of a nonzero number is always positive.

Absolute Value

*The **absolute value of a real number is its distance from 0**.*
In symbols:

For any real number a ,

If a is positive or 0, then $|a| = a.$

If a is negative, then $|a| = -a.$

Another form of the same definition is the following:

$$|a| = \begin{cases} a & \text{if } a \geq 0 \\ -a & \text{if } a < 0 \end{cases}.$$

Example 1: Absolute Value

a. $|5| = 5$

b. $|-6| = -(-6) = 6$

c. $|\pi| = \pi$

d. $|0| = 0$

e. $\left|-\dfrac{3}{4}\right| = -\left(-\dfrac{3}{4}\right) = \dfrac{3}{4}$

f. $-|-5.2| = -5.2$

g. If $|x| = 6$, what are the possible values for x?

> **Solution:** $x = 6$ or $x = -6$ because both $|6| = 6$ and $|-6| = 6$.
> We say that $\{-6, 6\}$ is the **solution set** of the equation.

h. If $|x| = -4$, what are the possible values for x?

> **Solution:** There are no values of x for which $|x| = -4$. The absolute value can never be negative. The solution set is the empty set, $\varnothing$.

Addition with Real Numbers

The rules for addition with positive and negative real numbers are stated here for easy reference. They illustrate the need for understanding absolute value.

Rules for Addition with Real Numbers

> *1. To add two real numbers with **like signs**,*
> *a. add their absolute values and*
> *b. use the common sign.*
>
> *2. To add two real numbers with **unlike signs**,*
> *a. subtract their absolute values (the smaller from the larger), and*
> *b. use the sign of the number with the larger absolute value.*

Study the following examples carefully to help in understanding the rules for addition.

Example 2: Addition with Like Signs

a. $(+10) + (+3) = +\big(|+10| + |+3|\big) = +(10 + 3) = 13$

b. $(-10) + (-3) = -\big(|-10| + |-3|\big) = -(10 + 3) = -13$

c. $(-1.4) + (-2.5) = -\big(|-1.4| + |-2.5|\big) = -(1.4 + 2.5) = -3.9$

Example 3: Addition with Unlike Signs

a. $(-10)+(+3)=-\left(\left|-10\right|-\left|+3\right|\right)=-\left(10-3\right)=-7$

b. $(+10)+(-3)=+\left(\left|+10\right|-\left|-3\right|\right)=+\left(10-3\right)=7$

c. $-\dfrac{7}{11}+\dfrac{5}{11}=-\left(\left|-\dfrac{7}{11}\right|-\left|\dfrac{5}{11}\right|\right)=-\left(\dfrac{7}{11}-\dfrac{5}{11}\right)=-\dfrac{2}{11}$

NOTES If a negative number occurs after an addition symbol then we must place the number in parentheses so that the two operation symbols are not next to each other.

Example 4: Application

Jeremy is planning a cookout. He wants to buy $3\dfrac{1}{2}$ lbs ground beef, 3 lbs pork chops, 2 lbs chicken, and $3\dfrac{1}{4}$ lbs steak. How much meat does Jeremy intend to buy?

[**Note:** If you have trouble adding fractions, see Appendix A.1 for a review.]

Solution: $3\dfrac{1}{2}+3+2+3\dfrac{1}{4}=\dfrac{7}{2}+3+2+\dfrac{13}{4}$

$$=\dfrac{14}{4}+\dfrac{12}{4}+\dfrac{8}{4}+\dfrac{13}{4}$$

$$=\dfrac{47}{4}$$

$$=11\dfrac{3}{4}$$

Jeremy will need to buy $11\dfrac{3}{4}$ lbs of meat.

Note: The improper fraction $\dfrac{47}{4}$ is a correct solution and may be preferred in many algebraic situations. However, in practical applications, the mixed number form $11\dfrac{3}{4}$ is more appropriate and understandable to most people.

Example 5: Application

Susan is a salesperson for a shoe store. Last week her sales of pairs of shoes were as follows:

Day	Sales	Returns	Daily Net Sales
Monday	7	1	6
Tuesday	3	0	3
Wednesday	2	4	–2
Thursday	6	1	5
Friday	8	3	5

What were Susan's net sales for last week?

Solution: $6 + 3 + (-2) + 5 + 5 = 17$

Susan's net sales for the week were 17 pairs of shoes.

Subtraction with Real Numbers

Note that subtraction with real numbers is defined in terms of addition.

Rule for Subtraction with Real Numbers

For real numbers a and b,

$$a - b = a + (-b).$$

To subtract b, add the opposite of b.

Thus we see that any subtraction problem can be thought of as an addition problem. Of course, as illustrated in Example 6, we must be careful with the signs.

Example 6: Subtraction with Real Numbers

a. $18 - 13 = 18 + (-13) = 5$ Subtracting 13 is the same as adding –13.

b. $-18 - 13 = -18 + (-13) = -31$

c. $14-(-6)=14+(+6)=20$ Subtracting –6 is the same as adding +6.

Remember: To subtract, add the opposite of the number being subtracted. If more than two numbers are involved, add or subtract from left to right.

d. $8-12-21=8+(-12)+(-21)=-4+(-21)=-25$

e. $\dfrac{3}{5}+\dfrac{4}{5}-\dfrac{7}{5}=\dfrac{7}{5}-\dfrac{7}{5}=0$

f. $8.2-3.1-0.6=5.1-0.6=4.5$

Example 7: Application

At noon on Tuesday the temperature was 34°F. By noon on Thursday the temperature had dropped to –5° F. How much did the temperature drop between Tuesday and Thursday?

12pm Tuesday   12pm Thursday

Solution: For change in value:

$$(\text{end value} - \text{beginning value})=(-5)-(+34)$$

$$=-5+(-34)$$

$$=-39$$

Between Tuesday and Thursday the temperature changed –39° F (or dropped 39° F).

Multiplication with Real Numbers

Multiplication can be indicated by any of the following conventions.

Symbols for Multiplication

Symbol	Description	Example
·	*raised dot*	$4 \cdot 7$
( )	*numbers inside or next to parentheses*	$5(10)$ *or* $(5)10$ *or* $(5)(10)$
×	*cross sign*	6×12 *or* $\begin{array}{r} 12 \\ \underline{\times\, 6} \end{array}$
	number written next to variable	$8x$
	variable written next to variable	xy

Table 1.3

From previous experience, we know that

the product of two positive real numbers is positive.

For the product of a positive number and a negative number, consider the product $5(-3)$. We can think of this as repeated addition,

$$5(-3) = (-3) + (-3) + (-3) + (-3) + (-3) = -15$$

and we see that this is the same as adding numbers that are all negative. The result is negative. Thus it is reasonable to conclude that

the product of a positive real number and a negative real number is negative.

By using the facts that $-a = -1 \cdot a$ and $-b = -1 \cdot b$, as well as the commutative and associative properties of multiplication, we can show that

the product of two negative real numbers is positive.

This discussion leads to the following statement about multiplying positive and negative real numbers.

Rules for Multiplying Positive and Negative Real Numbers

For positive real numbers a and b,

1. *The product of two positives is positive:* $(a)(b) = ab$.

2. *The product of two negatives is positive:* $(-a)(-b) = ab$.

3. *The product of a positive and a negative is negative:* $a(-b) = (-a)b = -ab$.

In summary:

The product of real numbers with like signs is positive.

The product of real numbers with unlike signs is negative.

Example 8: Multiplication with Positive and Negative Real Numbers

a. $8(-5) = -40$ Product of a positive and a negative is negative.

b. $-6\left(\dfrac{1}{2}\right) = -3$

c. $9(-4)(2) = -36(2) = -72$

d. $-3(-4) = 12$ Product of two negatives is positive.

e. $\left(-\dfrac{3}{4}\right)\left(-\dfrac{1}{2}\right) = \dfrac{3}{8}$

f. $(-2.1)(-0.03) = 0.063$

Division with Real Numbers

Because division is defined in terms of multiplication, the rules for dividing positive and negative real numbers are similar to those for multiplying. We know that $a \div b = \dfrac{a}{b}$, so we can use the fraction form $\dfrac{a}{b}$ to indicate division. Division can be stated as multiplication by a reciprocal as follows:

$$\frac{a}{b} = a \cdot \frac{1}{b}.$$

Thus a division problem can be treated as a multiplication problem.

Rules for Dividing Positive and Negative Real Numbers

For positive real numbers a and b,

1. *The quotient of two positives is positive:* $\dfrac{a}{b} = +\dfrac{a}{b}$.

2. *The quotient of two negatives is positive:* $\dfrac{-a}{-b} = +\dfrac{a}{b}$.

3. *The quotient of a positive and a negative is negative:* $\dfrac{-a}{b} = \dfrac{a}{-b} = -\dfrac{a}{b}$.

In summary:

> *The quotient of numbers with like signs is positive.*
> *The quotient of numbers with unlike signs is negative.*

Example 9: Division with Positive and Negative Real Numbers

a. $\dfrac{30.6}{-2} = -15.3$ **b.** $\dfrac{-18}{-6} = 3$ **c.** $-\dfrac{51}{3} = -17$

Division by 0 is Undefined

1. *Suppose that* $a \neq 0$ *and* $\dfrac{a}{0} = x$. *Then, since division is related to multiplication, we must have* $a = 0 \cdot x$. *But this is not possible because* $0 \cdot x = 0$ *for any value of x and we stated that* $a \neq 0$.

2. *Suppose that* $\dfrac{0}{0} = x$. *Then* $0 = 0 \cdot x$ *which is true for all values of x. But we must have a unique answer for x.*

Therefore, in any case, we conclude that division by 0 is undefined.

Thus $\dfrac{a}{0}$ *is* **undefined**, *but for* $b \neq 0, \dfrac{0}{b} = 0$.

Example 10: Division with 0

a. $\dfrac{0}{-7} = 0$ **b.** $\dfrac{9}{0}$ *is undefined.*

Order of Operations

Consider the problem of evaluating an expression with more than one operation, such as $5 + 2 \cdot 3$.

$$\text{Addition first gives: } 5 + 2 \cdot 3 = 7 \cdot 3 = 21.$$
$$\text{Multiplication first gives: } 5 + 2 \cdot 3 = 5 + 6 = 11.$$
$$\text{But } 21 \neq 11.$$

Only one answer can be right, and mathematicians have agreed on the following **rules for order of operations**.

Rules for Order of Operations

1. *Simplify within symbols of inclusion (parentheses, brackets, braces, fraction bar, absolute value bars) beginning with the innermost symbols.*

2. *Find any powers indicated by exponents or roots.*

3. *Multiply or divide from **left to right**.*

4. *Add or subtract from **left to right**.*

Using these rules, we find that the correct value for $5 + 2 \cdot 3$ is found by multiplying first. Thus

$$5 + 2 \cdot 3 = 5 + 6 = 11 \text{ is correct.}$$

A well-known mnemonic device for remembering these rules is the following.

Please	Excuse	My	Dear	Aunt	Sally
↑	↑	↑	↑	↑	↑
Parentheses	**Exponents**	**Multiplication**	**Division**	Addition	Subtraction

The mnemonic **PEMDAS** here shows multiplication and division in the same color and addition and subtraction in the same color. **These operations are to be done from left to right with neither operation having priority.** This means that the mnemonic could just as easily be **PEDMSA**. But this wouldn't be as much fun to say or as easy to remember.

Example 11: Order of Operations

Using the rules for order of operations, find the value of each of the following expressions.

a. $10 - 21 \div 3 + 2$

Solution: $10 - 21 \div 3 + 2$

$$= 10 - 7 + 2 \qquad \text{Perform the division first.}$$

$$= 3 + 2 \qquad \text{Add or subtract from left to right.}$$

$$= 5 \qquad \textit{Continued on the next page...}$$

b. $5(-2)+6\cdot4-2$

Solution: $5(-2)+6\cdot4-2$

$= -10 + 24 - 2$ Multiply from left to right first.

$= 14 - 2$ Add or subtract from left to right.

$= 12$

c. $12 \div 2 \cdot 3 - 4$

Solution: $12 \div 2 \cdot 3 - 4$

$= 6 \cdot 3 - 4$ Perform the division first.

$= 18 - 4$ Multiply.

$= 14$ Add or subtract from left to right.

d. $-5 - 3\big[4 + (16 - 10) \div 2\big]$

Solution: $-5 - 3\big[4 + (16 - 10) \div 2\big]$

$= -5 - 3\big[4 + 6 \div 2\big]$ Subtract within the innermost parentheses first.

$= -5 - 3\big[4 + 3\big]$ Divide within brackets.

$= -5 - 3\big[7\big]$ Add within brackets.

$= -5 - 21$ Multiply.

$= -26$ Add or subtract from left to right.

e. $16 \cdot 3 \div 2^3 - (18 + 20)$

Solution: $16 \cdot 3 \div 2^3 - (18 + 20)$

$= 16 \cdot 3 \div 2^3 - 38$ Add within the parentheses.

$= 16 \cdot 3 \div 8 - 38$ Evaluate the exponents.

$= 48 \div 8 - 38$ Multiply or divide from left to right.

$= 6 - 38$

$= -32$ Add or subtract from left to right.

Practice Problems

Find the value of each expression, if possible.

1. $-8.6 - 4.1 - 0.2$

2. $(-3)(-5)(-6)$

3. $\dfrac{8}{0}$

4. $4(-2) + 6 \cdot 3$

5. $2(-5 + 3) + 16 \div 8 \cdot 2$

6. $3(-8 + 2^3) + 6^2 \div 2^2 - 4$

Answers to Practice Problems: **1.** −12.9 **2.** −90 **3.** Undefined **4.** 10 **5.** 0 **6.** 5

1.3 Exercises

Find the value of each expression in Exercises 1 – 5.

1. $|7|$ **2.** $\left|-\dfrac{3}{4}\right|$ **3.** $\left|-\sqrt{5}\right|$ **4.** $|0|$ **5.** $-|-8|$

Find the set of values for x in Exercises 6 – 15 that make true statements. If a statement is never true, indicate this by writing the empty set, $\emptyset$.

6. $|x| = 4$ **7.** $|x| = 7$ **8.** $|x| = 0$ **9.** $|x| = 2$

10. $|x| = -3$ **11.** $|x| = \dfrac{4}{5}$ **12.** $|x| = 2.6$ **13.** $|x| = -2.8$

14. $|x| = -x$ **15.** $|x| = x$

Perform the indicated operations in Exercises 16 – 65.

16. $(-16) + 20$ **17.** $(-2) + (-9)$ **18.** $-5 + |-3|$

19. $(-8) + (-6) + 5$ **20.** $-3 + |7| + (-2)$ **21.** $\left(-\dfrac{3}{8}\right) + \dfrac{7}{8}$

22. $\dfrac{9}{16} + \left|-\dfrac{5}{16}\right|$ **23.** $12 - 15$ **24.** $-4 - (-8)$

25. $(-9) - (-9)$ **26.** $0 - (-12)$ **27.** $17 - |-4|$

28. $-\dfrac{4}{13} - \dfrac{3}{13}$ **29.** $\dfrac{3}{5} - \dfrac{9}{5}$ **30.** $\left|-\dfrac{8}{3}\right| - \left(-\dfrac{2}{3}\right)$

31. $(-1.7) + (-5.2)$ **32.** $(8.5) + (-7.9)$ **33.** $-7 - (-2) + 6$

34. $-18 - 22 - 41$ **35.** $-8 + (-7) - (-15)$ **36.** $9 - (-3) + (-2)$

37. $21 + |-3| - |-4|$ **38.** $|13| - |-9| + |-3|$ **39.** $-\dfrac{7}{6} + \left(-\dfrac{5}{6}\right) - \dfrac{1}{6}$

40. $\dfrac{4}{15} + \left|-\dfrac{7}{15}\right| - \left|\dfrac{16}{15}\right|$ **41.** $\left(-\dfrac{9}{16}\right) + \left(-\dfrac{7}{8}\right)$ **42.** $\dfrac{1}{8} - \left(-\dfrac{1}{2}\right) + \dfrac{1}{4}$

43. $\dfrac{4}{5} + \left(-\dfrac{2}{3}\right) - \dfrac{1}{6}$ **44.** $-\dfrac{3}{8} - \dfrac{5}{6} + \left(-\dfrac{1}{2}\right)$ **45.** $(-8)(-7)$

46. $(-3)(17)$ **47.** $(-8)(-1)(-5)(6)(-2)$

48. $(12)\left(-\dfrac{5}{6}\right)$

49. $\dfrac{3}{8} \cdot \dfrac{5}{2}$

50. $-\dfrac{5}{16} \cdot \dfrac{3}{4}$

51. $6(5.3)$

52. $\left(-\dfrac{3}{10}\right)\left(\dfrac{5}{6}\right)\left(-\dfrac{8}{7}\right)\left(\dfrac{1}{2}\right)\left(-\dfrac{1}{4}\right)$

53. $(-0.8)(4.9)$

54. $(11.7)(2.06)(-1.3)$

55. $(-20) \div (-10)$

56. $\dfrac{-39}{-13}$

57. $\dfrac{-91}{-7}$

58. $\dfrac{52}{13}$

59. $\dfrac{6}{16} \div 0$

60. $60 \div (-15)$

61. $0 \div \dfrac{11}{12}$

62. $\dfrac{28.7}{-7}$

63. $-68.05 \div 5$

64. $-88.64 \div (-8)$

65. $-6.084 \div (-9)$

Find the value of each expression in Exercises 66 – 87 by using the rules for order of operations.

66. $18 \div 3 \cdot 6 + 3$

67. $7(4-2) \div 7 + 3$

68. $10 \div 2 - 4 \cdot 3^2$

69. $2^2 \cdot 3 \div 3 + 6 \div 3$

70. $-6 \cdot 3 \div (-1) + 4 - 2$

71. $5(-2) \div (-5) + 5 - 3$

72. $(4^2 + 6) - 2 \cdot 19$

73. $(5^2 - 4^2)^2 - 11$

74. $[(4+14) \div (3 \cdot 3)] - 5$

75. $[8 - (5 \cdot 6 - 2)] + 3$

76. $(12 \cdot 4 \div 2^3) - [(3 \cdot 2^3) \div (4 \cdot 6)]$

77. $[(3 \cdot 0) \div (2 \cdot 1)] - (24 - 6^2) \div (4^2 - 3 \cdot 4)$

78. $(3 \cdot 2^3) \div (3 \cdot 4) + (2 \cdot 3 + 4) \div (6 - 1)$

79. $-6 + (-2)(12 \cdot 2 \div 3)4$

80. $14 - \left[11 \cdot 4 - (2 \cdot 3^2 + 1)\right]$

81. $6 + 3\left[-4 - 2(3 - 1)\right]$

82. $7 - \left[4 \cdot 3 - (4 - 3 \cdot 2)\right]$

83. $-2\left[6 + 4(1 + 7)\right] \div 4$

84. $\dfrac{(-3)(-6)}{5 - (-4)} - 2$

85. $\dfrac{4 - (-10)}{-2 - 5} \div (-2)$

86. $\dfrac{16 - (-4)}{-3 + 9} \div \dfrac{10^2 + 10}{-5 \cdot 11}$

87. $\dfrac{3^3 - (-27)}{2 \cdot 3^2} + \dfrac{-6 \cdot 5}{-2 \cdot 5}$

88. Stock market: During the first hour of trading, a stock trader has stock worth $1973.27. During the second hour, the trader loses $797.53. During the third hour, he gains $925.87. What was the net worth of the trader's stock after the first three hours of trading?

89. **Checking Account:** A college student opens a checking account with a deposit of $1000.00. She withdraws $252.68 to pay for textbooks. Later that evening, she writes a check for $116.89 for groceries. The next day, she deposits a graduation gift of $75.25 cash. What is her final account balance?

90. **Stock market:** In a 5-day week the NASDAQ stock market posted a gain of 38 points, a loss of 65 points, a loss of 32 points, a gain of 10 points, and a gain of 15 points. If the NASDAQ started the week at 2050 points, what was the market at the end of the week?

91. **Football:** In ten running plays in a football game, the tailback gained 5 yards, lost 3 yards, gained 15 yards, gained 7 yards, gained 12 yards, lost 4 yards, lost 2 yards, gained 20 yards, lost 5 yards, and gained 6 yards. What was his cumulative yardage in the game?

92. **Hiking:** A hiker, beginning at an altitude of 970 ft, ascends a peak 5260 ft. Next, he descends 3130 ft and climbs another peak 1570 ft. After a brief rest, he continues his ascent another 2190 ft. Finally he descends 4040 ft. What is his final altitude?

93. **Fishing:** A commercial fishing boat casts a net and brings in 258 fish. The fishermen find that 77 of the fish are too small to sell and throw them back. They cast their net again and bring in 401 more fish. Of these, 98 are too small to sell. How many fish do the fishermen have left to sell?

Use your graphing calculator to evaluate each expression in Exercises 94 – 99. Note that for a leading negative sign you must use the **(−)** *key next to* **ENTER** *. Also, brackets must be replaced with parentheses on the calculator and exponents can be indicated with the caret key* **^** *. To get the value of an expression, press* **ENTER** *after you have entered the expression. For example, the picture shows the evaluation of the expression* $5^2 \div 5 \cdot 2 + (18 - 20)^3$ *.*

94. $12^2 \div 3 \cdot 2 + (17 - 5)^3$

95. $5^3 - 7^3 + (5 - 7)^3$

96. $0.8 + 2.1(17 - 14.1 \div 2) \div 7$

97. $(140 - 20 \cdot 6 \div 2^3) \div 5^3$

98. $16 - \left[18 \cdot 4 - (11.1 \cdot 5^2 + 1) \right]$

99. $-10 \left[45 + 40(10 - 76.5) \right] \div 4$

Writing and Thinking About Mathematics

100. Determine a general statement related to positive and negative signs that can be made about the sign of a product for any number of values being multiplied.

101. Explain, in your own words, how the fraction bar works as an inclusion symbol.

HAWKES LEARNING SYSTEMS: INTERMEDIATE ALGEBRA SOFTWARE

- Introduction to Absolute Values
- Addition with Real Numbers
- Subtraction with Real Numbers
- Multiplication and Division with Real Numbers
- Order of Operations

1.4

Linear Equations in One Variable: $ax + b = c$

- *Combine like terms.*
- *Solve linear equations of the form $ax + b = c$.*
- *Solve absolute value equations of the form $|ax + b| = c$.*

Combining Like Terms

A **term** is an expression that involves only multiplication and/or division with constants and/or variables. A term consisting of only a real number is called a **constant** or a **constant term**. Exponents are used to indicate repeated factors in a term.

Examples of terms: $2x^5$, $\dfrac{1}{3}x^2y$, -14, $5.6a$

The numerical factor of a term containing variables is called the **numerical coefficient** (or the **coefficient**) of the variables. For example, in the term $4x^2y$, 4 is the coefficient. If no coefficient is written, then the coefficient is *understood to be* 1. If a negative sign is in front of a variable expression, then the coefficient is *understood to be* -1. For example,

$$x = 1 \cdot x, \quad n^2 = 1 \cdot n^2, \quad -y = -1 \cdot y, \quad \text{and} \quad -pq = -1 \cdot pq.$$

If terms contain variables, then **like terms** (or **similar terms**) are those terms that contain the same variables raised to the same powers. That is, whatever power a variable is raised to in one term, it is raised to that same power in other like terms. Constants are considered to be like terms.

Like Terms	
$-5, 1.3,$ and 144	are **like terms** because each term is a constant.
$x^2y, 9x^2y,$ and $-3x^2y$	are **like terms** because each term contains the same two variables with x having an exponent of 2 and y having an exponent of 1.

Table 1.4

Unlike Terms	
$9x$ and $5x^3$	are **unlike terms** (**not** like terms) because the variable x is not to the same power in both terms.
$8xy, 13x^2,$ and $17y$	are **unlike terms** because not all terms have the same variables and the variables are not to the same power in all terms.

Table 1.5

To simplify expressions that contain like terms, we want to **combine like terms**.

Combining Like Terms

> To **combine like terms**, add (or subtract) the coefficients and keep the common variable expression.

The procedure for combining like terms uses the distributive property in the form
$$ba + ca = (b + c)a.$$

Example 1: Combining Like Terms

Combine like terms in the following expressions.

a. $4x^2 + 11x^2$

   Solution: $4x^2 + 11x^2$

   $$= (4 + 11)x^2$$

   $$= 15x^2$$

b. $-6y + 4y$

   Solution: $-6y + 4y$

   $$= (-6 + 4)y$$

   $$= -2y$$

c. $3x^2 - x + 3 - (x^2 - 5x + 6)$

   Solution: $3x^2 - x + 3 - (x^2 - 5x + 6)$

   $$= 3x^2 - x + 3 + (-1)(x^2 - 5x + 6)$$

   $$= 3x^2 - x + 3 + (-1)(x^2) + (-1)(-5x) + (-1)(6)$$

   $$= 3x^2 - x + 3 - x^2 + 5x - 6$$

   $$= (3 - 1)x^2 + (-1 + 5)x + (3 - 6)$$

   $$= 2x^2 + 4x - 3$$

   The $-$ sign in front of $x^2 - 5x + 6$ can be interpreted as multiplication by -1. Thus each term in parentheses is multiplied by -1, and each term in parentheses is changed.

d. $\dfrac{3x + 5x}{4} + 9x$

   Solution: $\dfrac{3x + 5x}{4} + 9x$

   $$= \dfrac{8x}{4} + 9x$$

   $$= 2x + 9x$$

   $$= 11x$$

   Note that the fraction bar is treated as a symbol of inclusion, and $3x$ and $5x$ are added first.

e. $4x^2 - \left[3x - \left(x^2 + x\right)\right]$

Solution: $4x^2 - \left[3x - \left(x^2 + x\right)\right]$

$$= 4x^2 - \left[3x - x^2 - x\right]$$

$$= 4x^2 - \left[2x - x^2\right]$$

$$= 4x^2 - 2x + x^2$$

$$= 5x^2 - 2x$$

Remove the innermost symbol of inclusion first. The coefficient of x $\left(\text{and of } x^2\right)$ is 1. With practice, this step can be done mentally.

Linear Equations in One Variable: $ax + b = c$

An **algebraic expression** is a combination of variables and numbers using any of the operations of addition, subtraction, multiplication, or division as well as exponents. An **equation** is a statement that two algebraic expressions are equal. That is, both expressions represent the same number. If an equation contains a variable, any number that gives a true statement when substituted for the variable is called a **solution** to the equation. For example, replacing x with 5 in the equation $3x + 4 = 10$ gives the false statement $3 \cdot 5 + 4 = 10$. Therefore, **5 is not a solution** to the equation. However, replacing x with 2 gives the true statement $3 \cdot 2 + 4 = 10$. Therefore, **2 is a solution** to the equation.

The solutions to an equation are said to form a **solution set**. The process of finding the solution set is called **solving the equation**. In this course we will study various types of equations that have more than one solution. However, **equations of the form $ax + b = c$ (linear equations in one variable) have exactly one solution**.

Linear Equations in x

*If a, b, and c are constants and a ≠ 0, then a **linear equation in x** is an equation that can be written in the form*

$$ax + b = c.$$

*[**Note:** A linear equation in x is also called a **first-degree equation in x** because the variable x can be written with the exponent 1. That is, $x = x^1$.]*

To **solve** (or **find the solution set of**) a linear equation, we need the following two properties of equality.

Addition Property of Equality

If the same algebraic expression is added to both sides of an equation, the new equation has the same solutions as the original equation. Symbolically, if A, B, and C are algebraic expressions, then the equations

$$A = B$$
$$\text{and} \qquad A + C = B + C$$

have the same solutions. Equations with the same solutions are said to be **equivalent**.

Multiplication (or Division) Property of Equality

If both sides of an equation are multiplied by (or divided by) the same nonzero constant, the new equation has the same solutions as the original equation. Symbolically, if A and B are algebraic expressions and C is any nonzero constant, then the equations

$$A = B$$
$$\text{and} \qquad AC = BC \qquad \text{where } C \neq 0$$
$$\text{and} \qquad \frac{A}{C} = \frac{B}{C} \qquad \text{where } C \neq 0$$

have the same solutions and are **equivalent**.

The basic strategy in solving linear equations in one variable is to find equivalent equations until an equation is found with a single variable on one side and a constant on the other side. We use the Addition Property and the Multiplication Property of Equality in this process. Then a simplified equation such as $x = 5$ or $x = 7$, in which the variable has a coefficient of +1, gives the solution to the original equation.

Procedure for Solving Linear Equations

1. *Simplify each side of the equation by removing any grouping symbols and combining like terms. (In some cases, you may want to multiply both sides of the equation by a constant to clear fractional or decimal coefficients.)*

2. *Use the addition property of equality to add the opposites of constants or variable expressions so that variable expressions are on one side of the equation and constants on the other.*

Continued on the next page...

Procedure for Solving Linear Equations (cont.)

3. *Use the multiplication property of equality to multiply both sides by the reciprocal of the coefficient of the variable (that is, divide both sides by the coefficient) so that the new coefficient is 1.*

4. *Check your answer by substituting it into the **original** equation.*

Example 2: Solving a Linear Equation

Solve the linear equation $4x + 7 + x - 3x = -8 + 3$.

Solution:

$4x + 7 + x - 3x = -8 + 3$	Write the equation.
$2x + 7 = -5$	Combine like terms.
$2x + 7 - 7 = -5 - 7$	Add -7 to both sides of the equation.
$2x = -12$	Simplify.
$\dfrac{2x}{2} = \dfrac{-12}{2}$	Divide both sides of the equation by 2.
$x = -6$	Simplify.

Check:
$$4(-6) + 7 + (-6) - 3(-6) \overset{?}{=} -8 + 3$$
$$-24 + 7 - 6 + 18 \overset{?}{=} -5$$
$$-5 = -5$$

The solution is -6. We usually write just $x = -6$ to indicate the solution to the original equation. But, writing $\{-6\}$ as the **solution set** is also acceptable.

Many of the steps shown in Example 2 can be done mentally. Also, there is generally more than one correct way to proceed. In Example 2, you may choose to add $+5$ to both sides of the equation instead of adding -7 to both sides. In this case, the steps that follow will be different, too. However, the solution will be the same.

NOTES   To avoid errors and to help make your work easy to read and understand, try to align the = signs in a vertical format so that each new equation is directly below the previous equation.

Example 3: Solving Linear Equations

a. $7(x-3) = x + 3(x+5)$

Solution:

$$7(x-3) = x + 3(x+5) \quad \text{Write the equation.}$$
$$7x - 21 = x + 3x + 15 \quad \text{Use the distributive property (twice).}$$
$$7x - 21 = 4x + 15 \quad \text{Combine like terms.}$$
$$7x - 21 - 4x = 4x + 15 - 4x \quad \text{Add } -4x \text{ to both sides.}$$
$$3x - 21 = 15 \quad \text{Simplify.}$$
$$3x - 21 + 21 = 15 + 21 \quad \text{Add } 21 \text{ to both sides.}$$
$$3x = 36 \quad \text{Simplify.}$$
$$\frac{3x}{3} = \frac{36}{3} \quad \text{Divide both sides by } 3.$$
$$x = 12 \quad \text{Simplify.}$$

Check:
$$7(12-3) \overset{?}{=} 12 + 3(12+5)$$
$$7(9) \overset{?}{=} 12 + 3(17)$$
$$63 \overset{?}{=} 12 + 51$$
$$63 = 63$$

b. $\dfrac{x-5}{4} + \dfrac{3}{2} = \dfrac{x+2}{3}$

Solution:

$$\frac{x-5}{4} + \frac{3}{2} = \frac{x+2}{3} \quad \text{Write the equation.}$$
$$12\left(\frac{x-5}{4}\right) + 12\left(\frac{3}{2}\right) = 12\left(\frac{x+2}{3}\right) \quad \text{Multiply both sides by } 12, \text{ the LCM of the denominators.}$$
$$3(x-5) + 6(3) = 4(x+2)$$
$$3x - 15 + 18 = 4x + 8 \quad \text{Use the distributive property.}$$
$$3x + 3 = 4x + 8 \quad \text{Combine like terms.}$$
$$3x + 3 - 3x = 4x + 8 - 3x \quad \text{Add } -3x \text{ to both sides.}$$
$$3 = x + 8 \quad \text{Simplify.}$$
$$3 - 8 = x + 8 - 8 \quad \text{Add } -8 \text{ to both sides.}$$
$$-5 = x \quad \text{Simplify.}$$

Check:
$$\frac{-5-5}{4} + \frac{3}{2} \overset{?}{=} \frac{-5+2}{3}$$
$$\frac{-10}{4} + \frac{6}{4} \overset{?}{=} \frac{-3}{3}$$
$$\frac{-4}{4} \overset{?}{=} \frac{-3}{3}$$
$$-1 = -1$$

Conditional Equations, Identities, and Contradictions

When solving equations, there are times that we are concerned with the number of solutions that an equation has. If an equation has a finite number of solutions (the number of solutions is a countable number), the equation is said to be a **conditional equation**. As stated earlier, every linear equation has exactly one solution. Thus **every linear equation is a conditional equation**. However, in some cases, simplifying an equation will lead to a statement that is always true, such as $0 = 0$. In these cases the original equation is called an **identity** and has an infinite number of solutions which can be written as all real numbers or $\mathbb{R}$. If the equation simplifies to a statement that is never true, such as $0 = 2$, then the original equation is called a **contradiction** and its solution set is the empty set, $\varnothing$. Table 1.6 summarizes these ideas.

Type of Equation	Number of Solutions
Conditional	Finite number of solutions
Identity	Infinite number of solutions
Contradiction	No solutions

Table 1.6

Example 4: Solutions of Equations

Determine whether each of the following equations is a conditional equation, an identity, or a contradiction.

a. $0.3x + 15 = -1.2$

Solution:

$$0.3x + 15 = -1.2$$
$$0.3x = -16.2 \qquad \text{Add } -15 \text{ to both sides.}$$
$$x = -54 \qquad \text{Solve for } x.$$

The equation has one solution and it is a conditional equation.

b. $3(x - 25) + 3x = 6(x + 10)$

Solution:

$$3(x - 25) + 3x = 6(x + 10)$$
$$3x - 75 + 3x = 6x + 60 \qquad \text{Use the distributive property.}$$
$$6x - 75 = 6x + 60 \qquad \text{Simplify.}$$
$$-75 = 60 \qquad \text{Add } -6x \text{ to both sides.}$$

The last equation is never true. Therefore, the original equation is a contradiction and has no solution.

c. $-2(x - 7) + x = 14 - x$

Solution:

$$-2(x - 7) + x = 14 - x$$
$$-2x + 14 + x = 14 - x \qquad \text{Use the distributive property.}$$
$$14 - x = 14 - x \qquad \text{Simplify.}$$
$$14 = 14 \qquad \text{Add } x \text{ to both sides.}$$

The last equation is always true. Therefore, the original equation is an identity and has an infinite number of solutions. Every real number is a solution.

Absolute Value Equations

The definition of **absolute value** was given in Section 1.3 and is stated again here for easy reference. Remember, the absolute value of a number is its distance from 0 on a number line.

Absolute Value

For any real number x,
$$|x| = \begin{cases} x & \text{if } x \geq 0 \\ -x & \text{if } x < 0 \end{cases}.$$

Equations involving absolute value may have more than one solution (all of which must be included when giving an answer). For example, suppose that $|x| = 3$. Since $|3| = 3$ and $|-3| = -(-3) = 3$, we have either $x = 3$ or $x = -3$. We can say that the solution set is $\{3, -3\}$. In general, **any number and its opposite have the same absolute value**.

Solving Absolute Value Equations

For $c > 0$:
a. If $|x| = c$, then $x = c$ or $x = -c$.
b. If $|ax + b| = c$, then $ax + b = c$ or $ax + b = -c$.

Note: *If the absolute value expression is isolated on one side of the equation, we say that the expression is in **standard form**. You may need to manipulate the absolute value equation to get it into standard form before you can solve it. (See Example 5d.)*

Example 5: Solving Absolute Value Equations

Solve the following equations involving absolute value.

a. $|x| = 5$

　　Solution:　$x = 5$ or $x = -5$

b. $|3x - 4| = 5$

　　Solution:　
$3x - 4 = 5$	or	$3x - 4 = -5$
$3x = 9$		$3x = -1$
$x = 3$		$x = \dfrac{-1}{3}$

c. $|4x - 1| = -8$

Solution: There is no number that has a negative absolute value. Therefore, this equation has no solution. (The solution is $\varnothing$ and the equation is a contradiction.)

d. $5|3x + 17| - 4 = 51$

Solution:

$5\|3x + 17\| - 4 = 51$	Write the equation.
$5\|3x + 17\| = 55$	Add 4 to both sides.
$\|3x + 17\| = 11$	Divide both sides by 5. **We must have the absolute value expression by itself.**

$$3x + 17 = 11 \quad \text{or} \quad 3x + 17 = -11$$
$$3x = -6 \qquad\qquad\qquad 3x = -28$$
$$x = -2 \qquad\qquad\qquad x = -\frac{28}{3}$$

Equations with Two Absolute Value Expressions

If two numbers have the same absolute value, then either they are equal or they are opposites of each other. This fact can be used to solve equations that involve two absolute values.

Two Absolute Values

If $|a| = |b|$, then either $a = b$ or $a = -b$.

More generally,

if $|ax + b| = |cx + d|$, then either $ax + b = cx + d$ or $ax + b = -(cx + d)$.

Example 6: Solving Equations with Two Absolute Values

Solve $|x + 5| = |2x + 1|$.

Solution: In this case, the two expressions $(x + 5)$ and $(2x + 1)$ are equal to each other or are opposites of each other.

$$|x + 5| = |2x + 1|$$

$$x + 5 = 2x + 1 \quad \text{or} \quad x + 5 = -(2x + 1)$$

Note the use of parentheses. We want the opposite of the entire expression $(2x + 1)$.

$$5 = x + 1 \qquad\qquad x + 5 = -2x - 1$$
$$4 = x \qquad\qquad\quad 3x + 5 = -1$$
$$3x = -6$$
$$x = -2$$

Make sure to check that both 4 and −2 satisfy the original equation.

Practice Problems

1. *Combine like terms:* $3x^2 - \left[x^2 + \left(3x - 2x^2 \right) \right]$.

Solve the following equations.

2. $3x - 4 = 2x + 6 - x$ ***3.*** $6(x - 4) + x = 4(1 - x) + 4x$

4. $\dfrac{x+4}{2} + \dfrac{1}{4} = \dfrac{x+52}{12}$ ***5.*** $|2x - 1| = 8.2$

1.4 Exercises

In each of the expressions in Exercises 1 – 16, simplify by combining like terms.

1. $-2x + 5y + 6x - 2y$ **2.** $4x + 2x - 3y - x$

3. $4x - 3y + 2(x + 2y)$ **4.** $5(x - y) + 2x - 3y$

5. $\left(3x^2 + x \right) - \left(7x^2 - 2x \right)$ **6.** $-\left(4x^2 - 2x \right) - \left(5x^2 + 2x \right)$

7. $4x - [5x + 3 - (7x - 4)]$ **8.** $3x - [2y - (3x + 4y)]$

9. $\dfrac{-4x - 2x}{3} + 7x$ **10.** $\dfrac{2(4x - x)}{3} - \dfrac{3(6x - x)}{3}$

11. $\dfrac{8(5x + 2x)}{7} - \dfrac{2(3x + x)}{4}$ **12.** $\dfrac{6(5x - x)}{8} + \dfrac{5(x + 5x)}{3}$

13. $2x + \left[9x - 4(3x + 2) - 7 \right]$ **14.** $7x - 3\left[4 - (6x - 1) + x \right]$

15. $6x^2 - \left[9 - 2\left(3x^2 - 1 \right) + 7x^2 \right]$ **16.** $2x^2 + \left[4x^2 - \left(8x^2 - 3x \right) + \left(2x^2 + 7x \right) \right]$

Solve each of the equations in Exercises 17 – 60.

17. $7x - 4 = 17$ **18.** $9x + 6 = -21$ **19.** $4 - 3x = 19$

20. $18 + 11x = 23$ **21.** $6x - 2.5 = 1.1$ **22.** $5x + 4.06 = 2.31$

23. $7x + 3.4 = -1.5$ **24.** $8.2 = 2.6 + 8x$ **25.** $7x - 6 = 2x + 9$

26. $x - 7 = 4x + 11$ **27.** $\dfrac{x}{5} - 1 = -6$ **28.** $\dfrac{3x}{4} + 11 = 20$

Answers to Practice Problems: **1.** $4x^2 - 3x$ **2.** $x = 5$ **3.** $x = 4$ **4.** $x = 5$ **5.** $x = 4.6$ or $x = -3.6$

29. $\dfrac{2x}{3} - 4 = 8$ **30.** $\dfrac{5x}{4} + 1 = 11$ **31.** $2(x - 2) = 2x - 4$

32. $3x - 7 = 4(x + 3)$ **33.** $3(2x - 3) = 4x + 5$ **34.** $3(x - 1) = 4x + 6$

35. $4(3 - 2x) = 2(x - 4)$ **36.** $-3(x + 5) = 6(x + 2)$ **37.** $4x + 3 - x = 3x - 9$

38. $5x + 13 = x - 8 - 3x$ **39.** $x + 7 = 6x + 4 - x$ **40.** $x - 9 + 5x = 2x - 3$

41. $x - 2 = \dfrac{x + 2}{4} + 5$ **42.** $x + 8 = \dfrac{x}{3}$ **43.** $\dfrac{4x}{5} + 2 = 2x - 4$

44. $\dfrac{3x}{2} + 1 = x - 1$ **45.** $0.8x + 6.2 = 0.2x - 1.0$ **46.** $2.4x - 8.5 = 1.1x + 0.6$

47. $2.5x + 2.0 = 0.7x + 5.6$ **48.** $3.2x + 9.5 = 1.8x - 1.7$

49. $12x - (4x - 6) = 3x - (9x - 27)$ **50.** $5(x - 3) - 3 = 2x - 6(2 - x)$

51. $\dfrac{x}{4} + 2 = \dfrac{3x}{2} - 3$ **52.** $\dfrac{2x + 1}{8} - \dfrac{1}{4} = \dfrac{x - 3}{2} + 1$ **53.** $\dfrac{x}{2} - \dfrac{2x}{3} = \dfrac{3}{4} + \dfrac{x}{3}$

54. $\dfrac{1}{3}x + \dfrac{1}{4} = \dfrac{1}{5}x + \dfrac{1}{6}$ **55.** $5(x - 4) = 3(4x - 7) - 2(3x + 4)$

56. $5(x + 1) - 4(3 - x) = 2x - 7(1 - x)$ **57.** $2(4 - x) - (3x + 2) = 7 + 4(x - 7)$

58. $3(x - 1) - (4x + 2) = 3\big[(2x - 1) - 2(x + 3)\big]$

59. $3 - x = \dfrac{1}{4}(7 - x) - \dfrac{1}{3}(2x - 3)$ **60.** $\dfrac{3}{2}(x - 1) = \dfrac{1}{2}(x - 3) - \dfrac{7}{10}$

In Exercises 61 – 70, determine whether each of the equations is a conditional equation, an identity, or a contradiction.

61. $2x + 3x = 17.4 - x$ **62.** $2(3x - 1) + 5 = 3$

63. $7(x - 1) = -3(3 - x) + 4x$ **64.** $5x + 13 = -2(x - 7) + 3$

65. $5x + 12 - 9x = -4(x - 3) - x$ **66.** $5.2x + 3.4x = 0.2(x - 0.42)$

67. $\dfrac{1}{2}(x - 24) = \dfrac{1}{3}(x - 24)$ **68.** $4(3x - 5) = x + 3(x - 1) + 10$

69. $3(x - 2) + 4x = 6(x - 1) + x$ **70.** $\dfrac{1}{4}(2x + 1) - 7 = \dfrac{1}{2}(2x - 1)$

Solve each of the absolute value equations in Exercises 71 – 90.

71. $|x| = 8$

72. $|x| = 6$

73. $|z| = -\dfrac{1}{5}$

74. $|z| = \dfrac{1}{5}$

75. $|x + 3| = 2$

76. $|y + 5| = -7$

77. $|x - 4| = \dfrac{1}{2}$

78. $|3x + 1| = 8$

79. $|5x - 2| + 4 = 7$

80. $|2x - 7| - 1 = 0$

81. $3\left|\dfrac{x}{3} + 1\right| - 5 = -2$

82. $2\left|\dfrac{x}{4} - 3\right| + 6 = 10$

83. $|2x - 1| = |x + 2|$

84. $|2x - 5| = |x - 3|$

85. $|3x + 1| = |4 - x|$

86. $|5x + 4| = |1 - 3x|$

87. $\left|\dfrac{3x}{2} + 2\right| = \left|\dfrac{x}{4} + 3\right|$

88. $\left|\dfrac{x}{3} - 4\right| = \left|\dfrac{5x}{6} + 1\right|$

89. $\left|\dfrac{2x}{5} - 3\right| = \left|\dfrac{x}{2} - 1\right|$

90. $\left|\dfrac{4x}{3} + 7\right| = \left|\dfrac{x}{4} + 2\right|$

Writing and Thinking About Mathematics

91. Discuss the circumstances in which an equation would have an infinite number of solutions.

92. Discuss the circumstances in which an equation would have no solution.

 HAWKES LEARNING SYSTEMS: INTERMEDIATE ALGEBRA SOFTWARE

- Simplifying Expressions
- Solving Linear Equations
- Solving Absolute Value Equations

1.5   Evaluating and Solving Formulas

- Solve applied problems by using known formulas.
- Solve formulas for specified variables in terms of the other variables.

Using Formulas

A **formula** is an equation that represents a general relationship between two or more quantities or measurements. Several variables may appear in a formula, and the formula is not always in the most convenient form for application in some word problems. In such situations, we may want to solve a formula for a particular variable and use the formula in a different form. For example, the formula $d = rt$ (distance equals rate times time) is solved for d. Solving for r or t gives

$$r = \frac{d}{t} \quad \text{or} \quad t = \frac{d}{r}.$$

Formulas are useful in many fields of study, such as business, economics, medicine, physics, technology, and chemistry, as well as mathematics. Some formulas and their meanings are shown here and will be used with others in the exercises.

Formula	Meaning
1. $I = Prt$	The simple interest (I), earned by investing money, is equal to the product of the principal (P) times the rate of interest (r) times the time (t) in years.
2. $C = \frac{5}{9}(F - 32)$	Temperature in degrees Celsius (C) equals $\frac{5}{9}$ times the difference between the Fahrenheit temperature (F) and 32.
3. $IQ = \frac{100M}{C}$	Intelligence Quotient (IQ) is calculated by multiplying 100 times mental age (M), as measured by some test, and dividing by chronological age (C).
4. $\alpha + \beta + \gamma = 180°$	The sum of the measures of the angles (α, β, γ) of a triangle is 180°. (**Note:** α, β, and γ are the Greek letters alpha, beta, and gamma, respectively.)

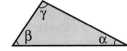

Table 1.7

Formulas for the perimeter (P) and the area (A) of geometric figures are as follows.

Formula	Figure Name	Figure
1. $P = 4s$ $A = s^2$	SQUARE s = side	
2. $P = 2l + 2w$ $A = lw$	RECTANGLE l = length w = width	
3. $P = 2a + 2b$ $A = bh$	PARALLELOGRAM b = base a = side h = height	
4. $P = a + b + c$ $A = \dfrac{1}{2}bh$	TRIANGLE b = base a, c = sides h = height	
5. $C = 2\pi r$ or $C = \pi d$ $A = \pi r^2$	CIRCLE r = radius d = diameter C = circumference or perimeter of a circle	
6. $P = a + b + c + d$ $A = \dfrac{1}{2}h(b + c)$	TRAPEZOID b, c = parallel sides a, d = other sides h = height	

Table 1.8

Evaluating Formulas

If the values for all but one variable in a formula are known, they can be substituted into the formula and the unknown value can be found by solving the equation as we did in Section 1.4. This is essentially how formulas are used in solving applications.

Example 1: Triangle

The perimeter of a triangle is 38 feet. One side is 5 feet long and a second side is 18 feet long. How long is the third side?

Solution 1:   Using the formula $P = a + b + c$, substitute $P = 38, a = 5$, and $b = 18$. Then solve for the third side.

$$38 = 5 + 18 + c$$
$$38 = 23 + c$$
$$15 = c$$

The third side is 15 feet long.

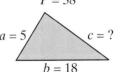

Solution 2:   First, solve for c in terms of P, a, and b. Then substitute for P, a, and b.

$$P = a + b + c$$
$$P - a - b = c \qquad \text{Treat } a \text{ and } b \text{ as constants.}$$
$$\text{Add } -a - b \text{ to both sides.}$$

or

$$c = P - a - b$$

Substituting gives

$$c = 38 - 5 - 18$$
$$= 33 - 18 = 15.$$

Solving Formulas

Many times, we simply want the formula solved for one of the variables without substituting in particular values. In this situation, we treat all the other variables as if they were constants and follow the same procedure for solving an equation as we did in Section 1.4. Thus, in Solution 2, we solved the formula $P = a + b + c$ for c to obtain $c = P - a - b$.

Example 2: Solving Formulas

a. Given $C = \dfrac{5}{9}(F-32)$, solve for F in terms of C.

Solution:

$C = \dfrac{5}{9}(F-32)$	Treat C as a constant.
$\dfrac{9}{5}C = \dfrac{9}{5} \cdot \dfrac{5}{9}(F-32)$	Multiply both sides by $\dfrac{9}{5}$.
$\dfrac{9}{5}C = F-32$	Simplify.
$\dfrac{9}{5}C + 32 = F-32+32$	Add 32 to both sides.
$\dfrac{9}{5}C + 32 = F$	Simplify.

Thus the formula for Celsius and Fahrenheit solved for C is

$$C = \frac{5}{9}(F-32),$$

and the same formula solved for F is

$$F = \frac{9}{5}C + 32.$$

These are two forms of the same formula.

b. Solve for l given $P = 2l + 2w$.

Solution:

$P = 2l + 2w$	
$P - 2w = 2l + 2w - 2w$	Add $-2w$ to both sides.
$P - 2w = 2l$	Simplify.
$\dfrac{P-2w}{2} = \dfrac{2l}{2}$	Divide both sides by 2.
$\dfrac{P-2w}{2} = l$	Simplify.

c. Solve the formula $y = mx + b$ for x.

Solution:

$y = mx + b$	
$y - b = mx$	Add $-b$ to both sides.
$\dfrac{y-b}{m} = \dfrac{mx}{m}$	Divide both sides by m.
$\dfrac{y-b}{m} = x$	Simplify.

d. Solve for R given the formula $F = \dfrac{1}{R+r}$.

Solution:

$$F = \frac{1}{R+r}$$

$$(R+r)F = (R+r)\frac{1}{R+r} \qquad \text{Multiply both sides by the denominator, } R+r.$$

$$RF + rF = 1 \qquad\qquad\qquad \text{Use the distributive property.}$$

$$RF = 1 - rF \qquad\qquad\quad \text{Add } -rF \text{ to both sides.}$$

$$\frac{RF}{F} = \frac{1-rF}{F} \qquad\qquad \text{Divide both sides by } F.$$

$$R = \frac{1-rF}{F}$$

This problem illustrates the importance of writing the correct form of a variable in a formula. Note that the uppercase R and the lowercase r represent completely different quantities.

Practice Problems

In each formula, solve for the indicated variable.

1. $P = a + 2b$; *solve for b.*

2. $y = mx + b$; *solve for m.*

3. $\alpha + \beta + \gamma = 180$; *solve for α.*

4. $F = \dfrac{1}{R+r}$; *solve for r.*

5. *The perimeter of a square is 12.8 cm. Find the length of the sides.*

1.5 Exercises

For Exercises 1 – 10, (a) determine the formula that relates the given information and solve it for the unknown quantity, and (b) substitute the given values in the formula to determine the value of the unknown quantity.

1. Investing: The interest earned in 2 years on an investment is $297. If the rate of interest is 9%, find the amount invested.

2. Temperature: The Celsius temperature is 45°. Find the Fahrenheit temperature.

Answers to Practice Problems: **1.** $b = \dfrac{P-a}{2}$ **2.** $m = \dfrac{y-b}{x}$ **3.** $\alpha = 180 - \beta - \gamma$ **4.** $r = \dfrac{1-FR}{F}$ **5.** 3.2 cm

3. **Triangles:** Two angles of a triangle measure 72° and 65°. Find the measure of the third angle.

4. **Squares:** The perimeter of a square is $10\frac{2}{3}$ meters. Find the length of the sides.

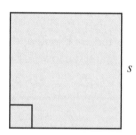

5. **Rectangles:** The perimeter of a rectangle is 88 feet. If the length is 31 feet, find the width.

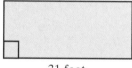

31 feet

6. **Parallelograms:** The area of a parallelogram is 1081 square inches. If the height is 23 inches, find the length of the base.

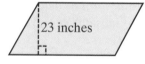

23 inches

7. **Triangles:** The perimeter of a triangle is 147 inches. Two of the sides measure 38 inches and 48 inches. Find the length of the third side.

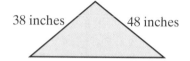

38 inches   48 inches

8. **Circles:** The circumference of a circle is 26π centimeters. Find the radius.

9. **Circles:** The radius of a circle is 14 feet. Find the area.

14 feet

10. **Trapezoids:** The area of a trapezoid is 51 square meters. One base is 7 meters long, and the other is 10 meters long. Find the height of the trapezoid.

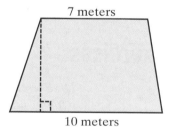

7 meters

10 meters

Solve for the indicated variables in Exercises 11 – 55.

11. $P = a + b + c$; solve for b.
12. $P = 3s$; solve for s.
13. $f = ma$; solve for m.

14. $C = \pi d$; solve for d.
15. $A = lw$; solve for w.
16. $P = R - C$; solve for C.

17. $R = np$; solve for n.   **18.** $v = k + gt$; solve for k.   **19.** $I = A - p$; solve for p.

20. $L = 2\pi rh$; solve for h.   **21.** $A = \dfrac{m+n}{2}$; solve for m.   **22.** $W = Rl^2t$; solve for R.

23. $P = 4s$; solve for s.   **24.** $C = 2\pi r$; solve for r.   **25.** $d = rt$; solve for t.

26. $P = a + 2b$; solve for a.   **27.** $I = Prt$; solve for t.   **28.** $R = \dfrac{E}{I}$; solve for E.

29. $P = a + 2b$; solve for b.   **30.** $c^2 = a^2 + b^2$; solve for b^2.

31. $S = \dfrac{a}{1-r}$; solve for a.   **32.** $A = \dfrac{h}{2}(a+b)$; solve for h.

33. $y = mx + b$; solve for x.   **34.** $V = lwh$; solve for h.

35. $A = 4\pi r^2$; solve for r^2.   **36.** $V = \pi r^2 h$; solve for h.

37. $IQ = \dfrac{100M}{C}$; solve for M.   **38.** $A = \dfrac{R}{2L}$; solve for R.

39. $V = \dfrac{1}{3}\pi r^2 h$; solve for h.   **40.** $A = \dfrac{1}{2}bh$; solve for b.

41. $R = \dfrac{E}{I}$; solve for I.   **42.** $IQ = \dfrac{100M}{C}$; solve for C.

43. $A = \dfrac{R}{2L}$; solve for L.   **44.** $K = \dfrac{mv^2}{2g}$; solve for g.

45. $A = \dfrac{h}{2}(a+b)$; solve for b.   **46.** $L = a + (n-1)d$; solve for d.

47. $Q = \dfrac{w_0 L}{R}$; solve for R.   **48.** $U_m = \dfrac{1}{2}LI^2$; solve for L.

49. $S = \dfrac{a}{1-r}$; solve for r.   **50.** $P = \dfrac{A}{1+ni}$; solve for n.

51. $W = \dfrac{2PR}{R-r}$; solve for P.   **52.** $V^2 = v^2 + 2gh$; solve for g.

53. $I = \dfrac{nE}{R+nr}$; solve for R.   **54.** $A = P + Prt$; solve for t.

55. $S = \dfrac{rL-a}{b-a}$; solve for r.

HAWKES LEARNING SYSTEMS: INTERMEDIATE ALGEBRA SOFTWARE

- Evaluating Formulas
- Solving Formulas

1.6 Applications

Solve the following by using first degree equations:

- *Number problems,*
- *Distance-rate-time problems,*
- *Cost-profit problems,*
- *Simple interest problems, and*
- *Average problems.*

Word problems (or applications) are designed to teach you to read carefully, to organize, and to think clearly. Whether or not a particular problem is easy for you depends a great deal on your personal experiences and general reasoning abilities. The problems generally do not give specific directions to add, subtract, multiply, or divide. You must decide what relationships are indicated through careful analysis of the problem.

George Pólya (1887-1985), a famous professor at Stanford University, studied the process of discovery learning. Among his many accomplishments, he developed the following four-step process as an approach to problem solving.

1. **Understand the problem.**
2. **Devise a plan.**
3. **Carry out the plan.**
4. **Look back over the results.**

For a complete discussion of these ideas, see *How to Solve It* by Pólya (Princeton University Press, 1945, 2nd edition, 1957). The following strategy is recommended for all word problems involving one variable and one equation.

Strategy for Solving Word Problems

1. *Understand the problem.*
   a. *Read the problem carefully. (Read it several times if necessary.)*
   b. *If it helps, restate the problem in your own words.*

2. *Devise a plan.*
   a. *Decide what is asked for. Assign a variable to the unknown quantity. Label this variable so you know exactly what it represents.*
   b. *Draw a diagram or set up a chart whenever possible.*
   c. *Write an equation that relates the information provided.*

Continued on the next page...

Strategy for Solving Word Problems (cont.)

  3. *Carry out the plan.*

   a. *Study your picture or diagram for insight into the solution.*

   b. *Solve the equation.*

  4. *Look back over the results.*

   a. *Does your solution make sense in terms of the wording of the problem?*

   b. *Check your solution in the equation.*

Problems involving numerical expressions will usually contain key words indicating the operations to be performed. Learn to look for words such as those in the following list.

Addition	Subtraction	Multiplication	Division	Equality
add	subtract	multiply	divide	gives
sum	difference	product	quotient	represents
plus	minus	times	ratio	amounts to
more than	less than	twice		is / was
increased by	decreased by	of (with fractions and percents)		is the same as

Table 1.9

Example 1: Number Problem

The sum of two numbers is 36. If $\frac{1}{2}$ of the smaller number is equal to $\frac{1}{4}$ of the larger number, find the two numbers.

Solution: Analyze the problem and identify the key words.

The key words are **sum** (indicating addition) and **of** (indicating multiplication when used with fractions).

Assign variables to the unknown quantities.

Let $\qquad\qquad\qquad\qquad x =$ smaller number.
Since $\quad x + ($larger number$) = 36,$
$\qquad\qquad\qquad 36 - x =$ larger number.

Continued on the next page...

Write an equation relating the given information.

$\frac{1}{2}$ of the smaller		$\frac{1}{4}$ of the larger
number	is equal to	number
$\downarrow$	$\downarrow$	$\downarrow$
$\frac{1}{2}x$	$=$	$\frac{1}{4}(36-x)$

Solve the equation.

$$\frac{1}{2}x = \frac{1}{4}(36-x)$$

$$4 \cdot \frac{1}{2}x = 4 \cdot \frac{1}{4}(36-x)$$ Multiplying both sides of the equation by the LCD 4 yields integer coefficients.

$$2x = 1(36-x)$$

$$2x = 36 - x$$

$$3x = 36$$

$$x = 12 \qquad \text{Smaller number}$$

$$36 - x = 24 \qquad \text{Larger number}$$

Check: $12 + 24 \overset{?}{=} 36$

$36 = 36$

$$\frac{1}{2}(12) \overset{?}{=} \frac{1}{4}(24)$$

$$6 = 6$$

The two numbers are 12 and 24.

Problems involving distance usually make use of the relationship indicated by the formula $d = rt$, where d = distance, r = rate, and t = time. A chart or table showing the known and unknown values is quite helpful and is illustrated in the next example.

Example 2: Distance-Rate-Time

A motorist averaged 45 mph for the first part of a trip and 54 mph for the last part of the trip. If the total trip of 303 miles took 6 hours, what was the time for each part?

Solution:

Analysis of Strategy:

What is being asked for?
Total time minus time for 1st part of trip gives time for 2nd part of trip.

Let t = time for 1st part of trip

$6 - t$ = time for 2nd part of trip

	rate	·	time	=	distance
1st Part	45		t		$45 \cdot t$
2nd Part	54		$6 - t$		$54(6 - t)$

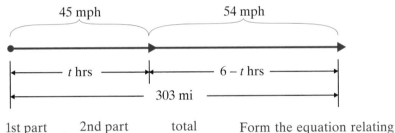

45 mph   54 mph

t hrs $6 - t$ hrs

303 mi

1st part distance	+	2nd part distance	=	total distance	Form the equation relating the given information.

$$45t + 54(6 - t) = 303 \qquad \text{Solve the equation.}$$
$$45t + 324 - 54t = 303$$
$$324 - 9t = 303$$
$$-9t = -21$$
$$t = \frac{-21}{-9} = \frac{7}{3} \qquad \text{1st part of the trip}$$
$$6 - t = 6 - \frac{7}{3} = \frac{11}{3} \qquad \text{2nd part of the trip}$$

Check: $45 \cdot \dfrac{7}{3} = 15 \cdot 7 = 105$ miles (1st part)

$54 \cdot \dfrac{11}{3} = 18 \cdot 11 = 198$ miles (2nd part)

$105 + 198 = 303$ miles total

The first part took $\dfrac{7}{3}$ hr or $2\dfrac{1}{3}$ hr. The second part took $\dfrac{11}{3}$ hr or $3\dfrac{2}{3}$ hr.

Problems involving cost come in a variety of forms. The next two examples illustrate more situations you may encounter in your own life.

Example 3: Cost

a. The Berrys sold their house. After paying the real estate agent a commission of 6% of the selling price and then paying $1486 in other costs and $90,000 on the mortgage, they received $49,514. What was the selling price of the house?

Solution: Use the relationship: selling price $-$ cost $=$ profit $(S - C = P)$.

Let $S =$ selling price
        cost $= 0.06s + 1486 + 90,000$

$$
\begin{array}{ccc}
\underbrace{\text{selling price}} & - & \underbrace{\text{cost}} & = & \underbrace{\text{profit}} \\
\downarrow & & \downarrow & & \downarrow
\end{array}
$$

$$
\begin{aligned}
S - (0.06S + 1486 + 90,000) &= 49{,}514 \\
S - 0.06S - 1486 - 90{,}000 &= 49{,}514 \\
0.94S &= 141{,}000 \\
S &= 150{,}000
\end{aligned}
$$

Check:

$$
\begin{array}{ll}
\$150{,}000 & \text{selling price} \\
\times \quad 0.06 & \text{commission \%} \\
\hline
\$9000 & \text{commission}
\end{array}
\qquad
\begin{array}{ll}
\$9000 & \text{commission} \\
1486 & \text{other costs} \\
+\,90{,}000 & \text{mortgage} \\
\hline
\$100{,}486 & \text{cost}
\end{array}
\qquad
\begin{array}{ll}
\$150{,}000 & \text{selling price} \\
-100{,}486 & \text{cost} \\
\hline
\$49{,}514 & \text{profit}
\end{array}
$$

The selling price was $150,000.

b. A jeweler paid $350 for a ring. He wants to price the ring for sale so that he can give a 30% discount on the marked selling price and still make a profit of 20% on his cost. What should be the marked selling price of the ring?

Solution: Again, we make use of the relationship $SP - C = P$.

Let x = marked selling price,
then $x - 0.30x$ = actual selling price
and 350 = cost.

$$\underbrace{\frac{\text{actual}}{\text{selling price}}}_{\downarrow} \quad - \quad \underbrace{\text{cost}}_{\downarrow} \quad = \quad \underbrace{\text{profit}}_{\downarrow}$$

$$x - 0.30x \quad - \quad 350 \quad = \quad 0.20(350)$$
$$0.7x \quad - \quad 350 \quad = \quad 70$$
$$0.7x \quad = \quad 420$$
$$x \quad = \quad 600$$

The actual selling price is the marked selling price minus the 30% discount on the ring.

The profit is 20% of what he paid originally.

Check: **Step 1:**
$600 marked selling price
$\times\,0.30$ discount %
─────────
$180 discount

Step 2:
$600 marked selling price
$-\,$180 discount
─────────
$420 actual selling price

Step 3:
$420 actual selling price
$-\,$350 cost
─────────
$70 profit

As a double check, $350 cost
$\times\,0.20$ profit %
─────────
$70 profit.

The jeweler should set the selling price at $600.

To work problems related to interest on money invested for one year, you need to know the basic relationship between the principal P (amount invested), the annual rate of interest r, and the amount of interest I (money earned). This relationship is described in the formula $P \cdot r = I$. (This is the formula for simple interest, $I = Prt$, with $t = 1$.) We use this relationship in Example 4.

Example 4: Interest

Kara has had \$40,000 invested for one year, some with a savings account which paid 7%, the rest in a high-risk stock which yielded 12% for the year. If her interest income last year was \$3550, how much did she have in the savings account and how much did she invest in the stock?

Solution: Let
$$x = \text{amount invested at } 7\%$$
$$40,000 - x = \text{amount invested at } 12\%$$

Total amount invested minus amount invested at 7% represents amount invested at 12%.

	principal	·	rate	=	interest
Savings Account	x		0.07		$0.07(x)$
Stock	$40,000 - x$		0.12		$0.12(40,000 - x)$

$$\underbrace{\text{interest at } 7\%}_{} + \underbrace{\text{interest at } 12\%}_{} = \underbrace{\text{total interest}}_{}$$

$$0.07(x) \; + \; 0.12(40,000 - x) \; = \quad 3550$$

Multiply both sides of the equation by 100 to eliminate the decimal.

$$7x + 12(40,000 - x) = \quad 355{,}000$$
$$7x + 480{,}000 - 12x = \quad 355{,}000$$
$$-5x = \quad -125{,}000$$
$$x = \quad 25{,}000 \quad \text{Amount invested at } 7\%$$
$$40{,}000 - x = \quad 15{,}000 \quad \text{Amount invested at } 12\%$$

Check: $25{,}000(0.07) = 1750$ and $15{,}000(0.12) = 1800$
and \$1750 + \$1800 = \$3550.
Kara had \$25,000 in the savings account at 7% interest and invested \$15,000 in the stock at 12% interest.

Average (or Mean)

You are probably already familiar with the concept of the **average** of a set of numbers. The average is also called the **arithmetic average** or **mean**. Your grade in most courses is related to an "average" of your exam scores. Magazines and newspapers report average income, average price of homes, average sales, batting averages, and so on. The mean is particularly important in the study of statistics. For example, traffic studies are interested in average speeds, census studies are concerned with the mean number of people living in a house, and universities study the mean test scores of incoming freshmen students.

Average

*The **average** (or **mean**) of a set of numbers is the value found by adding the numbers and then dividing the sum by the quantity of numbers in the set.*

Example 5: Average (or Mean)

Suppose that you have scores of 85, 92, 82 and 88 on four exams in your English class. What score will you need on the fifth exam to have an average of 90?

Solution: Let x = your score on the fifth exam.
The sum of all the scores, including the unknown fifth exam, divided by 5 must equal 90.

$$\frac{85 + 92 + 82 + 88 + x}{5} = 90$$

$$\frac{347 + x}{5} = 90$$

$$5 \cdot \frac{347 + x}{5} = 5 \cdot 90$$

$$347 + x = 450$$

$$x = 103$$

Assuming that each exam is worth 100 points, you cannot attain an average of 90 on the five exams.

Example 6: Bar Graphs

Given the enrollment at the main campuses of the following Big Ten Universities:
a. Find the average enrollment over the six schools. (Round to the nearest thousand.)
b. Find the university with the lowest enrollment.
c. Find the difference in enrollment between Ohio State and Penn State.

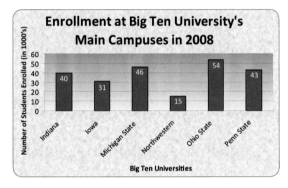

Solution: Note that the units on the graph are in thousands.

a. Find the sum: $40 + 31 + 46 + 15 + 54 + 43 = 229$.
Divide by 6: $229 \div 6 \approx 38$.
The average enrollment is 38,000 students.

b. Northwestern has the lowest enrollment: 15,000 students.

c. The difference is $54,000 - 43,000 = 11,000$ students.

1.6 Exercises

Refer to the formulas listed in Section 1.5 as necessary for Exercises 1 – 44.

1. If 15 is added to a number, the result is 56 less than twice the number. Find the number.

2. A number subtracted from 20 is equal to three times the number. Find the number.

3. Nine less than twice a number is equal to the number. What is the number?

4. What number gives a result of −2 when 5 is subtracted from the quotient of the number and 4?

5. Find a number such that −64 is 12 more than four times the number.

6. If 6 is added to the quotient of a number and 3, the result is 1. What is the number?

7. Seven times a certain number is equal to the sum of three times the number and 28. What is the number?

8. Twelve more than five times a number is equal to the difference between 5 and twice the number. Find the number.

9. Four added to the quotient of a number and 6 is equal to 11 less than the number. What is the number?

10. The quotient of twice a number and 8 is equal to 3 more than the number. What is the number?

11. One number is 6 more than three times another. If their sum is 38, what are the two numbers?

12. The sum of two numbers is 98 and their difference is 20. Find the two numbers.

13. **Fencing a yard:** The length of a rectangular-shaped backyard is 8 feet less than twice the width. If 260 feet of fencing is needed to enclose the yard, find the dimensions of the yard.

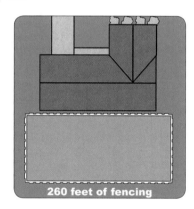

14. **Shopping:** The price of a pair of pants is reduced 15%. The sale price is $39.95. Find the original price.

15. **Salary:** After a raise of 8%, Juan's salary is $1620 per month. What was his salary before the increase?

16. **Rental cars:** The U-Drive Company charges $20 per day plus 22¢ per mile driven. For a one-day trip, Louis paid a rental fee of $66.20. How many miles did he drive?

17. **Telephone calls:** For a long-distance call, the telephone company charges 35¢ for each of the first three minutes and 15¢ for each additional minute. If the cost of a call was $12.30, how many minutes did the call last?

18. **Shopping:** Willis bought a shirt and necktie for $85. The shirt cost $15 more than the tie. Find the cost of each.

19. **Plane speeds:** Two planes which are 2475 miles apart fly toward each other. Their speeds differ by 75 mph. If they pass each other in 3 hours, what is the speed of each?

	rate ·	time =	distance
1st plane	r	3	
2nd plane	r + 75	3	

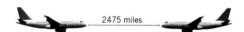
2475 miles

20. Biking: Jane rides her bike to Lake Junaluska. Going to the lake, she averages 12 mph. On the return trip, she averages 10 mph. If the round trip takes a total of 5.5 hours, how long does the return trip take?

	rate ·	time =	distance
Going	12	$5.5 - t$	
Returning	10	t	

21. Travel by car: Marcus drives from one town to another in 6 hours. On the return trip, his speed is increased by 10 mph and the trip takes 5 hours. Find his rate on the return trip. How far apart are the towns?

	rate ·	time =	distance
Going	r	6	
Returning	$r + 10$	5	

22. Travel time: The Reeds are moving across the state. Mr. Reed leaves $3\frac{1}{2}$ hours before Mrs. Reed. If he averages 40 mph and she averages 60 mph, how long will Mrs. Reed have to drive before she overtakes Mr. Reed?

23. Hiking: Carol has 8 hours to spend on a mountain hike. She can walk up the trail at an average of 2 mph and can walk down at an average of 3 mph. How long should she plan to hike uphill before turning around?

24. Car speed: After traveling for 40 minutes, Mr. Koole had to slow to $\frac{2}{3}$ his original speed for the rest of the trip due to heavy traffic. The total trip of 84 miles took 2 hours. Find his original speed.

25. Train speed: A train leaves Los Angeles at 2:00 PM. A second train leaves the same station in the same direction at 4:00 PM. The second train travels 24 mph faster than the first. If the second train overtakes the first at 7:00 PM, what is the speed of each of the two trains?

26. Running: Maria runs through the countryside at a rate of 10 mph. She returns along the same route at 6 mph. If the total trip took 1 hour 36 minutes, how far did she run in total?

27. Shoe sales: A particular style of shoe costs the dealer $81 per pair. At what price should the dealer mark them so he can sell them at a 10% discount off the selling price and still make a 25% profit?

28. Farming: Farmer McGregor raises strawberries. They cost him $0.80 a basket to produce. He is able to sell only 85% of those he produces. If he sells his strawberries at $2.40 a basket, how many baskets must he produce to make a profit of $2480?

29. Ice cream sales: A grocery store bought ice cream for $2.60 a half gallon and stored it in two freezers. During the night, one freezer malfunctioned and ruined 15 half gallons. If the remaining ice cream is sold for $3.98 per half gallon, how many half gallons did the store buy if it made a profit of $64.50?

30. Cabinetry: Craig builds cabinets in his spare time. Good quality cabinet plywood costs $4.00 per square foot. There is approximately a 10% waste of material due to cutting and fitting. He also figures $240 per month for finishing material, glue, tools, etc. If he charges $9.20 per square foot of finished cabinet, how many square feet of plywood would he have to buy to have a profit of $1129.60 in one month?

31. Farming: A citrus farmer figures that his fruit costs 96¢ a pound to grow. If he lost 20% of the crop he produced due to a frost and he sold the remaining 80% at $1.80 a pound, how many pounds did he produce to make a profit of $30,000?

32. Stock market: Mr. Wise bought $1950 worth of stock, some at $3.00 per share and some at $4.50 per share. If he bought a total of 450 shares of stock, how many of each did he buy?

33. Surfing: Last summer, Ernie sold surfboards. One style sold for $300 and the other sold for $250. He sold a total of 44 surfboards. How many of each style did he sell if the sales from each style were equal?

34. Golfing: The pro shop at the Divots Country Club ordered two brands of golf balls. Titleless balls cost $1.80 each and the Done Lob balls cost $1.50 each. The total cost of Titleless balls exceeded the total cost of the Done Lob balls by $108. If equal numbers of each brand were ordered, how many dozen of each brand were ordered?

35. Investing: Amanda invests $25,000, part at 5% and the rest at 6%. The annual return on the 5% investment exceeds the annual return on the 6% investment by $40. How much did she invest at each rate?

36. Investing: The annual interest earned on a $6000 investment was $120 less than the interest earned on $10,000 invested at 1% less interest per year. What was the rate of interest on each amount?

37. Investing: The annual interest on a $4000 investment exceeds the interest earned on a $3000 investment by $80. The $4000 is invested at a 0.5% higher rate of interest than the $3000. What is the interest rate of each investment?

38. Investing: Mr. Hill invests ten thousand dollars, part at 5.5% and part at 6%. The interest from the 5.5% investment exceeds the interest from the 6% investment by $251. How much did he invest at each rate?

39. Investing: Two investments totaling $16,000 produce an annual income of $1140. One investment yields 6% a year, while the other yields 8% per year. How much is invested at each rate?

40. Real estate: Sellit Realty Company gets a 6% fee for selling improved properties and 10% for selling unimproved land. Last week, the total sales were $220,000 and their total fees were $16,400. What were the sales from each of the two types of properties?

41. Temperature: Given the monthly temperatures over a year for Christchurch, New Zealand:

   a. Find the average temperature for the year. (Round to the nearest tenth.)

   b. Find the minimum temperature for the year.

   c. Find the difference in temperature between the months of June and December.

Source: http://www.weather.com

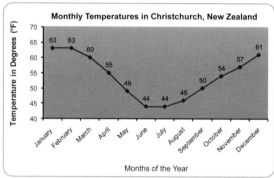

42. Average rainfall: Given the monthly rainfall averages over a year for Visakhapatnam, India:

   a. Find the average rainfall for the year. (Round to the nearest tenth.)

   b. Find the maximum rainfall for the year.

   c. Find the difference in rainfall between the months of October and December.

Source: http://www.weather.com

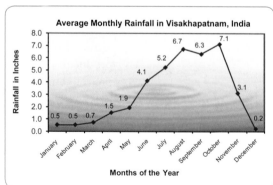

43. Company profits: Given the yearly profits of the world's largest companies in 2008:

a. Find the average profits of the companies in the year 2008 (round to the nearest tenth).

b. Find the yearly profits of Toyota Motor in the year 2008.

c. Which two companies had the same yearly profits in 2008?

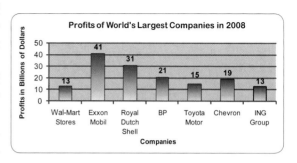

Source: http://money.cnn.com/magazines/fortune/global500/2008/full_list/

44. Airport usage: Given the number of passengers at the following airports:

a. Find the average number of passengers.

b. Find the difference in passengers between Atlanta, GA and Tokyo.

c. What was the total number of passengers to go through London?

Source: Airports Council International – North America

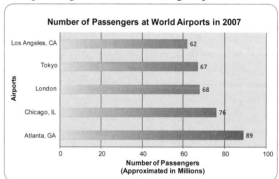

Writing and Thinking About Mathematics

45. List the four steps in Pólya's approach to problem solving. Then, in your own words, discuss how you used these steps in solving a "problem" you have had recently. Did you have trouble finding your car keys this morning? How did you decide what movie to see last weekend?

 HAWKES LEARNING SYSTEMS: INTERMEDIATE ALGEBRA SOFTWARE

- Applications

1.7 Solving Inequalities in One Variable

- *Understand and use **interval notation**.*
- *Solve linear inequalities.*
- *Solve compound inequalities.*
- *Solve absolute value inequalities.*

Intervals of Real Numbers

Suppose that a and b are two real numbers and that $a < b$. The set of all real numbers between a and b is called an **interval of real numbers**. Intervals of real numbers are used in relationship to everyday concepts such as the length of time you talk on your cell phone, the shelf life of chemicals or medicines, and the speed of an airplane.

As an aid in reading inequalities and graphing inequalities correctly, note that an inequality may be read either from left to right or right to left. Because we are concerned about which numbers satisfy an inequality, we **read the variable first**:

$$x > 7 \quad \text{is read from \textbf{left to right} as ``}x\text{ is greater than 7.''}$$
$$\text{and} \quad 7 < x \quad \text{is read from \textbf{right to left} as ``}x\text{ is greater than 7.''}$$

A compound interval such as $-3 < x < 6$ is read

"x is greater than -3 and x is less than 6."

Again, note that *the variable is read first*.

Graphs of real numbers on real number lines and inequalities with the symbols $<, >, \leq,$ and $\geq$ were discussed in Section 1.2. Various types of intervals and their corresponding **interval notation** are listed in Table 1.10.

Types of Intervals and Interval Notation

Type of Interval	Algebraic Notation	Interval Notation	Graph
Open Interval	$a < x < b$	(a, b)	
Closed Interval	$a \leq x \leq b$	$[a, b]$	

Half-open Interval	$\begin{cases} a \le x < b \\ a < x \le b \end{cases}$	$[a, b)$ $(a, b]$	
Open Interval	$\begin{cases} x > a \\ x < b \end{cases}$	(a, ∞) $(-\infty, b)$	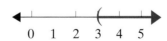
Half-open Interval	$\begin{cases} x \ge a \\ x \le b \end{cases}$	$[a, \infty)$ $(-\infty, b]$	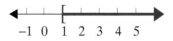

Table 1.10

NOTES

The symbol for infinity ∞ (or $-\infty$) is not a number. It is used to indicate that the interval is to include all real numbers from some point on (either in the positive direction or the negative direction) without end.

Example 1: Graphing Intervals

a. Graph the half-open interval $0 < x \le 4$.

Solution:

b. Graph the open interval $(3, \infty)$.

Solution:

c. Represent the following graph using algebraic notation, and state what kind of interval it is.

Solution: $x \ge 1$ is a half-open interval.

Continued on the next page...

d. Represent the following graph using interval notation, and state what kind of interval it is.

$$-4\,-3\,-2\,-1\ 0\ \ 1\ \ 2\ \ 3\ \ 4\ \ 5\ \ 6\ \ 7$$

Solution: $(-3,5)$ is an open interval.

Solving Linear Inequalities

In this section we will solve **linear inequalities**, such as $6x + 5 \le -7$, and write the solution in **interval notation**, such as $(-\infty, -2]$, to indicate all real numbers x less than or equal to –2. Note that this can also be written in set-builder notation as $\{x \mid x \in (-\infty, -2]\}$ or $\{x \mid x \le -2\}$.

Linear Inequalities

Inequalities of the given form, where a, b, and c are real numbers and $a \ne 0$,

$$ax + b < c \quad and \quad ax + b \le c$$

$$ax + b > c \quad and \quad ax + b \ge c$$

*are called **linear inequalities**.*

*The inequalities $c < ax + b < d$ and $c \le ax + b \le d$ are called **compound linear inequalities**. (This includes $c < ax + b \le d$ and $c \le ax + b < d$ as well.)*

The solutions to linear inequalities are intervals of real numbers, and the methods for solving linear inequalities are similar to those used to solve linear equations. There is only one important exception:

> **Multiplying or dividing both sides of an inequality by a negative number causes the "sense" of the inequality to be reversed.**

By the sense of the inequality, we mean "less than" or "greater than." Consider the following examples.

We know that $6 < 10$.

Add 5 to both sides	**Add –7 to both sides**	**Multiply both sides by 3**
$6 < 10$	$6 < 10$	$6 < 10$
$6 + 5 \ ? \ 10 + 5$	$6 + (-7) \ ? \ 10 + (-7)$	$3 \cdot 6 \ ? \ 3 \cdot 10$
$11 < 15$	$-1 < 3$	$18 < 30$

In the three cases just illustrated, addition, subtraction, and multiplication by a positive number, the sense of the inequality stayed the same. It remained <. Now we will see that multiplying or dividing each side by a negative number will **reverse the sense** of the inequality, from < to > or from > to <. This concept also applies to ≤ and ≥.

Multiply both sides by –3	**Divide both sides by –2**
$6 < 10$	$6 < 10$
$-3 \cdot 6 \ ? \ -3 \cdot 10$	$\dfrac{6}{-2} \ ? \ \dfrac{10}{-2}$
$-18 > -30$	$-3 > -5$

In each of these last two examples, the sense of the inequality is changed from < to >. While two examples do not prove a rule to be true, this particular rule is true and is included in the following rules for solving inequalities.

Rules for Solving Linear Inequalities

1.   *Simplify each side of the inequality by removing any grouping symbols and combining like terms.*

2.   *Use the addition property of equality to add the opposites of constants or variable expressions so that variable expressions are on one side of the inequality and constants on the other.*

3.   *Use the multiplication property of equality to multiply both sides by the reciprocal of the coefficient of the variable (that is, divide both sides by the coefficient) so that the new coefficient is 1.* **If this coefficient is negative, reverse the sense of the inequality.**

4.   *A quick (and generally satisfactory) check is to select any one number in your solution and substitute it into the original inequality.*

As with solving equations, the object of solving an inequality is to find equivalent inequalities of simpler form that have the same solution set. We want the variable with a coefficient of +1 on one side of the inequality and any constants on the other side. One key difference between solving linear equations and solving linear inequalities is that linear equations have only one solution while linear inequalities generally have an infinite number of solutions.

Example 2: Solving Linear Inequalities

Solve the following linear inequalities and graph the solutions.

a. $6x + 5 \le -1$

Solution:

$$6x + 5 \le -1$$

$6x + 5 - 5 \le -1 - 5$	Add –5 to both sides.
$6x \le -6$	Simplify.
$\dfrac{6x}{6} \le \dfrac{-6}{6}$	Divide both sides by 6.
$x \le -1$	Simplify.

x is in $(-\infty, -1]$	Use interval notation. Note that the interval $(-\infty, -1]$ is a half-open interval.

b. $x - 3 > 3x + 4$

Solution:

$$x - 3 > 3x + 4$$

$x - 3 - x > 3x + 4 - x$	Add –x to both sides.
$-3 > 2x + 4$	Simplify.
$-3 - 4 > 2x + 4 - 4$	Add –4 to both sides.
$-7 > 2x$	Simplify.
$\dfrac{-7}{2} > \dfrac{2x}{2}$	Divide both sides by 2.
$\dfrac{-7}{2} > x$	Simplify.

or $x < -\dfrac{7}{2}$

x is in $\left(-\infty, -\dfrac{7}{2}\right)$	Use interval notation. Note that the interval $\left(-\infty, -\dfrac{7}{2}\right)$ is an open interval.

c. $6 - 4x \le x + 1$

Solution:

$$6 - 4x \le x + 1$$

$6 - 4x - x \le x + 1 - x$	Add –x to both sides.
$6 - 5x \le 1$	Simplify.
$6 - 5x - 6 \le 1 - 6$	Add –6 to both sides.
$-5x \le -5$	Simplify.

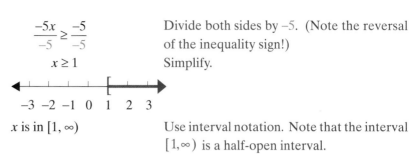

$$\frac{-5x}{-5} \geq \frac{-5}{-5}$$ Divide both sides by –5. (Note the reversal of the inequality sign!)

$$x \geq 1$$ Simplify.

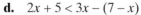

$$-3 \ -2 \ -1 \ \ 0 \ \ 1 \ \ 2 \ \ 3$$

x is in $[1, \infty)$ Use interval notation. Note that the interval $[1, \infty)$ is a half-open interval.

d. $2x + 5 < 3x - (7 - x)$

Solution:

$$2x + 5 < 3x - (7 - x)$$

$$2x + 5 < 3x - 7 + x$$ Distribute the negative sign.

$$2x + 5 < 4x - 7$$ Combine like terms.

$$2x + 5 - 2x < 4x - 7 - 2x$$ Add $-2x$ to both sides.

$$5 < 2x - 7$$ Simplify.

$$5 + 7 < 2x - 7 + 7$$ Add 7 to both sides.

$$12 < 2x$$ Simplify.

$$\frac{12}{2} < \frac{2x}{2}$$ Divide both sides by 2.

$$6 < x$$ Simplify.

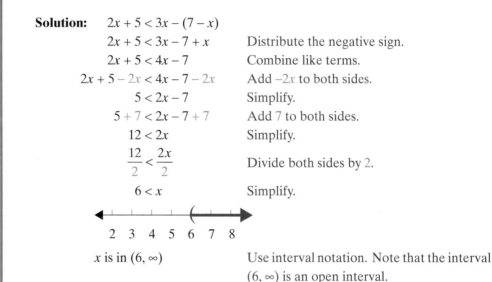

$$2 \ \ 3 \ \ 4 \ \ 5 \ \ 6 \ \ 7 \ \ 8$$

x is in $(6, \infty)$ Use interval notation. Note that the interval $(6, \infty)$ is an open interval.

Solving Compound Inequalities

Compound inequalities have three parts and can arise when a variable or variable expression is to be between two numbers. For example, the inequality

$$5 < x + 3 < 10$$

indicates that the values for the expression $x + 3$ are to be between 5 and 10. To solve this inequality, subtract 3 (or add -3) from each part.

$$5 < x + 3 < 10$$

$$5 - 3 < x + 3 - 3 < 10 - 3$$

$$2 < x < 7$$

Thus the variable x is isolated with coefficient +1, and we see that the solution set is the interval of real numbers $(2, 7)$. The graph of the solution set is the following.

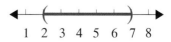

$$1 \ \ 2 \ \ 3 \ \ 4 \ \ 5 \ \ 6 \ \ 7 \ \ 8$$

Example 3: Solving Compound Inequalities

a. Solve the compound inequality $-5 \le 4x - 1 < 11$ and graph the solution set.

Solution:

$-5 \le \ 4x - 1 \ < 11$	Write the inequality.
$-5 + 1 \le 4x - 1 + 1 < 11 + 1$	Add 1 to each part.
$-4 \le \quad 4x \quad < 12$	Simplify.
$\dfrac{-4}{4} \le \ \dfrac{4x}{4} \ < \dfrac{12}{4}$	Divide each part by 4.
$-1 \le \quad x \quad < 3$	Simplify.

The solution set is the half-open interval $[-1, 3)$.

b. Solve the compound inequality $5 \le -3 - 2x \le 13$ and graph the solution set.

Solution:

$5 \le \ -3 - 2x \ \le 13$	Write the inequality.
$5 + 3 \le -3 - 2x + 3 \le 13 + 3$	Add 3 to each part.
$8 \le \quad -2x \quad \le 16$	Simplify.
$\dfrac{8}{-2} \ge \ \dfrac{-2x}{-2} \ \ge \dfrac{16}{-2}$	Divide each part by -2. Note that the inequalities change sense.
$-4 \ge \quad x \quad \ge -8$	Simplify.
$\left(\text{or} \ \ -8 \le \quad x \quad \le -4\right)$	

The solution set is the closed interval $[-8, -4]$.

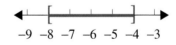

c. Solve the compound inequality $0 < \dfrac{3x - 5}{4} < 3$ and graph the solution set.

Solution:

$0 < \ \dfrac{3x - 5}{4} \ < 3$	Write the inequality.
$0 \cdot 4 < 4\left(\dfrac{3x - 5}{4}\right) < 3 \cdot 4$	Multiply each part by 4.
$0 < \quad 3x - 5 \quad < 12$	Simplify each part.

$$0+5 < 3x-5+5 < 12+5 \qquad \text{Add } 5 \text{ to each part.}$$

$$5 < \quad 3x \quad < 17 \qquad \text{Simplify.}$$

$$\frac{5}{3} < \quad \frac{3x}{3} \quad < \frac{17}{3} \qquad \text{Divide each part by } 3.$$

$$\frac{5}{3} < \quad x \quad < \frac{17}{3} \qquad \text{Simplify.}$$

The solution set is the open interval $\left(\dfrac{5}{3}, \dfrac{17}{3}\right)$.

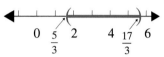

NOTES

You should note that a compound inequality is actually two linear inequalities in one expression and the solution set is the intersection $\cap$ of the two solution sets. Thus, in Example 3a, you found the intersection of the solutions sets of $-5 \le 4x-1$ and $4x-1 < 11$.

Graphs

For $-5 \le 4x-1$

$$-4 \le 4x$$

$$-1 \le x$$

For $4x-1 < 11$

$$4x < 12$$

$$x < 3$$

The solution set is the intersection.

Solving Absolute Value Inequalities

Now consider an inequality with absolute value such as $|x| < 3$. For a number to have an absolute value less than 3, it must be within 3 units of 0. That is, the numbers between -3 and 3 have their absolute values less than 3 because they are within 3 units of 0. Thus, for $|x| < 3$,

Algebraic Notation	Graph	Interval Notation
$|x| < 3$ $-3 < x < 3$ **(the intersection)**	3 units 3 units $-3 \qquad 0 \qquad 3$	$(-3, 3)$

Table 1.11

The inequality $|x - 5| < 3$ means that the distance between x and 5 is less than 3. That is, we want all the values of x that are within 3 units of 5. The inequality is solved algebraically as follows.

$$|x - 5| < 3$$

$$-3 < x - 5 < 3 \qquad x - 5 \text{ is between } -3 \text{ and } 3.$$

$$-3 + 5 < x - 5 + 5 < 3 + 5 \qquad \text{Add } 5 \text{ to each part of the expression, just as in solving linear inequalities.}$$

$$2 < x < 8 \qquad \text{Simplify each expression.}$$

$$x \text{ is in } (\,2, 8\,) \qquad \text{Use interval notation.}$$

The values for x are between 2 and 8 and are within 3 units of 5.

Solving Absolute Value Inequalities with < (or ≤)

For $c > 0$:

a.   If $|x| < c$, then $-c < x < c$.

b.   If $|ax + b| < c$, then $-c < ax + b < c$.

*The inequalities in **a.** and **b.** are also true if < is replaced by ≤.*

Note: *If the absolute value expression is isolated on one side of the inequality, we say that the expression is in **standard form**. You may need to manipulate the absolute value inequality to get it into standard form before you can solve it. (See Example 4c.)*

Example 4: Solving Absolute Value Inequalities

Solve the following absolute value inequalities and graph the solution sets.

a. $|x| \le 6$

Solution: $|x| \le 6$

$-6 \le x \le 6$

or x is in $[-6, 6]$

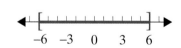

b. $|x + 3| < 2$

Solution:
$$|x + 3| < 2$$
$$-2 < x + 3 < 2$$
$$-2 - 3 < x + 3 - 3 < 2 - 3$$
$$-5 < x < -1$$

or x is in $(-5, -1)$

c. $3|2x - 7| < 15$

Solution:
$$3|2x - 7| < 15$$
$$|2x - 7| < 5 \qquad \text{Divide both sides by 3.}$$
$$-5 < 2x - 7 < 5$$
$$-5 + 7 < 2x - 7 + 7 < 5 + 7$$
$$2 < 2x < 12$$
$$1 < x < 6$$

or x is in $(1, 6)$

We have been discussing inequalities in which the absolute value is less than some positive number. Now consider an inequality where the absolute value is greater than some positive number, such as $|x| > 3$. For a number to have an absolute value greater than 3, its distance from 0 must be greater than 3. That is, numbers that are greater than 3 **or** less than −3 will have absolute values greater than 3. Thus, for $|x| > 3$,

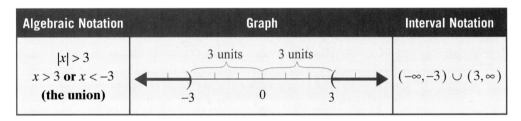

Algebraic Notation	Graph	Interval Notation
$|x| > 3$ $x > 3$ or $x < -3$ **(the union)**	3 units 3 units −3 0 3	$(-\infty, -3) \cup (3, \infty)$

Table 1.12

NOTES The expression $x > 3$ or $x < -3$ **cannot** be combined into one inequality expression. The word **or** must separate the inequalities since any number that satisfies one **or** the other is a solution to the absolute value inequality. There are **no** numbers that satisfy **both** inequalities.

The inequality $|x - 5| > 6$ means that the distance between x and 5 is more than 6. That is, we want all values of x that are more than 6 units from 5. The inequality is solved algebraically as follows.

$|x - 5| > 6$ indicates that

$x - 5 < -6$ **or** $x - 5 > 6$. $x - 5$ is less than −6 or greater than 6

Solving both inequalities gives

$x - 5 + 5 < -6 + 5$ **or** $x - 5 + 5 > 6 + 5$,   Add 5 to each side, just as in solving linear inequalities.

$x < -1$ **or** $x > 11$.   Simplify.

So x is in $(-\infty, -1) \cup (11, \infty)$.

Note: The values for x less than −1 or greater than 11 are more than 6 units from 5. Thus we can interpret the inequality $|x - 5| > 6$ to mean that the distance from x to 5 is greater than 6.

Solving Absolute Value Inequalities with > (or ≥)

For c > 0:

a.   If $|x| > c$, then $x < -c$ *or* $x > c$.

b.   If $|ax + b| > c$, then $ax + b < -c$ *or* $ax + b > c$.

Note: *The inequalities in **a.** and **b.** are true if > is replaced by ≥.*

Example 5: Solving Absolute Value Inequalities

Solve the following absolute value inequalities and graph the solution set.

a. $|x| \geq 5$

Solution: $|x| \geq 5$

$x \leq -5$ or $x \geq 5$

So x is in $(-\infty, -5] \cup [5, \infty)$.

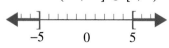

b. $|4x - 3| > 2$

Solution: $|4x - 3| > 2$

$$4x - 3 < -2 \quad \text{or} \quad 4x - 3 > 2$$

$$4x < 1 \quad \text{or} \quad 4x > 5$$

$$x < \frac{1}{4} \quad \text{or} \quad x > \frac{5}{4}$$

So x is in $\left(-\infty, \frac{1}{4}\right) \cup \left(\frac{5}{4}, \infty\right)$.

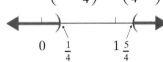

c. $|3x - 8| > -6$

Solution: There is nothing to do here except observe that no matter what is substituted for x, the absolute value will be greater than -6. Absolute value is always nonnegative (greater than or equal to 0). The solution to the inequality is **all real numbers**, so shade the entire number line. In interval notation, x is in $(-\infty, \infty)$.

d. $|x + 9| < -\frac{1}{2}$

Solution: Since absolute value is always nonnegative (greater than or equal to 0), no number has an absolute value less than $-\frac{1}{2}$. Thus there is **no solution**, $\varnothing$.

e. $|2x + 4| + 4 < 7$

Solution: $|2x + 4| + 4 < 7$

$$|2x + 4| < 3$$

$$-3 < 2x + 4 < 3$$

$$-7 < 2x < -1$$

$$-\frac{7}{2} < x < -\frac{1}{2}$$

Add -4 to both sides in order to get the expression in **standard form** with the absolute value expression by itself. Remember, we must get the expression in standard form before solving any absolute value inequality.

So x is in $\left(-\frac{7}{2}, -\frac{1}{2}\right)$.

1.7 Exercises

In Exercises 1 – 10, graph each interval on a real number line and tell what type of interval it is.

1. $x \le -3$ **2.** $x \ge -0.5$ **3.** $x > 4$ **4.** $x < -\dfrac{1}{10}$

5. $0 < x \le 2.5$ **6.** $-1.5 \le x < 3.2$ **7.** $-2 \le x \le 0$ **8.** $-1 \le x \le 1$

9. $4 > x \ge 2$ **10.** $0 > x \ge -5$

In Exercises 11 – 20, the graph of an interval of real numbers is given. Write the corresponding algebraic notation for the graph and tell what type of interval it is.

11.

12.

13.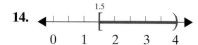

14.

15.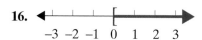

16.

17.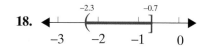

18.

19.

20.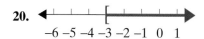

Solve the inequalities in Exercises 21 – 60 and graph the solution sets. Write each solution in interval notation. Assume that x is a real number.

21. $2x + 3 < 5$ **22.** $4x - 7 \ge 9$ **23.** $14 - 5x < 4$

24. $23 < 7x - 5$ **25.** $6x - 15 > 1$ **26.** $9 - 2x < 8$

27. $5.6 + 3x \ge 4.4$ **28.** $12x - 8.3 < 6.1$ **29.** $1.5x + 9.6 < 12.6$

30. $0.8x - 2.1 \ge 1.1$ **31.** $2 + 3x \ge x + 8$ **32.** $x - 6 \le 4 - x$

33. $3x - 1 \le 11 - 3x$ **34.** $5x + 6 \ge 2x - 2$ **35.** $4 - 2x < 5 + x$

36. $4 + x > 1 - x$ **37.** $x - 6 > 3x + 5$ **38.** $\dfrac{x}{2} - 1 \le \dfrac{5x}{2} - 3$

39. $\dfrac{x}{4} + 1 \le 5 - \dfrac{x}{4}$ **40.** $\dfrac{x}{3} - 2 > 1 - \dfrac{x}{3}$ **41.** $\dfrac{5x}{3} + 2 > \dfrac{x}{3} - 1$

42. $6x + 5.91 < 1.11 - 2x$ **43.** $4.3x + 21.5 \ge 1.7x + 0.7$

44. $6.2x - 5.9 > 4.8x + 3.2$ **45.** $0.9x - 11.3 < 3.1 - 0.7x$

46. $4(6 - x) < -2(3x + 1)$ **47.** $-3(2x - 5) \le 3(x - 1)$

48. $3x + 8 \le -3(2x - 3)$ **49.** $6(3x + 1) < 5(1 - 2x)$

50. $4 + 7x \le 3x - 8 + x$ **51.** $11x + 8 - 5x \ge 2x - (4 - x)$

52. $1 - (2x + 8) < (9 + x) - 4x$ **53.** $5 - 3(4 - x) + x \le -2(3 - 2x) - x$

54. $x - (2x + 5) \ge 7 - (4 - x) + 10$ **55.** $\dfrac{2(x - 1)}{3} < \dfrac{3(x + 1)}{4}$

56. $\dfrac{x + 2}{2} \ge \dfrac{2x}{3}$ **57.** $\dfrac{x - 2}{4} > \dfrac{x + 2}{2} + 6$

58. $\dfrac{x + 4}{9} \le \dfrac{x}{3} - 2$ **59.** $\dfrac{2x + 7}{4} \le \dfrac{x + 1}{3} - 1$ **60.** $\dfrac{4x}{7} - 3 > \dfrac{x - 6}{2} - 4$

Solve the compound inequalities in Exercises 61 – 70 and graph the solution sets. Write each solution in interval notation. Assume that x is a real number.

61. $-4 < x + 5 < 6$ **62.** $2 \le -x + 2 \le 6$ **63.** $3 \ge 4x - 3 \ge -1$

64. $13 > 3x + 4 > -2$ **65.** $1 \le \dfrac{2}{3}x - 1 \le 9$ **66.** $-2 \le \dfrac{1}{2}x - 5 \le -1$

67. $14 > -2x - 6 > 4$ **68.** $-11 \ge -3x + 2 > -20$ **69.** $-1.5 < 2x + 4.1 < 3.5$

70. $0.9 < 3x + 2.4 < 6.9$

71. Test scores: A statistics student has grades of 82, 95, 93, and 78 on four hourly exams. He must average 90 or higher to receive an A for the course. What scores can he receive on the final exam and earn an A if:
  a. The final is equivalent to a single hourly exam (100 points maximum)?
  b. The final is equivalent to two hourly exams (200 points maximum)?

72. **Test scores:** To receive a grade of B in a chemistry class, Melissa must average 80 or more but less than 90. If her five hourly exam scores were 75, 82, 90, 85, and 77, what score does she need on the final exam (100 points maximum) to earn a grade of B?

73. **Triangles:** The sum of the lengths of any two sides of a triangle must be greater than the third side. A triangle has sides as follows: the first side is 18 mm; and the second side is 3 mm more than twice the third side. What are the possible lengths of the second and third sides? [**Hint:** You will need to solve two inequalities.]

74. **Postage:**   Allison is going to the post office to buy 38¢ stamps and 2¢ adjustment stamps. Since the current postage rate is 44¢, she will need 3 times as many 2¢ adjustment stamps as 38¢ stamps. If she has $11 to spend, what is the largest number of 38¢ stamps she can buy?

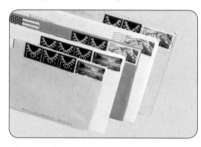

Solve each of the absolute value problems in Exercises 75 – 95 and graph the solution sets. Write each solution in interval notation. Assume that x is a real number.

75. $|x| \geq -2$

76. $|x| \geq 3$

77. $|x| \leq \dfrac{4}{5}$

78. $|x| \geq \dfrac{7}{2}$

79. $|x - 3| > 2$

80. $|y - 4| \leq 5$

81. $|x + 2| \leq -4$

82. $|3x + 4| > -8$

83. $|x + 6| \leq 4$

84. $|2x - 1| \geq 2$

85. $|3 - 2x| < -2$

86. $|3x + 4| - 1 < 0$

87. $\left|\dfrac{3x}{2} - 4\right| \geq 5$

88. $\left|\dfrac{3}{7}y + \dfrac{1}{2}\right| > 2$

89. $4|7x + 9| - 3 < 17$

90. $2|7x - 3| + 4 \geq 12$

91. $|2x - 9| - 7 \leq 4$

92. $5 > |4 - 2x| + 2$

93. $-4 < |6x - 1| + 4$

94. $7 > |8 - 5x| + 3$

95. $|3x - 7| + 4 \leq 4$

Writing and Thinking About Mathematics

In Exercises 96 – 100 a set of real numbers is described. **a.** *Sketch a graph of the set on a real number line.* **b.** *Represent each set by using absolute value notation.* **c.** *If the set is one interval, state what type of interval it is.*

96. the set of real numbers between −10 and 10, inclusive

97. the set of real numbers within 7 units of 4

98. the set of real numbers more than 6 units from 8

99. the set of real numbers greater than or equal to 3 units from −1

100. the set of real numbers within 2 units of −5

 HAWKES LEARNING SYSTEMS: INTERMEDIATE ALGEBRA SOFTWARE

- Solving Linear Inequalities
- Solving Absolute Value Inequalities

Chapter 1 Index of Key Ideas and Terms

Section 1.1   Prime Numbers, Exponents, and LCM

Natural Numbers (Counting Numbers)   page 2

$$\mathbb{N} = \{1, 2, 3, 4, 5, 6, 7, 8, 9, 10, 11, ...\}$$

Whole Numbers   page 2

$$\mathbb{W} = \{0, 1, 2, 3, 4, 5, 6, 7, 8, 9, 10, 11, ...\}$$

Factor   page 3

A **factor** (or **divisor**) divides into a product with a
remainder of 0.

Prime Numbers   page 3

A **prime number** has exactly two different factors, 1 and
the number itself. The prime numbers less than 50 are:
2, 3, 5, 7, 11, 13, 17, 19, 23, 29, 31, 37, 41, 43, 47.

Composite Numbers   page 3

A **composite number** has more than two different factors.

To Find the Prime Factorization of a Composite Number   page 4

1. Factor the composite number into any two factors.
2. Factor each factor that is not prime into two more
factors.
3. Continue this process until all factors are prime.
The **prime factorization** is the product of all the prime
factors.

Exponents   pages 4-5

An **exponent** is a number written to the right and slightly
above another number (called the **base**) that indicates
repeated multiplication of the base by itself.

$$\underbrace{a \cdot a \cdot a \cdot ... \cdot a}_{n \text{ factors}} = a^n.$$

exponent

base

Multiples   page 5

The **multiples** of a number are the products of that number
with the counting numbers.

Section 1.2 Introduction to Real Numbers (cont.)

Sets page 14

A **set** is a collection of objects or numbers. The items in the set are called **elements**. If the elements in a set can be counted, the set is said to be **finite**. If the elements cannot be counted, the set is said to be **infinite**.

Set-Builder Notation: $\{x \mid \ \}$ page 15

The notation $\{x \mid \ \}$ is read "the set of all x such that ..." and is called **set-builder notation**. A statement following the vertical bar (read "such that") gives a condition for x.

Inequality Symbols page 15

$<$	"less than"	$\leq$	"less than or equal to"
$>$	"greater than"	$\geq$	"greater than or equal to"

Properties of Addition and Multiplication page 18

For real numbers a, b, and c:

Addition	**Name of Property**	**Multiplication**
$a + b = b + a$	Commutative	$a \cdot b = b \cdot a$
$(a + b) + c = a + (b + c)$	Associative	$a \cdot (b \cdot c) = (a \cdot b) \cdot c$
$a + 0 = 0 + a = a$	Identity	$a \cdot 1 = 1 \cdot a = a$
$a + (-a) = 0$	Inverse	$a \cdot \dfrac{1}{a} = 1 \ (a \neq 0)$

Zero Factor Law page 18

$a \cdot 0 = 0 \cdot a = 0$

Distributive Property of Multiplication over Addition page 18

$a(b + c) = ab + ac$

Properties of Inequality (Order) page 20

For real numbers a, b, and c:

Trichotomy Property: Exactly one of the following is true:
$a < b$, $a = b$, or $a > b$.

Transitive Property: If $a < b$ and $b < c$, then $a < c$.
The transitive property applies to $>$, $\geq$, and $\leq$ as well.

Section 1.3 Operations with Real Numbers

Absolute Value page 25

For any real number a, $|a| = \begin{cases} a & \text{if } a \geq 0 \\ -a & \text{if } a < 0 \end{cases}$.

Operations with Real Numbers

Addition   page 26
1. To add two real numbers with like signs, add their absolute values and use the common sign.
2. To add two real numbers with unlike signs, subtract their absolute values (the smaller from the larger) and use the sign of the number with the larger absolute value.

Subtraction   page 28
For real numbers a and b, $a - b = a + (-b)$.

Multiplication   page 31
For positive real numbers a and b,
1. The product of two positives is positive: $(a)(b) = ab$.
2. The product of two negatives is positive: $(-a)(-b) = ab$.
3. The product of a positive and a negative is negative: $a(-b) = (-a)b = -ab$.

Division   page 32
For positive real numbers a and b,
1. The quotient of two positives is positive: $\dfrac{a}{b} = +\dfrac{a}{b}$.

2. The quotient of two negatives is positive: $\dfrac{-a}{-b} = +\dfrac{a}{b}$.

3. The quotient of a positive and a negative is negative:
$\dfrac{-a}{b} = \dfrac{a}{-b} = -\dfrac{a}{b}$.

Rules for Multiplication and Division   pages 31-32
1. If two nonzero numbers have like signs, then both their product and quotient will be positive.
2. If two nonzero numbers have unlike signs, then both their product and quotient will be negative.

Division by 0   page 32
Division by 0 is undefined.

Rules for Order of Operations   page 33
1. Simplify within symbols of inclusion (parentheses, brackets, braces, fraction bar, absolute value bars) beginning with the innermost symbols.
2. Find any powers indicated by exponents or roots.
3. Multiply or divide from left to right.
4. Add or subtract from left to right.

Section 1.4   Linear Equations in One Variable: $ax + b = c$

Terms   page 39
  A **term** is an expression that involves only multiplication
  and/or division with constants and/or variables.

Constant   page 39
  A term consisting of only a real number is called a **constant**.

Coefficient   page 39
  The numerical factor of a term containing variables is
  called the **coefficient** of the variables.

Like Terms   page 39
  Like terms (or similar terms) are those terms that contain
  the same variables raised to the same powers.

Combining Like Terms   page 40
  To **combine like terms**, add (or subtract) the coefficients
  and keep the common variable expression.

Linear Equations in x   page 41
  If a, b, and c are constants and $a \neq 0$, then a **linear equation
  in x** is an equation that can be written in the form $ax + b = c$.

Addition Property of Equality   page 42
  If the same algebraic expression is added to both sides of
  an equation, the new equation has the same solutions as the
  original equation. Symbolically, if A, B, and C are algebraic
  expressions, then the equations

$$A = B \qquad \text{and} \qquad A + C = B + C$$

  have the same solutions. Equations with the same solutions
  are said to be **equivalent**.

Multiplication (or Division) Property of Equality   page 42
  If both sides of an equation are multiplied by (or divided by)
  the same nonzero constant, the new equation has the same
  solutions as the original equation. Symbolically, if A and
  B are algebraic expressions and C is any nonzero constant,
  then the equations

$$A = B$$

  and $\qquad\qquad AC = BC \qquad\qquad$ where $C \neq 0$

  and $\qquad\qquad \dfrac{A}{C} = \dfrac{B}{C} \qquad\qquad$ where $C \neq 0$

  have the same solutions and are **equivalent**.

Section 1.4   Linear Equations in One Variable: $ax + b = c$ (cont.)

Procedure for Solving Linear Equations   pages 42-43
1. Simplify each side of the equation by removing any grouping symbols and combining like terms. (In some cases, you may want to multiply both sides of the equation by a constant to clear fractional or decimal coefficients.)
2. Use the addition property of equality to add the opposites of constants or variable expressions so that variable expressions are on one side of the equation and constants on the other.
3. Use the multiplication property of equality to multiply both sides by the reciprocal of the coefficient of the variable (that is, divide both sides by the coefficient) so that the new coefficient is 1.
4. Check your answer by substituting it into the original equation.

Types of Equations and their Solutions

Type of Equation	**Number of Solutions**	page 45
Conditional	Finite number of solutions	
Identity	Infinite number of solutions	
Contradiction	No solutions	

Solving Absolute Value Equations   page 46
For $c > 0$:
a. If $|x| = c$, then $x = c$ **or** $x = -c$.
b. If $|ax + b| = c$, then $ax + b = c$ **or** $ax + b = -c$.

Two Absolute Values   page 47
If $|a| = |b|$, then either $a = b$ **or** $a = -b$.
More generally, if $|ax + b| = |cx + d|$, then either
$ax + b = cx + d$ **or** $ax + b = -(cx + d)$.

Section 1.5   Evaluating and Solving Formulas

Formula   page 51
A **formula** is an equation that represents a general relationship between two or more quantities or measurements.

Section 1.5 Evaluating and Solving Formulas (cont.)

Evaluating Formulas page 53

If the values for all but one variable in a formula are known, they can be substituted into the formula and the unknown value can be found by solving the equation.

Solving Formulas page 53

Section 1.6 Applications

Strategy for Solving Word Problems pages 58-59
1. Understand the problem.
2. Devise a plan.
3. Carry out the plan.
4. Look back over the results.

Applications

Number problems	pages 59-60
Distance-rate-time problems	pages 60-61
Cost-profit problems	pages 62-63
Simple interest problems	page 64
Average problems	page 65
Bar graphs	page 66

Section 1.7 Solving Inequalities in One Variable

Interval Notation pages 72-73

Type of Interval	Interval Notation
Open	(a, b), (a, ∞), $(-\infty, b)$
Closed	$[a, b]$
Half-open	$[a, \infty)$, $(-\infty, b]$, $[a, b)$, $(a, b]$

Linear Inequalities page 74

Inequalities of the given form where a, b, and c are real numbers and $a \neq 0$,

$$ax + b < c \qquad \text{and} \qquad ax + b \leq c$$
$$ax + b > c \qquad \text{and} \qquad ax + b \geq c$$

are called **linear inequalities**.

The inequalities $c < ax + b < d$ and $c \leq ax + b \leq d$ are called **compound linear inequalities**. (This includes $c < ax + b \leq d$ and $c \leq ax + b < d$ as well.)

Section 1.7　Solving Inequalities in One Variable (cont.)

Rules for Solving Linear Inequalities　　　　　　　　page 75
1. Simplify each side of the inequality by removing any grouping symbols and combining like terms.
2. Use the addition property of equality to add the opposites of constants or variable expressions so that variable expressions are on one side of the inequality and constants on the other.
3. Use the multiplication property of equality to multiply both sides by the reciprocal of the coefficient of the variable (that is, divide both sides by the coefficient) so that the new coefficient is 1. If this coefficient is negative, reverse the sense of the inequality.
4. A quick check is to select any one number in your solution and substitute it into the original inequality.

Absolute Value Inequalities　　　　　　　　pages 80, 82

For $c > 0$:
1. If $|x| < c$, then $-c < x < c$.
2. If $|ax + b| < c$, then $-c < ax + b < c$.

For $c > 0$:
1. If $|x| > c$, then $x < -c$ **or** $x > c$.
2. If $|ax + b| > c$, then $ax + b < -c$ **or** $ax + b > c$.

HAWKES LEARNING SYSTEMS: INTERMEDIATE ALGEBRA SOFTWARE

For a review of the topics and problems from Chapter 1, look at the following lessons from *Hawkes Learning Systems: Intermediate Algebra*

- Prime Numbers, Exponents, and LCM
- Name that Real Number
- Properties of Real Numbers
- Introduction to Absolute Values
- Addition with Real Numbers
- Subtraction with Real Numbers
- Multiplication and Division with Real Numbers
- Order of Operations
- Simplifying Expressions

- Solving Linear Equations
- Solving Absolute Value Equations
- Evaluating Formulas
- Solving Formulas
- Applications
- Solving Linear Inequalities
- Solving Absolute Value Inequalities
- Chapter 1 Review and Test

Chapter 1 Review

1.1 Prime Numbers, Exponents, and LCM

In Exercises 1 – 8, find the prime factorization of each composite number. Use exponents to represent repeated factors.

1. 88   **2.** 95   **3.** 1000   **4.** 450

5. 660   **6.** 300   **7.** 150   **8.** 195

Find the least common multiple (LCM) of each set of whole numbers in Exercises 9 – 16.

9. $\{30, 70\}$ **10.** $\{20, 90\}$ **11.** $\{10, 25, 35\}$

12. $\{6, 30, 36\}$ **13.** $\{34, 51, 68\}$ **14.** $\{15, 21, 44\}$

15. $\{48, 120, 132, 144\}$ **16.** $\{35, 49, 63, 72\}$

1.2 Introduction to Real Numbers

17. Given the set of real numbers $A = \left\{ -\sqrt{6}, -1, -\dfrac{1}{3}, 0, \pi, 4.3, 5 \right\}$, list the numbers in A that are described by the following notation.

    a. $\{x \mid x \text{ is an integer}\}$ **b.** $\{x \mid x \text{ is an irrational number}\}$

18. Write the rational number $\dfrac{3}{7}$ in the form of an infinite repeating decimal.

19. Graph the set of real numbers $\{x \mid x \geq 0 \text{ and } x < 2\}$ on a real number line.

20. Graph the set of real numbers $\{x \mid x < -2 \text{ or } x > 7\}$ on a real number line.

In Exercises 21 – 26, state the property of real numbers that is illustrated.

21. $18 \cdot 0 = 0$ **22.** $73 + 6 = 6 + 73$ **23.** $89 + 0 = 89$

24. $2(y + 6) = 2 \cdot y + 2 \cdot 6$ **25.** $5 \cdot \dfrac{1}{5} = 1$

26. If $2 < x$ and $x < y$, then $2 < y$.

1.3   Operations with Real Numbers

Find the set of values for x in Exercises 27 and 28 that make true statements. If a statement is never true, indicate this by writing the empty set, ∅.

27. $|x| = -6$

28. $|x| = 1.4$

Perform the indicated operations in Exercises 29 – 38.

29. $(-3)(-4)(-6)$

30. $\dfrac{-8}{0}$

31. $\dfrac{0}{-3}$

32. $\left|-\dfrac{8}{3}\right| + \left|\dfrac{8}{3}\right|$

33. $6(-5.3)$

34. $\dfrac{-3.5}{-7}$

35. $-9 + (-7) - (-16)$

36. $75 \div (-15)$

37. $-\dfrac{3}{8} - \dfrac{1}{6} + \left(-\dfrac{1}{2}\right)$

38. $\dfrac{5}{8} - \left(-\dfrac{1}{2}\right) - \dfrac{3}{16}$

Find the value of each expression in Exercises 39 – 42 by using the rules for order of operations.

39. $\left(4^2 + 6\right) - \left(3^2 - 2^3\right)$

40. $2^2 \cdot 3 \div (-1) + 6 - 2$

41. $5 + 3\left[-5 - 2(3 - 4)\right]$

42. $17 - \left[4 \cdot 3 \div 2 + \left(24 - 6^2\right)\right]$

43. Stock market: During the first hour of trading, a stock trader has a profit of $601.97. During the second hour, he has a profit of $1490.21. During the third hour, he has a loss of $1252.05. What was the trader's net profit during the first three hours of trading?

1.4   Linear Equations in One Variable: $ax + b = c$

Combine like terms in the expressions in Exercises 44 – 47.

44. $4x^2 + 12x^2 - 10x$

45. $-6y + 4y - \left(y^2 + 3y\right)$

46. $\dfrac{2x + 6x}{2} + 19x$

47. $3x^2 - \left[3x - \left(x^2 + x - 2\right)\right]$

Solve the linear equations in Exercises 48 – 53.

48. $\dfrac{5x}{4} + 2 = 12$

49. $2(3x - 4) = 5x + 7$

50. $\dfrac{x - 2}{4} + 5 = \dfrac{x - 1}{3} + 2$

51. $5(3x - 10) + 7 = 2(7x - 3)$

52. $-2(x + 3) + 5(x - 7) = 4x + 21$

53. $2.7x + 6.5 = 1.4x - 1.3$

Determine whether each equation in Exercises 54 – 58 is conditional, an identity, or a contradiction.

54. $x = x + 1$ **55.** $5(x+2) = 4x + x + 10$ **56.** $3x + 14 = x - 2$

57. $3(x-15) + 4x = 7(x-5)$ **58.** $-2(x+8) + x = -16 - x$

Solve the absolute value equations in Exercises 59 – 63.

59. $|3x+1| = 8$ **60.** $2|x+3| - 4 = 10$ **61.** $|2x+1| = |3x-1|$

62. $|5x-2| = -10$ **63.** $3|x-7| - 15 = -3$

1.5 Evaluating and Solving Formulas

For Exercises 64 and 65, (a) determine the formula that relates the given information and solve it for the unknown quantity, and (b) substitute the given values in the formula to determine the value of the unknown quantity.

64. The sum of the measures of a triangle is 180°. If one angle equals 45° and another angle equals 63°, what is the measure of the third angle?

65. The area of a trapezoid is 70 square meters. One base is 3 meters long, and the height of the trapezoid is 14. Find the length of the second base.

Solve for the indicated variables in Exercises 66 – 75.

66. $V = \dfrac{k}{P}$; solve for P.   **67.** $A = \dfrac{1}{2}bh$; solve for h.

68. $C = 2\pi r$; solve for r.   **69.** $P = a + 2b$; solve for a.

70. $3x + y = 15$; solve for y.   **71.** $3x - y = 15$; solve for y.

72. $N = \dfrac{l}{d} + 1$; solve for d.   **73.** $P = 2l + 2w$; solve for w.

74. $2x + 3y = 6$; solve for x.   **75.** $A = \dfrac{1}{2}h(a+b)$; solve for a.

1.6 Applications

76. If a number is decreased by 30 and the result is 76 less than twice the number, what is the number?

77. Twice the difference between a number and 10 is equal to 5 times the number plus 13. Find the number.

78. **Travel:** Jerry drove from St. Louis, MO to Kansas City, MO in 5.5 hours. On the return trip he decided to increase his speed by 10 mph and the return took 4.5 hours. Find the rate on the return trip. What is the distance from St. Louis to Kansas City?

79. **Investing:** Ms. Clark has two investments totaling $20,000. One investment yields 6% a year and the other yields 8%. If her annual income from these investments is $1520, how much did she invest at each rate?

80. **Amusement parks:** Given the attendance at the top ten amusement park chains:
    a. Find the average attendance over the ten amusement park chains.
    b. Find the amusement park chain with the lowest attendance.
    c. Find the difference in attendance between Six Flags and Cedar Fair.

    Source: http://www.connectingindustry.com/downloads/pwteaerasupp.pdf page 6

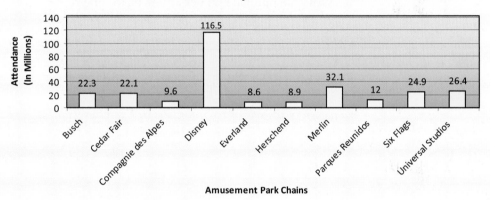

2007 Attendance at Top Ten Amusement Park Chains

1.7 Solving Inequalities in One Variable

Solve the inequalities in Exercises 81 – 90 and graph each solution set on a real number line. Write each solution set in interval notation.

81. $3x + 5 < -7$

82. $4x + 2 \geq 5x + 8$

83. $\dfrac{3x}{5} - 4 > \dfrac{x+2}{5} - 4$

84. $5 + 8x \leq 4x - 11 - 2x$

85. $-6 < 7x + 1 \leq 15$

86. $-2.5 < 2x + 5.1 < 4.5$

87. $|2 - 5x| < -3$

88. $2|x + 17| - 5 \leq 9$

89. $10 > |4 - 2x| + 6$

90. $|x - 8| \geq -4$

Chapter 1 Test

Find the prime factorization of each of the composite numbers in Exercises 1 and 2.

1. 1296

2. 575

In Exercises 3 and 4, find the least common multiple (LCM) of each set of whole numbers.

3. $\{9, 15, 20, 45\}$

4. $\{8, 24, 40, 48\}$

5. List the numbers in the set $A = \left\{ -\sqrt{11}, \ -2, \ -\dfrac{5}{3}, \ 0, \ \dfrac{1}{2}, \ \sqrt{3}, \ 2, \ \pi \right\}$ that are described below.

   a. $\{x \mid x \text{ is an integer}\}$

   b. $\{x \mid x \text{ is a rational number}\}$

6. Write the rational number $\dfrac{5}{12}$ as an infinite repeating decimal.

Graph each of the sets of numbers in Exercises 7 and 8 on a number line.

7. $\{x \mid x \leq 7, x \text{ is a whole number}\}$

8. $\{x \mid -8 < x \leq 0, x \text{ is an integer}\}$

Graph each of the sets of real numbers in Exercises 9 and 10 on a number line.

9. $\left\{ x \mid -2 \leq x < \dfrac{5}{8} \right\}$

10. $\left\{ x \mid -1.5 < x < 3 \text{ or } x \geq \sqrt{17} \right\}$

State the property that is illustrated in Exercises 11 and 12. All variables represent real numbers.

11. $3x + 15y = 3(x + 5y)$

12. $3 + (x + y) = (3 + x) + y$

Complete the expressions in Exercises 13 and 14 by using the given property.

13. $7y \cdot 0 =$ _____   Zero factor law

14. If $x < 17$ and $17 < a$, then _____.   Transitive property

15. Find the sets of values for x that make the following statements true. If a statement is never true, indicate this by writing the empty set, $\varnothing$.

   a. $|x| = 5$

   b. $|x| = -2$

Perform the indicated operations in Exercises 16 – 20.

16. $|-19| + 43 - (-8)$

17. $(-96) \div (-12)$

18. $\left(\dfrac{5}{8} \right)(-3)$

19. $\dfrac{3}{5} + \left(\dfrac{-7}{10}\right) - \dfrac{1}{6}$

20. $4(-6)(-2)(-3)$

Find the value of each expression in Exercises 21 and 22 by using the rules for order of operations.

21. $6 + \left(7^2 - 3^2\right) \div 5 \cdot 4$

22. $\left(2 \cdot 3 - 15 \div 3\right)^2 + 3 \cdot 4$

Simplify each expression in Exercises 23 and 24 by combining like terms.

23. $4(x + 6) - (8 - 2x)$

24. $-2x + 3\left[5x - (2x + 4)\right]$

Solve each of the equations in Exercises 25 – 29.

25. $5x - 3(x - 2) = 4$

26. $4x + (5 - x) = 3(x + 2)$

27. $\dfrac{3x - 2}{8} = \dfrac{x}{4} - 1$

28. $|2x + 1| = 2.8$

29. $|5 - 3x| + 1 = 4$

30. Determine whether each equation is a conditional equation, an identity, or a contradiction.

    a. $\dfrac{3}{4}x + 5 = \dfrac{2}{3}x - 1$ **b.** $5x + 7 = 3(x - 2) + 2x$ **c.** $-4(x + 2) = -2(2x + 4)$

For Exercises 31 and 32, (a) determine the formula that relates the given information and solve it for the unknown quantity, and (b) substitute the given values in the formula to determine the value of the unknown quantity.

31. Circles: The radius of a circle is 19 feet. Find the area of the circle.

32. Trapezoids: The area of a trapezoid is 84 square meters. The height of the trapezoid is 8 meters. One base is 9 meters long. Find the length of the other base.

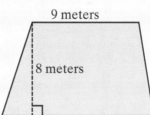

33. Solve the formula $P = 2l + 2w$ for w.

34. Solve the formula $U = \dfrac{1}{2}QV$ for Q.

35. Ten times a number equals ten less than twelve times a number. What is the number?

36. Traveling by bus: A bus leaves Kansas City headed for Phoenix traveling at a rate of 48 mph. Thirty minutes later, a second bus follows, traveling at 54 mph. How long will it take the second bus to overtake the first?

37. Selling Candy: The Candy Shack sells a particular candy in two different size packages. One size sells for \$1.25 and the other sells for \$1.75. If the store received \$65.50 for 42 packages of candy, how many of each size were sold?

38. Cell phone usage: Given the amount of U.S. cellular telephone subscribers:
    a. Find the average number of subscribers from 2002-2005.
    b. Find the difference in the number of subscribers from 1998-2007.
    c. How many subscribers were there in 2001?

Source: International Telecommunications Union

Number of US Cellular Telephone Subscribers From 1998 to 2007

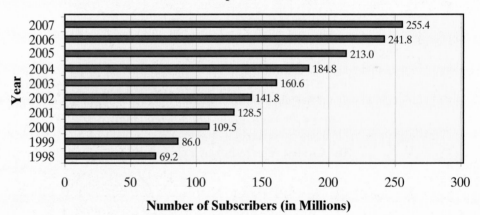

Number of Subscribers (in Millions)

In Exercises 39 and 40, the graph of an interval of real numbers is given. Write the corresponding algebraic notation for the graph and tell what type of interval it is.

39.

    −2 −1 0 1 2 3 4 5 6

40.

    −3 −2 −1 0 1 2 3 4 5

Solve the inequalities in Exercises 41 – 45 and graph the solution sets. Write each solution in interval notation. Assume that x is a real number.

41. $3x + 7 < 4(x + 3)$

42. $\dfrac{2x + 5}{4} > x + 3$

43. $-1 < 3x + 2 < 17$

44. $|7 - 2x| < 3$

45. $|2(x - 3) + 5| > 2.7$

Linear Equations and Functions

Did You Know?

One of the most difficult problems for students in algebra is to become comfortable with the idea that letters or symbols can be manipulated just like numbers in arithmetic. These symbols may be the cause of "math anxiety."

A great deal of publicity has recently been given to the concept that a large number of people suffer from math anxiety-a painful uneasiness caused by mathematical symbols or a problem-solving situation. Persons affected by math anxiety find it difficult to learn mathematics, or they may be able to learn but be unable to apply their knowledge or do well on tests. Persons suffering from math anxiety often develop math avoidance, and they avoid careers, majors, or classes that will require mathematics courses or skills. Sociologist Lucy Sells has determined that mathematics is a critical filter in the job market. Persons who lack quantitative skills are often channeled into high unemployment, low-paying, non-technical careers.

What causes math anxiety? Researchers are investigating the following hypotheses:

1. A lack of skills which leads to lack of confidence and, therefore, to anxiety;
2. An attitude that mathematics is not useful or significant to society;
3. Career goals that seem to preclude mathematics;
4. A self-concept that differs radically from the stereotype of a mathematician;
5. Perceptions that parents, peers, or teachers have low expectations for the person in mathematics;
6. Social conditioning to avoid mathematics.

We hope that you are finding your present experience with algebra successful and that the skills you are acquiring now will enable you to approach mathematical problems with confidence.

"The Science of Pure Mathematics, in its modern developments, may claim to be the most original creation of the human spirit."

Alfred North Whitehead (1861-1947)

This chapter uses the Cartesian coordinate system to develop relationships between algebra and geometry. These relationships have proven extremely important to the development of mathematics and mathematical problem-solving. In particular, you will develop graphing skills for linear equations and linear inequalities in two variables. The graphs of linear equations are straight lines, hence the term linear. The graphs of linear inequalities are half-planes separated by straight lines.

Three useful forms of linear equations are the **standard form**, the **slope-intercept form**, and the **point-slope form**. Each form is equally important and useful depending on the information given and the application of the equation. Your algebraic skills should allow you to change from one form to another and to recognize when the same equation is in a different form.

The concept of slope underlies all our work with straight lines. Horizontal lines (with slope 0) and vertical lines (with undefined slope) can be related to slope. Relationships between two or more lines can be discussed in terms of their slopes. Parallel lines have the same slope, and perpendicular lines have slopes that are negative reciprocals of each other.

Slope occurs as a part of our daily lives in many ways. The slope of a roof is particularly important in areas with lots of snow. Large trucks must be very careful on mountain roads with steep slopes. Airplanes must deal with their rate of descent or the slope of their paths in landing and taking off. We will see that slope can be related to the rate of change in prices of products that we purchase.

Cartesian Coordinate System and Linear Equations: $Ax + By = C$

2.1

- *Graph and label ordered pairs of real numbers as points on a plane.*
- *Find ordered pairs that satisfy a given linear equation.*
- *Recognize the **standard form** of a linear equation: $Ax + By = C$.*
- *Graph linear equations.*
- *Locate the **x-intercept** and the **y-intercept** of a linear equation.*

The linear (or first-degree) equations and inequalities we have discussed so far in this text have involved only one variable. In this chapter, we will discuss linear equations and inequalities that contain two variables. The context of the material should indicate clearly which type of linear equation is being discussed. As we will see in Section 2.4, except for special cases, linear equations in two variables are also called **linear functions**.

Cartesian Coordinate System

We begin with the concept of **ordered pairs** of real numbers. We write these in the form (x, y) and show how they are related to points on a graph and solutions of equations in two variables. For example, consider the equation $y = 3x + 1$ and the ordered pair $(2, 7)$. Substituting $x = 2$ in the equation gives $y = 3 \cdot 2 + 1 = 7$. We say that $(2, 7)$ is **a solution of** (or **satisfies**) the equation. The first number, 2, is called the **first coordinate** (or **x-coordinate**), and the second number, 7, is called the **second coordinate** (or **y-coordinate**). The ordered pair $(2, 7)$ is not the same as the ordered pair $(7, 2)$ because the order of the numbers is not the same. Note that $(7, 2)$ does not satisfy the equation because $3 \cdot 7 + 1 = 22 \neq 2$.

René Descartes

The **Cartesian coordinate system** (named after seventeenth century French mathematician René Descartes (1596-1650)) relates algebraic equations and ordered pairs of real numbers to geometry in a plane. In this system, two number lines intersect at a right angle and separate the plane into four **quadrants**. The **origin**, designated by the ordered pair $(0, 0)$, is the point of intersection of the two lines. The horizontal number line is called the **horizontal axis** or **x-axis**. The vertical number line is called the **vertical axis** or **y-axis**. All points that lie on the x-axis have a y-coordinate of 0 and all points that lie on the y-axis have an x-coordinate of 0. Points that lie on either axis are not in a quadrant. They are simply on an axis (Figure 2.1).

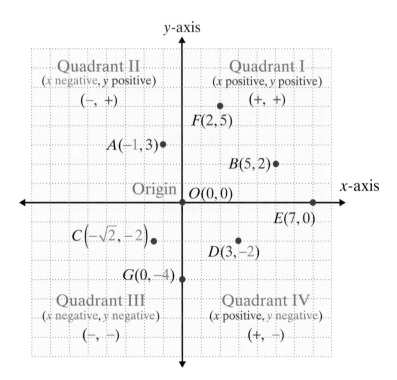

Figure 2.1

The following important relationship between ordered pairs of real numbers and points in a plane is the cornerstone of the Cartesian coordinate system.

One-to-One Correspondence

*There is a **one-to-one correspondence** between points in a plane and ordered pairs of real numbers.*

In other words, for each point in a plane there is one and only one corresponding ordered pair of real numbers, and for each ordered pair of real numbers there is one and only one corresponding point in the plane.

Example 1: Graphing Ordered Pairs

Graph the set of ordered pairs $\{A(-3,1), B(-1,-3), C(0,2), D(1,5), E(2,-4)\}$.

(**Note:** The listing of ordered pairs within the braces can be in any order.)

Solution: To locate each point, **start at the origin**, and:

For $A(-3,1)$, move 3 units left and 1 unit up.

For $B(-1,-3)$, move 1 unit left and 3 units down.

For $C(0,2)$, move 2 units up (do not move any units left or right.)

For $D(1,5)$, move 1 unit right and 5 units up.

For $E(2,-4)$, move 2 units right and 4 units down.

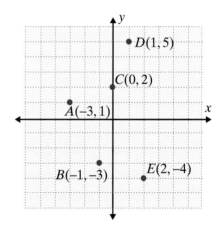

Example 2: Reading Points on a Graph

The graphs of two straight lines are given. Each line has an infinite number of points. Use the grid to help you locate (or estimate) three points on each line.

a.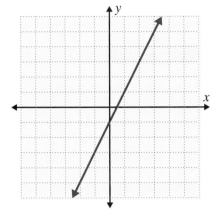

b.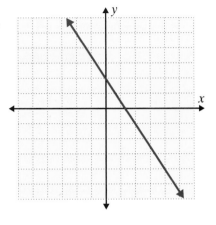

Solution:

a. Three points on this graph are $(-1,-3)$, $(0,-1)$, $(3,5)$. (Of course there are many different correct answers to this type of question. Use your own judgment.)

b. Three points on this graph are $(-2,5)$, $(2,-1)$, $(4,-4)$. (You may estimate with fractions. For example, one point appears to be $\left(\dfrac{1}{2}, \dfrac{5}{4}\right)$.)

Graphing Straight Lines

Now consider the equation in two variables

$$3x + y = 3$$

or, solved for y,

$$y = 3 - 3x.$$

The **solution set** to this equation consists of an infinite set of ordered pairs in the form (x, y). The variable x is called the **independent variable**, and the variable y is called the **dependent variable**.

Solution Set of an Equation in Two Variables

*The **solution set** of an equation in two variables, x and y, consists of all ordered pairs of real numbers (x, y) that satisfy the equation.*

To find some of the solutions of the equation $y = 3 - 3x$, we form a table by:

1. choosing arbitrary values for x, and
2. finding the corresponding values for y by substituting into the equation.

We say that these ordered pairs **satisfy** the equation. In Figure 2.2 we have found five ordered pairs that satisfy the equation and graphed the corresponding points.

Choices	Substitutions	Results
x	$3 - 3x = y$	(x, y)
-1	$3 - 3(-1) = 6$	$(-1, 6)$
0	$3 - 3 \cdot 0 = 3$	$(0, 3)$
$\dfrac{2}{3}$	$3 - 3\left(\dfrac{2}{3}\right) = 1$	$\left(\dfrac{2}{3}, 1\right)$
2	$3 - 3 \cdot 2 = -3$	$(2, -3)$
3	$3 - 3 \cdot 3 = -6$	$(3, -6)$

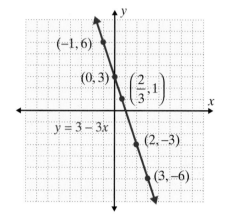

Figure 2.2

The five points in Figure 2.2 appear to lie on a straight line. They in fact do lie on a straight line, and any ordered pair that satisfies the equation $y = 3 - 3x$ will also lie on that same line.

Just as we use the terms **ordered pair** and **point** (the graph of an ordered pair) interchangeably, we use the terms **equation** and **graph of an equation** interchangeably. The equations

$$2x + 3y = 4, \quad y = -5, \quad x = 1.4, \quad \text{and} \quad y = 3x + 2$$

are called **linear equations**, and their graphs are straight lines on the Cartesian plane.

Standard Form of a Linear Equation

Any equation of the form

$$\mathbf{Ax + By = C,}$$

*where A, B, and C are real numbers and A and B are not both equal to 0, is called the **standard form** of a **linear equation**.*

NOTES   Note that in the standard form $Ax + By = C$, A and B may be positive, negative, or 0, but A and B **cannot both be 0**.

Every straight line corresponds to some linear equation, and the graph of every linear equation is a straight line. We know from geometry that **two points determine a line.** This means that the graph of a linear equation can be found by locating any two points that satisfy the equation.

To Graph a Linear Equation in Two Variables

1. *Locate any two points that satisfy the equation. (Choose values for x and y that lead to simple solutions. Remember that there are an infinite number of choices for either x or y. But, once a value for x or y is chosen, the corresponding value for the other variable is found by substituting into the equation.)*

2. *Plot these two points on a Cartesian coordinate system.*

3. *Draw a straight line through these two points. [**Note:** Every point on that line will satisfy the equation.]*

4. ***To check:*** *Locate a third point that satisfies the equation and check to see that it does indeed lie on the line.*

Example 3: Graphing a Linear Equation in Two Variables

Graph each of the following linear equations.

a. $2x + 3y = 6$

Solution: Make a table with headings x and y and, whenever possible, **choose values for *x* or *y* that lead to simple solutions for the other variable**. In this example, we have found four ordered pairs that satisfy this equation by choosing two x values and two y values that we felt would result in simple solutions.

x	$2x + 3y = 6$	y
0	$2(0) + 3y = 6$	2
-3	$2(-3) + 3y = 6$	4
3	$2x + 3(0) = 6$	0
$\dfrac{5}{2}$	$2x + 3\left(\dfrac{1}{3}\right) = 6$	$\dfrac{1}{3}$

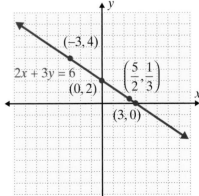

Continued on the next page...

b. $x - 2y = 1$

Solution: Solve for x and substitute $0, 1$, and 2 for y: $x = 2y + 1$.

Results	Substitutions	Choices
x	**x = 2y + 1**	**y**
1	$x = 2(0) + 1$	0
3	$x = 2(1) + 1$	1
5	$x = 2(2) + 1$	2

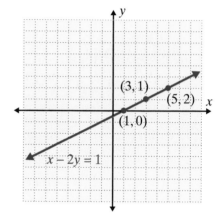

c. $y = 2x$

Solution: Substitute $-1, 0$, and 1 for x.

Choices	Substitutions	Results
x	**y = 2x**	**y**
−1	$y = 2(-1)$	−2
0	$y = 2(0)$	0
1	$y = 2(1)$	2

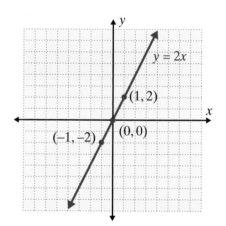

Locating the *y*-intercept and *x*-intercept

While the choice of the values for x or y can be arbitrary, letting $x = 0$ will locate the point on the graph where the line crosses (or intercepts) the *y*-axis. This point is called the **y-intercept** and is of the form $(0, y)$. The **x-intercept** is the point found by letting $y = 0$. This is the point where the line crosses (or intercepts) the *x*-axis and is of the form $(x, 0)$. These two points are generally easy to locate and are frequently used as the two points for drawing the graph of a linear equation. If the line passes through the point $(0, 0)$, then the *y*-intercept and the *x*-intercept are the same point, namely the origin. In this case you will need to locate some other point to draw the graph.

Intercepts

1. To find the **y-intercept** *(where the line crosses the y-axis), substitute x = 0 and solve for y.*

2. To find the **x-intercept** *(where the line crosses the x-axis), substitute y = 0 and solve for x.*

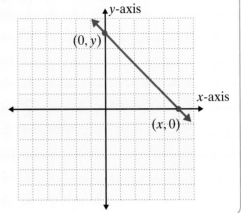

Example 4: *x*- and *y*-Intercepts

Graph the following linear equations by locating the *y*-intercept and the *x*-intercept.

a. $x + 3y = 9$

Solution: $x = 0 \rightarrow$ $(0) + 3y = 9$
$$3y = 9$$
$$y = 3$$
$(0, 3)$ is the *y*-intercept.
$y = 0 \rightarrow$ $x + 3(0) = 9$
$$x = 9$$
$(9, 0)$ is the *x*-intercept.

Plot the two intercepts and draw the line that contains them.

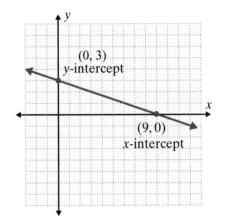

b. $3x - 2y = 12$

Solution: $x = 0 \rightarrow 3(0) - 2y = 12$
$$-2y = 12$$
$$y = -6$$
$(0, -6)$ is the *y*-intercept.
$y = 0 \rightarrow 3x - 2(0) = 12$
$$3x = 12$$
$$x = 4$$
$(4, 0)$ is the *x*-intercept.

Plot the two intercepts and draw the line that contains them.

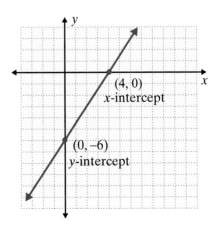

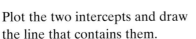

> **NOTES**
>
> In general, the intercepts are easy to find because substituting 0 for x or y leads to an easy solution for the other variable. However, when the intercepts result in a point with fractional (or decimal) coordinates and estimation is involved, then a third point that satisfies the equation should be found to verify that the line is graphed correctly.

Practice Problems

1. For x in the set $\{-1,\ 2,\ 3\}$, find the corresponding ordered pairs that satisfy the equation $x - 2y = 3$.

2. Find the missing coordinate of each ordered pair so that it belongs to the solution set of the equation $2x + y = 4$:
   $(0,\ \ ),\ (\ ,0),\ (\ ,8),\ (-1,\ \ )$.

3. Does the ordered pair $\left(1, \dfrac{3}{2}\right)$ satisfy the equation $3x + 2y = 6$?

4. Find the x-intercept and y-intercept of the equation $-3x + y = 9$.

2.1 Exercises

List the sets of ordered pairs that correspond to the graphs in Exercises 1 – 6. Assume that the grid lines are marked one unit apart.

1.
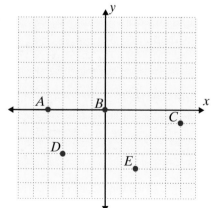

2.
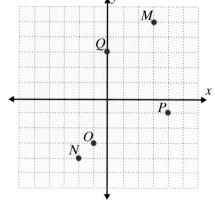

Answers to Practice Problems: 1. $(-1, -2),\ \left(2, -\dfrac{1}{2}\right),\ (3, 0)$ **2.** $(0, 4), (2, 0), (-2, 8), (-1, 6)$ **3.** Yes

4. x-intercept $= (-3, 0)$, y-intercept $= (0, 9)$

3.

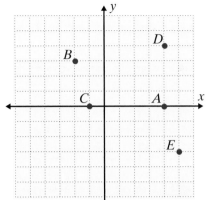

4.

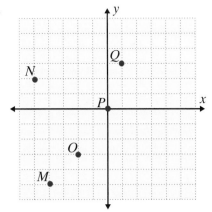

5.

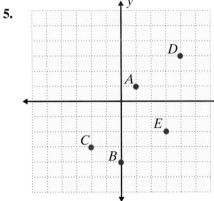

6.

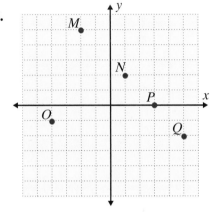

Graph the sets of ordered pairs and label the points in Exercises 7 – 12.

7. $\left\{ A(-5, 1),\, B(-3, 4),\, C(-1, 1),\, D(2, 2),\, E(2, -2) \right\}$

8. $\left\{ M(-4, 1),\, N(-2, 5),\, O(0, 3),\, P(1, 6),\, Q(3, 2) \right\}$

9. $\left\{ P(-3, 2),\, Q(-1, -1),\, R(1, 5),\, S(3, -2),\, T(6, 5) \right\}$

10. $\left\{ C(-5, -5),\, D(-3, 1),\, E(0, -2),\, F(3, 1),\, G(5, 0) \right\}$

11. $\left\{ A(-3, 4),\, B(-2, -1),\, C(-1, 6),\, D(2, 0),\, E(3, -3) \right\}$

12. $\left\{ M(-7, 2),\, N(-4, 5),\, O(0, -4),\, P(1, -2),\, Q(4, -4) \right\}$

In Exercises 13 – 18, the graph of a straight line is shown. Each line has an infinite number of points. List any three points on each line. (There is more than one correct answer.)

13.

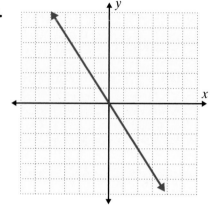

14.

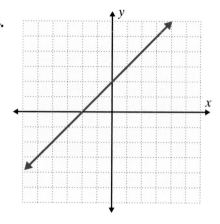

15.

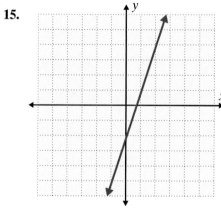

16.

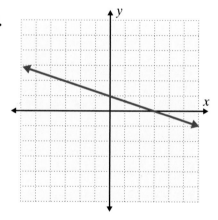

17.

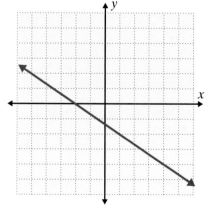

18.

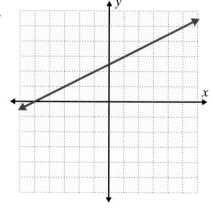

In Exercises 19 – 26, find the missing coordinate of each ordered pair so that the ordered pair belongs to the solution set of the given equation.

19. $2x + y = 5$

x	y
0	
	0
−2	
	3

20. $x + 2y = 6$

x	y
0	
	0
4	
	−2

21. $3x − y = 4$

x	y
0	
	0
2	
	5

22. $x − 3y = 9$

x	y
0	
	0
−3	
	−1

23. $y = 5 − 2x$
  a. (0,   )
  b. (, 0)
  c. (2,   )
  d. (, 7)

24. $y = 5x − 3$
  a. (0,   )
  b. (, 0)
  c. (−1,   )
  d. (, 7)

25. $3x − 2y = 6$
  a. (0,   )
  b. (, 0)
  c. (−2,   )
  d. (, 3)

26. $5x + 2y = 10$
  a. (0,   )
  b. (, 0)
  c. (4,   )
  d. (, 10)

Locate at least two ordered pairs of real numbers that satisfy each of the linear equations in Exercises 27 – 52 and graph the corresponding line in the Cartesian coordinate system.

27. $x + y = 3$ **28.** $x + y = 4$ **29.** $y = x$ **30.** $2y = x$

31. $2x + y = 0$ **32.** $3x + 2y = 0$ **33.** $2x + 3y = 7$ **34.** $4x + 3y = 11$

35. $3x − 4y = 12$ **36.** $2x − 5y = 10$ **37.** $−4x + y = 4$ **38.** $−3x + 3y = 6$

39. $3y = 2x − 4$ **40.** $4x = 3y + 8$ **41.** $3x + 5y = 6$ **42.** $2x + 7y = −4$

43. $2x + 3y = 1$ **44.** $5x − 3y = −1$ **45.** $5x − 2y = 7$ **46.** $3x + 4y = 7$

47. $\dfrac{2}{3}x − y = 4$ **48.** $x + \dfrac{3}{4}y = 6$ **49.** $2x + \dfrac{1}{2}y = 3$ **50.** $\dfrac{2}{5}x − 3y = 5$

51. $5x = y + 2$ **52.** $4x = 3y − 5$

Graph the linear equations in Exercises 53 – 60 by locating the y-intercept and the x-intercept.

53. $x - 2y = 8$

54. $x + y = 6$

55. $2x + 3y = 12$

56. $3x - 7y = -21$

57. $4x - y = 10$

58. $\frac{1}{2}x + 2y = 3$

59. $3x + 2y = 15$

60. $x - 4y = -6$

61. Interest rate: Given the equation $I = 0.06P$, where I is the interest earned on a principal P at the rate of 6%:

   a. Make a table of ordered pairs for the values of P and I if P has the values $1000, $2000, $4000, $5000, and $7000.

   b. Graph the points corresponding to the ordered pairs.

P	I
1000	
2000	
4000	
5000	
7000	

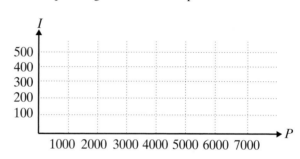

62. Volume of a box: Given the equation $V = 25h$, where V is the volume (in cm^3) of a box with a variable height h in cm and a fixed base of area 25 cm^2:

   a. Make a table of ordered pairs for the values of h and V with h as the values 2 cm, 3 cm, 8 cm, 10 cm, and 12 cm.

   b. Graph the points corresponding to the ordered pairs.

h	V
2	
3	
8	
10	
12	

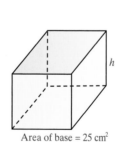

Area of base = 25 cm^2

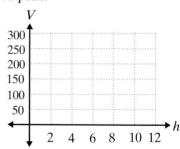

Writing and Thinking About Mathematics

In statistics, data is sometimes given in the form of ordered pairs where each ordered pair represents two pieces of information about one person. For example, ordered pairs might represent the height and weight of a person or the person's number of years of education and that person's annual income. The ordered pairs are plotted on a graph and the graph is called a **scatter diagram** (or **scatter plot**). Such scatter diagrams are used to see if there is any pattern to the data and, if there is, then the diagram is used to predict the value for one of the variables if the value of the other is known. For example, if you know that a person's height is 5 ft 6 in., then his or her weight might be predicted from information indicated in a scatter diagram that has several points of known information about height and weight.

63. Exercise: The following table of values gives the number of push-ups and pull-ups completed by ten students in a physical education class.
   a. Plot these points on a scatter diagram.

Person	#1	#2	#3	#4	#5	#6	#7	#8	#9	#10
Push-ups	20	15	25	23	35	30	42	40	25	35
Pull-ups	5	2	9	8	10	11	15	14	7	12

   b. Does there seem to be a pattern in the relationship between push-ups and pull-ups? If so, what is this pattern?
   c. Using the scatter diagram in Part **a**, predict the number of pull-ups that a student might be able to do if he or she has just done each of the following numbers of push-ups: 22, 32, 35, and 45. (**Note:** In each case, there is no one correct answer. The answers are only estimates based on the diagram.)

64. Explain in your own words why it is sufficient to find the x-intercept and y-intercept to graph a line (assuming that they are not the same point).

65. Explain in your own words how you can determine if an ordered pair is a solution to an equation.

 HAWKES LEARNING SYSTEMS: INTERMEDIATE ALGEBRA SOFTWARE

- Introduction to the Cartesian Coordinate System
- Graphing Linear Equations by Plotting Points

2.2 Slope-Intercept Form: $y = mx + b$

- *Interpret the slope of a line as a rate of change.*
- *Find the slopes of lines given two points.*
- *Find the slopes of and graph horizontal and vertical lines.*
- *Find the slopes and y-intercepts of lines and then graph the lines.*
- *Write the equations of lines given the slopes and y-intercepts.*

The Meaning of Slope

If you ride a bicycle up a mountain road, you certainly know when the **slope** (a measure of steepness) increases because you have to pedal harder. The contractor who built the road was aware of the **slope** because trucks traveling the road must be able to control their downhill speed and be able to stop in a safe manner. A carpenter given a set of house plans calling for a roof with a **pitch** of 7 : 12 knows that for every 7 feet of rise (vertical distance) there are 12 feet of run (horizontal distance). That is, the ratio of rise to run is $\dfrac{rise}{run} = \dfrac{7}{12}$.

Figure 2.3

Note that this ratio can be in units other than feet, such as inches or meters. (See Figure 2.4.)

$$\frac{rise}{run} = \frac{7}{12} = \frac{3.5}{6} = \frac{14}{24}$$

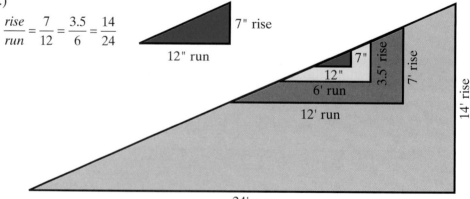

Figure 2.4

For a straight line, the **ratio of rise to run** is called the **slope of the line**. The graph of the linear equation $y = \dfrac{1}{3}x + 2$ is shown in Figure 2.5. What do you think is the slope of the line? Do you think that the slope is positive or negative? Do you think the slope might be $\dfrac{1}{3}$? $\dfrac{3}{1}$? 2?

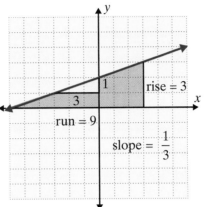

Figure 2.5

The concept of slope also relates to situations that involve **rate of change**. For example, the graphs in Figure 2.6 illustrate slope as miles per hour that a car can travel and as pages per minute that a printer can print.

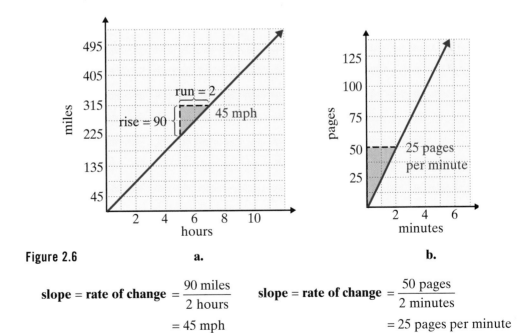

Figure 2.6   **a.**   **b.**

slope = rate of change $= \dfrac{90 \text{ miles}}{2 \text{ hours}}$ **slope = rate of change** $= \dfrac{50 \text{ pages}}{2 \text{ minutes}}$

$\qquad\qquad\qquad = 45 \text{ mph}$ $= 25 \text{ pages per minute}$

In general, the ratio of a change in one variable (say y) to a change in another variable (say x) is called the **rate of change of y with respect to x.** Figure 2.7 shows how the rate of change (the slope) can change over periods of time.

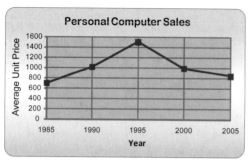

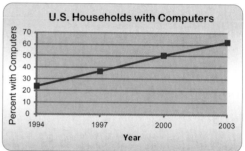

Source: Consumer Electonics Association

Source: U.S. Dept. of Commerce

Figure 2.7

Calculating the Slope: $m = \dfrac{y_2 - y_1}{x_2 - x_1}$

Consider the line $y = 2x + 3$ and two points on the line $P_1(-2, -1)$ and $P_2(2, 7)$ as shown in Figure 2.8.

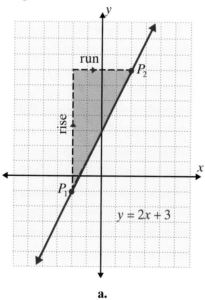

a.

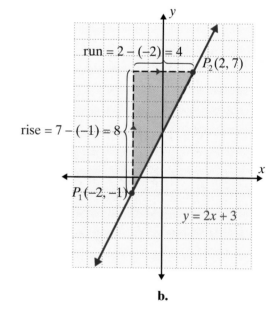

b.

Figure 2.8

NOTES   In the notation P_1, 1 is called a **subscript** and P_1 is read "P sub 1". Similarly, P_2 is read "P sub 2."

For the line $y = 2x + 3$ and using the points $(-2, -1)$ and $(2, 7)$ that are on the line,

$$\textbf{slope} = \frac{\text{rise}}{\text{run}} = \frac{\text{difference in } y\text{-values}}{\text{difference in } x\text{-values}} = \frac{7 - (-1)}{2 - (-2)} = \frac{8}{4} = 2.$$

From similar illustrations and the use of subscript notation, we can develop the following formula for the slope of any line.

Slope

Let $P_1(x_1, y_1)$ and $P_2(x_2, y_2)$ be two points on a line. The **slope** can be calculated as follows:

$$\textit{slope} = \textit{m} = \frac{\textit{rise}}{\textit{run}} = \frac{y_2 - y_1}{x_2 - x_1}.$$

Note: The letter m is standard notation for representing the slope of a line.

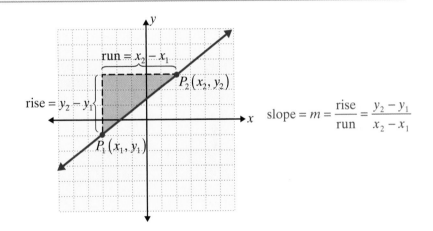

Figure 2.9

Example 1: Finding the Slope of a Line

Find the slope of the line that contains the points $(-1, 2)$ and $(3, 5)$, and then graph the line.

Solution: Using $(-1, 2)$ and $(3, 5)$,   slope $= m = \dfrac{y_2 - y_1}{x_2 - x_1}$

(x_1, y_1) (x_2, y_2)

$$= \frac{5 - 2}{3 - (-1)}$$

$$= \frac{3}{4}$$

Continued on the next page...

or using $(3, 5)$ and $(-1, 2)$,

(x_1, y_1) (x_2, y_2)

$$\text{slope} = m = \frac{y_2 - y_1}{x_2 - x_1}$$

$$= \frac{2 - 5}{-1 - 3}$$

$$= \frac{-3}{-4}$$

$$= \frac{3}{4}$$

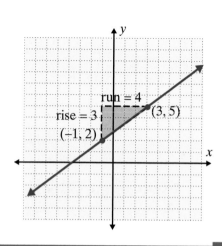

As we see in Example 1, **the slope is the same even if the order of the points is reversed**. The important part of the procedure is that **the coordinates must be subtracted in the same order in both the numerator and the denominator**.

In general,

$$\text{slope} = \frac{y_2 - y_1}{x_2 - x_1} = \frac{y_1 - y_2}{x_1 - x_2}.$$

Example 2: Finding the Slope of a Line

Find the slope of the line that contains the points $(1, 3)$ and $(5, 1)$, and then graph the line.

Solution: Using $(1, 3)$ and $(5, 1)$,

(x_1, y_1) (x_2, y_2)

$$\text{slope} = m = \frac{3 - 1}{1 - 5}$$

$$= \frac{2}{-4}$$

$$= -\frac{1}{2}$$

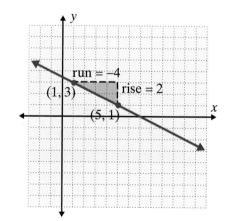

NOTES   Lines with **positive slope go up** as we move along the line from left to right.

Lines with **negative slope go down** as we move along the line from left to right.

Slopes of Horizontal and Vertical Lines

Suppose that two points on a line have the same *y*-coordinate, such as (–2, 3) and (5, 3). Then the line through these two points will be **horizontal** as shown in Figure 2.10. In this case, the *y*-coordinates of the horizontal line are all 3, and the equation of the line is simply *y* = 3. The slope is

$$m = \frac{3-3}{5-(-2)} = \frac{0}{7} = 0.$$

For any horizontal line, all of the *y*-values will be the same. Consequently, the formula for slope will always have 0 in the numerator. Therefore, **the slope of every horizontal line is 0**.

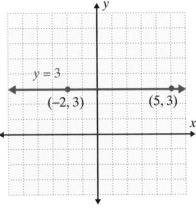

Figure 2.10

If two points have the same *x*-coordinates, such as (1, 3) and (1, –2), then the line through these two points will be **vertical** as in Figure 2.11. The *x*-coordinates for every point on the vertical line are all 1, and the equation of the line is simply *x* = 1. The slope is

$$m = \frac{-2-3}{1-1} = \frac{-5}{0}, \text{ which is } \textbf{undefined}.$$

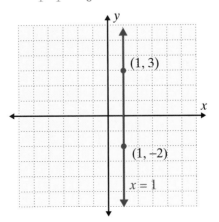

Figure 2.11

Horizontal and Vertical Lines

The following two general statements are true for horizontal and vertical lines:

1. For **horizontal lines** *(of the form $y = b$)*, the **slope is 0**.
2. For **vertical lines** *(of the form $x = a$)*, the **slope is undefined**.

Example 3: Slopes of Horizontal and Vertical Lines

a. Find the equation and slope of the horizontal line through the point $(-2, 5)$.

Solution: The equation is $y = 5$ and the slope is 0.

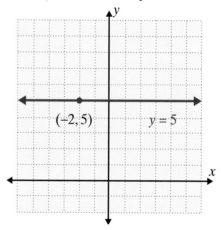

b. Find the equation and slope of the vertical line through the point $(3, 2)$.

Solution: The equation is $x = 3$ and the slope is undefined.

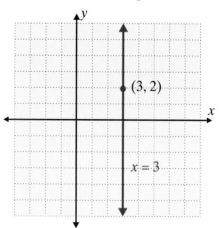

Slope-Intercept Form: $y = mx + b$

There are certain relationships between the coefficients in the equation of a line and the graph of that line. For example, consider the equation

$$y = 5x - 7.$$

First, find two points on the line and calculate the slope. $(0, -7)$ and $(2, 3)$ both satisfy the equation.

$$\text{slope} = m = \frac{-7 - 3}{0 - 2} = \frac{-10}{-2} = 5$$

$$\text{or } m = \frac{3 - (-7)}{2 - 0} = \frac{10}{2} = 5$$

Observe that the slope, $m = 5$, is the same as the coefficient of x in the equation $y = 5x - 7$. This is not just a coincidence. In fact, if a linear equation is solved for y, then the coefficient of x will always be the slope of the line.

For $y = mx + b$, m is the Slope

Statement: *Given an equation in the form $y = mx + b$, the slope of the line is m.*

Proof: *Suppose that the equation is solved for y and $y = mx + b$. Let (x_1, y_1) and (x_2, y_2) be two points on the line where $x_1 \neq x_2$. Then $y_1 = mx_1 + b$ and $y_2 = mx_2 + b$. The slope can be calculated as follows:*

$$slope = \frac{y_2 - y_1}{x_2 - x_1} = \frac{(mx_2 + b) - (mx_1 + b)}{x_2 - x_1}$$

$$= \frac{mx_2 + b - mx_1 - b}{x_2 - x_1}$$

$$= \frac{mx_2 - mx_1}{x_2 - x_1}$$

$$= \frac{m(x_2 - x_1)}{x_2 - x_1} = m.$$

Therefore, for an equation of the form $y = mx + b$, the slope of the line is m.

For the line $y = mx + b$, the point where $x = 0$ is the point where the line crosses the y-axis. This point is called the **y-intercept**. By letting $x = 0$, we get

$$y = mx + b$$

$$y = m \cdot 0 + b$$

$$y = b.$$

Thus the point $(0, b)$ is the y-intercept. The concepts of slope and y-intercept lead to the following definition.

Slope–Intercept Form

$y = mx + b$ is called the **slope-intercept form** for the equation of a line. m is the **slope** and $(0, b)$ is the **y-intercept**.

As illustrated in Example 4, an equation in the **standard form**

$$Ax + By = C \qquad \text{with } B \neq 0$$

can be written in the slope-intercept form by solving for y.

Example 4: Using the Form $y = mx + b$

a. Find the slope and y-intercept of $-2x + 3y = 6$ and graph the line.

Solution: Solve for y.

$$-2x + 3y = 6$$

$$3y = 2x + 6$$

$$\frac{3y}{3} = \frac{2x}{3} + \frac{6}{3}$$

$$y = \frac{2}{3}x + 2$$

Thus slope $= m = \dfrac{2}{3}$

and y-intercept $= (0, 2)$.

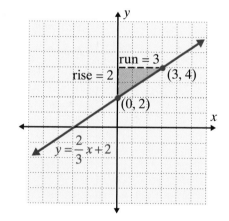

As shown in the graph, if we "rise" 2 units up and "run" 3 units to the right **from the y-intercept (0, 2)**, we locate another point $(3, 4)$. The line can be drawn through these two points.

Note: As shown in the graph on the right, we could also first "rise" 2 units down and "run" 3 units left from the y-intercept to locate the point $(-3, 0)$ on the graph. That is, we can interpret the slope $m = \dfrac{2}{3}$ as $m = \dfrac{-2}{-3}$.

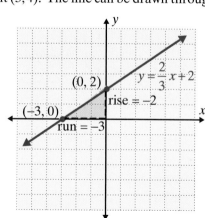

b. Find the slope and *y*-intercept of $x + 2y = -6$ and graph the line.

Solution: Solve for *y*.

$$x + 2y = -6$$

$$2y = -x - 6$$

$$\frac{2y}{2} = \frac{-x}{2} - \frac{6}{2}$$

$$y = -\frac{1}{2}x - 3$$

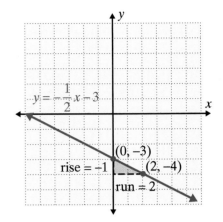

Thus slope $= m = -\dfrac{1}{2}$ and the *y*-intercept $= (0, -3)$

We can treat $m = -\dfrac{1}{2}$ as $m = \dfrac{-1}{2}$ and the "rise" as -1 and the "run" as 2. Moving from $(0, -3)$ as shown in the figure, we locate another point $(2, -4)$ on the graph and draw the line.

c. Find the equation of the line through the point $(0, -2)$ with slope $\dfrac{1}{2}$.

Solution: The point $(0, -2)$ is the *y*-intercept. So $b = -2$. The slope is $\dfrac{1}{2}$. So $m = \dfrac{1}{2}$. Substituting in slope-intercept form $y = mx + b$ gives the result:

$$y = \frac{1}{2}x - 2.$$

Practice Problems

1. Find the slope of the line through the two points (1, 3) and (4, 6). Graph the line.

2. Find the equation of the line through the point (0, 5) with slope $-\dfrac{1}{3}$.

3. Find the slope and y-intercept for the line $2x + y = 7$.

4. Write the equation for the horizontal line through the point (−1, 3). What is the slope of this line?

5. Write the equation for the vertical line through the point (−1, 3). What is the slope of this line?

2.2 Exercises

Find the slope of the line determined by each pair of points given in Exercises 1 – 12.

1. (2, 4); (1, −1)   **2.** (5, 1); (3, 0)   **3.** (−3, 7); (4, −1)

4. (−6, 3); (1, 2)   **5.** (−5, 8); (3, 8)   **6.** (0, 0); (−2, −3)

7. $\left(4, \dfrac{1}{2}\right)$; (−1, 2)   **8.** $\left(\dfrac{3}{4}, \dfrac{3}{2}\right)$; (1, 2)   **9.** (−2, 3); (−2, −1)

10. (1, −2); (1, 4)   **11.** $\left(\dfrac{3}{2}, \dfrac{4}{5}\right)$; $\left(-2, \dfrac{1}{10}\right)$ **12.** $\left(\dfrac{7}{2}, \dfrac{3}{4}\right)$; $\left(\dfrac{1}{2}, -3\right)$

Tell whether each equation in Exercises 13 – 20 represents a horizontal line or vertical line and give its slope. Graph the line.

13. $y = 5$ **14.** $y = -2$ **15.** $x = -3$ **16.** $x = 1.7$

17. $3y = -18$ **18.** $4x = 2.4$ **19.** $-3x + 21 = 0$ **20.** $2y + 5 = 0$

Answers to Practice Problems: **1.** $m = 1$ **2.** $y = -\dfrac{1}{3}x + 5$ **3.** $m = -2$, y-intercept $= (0, 7)$

4. $y = 3$, slope is 0

5. $x = -1$, slope is undefined

For Exercises 21 – 48, write the equation in slope-intercept form. Find the slope and the y-intercept, and then draw the graph.

21. $y = 2x - 1$ **22.** $y = 3x - 4$ **23.** $y = 5 - 4x$ **24.** $y = 4 - x$

25. $y = \dfrac{2}{3}x - 3$ **26.** $y = \dfrac{2}{5}x + 2$ **27.** $x + y = 5$ **28.** $x - 2y = 6$

29. $x + 5y = 10$ **30.** $4x + y + 3 = 0$ **31.** $2y - 8 = 0$ **32.** $2x + 7y + 7 = 0$

33. $4x + y = 0$ **34.** $3y - 9 = 0$ **35.** $2x = 3y + 6$ **36.** $4x = y + 2$

37. $3x + 9 = 0$ **38.** $3x + 6 = 6y$ **39.** $5x - 6y = 10$ **40.** $4x + 7 = 0$

41. $5 - 3x = 4y$ **42.** $5x = 11 - 2y$ **43.** $6x + 4y = -7$ **44.** $7x + 2y = 4$

45. $6y = 4 + 3x$ **46.** $6x + 5y = -15$ **47.** $5x - 2y + 5 = 0$ **48.** $4x = 3y - 7$

49. In reference to the equation $y = mx + b$, sketch the graph of three lines for each of the two characteristics listed below.

    a. $m > 0$ and $b > 0$ **b.** $m < 0$ and $b > 0$
    c. $m > 0$ and $b < 0$ **d.** $m < 0$ and $b < 0$

*In Exercises 50 – 57 the graph of a line is shown with two points highlighted. Find **a.** the slope, **b.** the y-intercept (if there is one), and **c.** the equation of the line.*

50.

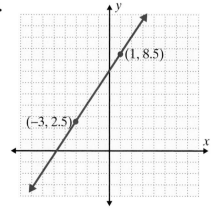

51.

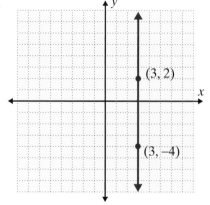

52.

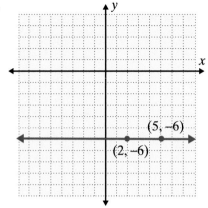

53.

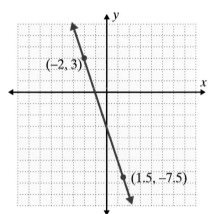

54.

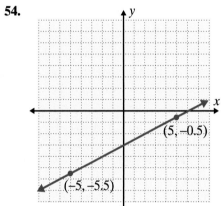

55.

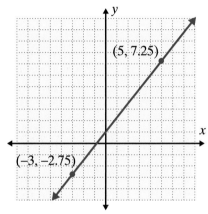

56.

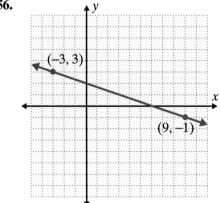

57.

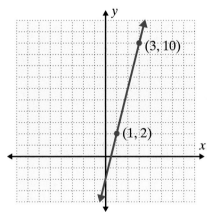

*Points are said to be **collinear** if they are on a straight line. If points are collinear, then the slope of the line through any two of them must be the same (because the line is the same line). Use this idea to determine whether or not the three points in each of the sets in Exercises 58 – 62 are collinear.*

58. $\{(-1, 3), (0, 1), (5, -9)\}$

59. $\{(-2, -4), (0, 2), (3, 11)\}$

60. $\{(-2, 0), (0, 30), (1.5, 5.25)\}$

61. $\left\{ \left(\dfrac{2}{3}, \dfrac{1}{2} \right), \left(0, \dfrac{5}{6} \right), \left(-\dfrac{3}{4}, \dfrac{29}{24} \right) \right\}$

62. $\{(-1, -7), (1, 1), (4, 12)\}$

63. Buying a new car: John bought his new car for $35,000 in the year 2004. He knows that the value of his car has depreciated linearly. If the value of the car in 2007 was $23,000, what was the annual rate of depreciation of his car? Show this information on a graph (graph years versus values of the cars).

64. Cell phone usage: The number of people in the United States with mobile cellular phones was about 142 million in 2002 and about 255 million in 2007. If the growth in mobile cellular phones was linear, what was the approximate rate of growth per year from 2002 to 2007. Show this information on a graph (graph years versus the number of users).

65. Internet usage: The given table shows the estimated number of internet users from 2003-2007. The number of users for each year is shown in millions.

Year	Internet Users
2003	162
2004	185
2005	198
2006	210
2007	220

a. Plot these points on a graph.
b. Connect the points with line segments.
c. Find the slope of each line segment.
d. Interpret the slope as a rate of change.

Source: International Telecommunications Union
Yearbook of Statistics

66. Urban growth: The following table shows the urban growth from 1850 to 2000 in New York, NY.

Year	Population
1850	515,547
1900	3,437,202
1950	7,891,957
2000	8,008,278

a. Plot these points on a graph.
b. Connect the points with line segments.
c. Find the slope of each line segment.
d. Interpret the slope as a rate of change.

Source: U.S. Census Bureau

67. Military: The following graph shows the number of female active duty military personnel over a span from 1945 to 2007. The number of women listed includes both officers and enlisted personnel from the Army, the Navy, the Marine Corps, and the Air Force.

a. Plot these points on a graph.
b. Connect the points with line segments.
c. Find the slope of each line segment.
d. Interpret the slope as a rate of change.

Source: U.S. Dept. of Defense

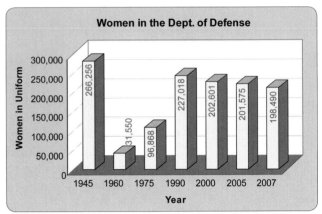

68. Marriage: The following graph shows the rates of marriage per 1000 people in the U.S., over a span from 1920 to 2007.

a. Plot these points on a graph.
b. Connect the points with line segments.
c. Find the slope of each line segment.
d. Interpret the slope as a rate of change.

Source: U.S. National Center for Health Statistics

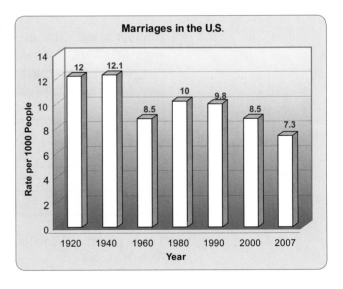

Writing and Thinking About Mathematics

Each of the following equations is not linear and the corresponding graph is not a straight line. Make a table and find several points, some where x is positive and some where x is negative, to determine the nature of the corresponding graph. After you have plotted your points and analyzed the nature of the graph, enter the expression for y in your graphing calculator to verify that you have a reasonably accurate graph.

69. $y = \dfrac{2}{x}$ **70.** $y = x^2$ **71.** $y = -2x^2$

72. $y = x^2 - 4$ **73.** $y = -x^3$

 HAWKES LEARNING SYSTEMS: INTERMEDIATE ALGEBRA SOFTWARE

- Graphing Linear Equations in Slope-Intercept Form

2.3

Point-Slope Form: $y - y_1 = m(x - x_1)$

- *Graph a line given its slope and one point on the line.*
- *Find the equation of a line given its slope and one point on the line by using the formula $y - y_1 = m(x - x_1)$.*
- *Find the equation of a line given two points on the line.*

Graphing a Line Given a Point and the Slope

Lines represented by equations in the **standard form** $Ax + By = C$ and in the **slope-intercept form** $y = mx + b$ have been discussed in Sections 2.1 and 2.2. In Section 2.2 we graphed lines by using the y-intercept $(0, b)$ and the slope by moving horizontally and then vertically (or by moving vertically and then horizontally) from the y-intercept. This same technique can be used to graph lines if the given point is on the line but is not the y-intercept. Consider the following example.

Example 1: Graph a Line Given a Point and the Slope

Graph the line with slope $m = -\dfrac{3}{4}$ which passes through the point $(2, 5)$.

Solution: Start from the point $(2, 5)$ and locate another point on the line by using the slope as $\dfrac{rise}{run} = \dfrac{-3}{4} = \dfrac{3}{-4}$. There are four ways to proceed. Here are two:

1. Move 4 units right and 3 units down, or
2. Move 3 units down and 4 units right.

Either way, you arrive at the same point $(6, 2)$.

This means that we can move from the given point either with the rise first or the run first.

(**Note:** Any numbers in the ratio of –3 to 4 can be used for the moves, such as –6 to 8 or 9 to –12.)

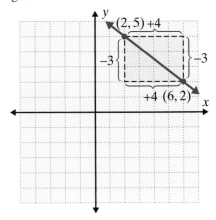

Point-Slope Form: $y - y_1 = m(x - x_1)$

Now consider finding the equation of the line given the point (x_1, y_1) on the line and the slope m. If (x, y) is **any other point** on the line, then the slope formula gives the equation

$$\frac{y - y_1}{x - x_1} = m.$$

Multiplying both sides of this equation by the denominator (assuming the denominator is not 0) gives

$$y - y_1 = m(x - x_1) \qquad \text{which is called the \textbf{point-slope form}.}$$

For example, suppose that a point $(x_1, y_1) = (8, 3)$ and the slope $m = -\dfrac{3}{4}$ are given.

If (x, y) represents any point on the line other than $(8, 3)$, then substituting into the formula for slope, gives

$$\frac{y - y_1}{x - x_1} = m \qquad\qquad \text{Formula for slope}$$

$$\frac{y - 3}{x - 8} = -\frac{3}{4} \qquad\qquad \text{Substitute the given information.}$$

$$(x - 8)\left(\frac{y - 3}{x - 8}\right) = -\frac{3}{4}(x - 8) \qquad \text{Multiply both sides by } (x - 8).$$

$$y - 3 = -\frac{3}{4}(x - 8). \qquad \text{Point-slope form: } y - y_1 = m(x - x_1)$$

From this point-slope form, we can manipulate the equation to get the other two forms:

$$y - 3 = -\frac{3}{4}(x - 8)$$

$$y - 3 = -\frac{3}{4}x + 6$$

$$\text{or} \qquad y = -\frac{3}{4}x + 9 \qquad \text{Slope-intercept form: } y = mx + b$$

$$\text{or} \qquad 3x + 4y = 36. \qquad \text{Standard form: } Ax + By = C$$

Point-Slope Form

An equation of the form

$$y - y_1 = m(x - x_1)$$

*is called the **point-slope form** for the equation of a line that contains the point (x_1, y_1) and has slope m.*

Example 2: Using the Point-Slope Form

Find the equation of the line with a slope of $-\dfrac{1}{2}$ and passing through the point $(2, 3)$. Graph the line using the point and slope.

Solution: Substitute into the point-slope form.

$$y - y_1 = m(x - x_1)$$

$$y - 3 = -\frac{1}{2}(x - 2)$$ Point-slope form

$$y - 3 = -\frac{1}{2}x + 1$$

or $$y = -\frac{1}{2}x + 4$$ Slope-intercept form

or $x + 2y = 8$ Standard form

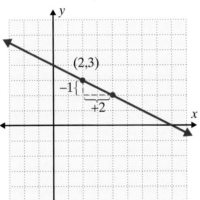

The point one unit down and two units right from $(2, 3)$ will be on the line because the slope is $m = \dfrac{\text{rise}}{\text{run}} = \dfrac{-1}{2} = -\dfrac{1}{2}$.

With a negative slope, either the rise is negative and the run is positive, or the rise is positive and the run is negative. In either case, as the previous figure and the following figure illustrate, the line is the same.

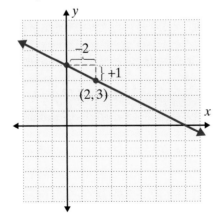

The point one unit up and two units to the left from $(2, 3)$ is on the line because the slope is $m = \dfrac{\text{rise}}{\text{run}} = \dfrac{1}{-2} = -\dfrac{1}{2}$.

In Example 2, the equation of the line is written in all three forms: point-slope form, slope-intercept form, and standard form. Generally any one of these forms is sufficient. However, there are situations in which one form is preferred over the others. Therefore, manipulation among the forms is an important skill. Also, if the answer in the text is in one form and your answer is in another form, you should be able to recognize that the answers are equivalent.

Finding the Equations of Lines Given Two Points

Given two points that lie on a line, the equation of the line can be found by using the following method.

Finding the Equation of a Line

If two points are given,

1. *Use the formula* $m = \dfrac{y_2 - y_1}{x_2 - x_1}$ *to find the slope.*

2. *Use this slope,* **m**, *and either point in the point-slope formula* $y - y_1 = m(x - x_1)$.

Example 3: Using Two Points to Find the Equation of a Line

Find the equation of the line containing the two points $(-1, 2)$ and $(4, -2)$.

Solution: First, find the slope.

$$m = \frac{y_2 - y_1}{x_2 - x_1}$$

$$= \frac{-2 - 2}{4 - (-1)}$$

$$= \frac{-4}{5}$$

$$= -\frac{4}{5}$$

Now use one of the given points and the point-slope form for the equation of a line. $[(-1, 2)$ and $(4, -2)$ are used here to illustrate that either point may be used.]

Continued on the next page...

<u>Using $(-1, 2)$</u>		<u>Using $(4, -2)$</u>
$y - y_1 = m(x - x_1)$	Point-slope form	$y - y_2 = m(x - x_2)$
$y - 2 = -\dfrac{4}{5}\left[x - (-1)\right]$	Substitute.	$y - (-2) = -\dfrac{4}{5}(x - 4)$
$y - 2 = -\dfrac{4}{5}x - \dfrac{4}{5}$	Distribute.	$y + 2 = -\dfrac{4}{5}x + \dfrac{16}{5}$
$y = -\dfrac{4}{5}x - \dfrac{4}{5} + 2$		$y = -\dfrac{4}{5}x + \dfrac{16}{5} - 2$
$y = -\dfrac{4}{5}x + \dfrac{6}{5}$	Slope-intercept form	$y = -\dfrac{4}{5}x + \dfrac{6}{5}$
or $4x + 5y = 6$	Standard form	$4x + 5y = 6$

Parallel Lines and Perpendicular Lines

Parallel and Perpendicular Lines

Parallel lines are lines that never intersect (cross each other) and these lines have the **same slope**. All vertical lines are parallel to one another.

Perpendicular lines are lines that intersect at 90° (right) angles and whose slopes are **negative reciprocals** of each other. Horizontal lines are perpendicular to vertical lines.

As illustrated in Figure 2.12, the lines $y = 2x + 1$ and $y = 2x - 3$ are **parallel**. They have the same slope. The lines $y = \dfrac{2}{3}x + 1$ and $y = -\dfrac{3}{2}x - 2$ are **perpendicular**. Their slopes are negative reciprocals of each other. In other words, the product of their slopes is -1 $\left(-\dfrac{3}{2} \cdot \dfrac{2}{3} = -1\right)$.

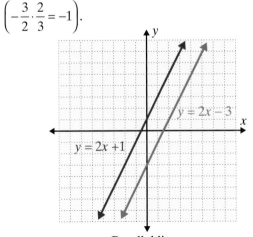

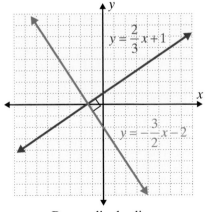

Figure 2.12   Parallel lines   Perpendicular lines

Example 4: Finding the Equations of Parallel Lines

Find the equation of the line through the point (2, 3) parallel to the line $5x + 3y = 1$. Graph both lines.

Solution: First, solve for y to find the slope of the given line.

$$5x + 3y = 1$$
$$3y = -5x + 1$$
$$y = -\frac{5}{3}x + \frac{1}{3}$$

Thus any line parallel to this line has slope $-\frac{5}{3}$.

Now use the point-slope form $y - y_1 = m(x - x_1)$ with $m = -\frac{5}{3}$ and $(x_1, y_1) = (2, 3)$.

$$y - 3 = -\frac{5}{3}(x - 2)$$ Point-slope form

$$3(y - 3) = -5(x - 2)$$ Multiply both sides by 3.

$$3y - 9 = -5x + 10$$ Simplify.

$$5x + 3y = 19$$ Standard form

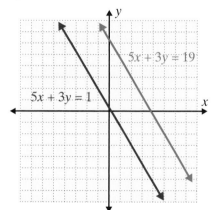

Example 5: Finding the Equations of Perpendicular Lines

Find the equation of the line through the point (2, 3) perpendicular to the line $5x + 3y = 1$. Graph both lines.

Solution: We know from Example 4 that the slope of the line $5x + 3y = 1$ is $-\frac{5}{3}$. Thus any line perpendicular to this line must have slope $m = \frac{3}{5}$.

Continued on the next page...

Now, using $m = \dfrac{3}{5}$, and $y - y_1 = m(x - x_1)$, we have

$$y - 3 = \frac{3}{5}(x - 2) \qquad \text{Point-slope form}$$

$$5(y - 3) = 3(x - 2) \qquad \text{Multiply both sides by 5.}$$

$$5y - 15 = 3x - 6 \qquad \text{Simplify.}$$

$$3x - 5y = -9 \qquad \text{Standard form}$$

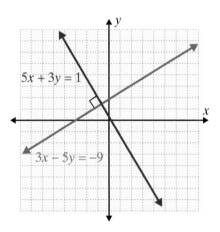

For easy reference, the following table summarizes what we know about straight lines.

Summary of Formulas and Properties of Straight Lines

1.	$Ax + By = C$	*Standard form*
2.	$m = \dfrac{y_2 - y_1}{x_2 - x_1}$	*Slope of a line*
3.	$y = mx + b$	*Slope-intercept form*
4.	$y - y_1 = m(x - x_1)$	*Point-slope form*
5.	$y = b$	*Horizontal line, slope 0*
6.	$x = a$	*Vertical line, undefined slope*
7.	*Parallel lines have the same slope.*	
8.	*Perpendicular lines have slopes that are negative reciprocals of each other.*	

Practice Problems

Find a linear equation in standard form that satisfies the given conditions.

1. Passes through the point (4, –1) with m = 2

2. Parallel to y = –3x + 4 and contains the point (–1, 5)

3. Perpendicular to 2x + y = 1 and passes through the origin (0, 0)

4. Contains the two points (6, –2) and (2, 0)

2.3 Exercises

In Exercises 1 – 6, find **a.** the slope, **b.** a point on the line, and **c.** the graph of the line for the given equations in point-slope form.

1. $y - 1 = 2(x - 3)$
2. $y - 4 = \dfrac{1}{2}(x - 1)$
3. $y + 2 = -5(x)$

4. $y - 3 = -\dfrac{1}{4}(x + 2)$
5. $y = -(x + 8)$
6. $y + 6 = \dfrac{1}{3}(x - 7)$

In Exercises 7 – 16, find an equation in standard form for the line passing through the given point with the given slope. Graph the line.

7. $(-2, 1)$; $m = -2$
8. $(3, 4)$; $m = 3$
9. $(5, -2)$; $m = 0$

10. $(-3, 6)$; $m = \dfrac{1}{2}$
11. $(-3, -1)$; m is undefined

12. $(7, 10)$; $m = \dfrac{3}{5}$
13. $(-1, -1)$; $m = -\dfrac{1}{4}$
14. $(0, 0)$; $m = -3$

15. $\left(-2, \dfrac{1}{3}\right)$; $m = \dfrac{2}{3}$
16. $\left(\dfrac{5}{2}, \dfrac{1}{2}\right)$; $m = -\dfrac{4}{3}$

In Exercises 17 – 26, find an equation in slope-intercept form for the line passing through the two given points.

17. $(-5, 2)$; $(3, 6)$
18. $(-3, 4)$; $(2, 1)$
19. $(-5, 1)$; $(2, 0)$
20. $(-4, -4)$; $(3, 1)$

21. $(0, 2)$; $\left(1, \dfrac{3}{4}\right)$
22. $\left(\dfrac{5}{2}, 0\right)$; $\left(2, -\dfrac{1}{3}\right)$
23. $(2, -5)$; $(4, -5)$

24. $(0, 4)$; $\left(1, \dfrac{1}{2}\right)$
25. $(-2, 6)$; $(3, 1)$
26. $(8, 2)$; $(0, 0)$

Answers to Practice Problems: 1. $2x - y = 9$ **2.** $3x + y = 2$ **3.** $x - 2y = 0$ **4.** $x + 2y = 2$

Find the slope-intercept form of the equations described in Exercises 27 – 42.

27. Find an equation for the horizontal line through the point $(-2, 6)$.

28. Find an equation for the vertical line through the point $(-1, -4)$.

29. Write an equation for the line parallel to the x-axis passing through $(2, 7)$.

30. Write an equation for the line parallel to the y-axis passing through $(-2, 0.5)$.

31. Find an equation for the line parallel to the y-axis containing the point $(2, -4)$.

32. Find an equation for the line parallel to the line $-6y = 1$ containing the point $(-3, 2)$.

33. Write an equation for the line parallel to the line $2x - y = 4$ containing the origin. Graph both lines.

34. Find an equation for the line parallel to $7x - 3y = 1$ containing the point $(1, 0)$. Graph both lines.

35. Write an equation for the line parallel to $5x = 7 + y$ through the point $(-1, -3)$. Graph both lines.

36. Write an equation for the line that contains the point $(2, 2)$ and is perpendicular to the line $4x + 3y = 4$. Graph both lines.

37. Find an equation for the line that passes through the point $(4, -1)$ and is perpendicular to the line $5x - 3y + 4 = 0$. Graph both lines.

38. Write an equation for the line perpendicular to $8 - 3x - 2y = 0$ through $(-4, -2)$.

39. Find an equation for the line perpendicular to $x = 4$ that passes through $(-1, 7)$.

40. Write an equation for the line through the origin perpendicular to $3x - y = 4$.

41. Write an equation for the line perpendicular to $2x - y = 7$ with the same y-intercept as $x - 3y = 6$.

42. Find an equation for the line with the same y-intercept as $5x + 4y = 12$ that is perpendicular to $3x - 2y = 4$.

43. Show that the points $A(-2, 4)$, $B(0, 0)$, $C(6, 3)$, and $D(4, 7)$ are the vertices of a rectangle. (Plot the points and show that opposite sides are parallel and that adjacent sides are perpendicular.)

44. Show that the points $A(0, -1)$, $B(3, -4)$, $C(6, 3)$, and $D(9, 0)$ are the vertices of a parallelogram. (Plot the points and show that opposite sides are parallel.)

*In Exercises 45 – 50, determine whether each pair of lines is (a) parallel, (b) perpendicular, or (c) neither. Graph both lines. (**Hint:** Write the equations in slope-intercept form and then compare slopes and y-intercepts.)*

45. $\begin{cases} y = -2x + 3 \\ y = -2x - 1 \end{cases}$

46. $\begin{cases} y = 3x + 2 \\ y = -\dfrac{1}{3}x + 6 \end{cases}$

47. $\begin{cases} 4x + y = 4 \\ x - 4y = 8 \end{cases}$

48. $\begin{cases} 2x + 3y = 5 \\ 3x + 2y = 10 \end{cases}$

49. $\begin{cases} 2x + 2y = 9 \\ 2x - y = 6 \end{cases}$

50. $\begin{cases} 3x - 4y = 16 \\ 4x + 3y = 15 \end{cases}$

Writing and Thinking About Mathematics

51. Handicapped access: Ramps for persons in wheelchairs or otherwise handicapped are now built into most buildings and walkways. (If ramps are not present in a building, then there must be elevators.) What do you think that the slope of a ramp should be for handicapped access? Look in your library or contact your local building permit office to find the recommended slope for such ramps.

52. Discuss the difference between the concepts of a line having slope 0 and a line having undefined slope.

HAWKES LEARNING SYSTEMS: INTERMEDIATE ALGEBRA SOFTWARE

- Graphing Linear Equations in Point-Slope Form
- Finding the Equation of a Line

<table>
<tr><td>

2.4

</td><td>

Introduction to Functions and Function Notation

- *Find the domain and range of a relation or function.*
- *Determine whether a relation is or is not a function.*
- *Use the vertical line test to determine whether a graph is or is not the graph of a function.*
- *Write functions using function notation.*
- *Use a TI-84 Plus graphing calculator to graph functions.*

</td></tr>
</table>

Everyday use of the term **function** is not far from the technical use in mathematics. For example, distance traveled is a function of time; profit is a function of sales; heart rate is a function of exertion; and interest earned is a function of principal invested. In this sense, one variable "depends on" (or "is a function of") another.

Mathematicians distinguish between graphs of ordered pairs of real numbers as those that represent **functions** and those that do not. Thus the concept of a function is one of the most important concepts in mathematics. For example, every equation of the form $y = mx + b$ represents a function and we say that y "is a function of " x. Thus straight lines that are not vertical are the graphs of functions. As the following discussion indicates, vertical lines do not represent functions.

 NOTES The ordered pairs discussed in this text are ordered pairs of real numbers. However, more generally, ordered pairs might be other types of pairs such as (parent, child), (city, state), or (name, batting average).

Relations and Functions

Relation, Domain, and Range

*A **relation** is a set of ordered pairs of real numbers.*

*The **domain**, **D**, of a relation is the set of all first coordinates in the relation.*

*The **range**, **R**, of a relation is the set of all second coordinates in the relation.*

In graphing relations, the horizontal axis is called the **domain axis**, and the vertical axis is called the **range axis**.

Example 1: Finding the Domain and Range

Find the domain and range for each of the following relations.

a. $g = \{(5,7),(6,2),(6,3),(-1,2)\}$

Solution: $D = \{5,6,-1\}$ All the first coordinates in g

$R = \{7,2,3\}$ All the second coordinates in g

Note that 6 is written only once in the domain and 2 is written only once in the range, even though each appears more than once in the relation.

b. $f = \{(-1,1),(1,5),(0,3)\}$

Solution: $D = \{-1,1,0\}$ All the first coordinates in f

$R = \{1,5,3\}$ All the second coordinates in f

The relation $f = \{(-1,1),(1,5),(0,3)\}$, used in Example 1b, meets a particular condition in that each first coordinate has a unique corresponding second coordinate. Such a relation is called a **function**. Notice that g in Example 1a is **not** a function because the first coordinate 6 has two corresponding second coordinates, 2 and 3. Also, for ease in discussion and understanding, the relations illustrated in Examples 1 and 2 have only a finite number of ordered pairs. The graphs of these relations are isolated dots or points. As we will see, the graphs of most relations and functions have an infinite number of points, and their graphs are smooth curves. [**Note:** Straight lines are also deemed to be curves in mathematics.]

Function

*A **function** is a relation in which each domain element has exactly one corresponding range element.*

The following statements are also true for functions:

1. A function is a relation in which each first coordinate appears only once.

2. A function is a relation in which no two ordered pairs have the same first coordinate.

Example 2: Functions

Determine whether or not each of the following relations is a function.

a. $s = \left\{(2,3),(1,6),\left(2,\sqrt{5}\right),(0,-1)\right\}$

Solution: *s* is not a function. The number 2 appears as a first coordinate more than once.

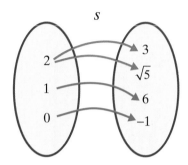

b. $t = \left\{(1,5),(3,5),\left(\sqrt{2},5\right),(-1,5),(-4,5)\right\}$

Solution: *t* is a function. Each first coordinate appears only once. The fact that the second coordinates are all the same has no effect on the concept of a function.

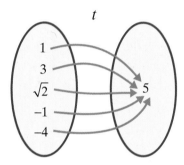

Vertical Line Test

If one point on the graph of a relation is directly above or below another point on the graph, then these points have the same first coordinate (or *x*-coordinate). Such a relation is **not** a function. Therefore, the **vertical line test** can be used to tell whether or not a graph represents a function (See Figure 2.13).

Vertical Line Test

*If **any** vertical line intersects the graph of a relation at more than one point, then the relation graphed is **not** a function.*

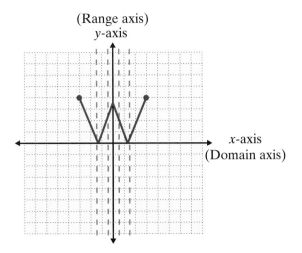

The vertical lines indicate that this graph represents a function. From the graph, we see that the domain of the function is the interval of real numbers [–3, 3] and the range of the function is the interval of real numbers [0, 4].

This relation is **not** a function because the vertical line drawn intersects the graph at more than one point. Thus, for that x-value, there is more than one corresponding y-value. Here $D = [-3, \infty)$ and $R = (-\infty, \infty)$.

Figure 2.13

Example 3: Vertical Line Test

Use the vertical line test to determine whether or not each of the following graphs represents a function. Then list the domain and range of each graph.

a.

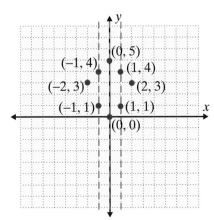

Solution:

The relation is **not** a function since a vertical line can be drawn that intersects the graph at more than one point. Listing the ordered pairs shows that several x-coordinates appear more than once.

$$s = \left\{ \begin{array}{l} (-2,3),(-1,1),(-1,4), \\ (0,0),(0,5),(1,1),(1,4),(2,3) \end{array} \right\}$$

Here $D = \{-2, -1, 0, 1, 2\}$
and $R = \{0, 1, 3, 4, 5\}$.

Continued on the next page...

b.

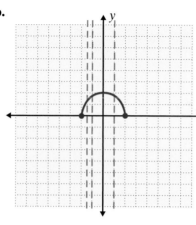

Solution:

The relation is a **function**. No vertical line will intersect the graph at more than one point. Several vertical lines are drawn to illustrate this. For this function, we see from the graph that $D = [-2, 2]$ and $R = [0, 2]$.

c.

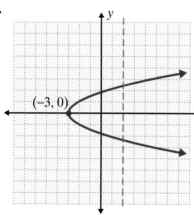

$(-3, 0)$

Solution:

The relation is **not** a function. At least one vertical line (drawn) intersects the graph at more than one point.

$D = [-3, \infty)$ and $R = (-\infty, \infty)$.

d.

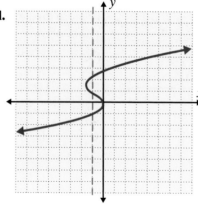

Solution:

The relation is **not** a function. At least one vertical line intersects the graph at more than one point.

$D = (-\infty, \infty)$ and $R = (-\infty, \infty)$.

Linear Functions

All non-vertical straight lines represent functions. Thus we have the following definition for a linear function.

Linear Function

A **linear function** is a function represented by an equation of the form

$$y = mx + b.$$

The domain of a linear function is the set of all real numbers: $D = (-\infty, \infty)$.

If the graph of a linear function is not a horizontal line, then the range is also the set of all real numbers. If the line is horizontal, then the domain is still all real numbers; however, the range is just a single number. For example, the graph of the linear equation $y = 5$ is a horizontal line. The domain of the function is all real numbers and the range is the number 5. Figure 2.14 shows two linear functions and the domain and range of each function.

a.

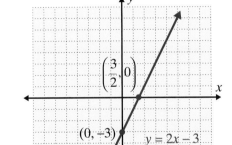

b.

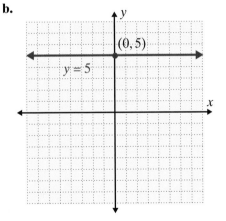

The graph of the linear function $y = 2x - 3$ is shown here.

$D = \{\text{all real numbers}\} = (-\infty, \infty)$

$R = \{\text{all real numbers}\} = (-\infty, \infty)$

The graph of the linear function $y = 5$ is shown here.

$D = \{\text{all real numbers}\} = (-\infty, \infty)$

$R = \{5\}$

Figure 2.14

Domains of Non-linear Functions

As we have seen for linear functions, the domain is the set of all real numbers. Now, for the nonlinear function

$$y = \frac{2}{x-1}$$

we say that the domain (all possible values for x) is every real number for which the expression $\frac{2}{x-1}$ is **defined**. Because the denominator cannot be 0, the domain consists of all real numbers except 1. That is, $D = (-\infty, 1) \cup (1, \infty)$ or simply $x \neq 1$. We adopt the following rule concerning equations and domains:

> **Unless a finite domain is explicitly stated, the domain will be implied to be the set of all real x-values for which the given function is defined. That is, the domain consists of all values of x that give real values for y.**

NOTES
In determining the domain of a function, two facts about real numbers are particularly important at this stage:
1. No denominator can equal 0, and
2. Square roots of negative numbers are not real numbers.

Example 4: Domain

Find the domain of each of the following functions.

a. $y = \frac{2x+1}{x-5}$

Solution: The domain is all real numbers for which the expression $\frac{2x+1}{x-5}$ is defined. Thus $D = (-\infty, 5) \cup (5, \infty)$, because the denominator is 0 when $x = 5$.

Note: Here interval notation tells us that x can be any real number except 5.

b. $y = \sqrt{x-2}$

Solution: For $\sqrt{x-2}$ to be a real number, the expression under the square root sign must be non-negative. So we have

$$x - 2 \geq 0 \qquad \text{or} \qquad x \geq 2.$$

Thus $D = [2, \infty)$.

Function Notation

We have used the ordered pair notation (x, y) to represent points in relations and functions. As the vertical line test will show, linear equations of the form

$$y = mx + b$$

where the equation is solved for y, represent **linear functions**. Another notation, called **function notation**, is more convenient for indicating calculations of values of a function and indicating operations performed with functions. In function notation,

instead of writing **y**,   write **f(x)**,   read "**f of x**."

The letter f is the name of the function. The letters f, g, h, F, G, and H are commonly used in mathematics, but any other letter than x will do. We have used r, s, and t in previous examples.

The linear equation $y = -3x + 2$ represents a linear function and we can replace y with $f(x)$ as follows:

$$f(x) = -3x + 2.$$

Now, in function notation, $f(4)$ means to replace x with 4 in the function.

$$f(4) = -3 \cdot (4) + 2 = -12 + 2 = -10$$

Thus the ordered pair $(4, -10)$ can be written as $(4, f(4))$.

Example 5: Function Evaluation

For the function $g(x) = 4x + 5$, find:

a. $g(2)$

   Solution: $g(2) = 4 \cdot (2) + 5 = 13$

b. $g(-1)$

   Solution: $g(-1) = 4(-1) + 5 = 1$

c. $g(0)$

   Solution: $g(0) = 4 \cdot (0) + 5 = 5$

Not all functions are linear functions, just as not all graphs are straight lines. Function notation is valid for a wide variety of types of functions. Example 6 illustrates the use of function notation with a nonlinear function.

Example 6: Nonlinear Function Evaluation

For the function $h(x) = x^2 - 3x + 2$, find:

a. $h(4)$

Solution: $h(4) = (4)^2 - 3 \cdot (4) + 2 = 16 - 12 + 2 = 6$

b. $h(0)$

Solution: $h(0) = (0)^2 - 3 \cdot (0) + 2 = 0 - 0 + 2 = 2$

c. $h(-3)$

Solution: $h(-3) = (-3)^2 - 3(-3) + 2 = 9 + 9 + 2 = 20$

Practice Problems

1. *State the domain and range of the relation {(5, 6), (7, 8), (9, 0.5), (11, 0.3)}. Is the relation a function? Explain briefly.*

2. *State the domain of the linear function f(x) = 5x − 2.*

3. *State the domain of the function $y = \sqrt{x+3}$.*

4. *State the domain of the function $F(x) = \dfrac{3}{x+6}$.*

Using a TI-84 Plus Graphing Calculator to Graph Functions

There are many types and brands of graphing calculators available. For convenience and so that directions can be specific, only the TI-84 Plus graphing calculator is used in the related discussions in this text. Other graphing calculators may be used, but the steps required may be different from those indicated in the text. If you choose to use another calculator, be sure to read the manual for your calculator and follow the relevant directions.

Answers to Practice Problems: 1. $D = \{5, 7, 9, 11\}; R = \{0.3, 0.5, 6, 8\}$ Yes, the relation is a function because each x-coordinate appears only once.
2. $D = (-\infty, \infty)$ 3. $D = [-3, \infty)$ 4. $D = (-\infty, -6) \cup (-6, \infty)$ or $x \neq -6$

In any case, remember that a calculator is just a tool to allow for fast calculations and to help in understanding some abstract concepts. A calculator does not replace your ability to think and reason or the need for algebraic knowledge and skills.

You should practice and experiment with your calculator until you feel comfortable with the results. **Do not be afraid of making mistakes. Note that** **CLEAR** **or** **2ND** QUIT **will get you out of most trouble and allow you to start over.**

Some Basics about the TI-84 Plus

1. **MODE** : Turn the calculator **ON** and press the **MODE** key. The screen should be highlighted as shown below. If it is not, use the arrow keys in the upper right corner of the keyboard to highlight the correct words and press **ENTER**. It is particularly important that **Func** is highlighted. This stands for function. See the manual for the meanings of the rest of the terms.

2. **WINDOW** : Press the **WINDOW** key and the standard window will be displayed. By default, the standard window displays *x*-values and *y*-values ranging from −10 to 10 with tic marks on the axis every 1 unit.

This window can be changed at any time by changing the individual numbers or pressing the ▇ZOOM▇ key and selecting an option from the menu displayed. Because of the shape of the display screen, the standard screen is not a square screen. Be aware that the slopes of lines are not truly depicted unless the screen is in a scale of about 3 : 2. A square screen can be attained by pressing zoom and 5: ZSquare or by pressing the window key and setting Xmin = −15 and Xmax = 15 to give the x-axis a length of 30 and the y-axis a length of 20 (a ratio of 3 : 2).

3. ▇Y=▇ : The ▇Y=▇ key is in the upper left corner of the keyboard. This key will allow ten different functions to be entered. These functions are labeled as $Y_1,\ldots,Y_{10}$. The variable x may be entered by using the ▇X,T,θ,n▇ key. The ▇^▇ key can be used to indicate exponents. (Also note that the negative sign (−) is next to the ▇ENTER▇ key.) For example, the equation $y = x^2 + 3x$ would be entered as:

$$Y_1 = X\wedge 2 + 3X.$$

To change an entry, practice with the keys ▇DEL▇ (delete), ▇CLEAR▇, and ▇2ND▇ **INS** (insert).

4. ▇GRAPH▇ : If this key is pressed, then the screen will display the graph of whatever functions are indicated in the ▇Y=▇ list with the = sign highlighted by using whatever scales are indicated in the current WINDOW. In many cases the WINDOW must be changed to accommodate the domain and range of the function or to show a point where two functions intersect.

5. ▇TRACE▇ : The ▇TRACE▇ key will display the current graph even if it is not already displayed and give the x- and y- coordinates of a point highlighted on the graph. The curve may be traced by pressing the left and right arrow keys. At each point on the graph, the corresponding x- and y- coordinates are indicated at the bottom of the screen. **(Remember that because of the limitations of the pixels (lighted dots) on the screen, these x- and y- coordinates are generally only approximations.)**

6. CALC: The CALC key (press 2ND   TRACE) gives a menu with seven items. Items 1 – 5 are used with graphs.

After displaying a graph, select CALC. Then press **2** and follow the steps outlined below to locate the point where the graph crosses the *x*-axis (the *x*-intercept). The graph must actually cross the axis. This point is called a **zero** of the function because the *y*-value will be 0.

Step 1: With the left arrow, move the cursor to the left of the *x*-intercept on the graph. Press ENTER in response to the question "`LeftBound?`".

Step 2: With the right arrow, move the cursor to the right of the *x*-intercept on the graph. Press ENTER in response to the question "`RightBound?`".

Step 3: With the left arrow, move the cursor near the *x*-intercept. Press **ENTER** in response to the question "**Guess?**". The calculator's estimate of the zero will appear at the bottom of the display.

Example 7: Graphing Functions with a TI-84 Plus

Use a TI-84 Plus graphing calculator to find the graphs of each of the following functions. Use the CALC key to find the point where each graph intersects the *x*-axis. Changing the WINDOW may help you get a "better" or "more complete" picture of the function. This is a judgment call on your part.

a. $3x + y = -1$

Solution: To have the calculator graph a nonvertical straight line, you must first solve the equation for *y*. Solving for *y* gives,

$$y = -3x - 1.$$

[It is important that the **(−)** key be used to indicate the negative sign in front of 3*x*. This is not the same as the subtraction key.]

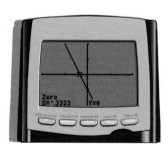

Note: Vertical lines are not functions and cannot be graphed by the calculator in function mode.

b. $y = \sqrt{x - 4}$

Solution: To graph a square root equation, we enter the $\sqrt{}$ sign by selecting

2ND **$.x^2$** .

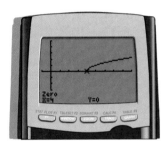

Note: Be sure to include the expression "$(x - 4)$" in parentheses after the $\sqrt{}$ sign.

c. $y = x^2 + 3x$

Solution: Since the graph of this function has two x-intercepts, we have shown the graph twice. Each graph shows the coordinates of a distinct x-intercept.

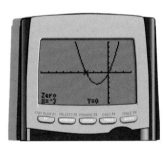   

d. $y = 2x - 1; y = 2x + 1; y = 2x + 3$

Solution:

$y = 2x - 1$ $y = 2x + 1$ $y = 2x + 3$

NOTES The standard window shows 96 pixels across the window and 64 pixels up and down the window. This gives a ratio of 3 to 2 and can give a slightly distorted view of the actual graph because the vertical pixels are squeezed into a smaller space. For Example 7d, the graphs of all three functions are in the standard window. Experiment by changing the window to a square window, say −9 to 9 for *x* and −6 to 6 for *y*. Then graph the functions and notice the slight differences (and better representation) in the appearances on the display. DO THIS!

2.4 Exercises

List the sets of ordered pairs that correspond to the points in Exercises 1 − 8. State the domain and range and indicate which of the relations are also functions.

1.

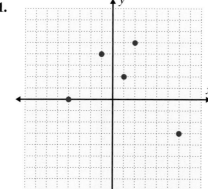

2.

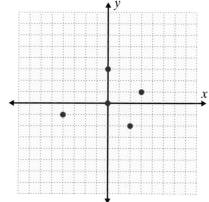

3.

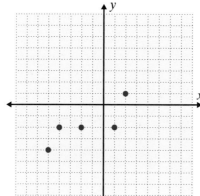

4.

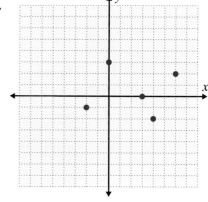

5.

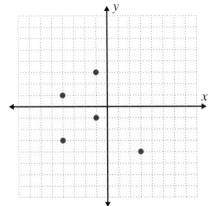

6.

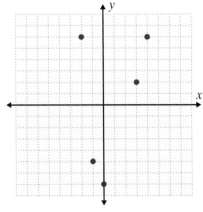

7.

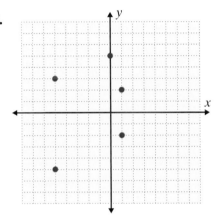

8.

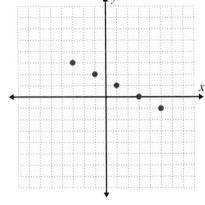

Graph the relations in Exercises 9 – 16. State the domain and range and indicate which of the relations are functions.

9. $a = \{(0,0),(1,6),(4,-2),(-3,5),(2,-1)\}$

10. $b = \{(1,-5),(2,-3),(-1,-3),(0,2),(4,3)\}$

11. $a = \{(-4,4),(-3,4),(1,4),(2,4),(3,4)\}$

12. $b = \{(-3,-3),(0,1),(-2,1),(3,1),(5,1)\}$

13. $a = \{(0,2),(-1,1),(2,4),(3,5),(-3,5)\}$

14. $b = \{(-1,-4),(0,-3),(2,-1),(4,1),(1,1)\}$

15. $a = \{(-1,4),(-1,2),(-1,0),(-1,6),(-1,-2)\}$

16. $b = \{(0,0),(-2,-5),(2,0),(4,-6),(5,2)\}$

Use the vertical line test to determine whether or not each graph in Exercises 17 – 28 represents a function. State the domain and range.

17.

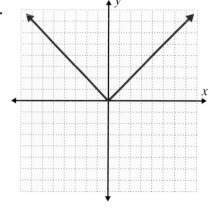

18.

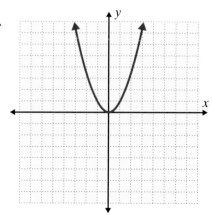

19.

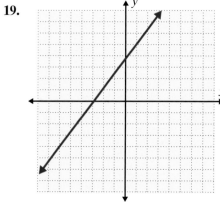

20.

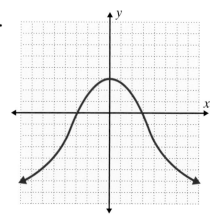

21.

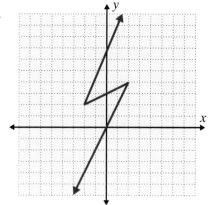

22.

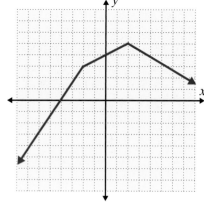

23.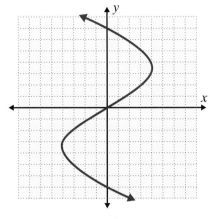

24.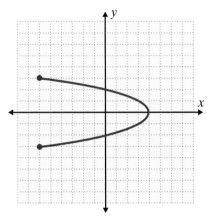

25.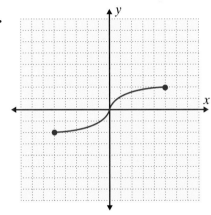

26.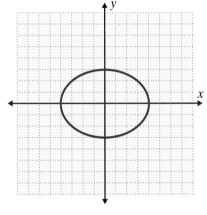

27.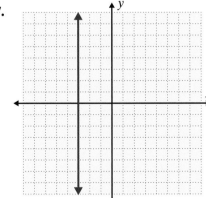

28.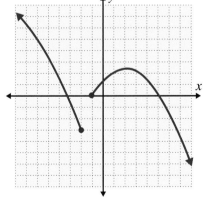

In Exercises 29 – 32, express the function as a set of ordered pairs for the given equation and given domain. [**Hint:** *Substitute each domain element for x and find the corresponding y-coordinate.*]

29. $y = 3x + 1$; $D = \left\{ -9, -\dfrac{1}{3}, 0, \dfrac{4}{3}, 2 \right\}$

30. $y = -\dfrac{3}{4}x + 2$; $D = \{ -4, -2, 0, 3, 4 \}$

31. $y = 1 - 3x^2$; $D = \{ -2, -1, 0, 1, 2 \}$

32. $y = x^3 - 4x$; $D = \left\{ -1, 0, \dfrac{1}{2}, 1, 2 \right\}$

In Exercises 33 – 38, evaluate each function as indicated.

33. $f(x) = 3x - 10$ Find **a.** $f(2)$ **b.** $f(-2)$ **c.** $f(0)$

34. $g(x) = -4x + 7$ Find **a.** $g(-3)$ **b.** $g(6)$ **c.** $g(0)$

35. $G(x) = x^2 + 5x + 6$ Find **a.** $G(-2)$ **b.** $G(1)$ **c.** $G(5)$

36. $F(x) = 6x^2 - 10$ Find **a.** $F(0)$ **b.** $F(-4)$ **c.** $F(4)$

37. $h(x) = x^3 - 8x$ Find **a.** $h(-3)$ **b.** $h(0)$ **c.** $h(3)$

38. $P(x) = x^2 + 4x + 4$ Find **a.** $P(-2)$ **b.** $P(10)$ **c.** $P(-5)$

State the domains of the functions given in Exercises 39 – 44.

39. $y = -5x + 10$ **40.** $2x + y = 14$

41. $f(x) = \dfrac{35}{2x - 1}$ **42.** $g(x) = \sqrt{2 - x}$

43. $h(x) = \sqrt{x^2 + 9}$ **44.** $y = \dfrac{13x^2 - 5x + 8}{x - 3}$

*Use a graphing calculator to graph the functions in Exercises 45 – 56. Use the CALC features of the calculator to find x-intercepts, if any. (Remember that the value of y will be 0 at those points.) For absolute value functions, select the MATH menu, then the NUM menu, and then **1: abs(**. Remember to press) after entering the absolute value.*

45. $y = 6$ **46.** $y = 4x$ **47.** $y = -2x + 3$

48. $y = x^2 - 4x$ **49.** $y = 1 + 2x - x^2$ **50.** $y = \sqrt{x + 5}$

51. $y = \sqrt{3 - x}$ **52.** $y = |x + 2|$ **53.** $y = |x^2 - 3x|$

54. $y = 2x^3 - 5x^2 + 1$ **55.** $y = -x^3 + 3x - 1$ **56.** $y = x^4 - 10x^2 + 9$

*In Exercises 57 and 58 use the CALC features of the calculator to find the coordinates of the highest point on the graph. [**Hint:** Item 4 on the CALC menu **4: maximum** will help in finding the highest point of a function, if there is one.]*

57. $y = 4x - x^2$ **58.** $y = 3 - 2x - x^2$

*In Exercises 59 and 60, use the CALC features of the calculator to find the coordinates of the lowest point on the graph. [**Hint:** Item 3 on the CALC menu **3: minimum** will help in finding the lowest point of a function, if there is one.]*

59. $y = 2x^2 - x + 1$ **60.** $y = 3(x - 1)^2 + 2$

*In Exercises 61 – 64, use the CALC features of the calculator to find the coordinates of any points of intersection of the graphs. [**Hint:** Item 5 on the CALC menu **5: intersect** will help in finding the point (or points) of intersection of two functions, if there is one.] In the* **Y** *= menu use both* **Y₁** *= and* **Y₂** *= to be able to graph both functions at the same time.*

61. $y = 3x + 2$
    $y = 4 - x$

62. $y = 2 - x$
    $y = x$

63. $y = 2x - 1$
    $y = x^2$

64. $y = x + 3$
    $y = -x^2 + x + 7$

In Exercises 65 – 69, the calculator display shows an incorrect graph for the corresponding equation. Explain how you know, just by looking at the graph, that a mistake has been made.

65. $y = 2x + 5$

66. $y = -3x + 4$

67. $y = \dfrac{2}{3}x - 2$

68. $y = -4x$

69. $y = -\dfrac{1}{3}x$

Writing and Thinking About Mathematics

70. Explain in your own words how to find the domain of a function
a. graphically
b. algebraically

71. Which, if any, of the following functions has no restrictions for the values in the domain? If the function's domain has restrictions, explain why.
a. $3x - 2y = -1$

b. $\sqrt{x+2} = y$

c. $\dfrac{1}{3x-1} = y$

72. JUST FOR FUN: Enter a variety of functions in your calculator, investigate your findings, and report these to your class. Certainly, interesting discussions will follow!

 HAWKES LEARNING SYSTEMS: INTERMEDIATE ALGEBRA SOFTWARE

- Introduction to Functions and Function Notation

2.5 Graphing Linear Inequalities: $y < mx + b$

- *Graph linear inequalities.*
- *Graph linear inequalities by using a TI-84 Plus graphing calculator.*

Graphing Linear Inequalities: $y < mx + b$

In this section, we will develop techniques for analyzing and graphing linear inequalities. We need the following terminology:

Half-plane	A straight line separates a plane into two **half-planes**. The points on one side of the line are in one of the half-planes, and the points on the other side of the line are in the other half-plane.
Boundary line	The line itself is called the **boundary line**.
Closed half-plane	If the boundary line is included, then the half-plane is said to be **closed**.
Open half-plane	If the boundary line is not included, then the half-plane is said to be **open**.

Figure 2.15 shows both **a.** an open half-plane and **b.** a closed half-plane with the line $2x - 3y = 10$ as the boundary line.

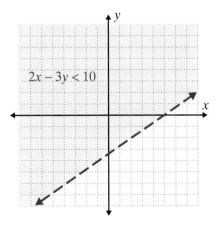

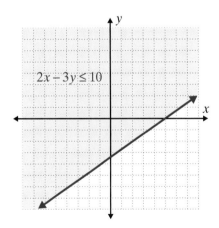

a. The points on the line $2x - 3y = 10$ are not included so the line is dashed. The half-plane is open.

b. The points on the line $2x - 3y = 10$ are included so the line is solid. The half-plane is closed.

Figure 2.15

Graphing Linear Inequalities

1. *First, graph the boundary line (dashed if the inequality is < or >, solid if the inequality is ≤ or ≥).*

2. *Next, determine which side of the line to shade using one of the following methods.*

Method 1
 a. *Test any one point obviously on one side of the line.*
 b. *If the test-point satisfies the inequality, shade the half-plane on that side of the line. Otherwise, shade the other half-plane.*
 [Note: The point (0, 0), if it is not on the boundary line, is usually the easiest point to test.]

Method 2
 a. *Solve the inequality for y (assuming that the line is not vertical).*
 b. *If the solution shows y < or y ≤ , then shade the half-plane below the line.*
 c. *If the solution shows y > or y ≥ , then shade the half-plane above the line.*
 [Note: If the boundary line is vertical, then solve for x. If the solution shows x > or x ≥ , then shade the half-plane to the right. If the solution shows x < or x ≤ , then shade the half-plane to the left.]

3. *The shaded half-plane (and the line if it is solid) is the solution to the inequality.*

Example 1: Graphing Linear Inequalities

a. Graph the half-plane that satisfies the inequality $2x + y \leq 6$.

Solution: Method 1 is used in this example.

Step 1: Graph the line $2x + y = 6$ as a solid line.

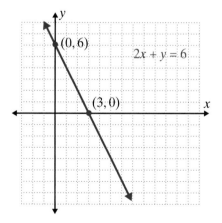

Step 2:   Test any point on one side of the line. In this example, we have chosen $(0, 0)$.

$$2 \cdot 0 + 0 \le 6$$
$$0 \le 6$$

This is a true statement.

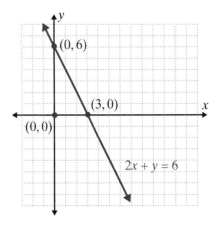

Step 3:   Shade the points on the same side as the point $(0, 0)$. (The shaded half-plane and the line are the solution to the inequality.)

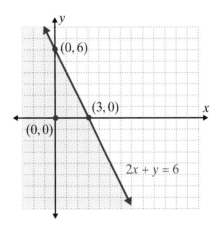

b. Graph the solution set to the inequality $y > 2x$.

Solution: Since the inequality is already solved for y, Method 2 is easy to apply.

Step 1:   Graph the line $y = 2x$ as a dashed line.

Step 2:   By Method 2, the graph consists of those points above the line. Shade the half-plane above the line.
[**Note:** As a check, we see that the point $(3, 0)$ gives $0 > 2 \cdot 3$, a false statement.]

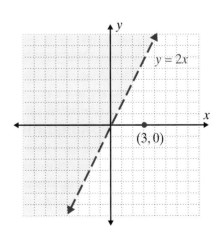

Continued on the next page...

c. Graph the half-plane that satisfies the inequality $y > 1$.

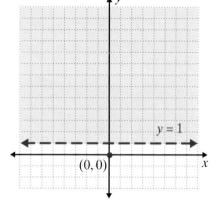

Solution: Again, the inequality is already solved for y and Method 2 is used.

Step 1:   Graph the horizontal line $y = 1$ as a dashed line.

Step 2:   By Method 2, shade the half-plane above the line.

d. Graph the solution set to the inequality $x \le 0$.

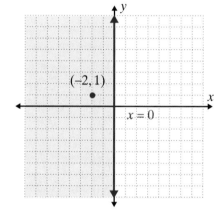

Solution: The boundary line is a vertical line and Method 1 is used.

Step 1:   Graph the line $x = 0$ as a solid line. Note that this is the y-axis.

Step 2:   Test the point $(-2, 1)$.
$$-2 \le 0$$
This statement is true.

Step 3:   Shade the half-plane on the same side of the line as $(-2, 1)$. This half-plane consists of the points with x-coordinate 0 or negative.

Using a TI-84 Plus Graphing Calculator to Graph Linear Inequalities

The first step in using the TI-84 Plus (or any other graphing calculator) to graph a linear inequality is to solve the inequality for y. This is necessary because this is the way that the boundary line equation can be graphed as a function. Thus, Method 2 for graphing the correct half-plane is appropriate.

Note that when you press the Y= key on the calculator, a slash (\ ) appears to the left of the Y expression as in $\backslash Y_1 =$. This slash is actually a command to the calculator to graph the corresponding function as a solid line or curve. If you move the cursor to position it over the slash and hit **ENTER** repeatedly, the following options will appear.

(1)   **(2)**   **(3)**

(4)   **(5)**   **(6)**

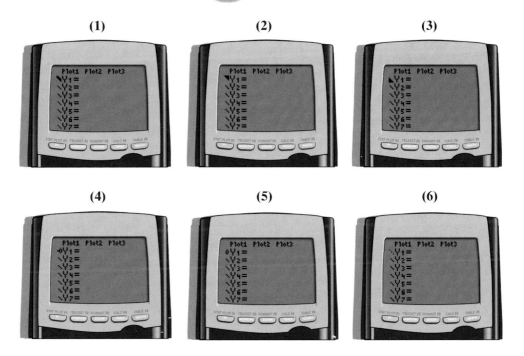

If the slash **(1)** (which is actually four dots if you look closely) becomes a set of three dots **(6)**, then the corresponding graph of the function will be dotted. By setting the shading above the slash **(2)**, the corresponding graph on the display will show shading above the line or curve. By setting the shading below the slash **(3)**, the corresponding graph on the display will show shading below the line or curve. (The solid line occurs only when the slash is four dots, so the calculator is not good for determining whether the boundary curve is included or not.) The following examples illustrate two situations.

Example 2:   Graphing Linear Inequalities Using a Calculator

a. Graph the linear inequality $2x + y \le 7$.

Solution:

Step 1: Solving the inequality for y gives: $y \le -2x + 7$.

Step 2: Press the key Y= and enter the function: $\backslash Y_1 = -2X + 7$.

Continued on the next page...

Step 3: Go to the \ and hit (ENTER) three times so that the display appears as follows:

Step 4: Press (GRAPH) and (using the standard WINDOW settings) the following graph should appear on the display.

b. Graph the linear inequality $-5x + 4y > -8$.

Solution:

Step 1: Solving the inequality for y gives: $y > \dfrac{5}{4}x - 2$.

Step 2: Press the key (Y=) and enter the function: $\backslash Y_1 = (5/4)X - 2$.

Step 3: Go to the \ and hit (ENTER) two times so that the display appears as follows:

Step 4: Press (GRAPH) and (using the standard WINDOW settings) the following graph should appear on the display. (**Note:** The boundary line should actually be dotted.)

1. Which of the following points satisfy the inequality $x + y < 3$?

    a. $(2, 1)$ **b.** $\left(\dfrac{1}{2}, 3\right)$ **c.** $(0, 5)$ **d.** $(-5, 2)$

2. Which of the following points satisfy the inequality $x - 2y \geq 0$?

    a. $(2, 1)$ **b.** $(1, 3)$ **c.** $(4, 2)$ **d.** $(3, 1)$

3. Which of the following points satisfy the inequality $x < 3$?

    a. $(1, 0)$ **b.** $(0, 1)$ **c.** $(4, -1)$ **d.** $(2, 3)$

2.5 Exercises

Graph the solution set of each of the linear inequalities in Exercises 1 – 30.

1. $x + y \leq 7$ **2.** $x - y > -2$ **3.** $x - y > 4$ **4.** $x + y \leq 6$

5. $y < 4x$ **6.** $y < -2x$ **7.** $y \geq -3x$ **8.** $y > x$

9. $x - 2y > 5$ **10.** $x + 3y \leq 7$ **11.** $4x + y \geq 3$ **12.** $5x - y < 4$

13. $y \leq 5 - 3x$ **14.** $y \geq 8 - 2x$ **15.** $2y - x \leq 0$ **16.** $x + y > 0$

17. $x + 4 \geq 0$ **18.** $x - 5 \leq 0$ **19.** $y \geq -2$ **20.** $y + 3 < 0$

21. $4x + 3y < 8$ **22.** $3x < 2y - 4$ **23.** $3y > 4x + 6$ **24.** $5x < 2y - 5$

25. $x + 3y < 7$ **26.** $3x + 4y > 11$ **27.** $\dfrac{1}{2}x - y > 1$ **28.** $\dfrac{1}{3}x + y \geq 3$

29. $\dfrac{2}{3}x + y \geq 4$ **30.** $2x - \dfrac{4}{3}y > 8$

Use your graphing calculator to graph each of the linear inequalities in Exercises 31 – 40.

31. $y > \dfrac{1}{2}x$ **32.** $x - y \leq 5$ **33.** $x + 2y > 8$ **34.** $3x + 2y \geq 12$

35. $2x + y \leq 6$ **36.** $y \geq -3$ **37.** $x - 3y \geq 9$ **38.** $y \leq -4$

39. $2x + 6y \geq 0$ **40.** $3x - 4y > 15$

Answers to Practice Problems: **1.** d   **2.** a, c, d   **3.** a, b, d

Writing and Thinking About Mathematics

41. Explain in your own words how to test to determine which side of the graph of an inequality should be shaded.

42. Describe the difference between a closed and an open half-plane.

 HAWKES LEARNING SYSTEMS: INTERMEDIATE ALGEBRA SOFTWARE

- Graphing Linear Inequalities

Chapter 2 Index of Key Ideas and Terms

Section 2.1 Cartesian Coordinate System and Linear Equations: $Ax + By = C$ **(cont.)**

***y*-intercept**   page 111
   The ***y*-intercept** is the point where the graph of a line crosses the *y*-axis. The *x*-coordinate will be 0.

***x*-intercept**   page 111
   The ***x*-intercept** is the point where the graph of a line crosses the *x*-axis. The *y*-coordinate will be 0.

Section 2.2 Slope-Intercept Form: $y = mx + b$

Slope as a Rate of Change   page 119

Slope of a line   page 121
   Let $P_1(x_1, y_1)$ and $P_2(x_2, y_2)$ be two points on a line. The slope can be calculated as follows:
   $$\text{slope} = m = \frac{rise}{run} = \frac{y_2 - y_1}{x_2 - x_1}.$$

Positive and Negative Slopes   page 122
   Lines with **positive slope** go up as we move along the line from left to right.
   Lines with **negative slope** go down as we move along the line from left to right.

Horizontal Lines   pages 123-124
   Any equation of the form $y = b$ represents a **horizontal line** with **slope 0**.

Vertical Lines   pages 123-124
   Any equation of the form $x = a$ represents a **vertical line** with **undefined slope**.

Slope-Intercept Form   page 125
   Any equation of the form $y = mx + b$ is called the **slope-intercept form** of the equation of a line. The slope is m and the *y*-intercept is b.

Section 2.3 Point-Slope Form: $y - y_1 = m(x - x_1)$

Point-Slope Form   page 135

An equation of the form $y - y_1 = m(x - x_1)$ is called the
point-slope form for the equation of a line that contains
the point (x_1, y_1) and has slope m.

Parallel Lines   page 138

Parallel lines are lines that never intersect (cross each
other) and these lines have the same slope. All vertical
lines are parallel.

Perpendicular Lines   page 138

Perpendicular lines are lines that intersect at 90° (right)
angles and whose slopes are negative reciprocals of each
other. Horizontal lines are perpendicular to vertical lines.

Section 2.4 Introduction to Functions and Function Notation

Relation, Domain, and Range   page 144

A **relation** is a set of ordered pairs of real numbers. The
domain, D, of a relation is the set of all first coordinates
in the relation. The **range, R,** of a relation is the set of all
second coordinates in the relation.

Functions   page 145

A **function** is a relation in which each domain element has
exactly one corresponding range element.

Vertical Line Test   page 146

If **any** vertical line intersects the graph of a relation at
more than one point, then the relation graphed is **not** a
function.

Linear Functions   page 149

A **linear function** is a function represented by an
equation of the form $y = mx + b$.

Function Notation   page 151

Instead of writing y, write $f(x)$, read "f of x." This is
function notation, where f is the name of the function. The
notation $f(4)$ means to replace x with 4 in the function.

Using a Graphing Calculator to Graph Linear Functions   pages 152-156

Section 2.5   Graphing Linear Inequalities: $y < mx + b$

1. First, graph the boundary line (dashed if the inequality is < or >, solid if the inequality is ≤ or ≥).
2. Next, determine which side of the line to shade using one of the following methods.

    Method 1
    a. Test any one point obviously on one side of the line.
    b. If the test-point satisfies the inequality, shade the half-plane on that side of the line. Otherwise, shade the other half-plane.

    Method 2
    a. Solve the inequality for y (assuming that the line is not vertical).
    b. If the solution shows $y <$ or $y \leq$, then shade the half-plane below the line.
    c. If the solution shows $y >$ or $y \geq$, then shade the half-plane above the line.
3. The shaded half-plane (and the line if it is solid) is the solution to the inequality.

 HAWKES LEARNING SYSTEMS: INTERMEDIATE ALGEBRA SOFTWARE

For a review of the topics and problems from Chapter 2, look at the following lessons from *Hawkes Learning Systems: Intermediate Algebra*

- Introduction to the Cartesian Coordinate System
- Graphing Linear Equations by Plotting Points
- Graphing Linear Equations in Slope-Intercept Form
- Graphing Linear Equations in Point-Slope Form
- Finding the Equation of a Line
- Introduction to Functions and Function Notation
- Graphing Linear Inequalities
- Chapter 2 Review and Test

Chapter 2 Review

2.1 Cartesian Coordinate System and Linear Equations: $Ax + By = C$

List the sets of ordered pairs that correspond to the graphs in Exercises 1 and 2. Assume that the grid lines are marked one unit apart.

1.

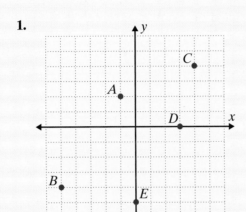

2.

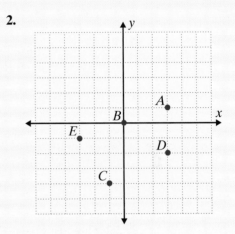

In Exercises 3 and 4, a graph of a straight line is shown. Each line has an infinite number of points. List any three points on each line. (There is more than one correct answer.)

3.

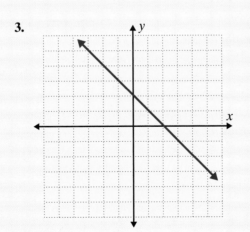

4.

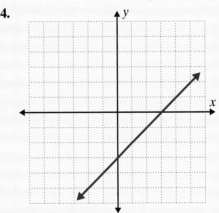

In Exercises 5 – 8, find the missing coordinate of each ordered pair so that the ordered pair belongs to the solution set of the given equation.

5. $x + 3y = 4$
    a. $(\ \ , 0)$
    b. $(1, \ \ )$
    c. $(2, \ \ )$
    d. $(\ \ , 2)$

6. $3x - y = 7$
    a. $(0, \ \ )$
    b. $(2, \ \ )$
    c. $(-2, \ \ )$
    d. $(\ \ , -2)$

7. $y = 8 - 2x$
    a. $(4, \ \ )$
    b. $(0, \ \ )$
    c. $(\ \ , 2)$
    d. $(\ \ , -2)$

8. $3x + 2y = 6$
    a. $(0, \ \ )$
    b. $(\ \ , 0)$
    c. $(-2, \ \ )$
    d. $(\ \ , -6)$

Locate at least two ordered pairs of real numbers that satisfy each of the linear equations in Exercises 9 – 12 and graph the corresponding line in a Cartesian coordinate system.

9. $x + y = 5$ **10.** $y = -2x$ **11.** $2x + 3y = 0$ **12.** $2x + \dfrac{1}{2}y = 5$

Graph the linear equations in Exercises 13 – 16 by locating the y-intercept and the x-intercept.

13. $x + 2y = 10$ **14.** $x + y = 2$ **15.** $4x - y = -6$ **16.** $3x + 2y = 10$

2.2 Slope-Intercept Form: $y = mx + b$

Find the slope of the line determined by each pair of points given in Exercises 17 – 20.

17. $(-1, 3); (0, 1)$ **18.** $(-2, -4); (0, 2)$ **19.** $\left(\dfrac{2}{3}, \dfrac{1}{2}\right); \left(0, \dfrac{5}{6}\right)$ **20.** $(-1, -7); (1, 1)$

Tell whether each equation in Exercises 21 – 24 represents a horizontal line or vertical line and give its slope. Graph the line.

21. $y = -6$ **22.** $x = 5$ **23.** $3x = -5.4$ **24.** $3y - 4 = 0$

For Exercises 25 – 34, write the equation in slope-intercept form. Find the slope and the y-intercept, and then draw the graph.

25. $y = -\dfrac{2}{5}x + 1$ **26.** $y = \dfrac{4}{3}x - 2$ **27.** $3x + y = 12$ **28.** $x - 2y = 4$

29. $x - y = -6$ **30.** $2x + y + 8 = 0$ **31.** $2y - 6 = 0$ **32.** $3x = -9$

33. $2x + 6 = 3y$ **34.** $4x = -2y + 7$

In Exercises 35 – 40, the graph of a line is shown with two points highlighted. Find ***a.*** *the slope,* ***b.*** *the y-intercept (if there is one), and* ***c.*** *the equation of the line in slope-intercept form.*

35.

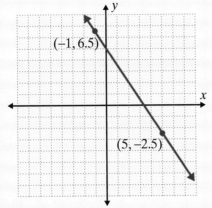

36.

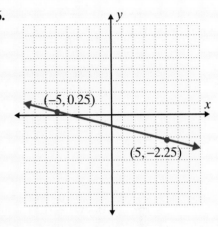

37.

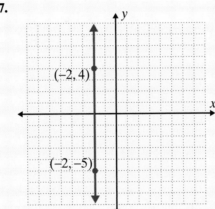

38.

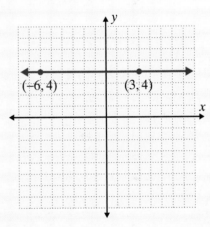

39.

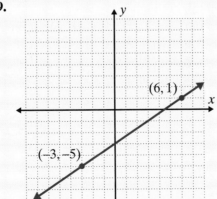

40.

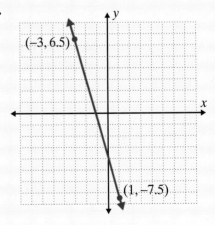

2.3 Point-Slope Form: $y - y_1 = m(x - x_1)$

*In Exercises 41 and 42, find **a.** the slope, **b.** a point on the line, and **c.** the graph of the line for the given equations in point-slope form.*

41. $y - 7 = 3(x + 1)$

42. $y + 3 = -(x - 1)$

In Exercises 43 – 46, find an equation in standard form for the line passing through the given point with the given slope. Graph the line.

43. $m = -3;\ (-4, 1)$

44. $m = \dfrac{5}{2};\ (3, -2)$

45. m is undefined; $(-4, -7)$

46. $m = 0;\ \left(-6, -\dfrac{3}{2}\right)$

In Exercises 47 – 50, find an equation in slope-intercept form for the line passing through the two given points.

47. $(-3, 5); (1, -3)$ **48.** $(4, 2); (6, 1)$ **49.** $\left(\dfrac{2}{3}, 0\right); \left(-2, \dfrac{1}{3}\right)$ **50.** $\left(0, -\dfrac{1}{2}\right); \left(\dfrac{3}{4}, \dfrac{1}{5}\right)$

Find the slope-intercept form of the equations described in Exercises 51 – 58.

51. Write an equation for the horizontal line through the point $(-3, 5)$.

52. Write an equation for the vertical line through the point $(5, 7)$.

53. Write an equation for the line through the point $(-6, -2)$ and parallel to the line $x - 4y = 4$. Graph both lines.

54. Write an equation for the line through the point $(-4, 1)$ and parallel to the line $x + y = -6$. Graph both lines.

55. Write an equation for the line through the point $(1, 1)$ and perpendicular to the line $2x + y = 9$. Graph both lines.

56. Write an equation for the line perpendicular to the line $x + 2y = 10$ and passing through $(-3, 0)$.

57. Write an equation for the line parallel to the line $2x + 3y = 5$ with the same y-intercept as the line $x - 2y = 8$.

58. Write an equation for the line perpendicular to the line $3x - y = 2$ with the same y-intercept as the line $2x + y = 5$.

*In Exercises 59 – 62, determine whether each pair of lines is **a.** parallel, **b.** perpendicular, or **c.** neither. Graph both lines.*

59. $\begin{cases} y = -3x + 4 \\ y = -3x - 1 \end{cases}$ **60.** $\begin{cases} 5x + y = 5 \\ x - 5y = 10 \end{cases}$ **61.** $\begin{cases} x + y = 6 \\ x - y = 4 \end{cases}$ **62.** $\begin{cases} 2x + y = 8 \\ y = -\dfrac{1}{2}x \end{cases}$

2.4 Introduction to Functions and Function Notation

Graph the relations in Exercises 63 – 66. State the domain and range and indicate which of the relations are functions.

63. $f = \{(0,3),(-1,2),(4,2),(5,3),(6,3)\}$ **64.** $g = \{(-5,1),(-3,-2),(-3,2),(0,0),(1,1)\}$

65. $f = \{(-4,3.2),(-3,0.1),(0,0.1),(0,-2),(2,3.2)\}$

66. $g = \left\{\left(-\dfrac{5}{2},2\right),\left(-\dfrac{1}{2},3\right),(0,4),(0,-4),\left(1,-\dfrac{1}{2}\right)\right\}$

In Exercises 67 – 72, use the vertical line test to determine whether or not each graph represents a function. State the domain and range.

67.

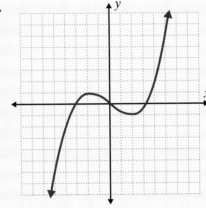

68.

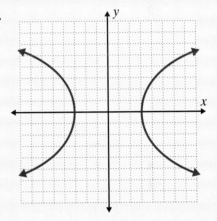

69.

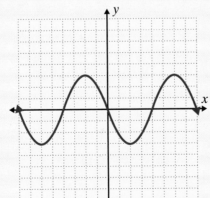

70.

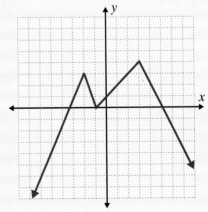

71.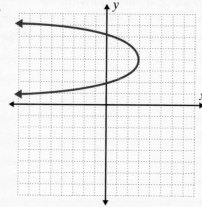

72.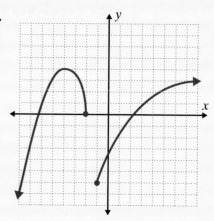

73. Given that $f(x) = -7x + 5$, find **a.** $f(0)$ **b.** $f(-3)$ and **c.** $f\left(\dfrac{2}{7}\right)$.

74. Given that $g(x) = 5x - 12$, find **a.** $g(6)$ **b.** $g(-1)$ and **c.** $g\left(\dfrac{3}{5}\right)$.

75. For $R(x) = x^2 - 5x + 6$, find **a.** $R(5)$ **b.** $R(-2)$ and **c.** $R(0)$.

76. For $h(x) = 6x^2 - 3x$, find **a.** $h(2)$ **b.** $h(-4)$ and **c.** $h(10)$.

State the domains of the functions given in Exercises 77 and 78.

77. $f(x) = \dfrac{x+5}{\sqrt{x-1}}$ **78.** $18x + 7y = 9$

*Use a graphing calculator to graph the functions in Exercises 79 – 84. Use the CALC features of the calculator to find the x-intercepts, if any. For absolute value functions, select the MATH menu, then the NUM menu, and then **1:abs(**.*

79. $y = 3x + 1$ **80.** $y = -2x - 5$ **81.** $y = |x + 3|$

82. $y = |x^2 - 2x|$ **83.** $y = x^3 - 4x - 1$ **84.** $y = \sqrt{2x - 1}$

2.5 Graphing Linear Inequalities: $y < mx + b$

Graph the solution set of each of the linear inequalities in Exercises 85 – 90.

85. $y \leq 2x - 1$ **86.** $y > -5$ **87.** $y < 3x$

88. $2x + 4 \geq 2y$ **89.** $\dfrac{3}{5}x + 6 \leq y$ **90.** $y > \dfrac{1}{4}x - 3$

Use your graphing calculator to graph each of the linear inequalities in Exercises 91 – 94.

91. $y > 2x$ **92.** $2x + 6y > 0$ **93.** $x + y \leq 5$ **94.** $x - y > -2$

Chapter 2 Test

In Exercises 1 and 2, find the missing coordinate of each ordered pair so that the ordered pair belongs to the solution set of the given equation.

1. $3x + y = 2$
   a. $(0, \quad)$
   b. $(\quad, 0)$
   c. $(-2, \quad)$
   d. $(\quad, -7)$

2. $x - 5y = 6$
   a. $(0, \quad)$
   b. $(\quad, 0)$
   c. $(11, \quad)$
   d. $(\quad, -2)$

Locate at least two ordered pairs of real numbers that satisfy each of the linear equations in Exercises 3 and 4 and graph the corresponding line in a Cartesian coordinate system.

3. $x + 4y = 5$

4. $2x - 5y = 1$

Find the slope of the line determined by each pair of points in Exercises 5 and 6. Graph the line.

5. $(1, -2), (9, 7)$

6. $(-2, 5), (8, 3)$

Write each equation in Exercises 7 and 8 in slope-intercept form. Find the slope and y-intercept, and then draw the graph.

7. $x - 3y = 4$

8. $4x + 3y = 3$

In Exercises 9 and 10, find the equation in standard form for the line determined by the given point and slope or the given two points.

9. $(3, 7), \quad m = -\dfrac{5}{3}$

10. $(-4, 6), (3, -2)$

In Exercises 11 – 13, find an equation in slope-intercept form for the line satisfying the given conditions.

11. Horizontal and passing through $(-1, 6)$
12. Parallel to $3x + 2y = -1$ and passing through $(2, 4)$
13. Perpendicular to the *y*-axis and passing through the point $(3, -2)$

14. Bicycling: The following table shows the number of miles Sam rode his bicycle from one hour to another.
   a. Plot the points indicated in the table.
   b. Calculate the slope of the line segments from point to point.
   c. Interpret each slope.

Hour	Miles
First	20
Second	31
Third	17

15. Write the definition of a function.

16. List the set of ordered pairs that corresponds to the points shown in the graph. State the domain and range of the relation and indicate whether or not the relation is a function.

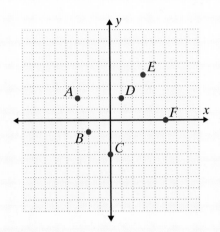

In Exercises 17 and 18, use the vertical line test to determine whether or not each graph represents a function. State the domain and range.

17.

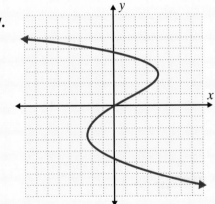

18.

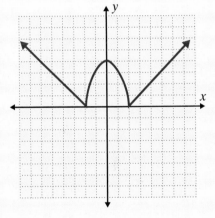

19. Given that $f(x) = x^2 - 2x + 5$, find **a.** $f(-2)$ **b.** $f(0)$ and **c.** $f(1)$.

20. State the domain of the function $y = \dfrac{x+3}{2x-1}$.

21. **a.** Use a graphing calculator to graph the functions $y = \sqrt{2x-3}$ and $y = 5 - x$.
   b. Use the **CALC** feature of the calculator to find the point of intersection of these two functions.

22. **a.** Use a graphing calculator to graph the function $y = 5 - 2x - 2x^2$.
   b. Use the **CALC** feature of the calculator to find the highest point on the graph.

Graph the linear inequalities in Exercises 23 and 24.

23. $3x - 5y \le 10$ **24.** $3x + 4y > 7$

Cumulative Review: Chapters 1–2

In Exercises 1 – 4, find the prime factorization of each of the composite numbers.

1. 93   **2.** 300   **3.** 188   **4.** 245

Find the LCM of each set of numbers given in Exercises 5 – 8.

5. $\{12,15,45\}$ **6.** $\{18,24,54\}$ **7.** $\{6,20,25,35\}$ **8.** $\{10,14,30,40\}$

9. Given the set of numbers $\left\{-10,-\sqrt{25},-1.6,-\sqrt{7},0,\dfrac{1}{5},\sqrt{9},\pi,\sqrt{12}\right\}$, list those numbers that belong to each of the following sets:

    a. $\{x|x\text{ is a natural number}\}$ **b.** $\{x|x\text{ is a whole number}\}$

    c. $\{x|x\text{ is a rational number}\}$ **d.** $\{x|x\text{ is an integer}\}$

    e. $\{x|x\text{ is an irrational number}\}$ **f.** $\{x|x\text{ is a real number}\}$

Graph each of the sets of real numbers described in Exercises 10 and 11 on a real number line.

10. $\{x|-1.8<x<5\text{ and }x\le 3\}$ **11.** $\{x|-5<x\le -4\text{ or }2<x\le 5\}$

State the properties of real numbers that are illustrated in Exercises 12 – 17. All variables represent real numbers.

12. $7+(x+3)=(7+x)+3$ **13.** $3(y+6)=3y+18$

14. $\dfrac{4}{5}+\left(-\dfrac{4}{5}\right)=0$ **15.** $x\cdot 1=x$

16. If $x<10$ and $10<y$, then $x<y$.   **17.** Either $x<-9$, $x=-9$, or $x>-9$.

Perform the indicated operations in Exercises 18 – 29.

18. $(-13)+(-7)$ **19.** $|-9|+2$ **20.** $17-(-5)$

21. $|9|-|-10|$ **22.** $6.5+(-4.2)-3.1$ **23.** $\dfrac{3}{4}-\dfrac{2}{3}+\left(-\dfrac{1}{6}\right)$

24. $(-7)\cdot(-12)$ **25.** $8(-5)$ **26.** $22\div(-11)$

27. $(-4)\div(-6)$ **28.** $8\div 0$ **29.** $0\div\dfrac{3}{5}$

Find the value of each expression in Exercises 30 – 33 by using the rules for order of operations.

30. $12 - 6 \div 2 \cdot 3 - 5$

31. $5 - (13 \cdot 5 - 5) \div 3 \cdot 2$

32. $3^2 + 5 \cdot 4 - 10 + |7|$

33. $6 \cdot 4 - 2^3 - (5 \cdot 10) - 5^2$

Simplify each expression in Exercises 34 – 37 by combining like terms.

34. $-4(x + 3) + 2x$

35. $x + \dfrac{x - 5x}{4}$

36. $\left(x^3 + 4x - 1\right) - \left(-2x^3 + x^2\right)$

37. $-2\left[7x - (2x + 5) + 3\right]$

Solve each of the equations in Exercises 38 – 43.

38. $9x - 11 = x + 5$

39. $5(1 - 2x) = 3x + 57$

40. $5(2x + 3) = 3(x - 4) - 1$

41. $\dfrac{7x}{8} + 5 = \dfrac{x}{4}$

42. $|2x + 1| = 5.6$

43. $|2(x - 4) + x| = 2$

44. Solve each equation for the indicated variable.

    a. Solve for n: $A = \dfrac{m + n}{2}$

    b. Solve for f: $\omega = 2\pi f$

45. The difference between twice a number and 3 is equal to the difference between five times the number and 2. Find the number.

46. **Rental cost:** The local supermarket charges a flat rate of \$5, plus \$3 per hour for rental of a carpet cleaner. If it cost Ron \$26 to rent the machine, how many hours did he keep it?

47. **Distance traveled:** Stephanie rode her new moped to Rod's house. Traveling the side streets, she averaged 20 mph. To save time on the return trip, they loaded the bike into Rod's truck and took the freeway, averaging 50 mph. The freeway distance is 2 miles less than the distance on the side streets and saves 24 minutes. Find the distance traveled on the return trip. [**Note:** You will need to convert 24 minutes into its equivalent value in hours before calculating the answer.]

48. **Entrance exams:** The mathematics component of the entrance exam at a certain Midwestern college consists of three parts: one part on geometry, one part on algebra, and one part on trigonometry. Prospective students must score at least 50 on each part and average at least 70 on the three parts. Beth learned that she had scored 60 and 66 on the first two parts of the exam. What minimum score did she need on the third part for her to pass this portion of the exam to gain entrance to the college?

Solve the inequalities in Exercises 49 – 54 and graph the solution sets. Write each solution in interval notation. Assume that x is a real number.

49. $5x - 7 > x + 9$ **50.** $5x + 10 \le 6(x + 3.8)$ **51.** $x + 8 - 5x \ge 2(x - 2)$

52. $\dfrac{2x + 1}{3} \le \dfrac{3x}{5}$ **53.** $|5x + 2| > 7$ **54.** $|3x + 2| + 4 < 10$

In Exercises 55 and 56, find the missing coordinate of each ordered pair so that the ordered pair belongs to the solution set of the given equation.

55. $2x - y = 4$
  a. $(0, \quad)$
  b. $(\quad, 0)$
  c. $(1, \quad)$
  d. $(\quad, 2)$

56. $x + 3y = 6$
  a. $(0, \quad)$
  b. $(\quad, 0)$
  c. $(2, \quad)$
  d. $(\quad, -1)$

Graph the linear equations in Exercises 57 – 59 by locating the y-intercept and the x-intercept.

57. $x + 2y = 6$ **58.** $2x - 5y = 10$ **59.** $3x - 4y = 6$

Write each equation in Exercises 60 – 62 in slope-intercept form. Find the slope and the y-intercept, and then draw the graph.

60. $x + 5y = 10$ **61.** $3x + y = 1$ **62.** $3x - 7y = 7$

Find the equation in standard form for the line determined by the given point and slope or two points in Exercises 63 – 68.

63. $(6, -1)$, $m = \dfrac{2}{5}$ **64.** $(-1, 2)$, $m = \dfrac{4}{3}$ **65.** $(0, 0)$, $m = 2$

66. $(5, 2)$, m is undefined   **67.** $(0, 3)$, $(5, -1)$ **68.** $(5, -2)$, $(1, 6)$

Find the equation in slope-intercept form for the line satisfying the conditions in Exercises 69 – 72.

69. Parallel to $3x + 2y - 6 = 0$, passing through $(2, 3)$

70. Parallel to the y-axis, passing through $(1, -7)$

71. Perpendicular to $4x + 3y = 5$, passing through $(4, 0)$

72. Perpendicular to $3x - 5y = 1$, passing through $(6, -2)$

73. List the ordered pairs corresponding to the points in the given graph. State the domain and range and whether or not the relation is a function.

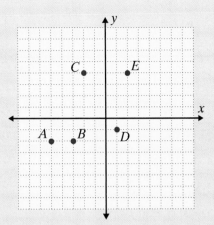

74. Use the vertical line test to determine whether or not the given graph represents a function. State the domain and range.

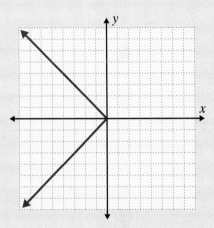

State the domains of the functions given in Exercises 75 – 77.

75. $y = \sqrt{3x+6}$ **76.** $y = -4x+1$ **77.** $f(x) = \dfrac{13}{x-9}$

78. For $f(x) = -3x+14$, find $f(5)$.

79. For $g(x) = x^2 - 4x + 7$, find $g(3)$.

80. For $F(x) = 3x^2 - 8x - 10$, find $F(0)$.

81. For $G(x) = 5x^3 - 4x$, find $G(-2)$.

Graph the linear inequalities in Exercises 82 and 83.

82. $y \geq 4x$ **83.** $3x + y < 2$

84. World Series of Poker: The following table shows the number of entrants in the main event of the World Series of Poker from 2003 to 2008.

Year	Number of Entrants
2003	839
2004	2576
2005	5619
2006	8773
2007	6358
2008	6844

a. Plot these points on a graph.
b. Connect the points with line segments.
c. Find the slope of each line segment.
d. Interpret the slope as a rate of change.

Source: Harrah's License Company

85. a. Use a graphing calculator to graph the linear functions $y = -2x + 7$ and $y = 3x$.
b. Use the **CALC** feature of the calculator to find the point of intersection of these two linear functions.

86. a. Use a graphing calculator to graph the function $y = \dfrac{1}{2}x^2 + x$.
b. Use the **CALC** feature of the calculator to find the lowest point on the graph.

87. a. Use a graphing calculator to graph the function $y = x^4 - 3x^3 + 1$.
b. Use the **CALC** feature of the calculator to find the values of the x-intercepts.

Systems of Linear Equations

Did You Know?

The subject of solutions to systems of linear equations, especially using determinants, received a great deal of attention in nineteenth-century mathematics. However, problems of this type are very old in the history of mathematics. As a matter of fact, determinants (or rules that were the equivalent of determinants) are found in pre-Christian, Chinese mathematical manuscripts. The solution to systems of equations was well known in China and carried to Japan also. The great Japanese mathematician Seki Kōwa (1642 – 1708) wrote a book on the subject in 1683 that was well in advance of European work on the subject. Seki Kōwa is probably the most distinguished of all Japanese mathematicians. Born into a samurai family, he showed great mathematical talent at an early age. He is credited with the independent development of calculus and is sometimes referred to as "the Newton of Japan." There is a traditional story that Seki Kōwa made a journey to the Buddhist shrines at Nara. There, ancient Chinese mathematical manuscripts were preserved, and Seki is supposed to have spent three years learning the contents of the manuscripts that previously no one had been able to understand. As was the custom, much of Seki Kōwa's work was done through finding solutions of very intricate problems. In 1907, the Emperor of Japan presented a posthumous award to the memory of Seki Kōwa, who did so much to awaken interest in scientific and mathematical research in Japan.

Native Japanese mathematics (the **wasan**) flourished until the nineteenth century when western mathematics and notation were completely adopted. In the 1940s, solutions to large systems of equations became a part of the new branch of mathematics called **operations research**. Operations research is concerned with deciding how best to design and operate man-machine systems, usually under conditions requiring the allocation of scarce resources. Although this new science initially had only military applications, recent applications have been made in the areas of business, industry, transportation, meteorology, and ecology. Computers now make it possible to solve extremely large systems of linear equations that are used to model the operating system being studied. Although computers do much of the work involved in solving systems of equations, it is necessary for you to understand the principles involved by studying small systems involving two or three unknowns, as presented in this chapter.

"The advancement and perfection of mathematics are intimately connected with the prosperity of the state."

Napoleon Bonaparte (1769 – 1821)

We saw in Chapter 2 that many applications involve two (or more) quantities. We will see in Chapter 3 that some of these applications can be solved by using two (or more) variables and a set of two (or more) equations. If the equations are linear, then the set of equations is called a **system of linear equations**. In this chapter we will develop techniques for solving systems of linear equations.

Graphing systems of two equations in two variables is helpful in visualizing the relationships between the equations. However, this approach is somewhat limited in finding solutions since numbers might be quite large, or solutions might involve fractions that must be estimated on the graph. Therefore, algebraic techniques are necessary to accurately solve systems of linear equations. Graphing in three dimensions will be left to later courses.

Two ideas probably new to you at this level are matrices and determinants. Matrices and determinants provide powerful general approaches to solving large systems of equations with many variables. Discussions in this chapter will be restricted to two linear equations in two variables and three linear equations in three variables.

3.1   Systems of Linear Equations in Two Variables

Solve systems of linear equations in two variables using three methods:
- *graphing,*
- *substitution, and*
- *addition.*

Two (or more) linear equations considered at one time are said to form a **system of equations** or a **set of simultaneous equations**. For example, consider the following system of two equations:

$$\begin{cases} 2x + y = 5 \\ x - y = 1 \end{cases}.$$

Each equation has an infinite number of solutions. That is, there are an infinite number of ordered pairs that satisfy each equation. But, the question of interest is, "Are there any ordered pairs that satisfy both equations at the same time?" In this example, the answer is "yes." The ordered pair $(2, 1)$ satisfies both equations.

$$2(2) + 1 = 5 \qquad \text{substituting into the first equation}$$
$$2 - 1 = 1 \qquad \text{substituting into the second equation}$$

There are several questions that need to be answered.

- How do we find the solution, if there is one?
- Will there always be a solution to a system of linear equations?
- Can there be more than one solution?

Graphing linear equations provides initial insight to the answers to these questions. However, as the chapter progresses we will see that other algebraic techniques are even more informative.

Table 3.1 illustrates the three possibilities for a system of two linear equations in two variables. Each system will be one of the following.

    1. Consistent   (has exactly one solution)

    2. Inconsistent   (has no solution)

    3. Dependent   (has an infinite number of solutions)

System	Graph	Intersection	Classification
$\begin{cases} 2x + y = 5 \\ x - y = 1 \end{cases}$		$(2, 1)$ or $x = 2$ and $y = 1$ (The lines intersect at one point.)	**Consistent**
$\begin{cases} 3x - 2y = 2 \\ 6x - 4y = -4 \end{cases}$		No Solution (The lines are parallel.)	**Inconsistent**
$\begin{cases} 2x - 4y = 6 \\ x - 2y = 3 \end{cases}$		Any ordered pair that satisfies $x - 2y = 3$. (The lines are the same line.) There are an infinite number of solutions: $(3 + 2y, y)$, where y is any real number.	**Dependent**

Table 3.1

The three basic methods for solving a system of two linear equations in two variables that will be discussed in this section are:

    1. graphing   (as illustrated in Table 3.1)

    2. substitution   (algebraically substituting from one equation into the other)

    3. addition   (combining like terms from both equations)

Solutions by Graphing

To Solve a System of Linear Equations by Graphing

1. *Graph both linear equations on the same set of axes.*

2. *Observe the point of intersection (if there is one).*

    a. *If the slopes of the two lines are different, then the lines intersect in one and only one point. The system is **consistent** and has a single point solution.*

    b. *If the lines are distinct and have the same slope, then the lines are parallel and the system is **inconsistent**. The system will have no solution.*

    c. *If the lines are the same line, the system is **dependent** and all the points on the line constitute the solution.*

Solving by graphing can involve estimating the solutions whenever the intersection of the two lines is at a point not represented by a pair of integers. (There is nothing wrong with this technique. Just be aware that at times it can lack accuracy. See Example 2.)

Example 1: A Consistent System

Solve the following system of linear equations by graphing: $\begin{cases} x+y=6 \\ \quad y=x+4 \end{cases}$.

Solution: The two lines intersect at the point $(1, 5)$. The system is **consistent** and the solution is $x = 1$ and $y = 5$.

Check: Substitution shows that $(1, 5)$ satisfies **both** of the equations in the system.

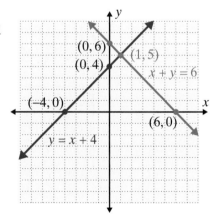

$$x+y=6 \qquad\qquad y=x+4$$
$$\overset{?}{(1)+(5)=6} \qquad \overset{?}{(5)=(1)+4}$$
$$6=6 \qquad\qquad 5=5$$

Example 2: A Consistent System with Estimation

Solve the following system of linear equations by graphing: $\begin{cases} x - 3y = 4 \\ 2x + y = 3 \end{cases}$.

Solution: The two lines intersect at one point, making this system **consistent**. However, we can only estimate the point of intersection as $\left(2, -\dfrac{1}{2}\right)$.

In this situation be aware that, although graphing gives a good "estimate," finding exact solutions to the system is not likely.

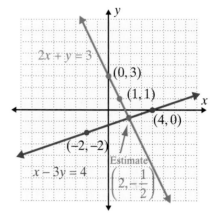

Check: Substituting $x = 2$ and $y = -\dfrac{1}{2}$ gives:

$$2 - 3\left(-\dfrac{1}{2}\right) \overset{?}{=} 4 \qquad \text{and} \qquad 2(2) + \left(-\dfrac{1}{2}\right) \overset{?}{=} 3$$

$$\dfrac{7}{2} \neq 4 \qquad\qquad\qquad \dfrac{7}{2} \neq 3.$$

Thus, checking shows that the estimated solution $\left(2, -\dfrac{1}{2}\right)$ does not satisfy either equation. The estimated point of intersection is just that – an estimate. The following discussion develops an algebraic technique that gives the exact solution as $\left(\dfrac{13}{7}, -\dfrac{5}{7}\right)$.

Solutions by Substitution

The objective in the substitution method is to eliminate one of the variables so that a new equation is formed with just one variable. If this new equation:

1. has one solution, the system is **consistent**.
2. is never true, the system is **inconsistent**.
3. is always true, the system is **dependent**.

To Solve a System of Linear Equations by Substitution

1. *Solve one of the equations for one of the variables.*

2. *Substitute the resulting expression into the other equation.*

3. *Solve this new equation, if possible, and then substitute back into one of the original equations to find the value of the other variable. (This is known as **back substitution**.)*

4. *Check the solution in both of the original equations.*

Example 3: A Consistent System

Use the method of substitution to solve the system: $\begin{cases} x - 3y = 4 \\ 2x + y = 3 \end{cases}$.

Solution: Note that this is the same problem we tried to solve by graphing in Example 2, but needed to estimate the solution.

$$x - 3y = 4 \qquad \text{First Equation}$$

$$x = 4 + 3y \qquad \text{Solve the first equation for } x.$$

$$2x + y = 3 \qquad \text{Second Equation}$$

$$2(4 + 3y) + y = 3 \qquad \text{Substitute } 4 + 3y \text{ for } x \text{ in the second equation.}$$

$$8 + 6y + y = 3 \qquad \text{Solve the new equation for } y.$$

$$7y = -5$$

$$y = -\frac{5}{7}$$

To find x, we "**back substitute**" $-\dfrac{5}{7}$ for y in one of the original equations.

$$x - 3\left(-\frac{5}{7}\right) = 4 \qquad\qquad 2x + \left(-\frac{5}{7}\right) = 3$$
$$\text{OR}$$
$$x + \frac{15}{7} = 4 \qquad\qquad 2x = 3 + \frac{5}{7}$$

$$x = 4 - \frac{15}{7} \qquad\qquad 2x = \frac{26}{7}$$

$$x = \frac{13}{7} \qquad\qquad x = \frac{13}{7}$$

The system is consistent, and the solution is

$$x = \frac{13}{7} \text{ and } y = -\frac{5}{7}, \text{ or } \left(\frac{13}{7}, -\frac{5}{7}\right).$$

In this example, the second equation could have been solved for y and the substitution made into the first equation. The solution would have been the same. (As a thorough check, substitute the solution in both of the original equations.)

 NOTES   In Example 3, we could have solved either equation for either variable and then substituted the result into the other equation. For simplicity, we generally solve for the variable that has a coefficient of 1 if there is such a variable.

Example 4: An Inconsistent System

Solve the following system by substitution: $\begin{cases} 6x + 3y = 14 \\ 2x + y = -3 \end{cases}$.

Solution:

$$2x + y = -3$$

$$y = -3 - 2x \qquad \text{Solve the second equation for } y.$$

$$6x + 3y = 14$$

$$6x + 3(-3 - 2x) = 14 \qquad \text{Substitute } -3 - 2x \text{ for } y \text{ in the first equation.}$$

$$6x - 9 - 6x = 14 \qquad \text{Simplify.}$$

$$-9 = 14 \qquad \text{False statement}$$

The variable x is eliminated, and this last equation is never true.
Therefore, the system is **inconsistent**.
There is **no solution** to this system of equations. (The lines are parallel.)

Solutions by Addition

A third method for solving a system of linear equations is the **method of addition** (or the **method of elimination**). As with the substitution method, the objective is to eliminate one of the variables so that a new equation is found with just one variable, if possible. If this new equation:

    1. has one solution, the system is **consistent**.
    2. is never true, the system is **inconsistent**.
    3. is always true, the system is **dependent**.

The procedure is outlined as follows:

To Solve a System of Linear Equations by Addition

1. *Write the equations one under the other so that **like terms are aligned**.*

2. *Multiply all terms of one equation by a constant (and possibly all terms of the other equation by another constant) so that **two like terms have opposite coefficients**.*

3. *Add the two equations by **combining like terms** and solve the resulting equation, if possible. Then, **back substitute into one of the original equations** to find the value of the other variable.*

4. *Check the solution in both of the original equations.*

Example 5: A Consistent System

Use the method of addition to solve the following system: $\begin{cases} 3x + 5y = -3 \\ -7x + 2y = 7 \end{cases}$.

Solution: Multiply each term in the first equation by 2 and each term in the second equation by -5. The y-coefficients will be opposites. Add the two equations by combining like terms which will eliminate y. Solve for x.

$$\begin{cases} 3x + 5y = -3 \\ -7x + 2y = 7 \end{cases}$$

$$\begin{cases} [2](3x + 5y = -3) \longrightarrow \quad 6x + 10y = -6 \\ [-5](-7x + 2y = 7) \longrightarrow \quad \underline{35x - 10y = -35} \end{cases}$$

$$41x \quad\quad = -41$$
$$x \quad\quad = -1$$

Substitute $x = -1$ into one of the original equations.

$$3x + 5y = -3 \quad\quad \text{OR} \quad\quad -7x + 2y = 7$$
$$3(-1) + 5y = -3 \quad\quad\quad\quad -7(-1) + 2y = 7$$
$$-3 + 5y = -3 \quad\quad\quad\quad 7 + 2y = 7$$
$$5y = 0 \quad\quad\quad\quad 2y = 0$$
$$y = 0 \quad\quad\quad\quad y = 0$$

The solution is $x = -1$ and $y = 0$, or $(-1, 0)$.

> **NOTES**
>
> In Example 5, we could have multiplied the terms in the first equation by 7 and the terms in the second equation by 3 and eliminated x instead of y. The solution would be the same.

Example 6: A Dependent System

Solve the following system by using the method of addition: $\begin{cases} 3x - \dfrac{1}{2}y = 6 \\ 6x - y = 12 \end{cases}$.

Solution: Multiply the first equation by -2 so that the y-coefficients will be opposites.

$$\begin{cases} [-2]\left(3x - \dfrac{1}{2}y = 6\right) & \longrightarrow & -6x + y = -12 \\ \left(6x - y = 12\right) & \longrightarrow & \dfrac{6x - y = 12}{0 = 0} \end{cases}$$

Because this last equation, $0 = 0$, is **always true**, the system is **dependent**. The solution consists of all points that satisfy the equation $6x - y = 12$. Solving for y gives $y = 6x - 12$, and we can write the solution in the general form $(x, 6x - 12)$. [Or, solving for x, $x = \dfrac{1}{6}y + 2$. The solution can be written in the form $\left(\dfrac{1}{6}y + 2, y\right)$.]

Using a TI-84 Plus Graphing Calculator to Solve a System of Linear Equations

A graphing calculator can be used to locate (or estimate) the point of intersection of two lines (and therefore the solution to the system). For example, consider the system

$$\begin{cases} x + y = 4 \\ 3x - 2y = 7 \end{cases}.$$

We can proceed as follows:

Step 1: To be able to use the calculator's function mode, solve each equation for y:

$$\begin{cases} y = 4 - x \\ y = \dfrac{3}{2}x - \dfrac{7}{2} \end{cases}.$$

Step 2: Press the ⬛ Y= key and enter the two functions for Y_1 and Y_2 as shown here:

Step 3: Press GRAPH. (You may need to check the WINDOW to be sure that both lines appear.)

Step 4: Press 2ND CALC.

Choose 5: intersect.

Move the cursor to one of the lines and press ENTER in response to the question First curve?.

Move the cursor to the second line and press ENTER in response to the question Second curve?.

Move the cursor near the point of intersection and press ENTER in response to the question Guess?.

We see that the solution is $x = 3$ and $y = 1$.

[**Note:** In this case the solution shown is exact. In many cases the solution shown will be only an estimate. Thus, even with a calculator, the graphing method is limited. Also, **if the lines are parallel (an inconsistent system), the calculator will give an error message when you try to find the intersection point**.]

Practice Problems

Solve each of the following systems of linear equations algebraically.

1. $\begin{cases} y = 3x + 4 \\ 2x + y = -1 \end{cases}$
    2. $\begin{cases} 2x - 3y = 0 \\ 6x + 3y = 4 \end{cases}$
    3. $\begin{cases} 4x + y = 3 \\ 4x + y = 2 \end{cases}$

3.1 Exercises

In Exercises 1 – 4, determine which of the given points, if any, lie on both of the lines in the systems of equations by substituting each point into both equations.

1. $\begin{cases} x - y = 6 \\ 2x + y = 0 \end{cases}$
    2. $\begin{cases} x + 3y = 8 \\ 3y = -x + 4 \end{cases}$
    3. $\begin{cases} 3x + 4y - 7 = 0 \\ 3x + 6y - 9 = 0 \end{cases}$
    4. $\begin{cases} 5x + 3y = 1 \\ 5x = -4y \end{cases}$

    a. $(0, 6)$
        a. $(2, -2)$
        a. $(1, 1)$
        a. $(-4, 5)$

    b. $(-2, 4)$
       b. $\left(0, \dfrac{4}{3}\right)$
    b. $(2, 3)$
        b. $(3, -1)$

    c. $(2, -4)$
       c. $(8, 0)$
        c. $(4, -3)$
      c. $(1, -1)$

    d. $(5, -1)$
       d. $\left(1, -\dfrac{7}{3}\right)$
    d. $(-2, 1)$
      d. $\left(\dfrac{4}{5}, -1\right)$

In Exercises 5 – 8, the graphs of the lines represented by each system of equations are given. Determine the solution of the system by looking at the graph. Check your solution by substituting into both equations.

5. $\begin{cases} x - 2y = -5 \\ 2x + y = 5 \end{cases}$
    6. $\begin{cases} x + 3y = 6 \\ 2x - 5y = -10 \end{cases}$
    7. $\begin{cases} 2x - 2y = 3 \\ x - y = -1 \end{cases}$
    8. $\begin{cases} 2x + y = -7 \\ x + 3y = -1 \end{cases}$

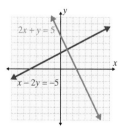
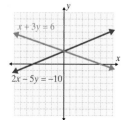
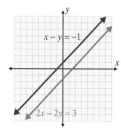
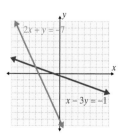

Answers to Practice Problems: **1.** $(-1, 1)$ **2.** $\left(\dfrac{1}{2}, \dfrac{1}{3}\right)$ **3.** No solution

In Exercises 9 – 12, show that each system of equations is inconsistent by determining the slope of each line and the y-intercept. That is, show that the lines are parallel (same slope, but different y-intercepts).

9. $\begin{cases} 2x + y = 4 \\ 4x + 2y = 1 \end{cases}$ **10.** $\begin{cases} 2x - 5y = 1 \\ 4x - 10y = 4 \end{cases}$ **11.** $\begin{cases} 4x - y = 8 \\ x - \dfrac{1}{4}y = 7 \end{cases}$ **12.** $\begin{cases} y = \dfrac{1}{3}x + 2 \\ x - 3y = 5 \end{cases}$

Solve each of the systems in Exercises 13 – 22 by graphing.

13. $\begin{cases} x + y = 5 \\ x - 4y = 5 \end{cases}$ **14.** $\begin{cases} 3x - y = 6 \\ 2x + y = -1 \end{cases}$ **15.** $\begin{cases} 2x - y = 8 \\ y = 2x \end{cases}$ **16.** $\begin{cases} y = \dfrac{5}{6}x + 1 \\ x - 2y = 2 \end{cases}$

17. $\begin{cases} 5x + 2y = 21 \\ x = y \end{cases}$ **18.** $\begin{cases} 2x + 3y = 4 \\ 4x - y = 1 \end{cases}$ **19.** $\begin{cases} 2x + y + 1 = 0 \\ 3x + 4y - 1 = 0 \end{cases}$ **20.** $\begin{cases} 4x - 2y = 10 \\ -6x + 3y = -15 \end{cases}$

21. $\begin{cases} x - 2y = 11 \\ 2x - 3y = 18 \end{cases}$ **22.** $\begin{cases} 4x + 3y + 7 = 0 \\ 5x = 2y - 3 \end{cases}$

Use the substitution method or the addition method to solve the systems of linear equations in Exercises 23 – 48. State whether each system is consistent, inconsistent, or dependent.

23. $\begin{cases} x + 4y = 6 \\ 2x + y = 5 \end{cases}$ **24.** $\begin{cases} 2x + y = 0 \\ x - 2y = -10 \end{cases}$ **25.** $\begin{cases} 5x - y = -2 \\ x + 2y = -7 \end{cases}$ **26.** $\begin{cases} 7x - y = 18 \\ x + 2y = 9 \end{cases}$

27. $\begin{cases} x + 2y = 3 \\ 4x + 8y = 8 \end{cases}$ **28.** $\begin{cases} 2x + 3y = 3 \\ x + 4y = 4 \end{cases}$ **29.** $\begin{cases} 6x + 2y = 16 \\ 3x + y = 8 \end{cases}$ **30.** $\begin{cases} 4x - y = 18 \\ 3x + 5y = 2 \end{cases}$

31. $\begin{cases} y = 3x + 3 \\ y = -2x + 8 \end{cases}$ **32.** $\begin{cases} x = -7 + 4y \\ 2x = 8y - 14 \end{cases}$ **33.** $\begin{cases} 2x + y = 4 \\ 4x + 5y = 11 \end{cases}$ **34.** $\begin{cases} 2x - 3y = 18 \\ 5x + 4y = -1 \end{cases}$

35. $\begin{cases} 3x + 4y = 6 \\ x - 8y = 9 \end{cases}$ **36.** $\begin{cases} 3x + 5y = 3 \\ 9x - y = -7 \end{cases}$ **37.** $\begin{cases} 2x = 5y - 1 \\ 4x - 10y = 0 \end{cases}$

38. $\begin{cases} 6x + 2y = 5 \\ 2x + y = 1 \end{cases}$ **39.** $\begin{cases} 4x + 12y = 5 \\ 5x - 6y = 1 \end{cases}$ **40.** $\begin{cases} 3x + y = 4 \\ 9x + 3y = 12 \end{cases}$

41. $\begin{cases} x + y = 7 \\ 2x + 3y = 16 \end{cases}$

42. $\begin{cases} 5x - 7y = 8 \\ 3x + 11y = -12 \end{cases}$

43. $\begin{cases} 6x - y = 15 \\ 1.2x - 0.2y = 3 \end{cases}$

44. $\begin{cases} 3x + y = 14 \\ 0.1x - 0.2y = 1.4 \end{cases}$

45. $\begin{cases} x + y = 12 \\ 0.05x + 0.25y = 1.6 \end{cases}$

46. $\begin{cases} x + y = 20 \\ 0.1x + 2.5y = 3.8 \end{cases}$

47. $\begin{cases} 0.6x + 0.5y = 5.9 \\ 0.8x + 0.4y = 6 \end{cases}$

48. $\begin{cases} 0.5x + 0.2y = 7 \\ 1.5x + 0.6y = 2 \end{cases}$

In Exercises 49 – 56, use a graphing calculator and the CALC *and* **5:intersect** *commands to find the solutions to the given systems of linear equations. If necessary, round values to four decimal places. (Remember to solve each equation for y. Use both* Y_1 *and* Y_2 *in the* Y= *menu.)*

49. $\begin{cases} 2x + y = 3 \\ x - y = 5 \end{cases}$

50. $\begin{cases} 3x + y = 6 \\ 2x + y = -1 \end{cases}$

51. $\begin{cases} 8x - 2y = 8 \\ y = -2x \end{cases}$

52. $\begin{cases} x + y = -5 \\ 4x - y = 5 \end{cases}$

53. $\begin{cases} x - 3y = 6 \\ -2x + y = -1 \end{cases}$

54. $\begin{cases} x + \dfrac{1}{2}y = 0 \\ 6x - y = 3 \end{cases}$

55. $\begin{cases} 2x + 3y = 2 \\ x + 2y = -3 \end{cases}$

56. $\begin{cases} x - 3y = 5 \\ 2x + 3y = 4 \end{cases}$

Writing and Thinking About Mathematics

57. Explain, in your own words, why the answer to a consistent system of linear equations is written as an ordered pair.

 HAWKES LEARNING SYSTEMS: INTERMEDIATE ALGEBRA SOFTWARE

- Solving Systems of Linear Equations by Graphing
- Solving Systems of Linear Equations by Substitution
- Solving Systems of Linear Equations by Addition

3.2 Applications

- *Solve applied problems using systems of two linear equations in two variables.*

Word problems were solved in Chapters 1 and 2 by using one variable and one equation. However, some of those problems and many other applications can be solved, and in fact are easier to solve, by using two variables and two equations. As we will see, by representing the information given in a problem with a system of two linear equations, the problem can be solved using the methods from Section 3.1.

Example 1: Mixture

A metallurgist needs to create 30 kg of a 40% copper alloy. He must do so by melting a 20% alloy and a 50% alloy and mixing them together. How much of each alloy will he need to make what he needs?

Solution: Let x = amount of 20% alloy in kg $\longrightarrow$ $0.20x$ is amount of copper
y = amount of 50% alloy in kg $\longrightarrow$ $0.50y$ is amount of copper

Form two equations based on the information given.

Phrase from problem	Equation formed
The total metal needed is 30 kg.	$x + y = 30$
The total amount of copper is 40% of 30.	$0.20x + 0.50y = 0.40(30)$

Now solve the system. We will use the addition method.

$$\begin{cases} [-2] & (x + y = 30) \\ [10] & (0.20x + 0.50y = 0.40(30)) \end{cases} \longrightarrow \begin{array}{rcrcr} -2x & - & 2y & = & -60 \\ 2x & + & 5y & = & 120 \\ \hline & & 3y & = & 60 \\ & & y & = & 20 \end{array}$$

Substituting $y = 20$ into the first equation yields:

$$x + (20) = 30$$
$$x = 10.$$

The metallurgist will need 10 kg of the 20% alloy and 20 kg of the 50% alloy.

Check: Substitute $x = 10$ and $y = 20$ into the original equations.
$(10) + (20) = 30$
$0.20(10) + 0.50(20) = 2 + 10 = 12 = 0.40(30)$

Example 2: Interest

A savings and loan company pays 7% interest on a long-term savings account, and a high-risk stock indicates that it should yield 12% interest. If a woman has $40,000 to invest and wants an annual income of $3550 from her investments, how much should she put in the savings account and how much in the stock?

Solution: Let x = amount invested at 7% $\longrightarrow$ $0.07x$ is amount of interest at 7%
y = amount invested at 12% $\longrightarrow$ $0.12y$ is amount of interest at 12%

Phrase from problem	Equation formed
The total invested is $40,000.	$\begin{cases} x + y = 40,000 \\ 0.07x + 0.12y = 3,550 \end{cases}$
The total interest is $3,550.	

$$\begin{cases} [-7] & (x & + & y & = & 40,000) \\ [100] & (0.07x & + & 0.12y & = & 3,550) \end{cases} \longrightarrow \begin{array}{rcrcr} -7x & - & 7y & = & -280,000 \\ 7x & + & 12y & = & 355,000 \\ \hline & & 5y & = & 75,000 \\ & & y & = & 15,000 \end{array}$$

Back substituting $y = 15,000$ gives:
$$x + (15,000) = 40,000$$
$$x = 25,000.$$

She should put $25,000 in the savings account at 7% and $15,000 in the stock at 12%.

Example 3: Work

Job A   Job B

Working his way through school, Richard works two part-time jobs for a total of 25 hours a week. Job A pays $7.50 per hour and job B pays $8.40 per hour. How many hours did he work at each job the week he made $200.10?

Solution: Let x = # of hrs at job A at $7.50 per hr $\longrightarrow$ $7.50x$ is earnings at $7.50
y = # of hrs at job B at $8.40 per hr $\longrightarrow$ $8.40y$ is earnings at $8.40

Phrase from problem	Equation formed
The total hours worked is 25.	$\begin{cases} x + y = 25 \\ 7.50x + 8.40y = 200.10 \end{cases}$
The total earnings were $200.10.	

$$\begin{cases} [-75] & (x & + & y & = & 25) \\ [10] & (7.50x & + & 8.40y & = & 200.10) \end{cases} \longrightarrow \begin{array}{rcrcr} -75x & - & 75y & = & -1875 \\ 75x & + & 84y & = & 2001 \\ \hline & & 9y & = & 126 \\ & & y & = & 14 \end{array}$$

Back substitute $y = 14$ to obtain:
$$x + (14) = 25$$
$$x = 11.$$

Richard worked 11 hours at job A and 14 hours at job B.

Example 4: Algebra

Determine the values of a and b such that the straight line $ax + by = 22$ passes through the point $(3, -1)$ and has slope $-\dfrac{5}{4}$.

Solution: Here, the unknown quantities are a and b, not x and y.

Since the point $(3, -1)$ is on the line, substitute $x = 3$ and $y = -1$ into the equation $ax + by = 22$.

$$3a - b = 22 \qquad \text{A linear equation in } a \text{ and } b$$

To find the slope in terms of a and b, write the equation $ax + by = 22$ in slope-intercept form.

$$ax + by = 22$$
$$by = -ax + 22$$
$$y = -\frac{a}{b}x + \frac{22}{b}$$

Thus the slope is $-\dfrac{a}{b}$. Therefore,

$$-\frac{a}{b} = -\frac{5}{4} \qquad \text{The problem tells us that the slope is } -\frac{5}{4}.$$

or $\qquad 4a = 5b$

or $\quad 4a - 5b = 0.$ Another linear equation in a and b

Now we have two linear equations in a and b, so we can use the addition method to solve the system of equations.

$$\begin{cases} [-5] & (3a - b = 22) \longrightarrow -15a + 5b = -110 \\ & (4a - 5b = 0) \longrightarrow \underline{\quad 4a - 5b = \quad 0} \\ & \qquad\qquad\qquad\qquad -11a \qquad\quad = -110 \\ & \qquad\qquad\qquad\qquad\quad a \qquad\qquad = 10 \end{cases}$$

Back substituting $a = 10$ yields:

$$3(10) - b = 22$$
$$-b = -8$$
$$b = 8.$$

Thus $a = 10$ and $b = 8$, so the line $10x + 8y = 22$ passes through the point $(3, -1)$ and has slope $-\dfrac{5}{4}$.

> **NOTES** Notice that the equation $-\dfrac{a}{b} = -\dfrac{5}{4}$ does not necessarily mean that $a = 5$ and $b = 4$. It means that the **ratio** of a to b is 5 to 4. In Example 4, $a = 10$ and $b = 8$, but the ratio of a to b is still 5 to 4.

3.2 Exercises

1. The sum of two integers is 102, and the larger number is 10 more than three times the smaller. Find the two integers.

2. The difference between two integers is 13, and their sum is 87. What are the two integers?

3. **Supplementary angles:** Two angles are supplementary if the sum of their measures is 180°. Find two supplementary angles such that the smaller is 30° more than one-half of the larger.

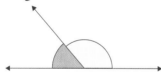

4. **Complementary angles:** Two angles are complementary if the sum of their measures is 90°. Find two complementary angles such that one is 15° less than six times the other.

5. **Soccer:** At present, the length of a rectangular soccer field is 55 yards longer than the width. The city council is thinking of rearranging the area containing the soccer field into two square playing fields. A math teacher on the council decided to test the council members' mathematical skills. (You know how math teachers are.) He told them that if the width of the current field were to be increased by 5 yards and the length cut in half, the resulting field would be a square. What are the dimensions of the field currently?

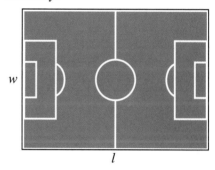

6. **Perimeter:** Consider a square and a regular hexagon (a six-sided figure with sides of equal length). One side of the square is 5 feet longer than a side of the hexagon, and the two figures have the same perimeter. What are the lengths of the sides of each figure?

7. **Chemistry:** How many liters each of a 12% iodine solution and a 30% iodine solution must be used to produce a total mixture of 90 liters of a 22% iodine solution?

8. **Food science:** A meat market has ground beef that is 40% fat and extra lean ground beef that is only 15% fat. How many pounds of each (ground beef and extra lean) must be ground together to get a total of 50 pounds of "lean" ground beef that is 25% fat?

9. **Food science:** A dairy needs 360 gallons of milk containing 4% butterfat. How many gallons each of milk containing 5% butterfat and milk containing 2% butterfat must be used to obtain the desired 360 gallons?

10. **Pharmacy:** A druggist has two solutions of alcohol. One is 25% alcohol. The other is 45% alcohol. He wants to mix these two solutions to get 36 ounces that will be 30% alcohol. How many ounces of each of these two solutions should he mix together?

11. **Investing:** Pam inherited $124,000 from her Uncle Harold. She invested a portion in bonds and the remainder in a long-term certificate account. The amount invested in bonds was $24,000 less than 3 times the amount invested in certificates. How much was invested in bonds and how much in certificates?

12. **Investing:** Sang has invested $48,000, part at 6% and the rest in a higher risk investment at 10%. How much did she invest at each rate to receive $4000 in interest after one year?

13. **Investments:** An investor bought 500 shares of stock, some at $3.50 per share and some at $6.00 per share. If the total cost was $2187.50, how many shares of each stock did the investor buy?

14. **Money:** Inez has 20 coins consisting of dimes and quarters. How many of each type does she have if all together she has $4.10?

15. **Candy:** A confectioner is going to mix candy worth $3.90 per pound with candy worth $2.50 per pound to obtain 70 pounds of candy worth $3.30 per pound. How many pounds of each kind should she use?

16. **Stamp collecting:** The postal service charges 42¢ for letters that weigh 1 ounce or less and 17¢ more for letters that weigh between 1 and 2 ounces. Jeff, testing his father's math skills, gave his father $42.10 and asked him to purchase 80 stamps for his stamp collection, some 42¢ stamps and some 59¢ stamps. How many of each type of stamp did his dad buy?

17. **Mixing nuts:** Mike wants to mix two kinds of nuts to be eaten at a party he and his fraternity brothers are hosting tonight. One kind sells for 70 cents per pound, and the other sells for $1.30 per pound. He wants to mix a total of 20 pounds and pay a total of 82 cents per pound. How many pounds of each kind should he use in the mix?

$0.70/lb $0.82/lb $1.30/lb

18. Manufacturing: A manufacturing plant is going to use two different stamping machines to complete an order of 975 units. One produces 100 units per hour, while the other produces 75 units per hour. How long must each machine operate to complete the order if, during the process, the faster machine is shut down for 2.5 hours for repairs?

19. Books: The bookstore can buy a popular book as a paperback or a hardback. A hardback book costs $3.50 more than the paperback book. What is the cost of each if 90 paperback books cost the same as 55 hardback books?

20. Elections: In an election, the winner received 430 votes more than twice as many votes as the loser. If there was a total of 2290 votes cast, how many did each candidate receive?

21. Government: A bill was defeated in the house of representatives after 25 more people voted against it than voted in favor of it. If one-tenth of those voting against the bill had voted in favor of it, then 21 more people would have voted in favor of it than against it. How many legislators voted in favor of the bill?

22. Car travel: Andrea made a car trip of 440 kilometers. She averaged 54 kilometers per hour for the first part of the trip and 80 kilometers per hour for the second part. If the total trip took 6 hours, how long was she traveling at 80 kilometers per hour?

23. Boating: A boat left Dana Point Marina at 11:00 am traveling at 10 knots (nautical miles per hour). Two hours later, a Coast Guard boat left the same marina traveling at 14 knots trying to catch the first boat. If both boats traveled the same course, at what time did the Coast Guard captain anticipate overtaking the first boat?

24. Car travel: Two cars are to start at the same place in Knoxville and travel in opposite directions (assume in straight lines). The drivers know that one driver drives an average of 5 mph faster than the other. (They have been married for 20 years.) They have agreed to stop and call each other after driving for 3 hours. In the telephone conversation they realize that they are 345 miles apart. What was the average speed of each driver?

25. Determine a and b such that the line with equation $ax + by = 7$ passes through the two points $(2, 1)$ and $(-1, 10)$.

26. Air travel: A private jet flies the same distance in 6 hours that a commercial jet flies in 2.5 hours. If the speed of the commercial jet was 75 mph less than three times the speed of the private jet, find the speed of each jet.

27. Determine a and b such that the line with equation $ax + by = 6$ passes through the points $(-6, -2)$ and $(3, 4)$.

28. Determine a and b such that the line with equation $ax + by = 4$ passes through the point $(5, 2)$ and has a slope of $\frac{2}{3}$.

29. Determine a and b such that the line with equation $ax + by = -4$ contains the point $(-1, -3)$ and has a slope of 5.

30. Toys: A manufacturer produces two new action figures, Ferocious Frank and Mighty Marcel. Ferocious Frank takes 4 hours to produce and costs $8 each. Mighty Marcel takes 3 hours to produce and costs $7 each. If the manufacturer allots a total of 5800 hours and $12,600 for production each week, how many of each model will be produced?

31. Manufacturing: A car parts company has begun manufacturing two new products. One requires 2.5 hours of labor, 3 pounds of raw materials, and costs $42.40 each to produce. The second product requires 4 hours of labor, 4 pounds of raw materials, and costs $64 each to produce. Find the cost of labor per hour and the cost of raw materials per pound.

32. Furniture: A furniture shop refinishes chairs. Employees use two methods to refinish a chair. Method I takes 1 hour, and the material costs $6. Method II takes an hour and a half, and the material costs $3. Next week, they plan to spend 144 hours in labor and $600 in material refinishing chairs. How many chairs should they plan to refinish by each method?

33. Cattle farming: A large feed lot uses two feed supplements, Ration I and Ration II. Each pound of Ration I contains 4 units of protein and 2 units of carbohydrates. Each pound of Ration II contains 3 units of protein and 6 units of carbohydrates. If the dietary requirement calls for 42 units of protein and 30 units of carbohydrates, how many pounds of each ration should be used to satisfy the requirements?

34. Triangles: The sum of the measures of the three angles of a triangle is 180°. In an isosceles triangle, two of the angles have the same measure. What are the measures of the angles of an isosceles triangle in which one angle measures 15° more than each of the other two equal angles?

35. Triangles: The sum of the measures of the three angles of a triangle is 180°. In an isosceles triangle, two of the angles have the same measure. What are the measures of the angles of an isosceles triangle in which each of the two equal angles measures 15° more than the third angle?

36. Triangles: The sum of the lengths of two equal sides of an isosceles triangle is 30 centimeters more than the third side. Find the length of each side of the triangle if its perimeter is 45 centimeters.

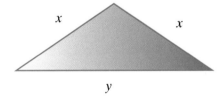

37. Age: Two years ago, Sue was half as old as Pat. Eight years from now, Sue will be two-thirds as old as Pat. How old is each of them now?

38. Age: George is 8 years older than his brother Kurt. Four years from now, he will be twice as old as Kurt. How old is each brother at this time?

39. Rectangles: The length of a rectangle is 10 meters more than one-half the width. If the perimeter is 50 meters, what is the length and what is the width of the rectangle?

40. Rectangles: The length of a rectangle is 2 feet less than twice the width. If each side is increased by 5 feet, the perimeter will be 118 feet. Find the length and width of the original rectangle.

 HAWKES LEARNING SYSTEMS: INTERMEDIATE ALGEBRA SOFTWARE

- Applications: Systems of Equations

<table>
<tr><td>**3.3**</td><td># Systems of Linear Equations in Three Variables</td></tr>
</table>

- *Solve systems of linear equations in three variables.*
- *Solve applied problems by using systems of linear equations in three variables.*

The equation $2x + 3y - z = 16$ is called a **linear equation in three variables**. The general form is

$$Ax + By + Cz = D \text{ where } A, B, \text{ and } C \text{ do not equal } 0.$$

The solutions to such equations are called **ordered triples** and are of the form (x_0, y_0, z_0) or $x = x_0$, $y = y_0$, and $z = z_0$. One ordered triple that satisfies the equation $2x + 3y - z = 16$ is $(1, 4, -2)$. To check this, substitute $x = 1$, $y = 4$, and $z = -2$ into the equation to see if the result is 16:

$$2(1) + 3(4) - (-2) = 2 + 12 + 2$$
$$= 16.$$

There are an infinite number of ordered triples that satisfy any linear equation in three variables in which at least two of the coefficients are nonzero. Any two values may be substituted for two of the variables, and then the value for the third variable can be calculated. For example, by letting $x = -1$ and $y = 5$, we find:

$$2(-1) + 3(5) - z = 16$$
$$-2 + 15 - z = 16$$
$$-z = 3$$
$$z = -3.$$

Hence, the ordered triple $(-1, 5, -3)$ satisfies the equation $2x + 3y - z = 16$.

Graphs can be drawn in three dimensions by using a coordinate system involving three mutually perpendicular number lines labeled as the x-axis, y-axis, and z-axis. Three planes are formed: the xy-plane, the xz-plane, and the yz-plane. The three axes separate space into eight regions called **octants**. You can "picture" the first octant as the region bounded by the floor of a room and two walls with the axes meeting in a corner. The floor is the xy-plane. The axes can be ordered in a "right-hand" or "left-hand" format. Figure 3.1 shows the point represented by the ordered triple $(2, 3, 1)$ in a right-hand system.

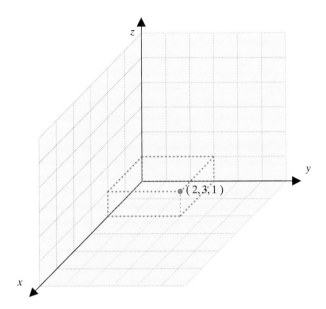

Figure 3.1

The graphs of linear equations in three variables are planes in three dimensions. A portion of the graph of $2x + 3y - z = 16$ appears in Figure 3.2.

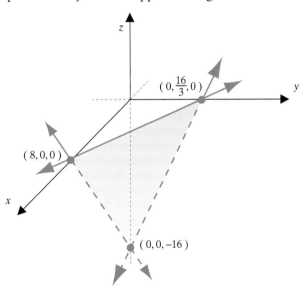

Figure 3.2

Two distinct planes will either be parallel or they will intersect. If they intersect, their intersection will be a straight line. If three distinct planes intersect, they will intersect in a straight line, or they will intersect in a single point represented by an ordered triple.

The graphs of systems of three linear equations in three variables can be both interesting and informative, but they can be difficult to sketch and points of intersection difficult to estimate. Also, most graphing calculators are limited to graphs in two dimensions, so they are not useful in graphically analyzing systems of linear equations in three variables.

Therefore, in this text, only algebraic techniques for solving these systems will be discussed. Figure 3.3 illustrates four different possibilities for the relative positions of three planes.

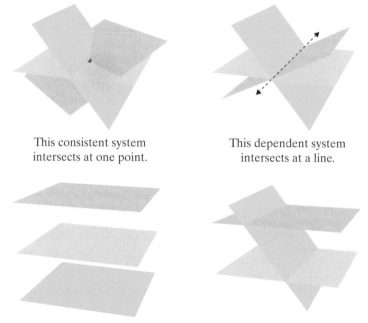

This consistent system intersects at one point.

This dependent system intersects at a line.

These two systems are both inconsistent. All three planes do not have a common intersection.

Figure 3.3

To Solve a System of Three Linear Equations in Three Variables

1. *Select two equations and eliminate one variable by using the addition method.*
2. *Select a different pair of equations and eliminate the **same** variable.*
3. *Steps 1 and 2 give **two** linear equations in **two** variables. Solve these equations by either addition or substitution as discussed in Section 3.1.*
4. *Back substitute the values found in Step 3 into any one of the original equations to find the value of the third variable.*
5. *Check the solution in all three of the original equations.*

The solution possibilities for a system of three equations in three variables are as follows:

1. There will be exactly one ordered triple solution.
   (Graphically, the three planes intersect at one point.)
2. There will be an infinite number of solutions.
   (Graphically, the three planes intersect in a line or are the same plane.)
3. There will be no solutions.
   (Graphically, there are no points common to all 3 planes.)

The technique is illustrated with the following system.

$$\begin{cases} 2x + 3y - z = 16 & \text{(I)} \\ x - y + 3z = -9 & \text{(II)} \\ 5x + 2y - z = 15 & \text{(III)} \end{cases}$$

Step 1:   Using equations (I) and (II), eliminate y.

[**Note:** We could just as easily have chosen to eliminate x or z. To be sure that you understand the process you might want to solve the system by first eliminating x and then again by first eliminating z. In any case, the answer will be the same.]

$$\begin{array}{ll} \text{(I)} & \\ \text{(II)} & [3] \end{array} \begin{cases} (2x + 3y - z = 16) \\ (x - y + 3z = -9) \end{cases} \longrightarrow \begin{array}{l} 2x + 3y - z = 16 \\ \underline{3x - 3y + 9z = -27} \\ 5x \quad\quad + 8z = -11 \ \text{(IV)} \end{array}$$

Step 2:   Using a different pair of equations, (II) and (III), eliminate the **same** variable, y.

$$\begin{array}{ll} \text{(II)} & [2] \\ \text{(III)} & \end{array} \begin{cases} (x - y + 3z = -9) \\ (5x + 2y - z = 15) \end{cases} \longrightarrow \begin{array}{l} 2x - 2y + 6z = -18 \\ \underline{5x + 2y - z = 15} \\ 7x \quad\quad + 5z = -3 \ \text{(V)} \end{array}$$

Step 3:   Using the results of Steps 1 and 2, solve the two equations for x and z.

$$\begin{array}{ll} \text{(IV)} & [-7] \\ \text{(V)} & [5] \end{array} \begin{cases} (5x + 8z = -11) \\ (7x + 5z = -3) \end{cases} \longrightarrow \begin{array}{l} -35x - 56z = 77 \\ \underline{35x + 25z = -15} \\ -31z = 62 \\ z = -2 \end{array}$$

Back substitute $z = -2$ into the equation $5x + 8z = -11$ to find x.

$$5x + 8(-2) = -11 \quad \text{Using equation (IV).}$$
$$5x = 5$$
$$x = 1$$

Step 4:   Using $x = 1$ and $z = -2$, back substitute to find y.

$$1 - y + 3(-2) = -9 \quad \text{Using equation (II).}$$
$$-y = -4$$
$$y = 4$$

The solution is $(1, 4, -2)$ or $x = 1$, $y = 4$, and $z = -2$. The solution can be checked by substituting the results into **all three** of the original equations.

$$\begin{cases} 2(1)+3(4)-(-2)=16 & \text{(I)} \\ (1)-(4)+3(-2)=-9 & \text{(II)} \\ 5(1)+2(4)-(-2)=15 & \text{(III)} \end{cases}$$

Example 1: Three Variables (Consistent System)

Solve the following system of linear equations.

$$\begin{cases} x-y+2z=-4 & \text{(I)} \\ 2x+3y+z=\dfrac{1}{2} & \text{(II)} \\ x+4y-2z=4 & \text{(III)} \end{cases}$$

Solution: Using equations (I) and (III), eliminate z.

$$\begin{array}{rrrrrr} \text{(I)} & x & - & y & + & 2z & = & -4 \\ \text{(III)} & x & + & 4y & - & 2z & = & 4 \\ \hline & 2x & + & 3y & & & = & 0 \ \text{(IV)} \end{array}$$

Using equations (I) and (II), eliminate z.

$$\begin{array}{l} \text{(I)} \\ \text{(II)} \end{array} \begin{cases} (x & - & y & + & 2z & = & -4) \\ [-2]\left(2x & + & 3y & + & z & = & \dfrac{1}{2}\right) \end{cases} \longrightarrow \begin{array}{rrrrrr} x & - & y & + & 2z & = & -4 \\ -4x & - & 6y & - & 2z & = & -1 \\ \hline -3x & - & 7y & & & = & -5 \ \text{(V)} \end{array}$$

Eliminate the variable x using the two equations in x and y.

$$\begin{array}{l} \text{(IV)} \\ \text{(V)} \end{array} \begin{cases} [3] & (2x & + & 3y & = & 0) \\ [2] & (-3x & - & 7y & = & -5) \end{cases} \longrightarrow \begin{array}{rrrrr} 6x & + & 9y & = & 0 \\ -6x & - & 14y & = & -10 \\ \hline & -5y & = & -10 \\ & y & = & 2 \end{array}$$

Back substituting to find x yields:

$$2x+3(2)=0$$
$$2x=-6$$
$$x=-3.$$

Finally, using $x = -3$ and $y = 2$, back substitute into (I).

$$(-3)-(2)+2z=-4$$
$$2z=1$$
$$z=\dfrac{1}{2}$$

The solution is $\left(-3,\ 2,\ \dfrac{1}{2}\right)$.

The solution can be checked by substituting $\left(-3,\ 2,\ \dfrac{1}{2}\right)$ into **all three** of the original equations.

$$\begin{cases} (-3)-(2)+2\left(\dfrac{1}{2}\right)=-4 & \text{(I)} \\[2mm] 2(-3)+3(2)+\left(\dfrac{1}{2}\right)=\dfrac{1}{2} & \text{(II)} \\[2mm] (-3)+4(2)-2\left(\dfrac{1}{2}\right)=4 & \text{(III)} \end{cases}$$

Example 2: Three Variables (Inconsistent System)

Solve the following system of linear equations.

$$\begin{cases} 3x-5y+z=6 & \text{(I)} \\ x-y+3z=-1 & \text{(II)} \\ 2x-2y+6z=5 & \text{(III)} \end{cases}$$

Solution: Using equations (I) and (II), eliminate z.

$$\begin{array}{ll} \text{(I)} & [-3] \quad (3x-5y+z=6) \longrightarrow -9x+15y-3z=-18 \\ \text{(II)} & \quad (x-y+3z=-1) \longrightarrow \underline{x-y+3z=-1} \\ & \qquad\qquad\qquad\qquad\qquad\quad -8x+14y=-19 \ \text{(IV)} \end{array}$$

Using equations (II) and (III), eliminate z.

$$\begin{array}{ll} \text{(II)} & [-2] \quad (x-y+3z=-1) \longrightarrow -2x+2y-6z=2 \\ \text{(III)} & \quad (2x-2y+6z=5) \longrightarrow \underline{2x-2y+6z=5} \\ & \qquad\qquad\qquad\qquad\qquad\qquad\quad 0=7 \ \text{(V)} \end{array}$$

This last equation is **false**. Thus, the system **does not have a solution**.

Example 3: Three Variables (Dependent System)

Solve the following system of linear equations.

$$\begin{cases} 3x - 2y + 4z = 1 & \text{(I)} \\ y + 3z = -2 & \text{(II)} \\ 6x - 5y + 5z = 4 & \text{(III)} \end{cases}$$

Solution: Using equations (I) and (III), eliminate x.

$$\begin{array}{ll} \text{(I)} \\ \text{(III)} \end{array} \begin{cases} [-2] & (3x - 2y + 4z = 1) \longrightarrow -6x + 4y - 8z = -2 \\ & (6x - 5y + 5z = 4) \longrightarrow \underline{6x - 5y + 5z = 4} \\ & -y - 3z = 2 \quad \text{(IV)} \end{cases}$$

Using (II) and the equation just found (IV), eliminate y.

$$\begin{array}{lrcr} \text{(II)} & y + 3z & = & -2 \\ \text{(IV)} & \underline{-y - 3z} & = & \underline{2} \\ & 0 & = & 0 \quad \text{(V)} \end{array}$$

Because this last equation, $0 = 0$, is **always true**, the system has an **infinite number of solutions**.

Example 4: Three Variables (Application)

A cash register contains $341 in $20, $5, and $2 bills. There are twenty-eight bills in all and three more $2 bills than $5 bills. How many bills of each kind are there?

Solution: Let x = number of $20 bills
   y = number of $5 bills
   z = number of $2 bills

$$\begin{cases} x + y + z = 28 & \text{(I)} \\ 20x + 5y + 2z = 341 & \text{(II)} \\ z = y + 3 & \text{(III)} \end{cases}$$

(I)	There are twenty-eight bills.
(II)	The total value is $341.
(III)	There are three more $2 bills than $5 bills.

Using equations (I) and (II), eliminate x.

$$\begin{array}{ll}\text{(I)} & [-20] \quad (x + y + z = 28) \longrightarrow \\ \text{(II)} & \qquad (20x + 5y + 2z = 341) \longrightarrow \end{array}$$

$$\begin{array}{rcrcrcr} -20x & - & 20y & - & 20z & = & -560 \\ 20x & + & 5y & + & 2z & = & 341 \\ \hline & - & 15y & - & 18z & = & -219 \ \text{(IV)} \end{array}$$

We rewrite equation (III) in the form $y - z = -3$ and use this equation along with the results just found.

$$\begin{array}{ll}\text{(III)} & [15] \quad (y - z = -3) \longrightarrow \\ \text{(IV)} & \qquad (-15y - 18z = -219) \longrightarrow \end{array}$$

$$\begin{array}{rcrcr} 15y & - & 15z & = & -45 \\ -15y & - & 18z & = & -219 \\ \hline & & -33z & = & -264 \ \text{(V)} \\ & & z & = & 8 \end{array}$$

Back substituting to solve for y gives:

$$y - 8 = -3$$
$$y = 5.$$

Now we can substitute the values $z = 8$ and $y = 5$ into equation (I).

$$x + 5 + 8 = 28$$
$$x = 15$$

There are fifteen \$20 bills, five \$5 bills, and eight \$2 bills.

Practice Problem

Solve the following system of linear equations: $\begin{cases} 2x + y + z = 4 \\ x + 2y + z = 1 \\ 3x + y - z = -3 \end{cases}$.

Answer to Practice Problem: $x = 1, y = -2, z = 4$

3.3 Exercises

Solve each system of equations in Exercises 1 – 20. State which systems, if any, have no solution or an infinite number of solutions.

1. $\begin{cases} x + y - z = 0 \\ 3x + 2y + z = 4 \\ x - 3y + 4z = 5 \end{cases}$

2. $\begin{cases} x - y + 2z = 3 \\ -6x + y + 3z = 7 \\ x + 2y - 5z = -4 \end{cases}$

3. $\begin{cases} 2x - y - z = 1 \\ 2x - 3y - 4z = 0 \\ x + y - z = 4 \end{cases}$

4. $\begin{cases} y + z = 6 \\ x + 5y - 4z = 4 \\ x - 3y + 5z = 7 \end{cases}$

5. $\begin{cases} x + y - 2z = 4 \\ 2x + y = 1 \\ 5x + 3y - 2z = 6 \end{cases}$

6. $\begin{cases} 2y + z = -4 \\ 3x + 4z = 11 \\ x + y = -2 \end{cases}$

7. $\begin{cases} x - y + 5z = -6 \\ x + 2z = 0 \\ 6x + y + 3z = 0 \end{cases}$

8. $\begin{cases} x - y + 2z = -3 \\ 2x + y - z = 5 \\ 3x - 2y + 2z = -3 \end{cases}$

9. $\begin{cases} y + z = 2 \\ x + z = 5 \\ x + y = 5 \end{cases}$

10. $\begin{cases} x - y - 2z = 3 \\ x + 2y + z = 1 \\ 3y + 3z = -2 \end{cases}$

11. $\begin{cases} 2x - y + 5z = -2 \\ x + 3y - z = 6 \\ 4x + y + 3z = -2 \end{cases}$

12. $\begin{cases} 2x - y + 5z = 5 \\ x - 2y + 3z = 0 \\ x + y + 4z = 7 \end{cases}$

13. $\begin{cases} 3x + y + 4z = -6 \\ 2x + 3y - z = 2 \\ 5x + 4y + 3z = 2 \end{cases}$

14. $\begin{cases} 2x + y - z = -3 \\ -x + 2y + z = 5 \\ 2x + 3y - 2z = -3 \end{cases}$

15. $\begin{cases} x - 2y + z = 4 \\ x - y - 4z = 1 \\ 2x - 4y + 2z = 8 \end{cases}$

16. $\begin{cases} 2x - 2y + 3z = 4 \\ x - 3y + 2z = 2 \\ x + y + z = 1 \end{cases}$

17. $\begin{cases} 2x - 3y + z = -1 \\ 6x - 9y - 4z = 4 \\ 4x + 6y - z = 5 \end{cases}$

18. $\begin{cases} x + y + z = 3 \\ 2x - y - 2z = -3 \\ 3x + 2y + z = 4 \end{cases}$

19. $\begin{cases} 2x + 3y + z = 4 \\ 3x - 5y + 2z = -5 \\ 4x - 6y + 3z = -7 \end{cases}$

20. $\begin{cases} x + 6y + z = 6 \\ 2x + 3y - 2z = 8 \\ 2x + 4z = 3 \end{cases}$

21. The sum of three integers is 67. The sum of the first and second integers exceeds the third by 13. The third integer is 7 less than the first. Find the three integers.

22. The sum of three integers is 189. The first integer is 28 less than the second. The second integer is 21 less than the sum of the first and third integers. Find the three integers.

23. Money: A wallet contains $218 in $10, $5, and $1 bills. There are forty-six bills in all and four more fives than tens. How many bills of each kind are there?

24. Money: Sally is trying to get her brother Robert to learn to think algebraically. She tells him that she has 23 coins in her purse, including nickels, dimes, and quarters. She has two more dimes than quarters, and the total value of the coins is $2.50. How many of each kind of coin does she have?

25. Find values for a, b, and c so that the points $(-1, -4)$, $(2, 8)$, and $(-2, -4)$ lie on the graph of the function $y = ax^2 + bx + c$.

26. Find values for a, b, and c so that the points $(1, 1)$, $(-3, 13)$, and $(0, -2)$ lie on the graph of the function $y = ax^2 + bx + c$.

27. Perimeter of a triangle: The perimeter of a triangle is 73 cm. The longest side is 13 cm less than the sum of the other two sides. The shortest side is 11 cm less than the longest side. Find the lengths of the three sides.

28. Fruit stand: At Steve's Fruit Stand, 4 pounds of bananas, 2 pounds of apples, and 3 pounds of grapes cost $16.40. Five pounds of bananas, 4 pounds of apples, and 2 pounds of grapes cost $16.60. Two pounds of bananas, 3 pounds of apples, and 1 pound of grapes cost $9.60. Find the price per pound of each kind of fruit.

29. Construction: The Tates are having a house built. The cost of building the house is $24,000 more than three times the cost of the lot. The cost of the landscaping, sidewalks, and upgrades is one-half the cost of the lot. If the total cost is $123,000, what is the cost of each part of the construction (the home, the lot, and the improvements)?

30. Fast food: At the Happy Burger Drive-In, you can buy 2 hamburgers, 1 chocolate shake, and 2 orders of fries, or 3 hamburgers and 1 order of fries, for $9.50. One hamburger, 2 chocolate shakes, and 1 order of fries cost $7.30. How much does a hamburger cost?

31. Investing: Kirk inherited $100,000 dollars from his aunt and decided to invest in three different accounts: savings, bonds, and stocks. The amount in his bond account was $10,000 more than three times the amount in his stock account. At the end of the first year, the savings account returned 5%, the bonds 8%, and the stocks 10% for total interest of $7400. How much did he invest in each account?

32. Stock market: Melissa has saved a total of $30,000 and wants to invest in three different stocks: PepsiCo, IBM, and Microsoft. She wants the PepsiCo amount to be $1000 less than twice the IBM amount and the Microsoft amount to be $2000 more than the total in the other two stocks. How much should she invest in each stock?

33. Triangles: The sum of the measures of the three angles of a triangle is 180°. In one particular triangle, the largest angle is 10° more than three times the smallest angle, and the third angle is one-half the largest angle. What are the measures of the three angles?

34. Theater: The local theater has three types of seats for Broadway plays: main floor, balcony, and mezzanine. Main floor tickets are $60, balcony tickets are $45, mezzanine tickets are $30. On one particular night the sales totaled $29,400. Main floor sales were 20 more than the total of balcony and mezzanine sales. Balcony sales were 40 more than two times mezzanine sales. How many of each type of ticket were sold?

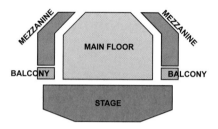

35. Chemistry: A chemist wants to mix 9 liters of a 25% acid solution. Because of limited amounts on hand, the mixture is to come from three different solutions, one with 10% acid, another with 30% acid, and a third with 40% acid. The amount of the 10% solution must be twice the amount of the 40% solution, and the amount of the 30% solution must equal the total amount of the other two solutions. How much of each solution must be used?

Writing and Thinking About Mathematics

36. Is it possible for three linear equations in three unknowns to have exactly two solutions? Explain your reasoning in some detail.

37. In geometry, we know that three non-collinear points determine a plane. (That is, if three points are not on a line, then there is a unique plane that contains all three points.) Find the values of A, B, and C (and therefore the equation of the plane) given $Ax + By + Cz = 3$ and the three points on the plane $(0, 3, 2)$, $(0, 0, 1)$ and $(-3, 0, 3)$. Sketch the plane in three dimensions as best you can by locating the three given points.

38. As stated in Exercise 37, three non-collinear points determine a plane. Find the values of A, B, and C (and therefore the equation of the plane) given $Ax + By + Cz = 10$ and the three points on the plane $(2, 0, -2)$, $(3, -1, 0)$ and $(-1, 5, -4)$. Sketch the plane in three dimensions as best you can by locating the three given points.

 HAWKES LEARNING SYSTEMS: INTERMEDIATE ALGEBRA SOFTWARE

- Solving Systems of Linear Equations with Three Variables

3.4 Matrices and Gaussian Elimination

- Write a system of equations as a coefficient matrix and an augmented matrix.
- Transform a matrix into triangular form by using elementary row operations.
- Solve systems of linear equations by using the Gaussian elimination method.

Matrices

A rectangular array of numbers is called a **matrix** (plural **matrices**). Matrices are usually named with capital letters, and each number in the matrix is called an **entry**. Entries written horizontally are said to form a **row**, and entries written vertically are said to form a **column**. The matrix A shown below has two rows and three columns and is a **2 × 3 matrix** (read "two by three matrix"). We say that the **dimension** of the matrix is the number of rows by the number of columns, which for this matrix is two by three (or 2×3). Similarly, if a matrix has three rows and two columns then its dimension is 3×2.

$$A = \begin{bmatrix} 3 & 4 & 0 \\ 7 & -2 & 5 \end{bmatrix} \qquad \begin{bmatrix} 3 & 4 & 0 \\ 7 & -2 & 5 \end{bmatrix} \begin{array}{l} \longleftarrow \text{row 1} \\ \longleftarrow \text{row 2} \end{array}$$

$$\begin{array}{ccc} & \text{column 2} & \\ \text{column 1} & \downarrow & \text{column 3} \\ \searrow & \downarrow & \swarrow \end{array}$$
$$\begin{bmatrix} 3 & 4 & 0 \\ 7 & -2 & 5 \end{bmatrix}$$

Three more examples are:

$$B = \begin{bmatrix} 5 & -1 \\ 2 & 3 \end{bmatrix} \qquad C = \begin{bmatrix} 5 & -1 & 0 & 7 \\ 2 & 3 & 2 & 8 \\ 1 & -3 & 0 & 6 \end{bmatrix} \qquad D = \begin{bmatrix} 0 & 4 \\ 1 & 6 \\ -1 & 3 \end{bmatrix}$$
$$2 \times 2 \text{ matrix} \qquad\qquad 3 \times 4 \text{ matrix} \qquad\qquad 3 \times 2 \text{ matrix}$$

A matrix with the same number of rows as columns is called a **square matrix**. Matrix B (shown above) is a square 2×2 matrix.

Elementary Row Operations

Matrices have many uses and are generated from various types of problems because they allow data to be presented in a systematic and orderly manner. (Business majors may want to look up a topic called Markov chains.)

Also, matrices can sometimes be added, subtracted, and multiplied. Some square matrices have inverses, much the same as multiplicative inverses for real numbers. Matrix methods of solving systems of linear equations can be done manually or with graphing calculators and computers. These topics are presented in courses such as finite mathematics and linear algebra.

In this text we will see that matrices can be used to solve systems of linear equations in which the equations are written in standard form. The two matrices derived from such a system are the **coefficient matrix** (made up of the coefficients of the variables) and the **augmented matrix** (including the coefficients and the constant terms). For example:

System	**Coefficient Matrix**	**Augmented Matrix**

$$\begin{cases} x - y + z = -6 \\ 2x + 3y \quad\;\; = 17 \\ x + 2y + 2z = 7 \end{cases}$$

$$\begin{bmatrix} 1 & -1 & 1 \\ 2 & 3 & 0 \\ 1 & 2 & 2 \end{bmatrix}$$

$$\left[\begin{array}{ccc|c} 1 & -1 & 1 & -6 \\ 2 & 3 & 0 & 17 \\ 1 & 2 & 2 & 7 \end{array}\right]$$

coefficients constants

Note that 0 is the entry in the second row, third column of both matrices. This 0 corresponds to the missing z-variable in the second equation. The second equation could have been written $2x + 3y + 0z = 17$.

As discussed in Section 2.2, notation with a small number to the right and below a variable is called **subscript** notation. For example, a_1 is read "a sub one," b_3 is read "b sub three" and R_2 is read "R sub two."

a_1 ← subscript b_3 ← subscript R_2 ← subscript

↑   ↑   ↑
variable   variable   variable

When working with matrices, R_1 represents row 1, R_2 represents row 2, and so on.

In Sections 3.1 and 3.3, systems of linear equations were solved by the addition method and back substitution. In solving these systems, we can make any of the following three manipulations **without changing the solution set of the system**.

The system $\begin{cases} x - y + z = -6 \\ 2x + 3y + 0z = 17 \\ x + 2y + 2z = 7 \end{cases}$ is used here to illustrate some possibilities.

1. Any two equations may be interchanged.

$$\begin{cases} x - y + z = -6 \\ 2x + 3y + 0z = 17 \\ x + 2y + 2z = 7 \end{cases} \xrightarrow{R_1 \leftrightarrow R_2} \begin{cases} 2x + 3y + 0z = 17 \\ x - y + z = -6 \\ x + 2y + 2z = 7 \end{cases}$$

Here we have interchanged the first two equations.

2. All terms of any equation may be multiplied by a constant.

$$\begin{cases} x - y + z = -6 \\ 2x + 3y + 0z = 17 \\ x + 2y + 2z = 7 \end{cases} \xrightarrow{-2R_1} \begin{cases} -2x + 2y - 2z = 12 \\ 2x + 3y + 0z = 17 \\ x + 2y + 2z = 7 \end{cases}$$

Here we have multiplied each term of the first equation by -2.

3. All terms of any equation may be multiplied by a constant and these new terms may be added to like terms of another equation. **(The original equation remains unchanged.)**

$$\begin{cases} x - y + z = -6 \\ 2x + 3y + 0z = 17 \\ x + 2y + 2z = 7 \end{cases} \xrightarrow{\ R_3 - R_1\ } \begin{cases} x - y + z = -6 \\ 2x + 3y + 0z = 17 \\ 0x + 3y + z = 13 \end{cases}$$

Here we have multiplied the first equation by −1 (mentally) and added the results to the third equation.

When dealing with matrices, the three corresponding operations are called **elementary row operations**. These operations are listed below and illustrated in Example 1. Follow the steps outlined in Example 1 carefully, and note how the row operations are indicated, such as $\frac{1}{2}R_3$ to indicate that all numbers in row 3 are multiplied by $\frac{1}{2}$. (Reasons for using these row operations are discussed under **Gaussian Elimination** on page 228.)

Elementary Row Operations

1. Interchange two rows.

2. Multiply a row by a nonzero constant.

3. Add a multiple of a row to another row.

*If any elementary row operation is applied to a matrix, the new matrix is said to be **row-equivalent** to the original matrix.*

Example 1: Coefficient and Augmented Matrices

a. For the system $\begin{cases} y + z = 6 \\ x + 5y - 4z = 4 \\ 2x - 6y + 10z = 14 \end{cases}$,

write the corresponding coefficient matrix and the corresponding augmented matrix.

Solution: Coefficient Matrix

$$\begin{bmatrix} 0 & 1 & 1 \\ 1 & 5 & -4 \\ 2 & -6 & 10 \end{bmatrix}$$

Augmented Matrix

$$\left[\begin{array}{ccc|c} 0 & 1 & 1 & 6 \\ 1 & 5 & -4 & 4 \\ 2 & -6 & 10 & 14 \end{array}\right]$$

b. In the augmented matrix in Example 1a, interchange rows 1 and 2 and multiply row 3 by $\dfrac{1}{2}$.

Solution:
$$\begin{bmatrix} 0 & 1 & 1 & | & 6 \\ 1 & 5 & -4 & | & 4 \\ 2 & -6 & 10 & | & 14 \end{bmatrix} \xrightarrow[\frac{1}{2}R_3]{R_1 \leftrightarrow R_2} \begin{bmatrix} 1 & 5 & -4 & | & 4 \\ 0 & 1 & 1 & | & 6 \\ 1 & -3 & 5 & | & 7 \end{bmatrix}$$

c. For the system $\begin{cases} x - y = 5 \\ 3x + 4y = 29 \end{cases}$, write the corresponding coefficient matrix and the corresponding augmented matrix.

Solution: Coefficient Matrix   Augmented Matrix

$$\begin{bmatrix} 1 & -1 \\ 3 & 4 \end{bmatrix} \qquad\qquad\qquad \begin{bmatrix} 1 & -1 & | & 5 \\ 3 & 4 & | & 29 \end{bmatrix}$$

d. In the augmented matrix in Example 1c, add -3 times row 1 to row 2.
(Note that row 1 is unchanged in the resulting matrix. Only row 2 is changed.)

Solution:

$$\begin{bmatrix} 1 & -1 & | & 5 \\ 3 & 4 & | & 29 \end{bmatrix} \xrightarrow{R_2 - 3R_1} \begin{bmatrix} 1 & -1 & | & 5 \\ 0 & 7 & | & 14 \end{bmatrix}$$

General Notation for a Matrix and a System of Equations

With matrices we use capital letters to name a matrix and **double subscript** notation with corresponding lower case letters to indicate both the row and column location of an entry. For example in a matrix A the entries will be designated as described below and as shown on the following page.

a_{11} is read "a sub one one" and indicates the entry in the first row and first column;
a_{12} is read "a sub one two" and indicates the entry in the first row and second column;
a_{13} is read "a sub one three" and indicates the entry in the first row and third column;
a_{21} is read "a sub two one" and indicates the entry in the second row and first column;
and so on.

NOTES

We will see in dealing with polynomials later that a_{11} can be read simply as "a sub eleven." However, with matrices, we need to indicate the row and column corresponding to the entry. If there are more than nine rows or columns, then commas are used to separate the numbers as $a_{10,10}$. You will see the commas in use on your calculator.

With double subscript notation we can write the general form of a 2×3 matrix A and a 3×3 matrix B as follows.

$$A = \begin{bmatrix} a_{11} & a_{12} & a_{13} \\ a_{21} & a_{22} & a_{23} \end{bmatrix} \qquad B = \begin{bmatrix} b_{11} & b_{12} & b_{13} \\ b_{21} & b_{22} & b_{23} \\ b_{31} & b_{32} & b_{33} \end{bmatrix}.$$

We will use this notation when discussing the use of calculators to define matrices and to operate with matrices. The general form of a system of three linear equations might use this notation in the following way.

$$\begin{cases} a_{11}x + a_{12}y + a_{13}z = k_1 \\ a_{21}x + a_{22}y + a_{23}z = k_2 \\ a_{31}x + a_{32}y + a_{33}z = k_3 \end{cases}$$

A matrix is in **upper triangular form** (or just **triangular form** for our purposes) if its entries in the lower left triangular region are all 0's. The **main diagonal** consists of the entries in the positions of b_{11}, b_{22}, and b_{33}. Note that in the main diagonal the column and row indices are equal. Thus, if all the entries below the main diagonal of a matrix are all 0's, the matrix is in triangular form as shown below.

$$B = \begin{bmatrix} b_{11} & b_{12} & b_{13} \\ 0 & b_{22} & b_{23} \\ 0 & 0 & b_{33} \end{bmatrix}$$

The upper triangular form of a matrix with all 1's in the main diagonal is called the **row echelon form** (or **ref**). We will see that a graphing calculator can be used to change a matrix into the row echelon form.

Gaussian Elimination

Another method of solving a system of linear equations is the **Gaussian elimination** method (named after the famous German mathematician Carl Friedrich Gauss, 1777 – 1855). This method makes use of augmented matrices and elementary row operations. The objective is to transform an augmented matrix into row echelon form and then use back substitution to find the values of the variables. The method is outlined as follows:

Strategy for Gaussian Elimination

1. Write the augmented matrix for the system.

2. Use elementary row operations to transform the matrix into row echelon form.

3. Solve the corresponding system of equations by using back substitution.

The following examples illustrate the method. Study the steps and the corresponding comments carefully.

Example 2: Gaussian Elimination

Solve the following system of linear equations by using the Gaussian elimination method with back substitution.

$$\begin{cases} 2x + 4y = -6 \\ 5x - y = 7 \end{cases}$$

Solution:

Step 1: Write the augmented matrix.
[The following steps show how to use elementary row operations to get the matrix in row echelon form with 0 in the lower left corner.]

$$\begin{bmatrix} 2 & 4 & | & -6 \\ 5 & -1 & | & 7 \end{bmatrix}$$

Step 2: Multiply row 1 by $\dfrac{1}{2}$ so that the entry in the upper left corner will be 1. This will help to get 0 below the 1 in the next step.

$$\begin{bmatrix} 2 & 4 & | & -6 \\ 5 & -1 & | & 7 \end{bmatrix} \xrightarrow{\frac{1}{2}R_1} \begin{bmatrix} 1 & 2 & | & -3 \\ 5 & -1 & | & 7 \end{bmatrix}$$

Step 3: To get 0 in the lower left corner, add -5 times row 1 to row 2.

$$\begin{bmatrix} 1 & 2 & | & -3 \\ 5 & -1 & | & 7 \end{bmatrix} \xrightarrow{R_2 - 5R_1} \begin{bmatrix} 1 & 2 & | & -3 \\ 0 & -11 & | & 22 \end{bmatrix}$$

Continued on the next page ...

Step 4: Now multiply row 2 by $-\dfrac{1}{11}$.

$$\begin{bmatrix} 1 & 2 & | & -3 \\ 0 & -11 & | & 22 \end{bmatrix} \quad \xrightarrow{-\frac{1}{11}R_2} \quad \begin{bmatrix} 1 & 2 & | & -3 \\ 0 & 1 & | & -2 \end{bmatrix}$$

Step 5: The last triangular matrix in Step 4 represents the following system of linear equations.

$$\begin{cases} x + 2y = -3 \\ 0x + y = -2 \end{cases}$$

The last equation gives $y = -2$.

Back substitute to find the value for x.

$$x + 2(-2) = -3$$
$$x - 4 = -3$$
$$x = 1$$

Thus the solution is $x = 1$ and $y = -2$. Or we can write $(1, -2)$.

Example 3: Gaussian Elimination

Solve the following system of linear equations by using the Gaussian elimination method with back substitution.

$$\begin{cases} 2x - 3y - z = -4 \\ -x + 2y + z = 6 \\ x - y + 2z = 14 \end{cases}$$

Solution:

Step 1: Write the augmented matrix.

$$\begin{bmatrix} 2 & -3 & -1 & | & -4 \\ -1 & 2 & 1 & | & 6 \\ 1 & -1 & 2 & | & 14 \end{bmatrix}$$

Step 2: Exchange row 1 and row 3 so that the entry in the upper left corner will be 1.

$$\begin{bmatrix} 2 & -3 & -1 & | & -4 \\ -1 & 2 & 1 & | & 6 \\ 1 & -1 & 2 & | & 14 \end{bmatrix} \quad \xrightarrow{R_1 \leftrightarrow R_3} \quad \begin{bmatrix} 1 & -1 & 2 & | & 14 \\ -1 & 2 & 1 & | & 6 \\ 2 & -3 & -1 & | & -4 \end{bmatrix}$$

Step 3: To get a 0 under the 1 in Column 1, add row 1 to row 2.

$$\begin{bmatrix} 1 & -1 & 2 & | & 14 \\ -1 & 2 & 1 & | & 6 \\ 2 & -3 & -1 & | & -4 \end{bmatrix} \xrightarrow{R_2 + R_1} \begin{bmatrix} 1 & -1 & 2 & | & 14 \\ 0 & 1 & 3 & | & 20 \\ 2 & -3 & -1 & | & -4 \end{bmatrix}$$

Step 4: To get a_{31} to be 0, add -2 times row 1 to row 3.

$$\begin{bmatrix} 1 & -1 & 2 & | & 14 \\ 0 & 1 & 3 & | & 20 \\ 2 & -3 & -1 & | & -4 \end{bmatrix} \xrightarrow{R_3 - 2R_1} \begin{bmatrix} 1 & -1 & 2 & | & 14 \\ 0 & 1 & 3 & | & 20 \\ 0 & -1 & -5 & | & -32 \end{bmatrix}$$

Step 5: Add row 2 to row 3 to arrive at triangular form.

$$\begin{bmatrix} 1 & -1 & 2 & | & 14 \\ 0 & 1 & 3 & | & 20 \\ 0 & -1 & -5 & | & -32 \end{bmatrix} \xrightarrow{R_3 + R_2} \begin{bmatrix} 1 & -1 & 2 & | & 14 \\ 0 & 1 & 3 & | & 20 \\ 0 & 0 & -2 & | & -12 \end{bmatrix}$$

Step 6: Then multiply row 3 by $-\dfrac{1}{2}$.

$$\begin{bmatrix} 1 & -1 & 2 & | & 14 \\ 0 & 1 & 3 & | & 20 \\ 0 & 0 & -2 & | & -12 \end{bmatrix} \xrightarrow{-\frac{1}{2}R_3} \begin{bmatrix} 1 & -1 & 2 & | & 14 \\ 0 & 1 & 3 & | & 20 \\ 0 & 0 & 1 & | & 6 \end{bmatrix}$$

Step 7: The triangular matrix in Step 6 represents the following system of linear equations.

$$\begin{cases} x - y + 2z = 14 \\ y + 3z = 20 \\ z = 6 \end{cases}$$

From the last equation, we have $z = 6$.
Back substitution into the equation $y + 3z = 20$ gives:

$$y + 3(6) = 20$$
$$y = 2.$$

Back substitution into the equation $x - y + 2z = 14$ gives:

$$x - 2 + 2(6) = 14$$
$$x = 4.$$

Thus the solution is $x = 4$, $y = 2$, and $z = 6$. Or, as an ordered triple, $(4, 2, 6)$.

If the final matrix, in triangular form, has a row with all entries 0, then the system has an infinite number of solutions.

For example, solving the system $\begin{cases} x+3y=8 \\ 2x+6y=16 \end{cases}$ will result in the matrix $\begin{bmatrix} 1 & 3 & | & 8 \\ 0 & 0 & | & 0 \end{bmatrix}$.

The last line indicates that $0x + 0y = 0$ which is always true. Therefore, the solution to the system is the set of all solutions of the equation $x + 3y = 8$. The system is **dependent**.

If the triangular form of the augmented matrix shows the coefficient entries in one or more rows to be all 0's and the constant not 0, then the system has no solution.

For example, the last row of the augmented matrix $\begin{bmatrix} 1 & 2 & 2 & | & 7 \\ 0 & 1 & 3 & | & 6 \\ 0 & 0 & 0 & | & 15 \end{bmatrix}$ indicates that

$0x + 0y + 0z = 15$. Since this is never true, the system has no solution. That is, the system is **inconsistent**.

Using the TI-84 Plus Graphing Calculator to Solve a System of Linear Equations

The TI-84 Plus calculator can be used to define a matrix and perform matrix operations. (Matrices can be added, subtracted, and multiplied under special restrictions. These operations are saved for another course.) In particular, the Gaussian elimination method is used by the calculator to solve a system of linear equations by reducing the corresponding augmented matrix to **row echelon form (ref)**.

Pressing the **MATRIX** key (found by pressing ⬛ **2ND** x^{-1}) will give the menu shown here.

Pressing the right arrow and moving to **MATH** will give the following choices.

Pressing the right arrow again and moving to **EDIT** will give the following choices.

Example 4 shows how to use the calculator to solve a system of three linear equations in three variables. Study each step carefully.

Example 4: Using a Graphing Calculator to Solve Systems of Equations

Use a TI-84 Plus calculator to solve the following system of linear equations.

$$\begin{cases} x + 2y + z = 1 \\ -x + y + z = -6 \\ 4x - y + 3z = -1 \end{cases}$$

Solution:

Step 1: Press **2ND** > **MATRIX** and move to the **EDIT** menu.

Press **ENTER**. The following display will appear.

Continued on the next page ...

Step 2: The augmented matrix is a 3 × 4 matrix. So, in the top line enter 3, press **ENTER**, enter 4, press **ENTER** and the display will appear as follows.

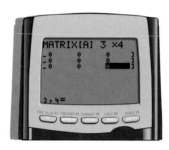

(**Note:** If other numbers are already present on the display, just type over them. The calculator will adjust automatically.)

Step 3: Move the cursor to the upper left entry position and enter the coefficients and constants in the matrix. As you enter each number press **ENTER** and the cursor will automatically move to the next position in the matrix. Note that the double subscripts appear at the bottom of the display as each number is entered. The final display for matrix [*A*] should appear as follows.

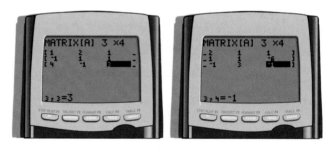

Note: The display only shows three columns at a time.

Step 4: Press **2ND** > **QUIT**, press **2ND** > **MATRIX** again, go to **MATH**; move the cursor down to **A:ref(**; press **ENTER**. The display will appear as follows.

Step 5: Press **2ND** > **MATRIX** again; press **ENTER** (this selects the matrix A); enter a right parenthesis) ; press the **MATH** key; and choose `1:>Frac` by pressing **ENTER**. The display will appear as follows.

[**Note:** You must select the matrix from the matrix menu. The calculator will not recognize the matrix if you manually type in [A].]

Step 6: Press **ENTER** and the row echelon form of matrix will appear as follows.

With back substitution we get the following solution: $x = 3, y = 1, z = -4$.

Practice Problem

Solve the following system of linear equations by using the Gaussian elimination method with back substitution.

$$\begin{cases} x - 2y + 3z = 4 \\ 2x + y = 0 \\ 3x + y - z = -4 \end{cases}$$

Answer to Practice Problem: $x = -1, y = 2, z = 3$

3.4 Exercises

In Exercises 1 – 6, write the coefficient matrix and the augmented matrix for the given systems of linear equations.

1. $\begin{cases} 2x + 2y = 13 \\ 5x - y = 10 \end{cases}$

2. $\begin{cases} x + 4y = -1 \\ 2x - 3y = 7 \end{cases}$

3. $\begin{cases} 7x - 2y + 7z = 2 \\ -5x + 3y = 2 \\ 4y + 11z = 8 \end{cases}$

4. $\begin{cases} -8x + 2y - z = 6 \\ 2x + 3z = -3 \\ -4x - 2y + 5z = 13 \end{cases}$

5. $\begin{cases} 3x + y - z + 2w = 6 \\ x - y + 2z - w = -8 \\ 2y + 5z + w = 2 \\ x + 3y + 3w = 14 \end{cases}$

6. $\begin{cases} 4x + y + 3z - 2w = 13 \\ x - 2y + z - 4w = -3 \\ x + y + 4z + 2w = 12 \\ -2x + 3y - z - 3w = 5 \end{cases}$

In Exercises 7 – 10, write the system of linear equations represented by each of the augmented matrices. Use x, y, and z as the variables.

7. $\begin{bmatrix} -3 & 5 & \vdots & 1 \\ -1 & 3 & \vdots & 2 \end{bmatrix}$

8. $\begin{bmatrix} 3 & -1 & \vdots & 5 \\ -2 & 10 & \vdots & 9 \end{bmatrix}$

9. $\begin{bmatrix} 1 & 3 & 4 & \vdots & 1 \\ 2 & -3 & -2 & \vdots & 0 \\ 1 & 1 & 0 & \vdots & -4 \end{bmatrix}$

10. $\begin{bmatrix} 2 & -9 & 14 & \vdots & 0 \\ -3 & 0 & -8 & \vdots & 5 \\ 2 & -6 & 1 & \vdots & 3 \end{bmatrix}$

In Exercises 11 – 28, use the Gaussian elimination method with back substitution to solve the given system of linear equations.

11. $\begin{cases} x + 2y = 3 \\ 2x - y = -4 \end{cases}$

12. $\begin{cases} 4x + 3y = 5 \\ -x - 2y = 0 \end{cases}$

13. $\begin{cases} -8x + 2y = 6 \\ x - 2y = 1 \end{cases}$

14. $\begin{cases} 2x + y = -2 \\ 4x + 3y = -2 \end{cases}$

15. $\begin{cases} x - 3y + 2z = 11 \\ -2x + 4y + z = -3 \\ x - 2y + 3z = 12 \end{cases}$

16. $\begin{cases} x + 2y - z = 6 \\ 3x - y + 2z = 9 \\ x + y + z = 6 \end{cases}$

17. $\begin{cases} x + 2y + 3z = 4 \\ x - y - z = 0 \\ 4x - 3y + z = 5 \end{cases}$

18. $\begin{cases} x + y - 2z = -1 \\ 3x + 4y - 2z = 0 \\ x - y + z = 4 \end{cases}$

19. $\begin{cases} x - y - 2z = 3 \\ x + 2y - z = 5 \\ 2x - 3y - 2z = 3 \end{cases}$

20. $\begin{cases} x + y + 3z = 2 \\ 2x - y + z = 1 \\ 4x + y + 7z = 5 \end{cases}$

21. $\begin{cases} x - y + 5z = -6 \\ x + 2z = 0 \\ 6x + y + 3z = 0 \end{cases}$

22. $\begin{cases} x - 3y - z = -4 \\ 3x - 2y + z = 1 \\ -2x + y + 2z = 13 \end{cases}$

23. $\begin{cases} x - y + 2z = 5 \\ 2x - 2y + 4z = 5 \\ 3x - 3y + 6z = 8 \end{cases}$　**24.** $\begin{cases} 2x - y - 5z = -9 \\ x - 3y + 2z = 0 \\ 3x + 2y + 10z = 4 \end{cases}$　**25.** $\begin{cases} x - 2y + 3z = 0 \\ x + y + 4z = 7 \\ 2x - y + 5z = 5 \end{cases}$

26. $\begin{cases} 2x - y + 5z = -2 \\ 4x + y + 3z = -2 \\ x + 3y - z = 6 \end{cases}$　**27.** $\begin{cases} 3x + 4z = 11 \\ x + y = -2 \\ 2y + z = -4 \end{cases}$　**28.** $\begin{cases} y + z = 2 \\ x + y = 5 \\ x + z = 5 \end{cases}$

For Exercises 29 – 32, set up a system of linear equations that represents the information and solve the system using Gaussian elimination.

29. The sum of three integers is 169. The first integer is twelve more than the second integer. The third integer is fifteen less than the sum of the first and second integers. What are the integers?

30. Pizza: A pizzeria sells three sizes of pizzas: small, medium, and large. The pizzas sell for $6.00, $8.00, and $9.50, respectively. One evening they sold 68 pizzas for a total of $528.00. If they sold twice as many medium-sized pizzas as large-sized pizzas, how many of each size did they sell?

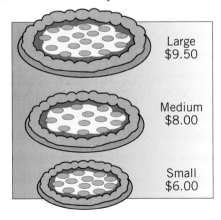

31. Grocery shopping: Caroline bought a pound of bacon, a dozen eggs, and a loaf of bread. The total cost was $8.52. The eggs cost $0.94 more than the bacon. The combined cost of the bread and eggs was $2.34 more than the cost of the bacon. Find the cost of each item.

32. Investments: An investment firm is responsible for investing $250,000 from an estate according to three conditions in the will of the deceased. The money is to be invested in three accounts paying 6%, 8%, and 11% interest. The amount invested in the 6% account is to be $5000 more than the total invested in the other two accounts, and the total annual interest for the first year is to be $19,250. How much is the firm supposed to invest in each account?

Use your graphing calculator to solve the systems of linear equations in Exercises 33 – 40.

33. $\begin{cases} x + y = -4 \\ 2x + 3y = -12 \end{cases}$　**34.** $\begin{cases} 2x + 3y = 1 \\ x - 5y = -19 \end{cases}$

35. $\begin{cases} x+y+z=10 \\ 2x-y+z=10 \\ -x+2y+2z=14 \end{cases}$

36. $\begin{cases} 2x+y+2z=-1 \\ x-y+4z=-3 \\ 3x-y+\dfrac{1}{2}z=-\dfrac{25}{4} \end{cases}$

37. $\begin{cases} x-3y+z=0 \\ 2x+2y-z=2 \\ x+y+z=5 \end{cases}$

38. $\begin{cases} x+5y=13 \\ 2x+z=6 \\ 4y-z=8 \end{cases}$

39. $\begin{cases} x-2y-2z=-13 \\ 2x+y-z=-5 \\ x+y+z=6 \end{cases}$

40. $\begin{cases} x+y+z+w=0 \\ x-y-z+w=-2 \\ 3x+3y-z-w=11 \\ y-2z=6 \end{cases}$

Writing and Thinking About Mathematics

41. Suppose that Gaussian elimination with a system of three linear equations in three unknowns results in the following triangular matrix. Discuss how you can use back substitution to find that the system has an infinite number of solutions. That is, the system is dependent. [**Hint:** Solve the second equation for z.]

$$\begin{bmatrix} 1 & 2 & -1 & 4 \\ 0 & 3 & 1 & 2 \\ 0 & 0 & 0 & 0 \end{bmatrix}$$

 HAWKES LEARNING SYSTEMS: INTERMEDIATE ALGEBRA SOFTWARE

- Matrices and Gaussian Elimination

3.5 Determinants

- *Evaluate 2 × 2 and 3 × 3 determinants.*
- *Solve equations involving determinants.*

As was discussed in Section 3.4, a rectangular array of numbers is called a matrix, and matrices arise in connection with solving systems of linear equations such as:

$$\begin{cases} a_{11}x + a_{12}y = k_1 \\ a_{21}x + a_{22}y = k_2 \end{cases}.$$

The **matrix of the coefficients** (or the **coefficient matrix**) is

$$A = \begin{bmatrix} a_{11} & a_{12} \\ a_{21} & a_{22} \end{bmatrix}.$$

If a matrix is **square** (the number of rows is equal to the number of columns), then there is a number associated with the matrix called its **determinant**. In this section, we will show how to evaluate determinants and, in the next section, we will show how determinants can be used to solve systems of linear equations by using a method called **Cramer's Rule**.

Determinant

*A **determinant** is a real number associated with a square array of real numbers and is indicated by enclosing the array between two vertical bars. For a matrix A, the corresponding determinant is designated as det(A) and is read "determinant of A."*

Examples of determinants are:

(a) For the matrix, $A = \begin{bmatrix} 3 & 4 \\ 7 & -2 \end{bmatrix}$, $\qquad \det(A) = \begin{vmatrix} 3 & 4 \\ 7 & -2 \end{vmatrix}$.

(b) For the matrix, $B = \begin{bmatrix} 1 & 6 & -3 \\ 4 & 5 & 5 \\ -1 & -1 & -1 \end{bmatrix}$, $\qquad \det(B) = \begin{vmatrix} 1 & 6 & -3 \\ 4 & 5 & 5 \\ -1 & -1 & -1 \end{vmatrix}$.

Example (a) is a 2 × 2 determinant and has two rows and two columns.
Example (b) is a 3 × 3 determinant and has three rows and three columns.

column 1 column 2

↓   ↓

row 1 → | 3   4 |
row 2 → | 7 −2 |

| 3   4 |
| 7 −2 |

| column 1 column 2 column 3
| ↓   ↓   ↓
row 1 → | 1   6   −3 |
row 2 → | 4   5   5 |
row 3 → | −1 −1 −1 |

| column
| column 1   2   column 3
| ↓   ↓   ↓
| 1   6   −3 |
| 4   5   5 |
| −1 −1 −1 |

A 4×4 determinant has four rows and four columns. A determinant may be of any size $n \times n$ where n is a positive integer and $n \geq 2$. In this text, the discussion will be restricted to 2×2 and 3×3 determinants, and the entries will be real numbers. [Huge matrices and determinants (1000×1000 or larger) are common in industry, and their values are calculated by computers. Even then, someone must understand the algebraic techniques to be able to write the necessary programs.]

Every determinant with real entries has a real value. The method for finding the value of 3×3 determinants involves finding the value of 2×2 determinants. Determinants of larger matrices can be evaluated by using techniques similar to those shown here. Their applications occur in higher mathematics such as linear algebra and differential equations.

Value of a 2×2 Determinant

For the square matrix $A = \begin{bmatrix} a_{11} & a_{12} \\ a_{21} & a_{22} \end{bmatrix}$, $\qquad det(A) = \begin{vmatrix} a_{11} & a_{12} \\ a_{21} & a_{22} \end{vmatrix} = a_{11}a_{22} - a_{21}a_{12}.$

As the definition indicates and the following examples illustrate, the value of a 2×2 determinant is the **product of the numbers in the diagonal containing the term in the first row, first column, minus the product of the numbers in the other diagonal.**

Example 1: 2×2 Determinant

Evaluate the following 2×2 determinants.

a. $\begin{vmatrix} 3 & 4 \\ 7 & -2 \end{vmatrix} = 3(-2) - 7(4) = -6 - 28 = -34$

b. $\begin{vmatrix} -5 & -\dfrac{1}{2} \\ 6 & 3 \end{vmatrix} = -5(3) - 6\left(-\dfrac{1}{2}\right) = -15 + 3 = -12$

c. $\begin{vmatrix} 1 & 7 \\ 2 & 14 \end{vmatrix} = 1(14) - 2(7) = 14 - 14 = 0$

One method of evaluating 3×3 determinants is called **expanding by minors**. In this method, **one row is chosen** and each entry in that row has a minor. Each minor is found by mentally crossing out both the row and column (shown here in the shaded regions) that contain that entry. The minors of the entries in the first row are illustrated here.

$$\begin{vmatrix} a_{11} & a_{12} & a_{13} \\ a_{21} & a_{22} & a_{23} \\ a_{31} & a_{32} & a_{33} \end{vmatrix} \longrightarrow \begin{vmatrix} a_{22} & a_{23} \\ a_{32} & a_{33} \end{vmatrix} \longleftarrow \text{minor of } a_{11}$$

$$\begin{vmatrix} a_{11} & a_{12} & a_{13} \\ a_{21} & a_{22} & a_{23} \\ a_{31} & a_{32} & a_{33} \end{vmatrix} \longrightarrow \begin{vmatrix} a_{21} & a_{23} \\ a_{31} & a_{33} \end{vmatrix} \longleftarrow \text{minor of } a_{12}$$

$$\begin{vmatrix} a_{11} & a_{12} & a_{13} \\ a_{21} & a_{22} & a_{23} \\ a_{31} & a_{32} & a_{33} \end{vmatrix} \longrightarrow \begin{vmatrix} a_{21} & a_{22} \\ a_{31} & a_{32} \end{vmatrix} \longleftarrow \text{minor of } a_{13}$$

To find the value of a determinant (of any dimension other than 2×2), first choose a row (or column) and find the product of each entry in that row (or column) with its corresponding minor. Then the value is determined by adding these products with appropriate adjustments of alternating signs of the minors. We say that the determinant has been expanded by that row (or column). The following illustrates how to find the value of a 3×3 determinant by expanding by the first row.

Value of a 3×3 Determinant

For the square matrix $A = \begin{bmatrix} a_{11} & a_{12} & a_{13} \\ a_{21} & a_{22} & a_{23} \\ a_{31} & a_{32} & a_{33} \end{bmatrix}$,

$$\boldsymbol{det(A)} = \begin{vmatrix} a_{11} & a_{12} & a_{13} \\ a_{21} & a_{22} & a_{23} \\ a_{31} & a_{32} & a_{33} \end{vmatrix} = \boldsymbol{a_{11}}(minor\ of\ a_{11}) - \boldsymbol{a_{12}}(minor\ of\ a_{12}) + \boldsymbol{a_{13}}(minor\ of\ a_{13})$$

$$= \boldsymbol{a_{11}}\begin{vmatrix} a_{22} & a_{23} \\ a_{32} & a_{33} \end{vmatrix} - \boldsymbol{a_{12}}\begin{vmatrix} a_{21} & a_{23} \\ a_{31} & a_{33} \end{vmatrix} + \boldsymbol{a_{13}}\begin{vmatrix} a_{21} & a_{22} \\ a_{31} & a_{32} \end{vmatrix}.$$

NOTES   **CAUTION:** The negative sign in the middle term of the expansion (representing -1 times a_{12}) is a critical part of the method and is a source of error for many students. **Be careful.**

Each minor is multiplied by its corresponding entry and +1 or −1 according to the pattern illustrated in Figure 3.4. [**Note:** The signs alternate and this pattern can be extended to apply to any $n \times n$ determinant.]

$$\begin{vmatrix} + & - & + \\ - & + & - \\ + & - & + \end{vmatrix}$$

Figure 3.4

For example, the value of a 3×3 determinant can be found by expanding by the minors of the second row as follows:

$$\det(A) = -a_{21}\left(\text{minor of } a_{21}\right) + a_{22}\left(\text{minor of } a_{22}\right) - a_{23}\left(\text{minor of } a_{23}\right).$$

Note the use of the alternating + and − signs from the pattern in Figure 3.4. You may want to try this for practice with some of the exercises.

NOTES There are methods other than expanding by minors for evaluating 3×3 determinants. Your instructor may wish to show you some of these, but they will not be discussed in the text. The advantage of learning to expand by minors is that this method can be used for evaluating higher-order determinants.

Example 2: 3×3 Determinant

Evaluate the following 3×3 determinants.

a.
$$\begin{vmatrix} 5 & 1 & -4 \\ 2 & 6 & 3 \\ 2 & 2 & 1 \end{vmatrix}$$

Using Row 1, mentally delete the shaded regions.

Solution:

$$\begin{vmatrix} 5 & 1 & -4 \\ 2 & 6 & 3 \\ 2 & 2 & 1 \end{vmatrix} = (5)\begin{vmatrix} 5 & 1 & -4 \\ 2 & 6 & 3 \\ 2 & 2 & 1 \end{vmatrix} - (1)\begin{vmatrix} 5 & 1 & -4 \\ 2 & 6 & 3 \\ 2 & 2 & 1 \end{vmatrix} + (-4)\begin{vmatrix} 5 & 1 & -4 \\ 2 & 6 & 3 \\ 2 & 2 & 1 \end{vmatrix}$$

$$= 5\begin{vmatrix} 6 & 3 \\ 2 & 1 \end{vmatrix} - 1\begin{vmatrix} 2 & 3 \\ 2 & 1 \end{vmatrix} - 4\begin{vmatrix} 2 & 6 \\ 2 & 2 \end{vmatrix}$$

$$= 5(6 \cdot 1 - 2 \cdot 3) - 1(2 \cdot 1 - 2 \cdot 3) - 4(2 \cdot 2 - 2 \cdot 6)$$

$$= 5(6 - 6) - 1(2 - 6) - 4(4 - 12)$$

$$= 5(0) - 1(-4) - 4(-8)$$

$$= 0 + 4 + 32$$

$$= 36$$

b. $\begin{vmatrix} 6 & -2 & 4 \\ 1 & 7 & 0 \\ -3 & 2 & -1 \end{vmatrix}$

Solution: $\begin{vmatrix} 6 & -2 & 4 \\ 1 & 7 & 0 \\ -3 & 2 & -1 \end{vmatrix} = (6)\begin{vmatrix} 7 & 0 \\ 2 & -1 \end{vmatrix} - (-2)\begin{vmatrix} 1 & 0 \\ -3 & -1 \end{vmatrix} + (4)\begin{vmatrix} 1 & 7 \\ -3 & 2 \end{vmatrix}$

$= 6(-7 - 0) + 2(-1 - 0) + 4(2 + 21)$

$= -42 - 2 + 92$

$= 48$ After some practice, many of these steps can be done mentally.

Example 3: Equations with Determinants

Solve the following equation for x: $\begin{vmatrix} 2 & 3 & 0 \\ 6 & x & 5 \\ 1 & -2 & 9 \end{vmatrix} = 53$.

Solution: First, evaluate the determinant.

$\begin{vmatrix} 2 & 3 & 0 \\ 6 & x & 5 \\ 1 & -2 & 9 \end{vmatrix} = (2)\begin{vmatrix} x & 5 \\ -2 & 9 \end{vmatrix} - (3)\begin{vmatrix} 6 & 5 \\ 1 & 9 \end{vmatrix} + (0)\begin{vmatrix} 6 & x \\ 1 & -2 \end{vmatrix}$

$= 2(9x + 10) - 3(54 - 5) + 0$

$= 2(9x + 10) - 3(49)$

$= 18x + 20 - 147$

$= 18x - 127$

Now solve the equation.

$18x - 127 = 53$

$18x = 180$

$x = 10$

The technique of expanding by minors may be used (with appropriate adjustments) to evaluate any $n \times n$ determinant. For example, in a 4×4 determinant, the minors of the entries in a particular row will be 3×3 determinants. Also, there are techniques for simplifying determinants and there are rules for arithmetic with determinants. The general rules governing these operations are discussed in courses like precalculus, finite mathematics, and linear algebra.

Using the TI-84 Plus Graphing Calculator to Evaluate a Determinant

A TI-84 Plus calculator (and other graphing calculators) can be used to find the value of the determinant of a square matrix. The determinant command, **1: det(** is found as the first entry in the MATRIX / MATH menu. Example 4 shows, in a step by step format, how to find the determinant of a given 3×3 matrix.

Example 4: Evaluating Determinants with a Calculator

Use a TI-84 Plus calculator to find the value of det(A) for the matrix

$$A = \begin{bmatrix} 2 & 5 & 7 \\ 3 & 1 & 0 \\ 4 & 0 & 3 \end{bmatrix}.$$

Solution:

Step 1: Press **2ND** > **MATRIX**, go to the **EDIT** menu and enter the appropriate dimensions and numbers in the matrix A. The display should appear as follows.

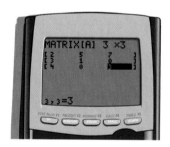

Step 2: Press **2ND** > **QUIT** then **2ND** > **MATRIX** again and go to the **MATH** menu. On the **MATH** menu choose **1: det (** and press **ENTER**. The display should appear as follows.

Step 3: Press 2ND > **MATRIX** again and on the **NAMES** menu choose
1:[A] 3×3. Press ENTER and type a right parenthesis). The display
should appear as follows.

Step 4: Press ENTER and the display should appear as follows with the answer.

Practice Problems

Evaluate each of the following determinants.

1. $\begin{vmatrix} -3 & 2 \\ 4 & 7 \end{vmatrix}$ **2.** $\begin{vmatrix} 6 & 3 \\ 4 & 2 \end{vmatrix}$ **3.** $\begin{vmatrix} 1 & 4 & 0 \\ 2 & -1 & 5 \\ 0 & 7 & -1 \end{vmatrix}$

Use a graphing calculator to find the value of the determinant.

4. $\begin{vmatrix} 5 & -1 & 3 \\ 0 & 4 & 2 \\ -3 & 1 & 3 \end{vmatrix}$

Use the method for evaluating determinants to solve the equation.

5. $\begin{vmatrix} 3 & 5 \\ 6 & x \end{vmatrix} = 18$

Answers to Practice Problems: **1.** −29 **2.** 0 **3.** −26 **4.** 92 **5.** 16

3.5 Exercises

In Exercises 1 – 4, the matrix A is given. Find det(A).

1. $A = \begin{bmatrix} 2 & 7 \\ 4 & 3 \end{bmatrix}$

2. $A = \begin{bmatrix} 7 & 3 \\ 8 & 5 \end{bmatrix}$

3. $A = \begin{bmatrix} -5 & 2 & 1 \\ 4 & 8 & 0 \\ -2 & 3 & 5 \end{bmatrix}$

4. $A = \begin{bmatrix} -6 & 5 & -3 \\ 4 & 0 & -1 \\ -2 & 7 & -2 \end{bmatrix}$

Evaluate the determinants in Exercises 5 – 20.

5. $\begin{vmatrix} 1 & 3 \\ -2 & 5 \end{vmatrix}$

6. $\begin{vmatrix} 7 & 2 \\ 3 & -6 \end{vmatrix}$

7. $\begin{vmatrix} 6 & 3 \\ -11 & -5 \end{vmatrix}$

8. $\begin{vmatrix} 2 & 3 \\ 3 & -4 \end{vmatrix}$

9. $\begin{vmatrix} 9 & 4 \\ 4 & 7 \end{vmatrix}$

10. $\begin{vmatrix} 3 & -4 \\ 8 & -6 \end{vmatrix}$

11. $\begin{vmatrix} 0 & -1 & 2 \\ 3 & 5 & -7 \\ -3 & 4 & 1 \end{vmatrix}$

12. $\begin{vmatrix} 1 & 0 & -1 \\ -2 & 3 & 5 \\ 6 & -3 & 4 \end{vmatrix}$

13. $\begin{vmatrix} 1 & -1 & 2 \\ -2 & 5 & -7 \\ 6 & 4 & 1 \end{vmatrix}$

14. $\begin{vmatrix} 2 & -1 & -3 \\ 5 & 9 & 4 \\ 7 & 6 & -2 \end{vmatrix}$

15. $\begin{vmatrix} 2 & 1 & 3 \\ 3 & 4 & 5 \\ 1 & 7 & 2 \end{vmatrix}$

16. $\begin{vmatrix} -3 & 2 & 1 \\ 1 & -4 & -1 \\ 2 & 5 & 3 \end{vmatrix}$

17. $\begin{vmatrix} 2 & 1 & -1 \\ 4 & 3 & 2 \\ 1 & 5 & 5 \end{vmatrix}$

18. $\begin{vmatrix} 6 & 7 & 1 \\ 0 & 3 & 3 \\ 4 & 1 & -5 \end{vmatrix}$

19. $\begin{vmatrix} 3 & -1 & -1 \\ 2 & 4 & 1 \\ -1 & 1 & 2 \end{vmatrix}$

20. $\begin{vmatrix} 2 & 3 & 2 \\ 1 & -1 & 5 \\ 0 & 5 & 1 \end{vmatrix}$

Use the method for evaluating determinants to solve the equations for x in Exercises 21 – 25.

21. $\begin{vmatrix} 1 & 3 & 4 \\ 2 & x & 3 \\ 1 & 3 & 5 \end{vmatrix} = 1$

22. $\begin{vmatrix} -2 & -1 & 1 \\ x & 1 & -1 \\ 4 & 3 & -2 \end{vmatrix} = 7$

23. $\begin{vmatrix} 1 & x & x \\ 2 & -2 & 1 \\ -1 & 3 & 2 \end{vmatrix} = 0$

24. $\begin{vmatrix} x & x & 1 \\ 1 & 5 & 0 \\ 0 & 1 & -2 \end{vmatrix} = -15$

25. $\begin{vmatrix} 3 & 1 & -2 \\ 1 & x & 4 \\ 2 & x & 0 \end{vmatrix} = 38$

The equation $\begin{vmatrix} x & y & 1 \\ x_1 & y_1 & 1 \\ x_2 & y_2 & 1 \end{vmatrix} = 0$ *is an equation of the line passing through the two points*

$P_1(x_1, y_1)$ *and* $P_2(x_2, y_2)$. *Find an equation for the line determined by the pairs of points given in Exercises 26 – 28.*

26. $(3, 2), (-1, 4)$ **27.** $(-2, 1), (5, 3)$ **28.** $(4, -4), (0, 6)$

The area of the triangle having the vertices $P_1(x_1, y_1)$, $P_2(x_2, y_2)$, *and* $P_3(x_3, y_3)$ *is given*

by the absolute value of the expression $\dfrac{1}{2} \begin{vmatrix} x_1 & y_1 & 1 \\ x_2 & y_2 & 1 \\ x_3 & y_3 & 1 \end{vmatrix}$. *In Exercises 29 – 31, draw the*

triangle with the given points as vertices and then find the area of the triangle.

29. $(3, 1), (1, -1), (5, 2)$ **30.** $(4, 0), (5, -2), (7, 1)$ **31.** $(-1, 3), (-4, -1), (3, -2)$

32. Explain, in your own words, the position of the three points $P_1(x_1, y_1)$, $P_2(x_2, y_2)$,

and $P_3(x_3, y_3)$ if the expression $\dfrac{1}{2} \begin{vmatrix} x_1 & y_1 & 1 \\ x_2 & y_2 & 1 \\ x_3 & y_3 & 1 \end{vmatrix}$ has a value of 0. [**Hint:** Refer to

the discussion before Exercises 29 – 31.]

In Exercises 33 – 35, use a graphing calculator to find the value of the determinant.

33. $\begin{vmatrix} 3 & -4 & 6 \\ 2 & 4 & -1 \\ 7 & 9 & -1 \end{vmatrix}$ **34.** $\begin{vmatrix} 2.1 & 3.5 & -3.4 \\ 2.6 & 5.0 & 1.2 \\ -1.0 & 3.4 & 6.3 \end{vmatrix}$ **35.** $\begin{vmatrix} 1.6 & \frac{1}{2} & -5.9 \\ 0.7 & \frac{3}{4} & 1.7 \\ 5.0 & 8.2 & -4.1 \end{vmatrix}$

Writing and Thinking About Mathematics

36. Suppose that in a 2 × 2 determinant two rows are identical. What will be the value of this determinant? Give two specific examples and a general example to back up your conclusion.

37. Suppose that in a 3 × 3 determinant one row is all 0's. What will be the value of this determinant? Give two specific examples and a general example to back up your conclusion.

Writing and Thinking About Mathematics (cont.)

38. In each part, give two specific examples and a general example to back up your conclusion.

   a. Suppose that in a 2×2 determinant two rows (or columns) are switched. How will the value of this new determinant relate to the value of the original determinant?

   b. Suppose that in a 3×3 determinant two rows (or columns) are switched. How will the value of this new determinant relate to the value of the original determinant?

 HAWKES LEARNING SYSTEMS: INTERMEDIATE ALGEBRA SOFTWARE

    ▪ Determinants

Cramer's Rule

3.6

- *Solve systems of linear equations using Cramer's Rule.*

Cramer's Rule is a method that uses determinants for solving systems of linear equations. To explain the method and how these determinants are generated, we first illustrate the solution to a system of linear equations by the addition method and do not simplify the fractional answers. We will see that these fractional answers can be represented as determinants.

Consider the following system of linear equations **with the equations in standard form**:

$$\begin{cases} 2x + 3y = -5 \\ 4x + y = 5 \end{cases}.$$

Eliminating y gives:

$$\begin{cases} 2x + 3y = -5 \\ [-3] \ (4x + y = 5) \end{cases} \longrightarrow$$

$$\begin{array}{rrr} 2x & +3y & = -5 \\ -3(4x) & -3(1y) & = -3(5) \end{array}$$

$$\overline{[2-3(4)]x + [3-3(1)]y = -5-3(5)}$$

$$(-10)x + (0)y = -20$$

$$x = \frac{-20}{-10}$$

Eliminating x gives:

$$\begin{cases} [-4] \ (2x + 3y = -5) \\ [2] \ \ (4x + y = 5) \end{cases} \longrightarrow$$

$$\begin{array}{rrr} -4(2x) & -4(3y) & = -4(-5) \\ 2(4x) & +2(1y) & = 2(5) \end{array}$$

$$\overline{[-4(2)+2(4)]x + [-4(3)+2(1)]y = -4(-5)+2(5)}$$

$$(0)x + (-10)y = 30$$

$$y = \frac{30}{-10}$$

Notice that the denominators for both x and y are the same number. This number is the value of the determinant of the coefficient matrix. (Remember that the equations are in standard form.)

Determinant of coefficient matrix $= D = \begin{vmatrix} 2 & 3 \\ 4 & 1 \end{vmatrix} = 2(1) - 4(3) = -10.$

In determinant form, the numerator D_x is found by replacing the x-coefficients with the constant terms

$$D_x = \begin{vmatrix} -5 & 3 \\ 5 & 1 \end{vmatrix} = -5(1) - 5(3) = -20$$

and the numerator D_y is found by replacing the y-coefficients with the constant terms:

$$D_y = \begin{vmatrix} 2 & -5 \\ 4 & 5 \end{vmatrix} = 2(5) - 4(-5) = 30.$$

Therefore, the values for x and y can be written in fraction form using determinants as follows:

$$x = \frac{D_x}{D} = \frac{-20}{-10} = 2 \qquad \text{and} \qquad y = \frac{D_y}{D} = \frac{30}{-10} = -3.$$

The determinant D_x is formed as follows:

1. Form D, the determinant of the coefficients.
2. Replace the coefficients of x with the corresponding constants on the right-hand side of the equations.

The determinant D_y is formed as follows:

1. Form D, the determinant of the coefficients.
2. Replace the coefficients of y with the corresponding constants on the right-hand side of the equations.

Cramer's Rule is stated here only for 2×2 systems (systems of two linear equations in two variables) and 3×3 systems (systems of three linear equations in three variables). However, Cramer's Rule applies to all $n \times n$ systems of linear equations.

Cramer's Rule for 2 × 2 Systems

For the system $\begin{cases} a_{11}x + a_{12}y = k_1 \\ a_{21}x + a_{22}y = k_2 \end{cases}$,

where

$$D = \begin{vmatrix} a_{11} & a_{12} \\ a_{21} & a_{22} \end{vmatrix}, \qquad D_x = \begin{vmatrix} k_1 & a_{12} \\ k_2 & a_{22} \end{vmatrix}, \qquad \text{and} \qquad D_y = \begin{vmatrix} a_{11} & k_1 \\ a_{21} & k_2 \end{vmatrix},$$

if $D \neq 0$, *then*

$$x = \frac{D_x}{D} \qquad \text{and} \qquad y = \frac{D_y}{D}$$

is the unique solution to the system.

Cramer's Rule for 3 × 3 Systems

For the system $\begin{cases} a_{11}x + a_{12}y + a_{13}z = \boldsymbol{k_1} \\ a_{21}x + a_{22}y + a_{23}z = \boldsymbol{k_2} \\ a_{31}x + a_{32}y + a_{33}z = \boldsymbol{k_3} \end{cases}$,

where $\quad D = \begin{vmatrix} a_{11} & a_{12} & a_{13} \\ a_{21} & a_{22} & a_{23} \\ a_{31} & a_{32} & a_{33} \end{vmatrix}$,

$$D_x = \begin{vmatrix} \boldsymbol{k_1} & a_{12} & a_{13} \\ \boldsymbol{k_2} & a_{22} & a_{23} \\ \boldsymbol{k_3} & a_{32} & a_{33} \end{vmatrix}, \qquad D_y = \begin{vmatrix} a_{11} & \boldsymbol{k_1} & a_{13} \\ a_{21} & \boldsymbol{k_2} & a_{23} \\ a_{31} & \boldsymbol{k_3} & a_{33} \end{vmatrix}, \qquad and \qquad D_z = \begin{vmatrix} a_{11} & a_{12} & \boldsymbol{k_1} \\ a_{21} & a_{22} & \boldsymbol{k_2} \\ a_{31} & a_{32} & \boldsymbol{k_3} \end{vmatrix},$$

if $D \neq 0$, then

$$x = \frac{D_x}{D}, \quad y = \frac{D_y}{D}, \quad and \quad z = \frac{D_z}{D}$$

is the unique solution to the system.

Cramer's Rule when $D = 0$

If $D = 0$, Cramer's Rule cannot be used. *In a case where $D = 0$ (either for a 2×2 or a 3×3 matrix), use the algebraic method of elimination or substitution. You will find that the system is either dependent (infinite solutions) or inconsistent (no solution).*

Example 1: Cramer's Rule

Using Cramer's Rule, solve the following systems of linear equations. The solutions are not checked here, but they can be checked by substituting them into all of the equations in the system.

a.
$$\begin{cases} 2x + y = 3 \\ 3x - 2y = 5 \end{cases}$$

Solution: $D = \begin{vmatrix} 2 & 1 \\ 3 & -2 \end{vmatrix} = -7,$ $\qquad D_x = \begin{vmatrix} 3 & 1 \\ 5 & -2 \end{vmatrix} = -11,$ $\qquad D_y = \begin{vmatrix} 2 & 3 \\ 3 & 5 \end{vmatrix} = 1$

$$x = \frac{D_x}{D} = \frac{-11}{-7} = \frac{11}{7} \qquad y = \frac{D_y}{D} = \frac{1}{-7} = -\frac{1}{7}$$

b.
$$\begin{cases} 2x + 2y = 8 \\ -x + 3y = -8 \end{cases}$$

Solution: $D = \begin{vmatrix} 2 & 2 \\ -1 & 3 \end{vmatrix} = 8,$ $\qquad D_x = \begin{vmatrix} 8 & 2 \\ -8 & 3 \end{vmatrix} = 40,$ $\qquad D_y = \begin{vmatrix} 2 & 8 \\ -1 & -8 \end{vmatrix} = -8$

$$x = \frac{D_x}{D} = \frac{40}{8} = 5 \qquad y = \frac{D_y}{D} = \frac{-8}{8} = -1$$

c.
$$\begin{cases} x + 2y + 3z = 3 \\ 4x + 5y + 6z = 1 \\ 7x + 8y + 9z = 0 \end{cases}$$

Solution: $D = \begin{vmatrix} 1 & 2 & 3 \\ 4 & 5 & 6 \\ 7 & 8 & 9 \end{vmatrix} = 1 \begin{vmatrix} 5 & 6 \\ 8 & 9 \end{vmatrix} - 2 \begin{vmatrix} 4 & 6 \\ 7 & 9 \end{vmatrix} + 3 \begin{vmatrix} 4 & 5 \\ 7 & 8 \end{vmatrix}$

$$= 1(-3) - 2(-6) + 3(-3) = 0$$

Because $D = 0$, Cramer's Rule cannot be used to solve the system. (You might try to solve the system by using the elimination method to see what happens in this case.)

d. $\begin{cases} x + y + 3z = 7 \\ 2x - y - 3z = -4 \\ 5x - 2y = -5 \end{cases}$

Solution: $D = \begin{vmatrix} 1 & 1 & 3 \\ 2 & -1 & -3 \\ 5 & -2 & 0 \end{vmatrix} = 1\begin{vmatrix} -1 & -3 \\ -2 & 0 \end{vmatrix} - 1\begin{vmatrix} 2 & -3 \\ 5 & 0 \end{vmatrix} + 3\begin{vmatrix} 2 & -1 \\ 5 & -2 \end{vmatrix} = -18$

$D_x = \begin{vmatrix} 7 & 1 & 3 \\ -4 & -1 & -3 \\ -5 & -2 & 0 \end{vmatrix} = 7\begin{vmatrix} -1 & -3 \\ -2 & 0 \end{vmatrix} - 1\begin{vmatrix} -4 & -3 \\ -5 & 0 \end{vmatrix} + 3\begin{vmatrix} -4 & -1 \\ -5 & -2 \end{vmatrix} = -18$

$D_y = \begin{vmatrix} 1 & 7 & 3 \\ 2 & -4 & -3 \\ 5 & -5 & 0 \end{vmatrix} = 1\begin{vmatrix} -4 & -3 \\ -5 & 0 \end{vmatrix} - 7\begin{vmatrix} 2 & -3 \\ 5 & 0 \end{vmatrix} + 3\begin{vmatrix} 2 & -4 \\ 5 & -5 \end{vmatrix} = -90$

$D_z = \begin{vmatrix} 1 & 1 & 7 \\ 2 & -1 & -4 \\ 5 & -2 & -5 \end{vmatrix} = 1\begin{vmatrix} -1 & -4 \\ -2 & -5 \end{vmatrix} - 1\begin{vmatrix} 2 & -4 \\ 5 & -5 \end{vmatrix} + 7\begin{vmatrix} 2 & -1 \\ 5 & -2 \end{vmatrix} = -6$

$x = \dfrac{-18}{-18} = 1 \qquad y = \dfrac{-90}{-18} = 5 \qquad z = \dfrac{-6}{-18} = \dfrac{1}{3}$

NOTES

The determinants shown in Examples 1c and 1d are expanded by the first row. However, you should remember that any row or column can be used in the expansion as long as the corresponding adjustments in the + and – signs are used with the minors. This may be particularly useful when a row or column has one or more 0's because multiplication by 0 will always give 0 and this will reduce the time needed for the expansion.

Practice Problems

1. Solve the following system using Cramer's Rule.

$\begin{cases} 2x - y = 11 \\ x + y = -2 \end{cases}$

2. Find D_x for the following system.

$\begin{cases} x + 2y + z = 0 \\ 2x + y - 2z = 5 \\ 3x - y + z = -3 \end{cases}$

Answers to Practice Problems: **1.** $x = 3$, $y = -5$ **2.** $D_x = 0$

3.6 Exercises

For Exercises 1 – 30, use Cramer's Rule to solve the following systems of linear equations, if possible. If the determinant of the coefficient matrix is zero, solve the system using addition or substitution to determine whether the system has no solution or infinite solutions.

1. $\begin{cases} 2x - 5y = -7 \\ 3x - 2y = 6 \end{cases}$ **2.** $\begin{cases} 3x + 5y = 17 \\ x + 3y = 15 \end{cases}$ **3.** $\begin{cases} 6x - 4y = 5 \\ 3x + 8y = 0 \end{cases}$ **4.** $\begin{cases} 3x + 4y = 24 \\ 2x + y = 11 \end{cases}$

5. $\begin{cases} 3x + y = 1 \\ -9x - 3y = 2 \end{cases}$ **6.** $\begin{cases} 4x + 8y = 12 \\ 3x + 6y = 9 \end{cases}$ **7.** $\begin{cases} 12x + 4y = 3 \\ -10x + 3y = 7 \end{cases}$ **8.** $\begin{cases} 4x - 9y = 2 \\ 8x - 15y = 3 \end{cases}$

9. $\begin{cases} 2x + 3y = 4 \\ 3x - 4y = 5 \end{cases}$ **10.** $\begin{cases} 5x + 2y = 7 \\ 2x - 3y = 4 \end{cases}$ **11.** $\begin{cases} 7x + 3y = 9 \\ 4x + 8y = 11 \end{cases}$ **12.** $\begin{cases} 5x - 9y = 3 \\ 11x + 6y = 12 \end{cases}$

13. $\begin{cases} 6x - 13y = 21 \\ 5x - 12y = 18 \end{cases}$ **14.** $\begin{cases} 10x + 7y = 15 \\ 13x - 4y = 11 \end{cases}$ **15.** $\begin{cases} 8x - 9y = -14 \\ 15x + 6y = 7 \end{cases}$ **16.** $\begin{cases} 17x - 5y = 21 \\ 4x + 3y = 6 \end{cases}$

17. $\begin{cases} 0.8x + 0.3y = 4 \\ 0.9x - 1.2y = 5 \end{cases}$ **18.** $\begin{cases} 0.4x + 0.7y = 3 \\ 0.5x + y = 6 \end{cases}$ **19.** $\begin{cases} 1.6x - 4.5y = 1.5 \\ 0.4x + 1.2y = 3.1 \end{cases}$

20. $\begin{cases} 2.3x + 1.8y = 4.6 \\ 0.8x - 1.4y = 3.2 \end{cases}$ **21.** $\begin{cases} x - 2y - z = -7 \\ 2x + y + z = 0 \\ 3x - 5y + 8z = 13 \end{cases}$ **22.** $\begin{cases} 2x + 3y + z = 0 \\ 5x + y - 2z = 9 \\ 10x - 5y + 3z = 4 \end{cases}$

23. $\begin{cases} 5x - 4y + z = 17 \\ x + y + z = 4 \\ -10x + 8y - 2z = 11 \end{cases}$ **24.** $\begin{cases} 9x + 10y = 2 \\ 2x + 6z = 4 \\ -3y + 3z = 1 \end{cases}$ **25.** $\begin{cases} 2x - 3y - z = -4 \\ -x + 2y + z = 6 \\ x - y + 2z = 14 \end{cases}$

26. $\begin{cases} 2x - 3y - z = 4 \\ x - 2y - z = 1 \\ x - y + 2z = 9 \end{cases}$ **27.** $\begin{cases} 3x + 2y + z = 5 \\ 2x + y - 2z = 4 \\ 5x + 3y - z = 9 \end{cases}$ **28.** $\begin{cases} 8x + 3y + 2z = 15 \\ 3x + 5y + z = -4 \\ 2x + 3y = -7 \end{cases}$

29. $\begin{cases} 2x - y + 3z = 1 \\ 5x + 2y - z = 2 \\ x - 2y + 5z = 2 \end{cases}$ **30.** $\begin{cases} 2x + 3y + 2z = -5 \\ 2x - 2y + z = -1 \\ 5x + y + z = 1 \end{cases}$

For Exercises 31 – 34, set up a system of linear equations that represents the information, then solve the system by using Cramer's Rule.

31. Triangles: The three sides of a triangle are related as follows: the perimeter is 43 feet, the second side is 5 feet more than twice the first side, and the third side is 3 feet less than the sum of the other two sides. Find the lengths of the three sides of the triangle.

32. Nutrition: Joel loves candy bars and ice cream, and they have fat and calories as follows: each candy bar contains 5 grams of fat and 280 calories; each serving of ice cream contains 10 grams of fat and 150 calories. How many candy bars and how many servings of ice cream did he eat the week that he consumed 85 grams of fat and 2300 calories from these two foods?

33. Investments: A financial advisor has $6 million to invest for her clients. She chooses, for one month, to invest in mutual funds and technology stocks. If the mutual funds earned 2% and the stocks earned 4% for a total of $170,000 in earnings for the month, how much money did she invest in each type of investment?

34. Farming: A farmer plants corn, wheat, and soybeans and rotates the planting each year on his 500-acre farm. In one particular year, the profits were: $120 per acre for corn, $100 per acre for wheat, and $80 per acre for soybeans. He planted twice as many acres with corn as with soybeans. How many acres did he plant with each crop that year, if he made a total profit of $51,800?

3.7 Graphing Systems of Linear Inequalities

- *Solve systems of linear inequalities graphically.*

In some branches of mathematics, in particular a topic called (interestingly enough) game theory, the solution to a very sophisticated problem can involve the set of points that satisfy a system of several **linear inequalities**. In business, these ideas relate to problems such as minimizing the cost of shipping goods from several warehouses to distribution outlets. In this section we will consider graphing the solution sets to only two inequalities. We will leave the problem solving techniques to another course.

First, we review the ideas related to systems of equations discussed in Section 3.1. Systems of two linear equations were solved by using three methods: graphing, substitution, and addition. We found that such systems can be:

a. consistent (one point satisfies both equations, and the lines intersect at one point),

b. inconsistent (no point satisfies both equations, and the lines are parallel), or

c. dependent (an infinite number of points satisfy both equations, and the lines are the same).

Table 3.2 shows an example of each case.

System	Graph	Intersection	Classification
$\begin{cases} y = 3x - 1 \\ y = -x + 3 \end{cases}$		One point: $(1, 2)$	**Consistent**
$\begin{cases} 2x + y = 4 \\ y = -2x + 1 \end{cases}$		No points; lines are parallel	**Inconsistent**

| $\begin{cases} y = -x + 5 \\ 2y + 2x = 10 \end{cases}$ | 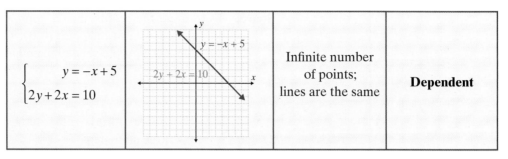 | Infinite number of points; lines are the same | **Dependent** |

Table 3.2

In this section, we will develop techniques for graphing (and therefore solving) **systems of two linear inequalities**. The solution set (if there are any solutions) to a system of two linear inequalities consists of the points in the intersection of two half-planes and possibly portions of the boundary lines indicated by the inequalities.

We know that a straight line separates a plane into two **half-planes**. The line itself is called the **boundary line**, and the boundary line may be included (the half-plane is **closed**) or the boundary line may not be included (the half-plane is **open**). The following procedure may be used to solve a system of linear inequalities.

To Solve a System of Two Linear Inequalities

1. *For each inequality, graph the boundary line and shade the appropriate half-plane. (Refer to Section 2.5 to review this process.)*

2. *Determine the region of the graph that is common to both half-planes (the region where the shading overlaps). (This region is called the **intersection** of the two half-planes.)*

3. *To check, pick one test-point in the intersection and verify that it satisfies both inequalities.*

Note: *If there is no intersection, then the system is inconsistent and has no solution.*

Example 1: Graphing Systems of Linear Inequalities

a.   Graph the points that satisfy the system of inequalities: $\begin{cases} x \le 2 \\ y \ge -x + 1 \end{cases}$.

Solution: **Step 1:** For $x \le 2$, the points are to the left of and on the line $x = 2$.

Step 2: For $y \ge -x + 1$, the points are above and on the line $y = -x + 1$.

Continued on the next page...

Step 3: Determine the region that is common to both half-planes. In this case, we test the point $(0, 3)$. On the graph below, the solution is the purple-shaded region and its boundary lines.

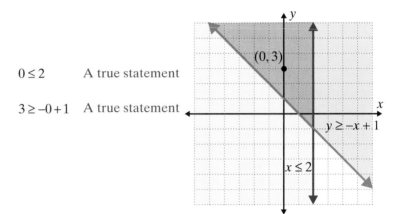

$0 \leq 2$ A true statement

$3 \geq -0 + 1$ A true statement

b. Solve the system of linear inequalities graphically: $\begin{cases} 2x + y \leq 6 \\ x + y < 4 \end{cases}$.

Solution: **Step 1:** Solve each inequality for y: $\begin{cases} y \leq -2x + 6 \\ y < -x + 4 \end{cases}$.

         Step 2: For $y \leq -2x + 6$, the points are below and on the line $y = -2x + 6$.

         Step 3: For $y < -x + 4$, the points are below but not on the line $y = -x + 4$.

         Step 4: Determine the region that is common to both half-planes. Note that the line $y = -x + 4$ is dashed to indicate that the points on the line are not included.
                 In this case, we test the point $(0, 0)$.

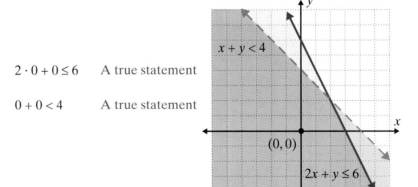

$2 \cdot 0 + 0 \leq 6$ A true statement

$0 + 0 < 4$ A true statement

c. Solve the system of linear inequalities graphically: $\begin{cases} y \geq x \\ y \leq x+2 \end{cases}$.

Solution: For $y \geq x$, the points are above and on the line $y = x$.

For $y \leq x + 2$, the points are below and on the line $y = x + 2$.

The solution set consists of the boundary lines and the region between them.

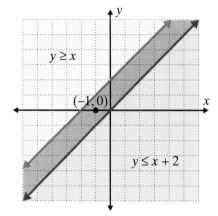

Note: When the boundary lines are parallel there are three possibilities:
1. The common region will be in the form of a strip between two lines (as in this example).
2. The common region will be a half-plane, as the solution to one inequality will be entirely contained within the solution of the other inequality.
3. There will be no common region and the solution set will be the empty set, $\varnothing$.

Using a TI-84 Plus Graphing Calculator to Graph Systems of Linear Inequalities

To graph a system of linear inequalities with a TI-84 Plus graphing calculator, first solve each inequality for y and then enter both of the corresponding functions after pressing the [Y=] key. By setting the graphing symbol to the left of Y_1 and Y_2 to the desired form and then pressing [GRAPH] (see section 2.5, Example 2), the desired region will be graphed as a cross-hatched area on the display (assuming that the window is set correctly). The following example shows how this can be done.

Example 2: Graphing Systems of Linear Inequalities

Use a TI-84 Plus graphing calculator to graph the following system of linear inequalities: $\begin{cases} 2x + y < 4 \\ 2x - y \leq 0 \end{cases}$.

Solution: **Step 1:** Solve each inequality for y: $\begin{cases} y < -2x + 4 \\ y \geq 2x \end{cases}$.

[**Note:** When solving $2x - y \leq 0$ for y, we write the inequality as $2x \leq y$ and then as $y \geq 2x$.]

Step 2: Press the 〔 Y= 〕 key and enter both functions and the corresponding symbols as they appear here.

[**Remember:** To shade your graphs, position the cursor over the slash next to Y_1 (or Y_2) and hit 〔ENTER〕 repeatedly until the appropriate shading is displayed.]

Step 3: Press 〔GRAPH〕. The display should appear as follows. The solution is the cross-hatched region.

3.7 Exercises

In Exercises 1 – 24, solve the systems of two linear inequalities graphically.

1. $\begin{cases} y > 2 \\ x \geq -3 \end{cases}$

2. $\begin{cases} 2x + 5 < 0 \\ y \geq 2 \end{cases}$

3. $\begin{cases} x < 3 \\ y > -x + 2 \end{cases}$

4. $\begin{cases} y \leq -5 \\ y \geq x - 5 \end{cases}$

5. $\begin{cases} x \leq 3 \\ 2x + y > 7 \end{cases}$

6. $\begin{cases} 2x - y > 4 \\ y < -1 \end{cases}$

7. $\begin{cases} x - 3y \leq 3 \\ x < 5 \end{cases}$

8. $\begin{cases} 3x - 2y \geq 8 \\ y \geq 0 \end{cases}$

9. $\begin{cases} x - y \geq 0 \\ 3x - 2y \geq 4 \end{cases}$

10. $\begin{cases} y \geq x - 2 \\ x + y \geq -2 \end{cases}$

11. $\begin{cases} 3x + y \leq 10 \\ 5x - y \geq 6 \end{cases}$

12. $\begin{cases} y \geq 2x - 5 \\ 3x + 2y > -3 \end{cases}$

13. $\begin{cases} 3x + 4y \geq -7 \\ y < 2x + 1 \end{cases}$

14. $\begin{cases} 2x - 3y \geq 0 \\ 8x - 3y < 36 \end{cases}$

15. $\begin{cases} x + y < 4 \\ 2x - 3y < 3 \end{cases}$

16. $\begin{cases} 2x + 3y < 12 \\ 3x + 2y > 13 \end{cases}$

17. $\begin{cases} x + y \geq 0 \\ x - 2y \geq 6 \end{cases}$

18. $\begin{cases} y \geq 3x + 1 \\ y \leq x - 2 \end{cases}$

19. $\begin{cases} x + 3y \leq 9 \\ x - y \geq 5 \end{cases}$

20. $\begin{cases} x - y \geq -2 \\ x + 2y < -1 \end{cases}$

21. $\begin{cases} y \leq -2x \\ y > -2x - 6 \end{cases}$

22. $\begin{cases} y > 3x + 1 \\ -3x + y < -1 \end{cases}$

23. $\begin{cases} y \leq x + 3 \\ x - y \leq -5 \end{cases}$

24. $\begin{cases} y > x - 4 \\ y < x + 2 \end{cases}$

Use a graphing calculator to solve the systems of linear inequalities in Exercises 25 – 35.

25. $\begin{cases} y \geq 0 \\ 3x - 5y \leq 10 \end{cases}$

26. $\begin{cases} 3x + 2y \leq 15 \\ 2x + 5y \geq 10 \end{cases}$

27. $\begin{cases} 4x - 3y \geq 6 \\ 3x - y \leq 3 \end{cases}$

28. $\begin{cases} y \leq 0 \\ 3x + y \leq 11 \end{cases}$

29. $\begin{cases} 3x - 4y \geq -6 \\ 3x + 2y \leq 12 \end{cases}$

30. $\begin{cases} 3y \leq 2x \\ x + 2y \leq 11 \end{cases}$

31. $\begin{cases} x + y \leq 8 \\ 3x - 2y \geq -6 \end{cases}$

32. $\begin{cases} x + y \leq 7 \\ 2x - y \leq 8 \end{cases}$

33. $\begin{cases} y \leq x \\ y < 2x + 1 \end{cases}$

34. $\begin{cases} x - y \geq -2 \\ 4x - y < 16 \end{cases}$

35. $\begin{cases} y \geq x \\ y \leq x + 7 \end{cases}$

Writing and Thinking About Mathematics

36. Example 1c discusses a system of two linear inequalities in which the boundary lines are parallel. Describe, in your own words, how you might test whether or not you have graphed the correct solution set. Solve the following systems graphically and indicate how your method of testing works in each case.

a. $\begin{cases} y \leq 2x - 5 \\ y \geq 2x + 3 \end{cases}$ b. $\begin{cases} y \leq -x + 2 \\ y \geq -x - 1 \end{cases}$ c. $\begin{cases} y \leq \dfrac{1}{2}x + 3 \\ y \geq \dfrac{1}{2}x - 3 \end{cases}$

 HAWKES LEARNING SYSTEMS: INTERMEDIATE ALGEBRA SOFTWARE

▪ Systems of Linear Inequalities

Chapter 3 Index of Key Ideas and Terms

Section 3.1   Systems of Linear Equations in Two Variables

Systems of Equations (Two Variables)   page 192
   Two (or more) linear equations considered at one time are
   said to form a **system of equations** or a **set of simultaneous
   equations**.

Consistent   page 193
   If a system of linear equations has exactly one solution,
   it is said to be **consistent**.

Inconsistent   page 193
   If a system of linear equations has no solution, it is said to
   be **inconsistent**.

Dependent   page 193
   If a system of linear equations has an infinite number of
   solutions, it is said to be **dependent**.

To Solve a System of Linear Equations by Graphing   page 194
   1. Graph both linear equations on the same set of axes.
   2. Observe the point of intersection (if there is one).

To Solve a System of Linear Equations by Substitution   page 196
   1. Solve one of the equations for one of the variables.
   2. Substitute the resulting expression into the other
           equation.
   3. Solve this new equation, if possible, and then substitute
           back into one of the original equations to find the
           value of the other variable. (This is known as **back
           substitution**.)

To Solve a System of Linear Equations by Addition   page 198
   1. Write the equations one under the other so that like
           terms are aligned.
   2. Multiply all terms of one equation by a constant (and
           possibly all terms of the other equation by another
           constant) so that two like terms have opposite coefficients.
   3. Add the two equations by combining like terms and solve
           the resulting equation, if possible. Then, back substitute
           into one of the original equations to find the value of the
           other variable.

Section 3.3 Systems of Linear Equations in Three Variables

Linear Equation in Three Variables page 212
The general form of an equation in three variables is
$Ax + By + Cz = D$ where A, B, and C do not equal 0.

Ordered Triples page 212
An **ordered triple** is an ordering of three real numbers in
the form (x_0, y_0, z_0) and is used to represent the solution to
a linear equation in three variables.

Graphs in Three Dimensions pages 212-213
Three mutually perpendicular number lines labeled as the
x-axis, y-axis, and z-axis are used to separate space into
eight regions called **octants**.

To Solve a System of Three Linear Equations in Three Variables page 214
1. Select two equations and eliminate one variable by
   using the addition method.
2. Select a different pair of equations and eliminate the
   same variable.
3. Steps 1 and 2 give two linear equations in two variables.
   Solve these equations by either addition or substitution
   as discussed in Section 3.1.
4. Back substitute the values found in Step 3 into any one
   of the original equations to find the value of the third
   variable.

Matrices   page 224

A rectangular array of numbers is called a **matrix**. Each number in the matrix is called an **entry**. Entries written horizontally are said to form a **row**, and entries written vertically are said to form a **column**. The number of rows by the number of columns, such as 2×3, is the **dimension** of the matrix. A matrix with the same number of rows as columns, such as 3×3, is called a **square matrix**.

Coefficient Matrix   page 225

A matrix formed from the coefficients of the variables in a system of linear equations is called a **coefficient matrix**.

Augmented Matrix   page 225

A matrix that includes the coefficients and the constant terms is called an **augmented matrix**.

Elementary Row Operations   page 226

There are three elementary row operations with matrices:
1. Interchange two rows.
2. Multiply a row by a nonzero constant.
3. Add a multiple of a row to another row.

If any elementary row operation is applied to a matrix, the new matrix is said to be **row-equivalent** to the original matrix.

Upper Triangular Form   page 228

If all the entries in the lower left triangular region are 0's, the matrix is said to be in **upper triangular form**.

Gaussian Elimination   page 229

To solve a system of linear equations using Gaussian elimination:
1. Write the augmented matrix for the system.
2. Use elementary row operations to transform the matrix into row echelon form.
3. Solve the corresponding system of equations by using back substitution.

Using a Graphing Calculator to Solve a System of Linear Equations   pages 232-233

Determinants   page 239

A **determinant** is a real number associated with a square array of real numbers and is indicated by enclosing the array between two vertical bars. For a matrix A, the corresponding determinant is designated as $\det(A)$.

Value of a 2 × 2 Determinant   page 240

For the square matrix $A = \begin{bmatrix} a_{11} & a_{12} \\ a_{21} & a_{22} \end{bmatrix}$,

$$\det(A) = \begin{vmatrix} a_{11} & a_{12} \\ a_{21} & a_{22} \end{vmatrix} = a_{11}a_{22} - a_{21}a_{12}.$$

Value of a 3 × 3 Determinant   page 241

For the square matrix $A = \begin{bmatrix} a_{11} & a_{12} & a_{13} \\ a_{21} & a_{22} & a_{23} \\ a_{31} & a_{32} & a_{33} \end{bmatrix}$,

$$\det(A) = \begin{vmatrix} a_{11} & a_{12} & a_{13} \\ a_{21} & a_{22} & a_{23} \\ a_{31} & a_{32} & a_{33} \end{vmatrix}$$

$$= a_{11}\left(\text{minor of } a_{11}\right) - a_{12}\left(\text{minor of } a_{12}\right) + a_{13}\left(\text{minor of } a_{13}\right)$$

$$= a_{11}\begin{vmatrix} a_{22} & a_{23} \\ a_{32} & a_{33} \end{vmatrix} - a_{12}\begin{vmatrix} a_{21} & a_{23} \\ a_{31} & a_{33} \end{vmatrix} + a_{13}\begin{vmatrix} a_{21} & a_{22} \\ a_{31} & a_{32} \end{vmatrix}.$$

Sign Table for Minors of a 3 × 3 Determinant   page 242

This pattern can be extended to apply to any $n \times n$ determinant.

$$\begin{vmatrix} + & - & + \\ - & + & - \\ + & - & + \end{vmatrix}$$

Using a Graphing Calculator to Evaluate a Determinant   page 244

Cramer's Rule

For 2 × 2 Systems   page 250

For the system $\begin{cases} a_{11}x + a_{12}y = k_1 \\ a_{21}x + a_{22}y = k_2 \end{cases}$, where

$$D = \begin{vmatrix} a_{11} & a_{12} \\ a_{21} & a_{22} \end{vmatrix}, \quad D_x = \begin{vmatrix} k_1 & a_{12} \\ k_2 & a_{22} \end{vmatrix}, \quad \text{and} \quad D_y = \begin{vmatrix} a_{11} & k_1 \\ a_{21} & k_2 \end{vmatrix},$$

if $D \neq 0$, then $x = \dfrac{D_x}{D}$ and $y = \dfrac{D_y}{D}$

is the unique solution to the system.

For 3 × 3 Systems   page 251

For the system $\begin{cases} a_{11}x + a_{12}y + a_{13}z = k_1 \\ a_{21}x + a_{22}y + a_{23}z = k_2 \\ a_{31}x + a_{32}y + a_{33}z = k_3 \end{cases}$, where

$$D = \begin{vmatrix} a_{11} & a_{12} & a_{13} \\ a_{21} & a_{22} & a_{23} \\ a_{31} & a_{32} & a_{33} \end{vmatrix}, \quad D_x = \begin{vmatrix} k_1 & a_{12} & a_{13} \\ k_2 & a_{22} & a_{23} \\ k_3 & a_{32} & a_{33} \end{vmatrix},$$

$$D_y = \begin{vmatrix} a_{11} & k_1 & a_{13} \\ a_{21} & k_2 & a_{23} \\ a_{31} & k_3 & a_{33} \end{vmatrix}, \quad \text{and} \quad D_z = \begin{vmatrix} a_{11} & a_{12} & k_1 \\ a_{21} & a_{22} & k_2 \\ a_{31} & a_{32} & k_3 \end{vmatrix}$$

if $D \neq 0$, then $x = \dfrac{D_x}{D}$, $y = \dfrac{D_y}{D}$, and $z = \dfrac{D_z}{D}$

is the unique solution to the system.

Cramer's Rule when $D = 0$ page 251

If $D = 0$, Cramer's Rule cannot be used. In a case
where $D = 0$, use the algebraic method of elimination
or substitution. The system will be either dependent or
inconsistent.

Section 3.7 Graphing Systems of Linear Inequalities

To Solve a System of Two Linear Inequalities   page 257
1. For each inequality, graph the boundary line and shade the appropriate half-plane.
2. Determine the region of the graph that is common to both half-planes (the region where the shading overlaps). (This region is called the **intersection** of the two half-planes.)
3. To check, pick one test-point in the intersection and verify that it satisfies both inequalities.
   Note: If there is no intersection, then the system is inconsistent and has no solution.

Using a Graphing Calculator to Graph Systems of Linear Inequalities   page 259

 ## HAWKES LEARNING SYSTEMS: INTERMEDIATE ALGEBRA SOFTWARE

For a review of the topics and problems from Chapter 3, look at the following lessons from *Hawkes Learning Systems: Intermediate Algebra*

- Solving Systems of Linear Equations by Graphing
- Solving Systems of Linear Equations by Substitution
- Solving Systems of Linear Equations by Addition
- Applications: Systems of Equations
- Solving Systems of Linear Equations with Three Variables
- Matrices and Gaussian Elimination
- Determinants
- Determinants and Systems of Linear Equations: Cramer's Rule
- Systems of Linear Inequalities
- Chapter 3 Review and Test

Chapter 3 Review

3.1 Systems of Linear Equations in Two Variables

1. For the following system of linear equations, find the slope and y-intercept of each line. What does this information indicate about the graphs of the two lines, and what does this tell you about the type of system?

$$\begin{cases} y = 3x + 2 \\ -6x + 2y = 14 \end{cases}$$

2. Determine which of the points in set A, if any, satisfy the given system of linear equations.

$$A = \{(-3, -4), (-2, 1), (0, 5), (1, 5)\} \qquad \begin{cases} x + y = 6 \\ 2x - y = -3 \end{cases}$$

Solve each of the systems in Exercises 3 – 6 by graphing.

3. $\begin{cases} x + y - 5 = 0 \\ x - 4y = 5 \end{cases}$
4. $\begin{cases} x - y = 5 \\ x = -2 \end{cases}$
5. $\begin{cases} 4x + 3y = -7 \\ 5x - 2y = -3 \end{cases}$
6. $\begin{cases} y = 2x \\ y = -2x + 4 \end{cases}$

Use the substitution method or the addition method to solve the systems of linear equations in Exercises 7 – 12. State whether each system is consistent, inconsistent, or dependent.

7. $\begin{cases} x = -2 \\ y = 2x + 5 \end{cases}$
8. $\begin{cases} y = \dfrac{1}{6}x - 2 \\ \dfrac{5}{6}x + y = 4 \end{cases}$
9. $\begin{cases} y = 4x \\ -4x + y = 13 \end{cases}$

10. $\begin{cases} x + 4y = 3 \\ 3x - 4y = 7 \end{cases}$
11. $\begin{cases} 3x + 2y = 4 \\ -6x - 4y = 8 \end{cases}$
12. $\begin{cases} 6x + 3y = 9 \\ y = 3 - 2x \end{cases}$

*In Exercises 13 – 16, use a graphing calculator and the **CALC** and `5:intersect` commands to find the solutions to the given systems of linear equations. If necessary, round values to four decimal places.*

13. $\begin{cases} y = 2 \\ -2x + 3y = 3 \end{cases}$
14. $\begin{cases} y = -4x + 0.5 \\ x + 2y = -8 \end{cases}$

15. $\begin{cases} x + 2y = 9 \\ 2x - y = -2 \end{cases}$
16. $\begin{cases} y = \dfrac{1}{3}x \\ 2x + y = 7 \end{cases}$

3.2 Applications

17. Rectangles: The perimeter of a rectangle is 100 meters and the length is 12 meters longer than the width. Find the dimensions of the rectangle.

18. School supplies: A student bought a calculator and a textbook for a course in algebra. She told her friend that the total cost was $150 and that the calculator cost $30 more than twice the cost of the textbook. What was the cost of each item?

19. Chemistry: A salt solution consists of ten gallons of 30% salt. This solution is the result of mixing a 25% solution with a 50% solution. How many gallons of each of the original salt solutions were used?

20. Investing: Jiao-long made two investments, one at 8% interest and the other at 6%. He decided to invest $1000 more at 8% than at 6%. What amount is invested at each rate, if the interest from both investments combined is $640 per year?

21. Determine a and b such that the line with equation $ax + by = 8$ passes through the points $(1, 1)$ and $(8, 0)$.

22. Chemistry: How many ounces each of a 40% acid solution and a 30% solution must be mixed to produce 100 ounces of a 36% solution?

23. The difference between two numbers is 16. Four times the smaller number is equal to 8 more than the larger. What are the numbers?

24. Car travel: George drove 190 miles on a trip that took him 3.5 hours. If he averaged 52 mph on the first part of the trip and 56 mph on the second part, how much time did he spend on each part of the trip?

3.3 Systems of Linear Equations in Three Variables

Solve each system of equations in Exercises 25 – 30. State which systems, if any, have no solution or an infinite number of solutions.

25. $\begin{cases} x + 3y = 6 \\ 6y + z = 12 \\ x - 3z = -15 \end{cases}$

26. $\begin{cases} x + y + z = 6 \\ x - 2y + z = 0 \\ 3x + y - 2z = -7 \end{cases}$

27. $\begin{cases} x + y - z = -4 \\ 2x + y + z = 6 \\ x + 4y - 2z = -6 \end{cases}$

28. $\begin{cases} x + y - 2z = -2 \\ 2x + y = 2 \\ 3y - 4z = -14 \end{cases}$

29. $\begin{cases} \dfrac{1}{3}x + \dfrac{1}{2}y = \dfrac{7}{2} \\ x - y + \dfrac{1}{2}z = -5 \\ \dfrac{2}{3}x - y + \dfrac{1}{2}z = -6 \end{cases}$

30. $\begin{cases} x - 2y + 3z = 0 \\ 3x - y + 3z = 2 \\ 5x - 5y + 9z = 6 \end{cases}$

31. The sum of three integers is 65. The sum of the second and third is 25 more than the first. If the third is subtracted from the sum of the first two, the difference is 17. Find the three integers.

32. Investing: Kathleen has three investments for a total of $50,000. She has invested in savings, stocks, and bonds. Her savings account paid 5% interest, stocks earned 6%, and bonds earned 4% for a total interest of $2400 last year. If her investment in bonds is equal to the total invested in the other two accounts, how much does she have invested in each type of investment?

33. Coin collecting: Opal has a small coin collection consisting only of nickels, dimes, and quarters with a total value of $5.40. She has 34 coins in the collection and twice as many quarters as nickels. How many of each kind of coin does she have?

34. Triangles: A triangle has a perimeter of 42 meters. The longest side is 2 meters less than the sum of the other two sides and four times the shortest side is 32 meters more than the difference between the two longer sides. What are the lengths of the three sides of the triangle?

3.4   Matrices and Gaussian Elimination

In Exercises 35 and 36, write the coefficient matrix and the augmented matrix for the given systems of linear equations.

35. $\begin{cases} 2x - y = 2 \\ 7x + y = 5 \end{cases}$

36. $\begin{cases} 2x - y + 3z = 5 \\ 4x - z = 1 \\ x - 9y + 2z = 3 \end{cases}$

Use the Gaussian elimination method with back substitution to solve the given system of equations in Exercises 37 – 42.

37. $\begin{cases} x + y = -2 \\ 3x + 2y = 0 \end{cases}$

38. $\begin{cases} 2x + y = -3 \\ x + 2y = 6 \end{cases}$

39. $\begin{cases} x + y - z = 13 \\ 2x - 2y + z = -4 \\ x + 2y - 2z = 15 \end{cases}$

40. $\begin{cases} 2x - 3y + z = 8 \\ x - 2y + 3z = 1 \\ x + 5y + 4z = 7 \end{cases}$

41. $\begin{cases} 2x + y - z = -2 \\ 5x - 2y + 4z = 1 \\ 3x + y + z = -13 \end{cases}$

42. $\begin{cases} x + y + z = 1 \\ x - 2y + 3z = 0 \\ x + 4y + 2z = -4 \end{cases}$

In Exercises 43 and 44, use your graphing calculator to solve the systems of linear equations.

43. $\begin{cases} 3x - 2y = 1 \\ 4x - 3y = -1 \end{cases}$

44. $\begin{cases} x + y + z = 1 \\ 3x + 2y + z = 5 \\ x - 4y + 2z = -5 \end{cases}$

3.5 Determinants

Evaluate the determinants in Exercises 45 – 50.

45. $\begin{vmatrix} 2 & 3 \\ -1 & -2 \end{vmatrix}$

46. $\begin{vmatrix} 5 & 1 \\ 3 & -2 \end{vmatrix}$

47. $\begin{vmatrix} 1 & 2 & 3 \\ 4 & 5 & 5 \\ 6 & -1 & 4 \end{vmatrix}$

48. $\begin{vmatrix} 4 & 1 & -2 \\ 2 & 3 & 3 \\ -6 & -2 & 1 \end{vmatrix}$

49. $\begin{vmatrix} 1 & 1 & -3 \\ 1 & -1 & \frac{1}{3} \\ 1 & 2 & 1 \end{vmatrix}$

50. $\begin{vmatrix} 0 & 6 & -3 \\ 4 & 0 & 6 \\ 0 & 5 & -10 \end{vmatrix}$

Use a graphing calculator to find the value of the determinants in Exercises 51 and 52.

51. $\begin{vmatrix} 5 & -8 \\ 3 & -7 \end{vmatrix}$

52. $\begin{vmatrix} \frac{1}{2} & \frac{2}{3} & 0 \\ \frac{1}{4} & \frac{3}{5} & 0 \\ 1 & -1 & 1 \end{vmatrix}$

Use the method for evaluating determinants to solve the equations for x in Exercises 53 and 54.

53. $\begin{vmatrix} 1 & 3 & -3 \\ 5 & x & 2 \\ 7 & 7 & -7 \end{vmatrix} = 28$

54. $\begin{vmatrix} 1 & 0 & -2 \\ 4 & -4 & 1 \\ 2 & x & -2 \end{vmatrix} = 37$

3.6 Cramer's Rule

Use Cramer's Rule to solve the systems of linear equations in Exercises 55 – 60.

55. $\begin{cases} 2x + y = 6 \\ 5x - y = 8 \end{cases}$

56. $\begin{cases} 2x - 3y = 8 \\ -x + 5y = -9 \end{cases}$

57. $\begin{cases} x + y = 7 \\ x + z = 9 \\ x - y + z = 1 \end{cases}$

58. $\begin{cases} x + y + 4z = 7 \\ 2x - y - 3z = -5 \\ 5x - 3y = -2 \end{cases}$

59. $\begin{cases} x + y + 2z = 1 \\ 2x - y - z = 0 \\ 3x + y + 2z = 2 \end{cases}$

60. $\begin{cases} 2x + y = -3 \\ y + z = -2 \\ 4x - z = -3 \end{cases}$

3.7   Graphing Systems of Linear Inequalities

In Exercises 61 – 66, solve the systems of two linear inequalities graphically.

61. $\begin{cases} x > 3 \\ y < -x + 4 \end{cases}$

62. $\begin{cases} y \le -1 \\ y \ge 2x + 1 \end{cases}$

63. $\begin{cases} 2x + 5y < 30 \\ x - 2y < -8 \end{cases}$

64. $\begin{cases} y > 5x + 1 \\ y < 2x + 1 \end{cases}$

65. $\begin{cases} x + y < 6 \\ 2x - 3y < 10 \end{cases}$

66. $\begin{cases} x + 3y < 7 \\ 5x - 2y \ge 4 \end{cases}$

In Exercises 67 – 70, use a graphing calculator to solve the systems of linear inequalities.

67. $\begin{cases} y \le 4x + 8 \\ y \le -x + 2 \end{cases}$

68. $\begin{cases} 3x + 2y < 12 \\ x + y < 6 \end{cases}$

69. $\begin{cases} y \ge 3x + 1 \\ y \le 2x + 1 \end{cases}$

70. $\begin{cases} y > x - 5 \\ y < x + 3 \end{cases}$

Chapter 3 Test

1. Solve the system by graphing: $\begin{cases} 4x - y = 13 \\ 2x - 3y = 9 \end{cases}$.

In Exercises 2 – 5, use the substitution method or the addition method to solve the systems of linear equations. State whether each system is consistent, inconsistent, or dependent.

2. $\begin{cases} x + y = 9 \\ x - y = 5 \end{cases}$ 3. $\begin{cases} 6x + 3y = 5 \\ 4x + 2y = -3 \end{cases}$ 4. $\begin{cases} 7x - 6y = 2 \\ 5x + 2y = 3 \end{cases}$ 5. $\begin{cases} 2x + 3y = 1 \\ 5x + 7y = 6 \end{cases}$

6. Determine a and b such that the line $ax + by = 17$ passes through the two points $(-3, 2)$ and $(1, 5)$.

7. **Rectangles:** The length of a rectangle is 7 ft more than twice its width. The perimeter is 62 ft. Find the dimensions of the rectangle.

8. Solve the following system of equations algebraically: $\begin{cases} x - 2y - 3z = 3 \\ x + y - z = 2 \\ 2x - 3y - 5z = 5 \end{cases}$.

9. Solve the following system of equations algebraically: $\begin{cases} x + 2y - 2z = 0 \\ x - y + z = 2 \\ -x + 4y - 4z = -8 \end{cases}$.

10. For the following system of equations:
    a. write the coefficient matrix and the augmented matrix and state the dimension of each, and
    b. solve the system using Gaussian elimination.
    $$\begin{cases} x + 2y - 3z = -11 \\ x - y - z = 2 \\ x + 3y + 2z = -4 \end{cases}$$

11. **Stamp collecting:** Kimberly bought 90 stamps in denominations of 41¢, 58¢, and 75¢. To test her daughter, who is taking an algebra class, she said that she bought three times as many 41¢ stamps as 58¢ stamps and that the total cost of the stamps was $43.70. How many stamps of each denomination did she buy?

Evaluate the determinants in Exercises 12 and 13.

12. $\begin{vmatrix} 6 & -3 \\ 4 & 5 \end{vmatrix}$

13. $\begin{vmatrix} 1 & 3 & 2 \\ 2 & 5 & 1 \\ 0 & 2 & 1 \end{vmatrix}$

Use the method for evaluating determinants to solve the equations for x in Exercises 14 and 15.

14. $\begin{vmatrix} 3 & x \\ 5 & 7 \end{vmatrix} = -9$

15. $\begin{vmatrix} -2 & 3 & 1 \\ 1 & 3 & x \\ 2 & 3 & x \end{vmatrix} = 14$

In Exercises 16 and 17, use Cramer's Rule to solve the given system of equations.

16. $\begin{cases} 3x + 8y = 14 \\ 2x + 7y = 22 \end{cases}$

17. $\begin{cases} x + 2y - z = 2 \\ x - 4y - 5z = -7 \\ x + 3y + 4z = 5 \end{cases}$

18. Triangles: The sum of the measures of the three angles of a triangle is $180°$. If the largest angle is $40°$ less than the sum of the other two, and the middle angle is $40°$ less than twice the smallest angle, find the measures of the angles.

19. Solve the following system of linear inequalities graphically.

$$\begin{cases} y < 3x + 4 \\ 2x + y \geq 1 \end{cases}$$

20. Solve the following system of equations using a graphing calculator.

$$\begin{cases} x - 3y - 4z = -12 \\ 3x + 4y + \dfrac{1}{2}z = -4 \\ -x - y + z = 2 \end{cases}$$

21. Use a graphing calculator to find the value of the determinant for the coefficient matrix of the system of equations given in Exercise 20.

Special Problem to Check Your Understanding

22. Use the method of expanding by minors to find the value of the following 4×4 determinant. Show all your steps in the calculation. Then, if you have time, use a graphing calculator to verify your result.

$$\begin{vmatrix} 5 & 3 & 1 & 0 \\ -2 & 0 & 1 & 2 \\ 0 & 6 & 7 & -5 \\ 0 & 8 & -1 & -3 \end{vmatrix}$$

Cumulative Review: Chapters 1–3

Find the prime factorization for each of the composite numbers in Exercises 1 – 3.

1. 82

2. 300

3. 5000

Find the LCM for each set of counting numbers in Exercises 4 – 6.

4. $\{12, 50, 100\}$

5. $\{8, 44, 56\}$

6. $\{6, 18, 96, 120\}$

7. Given the set of numbers, $A = \left\{-\sqrt{13}, -3, -\frac{1}{2}, 0, \frac{5}{8}, 1, \sqrt{2}, \pi\right\}$, list the numbers in A that are described by the following notation.

    a. integers
                           b. rational numbers

    c. irrational numbers
                   d. real numbers

State the property of real numbers that is illustrated in Exercises 8 and 9. All variables represent real numbers.

8. $4 + (-4) = 0$

9. $a(b + 2) = ab + 2a$

In Exercises 10 and 11, simplify each expression using the rules for order of operations.

10. $16 + \left(3^2 - 7^2\right) \div 5 \cdot 2$

11. $15 - 3\left(4^2 - 10^3 \div 2^3 + 6 \cdot 3\right) + 2^5$

Solve each equation in Exercises 12 – 14.

12. $4(x + 2) - (8 - 2x) = 12$

13. $\dfrac{3x - 2}{8} = \dfrac{x}{4} - 1$

14. $|2x + 1| = 28$

15. Solve the formula $K = \dfrac{mv^2}{2g}$ for m.

Solve the inequalities in Exercises 16 – 18. Write the solutions in interval notation then graph each solution on a real number line.

16. $5x - 13 > 7(x - 3)$

17. $\dfrac{x}{3} - 21 \le \dfrac{x}{2} + 3$

18. $|3x + 2| + 4 < 10$

Graph the linear equations in Exercises 19 and 20 by locating the y-intercept and the x-intercept.

19. $6x + 3y = 12$

20. $-2x + 5y = 10$

For Exercises 21 and 22, write the equation in slope-intercept form. Find the slope and the y-intercept, and then draw the graph.

21. $2x - y = 6$

22. $6x = 4$

23. Given the two points $(-5, 2)$ and $(5, -3)$:
   a. Find the slope of the line that passes through the two points.
   b. Find the equation of the line in slope-intercept form.
   c. Graph the line.

24. Find the equation of the line that is parallel to the line $2x - 5y = 1$ and passes through the origin.

25. Find the equation of the line that is perpendicular to the line $y = 7$ and passes through the point $(3, -5)$.

26. Graph the relation $r = \{(1,2),(-1,-2),(-3,2),(1,-1),(4,2)\}$. State the domain and range and indicate whether the relation is a function.

27. Given that $f(x) = -2x + 10$, find **a.** $f(0)$ **b.** $f(-6)$ **c.** $f(10)$.

28. For the function $H(x) = x^2 - 5x + 6$, find **a.** $H(2)$ **b.** $H(3)$ **c.** $H(-2)$.

29. Solve the system of linear equations by graphing.
$$\begin{cases} 2x + y = 6 \\ 3x - 2y = -5 \end{cases}$$

30. Solve the system of linear equations by the addition method.
$$\begin{cases} x + 3y = 10 \\ 5x - y = 2 \end{cases}$$

31. Solve the system of linear equations using Gaussian elimination.
$$\begin{cases} x - 3y + 2z = -1 \\ -2x + y + 3z = 1 \\ x - y + 4z = 9 \end{cases}$$

Evaluate the determinants in Exercises 32 and 33.

32. $\begin{vmatrix} 8 & -4 \\ -6 & 3 \end{vmatrix}$

33. $\begin{vmatrix} 7 & 5 \\ 5 & 2 \end{vmatrix}$

34. Use the method for evaluating determinants to solve the given equation for x.

$$\begin{vmatrix} 1 & -7 & x \\ 5 & 9 & x \\ 0 & 1 & 0 \end{vmatrix} = -12$$

In Exercises 35 and 36, solve the systems of linear equations using Cramer's Rule.

35. $\begin{cases} x - y = 0 \\ 2x + y = 3 \end{cases}$

36. $\begin{cases} 3x - 5y + 2z = 3 \\ 2x + 2z = 3 \\ -x + 5y - 4z = 2 \end{cases}$

In Exercises 37 and 38, solve the systems of two linear inequalities graphically.

37. $\begin{cases} x + 2y > 4 \\ x - y > 7 \end{cases}$

38. $\begin{cases} y \le x - 5 \\ 3x + y \ge 2 \end{cases}$

39. Baking: Karl makes two kinds of cookies. Choc-O-Nut requires 4 oz of peanuts for each 10 oz of chocolate chips. Chocolate Krunch requires 12 oz of peanuts per 8 oz of chocolate chips. How many batches of each can he make if he has 36 oz of peanuts and 46 oz of chocolate chips?

40. Investing: Alicia has $7000 invested, some at 7% and the remainder at 8%. After one year, the interest from the 7% investment exceeds the interest from the 8% investment by $70. How much is invested at each rate?

41. The points $(0, 4)$, $(-2, 6)$, and $(1, 9)$ lie on the curve described by the function $y = ax^2 + bx + c$. Find the values of $a, b,$ and c.

42. Graph the function $y = 5 - 2x - x^2$ on a graphing calculator. With the **CALC** features of the calculator, find the x-intercepts and the maximum value for y on the curve.

43. Use a graphing calculator and the **CALC** and **5:intersect** commands to find the solution to the following system of linear equations.

$$\begin{cases} 6x + y = 0 \\ -3x + 2y = -15 \end{cases}$$

In Exercises 44 and 45, solve the systems of equations using a graphing calculator and Gaussian elimination.

44. $\begin{cases} x + 2y = 7 \\ -x + 3y = \dfrac{13}{2} \end{cases}$

45. $\begin{cases} x + y - z = -8 \\ 2x - 3y = -6 \\ x + y + z = 2 \end{cases}$

Exponents and Polynomials

Did You Know?

Throughout history, teachers of mathematics have tried to develop calculation methods that were easy to use or to memorize. One of the more interesting of these techniques is the Rule of Double False Position. As a student of algebra, it may seem strange to you that such a complicated method would be developed to solve a simple first-degree equation of the form $ax + b = 0$. But remember that you have modern symbolism at your disposal. To use the rule, we will make two guesses as to the solution of the equation. We shall designate the guesses as g_1 and g_2. Now we will let $e_1 = ag_1 + b$ and $e_2 = ag_2 + b$, where e_1 and e_2 represent the amount of error in our guesses. Then the solution is

$$x = \frac{e_1 g_2 - e_2 g_1}{e_1 - e_2}.$$

We now illustrate the Rule of Double False Position with an example: $3x + 6 = 0$. Suppose we guess that the solution is either 1 or 2. That is, let $g_1 = 1$ and $g_2 = 2$. Then $e_1 = 3(1) + 6 = 9$ and $e_2 = 3(2) + 6 = 12$, and the solution to the equation would be

$$\frac{9(2) - 12(1)}{9 - 12} = \frac{18 - 12}{-3} = \frac{6}{-3} = -2.$$

This unnecessarily complicated method of solving first-degree equations was taught until the nineteenth century. One of the most popular English texts of the sixteenth century, *The Grounde of Artes* by Robert Recorde (1510 – 1558), gives the Rule of Double False Position in verse:

Gesse at this woorke as happe doth leade.
By chaunce to truthe you may procede.
And firste woorke by the question,
Although no truthe therein be don.
Suche falsehode is so good a grounde,
That truth by it will soone be founde.
From many bate to many mo,
From to fewe take to fewe also.
With to much ioyne to fewe againe,
To to fewe adde to manye plaine.
To crossewaies multiplye contrary kinde,
All truthe by falsehode for to fynde.

Recorde

Students memorized the poem as a method of remembering the rule, which they generally did not understand. Can you figure out why making two false guesses can lead to the correct answer? After studying Chapters 3 and 4, you may be able to verify the Rule of Double False Position.

"But it should always be required that a mathematical subject not be considered exhausted until it has become intuitively evident".

Felix Klein (1849 – 1925)

In Chapter 4 you will learn the rules of exponents and how, in scientific notation, exponents can be used in representing very large and very small decimal numbers. These ideas are very familiar to astronomers and chemists and are used in scientific calculators. (Multiply 9,000,000 by 9,000,000 on your calculator and read the result. What do you think the answer means?)

Polynomials and the operations with polynomials are used throughout all levels of mathematics from elementary and intermediate algebra through statistics, calculus, and beyond. You will see that the related techniques in dealing with polynomials are a basic part of operating with and simplifying rational expressions and solving equations in Chapter 5 and again in Chapter 7. **The skills you learn in factoring polynomials may be one of the determining factors in successfully completing this course and future courses, so work particularly hard at this early stage.**

4.1   Exponents and Scientific Notation

- *Simplify expressions by using the properties of integer exponents.*
- *Write decimal numbers in scientific notation.*
- *Write numbers given in scientific notation as decimal numbers.*

As we discussed in Section 1.1, **exponents** can be used to indicate repeated multiplication by the same number (called the **base**). In this section, we will discuss the rules of exponents including 0 and negative integer exponents. These rules are particularly useful in simplifying algebraic expressions.

In the expression x^5 (read "x to the fifth power"), x is the **base** and 5 is the **exponent**. Remember, if a variable or constant is written without an exponent, then the exponent is understood to be 1.

The Product Rule

Now consider the product $x^3 \cdot x^5$. We can write

$$x^3 \cdot x^5 = (x \cdot x \cdot x) \cdot (x \cdot x \cdot x \cdot x \cdot x) = x^8. \qquad \text{Note that } 3 + 5 = 8.$$

This example illustrates the following **Product Rule** for multiplying factors with the **same base**.

Product Rule for Exponents

If a is a nonzero real number and m and n are integers, then
$$a^m \cdot a^n = a^{m+n}.$$

Example 1: The Product Rule for Exponents

Use the Product Rule for Exponents to simplify each expression.

a. $3^2 \cdot 3^3 = 3^{2+3} = 3^5$ Note that the bases are **not** multiplied. The base, 3, stays the same.

b. $6^2 \cdot 6 = 6^{2+1} = 6^3$ Remember $6 = 6^1$.

c. $\left(7y^2\right)\left(-3y^5\right) = 7(-3) \cdot y^2 \cdot y^5 = -21y^{2+5} = -21y^7$ In **c.** and **d.**, the commutative and associative properties of multiplication have been used.

d. $3x^2 y^3 \cdot 2x^3 y^5 = 3 \cdot 2 \cdot x^2 \cdot x^3 \cdot y^3 \cdot y^5 = 6x^{2+3} y^{3+5} = 6x^5 y^8$

The Product Rule is stated for **integer exponents**. Applying this rule with the exponent 0 gives results such as the following:

$$5^0 \cdot 5^2 = 5^{0+2} = 5^2 \qquad \text{This implies that } 5^0 = 1.$$

$$4^0 \cdot 4^3 = 4^{0+3} = 4^3 \qquad \text{This implies that } 4^0 = 1.$$

These results lead to the following definition.

The Exponent 0

If a is a nonzero real number, then
$$a^0 = 1.$$
Note: *the expression 0^0 is undefined.*

NOTES   Throughout this text, unless specifically stated otherwise, we will assume that the bases of exponents are nonzero.

Example 2: The Exponent 0

Find the value of each expression.

a. $7^0 = 1$

b. $(-5)^0 = 1$

c. $(8x)^0 = 1$ for $x \neq 0$

d. $-3^0 = -1 \cdot 3^0 = -1 \cdot 1 = -1$

The Quotient Rule

Consider a fraction in which the numerator and the denominator are powers of the same base such as $\dfrac{4^5}{4^3}$ or $\dfrac{x^3}{x^2}$. In these cases, we can write and simplify the fractions as follows:

$$\frac{4^5}{4^3} = \frac{\cancel{4} \cdot \cancel{4} \cdot \cancel{4} \cdot 4 \cdot 4}{\cancel{4} \cdot \cancel{4} \cdot \cancel{4}} = \frac{4^2}{1} = 16 \qquad \text{or} \qquad \frac{4^5}{4^3} = 4^{5-3} = 4^2 = 16$$

$$\text{and} \qquad \frac{x^3}{x^2} = \frac{\cancel{x} \cdot \cancel{x} \cdot x}{\cancel{x} \cdot \cancel{x}} = \frac{x}{1} = x \qquad \text{or} \qquad \frac{x^3}{x^2} = x^{3-2} = x^1 = x.$$

We now have the following **Quotient Rule for Exponents**.

Quotient Rule for Exponents

If a is a nonzero real number and m and n are integers, then

$$\frac{a^m}{a^n} = a^{m-n}.$$

Example 3: The Quotient Rule for Exponents

Use the Quotient Rule for Exponents to simplify each expression.

a. $\dfrac{x^7}{x}$

   Solution: $\dfrac{x^7}{x} = x^{7-1} = x^6$

b. $\dfrac{y^2}{y^2}$

Solution: $\dfrac{y^2}{y^2} = y^{2-2} = y^0 = 1$

Note how this example shows another way to justify the idea that $a^0 = 1$ for any nonzero base.

c. $\dfrac{12x^{12}}{-3x^3}$

Solution: $\dfrac{12x^{12}}{-3x^3} = \dfrac{12}{-3} \cdot x^{12-3} = -4x^9$

Note that the **coefficients are divided** and the **exponents are subtracted**.

d. $\dfrac{14x^{10}y^6}{2xy^4}$

Solution: $\dfrac{14x^{10}y^6}{2xy^4} = \dfrac{14}{2} \cdot x^{10-1} \cdot y^{6-4} = 7x^9y^2$

Note, as illustrated in Examples 3c and 3d, in division with terms that have numerical coefficients, the **coefficients are divided** as usual and any **exponents are subtracted** by using the Quotient Rule.

Negative Exponents

In Example 3, the exponents in the numerators were greater than or equal to the exponents in the denominators in each case. But, what if the larger exponent is in the denominator and we still apply the Quotient Rule? (Remember that the Quotient Rule was stated for **integer** exponents.)

For example, applying the Quotient Rule to $\dfrac{7^3}{7^5}$ gives

$$\dfrac{7^3}{7^5} = 7^{3-5} = 7^{-2} \text{ which results in a negative exponent.}$$

But, simply reducing $\dfrac{7^3}{7^5}$ gives

$$\dfrac{7^3}{7^5} = \dfrac{\cancel{7} \cdot \cancel{7} \cdot \cancel{7}}{\cancel{7} \cdot \cancel{7} \cdot \cancel{7} \cdot 7 \cdot 7} = \dfrac{1}{7 \cdot 7} = \dfrac{1}{7^2}.$$

This means that $7^{-2} = \dfrac{1}{7^2}$.

The **Rule for Negative Exponents** follows on the next page.

Rule for Negative Exponents

If a is a nonzero real number and n is an integer, then

$$a^{-n} = \frac{1}{a^n}.$$

In words, a negative exponent indicates the reciprocal of the base.

Each expression in Example 4 is simplified by using the appropriate rules for exponents. Study each example carefully. In each case, the **expression is considered simplified if each base appears only once and each base has only positive exponents**.

> **NOTES**
>
> There is nothing wrong with negative exponents. In fact, negative exponents are preferred in some courses in mathematics and science. However, so that all answers are the same, in this course we will consider expressions to be simplified if:
>
> **1.** all exponents are positive, and
>
> **2.** each base appears only once.

Example 4: Negative Exponents

Use the rules for exponents that apply and simplify each expression so that it contains only positive exponents.

a. y^{-3}

Solution: $y^{-3} = \dfrac{1}{y^3}$ 　　　　　　Use the Rule for Negative Exponents.

b. $\dfrac{10^{-6}}{10^{-3}}$

Solution: $\dfrac{10^{-6}}{10^{-3}} = 10^{-6-(-3)}$ 　　　Use the Quotient Rule.

$\qquad\qquad = 10^{-6+3}$ 　　　Simplify.

$\qquad\qquad = 10^{-3}$

$\qquad\qquad = \dfrac{1}{10^3}$ or $\dfrac{1}{1000}$ 　　Use the Rule for Negative Exponents.

c. $x^{-10} \cdot x^7 \cdot x$

Solution: $x^{-10} \cdot x^7 \cdot x = x^{-10+7+1} = x^{-2} = \dfrac{1}{x^2}$ 　　Use the Product Rule and the Rule for Negative Exponents.

d. $\dfrac{18a^8 \cdot 2a^2}{3a^{15}}$

Solution: $\dfrac{18a^8 \cdot 2a^2}{3a^{15}} = \dfrac{36a^{8+2}}{3a^{15}}$ Use the Product Rule.

$= 12a^{10-15}$ Use the Quotient Rule.

$= 12a^{-5}$

$= \dfrac{12}{a^5}$ Use the Rule for Negative Exponents.

NOTES

Special Note about Using the Quotient Rule:

Regardless of the size of the exponents or whether they are positive or negative, the following single subtraction policy can be used with the Quotient Rule.

(numerator exponent) − (denominator exponent)

This subtraction will always lead to the correct answer.

Power Rules for Exponents

There are three power rules for exponents. The following examples illustrate these rules.

$$\left(3^5\right)^2 = 3^5 \cdot 3^5 = 3^{5+5} = 3^{10} \qquad \text{Note that } 5 \cdot 2 = 10.$$

$$(xy)^3 = xy \cdot xy \cdot xy = (x \cdot x \cdot x) \cdot (y \cdot y \cdot y) = x^3 y^3$$

$$\left(\frac{2}{y}\right)^3 = \frac{2}{y} \cdot \frac{2}{y} \cdot \frac{2}{y} = \frac{2 \cdot 2 \cdot 2}{y \cdot y \cdot y} = \frac{2^3}{y^3} = \frac{8}{y^3}$$

The three rules are included more formally in the following box.

Power Rules for Exponents

If a and b are nonzero real numbers and m and n are integers:

1. Power Rule: $\left(a^m\right)^n = a^{mn}$

To raise a power to a power, multiply the exponents.

2. Power Rule for Products: $(ab)^n = a^n b^n$

To raise a product to a power, raise each factor to that power.

3. Power Rule for Fractions: $\left(\dfrac{a}{b}\right)^n = \dfrac{a^n}{b^n}$

To raise a fraction to a power, raise the numerator and denominator to that power.

NOTES

Special Note about Negative Numbers and Exponents:

In an expression such as $-x^2$, we know that -1 is understood to be the coefficient of x^2. That is,

$$-x^2 = -1 \cdot x^2.$$

The same is true for expressions with numbers such as -4^2. That is,

$$-4^2 = -1 \cdot 4^2 = -1 \cdot 16 = -16.$$

We see that the exponent refers to 4 and **not** to -4. For the exponent to refer to -4 as the base, -4 **must be in parentheses** as follows:

$$(-4)^2 = (-4) \cdot (-4) = +16.$$

As another example,

$$-9^0 = -1 \cdot 9^0 = -1 \cdot 1 = -1 \text{ and } (-9)^0 = 1.$$

Example 5: Using Combinations of the Rules for Exponents

Simplify each expression by using the power rules and any other appropriate rules for exponents. (There may be more than one correct sequence of steps to arrive at the correct answer.)

a. $\left(y^{-6}\right)^2$

Solution: $\left(y^{-6}\right)^2 = y^{-6 \cdot 2} = y^{-12} = \dfrac{1}{y^{12}}$

Use the Power Rule and the Rule for Negative Exponents.

b. $\left(a^3 b^{-5} c\right)^4$

Solution: $\left(a^3 b^{-5} c\right)^4 = a^{3 \cdot 4} b^{-5 \cdot 4} c^{1 \cdot 4} = a^{12} b^{-20} c^4 = \dfrac{a^{12} c^4}{b^{20}}$

Use the Power Rule for Products and the Rule for Negative Exponents.

c. $\left(\dfrac{2x^{2k} y^k}{x^k y}\right)^3$

Solution:

Option 1: Simplify within the parentheses first.

$$\left(\frac{2x^{2k} y^k}{x^k y}\right)^3 = \left(2x^{2k-k} y^{k-1}\right)^3 = \left(2x^k y^{k-1}\right)^3 = 2^3 x^{3k} y^{3(k-1)} = 8x^{3k} y^{3k-3}$$

Option 2: Apply the Power Rule for Fractions first.

$$\left(\frac{2x^{2k} y^k}{x^k y}\right)^3 = \frac{\left(2x^{2k} y^k\right)^3}{\left(x^k y\right)^3} = \frac{2^3 x^{3 \cdot 2k} y^{3 \cdot k}}{x^{3 \cdot k} y^{3 \cdot 1}} = \frac{8x^{6k} y^{3k}}{x^{3k} y^3} = 8x^{6k-3k} y^{3k-3} = 8x^{3k} y^{3k-3}$$

Another general approach with fractions involving negative exponents is to note that

$$\left(\frac{a}{b}\right)^{-n} = \frac{a^{-n}}{b^{-n}} = \frac{\frac{1}{a^n}}{\frac{1}{b^n}} = \frac{1}{a^n} \cdot \frac{b^n}{1} = \frac{b^n}{a^n} = \left(\frac{b}{a}\right)^{n}.$$

NOTES

In effect there are two basic shortcuts with negative exponents and fractions:

1. Taking the reciprocal of a fraction changes the sign of any exponent in the fraction.
2. Moving any term from the numerator to the denominator, or vice versa, changes the sign of the corresponding exponent.

Example 6: Two Approaches with Fractional Expressions and Negative Exponents

Simplify $\left(\dfrac{-3a^2}{b^3}\right)^{-2}$.

Solution: **Option 1:**   Use the idea of reciprocal first.

$$\left(\frac{-3a^2}{b^3}\right)^{-2} = \left(\frac{b^3}{-3a^2}\right)^{2} = \frac{b^{3\cdot2}}{(-3)^2 a^{2\cdot2}} = \frac{b^6}{9a^4}$$

Option 2:   Apply the Power Rule for Fractions first.

$$\left(\frac{-3a^2}{b^3}\right)^{-2} = \frac{(-3)^{-2} a^{-2\cdot2}}{b^{-2\cdot3}} = \frac{(-3)^{-2} a^{-4}}{b^{-6}} = \frac{b^6}{9a^4}$$

Example 7: A More Complex Example

This example involves the application of a variety of steps. Study it carefully and see if you can get the same result by following a different sequence of steps.

Simplify $\left(\dfrac{x^2 y^3}{3xy^{-2}}\right)^{-2} \left(\dfrac{4x^2 y^{-1}}{x^{-5} y^3}\right)^{-1}$.

Continued on the next page...

Solution: $\left(\dfrac{x^2 y^3}{3xy^{-2}}\right)^{-2}\left(\dfrac{4x^2 y^{-1}}{x^{-5}y^3}\right)^{-1} = \left(\dfrac{3xy^{-2}}{x^2 y^3}\right)^{2}\left(\dfrac{x^{-5}y^3}{4x^2 y^{-1}}\right)$ Use the Rule for Negative Exponents.

$$= \left(3x^{1-2}y^{-2-3}\right)^{2}\left(\dfrac{x^{-5-2}y^{3-(-1)}}{4}\right) \quad \text{Use the Quotient Rule.}$$

$$= \left(3x^{-1}y^{-5}\right)^{2}\left(\dfrac{x^{-7}y^{4}}{4}\right)$$

$$= \left(3^2 x^{-1\cdot 2}y^{-5\cdot 2}\right)\left(\dfrac{x^{-7}y^{4}}{4}\right) \quad \begin{array}{l}\text{Use the Power Rule}\\\text{and the Power Rule for}\\\text{Products.}\end{array}$$

$$= 3^2 x^{-2}y^{-10}\cdot\dfrac{x^{-7}y^{4}}{4}$$

$$= \dfrac{9x^{-2+(-7)}y^{-10+4}}{4} \quad \text{Use the Product Rule.}$$

$$= \dfrac{9x^{-9}y^{-6}}{4} = \dfrac{9}{4x^{9}y^{6}} \quad \begin{array}{l}\text{Use the Rule for}\\\text{Negative Exponents.}\end{array}$$

Remember that the choice of steps in dealing with exponents is yours. As long as you correctly apply the properties of exponents, the answer will be the same regardless of the order of the steps.

Summary of Properties and Rules for Exponents

If a and b are nonzero real numbers and m and n are integers:

1. *The Exponent 1:* $a^1 = a$

2. *The Exponent 0:* $a^0 = 1$

3. *Product Rule:* $a^m \cdot a^n = a^{m+n}$

4. *Quotient Rule:* $\dfrac{a^m}{a^n} = a^{m-n}$

5. *Negative Exponents:* $a^{-n} = \dfrac{1}{a^n}$

6. *Power Rule:* $\left(a^m\right)^n = a^{mn}$

7. *Power of a Product:* $(ab)^n = a^n b^n$ }   ***The Power Rules***

8. *Power of a Quotient:* $\left(\dfrac{a}{b}\right)^n = \dfrac{a^n}{b^n}$

Scientific Notation and Calculators

A basic application of integer exponents occurs in scientific disciplines, such as astronomy and biology, when very large and very small numbers are involved. For example, the distance from the earth to the sun is approximately 93,000,000 miles, and the approximate radius of a carbon atom is 0.000 000 007 7 centimeters.

In **scientific notation** (an option in all scientific and graphing calculators), **decimal numbers are written as the product of a number greater than or equal to 1 and less than 10 and an integer power of 10**. In scientific notation there is just one nonzero digit to the left of the decimal point. For example,

$$93,000,000 = 9.3 \times 10^7 \quad \text{and} \quad 0.0000000077 = 7.7 \times 10^{-9}.$$

The exponent tells how many places the decimal point is to be moved and in what direction. If the exponent is positive, the decimal point is moved to the right.

$$2.7 \times 10^3 = 2700. \quad \text{3 places right}$$

A negative exponent indicates that the decimal point should move to the left.

$$3.92 \times 10^{-6} = 0.000\,003\,92 \quad \text{6 places left}$$

Scientific Notation

> *If N is a decimal number, then in **scientific notation***
>
> $$N = a \times 10^n \text{ where } 1 \le a < 10 \text{ and n is an integer.}$$

Example 8: Decimals in Scientific Notation

Write the following decimal numbers in scientific notation.

a. 867,000,000,000
   Solution: $867,000,000,000 = 8.67 \times 10^{11}$

b. 420,000
   Solution: $420,000 = 4.2 \times 10^5$

c. 0.0036
   Solution: $0.0036 = 3.6 \times 10^{-3}$

d. 0.000 000 025
   Solution: $0.000000025 = 2.5 \times 10^{-8}$

Example 9: Using Scientific Notation and the Properties of Exponents

Simplify the following expressions by first writing the decimal numbers in scientific notation and then using the properties of exponents.

a. $\dfrac{0.0023 \times 560,000}{0.0004}$

Solution: $\dfrac{0.0023 \times 560,000}{0.0004} = \dfrac{2.3 \times 10^{-3} \times 5.6 \times 10^{5}}{4.0 \times 10^{-4}}$

$$= \dfrac{2.3 \times \overset{1.4}{\cancel{5.6}}}{\cancel{4.0}} \times \dfrac{10^{-3} \times 10^{5}}{10^{-4}}$$

$$= 3.22 \times \dfrac{10^{-3+5}}{10^{-4}} = 3.22 \times \dfrac{10^{2}}{10^{-4}}$$

$$= 3.22 \times 10^{2-(-4)}$$

$$= 3.22 \times 10^{6}$$

b. $\dfrac{6.3 \times 8200}{3,000,000 \times 4.1}$

Solution: $\dfrac{6.3 \times 8200}{3,000,000 \times 4.1} = \dfrac{6.3 \times 8.2 \times 10^{3}}{3.0 \times 10^{6} \times 4.1}$

$$= \dfrac{\overset{2.1}{\cancel{6.3}} \times \overset{2}{\cancel{8.2}}}{\cancel{3.0} \times \cancel{4.1}} \times \dfrac{10^{3}}{10^{6}}$$

$$= 4.2 \times 10^{3-6} = 4.2 \times 10^{-3}$$

c. Light travels approximately 3×10^{8} meters per second. How many meters per minute does light travel?

Solution: Since there are 60 seconds in one minute, multiply by 60.

$$3 \times 10^{8} \times 60 = 3 \times 10^{8} \times 6 \times 10^{1} = 18 \times 10^{9} = 1.8 \times 10^{1} \times 10^{9} = 1.8 \times 10^{10}$$

Thus light travels 1.8×10^{10} meters per minute.

Example 10: Scientific Notation and Calculators

a. Use a TI-84 Plus graphing calculator to evaluate the expression $\dfrac{8600\left(4.5 \times 10^{4}\right)}{1.5 \times 10^{-3}}$. Leave the answer in scientific notation.

[**Note:** The caret key is used to indicate an exponent.]

Solution: With a TI-84 Plus calculator (set in scientific notation mode by pressing the **MODE** key and selecting **SCI**) the display should appear as shown below:

The E on the display indicates an exponent with base 10. Thus,

$$2.58 \, \text{E} \, 11 = 2.58 \times 10^{11}.$$

Note that the numerator and denominator must be set in parentheses and be sure to use the ⊖ key on the bottom row for −3.

b. Use a calculator to find the number of miles that light travels in one year (a light-year) if light travels 186,000 miles per second.

Solution: We know

$$60 \text{ sec} = 1 \text{ min}$$
$$60 \text{ min} = 1 \text{ hr}$$
$$24 \text{ hr} = 1 \text{ day}$$
$$365 \text{ days} = 1 \text{ yr}$$

$$\frac{186000 \text{ miles}}{1 \, \cancel{\text{sec}}} \cdot \frac{60 \, \cancel{\text{sec}}}{1 \, \cancel{\text{min}}} \cdot \frac{60 \, \cancel{\text{min}}}{1 \, \cancel{\text{hr}}} \cdot \frac{24 \, \cancel{\text{hr}}}{1 \, \cancel{\text{day}}} \cdot \frac{365 \, \cancel{\text{day}}}{1 \text{yr}}$$

Multiplication should give the following display on your calculator:

Thus a light-year is 5,865,696,000,000 miles (or approximately 5.87×10^{12} miles.

Practice Problems

Use the properties of exponents to simplify each expression.

1. 6^{-2}

2. $x^4 \cdot x^0 \cdot x$

3. $\left(x^2 x^{-3}\right)^4$

4. $\dfrac{y^{10} y^4}{y^6}$

5. $\left(\dfrac{-2^{-1} a}{3b^2}\right)^{-2}$

6. $\left(\dfrac{7x^{-2}y}{xy^4}\right)^2 \left(\dfrac{14xy^{-3}}{x^4 y^2}\right)^{-1}$

7. Write the following expression in decimal notation: 2.47×10^{-3}.

8. Write the number 186,000 in scientific notation. This is the speed of light in miles per second.

4.1 Exercises

Use the properties of exponents to simplify the expressions in Exercises 1 – 64. Answers should contain only positive exponents.

1. $\left(7^2\right)\left(7^0\right)$

2. 7^{-2}

3. $3 \cdot 2^2$

4. -5^{-2}

5. $\left(-8\right)^{-2}$

6. $x^3 \cdot x^5$

7. $x^2 \cdot x^{-1}$

8. $x^{-2} \cdot x^3 \cdot x^5$

9. $y^{-3} \cdot y^{-2} \cdot y^0$

10. $\dfrac{x^{12}}{x^4}$

11. $\dfrac{x^2}{x^{-1}}$

12. $\dfrac{y^2}{y^{-5}}$

13. $\dfrac{x^3 x^5}{x^4}$

14. $\dfrac{x^0 x^3}{x^6}$

15. $\dfrac{x \cdot x^3}{x^5}$

16. $\dfrac{x^{-1} x^3}{x^{-4}}$

17. $\dfrac{x \cdot x^{-2}}{x^2 x^{-3}}$

18. $\dfrac{x^{16}}{x^{-2} x^{-8}}$

19. $\left(x^4\right)^2$

20. $\left(x^2\right)^{-2}$

21. $\left(x^0\right)^{-1}$

22. $\left(-x^3\right)^0$

23. $\left(y^0 y^{-1}\right)^5$

24. $\left(x^3 x^{-3}\right)^0$

Answers to Practice Problems: **1.** $\dfrac{1}{36}$ **2.** x^5 **3.** $\dfrac{1}{x^4}$ **4.** y^8 **5.** $\dfrac{36b^4}{a^2}$ **6.** $\dfrac{7}{2x^3 y}$ **7.** 0.00247

8. 1.86×10^5

25. $\dfrac{y^2 y^4}{y}$ **26.** $\dfrac{y \cdot y^4}{y}$ **27.** $\dfrac{x^5 x^2}{\left(x^2\right)^2}$ **28.** $\dfrac{x^{10} x^{-3}}{x^3 x^{-1}}$

29. $\dfrac{x^8 x^{-2}}{\left(x^2\right)^3}$ **30.** $\dfrac{\left(x^{-2}\right)^3}{x \cdot x^{-3}}$ **31.** $\dfrac{\left(y^2\right)^4}{y^{-2} y^{-1}}$ **32.** $\left(\dfrac{y^2 y^{-1}}{y^5 y^2}\right)^{-2}$

33. $\left(\dfrac{x^2 x^0}{x^4 x^{-1}}\right)^{-3}$ **34.** $\left(\dfrac{x^{-3} x^0}{x^2 x}\right)^3$ **35.** $\left(\dfrac{x^5 x^{-2}}{x \cdot x^{-3}}\right)^2$ **36.** $x^k \cdot x$

37. $x^k \cdot x^3$ **38.** $x^k \cdot x^{2k}$ **39.** $x^{3k} \cdot x^4$ **40.** $\dfrac{x^k}{x^2}$

41. $\dfrac{x^{2k}}{x^k}$ **42.** $\dfrac{x^{k+1}}{x^3}$ **43.** $\left(x^k\right)^2$ **44.** $\left(x^5\right)^k$

45. $x\left(x^2\right)^k$ **46.** $\dfrac{x^2 x^k}{\left(x^2\right)^k}$ **47.** $\dfrac{x^{k+1} x^{-2}}{x^4}$ **48.** $\dfrac{x^{k+3} x}{x^{-2}}$

49. $\left(\dfrac{-3x^{-2}}{y^3}\right)^{-1}$ **50.** $\left(\dfrac{2ab^4}{3b^2}\right)^{-3}$ **51.** $\left(\dfrac{x^2 y^{-3}}{3x^{-1} y}\right)^{-1}$ **52.** $\left(x^k y^m\right)^2$

53. $\left(x^{4n} y^3\right)\left(x^n y^{-k}\right)$ **54.** $\left(x^{k+1} y^{3k}\right)\left(x^2 y^{-k}\right)$ **55.** $\left(\dfrac{a^2 b}{ab^{-2}}\right)\left(\dfrac{a^{-3} b}{b^{-3}}\right)$

56. $\left(\dfrac{x^2 y^{-3}}{y^{-1}}\right)^2\left(\dfrac{xy^2}{2y}\right)^{-1}$ **57.** $\left(\dfrac{x^4 y}{2}\right)^2\left(\dfrac{y^3}{x^2}\right)^{-1}$ **58.** $\left(\dfrac{3x}{2x^2 y^2}\right)^{-1}\left(\dfrac{y^3}{2x}\right)^2$

59. $\left(\dfrac{5x^3 y}{x^{-2} y^3}\right)^{-1}\left(\dfrac{4x^{-2} y^{-1}}{15xy^4}\right)^{-1}$ **60.** $\dfrac{\left(7x^3 y^4\right)^0}{\left(2x^2 y\right)\left(xy^{-3}\right)^{-1}}$ **61.** $\dfrac{\left(4^{-2} x^{-3} y\right)^{-1}}{\left(x^{-2} y^2\right)^3\left(5xy^{-2}\right)^{-1}}$

62. $\dfrac{\left(6x^2 y\right)\left(x^{-1} y^3\right)^2}{\left(x^{-1} y\right)^2\left(3x^2 y\right)^3}$ **63.** $\dfrac{\left(x^{-3} y^{-5}\right)^{-2}\left(x^2 y^{-3}\right)^3}{\left(x^3 y^{-4}\right)^2\left(x^{-1} y^{-2}\right)^{-2}}$ **64.** $\dfrac{\left(4xy\right)^2\left(x^{-2} y^2\right)^{-1}}{\left(3x^3 y\right)^{-2}\left(2x^2 y^{-2}\right)^3}$

Write each number in Exercises 65 – 70 in decimal notation.

65. 4.72×10^5 **66.** 6.91×10^{-4} **67.** 1.28×10^{-7}

68. 1.63×10^8 **69.** 9.23×10^{-3} **70.** 5.88×10^6

In Exercises 71 – 86, write each number in the following expressions in scientific notation and simplify (if applicable). Show the steps you use, as in Examples 9a and 9b in the text. Do not use a calculator. Leave all answers in scientific notation.

71. 479,000

72. 0.000 367

73. 0.000 000 871

74. $52,800 \times 1,000$

75. $143,000 \times 0.0003$

76. 0.007×0.00012

77. $0.036 \times 4,000,000$

78. $\dfrac{27,000}{0.0009}$

79. $\dfrac{1800 \times 0.00045}{1350}$

80. $\dfrac{0.0032 \times 120}{0.0096}$

81. $\dfrac{0.084 \times 0.0093}{0.21 \times 0.031}$

82. $\dfrac{0.0070 \times 50 \times 0.55}{1.4 \times 0.0011 \times 0.25}$

83. $\dfrac{0.36 \times 5200}{0.00052 \times 720}$

84. $\dfrac{0.0016 \times 0.09 \times 460}{0.00012 \times 0.023}$

85. $\dfrac{760 \times 84 \times 0.063}{900 \times 0.38 \times 210}$

86. $\dfrac{420 \times 0.016 \times 80}{0.028 \times 120 \times 0.2}$

For Exercises 87 – 92, use your calculator and leave all answers in scientific notation.

87. Speed of light: Light travels approximately 3×10^{10} centimeters per second. How many centimeters would this be per minute? per hour?

88. Astronomy: One light-year is approximately 9.46×10^{15} meters. The distance to a certain star is about 4.3 light years. How many meters is this?

89. Astronomy: One light-year is about 5.88×10^{12} miles. The mean distance from the sun to Pluto is 3.675×10^{9} miles. How many light years is this?

90. Atomic weight: An atom of gold weighs approximately 3.25×10^{-22} grams. What would be the weight of 3,000 atoms of gold?

91. Atomic weight: The weight of an atom is measured in atomic weight units (amu), where 1 amu = 1.6605×10^{-27} kilograms. The atomic weight of carbon-12 is 12 amu. Express the atomic weight of carbon 12 in kilograms.

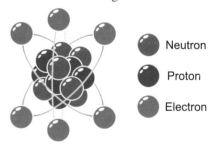

Neutron

Proton

Electron

92. Atomic weight: The atomic weight of argon is about 40 amu. Express this weight in kilograms. (See Exercise 91.)

In Exercises 93 – 98, use your calculator (set in scientific notation mode) to evaluate each expression. Leave all answers in scientific notation.

93. $\dfrac{5.4 \times 0.003 \times 5000}{15 \times 0.0027 \times 20}$

94. $\dfrac{0.0005 \times 650 \times 3.3}{0.00011 \times 2500}$

95. $\dfrac{\left(1.4 \times 10^{-3}\right)\left(922\right)}{\left(3.5 \times 10^{3}\right)\left(2.0 \times 10^{-6}\right)}$

96. $\dfrac{0.0084 \times 0.003}{0.21 \times 600}$

97. $\dfrac{0.02\left(3.9 \times 10^{3}\right)}{0.013\left(5.0 \times 10^{-3}\right)}$

98. $\dfrac{(43,000)\left(3.0 \times 10^{5}\right)}{\left(8.6 \times 10^{-2}\right)\left(1.5 \times 10^{-3}\right)}$

Writing and Thinking About Mathematics

99. Without looking at the text, show that $\left(\dfrac{x}{y}\right)^{-n} = \left(\dfrac{y}{x}\right)^{n}$ by using the Power Rules and the Rule for Negative Exponents. (Check to see if your method is similar to that on page 287.)

 HAWKES LEARNING SYSTEMS: INTERMEDIATE ALGEBRA SOFTWARE

- Simplifying Integer Exponents I
- Simplifying Integer Exponents II
- Scientific Notation

4.2

Addition and Subtraction with Polynomials

- *Identify polynomial expressions.*
- *Classify certain polynomials as monomials, binomials, or trinomials.*
- *Add and subtract polynomials.*
- *Evaluate polynomials for given values of the variables.*

Monomial

A **monomial** is a term that has no variable in its denominator, and its variables have only whole number exponents. Thus, a monomial does **not** have variables with negative exponents in the numerator, positive exponents in the denominator, or fractional exponents. For example:

Monomial terms: $\qquad 5x, \quad -7y^2, \quad 4, \quad \dfrac{1}{8}x^2y^2$

Not monomial terms: $\qquad \dfrac{2x^2}{y^2}, \quad -6x^{-1}, \quad 3y^{\frac{1}{2}}$

Monomial

> *A **monomial in x** is an expression of the form*
>
> $$kx^n$$
>
> *where n is a whole number and k is any real number.*
> *n is called the **degree** of the monomial, and k is the **coefficient**.*

A monomial may be in more than one variable. For example, $7x^2y$ is a monomial in x and y. The **degree of a monomial in more than one variable** is the sum of the exponents of its variables. Thus, $7x^2y$ is third-degree ($2 + 1 = 3$) in x and y, and $8a^2b^3$ is fifth-degree in a and b.

In the case of a constant monomial, such as 5, we can write $5 = 5 \cdot 1 = 5 \cdot x^0$. Therefore, a nonzero constant is considered to be a monomial of **degree 0**. Because there is more than one way of writing 0 with a variable, as $0 = 0x = 0x^3 = 0x^{16}$, 0 is considered to be a monomial of **no degree**.

Polynomials

Any monomial or algebraic sum or difference of monomials is a **polynomial**. For example:

Polynomials: $18,\ 3x^2 + 2,\ 5x^2y + 4xy^2,$ and $3x - 5.4w^2 - \dfrac{3}{2}xy$

Not polynomials: $2x^{\frac{1}{3}} + 1,\ \dfrac{x}{y},\ \dfrac{1}{x-2},$ and $x^{-1} + 5x$

In general, expressions with variables in the denominator, or variables with fractional or negative exponents are not polynomials.

Polynomial

*A **polynomial** is a monomial or the algebraic sum or difference of monomials.*

*The **degree of a polynomial** is the largest of the degrees of its terms after like terms have been combined.*

*The coefficient of the term of largest degree is called the **leading coefficient**.*

For consistency and easy identification, we will write polynomials in the generally accepted form with terms in **descending order of degree** from left to right. Thus the polynomial

$$3x + 5x^2 - 10 + \frac{1}{3}x^3 \text{ will be written as } \frac{1}{3}x^3 + 5x^2 + 3x - 10.$$

You will find this ordering technique very helpful when performing operations with polynomials.

Polynomials with one, two, or three terms (**after they have been simplified**) are classified as follows.

Classification of Polynomials

Term	Description	Example
Monomial:	*polynomial with one term*	$15x^3$ *(third-degree monomial)*
Binomial:	*polynomial with two terms*	$4x - 10$ *(first-degree binomial)*
Trinomial:	*polynomial with three terms*	$-x^4 + 2x - 1$ *(fourth-degree trinomial)*

No special name is given to polynomials with more than three terms. They are referred to simply as polynomials. Of course, monomials, binomials, and trinomials can be referred to as polynomials as well.

A polynomial in one variable can be classified in reference to its degree as follows:

If a polynomial is of
  a. degree 0 or 1, it is called a **linear** polynomial,
  b. degree 2, it is called a **quadratic** polynomial,
  c. degree 3, it is called a **cubic** polynomial.

For example,

$3x + 8$ is a linear polynomial (first-degree binomial, leading coefficient 3).

$5x^2 - 3x + 7$ is a quadratic polynomial (second-degree trinomial, leading coefficient 5).

$-10y^3$ is a cubic polynomial (third-degree monomial, leading coefficient -10).

Addition with Polynomials

The **sum** of two or more polynomials can be found by combining like terms. (See Section 1.4.)

Example 1: Addition with Polynomials

Simplify each of the following expressions.

a. $(x^3 - 2x + 1) + (3x^2 + 4x + 5)$

> **Solution:** $(x^3 - 2x + 1) + (3x^2 + 4x + 5) = x^3 - 2x + 1 + 3x^2 + 4x + 5$
> $$= x^3 + 3x^2 - 2x + 4x + 1 + 5$$
> $$= x^3 + 3x^2 + 2x + 6$$

b. $(2x^2 + 4x - 7) + (x^2 + 6x + 8)$

> **Solution:** $(2x^2 + 4x - 7) + (x^2 + 6x + 8) = 2x^2 + 4x - 7 + x^2 + 6x + 8$
> $$= 2x^2 + x^2 + 4x + 6x - 7 + 8$$
> $$= 3x^2 + 10x + 1$$

As illustrated in Example 2, polynomials can be added vertically with like terms aligned. The polynomials are written in descending order and a 0 is written as a placeholder for any missing powers of the variable.

Example 2: Addition with Polynomials in Vertical Format

Write the sum $(5x^3 - 9x^2 - 10x + 12) + (3x^3 + 6x^2 - 7)$ in a vertical format and evaluate.

Solution:

$$
\begin{array}{r}
5x^3 - 9x^2 - 10x + 12 \\
3x^3 + 6x^2 + \ 0x - \ 7 \\
\hline
8x^3 - 3x^2 - 10x + \ 5
\end{array}
$$

Note that $0x$ is written as a placeholder.

Subtraction with Polynomials

To find the **difference** of two polynomials, either

a. add the opposite of each term being subtracted, or equivalently,

b. use the distributive property and multiply each term being subtracted by –1, then add.

Example 3: Subtraction with Polynomials

Find the difference in simplest form: $(x^2y + 3y - 4x) - (2x^2y - 7x)$.

Solution:

a. Add the opposites of the terms being subtracted.

$$
\begin{aligned}
(x^2y + 3y - 4x) - (2x^2y - 7x) &= x^2y + 3y - 4x + (-2x^2y) + (7x) \\
&= \underbrace{x^2y - 2x^2y} + 3y \underbrace{- 4x + 7x} \\
&= \ -x^2y \ + \ 3y \ + \ 3x
\end{aligned}
$$

b. Multiply each term being subtracted by –1.

$$
\begin{aligned}
(x^2y + 3y - 4x) - (2x^2y - 7x) &= (x^2y + 3y - 4x) + (-1)(2x^2y - 7x) \\
&= x^2y + 3y - 4x + (-2x^2y + 7x) \\
&= \underbrace{x^2y - 2x^2y} + 3y \underbrace{- 4x + 7x} \\
&= \ -x^2y \ + \ 3y \ + \ 3x
\end{aligned}
$$

Adding the opposites of the terms or multiplying by –1 and then adding both have the effect of changing the sign of each term in the polynomial being subtracted. Subtraction can be performed using either the horizontal or vertical format.

Example 4: Subtraction with Polynomials

a. Find the difference $(x^2 + 12x - 23) - (2x^2 + 7x - 20)$.

Solution: Change the sign of each term in the polynomial being subtracted and combine like terms.

$$(x^2 + 12x - 23) - (2x^2 + 7x - 20) = x^2 + 12x - 23 - 2x^2 - 7x + 20$$
$$= -x^2 + 5x - 3$$

If the polynomials are written in a vertical format, one beneath the other, we change the signs of the terms of the polynomial being subtracted, and then combine like terms.

$$\begin{array}{r} x^2 + 12x - 23 \\ -(2x^2 + 7x - 20) \\ \hline \end{array} \quad \rightarrow \quad \begin{array}{r} x^2 + 12x - 23 \\ -2x^2 - 7x + 20 \\ \hline -x^2 + 5x - 3 \end{array}$$

b. Find the difference $(6x^4 + 2x^3 - 4x^2 - 8) - (3x^4 + 5x^3 - x^2 + 6x + 10)$ by writing the terms in a vertical format and changing the signs of the polynomial being subtracted.

Solution:

$$\begin{array}{r} 6x^4 + 2x^3 - 4x^2 + 0x - 8 \\ -(3x^4 + 5x^3 - x^2 + 6x + 10) \\ \hline \end{array} \quad \rightarrow \quad \begin{array}{r} 6x^4 + 2x^3 - 4x^2 + 0x - 8 \\ -3x^4 - 5x^3 + x^2 - 6x - 10 \\ \hline 3x^4 - 3x^3 - 3x^2 - 6x - 18 \end{array}$$

Note that $0x$ is written as a placeholder.

$P(x)$ Notation and Evaluation of Polynomials

Function notation, such as $f(x)$, introduced in Section 2.4, is convenient to use in evaluating polynomials. $P(x)$ [read "P of x"] indicates that $\boldsymbol{P}$ is the name of the polynomial and $\boldsymbol{x}$ is the variable used in the polynomial. With this notation, $P(3)$ [read "P of 3"] indicates that the value of the polynomial P is to be calculated by substituting 3 for x throughout the expression and then **following the rules for order of operations**.

If a polynomial is in more than one variable, such as x and y, then the notation can be written as $P(x, y)$ [read "P of x and y"]. **Be sure to put negative numbers in parentheses when substituting them into the equation as illustrated in Example 5b.**

Example 5: Evaluation of Polynomials

a. For the polynomial $P(x) = x^3 - 2x^2 + 3x + 5$, find $P(4)$.

Solution: Substitute 4 for x throughout the polynomial.

$$P(4) = 4^3 - 2 \cdot 4^2 + 3 \cdot 4 + 5 = 64 - 32 + 12 + 5$$
$$= 32 + 12 + 5$$
$$= 49$$

b. Evaluate the polynomial $P(x, y) = 2x^2y - xy + 3x - 4y + 15$ for $x = -1$ and $y = -6$.

Solution: Substitute -1 for x and -6 for y throughout the polynomial.

$$P(-1, -6) = 2(-1)^2(-6) - (-1)(-6) + 3(-1) - 4(-6) + 15$$
$$= -12 - 6 - 3 + 24 + 15$$
$$= -21 + 39$$
$$= 18$$

Practice Problems

Add or subtract as indicated and simplify the result.

1. $(3x^2 - 2x + 5) + (2x^2 - x + 3)$ **2.** $(x^3 - 2x^2) - (x^2 - 1)$

3. $(5x^2 - 9x - 11) - (x^2 - 3x + 1)$

4. For $P(x) = x^3 - 8x^2 - 5x + 10$, find $P(-2)$.

Answers to Practice Problems: **1.** $5x^2 - 3x + 8$ **2.** $x^3 - 3x^2 + 1$ **3.** $4x^2 - 6x - 12$ **4.** -20

4.2 Exercises

In Exercises 1 – 12, state whether the expression is or is not a polynomial. If the expression is a polynomial, state its degree, its classification as a monomial, binomial, or trinomial, and its leading coefficient.

1. $x^3 - x^2$

2. 9

3. $-3x^{\frac{1}{2}} + x$

4. $x^4 + 8x^3 - y^2$

5. $\frac{1}{2}y^2 + \frac{5}{4}y^3 - \frac{7}{4}y$

6. $x^2 + y^2 - \frac{1}{y}$

7. 0

8. $-\sqrt{2}$

9. $\left(x^5 - y^3\right)^{\frac{1}{2}}$

10. $\frac{3}{2}x^2 - \sqrt{3}x - 7$

11. $7x^2 - 6x + 9x^{\frac{2}{3}}$

12. $\frac{x^3 - 3y^2}{x}$

In Exercises 13 – 35, find the indicated sums and differences. Simplify each answer.

13. $\left(3x^2 - 5x + 1\right) + \left(x^2 + 2x - 7\right)$

14. $\left(5x^2 + 8x - 3\right) + \left(-2x^2 + 6x - 4\right)$

15. $\left(x^2 - 9x + 2\right) + \left(-x^2 + 2x - 8\right)$

16. $\left(7x^2 - 4x + 6\right) + \left(4x^2 - 2x + 5\right)$

17. $\left(x^2 + y^2\right) + \left(2x^2 - 5y^2\right)$

18. $\left(x^2 - 3xy + y^2\right) + \left(2x^2 - 5xy - y^2\right)$

19. $\left(2x^2 + 3x + 8\right) - \left(x^2 + 4x - 2\right)$

20. $\left(6x^3 - 5x + 1\right) - \left(2x^3 + 3x - 4\right)$

21. $\left(2x^4 + 3x\right) - \left(5x^3 + 4x + 3\right)$

22. $\left(2x^3 - 3x^2 + 6\right) - \left(x^4 + x + 1\right)$

23. $\left(5x^2 + 6x - 1\right) + \left(x^4 - 3x^2 + 2x\right)$

24. $\left(7x^2 - 2xy + 3y^2\right) + \left(-3x^2 - 2xy + 5y^2\right)$

25. $\left(4x^3 - 7x^2 + 3x + 2\right) - \left(-2x^3 - 5x - 1\right)$

26. $\left(4x^2 - 8xy - 2y^2\right) + \left(-9x^2 + 5xy - 6y^2\right)$

27. $\left(3x^2 - 2y^2\right) + \left(7xy + 4y^2\right) - \left(-6x^2 - 6xy + 8y^2\right)$

28. $\left(9xy + 8y^2\right) - \left(6x^2 - 8xy\right) + \left(5x^2 - 3xy + 7y^2\right)$

29. $\left(5x^3 - 14x^2\right) - \left(5x^2 + 2x + 1\right) - \left(-7x^3 + 2x^2 - 13\right)$

30. $\left(7x^3 + 4x^2 - x\right) + \left(3x^3 - 4x + 5\right) - \left(8x^3 + x^2 - x + 3\right)$

31. $x^3 - \left[3x^2 - 1 - \left(x^3 + 4x^2 + 1\right)\right] + \left(3x^3 - 3x^2 - 2\right)$

32. $3x - 4xy + \left[6y + \left(4x + 3xy + 2y\right)\right] - \left[-6x - \left(xy - 4y\right)\right]$

33. $x^2 - 2xy + \left[y^2 - \left(3xy + 2y^2\right) - \left(3x^2 - xy - 2y^2\right)\right]$

34. $\left[\left(4x^2 - 3x\right) - \left(2x^2 + 5x\right)\right] + \left[\left(x^2 - 6x\right) + \left(-3x^2 + x\right)\right]$

35. $\left[\left(2x + xy - y\right) + \left(x - 2xy + 4y\right)\right] - \left[\left(-3x + 5xy + y\right) - \left(2x + 3xy - 2y\right)\right]$

Find each sum in Exercises 36 – 40.

36. $\begin{array}{r} 2x^2 - 5x - 6 \\ -3x^2 + 2x - 1 \end{array}$

37. $\begin{array}{r} x^3 + 2x^2 + x - 2 \\ x^3 - 2x^2 - 3x - 1 \end{array}$

38. $\begin{array}{r} 5x^3 - 4x^2 \quad\ - 9 \\ 2x^3 - 3x^2 - 6x + 5 \end{array}$

39. $\begin{array}{r} 3x^4 + 3x^3 + x^2 + x + 2 \\ 7x^4 - x^3 - 5x^2 + x - 1 \end{array}$

40. $\begin{array}{r} 14x^3 + 13x^2 + 10x - 13 \\ 20x^3 \qquad\ - 18x + 25 \end{array}$

Find each difference in Exercises 41 – 45.

41. $\begin{array}{r} 9x^2 - 2x + 3 \\ -\left(4x^2 + 5x - 2\right) \end{array}$

42. $\begin{array}{r} -3x^2 + 7x - 6 \\ -\left(2x^2 - x + 6\right) \end{array}$

43. $\begin{array}{r} 5x^3 \qquad -10x + 15 \\ -\left(x^3 - 4x^2 - 3x - 9\right) \end{array}$

44. $\begin{array}{r} x^3 - 8x^2 + 12x + 5 \\ -\left(-3x^3 + 8x^2 + 2x + 5\right) \end{array}$

45. $\begin{array}{r} 2x^4 - 5x^3 - 6x^2 + 7x + 7 \\ -\left(x^4 \qquad + 2x^2 + 4x + 10\right) \end{array}$

Evaluate each polynomial in Exercises 46 – 55 for the specified value(s) of the variable(s).

46. Given $P(x) = 2x^2 - x + 3$; find $P(1)$.

47. Given $P(x) = 3x^2 - 2x + 5$; find $P(2)$.

48. Given $P(x) = 3 - x^2$; find $P(-2)$.

49. Given $P(x) = x^3 - 2x^2 + x - 1$; find $P(2)$.

50. Given $P(x) = x^3 + x^2 - 4$; find $P(-3)$.

51. Given $P(x) = 4x^3 - 2x^2 - 1$; find $P(-4)$.

52. Given $P(x, y) = 2x^2 - 3xy + y^2$; find $P(2, -2)$.

53. Given $P(x, y) = 4x - 2xy + 5y$; find $P(1, 1)$.

54. Given $P(x, y, z) = 3x + 4xy - 2yz + z$; find $P(1, 0, 2)$.

55. Given $P(x, y, z) = 2xyz - 3x + yz - xz$; find $P(2, -1, 2)$.

56. Use a graphing calculator to graph the following linear functions.

    a. $P(x) = -2x + 5$ **b.** $P(x) = \dfrac{1}{4}x$

57. Use a graphing calculator to graph the following quadratic functions.

    a. $P(x) = -x^2$ **b.** $P(x) = x^2 - 4x + 4$

58. Use a graphing calculator to graph the following cubic functions.

    a. $P(x) = x^3$ **b.** $P(x) = x^3 - 4x$

Writing and Thinking About Mathematics

59. Write the definition of a polynomial.

60. Explain, in your own words, how to subtract one polynomial from another.

61. Describe what is meant by the degree of a polynomial in x.

62. Give two examples that show how the sum of two binomials might not be a binomial.

HAWKES LEARNING SYSTEMS: INTERMEDIATE ALGEBRA SOFTWARE

- Identifying Polynomials
- Adding and Subtracting Polynomials

4.3

Multiplication with Polynomials

- *Multiply polynomials using the distributive property.*
- *Multiply two binomials using the FOIL method.*
- *Multiply binomials, finding products that are the difference of squares.*
- *Square binomials, finding products that are perfect square trinomials.*

Multiplication of a Polynomial by a Monomial

Using the distributive property $a(b+c) = ab + ac$ with multiplication indicated on the left, we can find the product of a monomial and a polynomial of two or more terms as follows:

$$5x(2x^3 + 3) = 5x \cdot 2x^3 + 5x \cdot 3$$

$$= 10x^4 + 15x$$

and

$$3x^2(x^2 - 5x + 1) = 3x^2 \cdot x^2 + 3x^2(-5x) + 3x^2 \cdot 1$$

$$= 3x^4 - 15x^3 + 3x^2.$$

Multiplication with Two Polynomials

To find the product of two binomials, such as $(x+3)(x+8)$, we can apply the distributive property in the following way:

Compare $(x+3)(x+8)$ to

$$(b+c)a = ba + ca$$

and think of $(x + 8)$ as taking the place of a as follows:

$$(b+c)a = ba + ca$$

$$(x+3)(x+8) = x(x+8) + 3(x+8)$$

Continue to apply the distributive property.

$$= x \cdot x + x \cdot 8 + 3 \cdot x + 3 \cdot 8$$

$$= x^2 + 8x + 3x + 24$$

$$= x^2 + 11x + 24.$$

Similarly,

$$(2x-1)(x^2 + x - 4) = 2x(x^2 + x - 4) - 1(x^2 + x - 4)$$

$$= 2x \cdot x^2 + 2x \cdot x + 2x(-4) - 1 \cdot x^2 - 1 \cdot x - 1(-4)$$

$$= 2x^3 + 2x^2 - 8x - x^2 - x + 4$$

$$= 2x^3 + x^2 - 9x + 4.$$

The product of two polynomials can also be found by writing one polynomial under the other. **The distributive property is applied by multiplying each term of one polynomial by each term of the other.** Consider the product $(2x^2 + 4x - 3)(3x + 5)$. Now, writing one polynomial under the other and applying the distributive property, we have:

Multiply by $+5$:   Multiply by $3x$:

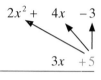   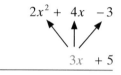

$$10x^2 + 20x - 15 \qquad\qquad 10x^2 + 20x - 15 \quad \text{Align the like terms so that}$$
$$6x^3 + 12x^2 - 9x \qquad \text{they can be easily combined.}$$

Finally, combine like terms:

$$
\begin{array}{r}
2x^2 + 4x - 3 \\
3x + 5 \\
\hline
10x^2 + 20x - 15 \\
6x^3 + 12x^2 - 9x \\
\hline
6x^3 + 22x^2 + 11x - 15
\end{array}
$$
    Combine like terms.

Example 1: Multiplication with Polynomials

Find each product.

a. $-6x(2x^2 + 3x - 11)$

     Solution: $-6x(2x^2 + 3x - 11) = -6x \cdot 2x^2 - 6x \cdot 3x - 6x(-11)$

$$= -12x^3 - 18x^2 + 66x$$

b. $x^{3k}(x^k + x)$

     Solution: $x^{3k}(x^k + x) = x^{3k} \cdot x^k + x^{3k} \cdot x$

$$= x^{3k+k} + x^{3k+1}$$
$$= x^{4k} + x^{3k+1}$$

c. $(2x + 1)(x - 5)$

     Solution: $(2x + 1)(x - 5) = 2x(x - 5) + 1(x - 5)$

$$= 2x \cdot x + 2x(-5) + 1x + 1(-5)$$
$$= 2x^2 - 10x + x - 5$$
$$= 2x^2 - 9x - 5$$

d. $(2x+3)(5x^2+4x-5)$

> **Solution:** We can arrange the polynomials in a vertical format and multiply by using the distributive property. Be sure to align like terms in the partial products.

$$
\begin{array}{r}
5x^2 + 4x - 5 \\
2x + 3 \\
\hline
15x^2 + 12x - 15 \\
10x^3 + 8x^2 - 10x \\
\hline
10x^3 + 23x^2 + 2x - 15
\end{array}
$$

Multiply by 3.

Multiply by $2x$.

Combine like terms.

e. $x^2 + 2x + 3$

$x^2 - 2x + 3$

> **Solution:**

$$
\begin{array}{r}
x^2 + 2x + 3 \\
x^2 - 2x + 3 \\
\hline
3x^2 + 6x + 9 \\
-2x^3 - 4x^2 - 6x \\
x^4 + 2x^3 + 3x^2 \\
\hline
x^4 \qquad + 2x^2 \qquad + 9
\end{array}
$$

Multiply by 3.

Multiply by $-2x$.

Multiply by x^2.

Combine like terms.

The FOIL Method

In the case of the **product of two binomials** such as $(2x+5)(3x-7)$, the **FOIL** method is useful. **F-O-I-L** is a mnemonic device (memory aid) to help in remembering which terms of the binomials to multiply together. First, by using the distributive property we can see how the terms are multiplied.

$$(2x+5)(3x-7) = 2x(3x-7) + 5(3x-7)$$

$$= 2x \cdot 3x + 2x \cdot (-7) + 5 \cdot 3x + 5 \cdot (-7)$$

First	Outside	Inside	Last
terms	terms	terms	terms
F	O	I	L

Now, we can use the FOIL method and then combine like terms to go directly to the answer.

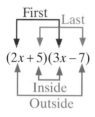

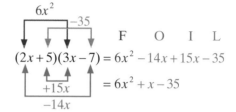

Example 2: FOIL Method

Use the FOIL method to find the products of the given binomials.

a. $(x+3)(2x+8)$

> **Solution:**
>
> $$(x+3)(2x+8) = 2x^2 + 8x + 6x + 24$$
> $$= 2x^2 + 14x + 24$$

b. $(2x-3)(3x-5)$

> **Solution:**
>
> $$(2x-3)(3x-5) = 6x^2 - 10x - 9x + 15$$
> $$= 6x^2 - 19x + 15$$

c. $(x+7)(x-7)$

> **Solution:**
>
> $$(x+7)(x-7) = x^2 - 7x + 7x - 49 \qquad \text{Apply the FOIL method mentally.}$$
>
> $$= x^2 - 49 \qquad \text{Note that in this special case the two middle terms are opposites and their sum is 0.}$$

The Difference of Two Squares: $(x + a)(x - a) = x^2 - a^2$

In Example 2c, the middle terms, $-7x$ and $+7x$, are opposites of each other and their sum is 0. Therefore, the resulting product has only two terms and each term is a square.

$$(x+7)(x-7) = x^2 - 49$$

In fact, when two binomials are in the form of the sum and difference of the same two terms, the product will always be the difference of the squares of the terms. The product is called the **difference of two squares**.

Difference of Two Squares

$$(x + a)(x - a) = x^2 - a^2$$

Example 3: Difference of Two Squares

Find each product.

a. $(x+4)(x-4)$

Solution: The two binomials represent the sum and difference of x and 4. So, the product is the difference of their squares.

$$(x+4)(x-4) = x^2 - 4^2 = x^2 - 16 \qquad \text{Difference of two squares}$$

b. $(3y+7)(3y-7)$

Solution:

$$(3y+7)(3y-7) = (3y)^2 - 7^2 = 9y^2 - 49 \qquad \text{Difference of two squares}$$

c. $(x^3 - 6)(x^3 + 6)$

Solution:

$$(x^3 - 6)(x^3 + 6) = (x^3)^2 - 6^2 = x^6 - 36 \qquad \text{Difference of two squares}$$

d. $(y^k + 3)(y^k - 3)$

Solution:

$$(y^k + 3)(y^k - 3) = (y^k)^2 - 3^2 = y^{2k} - 9 \qquad \text{Difference of two squares}$$

Perfect Square Trinomials: $\begin{cases} (x+a)^2 = x^2 + 2ax + a^2 \\ (x-a)^2 = x^2 - 2ax + a^2 \end{cases}$

Now we consider the case where the two binomials being multiplied are **the same**. That is, we want to consider the **square of a binomial**. The following examples, using the distributive property, illustrate two patterns that, after some practice, allow us to go directly to the products.

$$(x+3)^2 = (x+3)(x+3) = x^2 + 3x + 3x + 9$$

$$= x^2 + 2 \cdot 3x + 9 \qquad \text{The middle term is doubled.}$$

$$= x^2 + 6x + 9 \qquad \text{Perfect square trinomial}$$

$$(x-11)^2 = (x-11)(x-11) = x^2 - 11x - 11x + 121$$

$$= x^2 - 2 \cdot 11x + 121 \qquad \text{The middle term is doubled.}$$

$$= x^2 - 22x + 121 \qquad \text{Perfect square trinomial}$$

Note that in each case **the result of squaring the binomial is a trinomial**. These trinomials are called **perfect square trinomials**.

Squares of Binomials (Perfect Square Trinomials)

$$(x+a)^2 = x^2 + 2ax + a^2 \qquad \textit{Square of a binomial sum}$$

$$(x-a)^2 = x^2 - 2ax + a^2 \qquad \textit{Square of a binomial difference}$$

Example 4: Squares of Binomials (Perfect Square Trinomials)

Find each product.

a. $(5x+1)^2$

Solution: $(5x+1)^2 = (5x)^2 + 2 \cdot 5x \cdot 1 + 1^2 = 25x^2 + 10x + 1$

b. $(y^3 + 3)^2$

Solution: $(y^3 + 3)^2 = (y^3)^2 + 2 \cdot y^3 \cdot 3 + 3^2 = y^6 + 6y^3 + 9$

c. $(10 - x)^2$

Solution: $(10 - x)^2 = 10^2 - 2 \cdot 10 \cdot x + x^2 = 100 - 20x + x^2$ $\left(\text{or } x^2 - 20x + 100\right)$

d. $\left[(x + 3) - y\right]^2$

Solution: $\left[(x + 3) - y\right]^2 = (x + 3)^2 - 2 \cdot (x + 3) \cdot y + y^2$

$$= x^2 + 2 \cdot x \cdot 3 + 3^2 - 2xy - 6y + y^2$$

$$= x^2 + 6x + 9 - 2xy - 6y + y^2$$

NOTES

COMMON ERROR

For products raised to a power, we have $(ab)^n = a^n b^n$ and $(ab)^2 = a^2 b^2$. However, this rule does not apply to sums. In particular, it does **not** apply to binomials.

$$(a + b)^2 \neq a^2 + b^2$$

$$(a - b)^2 \neq a^2 - b^2$$

Remember, the squares of binomials are trinomials:

$$(a + b)^2 = (a + b)(a + b) = a^2 + 2ab + b^2$$

and

$$(a - b)^2 = (a - b)(a - b) = a^2 - 2ab + b^2.$$

Practice Problems

Find the indicated products and simplify if possible.

1. $4x^5 \left(2x^2 - 3x + 7\right)$ **2.** $a^2 \left(a^4 - 3a^2 - 5\right)$ **3.** $(x + 6)(x - 3)$

4. $\left(a^2 - 9\right)\left(a^2 + 9\right)$ **5.** $(2y + 5)^2$ **6.** $\left(x - \dfrac{1}{3}\right)^2$

7. $(x + 5)\left(x^2 - 5x + 25\right)$ **8.** $x^k \left(x^k - 1\right)$ **9.** $\quad 8x^2 + 3x + 2$

$$\underline{\qquad\qquad 2x + 7}$$

Answers to Practice Problems: **1.** $8x^7 - 12x^6 + 28x^5$ **2.** $a^6 - 3a^4 - 5a^2$ **3.** $x^2 + 3x - 18$ **4.** $a^4 - 81$

5. $4y^2 + 20y + 25$ **6.** $x^2 - \dfrac{2x}{3} + \dfrac{1}{9}$ **7.** $x^3 + 125$ **8.** $x^{2k} - x^k$

9. $16x^3 + 62x^2 + 25x + 14$

4.3 Exercises

In Exercises 1 – 62, find the indicated products and simplify, if possible.

1. $5x\left(x^2 - 2x + 3\right)$ **2.** $2x^2\left(3x^2 + 5x - 1\right)$ **3.** $xy^2\left(x^2 + 4y\right)$

4. $x^2z\left(x - 4y + z\right)$ **5.** $(x + 3)(x - 6)$ **6.** $(x - 2)(x - 5)$

7. $(x - 8)(x - 1)$ **8.** $(x + 2)(x + 4)$ **9.** $(2y + 1)(y - 6)$

10. $(y + 5)(3y + 2)$ **11.** $(3x - 4)(x - 5)$ **12.** $(2x - 1)(x - 2)$

13. $(2y + 3)(3y + 2)$ **14.** $(5y - 2)(3y + 1)$ **15.** $(8x + 3)(x - 5)$

16. $(7x + 6)(2x - 3)$ **17.** $(9x + 1)(3x - 2)$ **18.** $(5x - 11)(3x + 4)$

19. $(3x + 1)^2$ **20.** $(4x - 3)^2$ **21.** $(5x - 2y)^2$

22. $(7x + 4y)^2$ **23.** $(4x + 7)(4x - 7)$ **24.** $(3x + 5)(3x - 5)$

25. $(2x - 3y)(2x + 3y)$ **26.** $(6x - y)(6x + y)$ **27.** $x\left(3x^2 - 4\right)\left(3x^2 + 4\right)$

28. $3x\left(7x^2 + 8\right)\left(7x^2 - 8\right)$ **29.** $(x - 1)\left(x^2 + x + 1\right)$ **30.** $(y + 4)\left(y^2 - 4y + 16\right)$

31. $(x + 3)\left(x^2 + 6x + 9\right)$ **32.** $(y - 5)\left(y^2 + 3y + 2\right)$ **33.** $\left(x^3 + 2\right)^2$

34. $\left(2x^3 - 3\right)^2$ **35.** $\left(2x^3 - 7\right)\left(2x^3 + 7\right)$

36. $(x + 2y)\left(x^2 - 2xy + 4y^2\right)$ **37.** $(x - 3y)\left(x^2 + 3xy + 9y^2\right)$

38. $\left(8y^2 - 7\right)\left(3y^2 + 2\right)$ **39.** $\left(x^2 + 6y^2\right)\left(x^2 - 6y^2\right)$

40. $\left(x^2 - 6y^2\right)\left(x^2 + 3y^2\right)$ **41.** $\left(5x^2 + y^2\right)\left(2x^2 - 3y^2\right)$

42. $(x - 2y)\left(x^2 + 2xy + 4y^2\right)$ **43.** $\big[(x + y) + 2\big]\big[(x + y) - 2\big]$

44. $\big[(x + 1) + y\big]\big[(x + 1) - y\big]$ **45.** $\big[(5x - y) + 3\big]^2$

46. $\big[(2x + 1) - y\big]^2$ **47.** $\big[(x + 4) - 2y\big]^2$ **48.** $\big[(x - 3y) + 5\big]^2$

49. $x^2(x^k + 3)$

50. $x^3(x^{2k} + x)$

51. $(x^k + 3)(x^k - 5)$

52. $(x^k + 6)(x^k - 6)$

53. $(x^k + 1)(x^k + 4)$

54. $(2x^k - 3)(x^k + 2)$

55. $(3x^k + 2)(x^k + 5)$

56. $\left(x + \dfrac{1}{4}\right)\left(x - \dfrac{1}{4}\right)$

57. $\left(x + \dfrac{5}{8}\right)\left(x - \dfrac{5}{8}\right)$

58. $\left(x + \dfrac{2}{3}\right)^2$

59. $\left(y - \dfrac{1}{5}\right)^2$

60. $\left(y + \dfrac{1}{4}\right)\left(y - \dfrac{3}{4}\right)$

61. $(x + 2.5)(x - 2.5)$

62. $(x + 2.1)^2$

Find the indicated products and simplify in Exercises 63 – 68.

63. $2x^2 - 5x - 6$
$\underline{\ 3x + 1}$

64. $x^2 + 2x + 1$
$\underline{x^2 + 2x + 1}$

65. $x^3 - 3x + 4$
$\underline{\ 2x - 3}$

66. $2x^3 + 6x^2 + 5$
$\underline{\ x^2\ + 5}$

67. $x^3 - 7x - 4$
$\underline{\ 4x - 6}$

68. $x^3 - 5x + 14$
$\underline{\ 2x - 3}$

69. Probability: In the case of binomial probabilities, if x is the probability of success in one trial of an event, then the expression $P(x) = 10x^3(1 - x)^2$ is the probability of 3 successes in 5 trials where $0 \le x \le 1$.

a. Represent the expression $P(x)$ as a single polynomial function.

b. If a fair coin is tossed, the probability of heads occurring is $\dfrac{1}{2}$. That is, $x = \dfrac{1}{2}$. Find the probability of exactly 3 heads occurring in 5 tosses.

c. A basketball player is known to make 80% of his free throws. What is the probability that he will make exactly 3 of his next 5 attempts?

70. Geometry: A square is 30 inches on each side. A small square, x inches on each side, is cut from each corner of the original square.

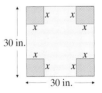

a. Represent the area of the remaining portion of the square in the form of a polynomial function $A(x)$.

b. Represent the perimeter of the remaining portion of the square in the form of a polynomial function $P(x)$.

71. Swimming Pools: A swimming pool, 25 meters by 50 meters, is surrounded by a concrete deck that is x meters wide.

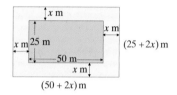

a. Represent the area covered by the deck and the pool in the form of a polynomial function.

b. Represent the area covered by the deck only in the form of a polynomial function.

72. Geometry: A rectangle has sides $(x + 5)$ ft and $(x + 10)$ ft. A square x feet on each side is cut from one corner of the rectangle.

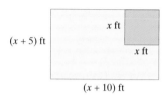

a. Represent the remaining area (light area in the figure shown) in the form of a polynomial function $A(x)$.

b. Represent the perimeter of the remaining figure (after the square in the corner has been removed) in the form of a polynomial function $P(x)$.

Writing and Thinking About Mathematics

73. A square with sides of length $(x + 5)$ can be broken up as shown in the diagram. The sums of the areas of the interior rectangles and squares is equal to the total area of the square: $(x+5)^2$. Show how this fits with the formula for the square of a sum.

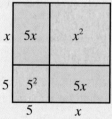

HAWKES LEARNING SYSTEMS: INTERMEDIATE ALGEBRA SOFTWARE

- Multiplying Polynomials
- The FOIL Method

Division with Polynomials

4.4

- *Divide a polynomial by a monomial.*
- *Divide polynomials using the division algorithm.*

Fractions, such as $\dfrac{127}{2}$ and $\dfrac{1}{8}$, in which the numerator and denominator are integers are called rational numbers. Fractions in which the numerator and denominator are polynomials are called **rational expressions**. (No denominator can be 0.)

Rational Expressions: $\dfrac{x^3 - 6x^2 + 2x}{3x}$, $\dfrac{6x^2 - 7x - 2}{2x - 1}$, $\dfrac{5x^3 - 3x + 1}{x + 3}$, and $\dfrac{1}{10x}$.

In this section, we will treat a rational expression as a division problem. With this basis, there are two situations to consider:

1. the denominator (divisor) is a monomial, or
2. the denominator (divisor) is not a monomial.

Division by a Monomial

We know that the sum of fractions with the same denominator can be written as a single fraction by adding the numerators and using the common denominator. For example,

$$\frac{5}{a} + \frac{3b}{a} + \frac{2c}{a} = \frac{5 + 3b + 2c}{a}.$$

If instead of adding the fractions, we start with the sum and want to divide the numerator by the denominator (with a monomial in the denominator), we divide each term in the numerator by the monomial denominator and simplify each fraction.

$$\frac{4x^3 + 8x^2 - 12x}{4x} = \frac{4x^3}{4x} + \frac{8x^2}{4x} - \frac{12x}{4x} = x^2 + 2x - 3$$

Example 1: Division by a Monomial

Divide each polynomial by the **monomial denominator** by writing each fraction as the sum (or difference) of fractions. Simplify each fraction, if possible.

a. $\dfrac{x^3 - 6x^2 + 2x}{3x}$

Solution: $\dfrac{x^3 - 6x^2 + 2x}{3x} = \dfrac{x^3}{3x} - \dfrac{6x^2}{3x} + \dfrac{2x}{3x} = \dfrac{x^2}{3} - 2x + \dfrac{2}{3}$

b. $\dfrac{15y^3 - 20y^2 + 5y}{5y^2}$

Solution: $\dfrac{15y^3 - 20y^2 + 5y}{5y^2} = \dfrac{15y^3}{5y^2} - \dfrac{20y^2}{5y^2} + \dfrac{5y}{5y^2} = 3y - 4 + \dfrac{1}{y}$

The Division Algorithm

In arithmetic, the **division algorithm** (called **long division**) is the process (or series of steps) that we follow when dividing two numbers. By this division algorithm, we can find $64 \div 5$ as follows.

$$
\begin{array}{r}
12 \quad \longleftarrow \text{ Quotient} \\
5\overline{)64} \quad \longleftarrow \text{ Dividend} \\
\underline{5} \quad \longleftarrow \text{ Subtract} \\
14 \\
\underline{10} \quad \longleftarrow \text{ Subtract} \\
4 \quad \longleftarrow \text{ 4 is the remainder (The remainder is} \\
\text{always smaller than the divisor.)}
\end{array}
$$

Divisor $\longrightarrow$

Check: $5 \cdot 12 + 4 = 60 + 4 = 64$ (Multiply the divisor times the quotient and add the remainder. The result should be the original dividend.)

We can also write the division in fraction form with the remainder over the divisor, giving a mixed number.

$$64 \div 5 = \frac{64}{5} = 12 + \frac{4}{5} = 12\frac{4}{5}$$

In algebra, **the division algorithm with polynomials** is quite similar. In dividing one polynomial by another, with the degree of the divisor smaller than the degree of the dividend, the quotient will be another polynomial with a remainder. Symbolically, if we have $P \div D$, then

$$P = Q \cdot D + R$$

or

$$\frac{P}{D} = Q + \frac{R}{D}$$

where
P is the dividend,
D is the divisor,
Q is the quotient, and
R is the remainder.

The remainder must be of smaller degree than the divisor. If the remainder is 0, then the divisor and quotient are factors of the dividend.

The Division Algorithm

*For polynomials P and D, the **division algorithm** gives*

$$\frac{P}{D} = Q + \frac{R}{D}, \ D \neq 0 \ \left[\text{or, in function notation, } \frac{P(x)}{D(x)} = Q(x) + \frac{R(x)}{D(x)}, \ D(x) \neq 0 \right]$$

*where Q and R are polynomials and the **degree of R** < degree of D.*

The actual process of long division is not clear from this abstract definition. Although you are familiar with the process of long division with decimal numbers, this same procedure appears more complicated with polynomials. The **division algorithm** (**long division**) is illustrated in a step-by-step form in the following example. Study it carefully.

Example 2: The Division Algorithm

Simplify $\dfrac{6x^2 - 7x - 2}{2x - 1}$ by using long division.

Solution:

	Calculation	**Explanation**
Step 1:	$2x-1\overline{\smash)6x^2 - 7x - 2}$	Write both polynomials in order of descending powers. **If any powers are missing, fill in with 0's.**

Step 2:
$$\begin{array}{r} 3x \\ 2x-1\overline{\smash)6x^2 - 7x - 2} \end{array}$$

Mentally divide $6x^2$ by $2x$:
$\dfrac{6x^2}{2x} = 3x$. Write $3x$ above $6x^2$.

Step 3:
$$\begin{array}{r} 3x \\ 2x-1\overline{\smash)6x^2 - 7x - 2} \\ \underline{-(6x^2 - 3x)} \end{array}$$

Multiply $3x$ times $(2x - 1)$ and write the terms under the like terms in the dividend. Use a '–' sign to indicate that the product is to be subtracted.

Step 4:
$$\begin{array}{r} 3x \\ 2x-1\overline{\smash)6x^2 - 7x - 2} \\ \underline{-6x^2 + 3x} \\ -4x \end{array}$$

Subtract $6x^2 - 3x$ by changing signs and adding.

Step 5:
$$\begin{array}{r} 3x \\ 2x-1\overline{\smash)6x^2 - 7x - 2} \\ \underline{-6x^2 + 3x}\big\downarrow \\ -4x - 2 \end{array}$$

Bring down the –2.

Step 6:
$$\begin{array}{r} 3x -2 \\ 2x-1\overline{\smash)6x^2 - 7x - 2} \\ \underline{-6x^2 + 3x} \\ -4x - 2 \end{array}$$

Mentally divide $-4x$ by $2x$:
$\dfrac{-4x}{2x} = -2$.
Write -2 in the quotient.

Continued on the next page...

Step 7:

$$2x-1 \overline{\smash{)}\ \begin{aligned} 3x\ -\ \ 2\\[-2pt] 6x^2\ -\ 7x-2 \end{aligned}}$$

$$\underline{-6x^2+3x}$$

$$-4x-2$$

$$\underline{-\left(-4x+2\right)}$$

Multiply –2 times $(2x-1)$ and write the terms under the like terms in the expression $-4x-2$. Use a '–' sign to indicate that the product is to be subtracted.

Step 8:

$$2x-1 \overline{\smash{)}\ \begin{aligned} 3x\ -\ \ 2\\[-2pt] 6x^2\ -\ 7x-2 \end{aligned}}$$

$$\underline{-6x^2+3x}$$

$$-4x-2$$

$$\underline{+4x-2}$$

$$-4$$

Subtract $-4x+2$ by changing signs and adding.

Thus the quotient is $3x-2$ and the remainder is -4.

In the form $Q+\dfrac{R}{D}$ we can write $\dfrac{6x^2-7x-2}{2x-1}=3x-2-\dfrac{4}{2x-1}$.

Check: Show $Q\cdot D+R=P$.

$$(3x-2)(2x-1)-4=6x^2-3x-4x+2-4=6x^2-7x-2$$

Example 3: Long Division (Remainder 0)

Divide $\left(25x^3-5x^2+3x+1\right)\div\left(5x+1\right)$ by using long division.

Solution:

$$5x+1 \overline{\smash{)}\ \begin{aligned} 5x^2\ -2x\ +\ 1\\[-2pt] 25x^3-5x^2+3x+1 \end{aligned}}$$

$$\underline{-\left(25x^3+5x^2\right)}$$

$$-10x^2+3x$$

$$\underline{-\left(-10x^2-2x\right)}$$

$$5x+1$$

$$\underline{-\left(5x+1\right)}$$

$$0$$

There is no remainder, thus the quotient is simply $5x^2-2x+1$.

In Example 3, because the remainder is 0, both $(5x+1)$ and $(5x^2-2x+1)$ are **factors** of $25x^3 - 5x^2 + 3x + 1$. That is,

$$(5x^2 - 2x + 1)(5x+1) = 25x^3 - 5x^2 + 3x + 1 \quad \text{or} \quad Q \cdot D = P.$$

Factoring polynomials will be discussed in detail in Sections 4.5, 4.6, and 4.7.

Example 4: Long Division (Terms Missing)

Simplify $\dfrac{x^4 + 9x^2 - 3x + 5}{x^2 - x + 2}$ by using long division.

Solution: Note that 0 is written as a placeholder for any missing powers of the variable. In this way, like terms are easily aligned vertically.

$$
\begin{array}{r}
x^2 + x + 8 \\
x^2 - x + 2 \overline{\smash{\big)}\ x^4 + 0x^3 + 9x^2 - 3x + 5} \\
-(x^4 - x^3 + 2x^2) \\
\hline
x^3 + 7x^2 - 3x \\
-(x^3 - x^2 + 2x) \\
\hline
8x^2 - 5x + 5 \\
-(8x^2 - 8x + 16) \\
\hline
3x - 11
\end{array}
$$

Note that the remainder is of smaller degree than the divisor.

Thus the quotient is $x^2 + x + 8$ and the remainder is $3x - 11$.

In the form $Q + \dfrac{R}{D}$ we can write $x^2 + x + 8 + \dfrac{3x - 11}{x^2 - x + 2}$.

Practice Problems

1. Express the quotient as a sum of fractions in simplified form: $\dfrac{8x^2 + 6x + 1}{2x}$.

Use the division algorithm to divide. Write the answer in the form $Q + \dfrac{R}{D}$.

2. $\dfrac{3x^2 - 8x + 5}{x + 2}$

3. $\left(x^3 + 4x^2 - 10\right) \div \left(x^2 + x - 1\right)$

Answers to Practice Problems: **1.** $4x + 3 + \dfrac{1}{2x}$ **2.** $3x - 14 + \dfrac{33}{x + 2}$ **3.** $x + 3 + \dfrac{-2x - 7}{x^2 + x - 1}$

4.4 Exercises

In Exercises 1 – 10, write each quotient as a sum (or difference) of fractions in simplified form.

1. $\dfrac{8y^3 - 16y^2 + 24y}{8y}$

2. $\dfrac{18x^4 + 24x^3 + 36x^2}{6x^2}$

3. $\dfrac{34x^5 - 51x^4 + 17x^3}{17x^3}$

4. $\dfrac{14y^4 + 28y^3 + 12y^2}{2y^2}$

5. $\dfrac{110x^4 - 121x^3 + 11x^2}{11x}$

6. $\dfrac{15x^7 + 36x^6 - 25x^3}{15x^3}$

7. $\dfrac{-56x^4 + 98x^3 - 35x^2}{14x^2}$

8. $\dfrac{108x^6 - 72x^5 + 63x^4}{18x^4}$

9. $\dfrac{16y^6 - 56y^5 - 120y^4 + 64y^3}{16y^3}$

10. $\dfrac{20y^5 - 14y^4 + 21y^3 + 42y^2}{4y^2}$

In Exercises 11 – 60, divide by using the division algorithm. Write the answers in the form
$Q + \dfrac{R}{D}$ *where the degree of R < the degree of D.*

11. $\dfrac{21x^2 + 25x - 3}{7x - 1}$

12. $\dfrac{15x^2 - 14x - 11}{3x - 4}$

13. $\dfrac{x^2 - 12x + 27}{x - 3}$

14. $\dfrac{x^2 - 12x + 35}{x - 5}$

15. $\dfrac{x^3 + 4x^2 + x - 1}{x + 8}$

16. $\dfrac{x^3 - 6x^2 + 8x - 5}{x - 2}$

17. $\dfrac{4x^3 + 2x^2 - 3x + 1}{x + 2}$

18. $\dfrac{3x^3 + 6x^2 + 8x - 5}{x + 1}$

19. $\dfrac{x^3 + 6x + 3}{x - 7}$

20. $\dfrac{2x^3 + 3x - 2}{x - 1}$

21. $\dfrac{2x^3 - 5x^2 + 6}{x + 2}$

22. $\dfrac{4x^3 - x^2 + 13}{x - 1}$

23. $\dfrac{2x^3 + 7x^2 + 10x - 6}{2x + 3}$

24. $\dfrac{6x^3 - 7x^2 + 14x - 8}{3x - 2}$

25. $\dfrac{21x^3 + 41x^2 + 13x + 5}{3x + 5}$

26. $\dfrac{6x^3 - 4x^2 + 5x - 7}{x - 2}$

27. $\dfrac{x^3 - x^2 - 10x - 10}{x - 4}$

28. $\dfrac{2x^3 - 3x^2 + 7x + 4}{2x - 1}$

29. $\dfrac{10x^3 + 11x^2 - 12x + 9}{5x + 3}$

30. $\dfrac{6x^3 + 19x^2 - 3x - 7}{6x + 1}$

31. $\dfrac{2x^3 - 7x + 2}{x + 4}$

32. $\dfrac{2x^3 + 4x^2 - 9}{x + 3}$

33. $\dfrac{9x^3 - 19x + 9}{3x - 2}$

34. $\dfrac{16x^3 + 7x + 12}{4x + 3}$

35. $\dfrac{6x^3 + 11x^2 + 25}{2x + 5}$

36. $\dfrac{4x^3 - 8x^2 - 9x}{2x - 3}$

37. $\dfrac{x^4 - 3x^3 + 2x^2 - x + 2}{x - 3}$

38. $\dfrac{x^4 + x^3 - 4x^2 + x - 3}{x + 6}$

39. $\dfrac{x^4 + 2x^2 - 3x + 5}{x - 2}$

40. $\dfrac{3x^4 + 2x^3 + 2x^2 + x - 1}{x + 1}$

41. $\dfrac{x^4 - x^2 + 3}{x - \dfrac{1}{2}}$

42. $\dfrac{x^3 + 2x^2 + 1}{x - \dfrac{2}{3}}$

43. $\dfrac{3x^3 + 5x^2 + 7x + 9}{x^2 + 2}$

44. $\dfrac{2x^4 + 2x^3 + 3x^2 + 6x - 1}{2x^2 + 3}$

45. $\dfrac{x^4 + x^3 - 4x + 1}{x^2 + 4}$

46. $\dfrac{2x^4 + x^3 - 8x^2 + 3x - 2}{x^2 - 5}$

47. $\dfrac{6x^3 + 5x^2 - 8x + 3}{3x^2 - 2x - 1}$

48. $\dfrac{x^3 - 9x^2 + 20x - 38}{x^2 - 3x + 5}$

49. $\dfrac{3x^4 - 7x^3 + 5x^2 + x - 2}{x^2 + x + 1}$

50. $\dfrac{2x^4 + 9x^3 - x^2 + 6x + 9}{x^2 - 3x + 1}$

51. $\dfrac{x^4 + 3x - 7}{x^2 + 2x - 3}$

52. $\dfrac{3x^4 - 2x^3 + 4x^2 - x + 3}{3x^2 + x - 1}$

53. $\dfrac{x^3 - 27}{x - 3}$

54. $\dfrac{x^3 + 125}{x + 5}$

55. $\dfrac{x^5 - 1}{x^2 + 1}$

56. $\dfrac{x^6 - 1}{x^3 - 1}$

57. $\dfrac{x^5 - 1}{x - 1}$

58. $\dfrac{x^5 - x^3 + x}{x + \dfrac{1}{2}}$

59. $\dfrac{x^4 - 2x^3 + 4}{x + \dfrac{4}{5}}$

60. $\dfrac{x^6 + 1}{x + 1}$

Writing and Thinking About Mathematics

61. Suppose that a polynomial is divided by $(3x - 2)$ and the answer is given as $x^2 + 2x + 4 + \dfrac{20}{3x - 2}$. What was the original polynomial? Explain how you arrived at this conclusion.

62. Suppose that a polynomial is divided by $(x + 5)$ and the answer is given as $x^2 - 3x + 2 - \dfrac{6}{x + 5}$. What was the original polynomial? Explain how you arrived at this conclusion.

63. Given that $P(x) = 2x^3 - 8x^2 + 10x + 15$.
   a. Find $P(2)$ then divide $P(x)$ by $x - 2$.
   b. Find $P(-1)$ then divide $P(x)$ by $x + 1$.
   c. Find $P(4)$ then divide $P(x)$ by $x - 4$.
   Do you see any pattern in the values of $P(a)$ for $x = a$ and the remainders you found in the division process? [**Hint:** Check the appendix section on Synthetic Division and the Remainder Theorem.]

 HAWKES LEARNING SYSTEMS: INTERMEDIATE ALGEBRA SOFTWARE

   ▪ Division by a Monomial
   ▪ The Division Algorithm

4.5 Introduction to Factoring Polynomials

- *Factor polynomials by finding the **greatest common factor**.*
- *Factor polynomials by **grouping**.*

The result of multiplication is called the **product** and the numbers or expressions being multiplied are called **factors** of the product. The reverse of multiplication is called **factoring**. That is, given a product, we want to find the factors.

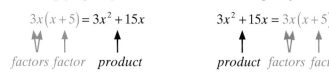

Multiplying Polynomials

$$3x(x+5) = 3x^2 + 15x$$

factors factor   product

Factoring Polynomials

$$3x^2 + 15x = 3x(x+5)$$

product   factors factor

Factoring polynomials relies heavily on the multiplication techniques developed in Section 4.3. You must remember how to multiply in order to be able to factor. Furthermore, you will find that the skills used in factoring polynomials are necessary when simplifying rational expressions (Chapter 5) and when solving equations. In other words, study this section and Sections 4.6 and 4.7 with extra care.

Greatest Common Factor

The first step in factoring polynomials is to factor out the monomial that is the **greatest common factor (GCF)** as follows.

Factoring Out the GCF

*To find a monomial that is the **greatest common factor (GCF)** of a polynomial:*

1. *Find the variable(s) of highest degree and the largest integer coefficient that is a factor of each term of the polynomial. (This is one factor.)*

2. *Divide this monomial factor into each term of the polynomial resulting in another polynomial factor.*

For example, the factor $3x$ can be factored out of the polynomial $6x^4 + 3x^3 - 21x^2$ by using the distributive property in a reverse sense. This gives:

$$6x^4 + 3x^3 - 21x^2 = 3x(2x^3) + 3x(x^2) + 3x(-7x)$$

$$= 3x(2x^3 + x^2 - 7x).$$

Note that the polynomial $2x^3 + x^2 - 7x$ has x as a common factor. Therefore, we have not factored **completely**. **An expression is factored completely if none of its factors can be factored.** Thus $3x$ is not the **greatest** common factor. The greatest common factor is $3x^2$. Factoring out $3x^2$ gives the factored form

$$6x^4 + 3x^3 - 21x^2 = 3x^2\left(2x^2\right) + 3x^2\left(x\right) + 3x^2\left(-7\right)$$
$$= 3x^2\left(2x^2 + x - 7\right).$$

A common factor may be an expression other than a monomial. It might be a binomial or a polynomial with three or more terms. In the following example, the binomial $x^2 + 1$ is the common factor and is factored out.

$$x^2(x^2 + 1) + 5x(x^2 + 1) + 2(x^2 + 1) = (x^2 + 1)(x^2 + 5x + 2)$$

In this text, factoring polynomials means to find factors that are integers or polynomials with integer coefficients. If this cannot be done, we say that the polynomial is **not factorable** (or **irreducible** or **prime**). For example, the polynomials $x^2 + 36$ and $x^2 + 5x + 2$ are not factorable.

Example 1: Finding the GCF of a Polynomial

Factor out the greatest common factor in each polynomial.

a. $4x^2 + 24x + 8$

Solution: By inspection, we see that the constant 4 is the GCF:

$$4x^2 + 24x + 8 = 4 \cdot x^2 + 4 \cdot 6x + 4 \cdot 2 = 4\left(x^2 + 6x + 2\right).$$

b. $6x^2y + 12xy^2$

Solution: Similarly, by looking at each term, we see that $6xy$ is the GCF:

$$6x^2y + 12xy^2 = 6xy \cdot x + 6xy \cdot 2y = 6xy(x + 2y).$$

By definition, the GCF of a polynomial will have a positive coefficient. However, if the leading coefficient of a polynomial is negative, we may choose to factor out the negative of the GCF (or $-1 \cdot$ GCF). This technique will leave a positive coefficient for the first term of the other polynomial factor.

Example 2: Factoring Out −1·GCF

Factor $-14x^4 + 21x^3 - 84x^2$ completely.

Solution: The GCF is $7x^2$ and we can factor as follows:

$$-14x^4 + 21x^3 - 84x^2 = 7x^2\left(-2x^2 + 3x - 12\right).$$

However, the leading coefficient is negative and we can choose to factor out $-7x^2$ as follows:

$$-14x^4 + 21x^3 - 84x^2 = -7x^2\left(2x^2 - 3x + 12\right).$$

Both answers are correct. But, we will see later that having a positive leading coefficient for the polynomial in parentheses may make that polynomial easier to factor. Therefore, the second answer is **generally preferred**.

Factoring by Grouping

Consider the expression

$$y(x+5) + 3(x+5)$$

where the common factor is the **binomial** $(x+5)$. Factoring out this common binomial factor by using the distributive property gives

$$y(x+5) + 3(x+5) = (x+5)(y+3).$$

If we multiply by using the distributive property twice,

$$y(x+5) + 3(x+5) = xy + 5y + 3x + 15,$$

we see that there are **four terms** in the product and no common monomial factor. Yet we know that the product has two factors, namely $(x+5)$ and $(y+3)$. Factoring polynomials with four or more terms can sometimes be accomplished by **grouping the terms** in such a way that a common binomial factor or some other form of factors can be recognized. The common factor can be a binomial or other polynomial.

Example 3: Factoring by Grouping

Factor each polynomial by grouping.

a. $ax - ay + bx - by$

Solution:

$ax - ay + bx - by = (ax - ay) + (bx - by)$ The first two terms have a in common.

$= a(x - y) + b(x - y)$ The second two terms have b in common. $(x - y)$ is a common factor.

$= (x - y)(a + b)$

b. $xy + 16 - 8x - 2y$

Solution:

$xy - 8x - 2y + 16 = (xy - 8x) + (-2y + 16)$

$= x(y - 8) + 2(-y + 8)$ Since $(y - 8)$ and $(-y + 8)$ are not the

$= x(y - 8) - 2(y - 8)$ same factor, we factor -2 instead of 2 from the last two terms.

$= (y - 8)(x - 2)$

c. $4xy - 28x + 3y - 15$

Solution: $4xy - 28x + 3y - 15 = 4x(y - 7) + 3(y - 5)$

But $y - 7$ and $y - 5$ are not the same factor. In fact, $4xy - 28x + 3y - 15$ is **not factorable**. That is, **the fact that some of the terms are factorable does not necessarily imply that the entire expression is factorable.** To be factorable, the entire expression must be a product of factors.

d. $4ax + 4ay + x + y$

Solution:

$4ax + 4ay + x + y = (4ax + 4ay) + (x + y)$

$= 4a(x + y) + 1(x + y)$ Note that a 1 is factored out of the last

$= (x + y)(4a + 1)$ two terms.

e. $5xy + 4uv - vy - 20ux$

Solution: In the polynomial $5xy + 4uv - vy - 20ux$ there is no common factor in the first two terms. In this case we try a rearrangement of the terms as follows:

$5xy + 4uv - vy - 20ux = 5xy - vy + 4uv - 20ux$

$= y(5x - v) + 4u(v - 5x)$ Note that $(5x - v) = -(v - 5x)$.

$= y(5x - v) - 4u(5x - v)$ Factor $-4u$ from the last two terms.

$= (5x - v)(y - 4u)$ Now $(5x - v)$ is a common factor.

Practice Problems

Completely factor each polynomial.

1. $5x - 20$

2. $10x^2 + 100x$

3. $-2x^3 - 4x$

4. $x^2 + xy - 2x - 2y$

5. $3xy - 27x + 2y - 18$

4.5 Exercises

In Exercises 1 – 16, completely factor each polynomial by factoring out the greatest common monomial factor.

1. $3x + 15$

2. $25x - 30$

3. $-5x^2 + 15x$

4. $7x^2 - 21x$

5. $-9x^2 + 27x$

6. $8x^3 - 24x^2$

7. $35x^4 + 49x^3$

8. $9y^4 + 9y^3$

9. $x^2y - 2xy + xy^2$

10. $4xy^2 + 10x^2y - 2xy$

11. $8x^3y - 4x^2y$

12. $5x^2y^2 + 20x^2y$

13. $6x^2y + 3xy - 9xy^2$

14. $4xy^2 - 16xy - 8x$

15. $-2x^4 - 30x^3 - 12x^2$

16. $-5x^2 - 25x + 35$

Use the method of grouping to factor the polynomials in Exercises 17–36. If a polynomial cannot be factored, write not factorable.

17. $ax + ay + 4x + 4y$

18. $ax + bx - 5a - 5b$

19. $xy + 4x - 3y - 12$

20. $xy + 5y + 2x + 10$

21. $7x - 3y + xy - 21$

22. $ay + by + a + b$

23. $a^2b + a^2c - b - c$

24. $5xy + 5y^2 - x - y$

25. $3x^2 + 3xy + x + y$

26. $13x^2 + 13y^2 + ax^2 + ay^2$

27. $3xy + 6y - x - 1$

28. $3xy + 5y + 6x + 10$

29. $4xy - 21 - 28x + 3y$

30. $5x^2 + 10x + 3y + 15$

31. $2xy + 10x + 3y + 15$

Answers to Practice Problems: **1.** $5(x - 4)$ **2.** $10x(x + 10)$ **3.** $-2x(x^2 + 2)$ **4.** $(x - 2)(x + y)$
5. $(3x + 2)(y - 9)$

32. $6x^2 + 36x + 7x + 42$ **33.** $9x^3 - 45x^2 - 5x + 25$ **34.** $2x^3y - 20x^2y - 3x + 30$

35. $4xy^2 + 4xy + 3y + 6$ **36.** $4x^3y^2 + 5x^2 - 2 - 8xy^2$

37. Area of a rectangle: The area (in square inches) of the rectangle shown is given by the polynomial function $A(x) = 8x^2 + 120x$. If the width of the rectangle is $4x$ inches, what is the representation of the length?

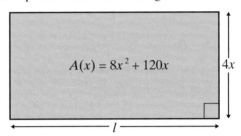

38. Area of a triangle: The area of a triangle is $\frac{1}{2}$ the product of its base and its height. The area (in square feet) of the triangle shown is given by the function $A(x) = \frac{1}{2}\left(x^2 + 32x\right)$.

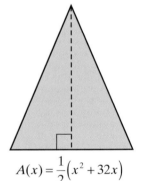

$A(x) = \frac{1}{2}\left(x^2 + 32x\right)$

a. Find representations for the lengths of its base and its height.

b. What is the value of $A(2)$?

c. What is the value of $A(3)$?

Writing and Thinking About Mathematics

39. Show two different ways to factor the expression $5xy + 4uv - vy - 20ux$. (Both approaches should arrive at at the same factorization.)

HAWKES LEARNING SYSTEMS: INTERMEDIATE ALGEBRA SOFTWARE

- GCF of a Polynomial
- Factoring by Grouping

Factoring Trinomials

4.6

- *Factor trinomials using the **trial-and-error method**.*
- *Factor trinomials using the **ac-method**.*

Second-degree trinomials in the variable *x* are of the form

$$ax^2 + bx + c \qquad \text{where } a, b, \text{ and } c \text{ are real constants.}$$

In this section we discuss two methods of factoring trinomials of the form $ax^2 + bx + c$ in which the coefficients *a*, *b*, and *c* are **restricted to integers**. These two methods are the **trial-and-error method** (a reverse of the **FOIL** method of multiplication) and the *ac*-**method** (a form of **grouping**).

The Trial-and-Error Method of Factoring

We consider the following two basic forms:

    1. The leading coefficient is 1:   → $x^2 + bx + c$
    2. The leading coefficient is not 1:   → $ax^2 + bx + c$

First, reviewing multiplication using the **FOIL** method to multiply $(x + 3)$ and $(x + 15)$, we find

$$\overset{\text{F}\quad\text{O}\quad\text{I}\quad\text{L}}{(x+3)(x+15)} = x^2 + 15x + 3x + 3\cdot15 = x^2 + \underset{\underset{3+15}{\uparrow}}{18x} + \underset{\underset{3\cdot15}{\uparrow}}{45}$$

More generally,

$$(x+a)(x+b) = x^2 + bx + ax + ab = x^2 + \underset{\underset{\substack{\text{sum} \\ \text{of constants}}}{\uparrow}}{(a+b)}x + \underset{\underset{\substack{\text{product} \\ \text{of constants}}}{\uparrow}}{ab}$$

Now, reversing this process, suppose that we are given the product $x^2 + 11x + 30$ and want to find two binomial factors (if they exist). Analyzing the FOIL method just shown, we realize that we need positive factors of +30 whose sum is +11.

Positive Factors of 30			Sums of These Factors
1	30	$\longrightarrow$	$1 + 30 = 31$
2	15	$\longrightarrow$	$2 + 15 = 17$
3	10	$\longrightarrow$	$3 + 10 = 13$
5	6	$\longrightarrow$	$5 + 6 = 11$

Now, because $5 \cdot 6 = 30$ and $5 + 6 = 11$, we have

$$x^2 + 11x + 30 = (x + 5)(x + 6).$$

If the middle term had been $-11x$, then we would have wanted pairs of negative integer factors to find a sum of -11. The following statement works for factoring trinomials with leading coefficient 1.

To factor a trinomial with leading coefficient 1, find two factors of the constant term whose sum is the coefficient of the middle term. (If these factors do not exist, the trinomial is **not factorable**.)

Example 1: Factoring Trinomials with Leading Coefficient 1

a. Factor $x^2 - 10x + 16$.

Solution: We need to find two negative factors of 16 whose sum is -10.

Negative Factors of 16			Sums of These Factors
-1	-16	$\longrightarrow$	$-1 + (-16) = -17$
-2	-8	$\longrightarrow$	$-2 + (-8) = -10$
-4	-4	$\longrightarrow$	$-4 + (-4) = -8$

Thus,

$$x^2 - 10x + 16 = (x - 2)(x - 8).$$

b. Factor $-3x^2 - 3x + 36$.

Solution: $-3x^2 - 3x + 36 = -3(x^2 + x - 12)$ -3 is a common factor.

$$= -3(x - 3)(x + 4)   (+4)(-3) = -12 \text{ and } 4x - 3x = x$$

In this case the constant term is negative (-12) so one factor is positive and the other negative.

NOTES

As illustrated in Example 1b, the first step in factoring should always be to factor out any common monomial factor. Also, if the leading coefficient is negative, factor out a negative monomial even if it is just −1. A polynomial with a positive leading coefficient is easier to factor.

To factor a trinomial with leading coefficient other than 1, we use the FOIL method also, but with more of a trial-and-error approach. For example, consider the problem of factoring

$$6x^2 + 23x + 7$$

as the product of two binomials.

$$6x^2 + 23x + 7 = (\qquad)(\qquad)$$

with $F = 6x^2$ and $L = +7$.

For $F = 6x^2$, we know that $6x^2 = 6x \cdot x$ and $6x^2 = 3x \cdot 2x$.

For $L = +7$, we know that $+7 = (+1)(+7)$ and $+7 = (-1)(-7)$.

Now we use various combinations for **F** and **L** in the **trial-and-error method** as follows:

1. List all the possible combinations of factors of $6x^2$ and $+7$ in their respective **F** and **L** positions. (See the following list.)
2. Check the sum of the products in the **O** and **I** positions until you find the sum to be $+23x$.
3. If none of these sums is $+23x$, the trinomial is not factorable.

F
L

a. $(6x+1)(x+7)$

b. $(6x+7)(x+1)$

c. $(3x+1)(2x+7)$

d. $(3x+7)(2x+1)$

e. $(6x-1)(x-7)$

f. $(6x-7)(x-1)$

g. $(3x-1)(2x-7)$

h. $(3x-7)(2x-1)$

We really don't need to check these last four because the **O** and **I** would have to be negative, and we are looking for $+23x$. In this manner, the trial-and-error method is more efficient than it first appears to be.

Now, investigating only the possibilities in the list with positive constants, we need to check the sums of the outer (**O**) and inner (**I**) products to find $+23x$.

a. $(6x+1)(x+7)$: $\mathbf{O} + \mathbf{I} = 42x + x = 43x$

x

$42x$

b. $(6x+7)(x+1)$: $\mathbf{O} + \mathbf{I} = 6x + 7x = 13x$

$7x$

$6x$

c. $(3x+1)(2x+7)$: $\mathbf{O} + \mathbf{I} = 21x + 2x = \boxed{23x}$ ← We found $23x$! With a little luck we could have found this first.

$2x$

$21x$

Example 2: Using the Trial-and-Error Method: Leading Coefficient Not 1

a. Factor $5x^2 + 23x - 10$.

Solution: For $\mathbf{F} = 5x^2$ we know that $5x^2 = 5x \cdot x$.

For $\mathbf{L} = -10$, we know that $-10 = (+1)(-10)$, $-10 = (-1)(+10)$, $-10 = (+2)(-5)$, and $-10 = (-2)(+5)$.

$(5x-10)(x+1)$ $5x - 10x = -5x \neq 23x$

$-10x$

$5x$ In fact, $(5x - 10)$ should not even be tried because it has a common factor of 5 and 5 is not a common factor of the trinomial.

$(5x-1)(x+10)$ $50x - x = 49x \neq 23x$

$-x$

$50x$

$(5x+2)(x-5)$ $-25x + 2x = -23x \neq 23x$

$2x$

$-25x$

$(5x-2)(x+5)$ $25x - 2x = 23x$

$-2x$

$25x$

Therefore $5x^2 + 23x - 10 = (5x - 2)(x + 5)$.

b. Factor $2x^2 + 6x + 10$.

Solution: $2x^2 + 6x + 10 = 2\left(x^2 + 3x + 5\right)$

Note that while $2x^2 + 6x + 10$ **is factorable** (because 2 is a common monomial factor), the trinomial $x^2 + 3x + 5$ is **not factorable**.

c. Factor $4x^2 - x - 5$.

Solution: The product of the first two terms of any binomial factors is to be $4x^2$. So, we might have:

$$\mathbf{F} = 4x^2 = 2x \cdot 2x \qquad\qquad \mathbf{F} = 4x^2 = 4x \cdot x$$

$(2x\quad)(2x\quad)$ **OR** $(4x\quad)(x\quad)$.

The product of the last terms, **L**, is to be -5. The factors could be -5 and $+1$ or $+5$ and -1. We try all possible pairings until we find the right product. (If none of the pairs gives the correct product, then the trinomial is **not factorable**.)

$$\left(2x + 1\right)\left(2x - 5\right) = 4x^2 - 10x + 2x - 5 = 4x^2 - 8x - 5$$

$$\left(2x + 5\right)\left(2x - 1\right) = 4x^2 - 2x + 10x - 5 = 4x^2 + 8x - 5$$

$$\left(4x + 1\right)\left(x - 5\right) = 4x^2 - 20x + x - 5 = 4x^2 - 19x - 5$$

$$\left(4x - 5\right)\left(x + 1\right) = 4x^2 + 4x - 5x - 5 = 4x^2 - x - 5 \quad \text{The desired product.}$$

Thus $4x^2 - x - 5 = \left(4x - 5\right)\left(x + 1\right)$.

With practice, most of the steps shown in the examples can be done mentally, and the final form can be found quickly. As noted in Example 2a, the binomial $(5x - 10)$ has a monomial factor and should not have been considered. In Example 2c, the second try could have been eliminated because only the sign of the middle term is affected. This type of reasoning can make the trial-and-error method relatively efficient.

Example 3:   Factoring Involving Binomial Expressions

a.   Factor $x^2(x-2)+6x(x-2)+5(x-2)$.

  Solution: In this case the binomial $(x-2)$ is the GCF.

$$x^2(x-2)+6x(x-2)+5(x-2)=(x-2)(x^2+6x+5)$$
$$=(x-2)(x+5)(x+1) \qquad x+5x=6x$$

b.   Factor $(5a+b)^2-4(5a+b)-12$.

  Solution: In this case if we substitute $u = 5a + b$ the resulting trinomial takes a more familiar look in one variable and is easily factored:

$$u^2-4u-12=(u-6)(u+2) \qquad (-6)(2)=-12 \quad \text{and} \quad -6+2=-4$$

  Now reversing the substitution, we have

$$(5a+b)^2-4(5a+b)-12=(5a+b-6)(5a+b+2)$$

ac-Method of Factoring

The ***ac*-method** of factoring trinomials is very systematic and involves the method of factoring by grouping discussed in Section 4.5.

Consider the problem of factoring $2x^2 + 9x + 10$ where $a = 2$, $b = 9$, and $c = 10$.

Analysis of Factoring by the *ac*-Method

General Method	*Example*
ax^2+bx+c	$2x^2+9x+10$

Step 1:   *Multiply $a \cdot c$.*   Multiply $2 \cdot 10 = 20$.

Step 2:   *Find two integers whose product is ac and whose sum is b. If this is not possible, then the trinomial is **not factorable**.*   Find two integers whose product is 20 and whose sum is 9. (In this case, $4 \cdot 5 = 20$ and $4+5=9$.)

Step 3:   *Rewrite the middle term (bx) using the two numbers found in step 2 as coefficients.*   Rewrite the middle term $(+9x)$ using $+4$ and $+5$ as coefficients.
$2x^2+9x+10=2x^2+4x+5x+10$

Continued on the next page...

Analysis of Factoring by the *ac*-Method (cont.)

	General Method	**_Example_**
Step 4:	_Factor by grouping the first two terms and the last two terms._	_Factor by grouping the first two terms and the last two terms._ $2x^2 + 4x + 5x + 10 = 2x(x+2) + 5(x+2)$
Step 5:	_Factor out the common binomial factor. This will give two binomial factors of the trinomial_ $ax^2 + bx + c.$	_Factor out the common binomial factor_ $(x+2)$. _Thus,_ $2x^2 + 9x + 10 = 2x^2 + 4x + 5x + 10$ $= 2x(x+2) + 5(x+2)$ $= (x+2)(2x+5)$

Example 4: Using the *ac*-Method

a.   Factor $x^2 - 2x - 15$ by using the *ac*-method.

Solution: $a = 1, b = -2,$ and $c = -15$

Step 1: Find the product $a \cdot c$: $1(-15) = -15$.

Step 2: Find two integers whose product is -15 and whose sum is -2.
$(-5)(+3) = -15$ and $-5 + 3 = -2$

Step 3: Rewrite $-2x$ as $-5x + 3x$ to obtain
$x^2 - 2x - 15 = x^2 - 5x + 3x - 15$.

Step 4: Factor by grouping.
$x^2 - 5x + 3x - 15 = x(x-5) + 3(x-5)$

Step 5: Factor out the common binomial factor $(x-5)$.
$x(x-5) + 3(x-5) = (x-5)(x+3)$

Continued on the next page...

b.   Factor $18x^3 - 39x^2 + 18x$ using the *ac*–method.

Solution: First factor out the greatest common factor $3x$.
$$18x^3 - 39x^2 + 18x = 3x\left(6x^2 - 13x + 6\right)$$

Now factor the trinomial $6x^2 - 13x + 6$ with $a = 6, b = -13,$ and $c = 6$.

Step 1: Find the product $a \cdot c$: $6(6) = 36$.

Step 2: Find two integers whose product is 36 and whose sum is −13.

>**Note:** This may take some time and experimentation. We do know that both numbers must be negative since the product is positive and the sum is negative.

$$(-9)(-4) = +36 \quad \text{and} \quad -9 + (-4) = -13$$

Step 3: Rewrite $-13x$ as $-9x - 4x$ to obtain
$$6x^2 - 13x + 6 = 6x^2 - 9x - 4x + 6.$$

Step 4: Factor by grouping.
$$6x^2 - 13x + 6 = 6x^2 - 9x - 4x + 6$$
$$= \left(6x^2 - 9x\right) + \left(-4x + 6\right)$$
$$= 3x(2x - 3) - 2(2x - 3)$$

>**Note:** −2 is factored from the last two terms so that there will be a common binomial factor $(2x - 3)$.

Step 5: Factor out the common binomial factor $(2x - 3)$.
$$6x^2 - 13x + 6 = 6x^2 - 9x - 4x + 6$$
$$= 3x(2x - 3) - 2(2x - 3)$$
$$= (2x - 3)(3x - 2)$$

Thus, for the original expression,
$$18x^3 - 39x^2 + 18x = 3x\left(6x^2 - 13x + 6\right)$$
$$= 3x(2x - 3)(3x - 2)$$

Remember to include the original GCF in the final product.

Summary Note:

a. When factoring polynomials, always look for a common factor first. Then, if there is one, remember to include this common factor as part of the answer.

b. **To factor completely** means to find factors of the polynomial such that none of the factors are themselves factorable.

c. Not all polynomials are factorable. (See $x^2 + 3x + 5$ in Example 2b.) **Any polynomial that cannot be factored as the product of polynomials with integer coefficients is not factorable.**

Practice Problems

Completely factor each polynomial.

1. $x^2 - 2x - 35$

2. $-2x^2 - 6x - 4$

3. $4x^2 - 31x - 8$

4. $8x^2 + 13x - 6$

5. $2x^2(x+1) + 9x(x+1) - 18(x+1)$

6. $5(x+3y)^2 + 5(x+3y) - 30$

4.6 Exercises

In Exercises 1 – 42, completely factor each polynomial. If a polynomial cannot be factored, write not factorable.

1. $x^2 + 9x + 18$

2. $y^2 - 7y - 30$

3. $x^2 - 6x - 27$

4. $y^2 - 5y - 14$

5. $x^2 - 27x + 50$

6. $x^2 - 15x + 36$

7. $2x^2 + 15x - 8$

8. $2x^2 + 9x - 35$

9. $6x^2 + 13x + 6$

10. $8y^2 + 10y - 25$

11. $2x^2 - 7x + 4$

12. $6x^2 - 35x - 5$

13. $2x^2 - 7x - 4$

14. $35y^2 + 9y - 18$

15. $18x^2 - 7x - 1$

16. $25x^2 + 5x - 6$

17. $12x^3 + 38x^2 + 20x$

18. $8x^3 - 6x^2 - 2x$

19. $2x^4 + 11x^2y - 15y^2$

20. $2x^4 + 11x^2y - 21y^2$

21. $2x^6 + 9x^3y^2 + 4y^4$

22. $5x^4 + 17x^2y^2 + 6y^4$

23. $-18x^2 + 72x - 8$

24. $-45y^2 + 30y + 120$

Answers to Practice Problems: **1.** $(x - 7)(x + 5)$ **2.** $-2(x + 1)(x + 2)$ **3.** $(x - 8)(4x + 1)$ **4.** $(x + 2)(8x - 3)$ **5.** $(x + 1)(x + 6)(2x - 3)$ **6.** $5(x + 3y + 3)(x + 3y - 2)$

25. $-5y^2 + 40y - 60$ **26.** $-12x^2 + 22x + 4$ **27.** $21x^4 - 4x^3 - 32x^2$

28. $2x^4 y^3 - 5x^3 y^3 - 18x^2 y^3$ **29.** $5x^4 + 17x^2 y^2 + 6y^4$ **30.** $2x^6 + 9x^3 y^2 + 4y^4$

31. $x^2(x-5) + 4x(x-5) - 21(x-5)$ **32.** $x^2(x+4) - 6x(x+4) - 15(x+4)$

33. $2x^2(2x+1) - 9x(2x+1) - 18(2x+1)$ **34.** $3x^2(4x-1) + 19x(4x-1) + 28(4x-1)$

35. $(2x+y)^2 - 9(2x+y) + 20$ **36.** $(x-2y)^2 + 10(x-2y) + 16$

37. $(x+3y)^2 - 14(x+3y) - 32$ **38.** $(x+5y)^2 + 8(x+5y) + 12$

39. $(6x-y)^2 + 8(6x-y) + 7$ **40.** $5(3x-y)^2 + 15(3x-y) - 20$

41. $4(2a+b)^2 + 4(2a+b) - 24$ **42.** $-6(a-b)^2 + 6(a-b) + 180$

43. Area of a rectangle: The area (in square inches) of the rectangle shown is given by the polynomial function $A(x) = 8x^2 + 22x + 5$. If the width of the rectangle is $4x + 1$ inches, what is the representation of the length?

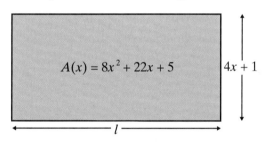

44. Area of a rectangle: The area (in square meters) of the rectangle shown is given by the polynomial function $A(x) = x^2 + 13x + 30$. If the length of the rectangle is $(x + 10)$ meters, what is the representation of the width?

45. Volume of a box: The volume of an open box is found by cutting equal squares (x units on a side) from a sheet of cardboard that is 12 inches by 36 inches. The function representing the volume is $V(x) = 4x^3 - 96x^2 + 432x,$ where $0 < x < 6.$

    a. Factor this function in such a way that the factors represent the lengths of the sides of the box.

    b. What is the value of $V(2)$?

    c. What is the value of $V(4)$?

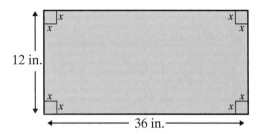

Writing and Thinking About Mathematics

46. The following statement is true:

$$4x^2 + 24x + 20 = (4x + 20)(x + 1) = (x + 5)(4x + 4) = (2x + 10)(2x + 2).$$

Explain how the trinomial can be factored in three ways. Is there some kind of error?

47. The following statement is true: $5x^2 + 5x - 60 = (5x + 20)(x - 3).$

Explain why this is not the completely factored form of the trinomial.

48. Explain, in your own words, what is meant by factoring a polynomial.

HAWKES LEARNING SYSTEMS: INTERMEDIATE ALGEBRA SOFTWARE

- Factoring Trinomials by Trial and Error
- Factoring Trinomials by the *ac*-Method

4.7   Special Factoring Techniques

- *Factor the difference of squares.*
- *Factor perfect square trinomials.*
- *Factor the sums and differences of two cubes.*

In Section 4.3 we discussed the following three products.

I. $(x+a)(x-a) = x^2 - a^2$ Difference of two squares

II. $(x+a)^2 = x^2 + 2ax + a^2$ Square of a binomial sum

III. $(x-a)^2 = x^2 - 2ax + a^2$ Square of a binomial difference

In this section you will learn to factor products of these types (and two others) without referring to the **ac-method** or the **trial-and-error method**. That is, with practice you will learn to be able to recognize the "form" of the special product and go directly to the factors.

Table 4.1 lists the squares of the integers from 1 to 20.

Perfect Squares from 1 to 400									
Integer n	1	2	3	4	5	6	7	8	9	10
Square n^2	1	4	9	16	25	36	49	64	81	100
Integer n	11	12	13	14	15	16	17	18	19	20
Square n^2	121	144	169	196	225	256	289	324	361	400

Table 4.1

Difference of Two Squares: $x^2 - a^2 = (x + a)(x - a)$

Consider the polynomial $x^2 - 25$. By recognizing this expression as the **difference of two squares**, we can go directly to the factors:

$$x^2 - 25 = (x)^2 - (5)^2 = (x+5)(x-5).$$

Similarly, we have

$$9 - y^2 = (3)^2 - (y)^2 = (3+y)(3-y) \quad \text{and} \quad 49x^2 - 36 = (7x)^2 - (6)^2 = (7x+6)(7x-6).$$

Remember to **look for a common monomial factor first**. For example,

$$6x^2y - 24y = 6y(x^2 - 4) = 6y(x+2)(x-2).$$

Example 1: Factoring the Difference of Two Squares

Factor completely.

a. $3a^2b - 3b$

Solution: $3a^2b - 3b = 3b(a^2 - 1)$ Factor out the GCF, $3b$.

$= 3b(a + 1)(a - 1)$ Difference of two squares

Don't forget that $3b$ is a factor.

b. $x^6 - 400$ Even powers, such as x^6, can always be treated as squares: $x^6 = (x^3)^2$.

Solution: $x^6 - 400 = (x^3 + 20)(x^3 - 20)$ Difference of two squares

Sum of Two Squares

*The **sum of two squares** is an expression of the form $x^2 + a^2$ and is **not factorable**. For example, $x^2 + 36$ is the sum of two squares and is not factorable. There are no factors with integer coefficients whose product is $x^2 + 36$. To understand this situation, write*

$$x^2 + 36 = x^2 + 0x + 36$$

and note that there are no factors of $+36$ that will add to 0.

Perfect Square Trinomials: $\begin{cases} x^2 + 2ax + a^2 = (x + a)^2 \\ x^2 - 2ax + a^2 = (x - a)^2 \end{cases}$

We know that **squaring a binomial** leads to a **perfect square trinomial**. Therefore, factoring a perfect square trinomial gives the square of a binomial. We simply need to recognize whether or not a trinomial fits the "form."

In a perfect square trinomial, both the first and last terms of the trinomial must be perfect squares. If the first term is of the form x^2 and the last term is of the form a^2, then the middle term must be of the form $2ax$ or $-2ax$. For example,

$x^2 + 8x + 16 = (x + 4)^2$ Here $16 = 4^2 = a^2$ and $8x = 2 \cdot 4 \cdot x = 2ax$.

$x^2 - 6x + 9 = (x - 3)^2$ Here $9 = 3^2 = a^2$ and $-6x = -2 \cdot 3 \cdot x = -2ax$.

Example 2: Factoring Perfect Square Trinomials

Factor completely.

a. $4y^2 + 12y + 9$

Solution: In the form $x^2 + 2ax + a^2$ we have $x = 2y$ and $a = 3$.

$$4y^2 + 12y + 9 = (2y)^2 + 2 \cdot 3 \cdot 2y + 3^2$$
$$= (2y + 3)^2$$

b. $(z + 2)^2 - 12(z + 2) + 36$

Solution: In the form $x^2 - 2ax + a^2$ we have $x = z + 2$ and $a = 6$.

$$(z + 2)^2 - 12(z + 2) + 36 = (z + 2)^2 - 2 \cdot 6(z + 2) + 6^2$$
$$= \left[(z + 2) - 6 \right]^2$$
$$= (z - 4)^2$$

c. $2x^3 - 8x^2y + 8xy^2$

Solution: Factor out the GCF first. Then factor the **perfect square trinomial**.

$$2x^3 - 8x^2y + 8xy^2 = 2x(x^2 - 4xy + 4y^2)$$
$$= 2x(x - 2y)^2$$

d. $(x^2 + 6x + 9) - y^2$

Solution: Treat $x^2 + 6x + 9$ as a perfect square trinomial, then factor the **difference of two squares**.

$$(x^2 + 6x + 9) - y^2 = (x + 3)^2 - y^2$$
$$= (x + 3 + y)(x + 3 - y)$$

Sums and Differences of Two Cubes:

$$\begin{cases} \boldsymbol{x^3 + a^3 = (x + a)(x^2 - ax + a^2)} \\ \boldsymbol{x^3 - a^3 = (x - a)(x^2 + ax + a^2)} \end{cases}$$

The formulas for the sums and differences of two cubes are new, and we can proceed to show that they are indeed true as follows:

$$(x+a)(x^2 - ax + a^2) = x \cdot x^2 - x \cdot ax + x \cdot a^2 + a \cdot x^2 - a \cdot ax + a \cdot a^2$$

$$= x^3 - ax^2 + a^2 x + ax^2 - a^2 x + a \cdot a^2$$

$$= x^3 + a^3 \qquad \text{Sum of two cubes}$$

$$(x-a)(x^2 + ax + a^2) = x \cdot x^2 + x \cdot ax + x \cdot a^2 - a \cdot x^2 - a \cdot ax - a^3$$

$$= x^3 + ax^2 + a^2 x - ax^2 - a^2 x - a^3$$

$$= x^3 - a^3 \qquad \text{Difference of two cubes}$$

NOTES

Important Notes about the Sum and Difference of Two Cubes:
1. In each case, the middle terms drop out and only two terms are left.
2. The expressions in parentheses are not perfect square trinomials. These trinomials are not factorable.
3. The sign in the binomial agrees with the sign in the result.

Because we know that factoring is the reverse of multiplication, we use these formulas to factor the sums and differences of two cubes. For example,

$$x^3 + 27 = (x)^3 + (3)^3$$

$$= (x+3)((x)^2 - (3)(x) + (3)^2)$$

$$= (x+3)(x^2 - 3x + 9)$$

and

$$x^6 - 125 = (x^2)^3 - (5)^3$$

$$= (x^2 - 5)((x^2)^2 + (5)(x^2) + (5)^2)$$

$$= (x^2 - 5)(x^4 + 5x^2 + 25).$$

As an aid in factoring sums and differences of two cubes, Table 4.2 contains the cubes of the integers from 1 – 10. These cubes are called perfect cubes.

Perfect Cubes from 1 to 1000									
Integer n	1	2	3	4	5	6	7	8	9	10
Cube n^3	1	8	27	64	125	216	343	512	729	1000

Table 4.2

Example 3: Factoring Sums and Differences of Two Cubes

Factor completely.

a. $x^6 + 64$

    Solution: $x^6 + 64 = \left(x^2\right)^3 + 4^3$

                                $= \left(x^2 + 4\right)\left(x^4 - 4x^2 + 16\right)$

b. $16y^{12} - 250$

    Solution: Factor out the GCF first. Then factor the **difference of two cubes**.

$$16y^{12} - 250 = 2\left(8y^{12} - 125\right)$$

$$= 2\left[\left(2y^4\right)^3 - 5^3\right]$$

$$= 2\left[\left(2y^4\right) - 5\right]\left[\left(2y^4\right)^2 + 2y^4 \cdot 5 + 5^2\right]$$

$$= 2\left(2y^4 - 5\right)\left(4y^8 + 10y^4 + 25\right)$$

c. $x^{3k} + y^{6k}$

    Solution: $x^{3k} + y^{6k} = \left(x^k\right)^3 + \left(y^{2k}\right)^3$

                              $= \left(x^k + y^{2k}\right)\left(x^{2k} - x^k y^{2k} + y^{4k}\right)$

Note: Remember that the second polynomial is not a perfect square trinomial and cannot be factored.

Practice Problems

Completely factor each of the following polynomials. If a polynomial cannot be factored, write not factorable.

1. $5x^2 - 80$

2. $9x^2 - 12x + 4$

3. $x^2 - 3x + 4$

4. $2x^3 - 250$

5. $\left(y^2 - 4y + 4\right) - x^4$

6. $x^3 + y^{3k}$

Answers to Practice Problems: **1.** $5(x - 4)(x + 4)$ **2.** $(3x - 2)^2$ **3.** $(x - 4)(x + 1)$ **4.** $2(x - 5)\left(x^2 + 5x + 25\right)$
    5. $\left(y - 2 - x^2\right)\left(y - 2 + x^2\right)$ **6.** $\left(x + y^k\right)\left(x^2 - xy^k + y^{2k}\right)$

4.7 Exercises

For Exercises 1 – 80 completely factor each polynomial. If a polynomial cannot be factored, write not factorable.

1. $x^2 - 25$ **2.** $y^2 - 121$ **3.** $2x^2 - 128$ **4.** $3x^2 - 147$ **5.** $4x^4 - 64$

6. $5x^4 - 125$ **7.** $y^2 + 100$ **8.** $4x^2 + 49$ **9.** $-4x^2 - 24x + 8$

10. $-12x^4 + 18x^3 - 72x^2$ **11.** $9x^2 - 25$ **12.** $4x^2 - 49$

13. $y^2 - 10y + 25$ **14.** $x^2 + 12x + 36$ **15.** $9y^2 + 12y + 4$

16. $49x^2 - 14x + 1$ **17.** $16x^2 - 40x + 25$ **18.** $9x^2 - 12x + 4$

19. $4x^3 - 64x$ **20.** $50x^3 - 8x$ **21.** $2x^3y + 32x^2y + 128xy$

22. $3x^2y - 30xy + 75y$ **23.** $(x+y)^2 + 6(x+y) + 9$ **24.** $(4x+y)^2 + 4(4x+y) + 4$

25. $(x-y)^2 - 81$ **26.** $(x+2y)^2 - 25$ **27.** $x^{2k} - 20x^k + 100$

28. $25x^{2k} - 10x^k + 1$ **29.** $x^4 + 10x^2y + 25y^2$ **30.** $16x^4 + 8x^2y + y^2$

31. $x^3 - 125$ **32.** $x^3 - 64$ **33.** $y^3 + 216$ **34.** $y^3 + 1$

35. $x^3 + 27y^3$ **36.** $8x^3 + 1$ **37.** $3x^3 + 81$ **38.** $4x^3 - 32$

39. $64x^3 + 27y^3$ **40.** $54x^3 - 2y^3$ **41.** $3x^4 + 375xy^3$ **42.** $x^3y + y^4$

43. $x^4y^3 - x$ **44.** $x^2y^2 - x^2y^5$ **45.** $2x^2 - 16x^2y^3$ **46.** $24x^4y + 81xy^4$

47. $x^6 - 64y^3$ **48.** $x^6 - y^9$ **49.** $x^3 + (x-y)^3$ **50.** $27x^3 + (y^2-1)^3$

51. $(x-3y)^3 - 64z^3$ **52.** $(x+2)^3 + (y-3)^3$ **53.** $(x+y)^3 + (y-4)^3$

54. $(x+2y)^3 - (y+4)^3$ **55.** $(3x+2y)^3 + (y-3)^3$ **56.** $xy + 4x - 3y - 12$

57. $xy + 5y + 2x + 10$ **58.** $6x + 4y + xy + 24$ **59.** $7x - 3y + xy - 21$

60. $3xy + 6y - x - 1$ **61.** $2xy + 10x + 3y + 15$ **62.** $4xy - 28x + 3y - 21$

63. $6xy - 9x + 4y - 6$ **64.** $5x^2 + 10x + 3y + 15$ **65.** $(x^2 + 2x + 1) - y^2$

66. $\left(x^2 - 2xy + y^2\right) - 36$ **67.** $\left(x^2 + 4xy + 4y^2\right) - 25$ **68.** $\left(16x^2 + 8x + 1\right) - y^2$

69. $x^2 - \left(y^2 + 6y + 9\right)$ **70.** $x^3 - 5x^2 + 6x - 30$ **71.** $4x^3 - 6x^2 - 14x + 21$

72. $x^3 + 12x^2 - 4x - 48$ **73.** $x^3 + 3x^2 - 9x - 27$ **74.** $x^3 + 2x^2 - 4x - 8$

75. $x^3 + 7x^2 - 4x - 28$ **76.** $x^2 y + 5x^2 - 9y - 45$ **77.** $x^{2k} - 4y^2$

78. $x^{3k} + 8$ **79.** $x^{3k} + 27y^{3k}$ **80.** $3x^{3k} - 24x^{3k}y^{3k}$

81. a. Represent the shaded region of the square shown below as the difference of two squares.
  b. Use the factors of the expression in part a to draw (and label the sides) of a rectangle that has the same area as the shaded region.

82. a. Use a polynomial function to represent the shaded region of the square.
  b. Use a polynomial function to represent the perimeter of the shaded figure.

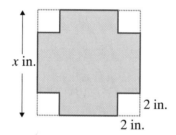

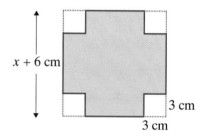

Writing and Thinking About Mathematics

83. a. Show that the sum of the areas of the rectangles and squares in the figure is a perfect square trinomial.

  b. Rearrange the rectangles and squares in the form of a square and represent its area as the square of a binomial.

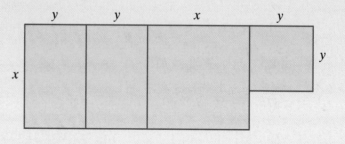

Writing and Thinking About Mathematics (cont.)

84. Compound interest is interest earned on interest. If a principal, P, is invested and compounded annually (once a year) at a rate of r, then the amount, A_1, accumulated in one year is $A_1 = P + Pr$.
In factored form, we have: $A_1 = P + Pr = P(1 + r)$.
At the end of the second year the amount accumulated is

$$A_2 = P + Pr + (P + Pr)r.$$

  a. Write the expression for A_2 in factored form similar to that for A_1.
  b. Write an expression for the amount accumulated in three years, A_3, in factored form.
  c. Write an expression for A_n the amount accumulated in n years.
  d. Use the formula you developed in part c and your calculator to find the amount accumulated if $10,000 is invested at 6% and compounded annually for 20 years.

85. You may have heard of (or studied) the following rules for division of an integer by 3 and 9:

  I. An integer is divisible by 3 if the sum of its digits is divisible by 3.
  II. An integer is divisible by 9 if the sum of its digits is divisible by 9.
    The proofs of both I and II can be started as follows:
    Let abc represent a three-digit integer.

    Then $abc = 100a + 10b + c$

$$= (99 + 1)a + (9 + 1)b + c$$

$$= (\text{now you finish the proofs}).$$

Use the pattern just shown and prove both I and II for a four-digit integer.

 HAWKES LEARNING SYSTEMS: INTERMEDIATE ALGEBRA SOFTWARE

- Special Factorizations - Squares
- Special Factorizations - Cubes

Polynomial Equations and Applications

4.8

- *Solve equations by factoring.*
- *Write equations given the roots.*
- *Solve problems related to consecutive integers.*
- *Solve problems related to the **Pythagorean Theorem**.*

The equations that we have solved to this point have been linear equations, identities, or contradictions. Each linear equation is first-degree and has exactly one solution. In order to solve equations of higher degree with more than one solution, new methods must be developed. These methods include factoring (*ac*-method, trial-and-error method, the difference of two squares, and the difference of two cubes). The solutions to an equation are also called the **roots** of the equation, and finding the solutions can be termed **finding the roots**.

Solving Equations by Factoring

Second-degree polynomials are called **quadratic polynomials** (or **quadratics**), and they play a major role in many applications in mathematics and physics. For example, the paths of thrown objects (or projectiles) affected by gravity, area (area of a circle, area of a rectangle, square, and triangle), and various number relationships are all related to quadratics.

Quadratic Equations

Quadratic equations are equations of the form

$$ax^2 + bx + c = 0 \quad \text{where a, b, and c are constants and } a \neq 0.$$

The form $ax^2 + bx + c = 0$ is called the **standard form** (or **general form**) of a **quadratic equation**. In standard form, a quadratic polynomial is on one side of the equation and 0 is on the other side. For example,

$$3x^2 + 5x + 2 = 0 \text{ is a quadratic equation in standard form,}$$

while,

$$x^2 - 8x = 12 \text{ and } x^2 = 36x \text{ are both quadratic equations, just not in standard form.}$$

These last two equations can be manipulated algebraically so that 0 is on one side. In solving equations by factoring, having 0 on one side is necessary because of the **zero-factor property**.

Zero-Factor Property

If a product is 0, then at least one of the factors must be 0. That is, for real numbers a and b,

if $a \cdot b = 0$, then $a = 0$ or $b = 0$ or both.

Example 1: Solving Factored Equations

Use the zero-factor property to solve each equation.

a. $(x-5)(2x-3) = 0$

Solution: The quadratic on the left of the equation is already in factored form and the product is 0. Set each factor equal to 0. This process yields two linear equations, which can, in turn, be solved.

$$x - 5 = 0 \qquad \text{or} \qquad 2x - 3 = 0$$
$$x = 5 \qquad\qquad\qquad 2x = 3$$
$$x = \frac{3}{2}$$

The solutions can be **checked** by substituting them, one at a time, for x in the equation. That is, there will be two "checks."

Substituting $x = 5$ gives

$$(x-5)(2x-3) = (5-5)(2 \cdot 5 - 3) = (0)(7) = 0.$$

Substituting $x = \frac{3}{2}$ gives

$$(x-5)(2x-3) = \left(\frac{3}{2} - 5\right)\left(\cancel{2} \cdot \frac{3}{\cancel{2}} - 3\right) = \left(-\frac{7}{2}\right)(0) = 0.$$

Thus the two solutions (or roots) to the original equation are $x = 5$ and $x = \frac{3}{2}$. Or, we can say that the solution set is $\left\{5, \frac{3}{2}\right\}$.

b. $(x+3)(x-2)(3x+5) = 0$

Solution: Again the polynomial is factored. But, in this case there are three factors and there are three solutions. Setting each factor equal to 0 and solving gives the three solutions (or roots):

Continued on the next page...

$$x + 3 = 0 \quad \text{or} \quad x - 2 = 0 \quad \text{or} \quad 3x + 5 = 0$$

$$x = -3 \qquad\qquad x = 2 \qquad\qquad 3x = -5$$

$$x = -\frac{5}{3}$$

The three solutions are $-3, 2,$ and $-\dfrac{5}{3}$.

(These can be checked in the original equation.)

In Example 1, the polynomials were given in factored form and the zero-factor property was applied directly. As we will see in Example 2, we may need to first perform algebraic manipulations so that one side of the equation is 0 and then factor before applying the zero-factor property. **One side of the equation must be 0.**

Example 2: Solving Equations By Factoring

Solve each polynomial equation by factoring.

a. $y^2 - 6y = 27$

> **Solution:** $\quad y^2 - 6y = 27$
>
> $\qquad\qquad y^2 - 6y - 27 = 0$ Add -27 to both sides.
> **One side must be 0.**
>
> $\qquad\qquad (y - 9)(y + 3) = 0$ Factor the left-hand side.
>
> $\qquad\qquad y - 9 = 0 \text{ or } y + 3 = 0$ Set each factor equal to 0.
>
> $\qquad\qquad y = 9 \qquad\quad y = -3$ Solve each linear equation.
>
> The solutions are -3 and 9.

b. $5x^2 + 10x + 5 = 0$

> **Solution:** $\;\; 5x^2 + 10x + 5 = 0$
>
> $\qquad\qquad 5(x^2 + 2x + 1) = 0$ Factor out the GCF, 5.
>
> $\qquad\qquad 5(x + 1)^2 = 0$ Factor the perfect square trinomial.
>
> $\qquad\qquad x + 1 = 0$ Set each factor equal to 0. (Since $5 \neq 0$ and $(x + 1)$ is a repeated factor, we need
>
> $\qquad\qquad x = -1$ to solve only one equation, $x + 1 = 0$.)

Because $(x + 1)$ is a double factor, the solution -1 is called a **double root** or a **root of multiplicity two**. (Note that the constant factor 5 does not affect the solution.)

c. $3x(x-1) = 2(5-x)$

> **Solution:**
> $$3x(x-1) = 2(5-x)$$
>
> | | |
> |---|---|
> | $3x^2 - 3x = 10 - 2x$ | Use the distributive property. |
> | $3x^2 - 3x + 2x - 10 = 0$ | Arrange terms so that 0 is on one side. |
> | $3x^2 - x - 10 = 0$ | Simplify. |
> | $(3x+5)(x-2) = 0$ | Factor. |
> | $3x + 5 = 0$ or $x - 2 = 0$ | Set each factor equal to 0. |
>
> $$3x = -5 \qquad\qquad x = 2$$
>
> $$x = -\frac{5}{3}$$
>
> The solutions are $-\dfrac{5}{3}$ and 2.

d. $100x = 4x^3$

> **Solution:** $100x = 4x^3$
>
> **Important Note : Do not divide by 4x. You will lose a solution if you divide by a variable.**
>
> | | |
> |---|---|
> | $0 = 4x^3 - 100x$ | Set one side, either the right side or the left side, equal to 0. |
> | $0 = 4x(x^2 - 25)$ | Factor out the GCF, $4x$. |
> | $0 = 4x(x+5)(x-5)$ | Factor the difference of two squares. |
> | $4x = 0$ or $x + 5 = 0$ or $x - 5 = 0$ | Set each factor equal to 0. |
>
> $$x = 0 \qquad\qquad x = -5 \qquad\qquad x = 5$$
>
> The solutions are $0, -5,$ and 5.

NOTES

All of the polynomial equations in this section can be solved by factoring. That is, all of the polynomials are factorable. However, as we have seen in some of the previous sections not all polynomials are factorable. In Chapter 7 we will develop techniques (other than factoring) for solving quadratic equations whether the quadratic polynomial is factorable or not.

NOTES

COMMON ERROR

A **common error** is to divide both sides of an equation by the variable x. This error can be illustrated by using the equation in Example 2d.

$$100x = 4x^3$$

$$\frac{100x}{x} = \frac{4x^3}{x}$$

INCORRECT
Do not divide by x because you lose the solution $x = 0$.

$$100 = 4x^2$$

$$25 = x^2$$

$$x = -5, 5$$

Factoring is the method to use. By factoring, you will find all solutions shown in the previous examples.

Note carefully that in Example 2d, had we divided by $4x$, we would have lost the solution 0.

Finding an Equation Given the Roots

To help develop a complete understanding of the relationships between factors, factoring, and solving equations, we reverse the process of finding solutions. That is, we want to **find an equation that has certain given solutions (or roots)**. For example, to find an equation that has the roots

$$x = 4 \quad \text{and} \quad x = -7$$

we proceed as follows:

1. Rewrite the linear equations with 0 on one side:
$$x - 4 = 0 \quad \text{and} \quad x + 7 = 0.$$

2. Form the product of the factors and set this product equal to 0:
$$(x - 4)(x + 7) = 0.$$

3. Multiply the factors. The resulting quadratic equation must have the two given roots (because we know the factors):
$$x^2 + 3x - 28 = 0.$$

NOTES

This equation can be multiplied by any nonzero constant and a new equation will be formed, but it will still have the same solutions, namely 4 and −7. Thus, technically, there are many equations with these two roots.

The formal reasoning is based on the following theorem called the **Factor Theorem**.

Factor Theorem

If $x = c$ is a root of a polynomial equation in the form $P(x) = 0$, then $x - c$ is a factor of the polynomial $P(x)$.

Proof:

The division algorithm says that

$$P(x) = (x - c) \cdot Q(x) + r.$$

But, because c is a root we must have $P(c) = 0$.

We also know that $P(c) = (c - c) \cdot Q(c) + r = 0 \cdot Q(c) + r = 0 + r = r$.

Now, since $P(c) = 0$ and $P(c) = r$, this gives the result that $r = 0$. Thus when c is a root of $P(x)$, $P(x) = (x - c) \cdot Q(x)$ and $(x - c)$ is a factor of $P(x)$.

Example 3: Using the Factor Theorem to Find Equations with Given Roots

Find a polynomial equation with integer coefficients that has the given roots:

$$x = 3 \text{ and } x = -\frac{2}{3}.$$

Solution: Form the linear equations and then the product of the factors.

$$x = 3 \qquad x = -\frac{2}{3}$$

$$x - 3 = 0 \qquad x + \frac{2}{3} = 0 \qquad \text{Set each equation equal to 0.}$$

$$3x + 2 = 0 \qquad \text{Multiply by 3 to get integer coefficients.}$$

Form the equation by setting the product of the factors equal to 0 and simplifying.

$$(x - 3)(3x + 2) = 0 \qquad \text{The equation has integer coefficients.}$$

$$3x^2 - 7x - 6 = 0 \qquad \text{This quadratic equation has the two given roots.}$$

Consecutive Integers

Applications related to integers often involve one of the following three categories: **consecutive integers**, **consecutive even integers**, or **consecutive odd integers**.

Consecutive Integers

*Integers are **consecutive** if each is 1 more than the previous integer.*
*Three consecutive integers can be represented as **n, n + 1**, and **n + 2**.*
For example: 5, 6, 7

Consecutive Even Integers

Even integers are consecutive if each is 2 more than the previous even integer.
*Three consecutive even integers can be represented as **n, n + 2**, and **n + 4**.*
For example: 24, 26, 28

Consecutive Odd Integers

Odd integers are consecutive if each is 2 more than the previous odd integer.
*Three consecutive odd integers can be represented as **n, n + 2**, and **n + 4**.*
For example: 41, 43, 45

Note that consecutive even and consecutive odd integers are represented in the same way. The value of the first integer n determines whether the remaining integers are even or odd.

Example 4: Consecutive Integers

a. Find two consecutive positive integers such that the sum of their squares is 265.

Solution: Let n = first integer
$n + 1$ = next consecutive integer
Set up and solve the related equation.

$$n^2 + (n+1)^2 = 265$$

$$n^2 + n^2 + 2n + 1 = 265$$

$$2n^2 + 2n - 264 = 0$$

$$2(n^2 + n - 132) = 0$$

$$2(n + 12)(n - 11) = 0$$

$$n + 12 = 0 \quad \text{or} \quad n - 11 = 0$$
$$n = -12 \qquad\qquad n = 11$$
$$n + 1 = -11 \qquad\quad n + 1 = 12$$

Consider the solution $n = -12$. The next consecutive integer, $n + 1$, is -11. While it is true that the sum of their squares is 265, we must remember that the problem calls for **positive** consecutive integers. Therefore, we can only consider positive solutions. Hence, the two integers are 11 and 12.

b. Find three consecutive odd integers such that the product of the first and second is 68 more than the third.

Solution: Let n = first odd integer
and $n + 2$ = second consecutive odd integer
and $n + 4$ = third consecutive odd integer.
Set up and solve the related equation.

$$n(n+2) = n + 4 + 68$$

$$n^2 + 2n = n + 72$$

$$n^2 + n - 72 = 0$$

$$(n+9)(n-8) = 0$$

$$n + 9 = 0 \quad \text{or} \quad n - 8 = 0$$

$$n = -9 \qquad\qquad n = 8$$

$$n + 2 = -7 \qquad n + 2 = 10$$

$$n + 4 = -5 \qquad n + 4 = 12$$

The three consecutive odd integers are -9, -7, and -5. Note that 8, 10 and 12 are even and therefore cannot be considered a solution to the problem.

The Pythagorean Theorem

Pythagoras

A geometric topic that often generates quadratic equations is right triangles. In a **right triangle**, one of the angles is a right angle (measures 90°), and the side opposite this angle (the longest side) is called the **hypotenuse**. The other two sides are called **legs**. Pythagoras (c. 585 – 501 B.C.), a famous Greek mathematician, is given credit for proving the following very important and useful theorem (even though history indicates that the Chinese knew of this theorem centuries before Pythagoras). Now, there are entire books written that contain only proofs of the Pythagorean Theorem developed by mathematicians since the time of Pythagoras. (You might want to visit the library!)

You will see the Pythagorean Theorem stated again in Chapter 7 and used throughout your studies in mathematics.

The Pythagorean Theorem

In a right triangle, the square of the hypotenuse is equal to the sum of the squares of the legs.

$$c^2 = a^2 + b^2$$

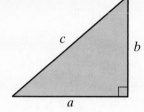

Example 5: The Pythagorean Theorem

A support wire is 25 feet long and stretches from a tree to a point on the ground. The point of attachment on the tree is 5 feet higher than the distance from the base of the tree to the point on the ground. How far up the tree is the point of attachment?

Solution: Let x = distance from base of tree to point on ground (see diagram), then $x + 5$ = height of point of attachment.

By the Pythagorean Theorem, we have:

$$(x+5)^2 + x^2 = 25^2$$

$$x^2 + 10x + 25 + x^2 = 625$$

$$2x^2 + 10x - 600 = 0$$

$$2(x^2 + 5x - 300) = 0$$

$$2(x - 15)(x + 20) = 0$$

$$x = 15 \text{ or } x = -20.$$

Because distance is positive, -20 is not a possible solution. The solution is

$$x = 15$$

and $x + 5 = 20.$

Thus the point of attachment is 20 feet up the tree.

Practice Problems

Solve the following equations by factoring.

1. $y^2 - 4y = 21$ **2.** $3x^2 - 16x + 5 = 0$

3. $z^2 + 6z = -9$ **4.** $x^3 = 25x$

5. Find a quadratic equation with integer coefficients that has the roots $\dfrac{2}{3}$ and $\dfrac{1}{2}$.

4.8 Exercises

Find the solution sets for the polynomial equations in Exercises 1 – 6.

1. $x(2x-1)(3x+1) = 0$ **2.** $2x(x-5)(2x+3) = 0$

3. $(x+1)(x-2)(x-6) = 0$ **4.** $(x-3)(2x-5)(x-4)(4x+9) = 0$

5. $x^3(5x-1)(2x+3) = 0$ **6.** $(4x+1)(3x-2)(x-8.5)(6x-1) = 0$

Solve the equations in Exercises 7 – 48 by factoring.

7. $x^2 + 13x + 36 = 0$ **8.** $x^2 + 17x + 72 = 0$ **9.** $5x^2 - 70x + 240 = 0$

10. $2y^2 - 24y + 70 = 0$ **11.** $4x^2 = 20x + 200$ **12.** $7n^2 + 14n = 168$

13. $3x^2 = 147$ **14.** $64 - 49x^2 = 0$ **15.** $3x^2 + 10 = 17x$

16. $2x^2 = 3x - 1$ **17.** $6x^2 - 11x + 4 = 0$ **18.** $4y^2 = 14y - 6$

19. $2n^2 - 72 = 0$ **20.** $3x^2 - 27 = 0$ **21.** $4z^2 - 49 = 0$

22. $9x^2 - 16 = 0$ **23.** $2z^2 + 3 = 7z$ **24.** $34x + 6 = 12x^2$

25. $12x^2 = 6 - x$ **26.** $12x^2 + 5x = 3$ **27.** $6n^2 + n = 35$

28. $50y^2 - 98 = 0$ **29.** $150y^2 - 96 = 0$ **30.** $8y^2 + 6y = 35$

Answers to Practice Problems: **1.** $y = 7, y = -3$ **2.** $x = \dfrac{1}{3}, x = 5$ **3.** $z = -3$ **4.** $x = 0,\ x = 5,\ x = -5$

            5. $6x^2 - 7x + 2 = 0$

31. $(x+5)(x-7)=13$

32. $(2x+3)(x-1)=-2$

33. $x(x-5)+9=3(x-1)$

34. $x(2x+3)-2=2(x+4)$

35. $2x(x+3)-14=x(x-2)+19$

36. $x(3x+5)=x(x+2)+14$

37. $18y^2-15y+2=0$

38. $14+11y=15y^2$

39. $63x^2=40x+12$

40. $12z^2-47z+11=0$

41. $3x^3+15x^2+18x=0$

42. $x^3=4x^2+12x$

43. $16x^3-100x=0$

44. $112x-2x^2=2x^3$

45. $12n^3+2n^2=70n$

46. $21n^3=13n^2-2n$

47. $63x=3x^2+30x^3$

48. $14x^3+60x^2=50x$

In Exercises 49 – 58, write a polynomial equation with integer coefficients that has the given roots.

49. $y=3, y=-2$

50. $x=5, x=7$

51. $x=\dfrac{1}{2}, x=\dfrac{3}{4}$

52. $y=\dfrac{2}{3}, y=\dfrac{1}{6}$

53. $x=0, x=3, x=-2$

54. $y=0, y=-4, y=1$

55. $x=-5, x=-3$

56. $x=\dfrac{1}{4}, x=-1$

57. $y=-2, y=3, y=3$ (3 is a double root.)

58. $x=-1, x=-1, x=-1$ (−1 is a triple root.)

59. Find two consecutive positive integers such that the sum of their squares is 113.

60. The product of two consecutive even integers is 168. Find the integers.

61. The product of two consecutive odd integers is 420 more than three times the smaller integer. Find the integers.

62. Find three consecutive positive integers such that twice the product of the two smaller integers is 88 more than the product of the two larger integers.

63. Find three consecutive even integers such that the sum of their squares is 440.

64. Find three consecutive odd integers such that the product of the first and third is 71 more than 10 times the second.

65. Four consecutive integers are such that if the product of the first and third is multiplied by 6, the result is equal to the sum of the second and the square of the fourth. What are the integers?

66. Find four consecutive even integers such that the square of the sum of the first and second is equal to 516 more than twice the product of the third and fourth.

67. Architecture: An architect wants to draw a rectangle with a diagonal of 13 inches. The length of the rectangle is to be 2 inches more than twice the width. What dimensions should she make the rectangle?

68. Holiday decorating: A Christmas tree is supported by a wire that is 1 foot longer than the height of the tree. The wire is anchored at a point whose distance from the base of the tree is 49 feet shorter than the height of the tree. What is the height of the tree?

69. Mountain climbing: Two mountain peaks are known to be 29 miles apart. A person located at a vista point, such that a right angle is spanned when he looks from one peak to the other, is told that the distances from the vista point to each peak differ by 1 mile. How far is the vista point from each mountain peak?

70. Communication: A telephone pole is to have a guy wire attached to its top and anchored to the ground at a point that is 34 feet less than the height of the pole. If the wire is to be 2 feet longer than the height of the pole, what is the height of the pole?

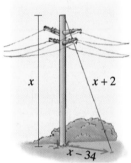

71. Gravity: A ball is dropped from the top of a building that is known to be 256 feet high. The formula for finding the height of the ball at any time is $h = 256 - 16t^2$ where t is measured in seconds.

   a. How many seconds will it take for the ball to hit the ground?

   b. How many seconds did it take for the ball to reach a height of 192 feet above the ground?

   c. How many seconds did it take for the ball to fall 64 feet?

72. If three positive integers satisfy the Pythagorean Theorem, they are called a **Pythagorean Triple**. For example, 3, 4, and 5 are a Pythagorean Triple because $3^2 + 4^2 = 5^2$. There are an infinite number of such triples. To see how some triples can be found, fill out the following table and verify that the numbers in the rightmost three columns are indeed Pythagorean Triples.

u	v	$2uv$	$u^2 - v^2$	$u^2 + v^2$
2	1	4	3	5
3	2			
5	2			
4	3			
7	1			
6	5			

73. The pattern in Kara's linoleum flooring is in the shape of a square 8 inches on a side with right triangles of sides x inches placed on each side of the original square so that a new larger square is formed. What is the area of the new square? Explain why you do not need to find the value of x.

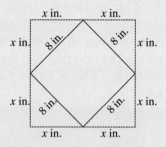

74. Write and prove the Factor Theorem.

 HAWKES LEARNING SYSTEMS: INTERMEDIATE ALGEBRA SOFTWARE

- Solving Equations by Factoring

4.9 Using a Graphing Calculator to Solve Equations

- *Use a TI-84 Plus graphing calculator to solve (or estimate the solutions of) polynomial equations by using one of the following strategies:*
  - *a. graph one function and find the zeros of the function, or*
  - *b. graph two functions and find the points of intersection of the functions.*
- *Use a TI-84 Plus graphing calculator to solve absolute value equations and inequalities.*

In Section 2.4, we introduced functions, showed how to use a graphing calculator to graph functions, and listed steps for finding the zeros of a function. **Remember that the zeros of a function are the values of *x* at the points, if any, where the graph of the function crosses the *x*-axis.** These are the points where $y = 0$. In this section we will simply expand on these ideas and use the related concepts to solve equations (or to estimate the solutions of equations).

NOTES

When a polynomial is second-degree or higher, the same factor may occur multiple times. This means that the corresponding zero may appear more than once. That is, if a binomial factor is squared, then the corresponding zero is said to be of multiplicity 2. If the factor is cubed, then the corresponding zero is of multiplicity 3, and so on. For example, $P(x) = x^3 - 3x^2 - 24x + 80 = (x + 5)(x - 4)^2$ and there are technically three zeros, $-5, 4$, and 4. But 4 appears twice, so 4 is a zero of multiplicity 2 and the only distinct zeros are -5 and 4. As we will see, this has a major effect on the nature of the corresponding graph of the function.

With the calculator as a tool, we can solve (or estimate the solutions of) linear, quadratic, or higher degree polynomial equations. But, we must be careful with the ⬭WINDOW⬭ settings or the display may not show enough of the graph to illustrate all of the solutions. The following information will help in determining whether or not your ⬭WINDOW⬭ setting is appropriate.

Zeros of Polynomial Functions

1. *Nonconstant linear functions have 1 zero. (The graph crosses the x-axis once.)*
2. *Quadratic functions have 2 zeros or none. (The graph crosses the x-axis twice, just touches the x-axis, or doesn't cross at all.)*
3. *Cubic functions have 3 zeros or 1 zero. (The graph crosses the x-axis three times, crosses once and just touches once, or crosses just once.)*

Note: A zero of multiplicity n is considered to be n zeros.

If you are graphing a cubic function and the display shows only 1 zero (the graph crosses the x-axis only once), you may need a larger [WINDOW] setting to determine whether or not there are more zeros. If you see 3 zeros, then you know that there are no more and the [WINDOW] setting is sufficient.

> **NOTES**
>
> **Important Note about the Graphs of Polynomial Functions:**
> The graph of every polynomial function is a smooth continuous graph. (That is, there are no holes, jumps from one point to another, or sharp points in the graph of a polynomial function.)

Now, there are two basic strategies to solving equations using the graphing calculator. We need to either:

> **a.** graph one function and find the zeros of the function, or
> **b.** graph two functions and find the points of intersection of the functions.

The following examples illustrate both of these strategies and the steps to use with the graphing calculator.

Example 1: Using One Graph to Solve a Polynomial Equation

Use a TI-84 Plus graphing calculator to solve the equation $x^3 - 3x^2 = 13x - 15$.

Solution:

Strategy: Manipulate the equation so that one side is 0. Graph the indicated function on the nonzero side. The zeros of this function are the roots of the original equation.

$$x^3 - 3x^2 = 13x - 15$$

$$x^3 - 3x^2 - 13x + 15 = 0$$

Enter the function as follows:

With the standard window the graph will appear as follows:

Note: You may want to increase the *y*-values on the window to see a more complete graph. This will not change the zeros.

With the **2ND** > **CALC** > **2:zero** sequence of commands you will find the following zeros (and therefore solutions to the equation):

$$x = -3, x = 1, \text{ and } x = 5.$$

[**Note:** See Section 2.4 for a more in-depth explanation of the zero function. With the **TRACE** command you will find only approximations of the zeros.]

Example 2: Using Two Graphs to Solve a Polynomial Equation

Solve the polynomial equation $2x^2 = 3x + 1$.

Solution:

Strategy: Graph the function indicated on each side of the equation. Find the points of intersection of these two graphs. The *x*-values of these points are the roots of the original equation.

Continued on the next page...

Enter the functions as follows:

With the standard window the graphs will appear as follows:

With the **2ND** > **CALC** > **5:intersect** sequence of commands you will find the following approximate x-values of the points of intersection (and therefore approximate solutions to the equation):

$$x \approx -.28 \quad \text{and} \quad x \approx 1.78 \quad \text{(accurate to two decimal places)}.$$

[**Note:** See Section 3.1 for an explanation of the intersect function.]

Example 3: Using Two Graphs to Solve an Absolute Value Equation

Solve the equation $|2x - 5| = 8$.

Solution:

Strategy: Graph the functions indicated on each side of the equation. This includes the constant function. Find the points of intersection of these two graphs. The x-values of these points are the roots of the original equation. Remember that the absolute value command can be found in the **MATH** > **NUM** menu.

Enter the functions as follows:

With the standard window the graphs will appear as follows:

With the **2ND** > **CALC** > **5:intersect** sequence of commands you will find the following x-values at the points of intersection (and therefore the solutions to the equation):

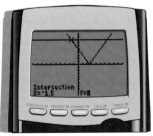

$$x = -1.5 \quad \text{and} \quad x = 6.5$$

[**Note:** These values of x are exact. The **TRACE** command will not give these exact values.]

Example 4: Using a Graphing Calculator to Solve Absolute Value Inequalities

Use a graphing calculator to solve the inequalities:

a. $|2x-5| < 8$ **b.** $|2x-5| > 8$

Both inequalities can be solved by using the graphs from Example 3.

Solution: We change the window for a clearer view: Use the interval $[-10, 10]$ for x and the interval $[-1, 15]$ for y. This gives:

a. From Example 3, we know that the intersections occur at $x = -1.5$ and $x = 6.5$. Looking at the graph we see that the graph of the absolute value is below the line $y = 8$ on the interval $(-1.5, 6.5)$. Thus the interval $(-1.5, 6.5)$ is the solution set for $|2x-5| < 8$.

b. Looking at the graph we see that the graph of the absolute value is above the line $y = 8$ on the intervals $(-\infty, -1.5)$ and $(6.5, \infty)$. Thus the solution set is $(-\infty, -1.5) \cup (6.5, \infty)$ for $|2x-5| > 8$.

4.9 Exercises

Use a graphing calculator to solve (or estimate the solutions of) the equations in Exercises 1 – 35. Find any approximations accurate to two decimal places.

1. $x^2 - 4 = 0$ **2.** $x^2 - 9 = 0$ **3.** $x^2 - 2 = 0$

4. $x^2 - 15 = 0$ **5.** $x^2 - 4x = 12$ **6.** $x^2 + 6x = 7$

7. $x^2 + 2x - 11 = 0$ **8.** $3x^2 - x - 6 = 0$ **9.** $-x^2 + 3x + 8 = 0$

10. $-2x^2 + 4x - 5 = 0$ **11.** $2x^2 + x + 2 = 0$ **12.** $3x + 15 = x^2$

13. $3x^2 - 9 = x$ **14.** $x^3 = 2x^2 - 5$ **15.** $5x - 3 = x^3$

16. $x^3 + 2x^2 = 4x + 6$

17. $x(x-1)(x-3) = 0$

18. $(x-2)(x+1)(x+4) = 0$

19. $(x+3)(x+1)(x-5) = 0$

20. $(x+2)(x-1)(x-6) = 0$

21. $2x^3 - 8x^2 + 7x - 1 = 9$

22. $3x^3 - x^2 + 4 = 10$

23. $-x^3 + 4x^2 - x = 5$

24. $x^4 - 10x^2 = 0$

25. $x^4 = 3x^2$

26. $x^4 - x^3 + 2x = 0$

27. $|2x - 3| = 11$

28. $|2x + 1| = 7$

29. $|3x - 2| = 7$

30. $|4x + 1| = 19$

31. $|2x + 1| = |x - 1|$

32. $|x - 3| = |x + 2|$

33. $\left|\dfrac{x}{2} + 1\right| = \dfrac{3}{2}$

34. $\left|\dfrac{x}{5} - 1\right| = |x|$

35. $|x - 2| = |5 - x|$

Use a graphing calculator to solve (or estimate the solutions of) the inequalities in Exercises 36 – 45. Write your answers in interval notation.

36. $|x| > 6$

37. $|x| \le 3$

38. $|x - 3| \le 1$

39. $|x - 5| > 2$

40. $|x - 4| \ge 2$

41. $|3x - 8| > 4$

42. $|x + 2| - 10 \le 17$

43. $|x - 4| - 2 > 10$

44. $\left|\dfrac{x}{3} - 1\right| < 2$

45. $\left|\dfrac{x}{4} + 3\right| \ge 1$

Writing and Thinking About Mathematics

Use a graphing calculator and three graphs to solve the inequality. Write the answer in interval notation. Explain how you might solve this inequality algebraically.

46. $1 \le |x - 4| \le 5$

 HAWKES LEARNING SYSTEMS: INTERMEDIATE ALGEBRA SOFTWARE

- Using a Graphing Calculator to Solve Equations

Chapter 4 Index of Key Ideas and Terms

Section 4.1   Exponents and Scientific Notation

Properties of Exponents   page 288

If a and b are nonzero real numbers and m and n are integers:

The Exponent 1: $a^1 = a$

The Exponent 0: $a^0 = 1$

Product Rule: $a^m \cdot a^n = a^{m+n}$

Quotient Rule: $\dfrac{a^m}{a^n} = a^{m-n}$

Negative Exponents: $a^{-n} = \dfrac{1}{a^n}$

Power Rule: $\left(a^m\right)^n = a^{mn}$

Power of a Product: $\left(ab\right)^n = a^n b^n$ $\Big\}$ The Power Rules

Power of a Quotient: $\left(\dfrac{a}{b}\right)^n = \dfrac{a^n}{b^n}$

Scientific Notation   page 289

If N is a decimal number, then in **scientific notation** $N = a \times 10^n$ where $1 \le a < 10$ and n is an integer.

Section 4.2   Addition and Subtraction with Polynomials

Monomials   page 296

A **monomial in x** is an expression of the form kx^n where n is a whole number and k is any real number. n is called the **degree** of the monomial, and k is the **coefficient**. The **degree of a monomial in more than one variable** is the sum of the exponents of its variables. A nonzero constant, such as 5, is considered to be a monomial of **degree 0**. Zero itself is considered to be a monomial of **no degree**.

Polynomials   page 297

A **polynomial** is a monomial or the algebraic sum or difference of monomials. The **degree of a polynomial** is the largest of the degrees of its terms after like terms have been combined. The coefficient of the term of largest degree is called the **leading coefficient**. Generally, we write polynomials in **descending order of degree** from left to right.

Section 4.2   Addition and Subtraction with Polynomials (cont.)

Classification of Polynomials   page 297
    Monomial: polynomial with one term
    Binomial: polynomial with two terms
    Trinomial: polynomial with three terms

Addition of Polynomials   page 298
    The sum of two or more polynomials can be found by
    combining like terms.

Subtraction of Polynomials   page 299
    To find the **difference** of two polynomials, either
    a. add the opposite of each term being subtracted, or
    b. use the distributive property and multiply each term
       being subtracted by −1, then add.

$P(x)$ Notation   page 300
    $P(x)$ [read "P of x"] is **function notation** and indicates
    that **P** is the name of the polynomial and **x** is the variable
    used in the polynomial.

Section 4.3   Multiplication with Polynomials

Multiplication by Monomials   page 305

Multiplication of Two Polynomials   pages 305-306
    The distributive property is applied by multiplying each
    term of one polynomial by each term of the other.

FOIL Method   page 307
    When finding the product of two binomials, F-O-I-L is a
    mnemonic device to help remember which terms of the
    binomials to multiply together:
    First terms, **O**utside terms, **I**nside terms, and **L**ast terms.

Special Products   pages 309-310
    I. $(x+a)(x-a) = x^2 - a^2$: Difference of two squares

    II. $(x+a)^2 = x^2 + 2ax + a^2$: Square of a binomial sum

    III. $(x-a)^2 = x^2 - 2ax + a^2$: Square of a binomial difference

Section 4.4 Division with Polynomials

Division by a Monomial   page 315

To divide a polynomial by a monomial, divide each term in
the numerator by the monomial denominator and simplify
each fraction.

The Division Algorithm   page 316

For polynomials P and D, the **division algorithm** gives

$$\frac{P}{D} = Q + \frac{R}{D}, \ D \neq 0$$

where Q and R are polynomials and the
degree of $R <$ degree of D.

Section 4.5 Introduction to Factoring Polynomials

Factoring out the GCF   page 323

To find a monomial that is the **greatest common factor
(GCF)** of a polynomial:
1. Find the variable(s) of highest degree and the largest
    integer coefficient that is a factor of each term of the
    polynomial. (This is one factor.)
2. Divide this monomial factor into each term of the
    polynomial resulting in another polynomial factor.

Not Factorable (or Irreducible or Prime)   page 324

Factoring polynomials means to find factors that are
integers or polynomials with integer coefficients. If
this cannot be done, we say that the polynomial is **not
factorable** (or **irreducible** or **prime**).

Factoring by Grouping   page 325

Factoring polynomials with four or more terms by
grouping the terms in such a way that a common binomial
factor or some other form of factors can be recognized is
called **factoring by grouping**.

Section 4.6 Factoring Trinomials

Trial-and-Error Method of Factoring (FOIL Method)   page 331

For a trinomial $ax^2 + bx + c$ where a, b, and c are real constants:

1. List all the possible combinations of factors of ax^2 and c in their respective F and L positions.
2. Check the sum of the products in the O and I positions until you find the sum to be bx.
3. If none of the sums is bx, the trinomial is not factorable.

***ac*-Method of Factoring**   pages 334-335

For a trinomial $ax^2 + bx + c$ where a, b, and c are real constants:

1. Multiply $a \cdot c$.
2. Find two integers whose product is ac and whose sum is b. If this is not possible, then the trinomial is not factorable.
3. Rewrite the middle term (bx) using the two numbers found in step 2 as coefficients.
4. Factor by grouping the first two terms and the last two terms.
5. Factor out the common binomial factor. This will give two binomial factors of the trinomial $ax^2 + bx + c$.

Section 4.7 Special Factoring Techniques

Special Factoring Techniques

I. $x^2 - a^2 = (x+a)(x-a)$: Difference of two squares   page 340

II. $x^2 + 2ax + a^2 = (x+a)^2$: Perfect square trinomial   page 341

III. $x^2 - 2ax + a^2 = (x-a)^2$: Perfect square trinomial   page 341

IV. $x^3 + a^3 = (x+a)(x^2 - ax + a^2)$: Sum of two cubes   page 343

V. $x^3 - a^3 = (x-a)(x^2 + ax + a^2)$: Difference of two cubes   page 343

Sum of Two Squares   page 341

The **sum of two squares** is an expression of the form $x^2 + a^2$ and is **not factorable**.

Section 4.8 Polynomial Equations and Applications

Section 4.9 Using a Graphing Calculator to Solve Equations

Zeros of Polynomial Functions   page 361
   1. Nonconstant linear functions have 1 zero.
   2. Quadratic functions have 2 zeros or none.
   3. Cubic functions have 3 zeros or 1 zero.

Solving Polynomial Equations with a Graphing Calculator   pages 362-364

Solving Absolute Value Equations with a Graphing Calculator   page 365

Solving Absolute Value Inequalities with a Graphing Calculator page 366

 HAWKES LEARNING SYSTEMS: INTERMEDIATE ALGEBRA SOFTWARE

For a review of the topics and problems from Chapter 4, look at the following lessons from *Hawkes Learning Systems: Intermediate Algebra*.

- Simplifying Integer Exponents I
- Simplifying Integer Exponents II
- Scientific Notation
- Identifying Polynomials
- Adding and Subtracting Polynomials
- Multiplying Polynomials
- The FOIL Method
- Division by a Monomial
- The Division Algorithm
- GCF of a Polynomial
- Factoring by Grouping
- Factoring Trinomials by Trial and Error
- Factoring Trinomials by the *ac*-Method
- Special Factorizations - Squares
- Special Factorization - Cubes
- Solving Equations by Factoring
- Using a Graphing Calculator to Solve Equations
- Chapter 4 Review and Test

Chapter 4 Review

4.1 Exponents and Scientific Notation

Use the properties of exponents to simplify the expressions in Exercises 1 – 10. Answers should contain only positive exponents.

1. $-2x^3 \cdot 3x^7$

2. $\dfrac{x}{x^{-1}}$

3. $x^{2k} \cdot x$

4. $\left(\dfrac{y^2 y^{-1}}{y^4 y}\right)^2$

5. $\dfrac{30y^3}{5y^{-3}}$

6. $\left(5x^4\right)^{-2}$

7. $\left(\dfrac{x^5 y}{xy^2}\right)\left(\dfrac{x^{-2}}{y^3}\right)^0$

8. $\left(\dfrac{a^3 b}{6b^2}\right)^{-2}$

9. $\left(\dfrac{a^2 b^{-3}}{a^{-1}}\right)^3 \left(\dfrac{a^{-1} b^2}{ab}\right)^2$

10. $\dfrac{\left(xy^4\right)^2 \left(x^{-2} y\right)^3}{\left(x^2 y^2\right)^{-3}}$

In Exercises 11 – 14, write each number in decimal notation.

11. 4.56×10^3

12. 5.91×10^{-5}

13. 6.7×10^{-8}

14. 8.235×10^6

In Exercises 15 – 20, write each number in the following expressions in scientific notation and simplify. Show the steps you use. Do not use a calculator. Leave the answers in scientific notation.

15. 896,000

16. 0.000 000 992

17. $0.003 \times 0.000\ 012$

18. $\dfrac{0.036 \times 0.0057}{0.18 \times 0.0019}$

19. $\dfrac{933 \times 1.4 \times 10^{-3}}{3.11 \times 70,000}$

20. $\dfrac{43,000 \times 1.5 \times 10^{-4}}{3.0 \times 10 \times 86}$

4.2 Addition and Subtraction with Polynomials

In Exercises 21 and 22, state whether the expression is or is not a polynomial. If the expression is a polynomial, state its degree, its classification as a monomial, binomial, or trinomial, and its leading coefficient.

21. $2x^2 + 3y^{\frac{1}{2}} + 3$

22. $\dfrac{3y + 5y^4 + y^2}{y}$

In Exercises 23 – 31, find the indicated sums and differences. Simplify each answer.

23. $\left(5x^2 - 8x + 3\right) + \left(2x^2 + 8x + 5\right)$

24. $\left(x^3 + 3x - 10\right) + \left(4x^2 + 7\right)$

25. $\left(16x^2 - 20x - 14\right) - \left(16x^2 + 20x + 14\right)$

26. $\left(-2x^2 + 6x + 23\right) - \left(3x^2 - 2x + 5\right)$

27. $y^2 - (4xy + 2y^2) + (3x^2 - xy + y^2)$

28. $9x^2 - 2x + 3$
$\underline{5x^2 + 5x - 10}$

29. $x^3 + 3x^2 + x - 4$
$\underline{x^3 - 2x^2 + x - 5}$

30. $4x^3 \qquad - 20x + 10$
$\underline{-(x^3 - 2x^2 + 13x - 14)}$

31. $-5x^2 + 9x - 4$
$\underline{-(-2x^2 - x + 5)}$

Evaluate each polynomial in Exercises 32 – 35 for the specified value(s) of the variable(s).

32. Given $P(x) = 3x^2 - x + 5$; find **a.** $P(-2)$ **b.** $P(0)$.

33. Given $P(x) = x^3 - 3x^2 + x - 1$; find **a.** $P(3)$ **b.** $P(-3)$.

34. Given $P(x, y) = 4x - 2xy + y^2$; find **a.** $P(1, -1)$ **b.** $P(3, 1)$.

35. Given $P(x, y, z) = x^2 - 2xyz + z^3$; find **a.** $P(-2, 3, -1)$ **b.** $P(1, 2, 3)$.

4.3 Multiplication with Polynomials

Find the indicated products and simplify in Exercises 36 – 50.

36. $-2x^2(3x^3 - 5x)$

37. $7x^3(-5x^2 + 4x + 1)$

38. $(2y + 1)(y^2 - 4y - 5)$

39. $(y + 6)(y^3 - 2y - 3)$

40. $(2x + 5)(x + 1)$

41. $(3x - 7)(2x + 9)$

42. $(3x + 2)^2$

43. $(2x + 9)^2$

44. $(7y - 8)^2$

45. $(x + 14)(x - 14)$

46. $(3y + 5)(3y - 5)$

47. $(8y + 1)(8y - 1)$

48. $x^2 + 4x + 3$
$\underline{\qquad 3x - 7}$

49. $2x^2 - x - 3$
$\underline{x^2 + 5x + 6}$

50. $x^2 + 3x + 1$
$\underline{x^2 + 3x + 1}$

4.4 Division with Polynomials

Write each quotient in Exercises 51 – 58 as a sum of fractions in simplified form.

51. $\dfrac{9y^2 + 6y + 2}{3}$

52. $\dfrac{-2x^2 - 10x + 4}{-2}$

53. $\dfrac{9x^3 - 12x^2 + 4x}{x^2}$

54. $\dfrac{4x^3 + 8x^2 - 5x}{x^2}$

55. $\dfrac{4y^2 + 16y + 20}{2y^2}$

56. $\dfrac{90x^6 + 72x^5 - 108x^3}{18x^4}$

57. $\dfrac{2x^4 + 8x^3 - 12x^2 + 4x}{4x}$ **58.** $\dfrac{12y^4 + 6y^3 - 3y^2 + 11y}{3y^2}$

In Exercises 59 – 66, divide by using the division algorithm. Write the answers in the form $Q + \dfrac{R}{D}$*, where the degree of R < the degree of D.*

59. $\dfrac{x^2 - 2x - 15}{x - 5}$

60. $\dfrac{x^2 + 9x + 20}{x + 5}$

61. $\dfrac{4x^2 - 20x + 3}{x - 2}$

62. $\dfrac{8y^2 + 10y + 5}{y + 3}$

63. $\dfrac{64x^3 - 125}{4x - 5}$

64. $\dfrac{8x^3 + 27}{2x + 3}$

65. $\dfrac{3x^3 + 4x^2 - 7x + 1}{x^2 + 2}$

66. $\dfrac{x^3 - 8x^2 + 20x - 50}{x^2 - 2x + 3}$

4.5 Introduction to Factoring Polynomials

In Exercises 67 – 76, completely factor each polynomial. If a polynomial cannot be factored, write not factorable.

67. $6ax^2 + 42ax$

68. $4ax^3 + 8ax$

69. $-6x^4 + 12x^3 - 24x^2$

70. $-7x^4 + 14x^3 + 28x^2$

71. $xy + 3y + 5x + 15$

72. $x^2 - xy - 6x + 6y$

73. $x^2 + ax + 4x + 4y$

74. $5x^2 - 5x + y^2 - 5y$

75. $xy - 10x - 2y + 20$

76. $4xy - 28x + 5y - 35$

4.6 Factoring Trinomials

In Exercises 77 – 88, completely factor each polynomial. If the polynomial cannot be factored, write not factorable.

77. $4x^2 - 5x - 6$

78. $4x^2 + 33x + 35$

79. $5x^2 + 20x + 5$

80. $12x^3 - 26x^2 + 12x$

81. $6x^2 + 17x + 5$

82. $2x^2 + 12x + 10$

83. $5x^2 - 23x - 10$

84. $2x^6 - 9x^3y^2 + 4y^4$

85. $5x^4 - 17x^2y^2 + 6y^4$

86. $-14by^3 - 7by + 21by^2$

87. $2x^2(x - 3) - 9x(x - 3) - 18(x - 3)$

88. $(3x + y)^2 - 8(3x + y) + 7$

4.7 Special Factoring Techniques

In Exercises 89 – 105, completely factor each polynomial. If a polynomial cannot be factored, write not factorable.

89. $x^6 - 100$ **90.** $16x^2 - 25$ **91.** $x^2 + 144$ **92.** $4y^2 + 169$

93. $8x^2 - 242$ **94.** $3x^3 - 48x$ **95.** $-3x^2 - 12x - 12$ **96.** $4x^2 - 70x + 245$

97. $x^{2k} + 10x^k + 25$ **98.** $2x^{2k} - 24x^k + 72$ **99.** $\left(x^2 + 6x + 9\right) - y^2$

100. $\left(9y^2 - 6y + 1\right) - 16x^2$ **101.** $8x^9 - y^6$ **102.** $3x^3 + 81$

103. $4x^4 + 500x$ **104.** $x^{3k} - 27y^{6k}$ **105.** $x^3 - \left(x - y\right)^3$

4.8 Polynomial Equations and Applications

Solve the equations in Exercises 106 – 115 by factoring.

106. $x^2 - 17x + 72 = 0$ **107.** $x^2 + 8x + 12 = 0$ **108.** $5x^2 - 10x = -5$

109. $7x^2 = 14x + 168$ **110.** $8x^2 - 32 = 0$ **111.** $9x^2 - 25 = 0$

112. $(x - 2)(x - 6) = 5$ **113.** $(x + 3)(2x + 1) = -3$ **114.** $12x^3 - 58x^2 - 10x = 0$

115. $9x^3 + 30x^2 = 24x$

In Exercises 116 – 123, write a polynomial equation with integer coefficients that has the given roots.

116. $x = 4, x = -5$ **117.** $x = -3, x = -7$ **118.** $y = -1, y = -8$

119. $y = \dfrac{2}{3}, y = \dfrac{1}{3}$ **120.** $x = -\dfrac{3}{4}, x = \dfrac{5}{8}$ **121.** $x = 0, x = -2, x = \dfrac{1}{6}$

122. $y = 1, \; y = -4, \; y = -4$ (−4 is a double root.)

123. $y = 5, \; y = 5, \; y = 5$ (5 is a triple root.)

124. Find two consecutive positive integers such that the sum of their squares is 145.

125. The product of two consecutive even integers is 72 more than 4 times the larger integer. Find the integers.

126. Find three consecutive odd integers such that the product of the first and second is 26 more than the third.

127. Find four consecutive integers such that the sum of the squares of the first and second is 65 more than the product of the third and fourth.

128. The diagonal of a rectangle is 9 meters longer than the width. The length is 21 meters. Find the width of the rectangle.

129. Distance: Hector's home is "around the corner" from Paul's home. Hector's home is 100 yards from the corner and Paul's home is 75 yards from the same corner. What is the distance between the two homes?

130. Gravity: A building is known to be 144 feet tall. A ball is dropped from the top of the building. In how many seconds will the ball hit the ground if the formula for the height of the ball at t seconds is given by $h = 144 - 16t^2$?

4.9 Using a Graphing Calculator to Solve Equations

Use a graphing calculator to solve (or estimate the solutions of) the equations in Exercises 131 – 138. Find any approximations accurate to two decimal places.

131. $2x^2 = x + 1$

132. $x^2 = 15$

133. $3x^2 - x - 5 = 0$

134. $x^3 + x^2 = x - 6$

135. $x^4 = 16x^2$

136. $x(x+2)(x-5) = 0$

137. $|3x - 2| = 1$

138. $|2x + 3| = 5$

In Exercises 139 – 142, use a graphing calculator to solve (or estimate the solutions of) the inequalities. Write your answers in interval notation.

139. $|x + 7| < 2$

140. $|x - 2| \le 4$

141. $\left|\dfrac{x}{5} - 6\right| \ge 1$

142. $|2x - 1| > 3$

Chapter 4 Test

Use the properties of exponents to simplify each of the expressions in Exercises 1 – 4. Answers should contain only positive exponents.

1. $\left(4x^2y\right)\left(-5xy^{-3}\right)$

2. $\dfrac{\left(xy^{-3}\right)^3}{\left(x^0y^{-1}\right)^2}$

3. $\left(\dfrac{x^2y^{-2}}{3y^5}\right)^{-2}\left(\dfrac{2y}{3x^2}\right)^{-1}$

4. $\dfrac{x^{k+3}x^{2k-1}}{x^{3k+1}}$

5. Write each number in the following expression in scientific notation and simplify. Show the steps you use. Do not use a calculator. Leave the answers in scientific notation.

a. $\dfrac{-27\times0.0016}{120}$

b. $\dfrac{650\times35,000}{0.0025}$

In Exercises 6 and 7, state whether the expression is or is not a polynomial. If the expression is a polynomial, state its degree, its classification as a monomial, binomial, or trinomial, and its leading coefficient.

6. $6y^{\frac{1}{3}}+2y^2+1$

7. $\dfrac{1}{4}x+2x^3$

In Exercises 8 – 11, find the indicated sums and differences and simplify the results.

8. $\left(6x^3+x-10\right)-\left(x^3+x^2+x-4\right)$

9. $\left(x^3+4x^2-x\right)+\left(-2x^3+6x-3\right)$

10. $\quad 5x^2-3x+7$
$\quad\ \underline{-2x^2+5x+6}$

11. $\quad 7x^3+3x^2-8x+10$
$\quad\ \underline{-\left(4x^3-5x^2-3x+15\right)}$

12. Evaluate the polynomial for the specified values of the variable:
Given $P(x)=x^2+6x-7$; find **a.** $P(2)$ and **b.** $P(-1)$

Find the indicated products and simplify in Exercises 13 – 16.

13. $(7x+3)(4x+5)$

14. $(8x+3)(8x-3)$

15. $\left[(x+1)-y\right]\left[(x+1)+y\right]$

16. $\quad 3x^2+x-1$
$\quad\ \underline{\quad\ 2x+1}$

In Exercises 17 and 18, divide by using long division. Write the answers in the form $Q + \dfrac{R}{D}$, where the degree of $R <$ the degree of D.

17. $\dfrac{2x^3 + 11x - 6}{x - 4}$

18. $\dfrac{4x^3 - 6x^2 + x - 3}{2x^2 + 1}$

Completely factor each polynomial in Exercises 19 – 28.

19. $30x^2y - 18xy^2 + 24xy$ **20.** $2x^2 + 4xy + x + 2y$ **21.** $x^2 + x - 42$

22. $7x^2 - 26x - 8$ **23.** $6x^2 + 7x - 5$ **24.** $x^2 - 9x - 36$

25. $x^2 + 81y^2$ **26.** $x^{2k} - y^{2k}$ **27.** $3x^2y - 18xy^2 + 27y^3$

28. $3x^3 + 81y^3$

Solve the equations in Exercises 29 – 32 by factoring.

29. $(x - 2)(3x + 5) = 0$ **30.** $x^2 - 7x - 8 = 0$

31. $2x(x + 2) = 3(x + 5)$ **32.** $3x^2 + 14x = 5$

33. Find a polynomial equation with integer coefficients that has $x = 3$ and $x = -8$ as solutions.

34. Find a polynomial equation with integer coefficients that has $x = \dfrac{1}{2}$ as a double root.

35. Find three consecutive odd integers such that the sum of the first and second is 201 less than the square of the third.

36. In a right triangle, the hypotenuse is 1 meter longer than the length of the first leg and the first leg is 17 meters longer than the length of the second leg. What are the lengths of the three sides of the right triangle? Sketch a figure and label the sides as an aid in understanding the problem.

37. Use a graphing calculator to solve the equation $x^3 - 5x^2 + 4x = 0$.

38. Use a graphing calculator to find the solution set for the inequality $\left| \dfrac{x}{3} - 1 \right| < 3$. Write the answer in interval notation.

Cumulative Review: Chapters 1–4

1. Define "*prime number.*" List the prime numbers less than 50.

2. List all of the even prime numbers.

In Exercises 3 – 5, find the prime factorization of each composite number.

3. 504
4. 544
5. 5000

Find the LCM of each set of integers given in Exercises 6 – 8.

6. $\{6, 8, 10\}$
7. $\{9, 20, 45\}$
8. $\{25, 75, 150, 300\}$

In Exercises 9 and 10, perform the indicated operations.

9. $\dfrac{1}{3} + \dfrac{11}{15} + \left(-\dfrac{7}{30}\right)$

10. $\dfrac{2}{3} \cdot \dfrac{7}{5} \div \dfrac{7}{12}$

Find the value of each expression in Exercises 11 and 12 by using the rules for order of operations.

11. $2 \cdot 3^2 \div 6 \cdot 3 - 3$

12. $6 + 3\left[4 - 2\left(3^3 - 1\right)\right]$

Simplify by combining like terms in Exercises 13 and 14.

13. $8x - \left[2x + 4(x - 3) - 5\right]$

14. $9x + \left[8 - 5(3 - 2x) - 7x\right]$

Solve each of the equations in Exercises 15 – 18.

15. $4(2x - 3) + 2 = 5 - (2x + 6)$

16. $\dfrac{4x}{7} - 3 = 9$

17. $\dfrac{2x + 3}{6} - \dfrac{x + 1}{4} = 2$

18. $\left|\dfrac{2x}{5} - 1\right| = 3$

Solve the inequalities in Exercises 19 and 20 and graph the solution sets on real number lines. Write each solution in interval notation. Assume that x is a real number.

19. $2x - 3 \geq 5x + 12$

20. $|3x + 2| < 5$

21. Graph the linear equation $5x - 2y = 10$ by locating the y-intercept and the x-intercept.

22. Write the equation $4x + 3y = 7$ in slope-intercept form and then graph the line.

23. Find an equation in standard form for the line determined by the two points $(-2,1)$ and $(5,3)$.

24. Find an equation for the horizontal line through the point $(8,3)$.

25. Find the equation in slope-intercept form for the line parallel to the line $3x+2y=-4$ and passing through the point $(2,-2)$. Graph both lines.

26. The graphs of three curves are shown. Use the vertical line test to determine whether or not each graph represents a function. State the domain and its range.

a. 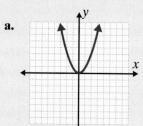   **b.** 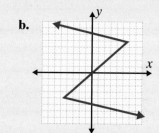   **c.**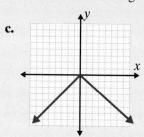

27. Given $f(x)=3x^2-8x-10$; find **a.** $f(-2)$ **b.** $f(0)$ and **c.** $f(1)$.

28. Given $g(x)=x^3-6x^2+3x$; find **a.** $g(0)$ **b.** $g(3)$ and **c.** $g(-2)$.

29. Graph the solution set of the following linear inequality: $3x+y\leq10$.

Solve the systems of equations in Exercises 30 and 31.

30. $\begin{cases} 3x+y=10 \\ 5x-y=6 \end{cases}$ **31.** $\begin{cases} y=2x-5 \\ 3x+2y=-3 \end{cases}$

32. Solve the following system of equations using Gaussian elimination.
$$\begin{cases} 2x+y-z=5 \\ x+2y-2z=4 \\ 4x+5y-5z=13 \end{cases}$$

33. Solve the following system of equations using Cramer's Rule.
$$\begin{cases} 3x-y=15 \\ 2x+y=5 \end{cases}$$

34. Find the value of $\det(A)$ for the matrix $A=\begin{bmatrix} 1 & 4 & 7 \\ 2 & 1 & 0 \\ -3 & 1 & 5 \end{bmatrix}$.

35. Graph the solution set to the following system of inequalities: $\begin{cases} y>-5x+1 \\ y<2x-3 \end{cases}$.

36. Use a graphing calculator to find the solution set to the following system of linear inequalities: $\begin{cases} 2x - y > 5 \\ 3x + y > 6 \end{cases}$.

Use the properties of exponents to simplify each of the expressions in Exercises 37 – 44. Answers should contain only positive exponents.

37. $\left(4x^2 y\right)^3$ **38.** $\left(7x^5 y^2\right)^2$ **39.** $\left(-2x^3 y^2\right)^{-3}$ **40.** $\left(\dfrac{6x^2}{y^5}\right)^2$

41. $\left(4x^2 y\right)\left(-5.2xy^{-3}\right)$ **42.** $\dfrac{\left(xy^0\right)^3}{\left(x^3 y^{-1}\right)^2}$

43. $\left(\dfrac{x^2 y^{-2}}{3y^5}\right)^{-2}$ **44.** $\dfrac{x^{k-2} \cdot x^{4k+3}}{x^{5k-4}}$

45. Write each number in the following expression in scientific notation and simplify. Show the steps you use. Do not use a calculator. Leave the answers in scientific notation.
   a. $0.000\,000\,56 \times 0.0003$ **b.** $\dfrac{81{,}000 \times 6200}{0.003 \times 0.2}$

In Exercises 46 and 47, state whether the expression is or is not a polynomial. If the expression is a polynomial, state its degree, its leading coefficient, and its classification as a monomial, binomial, or trinomial.

46. $5y^4$ **47.** $x^3 + x^{-3} - 2$

Perform the indicated operations in Exercises 48 – 52.

48. $\left(2x^2 + 9x - 3\right) + \left(5x^2 - 2x + 1\right)$ **49.** $\left(3x^2 + x - 9\right) - \left(-2x^2 + 5x - 3\right)$

50. $\left(5x + 3\right)\left(x - 8\right)$ **51.** $\left(2x + 5\right)\left(2x + 5\right)$ **52.** $\left(4x + 1\right)\left(2x + 3\right)$

53. Divide by using the division algorithm. Write the answer in the form $Q + \dfrac{R}{D}$, where the degree of $R <$ the degree of D.

$$\dfrac{4x^3 - 5x^2 + 7x - 13}{x^2 - 3}$$

Completely factor each polynomial in Exercises 54 – 59.

54. $3xy + y^2 + 3x + y$ **55.** $4x^2 - 4x - 15$ **56.** $6x^2 - 7x + 2$

57. $6x^3 - 22x^2 - 8x$ **58.** $9x^6 - 4y^2$ **59.** $8x^3 + 125$

Solve the equations in Exercises 60 – 63 by factoring.

60. $x^2 - 10x + 21 = 0$ **61.** $4x^2 + 20x + 25 = 0$

62. $3x(x-2) = x^2 - 2x + 16$ **63.** $x^3 - x^2 = 20x$

64. Find an equation that has $x = -10$ and $x = -5$ as roots.

65. Find an equation that has $x = 0$, $x = 4$ and $x = 13$ as roots.

66. Investments: Olivia invested $7000, some at 7%, and the remainder at 8%. During the first year, the interest from the 7% investment exceeded the interest from the 8% investment by $70. How much did she invest at each rate?

67. One number is 4 more than another and the sum of their squares is 976. What are the numbers?

68. Find the values for a, b, and c so that the points $(-1, 0)$, $(0, 1)$, and $(1, 0)$ lie on the graph of the function $y = ax^2 + bx + c$.

69. Chemistry: How many liters each of a 6% iodine solution and a 15% iodine solution must be used to produce 30 liters of a 12% iodine solution?

70. The following table shows the number of miles Kim drove from one hour to another:
   a. Plot the points indicated in the table.
   b. Calculate the slope of the line segments from point to point.
   c. Interpret each slope.

Hours	Miles
1	70
2	85
3	60
4	72
5	68

Use a graphing calculator to solve (or estimate the solutions of) the equations in Exercises 71 – 75. Find any approximations accurate to two decimal places.

71. $2x^2 - 8 = x$ **72.** $x^2 - 4x = 1$ **73.** $-x^2 + 3x + 5 = 0$

74. $x^3 - x^2 - 2x + 1 = 0$ **75.** $2x^2 = 5x - 4$

In Exercises 76 – 78, use a graphing calculator to solve the inequalities. Write your answers in interval notation.

76. $|x - 5| < 6$ **77.** $|2x - 4| \geq 4$ **78.** $\left| \dfrac{x}{2} + 1 \right| \leq 5$

Rational Expressions and Rational Equations

Did You Know?

Chapter 5 deals with rational expressions. Historically, one of the first properties listed for such expressions is the **cross-multiplication property**, which states that two rational expressions are equal if and only if the cross products are equal. Symbolically, this property is stated as $\frac{a}{b} = \frac{c}{d}$ if and only if $a \cdot d = b \cdot c$. This property is the key to understanding how a missing term of a proportion can be found if the remaining three terms are known. For example, using the cross-multiplication property to solve the proportion $\frac{x}{6} = \frac{3}{4}$ yields $4x = 18$, and the solution of this simple linear equation is $x = \frac{9}{2}$, the missing term of the proportion. This type of solution illustrates the so-called Rule of Three, which has been known for over 3000 years. The Hindu mathematician Brahmagupta (c. 628) called the rule by that name in his writings, although problems of this type also exist in ancient Egyptian and Chinese writings. Brahmagupta taught and wrote in the town of Ujjain in Central India, a center of Hindu science in the seventh century. Brahmagupta stated the Rule of Three as follows: "In the Rule of Three, argument, fruit and requisition are the names of the terms. The first and the last terms must be similar. Requisition multiplied by fruit and divided by argument is the produce;" or, as in our example, $x = \frac{3 \cdot 6}{4}$. In this case, the terms of the proportion have been given very fanciful names, and the cross-multiplication property has been concealed in an arbitrary rule.

The Rule of Three appears in Arabic and Latin works without explanation until the Renaissance. It was used in commercial arithmetic and occasionally was called the Merchant's Key or the Golden Rule. A popular seventeenth-century English arithmetician states: "The Rule of Three is commonly called The Golden Rule; and indeed it might be so termed; for as gold transcends all other mettals, so doth this rule all others in Arithmetick." The Rule of Three often appeared in verse as a memory aid.

Arithmetic texts of the 16th and 17th centuries had pages or chapters called Practice. At that time, the term "practice" was used to mean commercial arithmetic usually involving the Rule of Three and other short processes for solving applied problems. Sometimes such problems were called Italian Practice because the problems often related to the methods developed in Italian commercial arithmetic.

"Multiplication is vexation,
Division is as bad;
The Rule of Three doth puzzle me,
And Practice drives me mad."

Mother Goose Rhyme

In arithmetic, **rational numbers** are numbers that can be written in fraction form with **integers** for the numerator and denominator, the denominator not equal to 0. In algebra, **rational expressions** are expressions that can be written in fraction form with **polynomials** for the numerator and denominator, the denominator not equal to 0. As we will see in this section, all of the rules about operating with fractions in arithmetic can be applied to operating with rational expressions in algebra. For example, to add or subtract rational expressions, we need common denominators. To reduce rational expressions, we need to be able to factor both the numerator and denominator. Thus, all of the factoring skills learned in Chapter 4 are now going to be applied in dealing with rational expressions and in solving rational equations.

5.1 Multiplication and Division with Rational Expressions

- *Determine any restrictions on the variable in a rational expression.*
- *Reduce rational expressions to lowest terms.*
- *Multiply rational expressions.*
- *Divide rational expressions.*

Introduction to Rational Expressions

The term **rational number** is the technical name for a fraction in which both the numerator and denominator are integers. Similarly, the term **rational expression** is the technical name for a fraction in which both the numerator and denominator are polynomials.

Rational Expression

A **rational expression** is an algebraic expression that can be written in the form

$$\frac{P}{Q} \quad \text{where } P \text{ and } Q \text{ are polynomials and } Q \neq 0.$$

(In function notation, we can write $\frac{P(x)}{Q(x)}$ where $Q(x) \neq 0$.)

Examples of rational expressions are

$$\frac{4x^2}{9}, \qquad \frac{y^2 - 25}{y^2 + 25}, \qquad \text{and} \qquad \frac{x^2 + 7x - 6}{x^2 - 5x - 14}.$$

As with rational numbers, the denominators of rational expressions cannot be 0. If a numerical value is substituted for a variable in a rational expression and the denominator assumes a value of 0, we say that the expression is **undefined** for that value of the variable.

> **NOTES** **Remember, the denominator of a rational expression can never be 0.** Division by 0 is undefined.

Example 1: Finding Restrictions on the Variable

Determine what values of the variable, if any, will make the rational expression undefined. (These values are called **restrictions** on the variable.)

a. $\dfrac{5}{3x-1}$

    Solution: $3x-1=0$ Set the denominator equal to 0.

                $3x=1$ Solve the equation.

                $x=\dfrac{1}{3}$

Thus the expression $\dfrac{5}{3x-1}$ is undefined for $x=\dfrac{1}{3}$. Any other real number may be substituted for x in the expression. We write $x\neq\dfrac{1}{3}$ to indicate the restriction on the variable.

b. $\dfrac{x^2-4}{x^2-5x-6}$

    Solution: $x^2-5x-6=0$ Set the denominator equal to 0.

               $(x-6)(x+1)=0$ Solve the equation by factoring.

       $x-6=0$ or $x+1=0$

           $x=6$ $x=-1$

Thus there are two restrictions on the variable: 6 and –1. We write $x\neq 6,-1$.

c. $\dfrac{x+3}{x^2+36}$

    Solution: $x^2+36=0$ Set the denominator equal to 0.

              $x^2=-36$ Solve the equation.

However there is no real number whose square is –36. Thus there are **no restrictions** on the variable.

NOTES

Special Comment About the Numerator Being 0:

If the numerator of a rational expression has a value of 0 and the denominator is not 0 for that value of the variable, then the expression is defined and has a value of 0. If both numerator and denominator are 0, then the expression is **undefined** just as in the case where only the denominator is 0.

The rules for operating with rational expressions are essentially the same as those for operating with fractions in arithmetic. That is, simplifying, multiplying, and dividing rational expressions involve factoring and canceling. Addition and subtraction of rational expressions require common denominators. The basic rules for fractions are summarized here for easy reference.

Summary of Arithmetic Rules for Rational Numbers (or Fractions)

*A **fraction** (or **rational number**) is a number that can be written in the form $\dfrac{a}{b}$ where a and b are integers and $b \neq 0$. (Remember, no denominator can be 0.)*

The Fundamental Principle: $\dfrac{a}{b} = \dfrac{a \cdot k}{b \cdot k}$ *where b, k ≠ 0.*

*The **reciprocal** of* $\dfrac{a}{b}$ *is* $\dfrac{b}{a}$ *and* $\dfrac{a}{b} \cdot \dfrac{b}{a} = 1$ *where a, b ≠ 0.*

Multiplication: $\dfrac{a}{b} \cdot \dfrac{c}{d} = \dfrac{a \cdot c}{b \cdot d}$ *where b, d ≠ 0.*

Division: $\dfrac{a}{b} \div \dfrac{c}{d} = \dfrac{a}{b} \cdot \dfrac{d}{c}$ *where b, c, d ≠ 0.*

Addition: $\dfrac{a}{b} + \dfrac{c}{b} = \dfrac{a+c}{b}$ *where b ≠ 0.*

Subtraction: $\dfrac{a}{b} - \dfrac{c}{b} = \dfrac{a-c}{b}$ *where b ≠ 0.*

For rational expressions, each rule can be restated by replacing a and b with P and Q where P and Q represent polynomials. In particular, the Fundamental Principle can be restated as follows:

The Fundamental Principle of Rational Expressions

If $\dfrac{P}{Q}$ *is a rational expression and P, Q, and K are polynomials where Q, K ≠ 0, then*

$$\frac{P}{Q} = \frac{P \cdot K}{Q \cdot K}.$$

Reducing (or Simplifying) Rational Expressions

The Fundamental Principle can be used to **reduce** (or **simplify**) a rational expression to **lower terms** (for multiplication or division) and to **build** a rational expression to **higher terms** (for addition or subtraction). (Just as with rational numbers, a rational expression is said to be **reduced to lowest terms** if the numerator and denominator have no common factors other than 1 and –1.)

Example 2: Reducing Rational Expressions

Use the Fundamental Principle to reduce each expression to lowest terms. State any restrictions on the variable using the fact that no denominator can be 0. This restriction applies to denominators **before and after** a rational expression is reduced.

a. $\dfrac{2x-10}{3x-15}$

Solution: $\dfrac{2x-10}{3x-15} = \dfrac{2(x-5)}{3(x-5)} = \dfrac{2}{3}\ (x \neq 5)$

Note that $x - 5$ is a common **factor**. The key word here is **factor**. We reduce using **factors** only.

b. $\dfrac{x^3-64}{x^2-16}$

Solution: $\dfrac{x^3-64}{x^2-16} = \dfrac{(x-4)(x^2+4x+16)}{(x+4)(x-4)}$

$= \dfrac{x^2+4x+16}{x+4}\ (x \neq -4,4)$

Reduce. The common **factor** is $x - 4$. Note that $x^3 - 64$ is the difference of two cubes. Also, note that $x^2 + 4x + 16$ is not factorable.

c. $\dfrac{y-10}{10-y}$

Solution: $\dfrac{y-10}{10-y} = \dfrac{y-10}{-y+10}$

$= \dfrac{y-10}{-1(y-10)} = \dfrac{1}{-1} = -1\ (y \neq 10)$

Note that the expression $10 - y$ is the opposite of $y - 10$. When **nonzero opposites** are divided, the quotient is always –1.

Opposites in Rational Expressions

For a polynomial P, $\dfrac{-P}{P} = -1$ where $P \neq 0$.

In particular, $\dfrac{a-x}{x-a} = \dfrac{-(x-a)}{x-a} = -1$ where $x \neq a$.

Remember that the key word when reducing is **factor**. Many students make mistakes similar to the following when working with rational expressions.

> **NOTES**
>
> **COMMON ERRORS**
>
> **"Divide out" only common factors**.
>
> INCORRECT
>
> $$\dfrac{4x + \cancel{8}}{\cancel{8}}$$
>
> 8 is not a common factor.
>
> INCORRECT
>
> $$\dfrac{\cancel{x}^2 - \cancel{9}}{\cancel{x} - \cancel{3}}$$
>
> 3 and x are not common factors.
>
> CORRECT
>
> $$\dfrac{4x + 8}{8} = \dfrac{\cancel{4}\,(x + 2)}{\underset{2}{\cancel{8}}}$$
>
> 4 is a common factor.
>
> CORRECT
>
> $$\dfrac{x^2 - 9}{x - 3} = \dfrac{(x + 3)\,\cancel{(x - 3)}}{\cancel{(x - 3)}}$$
>
> $x - 3$ is a common factor.

Multiplication with Rational Expressions

Multiplying Rational Expressions

> *To multiply any two (or more) rational expressions,*
>
> 1. *Completely factor each numerator and denominator.*
> 2. *Multiply the numerators and multiply the denominators, keeping the expressions in factored form.*
> 3. *"Divide out" any common factors from the numerators and denominators.*
>
> *Remember that no denominator can have a value of 0.*

For example, here no factoring is necessary,

$$\frac{2x}{x - 6} \cdot \frac{x + 5}{x - 4} = \frac{2x(x + 5)}{(x - 6)(x - 4)} = \frac{2x^2 + 10x}{(x - 6)(x - 4)}. \qquad (x \neq 6, 4)$$

However, in this case the numerators and denominators must be factored.

$$\frac{y^2-4}{y^3}\cdot\frac{y^2-3y}{y^2-y-6}=\frac{(y+2)(y-2)(y)(y-3)}{y^3\ (y-3)(y+2)}=\frac{y-2}{y^2} \qquad (y\neq 0,3,-2)$$

Multiplication with Rational Expressions

If P, Q, R, and S are polynomials and Q, S ≠ 0, then

$$\frac{P}{Q}\cdot\frac{R}{S}=\frac{P\cdot R}{Q\cdot S}.$$

Example 3: Multiplication with Rational Expressions

Multiply and reduce, if possible. Use the rules of exponents when they apply. State any restrictions on the variable(s).

a. $\dfrac{5x^2y}{9xy^3}\cdot\dfrac{6x^3y^2}{15xy^4}=\dfrac{5\cdot 2\cdot 3\cdot x^5\cdot y^3}{3\cdot 3\cdot 3\cdot 5\cdot x^2\cdot y^7}=\dfrac{2x^{5-2}y^{3-7}}{9}=\dfrac{2x^3y^{-4}}{9}=\dfrac{2x^3}{9y^4}$ $(x\neq 0,y\neq 0)$

b. $\dfrac{x}{x-2}\cdot\dfrac{x^2-4}{x^2}=\dfrac{x(x+2)(x-2)}{(x-2)x^2}=\dfrac{x+2}{x}$ $(x\neq 2,0)$

c. $\dfrac{3x-3}{x^2+x}\cdot\dfrac{x^2+2x+1}{3x^2-6x+3}=\dfrac{3(x-1)(x+1)^2}{x(x+1)\cdot 3(x-1)^2}=\dfrac{x+1}{x(x-1)}$ $(x\neq 0,-1,1)$

Division with Rational Expressions

To divide any two rational expressions, multiply by the **reciprocal** of the divisor.

Division with Rational Expressions

If P, Q, R, and S are polynomials with Q, R, S ≠ 0, then

$$\frac{P}{Q}\div\frac{R}{S}=\frac{P}{Q}\cdot\frac{S}{R}.$$

Note that $\dfrac{S}{R}$ *is the reciprocal of* $\dfrac{R}{S}.$

Example 4: Division with Rational Expressions

Divide and reduce, if possible. Assume that no denominator has a value of 0.

a. $\dfrac{12x^2y}{10xy^2} \div \dfrac{3x^4y}{xy^3}$

Solution: $\dfrac{12x^2y}{10xy^2} \div \dfrac{3x^4y}{xy^3} = \dfrac{12x^2y}{10xy^2} \cdot \dfrac{xy^3}{3x^4y}$

$$= \dfrac{\cancel{2} \cdot 2 \cdot \cancel{3} \cdot x^3 \cdot y^4}{\cancel{2} \cdot 5 \cdot \cancel{3} \cdot x^5 \cdot y^3}$$

Note that in this example we have used the quotient rule for exponents.

$$= \dfrac{2x^{3-5}y^{4-3}}{5}$$

$$= \dfrac{2x^{-2}y}{5} = \dfrac{2y}{5x^2}$$

b. $\dfrac{x^3 - y^3}{x^3} \div \dfrac{y-x}{xy}$

Solution: $\dfrac{x^3 - y^3}{x^3} \div \dfrac{y-x}{xy} = \dfrac{x^3 - y^3}{x^3} \cdot \dfrac{xy}{y-x}$

$$= \dfrac{\overset{-1}{\cancel{(x-y)}}\left(x^2 + xy + y^2\right)\cancel{x}y}{\underset{x^2}{\cancel{x^3}}\,\cancel{(y-x)}}$$

Note that $\dfrac{x-y}{y-x} = -1$.

$$= \dfrac{-y\left(x^2 + xy + y^2\right)}{x^2} = \dfrac{-x^2y - xy^2 - y^3}{x^2}$$

c. $\dfrac{x^2 - 8x + 15}{2x^2 + 11x + 5} \div \dfrac{2x^2 - 5x - 3}{4x^2 - 1}$

Solution: $\dfrac{x^2 - 8x + 15}{2x^2 + 11x + 5} \div \dfrac{2x^2 - 5x - 3}{4x^2 - 1} = \dfrac{x^2 - 8x + 15}{2x^2 + 11x + 5} \cdot \dfrac{4x^2 - 1}{2x^2 - 5x - 3}$

$$= \dfrac{\cancel{(x-3)}(x-5)(2x-1)\cancel{(2x+1)}}{(2x+1)(x+5)\cancel{(x-3)}\cancel{(2x+1)}}$$

$$= \dfrac{(x-5)(2x-1)}{(2x+1)(x+5)} = \dfrac{2x^2 - 11x + 5}{(2x+1)(x+5)}$$

Remember that you have the option of leaving the numerator and/or denominator in factored form.

NOTES

As illustrated in the answer in Example 4c, generally the denominator will be left in factored form and the numerator will be multiplied out. This form makes the results easier to add or subtract, as we will see in the next section. However, be aware that leaving the denominator in factored form is just an option, and multiplying out the denominator is not an error. Thus in Example 4c we can write the answer either as

$$\frac{2x^2 - 11x + 5}{(2x+1)(x+5)} \text{ or as } \frac{2x^2 - 11x + 5}{2x^2 + 11x + 5}.$$

Practice Problems

Reduce to lowest terms. State any restrictions on the variables.

1. $\dfrac{5x + 20}{7x + 28}$

2. $\dfrac{4 - 2x}{2x - 4}$

3. $\dfrac{x^2 + x - 2}{x^2 + 3x + 2}$

Perform the following operations and simplify the results. Assume that no denominator is 0.

4. $\dfrac{x - 7}{x^3} \cdot \dfrac{x^2}{49 - x^2}$

5. $\dfrac{y^2 - y - 6}{y^2 - 5y + 6} \cdot \dfrac{y^2 - 4}{y^2 + 4y + 4}$

6. $\dfrac{x^3 + 3x}{2x + 1} \div \dfrac{x^2 + 3}{x + 1}$

7. $\dfrac{x^2 + 2x - 3}{x^2 - 3x - 10} \cdot \dfrac{2x^2 - 9x - 5}{x^2 - 2x + 1} \div \dfrac{4x + 2}{x^2 - x}$

Answers to Practice Problems: **1.** $\dfrac{5}{7}$, $x \neq -4$ **2.** -1, $x \neq 2$ **3.** $\dfrac{x - 1}{x + 1}$, $x \neq -1, -2$ **4.** $\dfrac{-1}{x(x + 7)}$ **5.** 1

6. $\dfrac{x^2 + x}{2x + 1}$ **7.** $\dfrac{x^2 + 3x}{2(x + 2)}$

5.1 Exercises

In Exercises 1 – 20, reduce to lowest terms. State any restrictions on the variable(s).

1. $\dfrac{9x^2y^3}{12xy^4}$

2. $\dfrac{18xy^4}{27x^2y}$

3. $\dfrac{20x^5}{30x^2y^3}$

4. $\dfrac{15y^4}{20x^3y^2}$

5. $\dfrac{x}{x^2-3x}$

6. $\dfrac{3x}{x^2+5x}$

7. $\dfrac{7x-14}{x-2}$

8. $\dfrac{4-2x}{2x-4}$

9. $\dfrac{9-3x}{4x-12}$

10. $\dfrac{6x^2+4x}{3xy+2y}$

11. $\dfrac{2x-8}{16-4x}$

12. $\dfrac{x+3y}{4x^2+12xy}$

13. $\dfrac{x^2+6x}{x^2+5x-6}$

14. $\dfrac{x^2-5x+6}{x^2-x-2}$

15. $\dfrac{x^2-y^2}{3x^2+3xy}$

16. $\dfrac{x^3-8}{x^2-4}$

17. $\dfrac{x^3+64}{2x^2+x-28}$

18. $\dfrac{3x^2+14x-24}{18-9x-2x^2}$

19. $\dfrac{x^3-8}{-2y+xy+5x-10}$

20. $\dfrac{xy-3y+2x-6}{y^2-4}$

Perform the indicated operations in Exercises 21 – 74 and reduce to lowest terms. Assume that no denominator has a value of 0.

21. $\dfrac{ax^2}{b}\cdot\dfrac{b^2}{x^2y}$

22. $\dfrac{18x^3}{5y^2}\cdot\dfrac{30y^3}{9x^4}$

23. $\dfrac{24x^3}{25y^2}\cdot\dfrac{10y^5}{18x}$

24. $\dfrac{16x^8}{3y^{11}}\cdot\dfrac{-21y^9}{10x^7}$

25. $\dfrac{x^2-9}{x^2+2x}\cdot\dfrac{x+2}{x-3}$

26. $\dfrac{16x^2-9}{3x^2-15x}\cdot\dfrac{6}{4x+3}$

27. $\dfrac{x^2+2x-3}{x^2+3x}\cdot\dfrac{x}{x+1}$

28. $\dfrac{4x+16}{x^2-16}\cdot\dfrac{x-4}{x}$

29. $\dfrac{x^2+6x-16}{x^2-64}\cdot\dfrac{1}{2-x}$

30. $\dfrac{4-x^2}{x^2-4x+4}\cdot\dfrac{3}{x+2}$

31. $\dfrac{x^2-5x+6}{x^2-4x}\cdot\dfrac{x-4}{x-3}$

32. $\dfrac{2x^2+x-3}{x^2+4x}\cdot\dfrac{2x+8}{x-1}$

33. $\dfrac{2x^2+10x}{3x^2+5x+2}\cdot\dfrac{6x+4}{x^2}$

34. $\dfrac{x+3}{x^2-16}\cdot\dfrac{x^2-3x-4}{x^2-1}$

35. $\dfrac{x}{x^2+7x+12}\cdot\dfrac{x^2-2x-24}{x^2-7x+6}$

36. $\dfrac{x^2-2x-3}{x+5}\cdot\dfrac{x^2-5x-14}{x^2-x-6}$

37. $\dfrac{8-2x-x^2}{x^2-2x}\cdot\dfrac{x-4}{x^2-3x-4}$

38. $\dfrac{3x^2+21x}{x^2-49}\cdot\dfrac{x^2-5x+4}{x^2+3x-4}$

39. $\dfrac{(x-2y)^2}{x^2-5xy+6y^2} \cdot \dfrac{x+2y}{x^2-4xy+4y^2}$

40. $\dfrac{4x^2+6x}{x^2+3x-10} \cdot \dfrac{x^2+4x-12}{x^2+5x-6}$

41. $\dfrac{2x^2+5x+2}{3x^2+8x+4} \cdot \dfrac{3x^2-x-2}{4x^3-x}$

42. $\dfrac{x^2+5x}{4x^2+12x+9} \cdot \dfrac{6x^2+7x-3}{x^2+10x+25}$

43. $\dfrac{x^2+x+1}{x^2-1} \cdot \dfrac{x^2-2x+1}{x^3-1}$

44. $\dfrac{x-2}{x^2-2x+4} \cdot \dfrac{x^3+8}{x^2-4x+4}$

45. $\dfrac{2x^2-7x+3}{x^2-9} \cdot \dfrac{3x^2+8x-3}{6x^2+x-1}$

46. $\dfrac{12x^2y}{9xy^9} \div \dfrac{4x^4y}{x^2y^3}$

47. $\dfrac{35xy^3}{24x^3y} \div \dfrac{15x^4y^3}{84xy^4}$

48. $\dfrac{45xy^4}{21x^2y^2} \div \dfrac{40x^4}{112xy^5}$

49. $\dfrac{x-3}{15x} \div \dfrac{4x-12}{5}$

50. $\dfrac{x-1}{6x+6} \div \dfrac{2x-2}{x^2+x}$

51. $\dfrac{7x-14}{x^2} \div \dfrac{x^2-4}{x^3}$

52. $\dfrac{6x^2-54}{x^4} \div \dfrac{x-3}{x^2}$

53. $\dfrac{x^2-25}{6x+30} \div \dfrac{x-5}{x}$

54. $\dfrac{2x-1}{x^2+2x} \div \dfrac{10x^2-5x}{6x^2+12x}$

55. $\dfrac{x+3}{x^2+3x-4} \div \dfrac{x+2}{x^2+x-2}$

56. $\dfrac{6x^2-7x-3}{x^2-1} \div \dfrac{2x-3}{x-1}$

57. $\dfrac{x^2-9}{2x^2+7x+3} \div \dfrac{x^2-3x}{2x^2+11x+5}$

58. $\dfrac{x^2-8x+15}{x^2-9x+14} \div \dfrac{x^2+4x-21}{x-1}$

59. $\dfrac{2x+1}{4x-x^2} \div \dfrac{4x^2-1}{x^2-16}$

60. $\dfrac{x^2-6x+9}{x^2-4x+3} \div \dfrac{2x^2-7x+3}{x^2-3x+2}$

61. $\dfrac{x^2-4x+4}{x^2+5x+6} \div \dfrac{x^2+2x-8}{x^2+7x+12}$

62. $\dfrac{x^2-x-6}{x^2+6x+8} \div \dfrac{x^2-4x+3}{x^2+5x+4}$

63. $\dfrac{x^2-x-12}{6x^2+x-9} \div \dfrac{x^2-6x+8}{3x^2-x-6}$

64. $\dfrac{6x^2+5x+1}{4x^3-3x^2} \div \dfrac{3x^2-2x-1}{3x^2-2x+1}$

65. $\dfrac{8x^2+2x-15}{3x^2+13x+4} \div \dfrac{2x^2+5x+3}{6x^2-x-1}$

66. $\dfrac{3x^2+13x+14}{4x^3-3x^2} \div \dfrac{6x^2-x-35}{4x^2+5x-6}$

67. $\dfrac{3x^2+2x}{9x^2-4} \div \dfrac{27x^3-8}{9x^2-6x+4}$

68. $\dfrac{x^3+2x^2}{x^3+64} \div \dfrac{4x^2}{x^2-4x+16}$

69. $\dfrac{6-11x-10x^2}{2x^2+x-3} \div \dfrac{5x^3-2x^2}{3x^2-5x+2}$

70. $\dfrac{x-6}{x^2-7x+6} \cdot \dfrac{x^2-3x}{x+3} \cdot \dfrac{x^2-9}{x^2-4x+3}$

71. $\dfrac{3x^2+11x+10}{2x^2+x-6} \cdot \dfrac{x^2+2x-3}{2x-1} \cdot \dfrac{2x-3}{3x^2+2x-5}$

72. $\dfrac{x^3+3x^2}{x^2+7x+12} \cdot \dfrac{2x^2+7x-4}{2x^2-x} \div \dfrac{2x^2-x-1}{x^2+4x-5}$

73. $\dfrac{x^2+2x-3}{x^2+10x+21} \div \dfrac{x^2-7x-8}{x^2+6x+5} \cdot \dfrac{x^2-x-56}{x^2-3x-40}$

74. $\dfrac{2x^2-5x+2}{4xy-2y+6x-3} \div \dfrac{xy-2y+3x-6}{2y^2+9y+9}$

Writing and Thinking About Mathematics

75. a. Define rational expression.

    b. Give an example of a rational expression that is undefined for $x = -2$ and $x = 3$ and has a value of 0 for $x = 1$. Explain how you determined this expression.

    c. Give an example of a rational expression that is undefined for $x = -5$ and never has a value of 0. Explain how you determined this expression.

76. Write the opposite of each of the following expressions.

    a. $3-x$ **b.** $2x-7$ **c.** $x+5$ **d.** $-3x-2$

77. Given the rational function $f(x)=\dfrac{x-4}{x^2-100}$:

    a. For what values, if any, will $f(x)=0$?

    b. For what values, if any, is $f(x)$ undefined?

HAWKES LEARNING SYSTEMS: INTERMEDIATE ALGEBRA SOFTWARE

- Defining Rational Expressions
- Multiplication and Division with Rational Expressions

5.2 Addition and Subtraction with Rational Expressions

- *Add rational expressions.*
- *Subtract rational expressions.*

Addition with Rational Expressions

To add rational expressions with a common denominator, proceed just as with fractions: add the numerators and keep the common denominator. For example,

$$\frac{5}{x+2}+\frac{6}{x+2}=\frac{5+6}{x+2}=\frac{11}{x+2}. \qquad (x\neq -2)$$

In some cases the sum can be reduced:

$$\frac{x^2+6}{x+2}+\frac{5x}{x+2}=\frac{x^2+5x+6}{x+2}=\frac{(x+2)(x+3)}{x+2}=x+3. \qquad (x\neq -2)$$

Adding Rational Expressions

For polynomials P, Q, and R, with Q ≠ 0,

$$\frac{P}{Q}+\frac{R}{Q}=\frac{P+R}{Q}.$$

Example 1: Adding Rational Expressions with a Common Denominator

Find each sum and reduce if possible. (Note the importance of the factoring techniques we studied in Chapter 4.)

a. $\dfrac{x}{x^2-1}+\dfrac{1}{x^2-1}$

Solution: $\dfrac{x}{x^2-1}+\dfrac{1}{x^2-1}=\dfrac{x+1}{x^2-1}=\dfrac{x+1}{(x+1)(x-1)}=\dfrac{1}{x-1}$ $(x\neq -1,1)$

Remember, if we cancel the entire expression in the numerator (or denominator) we are left with a factor of 1.

b. $\dfrac{1}{x^2+7x+10}+\dfrac{2x+3}{x^2+7x+10}$

Solution: $\dfrac{1}{x^2+7x+10}+\dfrac{2x+3}{x^2+7x+10}=\dfrac{2x+4}{x^2+7x+10}=\dfrac{2(x+2)}{(x+5)(x+2)}=\dfrac{2}{x+5}$

$(x\neq -5,-2)$

The rational expressions in Example 1 had common denominators. To add expressions with different denominators, we need to find the least common multiple (LCM) of the denominators. The LCM was discussed in Section 1.1. The procedure is stated here for polynomials.

To Find the LCM for a Set of Polynomials

1. Completely factor each polynomial (including prime factors for numerical factors).

2. Form the product of all factors that appear, using each factor the most number of times it appears in any one polynomial.

The LCM of a set of denominators is called the **least common denominator (LCD)**. To add fractions with different denominators, use the Fundamental Principle to change each fraction to an equivalent fraction with the LCD as denominator. This is called **building the fraction to higher terms**.

Use the following procedure when adding rational expressions with different denominators.

Procedure for Adding Rational Expressions with Different Denominators

1. Find the LCD (the LCM of the denominators).

2. Rewrite each fraction in an equivalent form with the LCD as the denominator.

3. Add the numerators and keep the common denominator.

4. Reduce if possible.

Example 2: Adding Rational Expressions with Different Denominators

Find each sum and reduce if possible. Assume that no denominator is equal to 0.

a. $\dfrac{y}{y-3}+\dfrac{6}{y+4}$

Solution: In this case, neither denominator can be factored so the LCD is the product of these factors. That is, $\text{LCD} = (y-3)(y+4)$.

Now, using the Fundamental Principle, we have

$$\frac{y}{y-3} + \frac{6}{y+4} = \frac{y(y+4)}{(y-3)(y+4)} + \frac{6(y-3)}{(y+4)(y-3)}$$

$$= \frac{(y^2+4y)+(6y-18)}{(y-3)(y+4)}$$

$$= \frac{y^2+10y-18}{(y-3)(y+4)}.$$

b. $\dfrac{1}{x^2+6x+9} + \dfrac{1}{x^2-9} + \dfrac{1}{2x+6}$

Solution: First, find the LCM for the polynomial denominators.

Step 1: Factor each expression completely:

$$x^2+6x+9 = (x+3)^2$$

$$x^2-9 = (x+3)(x-3)$$

$$2x+6 = 2(x+3).$$

Step 2: Form the product of 2, $(x+3)^2$, and $(x-3)$. That is, use each factor the most number of times it appears in any one factorization.

$$\text{LCM} = 2(x+3)^2(x-3)$$

Now use the LCM as the LCD of the fractions and add as follows:

$$\frac{1}{x^2+6x+9} + \frac{1}{x^2-9} + \frac{1}{2x+6}$$

$$= \frac{1}{(x+3)^2} + \frac{1}{(x+3)(x-3)} + \frac{1}{2(x+3)}$$

$$= \frac{1\cdot 2(x-3)}{(x+3)^2\cdot 2(x-3)} + \frac{1\cdot 2(x+3)}{(x+3)(x-3)\cdot 2(x+3)} + \frac{1\cdot(x+3)(x-3)}{2(x+3)\cdot(x+3)(x-3)}$$

$$= \frac{(2x-6)+(2x+6)+(x^2-9)}{2(x+3)^2(x-3)}$$

$$= \frac{x^2+4x-9}{2(x+3)^2(x-3)}.$$

Important Note About the Form of Answers:
In Examples 2a and 2b each denominator is left in factored form as a convenience for possibly reducing or adding to some other expression later. You may choose to multiply out these factors. Either form is correct. For consistency, denominators are left in factored form in the answers in the back of the text.

Subtraction with Rational Expressions

When subtracting fractions, the placement of negative signs can be critical. For example, note how -2 can be indicated in three different forms:

$$-\frac{6}{3} = -2, \quad \frac{-6}{3} = -2, \text{ and } \frac{6}{-3} = -2.$$

Thus, we have

$$-\frac{6}{3} = \frac{-6}{3} = \frac{6}{-3}.$$

With polynomials we have the following statement about the placement of negative signs which can be **very useful in subtraction**. We seldom leave the negative sign in the denominator.

Placement of Negative Signs

If P and Q are polynomials and $Q \neq 0$, then

$$-\frac{P}{Q} = \frac{-P}{Q} = \frac{P}{-Q}.$$

To subtract rational expressions with a common denominator, proceed just as with fractions: subtract the numerators and keep the common denominator. For example,

$$\frac{17}{x+7} - \frac{23}{x+7} = \frac{17-23}{x+7} = \frac{-6}{x+7} \quad \left(\text{or } -\frac{6}{x+7} \right).$$

Subtracting Rational Expressions

For polynomials P, Q, and R, with $Q \neq 0$,

$$\frac{P}{Q} - \frac{R}{Q} = \frac{P-R}{Q}.$$

Example 3: Subtracting Rational Expressions with a Common Denominator

Find each difference and reduce if possible. Assume that no denominator is equal to 0.

a. $\dfrac{2x-5y}{x+y} - \dfrac{3x-7y}{x+y}$

Solution: $\dfrac{2x-5y}{x+y} - \dfrac{3x-7y}{x+y} = \dfrac{2x-5y-(3x-7y)}{x+y}$ Subtract the entire numerator.

$$= \dfrac{2x-5y-3x+7y}{x+y}$$

$$= \dfrac{-x+2y}{x+y}$$

b. $\dfrac{x^2}{x^2+4x+4} - \dfrac{2x+8}{x^2+4x+4}$

Solution:

$$\dfrac{x^2}{x^2+4x+4} - \dfrac{2x+8}{x^2+4x+4} = \dfrac{x^2-(2x+8)}{x^2+4x+4}$$ Subtract the entire numerator.

$$= \dfrac{x^2-2x-8}{x^2+4x+4}$$

$$= \dfrac{(x-4)\,\cancel{(x+2)}}{(x+2)\,\cancel{(x+2)}}$$ Factor and reduce.

$$= \dfrac{x-4}{x+2}$$

c. $\dfrac{x}{x-5} - \dfrac{3}{5-x}$

Solution: Each denominator is the **opposite** of the other. Multiply both the numerator and denominator of the second fraction by –1 so that both denominators will be the same, in this case $x-5$.

$$\dfrac{x}{x-5} - \dfrac{3}{5-x} = \dfrac{x}{x-5} - \dfrac{3}{(5-x)} \cdot \dfrac{(-1)}{(-1)}$$

$$= \dfrac{x}{x-5} - \dfrac{-3}{x-5}$$

$$= \dfrac{x-(-3)}{x-5} = \dfrac{x+3}{x-5}$$

COMMON ERROR

Many beginning students make a mistake when subtracting rational expressions by not subtracting the entire numerator. They make a mistake similar to the following.

INCORRECT $\dfrac{10}{x+5} - \dfrac{3-x}{x+5} = \dfrac{10-3-x}{x+5} = \dfrac{7-x}{x+5}$

By using parentheses, you can avoid such mistakes.

CORRECT $\dfrac{10}{x+5} - \dfrac{3-x}{x+5} = \dfrac{10-(3-x)}{x+5} = \dfrac{10-3+x}{x+5} = \dfrac{7+x}{x+5} = \dfrac{x+7}{x+5}$

As with addition, if the rational expressions do not have the same denominator, find the LCM of the denominators (the LCD) and use the Fundamental Principle to **build each fraction to higher terms, if necessary**, so that each has the LCD as the denominator.

Example 4: Subtracting Rational Expressions with Different Denominators

Find each difference and reduce if possible. Assume that no denominator is equal to 0.

a. $\dfrac{x+5}{x-5} - \dfrac{100}{x^2-25}$

Solution: $\left.\begin{array}{l} x-5 = x-5 \\ x^2-25 = (x+5)(x-5) \end{array}\right\rangle$ LCD $= (x+5)(x-5)$

$\dfrac{x+5}{x-5} - \dfrac{100}{x^2-25} = \dfrac{(x+5)(x+5)}{(x-5)(x+5)} - \dfrac{100}{(x+5)(x-5)}$

$= \dfrac{(x^2+10x+25)-100}{(x+5)(x-5)}$

$= \dfrac{x^2+10x+25-100}{(x+5)(x-5)}$

$= \dfrac{x^2+10x-75}{(x+5)(x-5)}$

$= \dfrac{(x+15)\,(x-5)}{(x+5)\,(x-5)}$

$= \dfrac{x+15}{x+5}$

b. $\dfrac{x+y}{(x-y)^2} - \dfrac{x}{2x^2-2y^2}$

Solution: $(x-y)^2 = (x-y)^2$

$2x^2-2y^2 = 2(x-y)(x+y)$ $\Big\}$ LCD $= 2(x-y)^2(x+y)$

$\dfrac{x+y}{(x-y)^2} - \dfrac{x}{2x^2-2y^2}$

$= \dfrac{(x+y)\cdot 2(x+y)}{(x-y)^2\cdot 2(x+y)} - \dfrac{x(x-y)}{2(x-y)(x+y)(x-y)}$

$= \dfrac{2x^2+4xy+2y^2-\left(x^2-xy\right)}{2(x-y)^2(x+y)}$

$= \dfrac{2x^2+4xy+2y^2-x^2+xy}{2(x-y)^2(x+y)}$

$= \dfrac{x^2+5xy+2y^2}{2(x-y)^2(x+y)}$

c. $\dfrac{3x-12}{x^2+x-20} - \dfrac{x^2+5x}{x^2+9x+20}$

Hint: In this problem, both expressions can be reduced before looking for the LCD.

Solution: $\dfrac{3x-12}{x^2+x-20} - \dfrac{x^2+5x}{x^2+9x+20}$

$= \dfrac{3\cancel{(x-4)}}{(x+5)\cancel{(x-4)}} - \dfrac{x\cancel{(x+5)}}{\cancel{(x+5)}(x+4)}$

$= \dfrac{3}{x+5} - \dfrac{x}{x+4}$

Now subtract these two expressions with LCD $=(x+5)(x+4)$.

$\dfrac{3}{x+5} - \dfrac{x}{x+4} = \dfrac{3(x+4)}{(x+5)(x+4)} - \dfrac{x(x+5)}{(x+4)(x+5)}$

$= \dfrac{(3x+12)-\left(x^2+5x\right)}{(x+5)(x+4)}$

$= \dfrac{3x+12-x^2-5x}{(x+5)(x+4)}$

$= \dfrac{-x^2-2x+12}{(x+5)(x+4)}$

Continued on the next page...

d. $\dfrac{x+1}{xy-3y+4x-12} - \dfrac{x-3}{xy+6y+4x+24}$

Solution: $xy-3y+4x-12 = y(x-3)+4(x-3)$

$$= (x-3)(y+4)$$

$$xy+6y+4x+24 = y(x+6)+4(x+6)$$

$$= (x+6)(y+4)$$

$$\text{LCD} = (x-3)(y+4)(x+6)$$

$$\dfrac{x+1}{xy-3y+4x-12} + \dfrac{x-3}{xy+6y+4x+24} = \dfrac{(x+1)(x+6)}{(y+4)(x-3)(x+6)} - \dfrac{(x-3)(x-3)}{(y+4)(x+6)(x-3)}$$

$$= \dfrac{x^2+7x+6-\left(x^2-6x+9\right)}{(y+4)(x-3)(x+6)}$$

$$= \dfrac{x^2+7x+6-x^2+6x-9}{(y+4)(x-3)(x+6)}$$

$$= \dfrac{13x-3}{(y+4)(x-3)(x+6)}$$

Practice Problems

Perform the indicated operations and reduce if possible. Assume that no denominator is 0.

1. $\dfrac{1}{1-y} + \dfrac{2}{y^2-1}$

2. $\dfrac{x+3}{x^2+x-6} + \dfrac{x-2}{x^2+4x-12}$

3. $\dfrac{1}{y+2} - \dfrac{1}{y^3+8}$

4. $\dfrac{x}{x^2-1} - \dfrac{1}{x-1}$

Answers to Practice Problems: **1.** $\dfrac{-1}{y+1}$ **2.** $\dfrac{2x+4}{(x-2)(x+6)}$ **3.** $\dfrac{y^2-2y+3}{(y+2)\left(y^2-2y+4\right)}$ **4.** $\dfrac{-1}{(x+1)(x-1)}$

5.2 Exercises

In Exercises 1 – 60, perform the indicated operations and reduce if possible. Assume that no denominator has a value of 0.

1. $\dfrac{3x}{x+4}+\dfrac{12}{x+4}$

2. $\dfrac{7x}{x+5}+\dfrac{35}{x+5}$

3. $\dfrac{x-1}{x+6}+\dfrac{x+13}{x+6}$

4. $\dfrac{3x-1}{2x-6}+\dfrac{x-11}{2x-6}$

5. $\dfrac{3x+1}{5x+2}+\dfrac{2x+1}{5x+2}$

6. $\dfrac{x^2+3}{x+1}+\dfrac{4x}{x+1}$

7. $\dfrac{x-5}{x^2-2x+1}+\dfrac{x+3}{x^2-2x+1}$

8. $\dfrac{2x^2+5}{x^2-4}+\dfrac{3x-1}{x^2-4}$

9. $\dfrac{13}{7-x}-\dfrac{1}{x-7}$

10. $\dfrac{6x}{x-6}+\dfrac{36}{6-x}$

11. $\dfrac{3x}{x-4}+\dfrac{16-x}{4-x}$

12. $\dfrac{20}{x-10}-\dfrac{3}{10-x}$

13. $\dfrac{x^2+2}{x^2+x-12}+\dfrac{x+1}{12-x-x^2}$

14. $\dfrac{10}{x^2-x-6}-\dfrac{5x}{6+x-x^2}$

15. $\dfrac{x^2+2}{x^2-4}-\dfrac{4x-2}{x^2-4}$

16. $\dfrac{2x+5}{2x^2-x-1}-\dfrac{4x+2}{2x^2-x-1}$

17. $\dfrac{x+3}{7x-2}+\dfrac{2x-1}{14x-4}$

18. $\dfrac{3x+1}{4x+10}+\dfrac{4-x}{2x+5}$

19. $\dfrac{5}{x-3}+\dfrac{x}{x^2-9}$

20. $\dfrac{x+1}{x^2-3x-10}+\dfrac{x}{x-5}$

21. $\dfrac{x}{x-1}-\dfrac{4}{x+2}$

22. $\dfrac{x-1}{3x-1}-\dfrac{8+4x}{x+2}$

23. $\dfrac{x+2}{x+3}-\dfrac{4}{3-x}$

24. $\dfrac{x-1}{4-x}+\dfrac{3x}{x+5}$

25. $\dfrac{x+2}{3x+9}+\dfrac{2x-1}{2x-6}$

26. $\dfrac{x}{4x-8}-\dfrac{3x+2}{3x+6}$

27. $\dfrac{3x}{6+x}-\dfrac{2x}{x^2-36}$

28. $\dfrac{3x-4}{x^2-x-20}-\dfrac{2}{5-x}$

29. $\dfrac{4x+1}{7-x}+\dfrac{x-1}{x^2-8x+7}$

30. $\dfrac{4}{x+5}-\dfrac{2x+3}{x^2+4x-5}$

31. $\dfrac{4x}{x^2+3x-28}+\dfrac{3}{x^2+6x-7}$

32. $\dfrac{3x}{x^2+2x+1}-\dfrac{x}{x^2+4x+4}$

33. $\dfrac{x+1}{x^2+4x+4}-\dfrac{x-3}{x^2-4}$

34. $\dfrac{x-4}{x^2-5x+6}+\dfrac{2x}{x^2-2x-3}$

35. $\dfrac{3x}{9-x^2}+\dfrac{5}{x^2-7x+12}$

36. $\dfrac{4x}{3x^2+4x+1}-\dfrac{x+4}{x^2+7x+6}$

37. $\dfrac{x-6}{7x^2-3x-4}+\dfrac{7-x}{7x^2+18x+8}$

38. $\dfrac{x+5}{9x^2-26x-3}-\dfrac{8x}{9x^2+11x+3}$

39. $\dfrac{x-3}{4x^2-5x-6} - \dfrac{4x+10}{2x^2+x-10}$

40. $\dfrac{2x+1}{8x^2-37x-15} + \dfrac{2-x}{8x^2+11x+3}$

41. $\dfrac{3x}{4-x} + \dfrac{7x}{x+4} - \dfrac{x-3}{x^2-16}$

42. $\dfrac{x}{x+3} + \dfrac{x+1}{3-x} + \dfrac{x^2+4}{x^2-9}$

43. $2 - \dfrac{4x+1}{x-4} + \dfrac{x-3}{x^2-6x+8}$

44. $-4 + \dfrac{1-2x}{x+6} + \dfrac{x^2+1}{x^2+4x-12}$

45. $\dfrac{2}{x^2-4} - \dfrac{3}{x^2-3x+2} + \dfrac{x-1}{x^2+x-2}$

46. $\dfrac{4}{x^2+3x-10} + \dfrac{3}{x^2-25} - \dfrac{5}{x^2-7x+10}$

47. $\dfrac{x}{x^2+4x-21} + \dfrac{1-x}{x^2+8x+7} + \dfrac{3x}{x^2-2x-3}$

48. $\dfrac{3x+9}{x^2-5x+4} + \dfrac{49}{12+x-x^2} + \dfrac{3x+21}{x^2+2x-3}$

49. $\dfrac{5x+22}{x^2+8x+15} + \dfrac{4}{x^2+4x+3} + \dfrac{6}{x^2+6x+5}$

50. $\dfrac{x+1}{2x^2-x-1} + \dfrac{2x}{2x^2+5x+2} - \dfrac{2x}{3x^2+4x-4}$

51. $\dfrac{x-6}{3x^2+10x+3} - \dfrac{2x}{5x^2-3x-2} + \dfrac{2x}{3x^2-2x-1}$

52. $\dfrac{x}{xy+x-2y-2} + \dfrac{x+2}{xy+x+y+1}$

53. $\dfrac{4x}{xy-3x+y-3} + \dfrac{x+2}{xy+2y-3x-6}$

54. $\dfrac{3y}{xy+2x+3y+6} + \dfrac{x}{x^2-2x-15}$

55. $\dfrac{2}{xy-4x-2y+8} + \dfrac{5y}{y^2-3y-4}$

56. $\dfrac{x+6}{x^2+x+1} - \dfrac{3x^2+x-4}{x^3-1}$

57. $\dfrac{2x-5}{8x^2-4x+2} + \dfrac{x^2-2x+5}{8x^3+1}$

58. $\dfrac{x+1}{x^3-3x^2+x-3} + \dfrac{x^2-5x-8}{x^4-8x^2-9}$

59. $\dfrac{x+4}{x^3-5x^2+6x-30} - \dfrac{x-7}{x^3-2x^2+6x-12}$

60. $\dfrac{x+2}{9x^2-6x+4} + \dfrac{10x-5x^2}{27x^3+8} - \dfrac{2}{3x+2}$

Writing and Thinking About Mathematics

61. Discuss the steps in the process you go through when adding two rational expressions with different denominators. That is, discuss how you find the least common denominator when adding rational expressions and how you use this LCD to find equivalent rational expressions that you can add.

 HAWKES LEARNING SYSTEMS: INTERMEDIATE ALGEBRA SOFTWARE

- Addition and Subtraction with Rational Expressions

5.3 Complex Fractions

- *Simplify complex fractions.*

Simplifying Complex Fractions (First Method)

A **complex fraction** is a fraction in which the numerator and/or denominator are themselves fractions or the sum or difference of fractions. Examples of complex fractions are

$$\frac{x+y}{x^{-1}+y^{-1}} = \frac{x+y}{\dfrac{1}{x}+\dfrac{1}{y}} \qquad \text{and} \qquad \frac{\dfrac{1}{x+3}-\dfrac{1}{x}}{1+\dfrac{3}{x}}.$$

The objective here is to develop techniques for simplifying complex fractions so that they are written in the form of a single reduced rational expression.

In a complex fraction such as $\dfrac{\dfrac{1}{x+3}-\dfrac{1}{x}}{1+\dfrac{3}{x}}$, the large fraction bar is a symbol of inclusion.

The expression could also be written as follows.

$$\frac{\dfrac{1}{x+3}-\dfrac{1}{x}}{1+\dfrac{3}{x}} = \left(\frac{1}{x+3}-\frac{1}{x}\right) \div \left(1+\frac{3}{x}\right)$$

Thus a complex fraction indicates that the numerator is to be divided by the denominator.

Simplifying Complex Fractions (First Method)

To simplify complex fractions:

1. *Simplify the numerator so that it is a single rational expression.*
2. *Simplify the denominator so that it is a single rational expression.*
3. *Divide the numerator by the denominator and reduce to lowest terms.*

This method is used to simplify the complex fractions in Examples 1 and 2. Study the examples closely so that you understand what happens at each step.

Example 1: First Method for Simplifying Complex Fractions

Simplify the complex fraction $\dfrac{\dfrac{3x}{y^2}}{\dfrac{12x}{7y}}$.

Solution: $\dfrac{\dfrac{3x}{y^2}}{\dfrac{12x}{7y}} = \dfrac{3x}{y^2} \cdot \dfrac{7y}{12x} = \dfrac{7}{4y}$

Multiply by the reciprocal of the denominator.

Example 2: First Method for Simplifying Complex Fractions

Simplify the following complex fractions.

a. $\dfrac{x+y}{x^{-1}+y^{-1}}$

Solution: $\dfrac{x+y}{x^{-1}+y^{-1}} = \dfrac{x+y}{\dfrac{1}{x}+\dfrac{1}{y}}$

Recall that $x^{-1} = \dfrac{1}{x}$ and $y^{-1} = \dfrac{1}{y}$.

$= \dfrac{\dfrac{x+y}{1}}{\dfrac{1}{x}\cdot\dfrac{y}{y}+\dfrac{1}{y}\cdot\dfrac{x}{x}}$

Add the two fractions in the denominator.

$= \dfrac{\dfrac{x+y}{1}}{\dfrac{y}{xy}+\dfrac{x}{xy}} = \dfrac{\dfrac{x+y}{1}}{\dfrac{y+x}{xy}}$

$= \dfrac{x+y}{1} \cdot \dfrac{xy}{y+x} = \dfrac{xy}{1} = xy$

Multiply by the reciprocal of the denominator.

b. $\dfrac{\dfrac{1}{x+3}-\dfrac{1}{x}}{1+\dfrac{3}{x}}$

Solution: $\dfrac{\dfrac{1}{x+3}-\dfrac{1}{x}}{1+\dfrac{3}{x}} = \dfrac{\dfrac{1\cdot x}{(x+3)\cdot x}-\dfrac{1(x+3)}{x(x+3)}}{\dfrac{x}{x}+\dfrac{3}{x}}$

Combine the fractions in the numerator and in the denominator separately.

Note that $1 = \dfrac{x}{x}$.

Continued on the next page...

$$= \frac{\dfrac{x-(x+3)}{x(x+3)}}{\dfrac{x+3}{x}} = \frac{\dfrac{x-x-3}{x(x+3)}}{\dfrac{x+3}{x}}$$

$$= \frac{-3}{x(x+3)} \cdot \frac{x}{x+3} = \frac{-3}{(x+3)^2}$$ Multiply by the reciprocal of the denominator.

Simplifying Complex Fractions (Second Method)

A second method is to find the LCM of the denominators in the fractions in both the original numerator and the original denominator, and then multiply both the numerator and denominator by this LCM.

Simplifying Complex Fractions (Second Method)

To simplify complex fractions:

1. *Find the LCM of all the denominators in the numerator and denominator of the complex fraction.*

2. *Multiply both the numerator and denominator of the complex fraction by this LCM.*

3. *Simplify both the numerator and denominator and reduce to lowest terms.*

Example 3: Second Method for Simplifying Complex Fractions

Simplify the following complex fractions.

a. $\dfrac{x+y}{x^{-1}+y^{-1}}$

Solution: $\dfrac{x+y}{x^{-1}+y^{-1}} = \dfrac{\dfrac{x+y}{1}}{\dfrac{1}{x}+\dfrac{1}{y}}$

$$= \frac{\left(\dfrac{x+y}{1}\right)xy}{\left(\dfrac{1}{x}+\dfrac{1}{y}\right)xy} = \frac{(x+y)xy}{\dfrac{1}{x}\cdot xy+\dfrac{1}{y}\cdot xy}$$

Multiply by xy, the LCM of $\{x, y, 1\}$. This multiplication can be done because the net effect is that the fraction is multiplied by 1.

$$= \frac{(x+y)xy}{y+x} = xy$$

b. $\dfrac{\dfrac{1}{x+3}-\dfrac{1}{x}}{1+\dfrac{3}{x}}$

Solution: $\dfrac{\dfrac{1}{x+3}-\dfrac{1}{x}}{1+\dfrac{3}{x}}=\dfrac{\left(\dfrac{1}{x+3}-\dfrac{1}{x}\right)\cdot x(x+3)}{\left(1+\dfrac{3}{x}\right)\cdot x(x+3)}$ Multiply by $x(x+3)$, the LCM of $\{x, x+3\}$.

$=\dfrac{\dfrac{1}{x+3}\cdot x(x+3)-\dfrac{1}{x}\cdot x(x+3)}{1\cdot x(x+3)+\dfrac{3}{x}\cdot x(x+3)}$

$=\dfrac{x-(x+3)}{x(x+3)+3(x+3)}=\dfrac{x-x-3}{(x+3)(x+3)}$

$=\dfrac{-3}{(x+3)^2}$

Practice Problems

Simplify each complex fraction. Use both methods to solve each problem. In this way you will find which method seems easier to you, and you will see that both methods result in the same answer.

1. $\dfrac{\dfrac{1}{x+2}-\dfrac{1}{x}}{1+\dfrac{2}{x}}$

2. $\dfrac{\dfrac{1}{x+y}-\dfrac{1}{x-y}}{\dfrac{2y}{x^2-y^2}}$

5.3 Exercises

Simplify the complex fractions given in Exercises 1 – 36.

1. $\dfrac{\dfrac{2x}{3y^2}}{\dfrac{5x^2}{6y}}$ **2.** $\dfrac{\dfrac{6x^2}{5y}}{\dfrac{x}{10y^2}}$ **3.** $\dfrac{\dfrac{12x^3}{7y^2}}{\dfrac{3x^5}{2y}}$ **4.** $\dfrac{\dfrac{9x^2}{7y^3}}{\dfrac{3xy}{14}}$ **5.** $\dfrac{\dfrac{x+3}{2x}}{\dfrac{2x-1}{4x^2}}$

6. $\dfrac{\dfrac{x-2}{6x}}{\dfrac{x+3}{3x^2}}$ **7.** $\dfrac{\dfrac{3}{x}+\dfrac{1}{2x}}{1+\dfrac{2}{x}}$ **8.** $\dfrac{\dfrac{2x-1}{x}}{\dfrac{2}{x}+3}$ **9.** $\dfrac{1+\dfrac{1}{x}}{1-\dfrac{1}{x^2}}$ **10.** $\dfrac{\dfrac{2}{y}+1}{\dfrac{4}{y^2}-1}$

Answers to Practice Problems: **1.** $\dfrac{-2}{(x+2)^2}$ **2.** -1

11. $\dfrac{\dfrac{1}{x}+\dfrac{1}{3x}}{\dfrac{x+6}{x^2}}$

12. $\dfrac{\dfrac{3}{x}-\dfrac{6}{x^2}}{\dfrac{x-2}{x^2}}$

13. $\dfrac{\dfrac{7}{x}-\dfrac{14}{x^2}}{\dfrac{1}{x}-\dfrac{4}{x^3}}$

14. $\dfrac{\dfrac{3}{x}-\dfrac{6}{x^2}}{\dfrac{1}{x}-\dfrac{2}{x^2}}$

15. $\dfrac{\dfrac{x}{y}-\dfrac{1}{3}}{\dfrac{6}{y}-\dfrac{2}{x}}$

16. $\dfrac{\dfrac{3}{x}+\dfrac{5}{2x}}{\dfrac{1}{x}+4}$

17. $\dfrac{\dfrac{2}{x}+\dfrac{3}{4y}}{\dfrac{3}{2x}-\dfrac{5}{3y}}$

18. $\dfrac{1+x^{-1}}{1-x^{-2}}$

19. $\dfrac{1}{x^{-1}+y^{-1}}$

20. $\dfrac{x^{-1}+y^{-1}}{x+y}$

21. $\dfrac{x^{-1}+y^{-1}}{x^{-1}-y^{-1}}$

22. $\dfrac{x^{-1}+y^{-1}}{x^{-2}-y^{-2}}$

23. $\dfrac{2-\dfrac{4}{x}}{\dfrac{x^2-4}{x^2+x}}$

24. $\dfrac{\dfrac{1}{x}}{1-\dfrac{1}{x-2}}$

25. $\dfrac{x+\dfrac{3}{x-4}}{1-\dfrac{1}{x}}$

26. $\dfrac{1-\dfrac{4}{x+3}}{1-\dfrac{2}{x+1}}$

27. $\dfrac{1+\dfrac{4}{2x-3}}{1+\dfrac{x}{x+1}}$

28. $\dfrac{\dfrac{1}{x+h}-\dfrac{1}{x}}{h}$

29. $\dfrac{\dfrac{1}{(x+h)^2}-\dfrac{1}{x^2}}{h}$

30. $\dfrac{\left(2+\dfrac{1}{x+h}\right)-\left(2+\dfrac{1}{x}\right)}{h}$

31. $\dfrac{x^2-4y^2}{1-\dfrac{2x+y}{x-y}}$

32. $\dfrac{\dfrac{x+1}{x-1}-\dfrac{x-1}{x+1}}{\dfrac{x+1}{x-1}+\dfrac{x-1}{x+1}}$

33. $\dfrac{\dfrac{1}{x^2-1}-\dfrac{1}{x+1}}{\dfrac{1}{x-1}+\dfrac{1}{x^2-1}}$

34. $\dfrac{\dfrac{x}{x-4}-\dfrac{1}{x-1}}{\dfrac{x}{x-1}+\dfrac{2}{x-3}}$

35. $\dfrac{x-y}{x^{-2}-y^{-2}}$

36. $\dfrac{y^{-2}-x^{-2}}{x+y}$

Writing and Thinking About Mathematics

37. Some complex fractions involve the sum (or difference) of complex fractions. Beginning with the "farthest" denominator, simplify each of the following expressions.

a. $1+\dfrac{1}{1+\dfrac{1}{1+\dfrac{1}{1+1}}}$

b. $2-\dfrac{1}{2-\dfrac{1}{2-\dfrac{1}{2-1}}}$

c. $x+\dfrac{1}{x+\dfrac{1}{x+\dfrac{1}{x+1}}}$

 HAWKES LEARNING SYSTEMS: INTERMEDIATE ALGEBRA SOFTWARE

- Complex Fractions

5.4

Solving Equations with Rational Expressions

- *Solve proportions.*
- *Solve other equations with rational expressions.*

Proportions

A **ratio** is a comparison of two numbers by division. Ratios are written in the form

$$a : b \quad \text{or} \quad \frac{a}{b} \quad \text{or} \quad a \text{ to } b.$$

For example, suppose the ratio of pages in the first two chapters of a text is 4 to 3. We can also write this ratio in the form $4 : 3$ or in the fraction form $\frac{4}{3}$. This ratio does not mean that there are only 4 pages in one chapter and 3 pages in the other. There are many fractions that reduce to $\frac{4}{3}$. If there are 105 pages total in both chapters, then there are 60 pages in the first chapter and 45 pages in the second because $60 + 45 = 105$ and $\frac{60}{45} = \frac{4}{3}$.

Proportion

*A **proportion** is an equation stating that two ratios are equal.*

Proportions may involve only numbers as in $\frac{2}{3} = \frac{10}{15}$. However, proportions can also be used to find unknown quantities in applications, and in such cases, will involve variables.

One method of solving proportions with variables is to "clear" the equation of fractions by first multiplying both sides of the equation by the LCM of the denominators. This method is illustrated in Example 1.

Example 1: Proportions

Solve the following proportions.

a. $\dfrac{x-5}{2x} = \dfrac{6}{3x}$ LCM $= 6x$

Solution: $6x \cdot \left(\dfrac{x-5}{2x}\right) = 6x \cdot \left(\dfrac{6}{3x}\right)$ $(x \neq 0)$

$$3(x-5) = 2(6)$$
$$3x - 15 = 12$$
$$3x = 27$$
$$x = 9$$

Check: $\dfrac{9-5}{2 \cdot 9} \overset{?}{=} \dfrac{6}{3 \cdot 9}$

$$\dfrac{4}{18} \overset{?}{=} \dfrac{6}{27}$$

$$\dfrac{2}{9} = \dfrac{2}{9}$$ Thus the solution is $x = 9$.

b. $\dfrac{3}{x-6} = \dfrac{5}{x}$ LCM $= x\,(x-6)$

Solution: $x\,(x-6) \cdot \dfrac{3}{x-6} = x\,(x-6) \cdot \dfrac{5}{x}$ $(x \neq 0, 6)$

$$3x = 5x - 30$$
$$30 = 2x$$
$$15 = x$$

Check: $\dfrac{3}{15-6} \overset{?}{=} \dfrac{5}{15}$

$$\dfrac{3}{9} \overset{?}{=} \dfrac{5}{15}$$

$$\dfrac{1}{3} = \dfrac{1}{3}$$ Thus the solution is $x = 15$.

Proportions can be used to solve many everyday types of word problems. Using the correct ratios of units on both sides of the proportion is critical. One of the following conditions must be true:

1. The numerators agree in type and the denominators agree in type.
2. The numerators correspond and the denominators correspond.

Example 2 illustrates one situation.

Example 2: Application of Proportions

On an architect's scale drawing of a home, $\frac{1}{2}$ inch represents 10 feet. What length does a measure of $2\frac{1}{2}$ inches represent?

Solution: Set up a proportion representing the information. In this example the numerators are the same type and the denominators are the same type.

$$\frac{\frac{1}{2} \text{ inch}}{10 \text{ feet}} = \frac{2\frac{1}{2} \text{ inches}}{x \text{ feet}} \qquad \text{LCM} = 10x$$

$$10x \cdot \frac{\frac{1}{2}}{10} = 10x \cdot \frac{\frac{5}{2}}{x} \qquad \text{Multiply both sides by } 10x.$$

$$\frac{1}{2}x = 25 \qquad \text{Simplify.}$$

$$2 \cdot \frac{1}{2}x = 2 \cdot 25 \qquad \text{Multiply both sides by 2.}$$

$$x = 50 \qquad \text{Simplify.}$$

On this drawing, $2\frac{1}{2}$ inches represent 50 feet.

NOTES

Any of the following four equations could have been used to solve the problem in Example 2:

$$\frac{\frac{1}{2} \text{ inch}}{10 \text{ feet}} = \frac{2\frac{1}{2} \text{ inches}}{x \text{ feet}} \qquad\qquad \frac{10 \text{ feet}}{\frac{1}{2} \text{ inch}} = \frac{x \text{ feet}}{2\frac{1}{2} \text{ inches}}$$

$$\frac{\frac{1}{2} \text{ inch}}{2\frac{1}{2} \text{ inches}} = \frac{10 \text{ feet}}{x \text{ feet}} \qquad\qquad \frac{2\frac{1}{2} \text{ inches}}{\frac{1}{2} \text{ inch}} = \frac{x \text{ feet}}{10 \text{ feet}}$$

Proportions and Similar Triangles

Proportions are used when working with similar geometric figures. Similar figures are figures that meet the following two conditions:

1. The corresponding angles are equal.
2. The corresponding sides are proportional.

In **similar triangles**, corresponding sides are those sides opposite the equal angles. (See Figure 5.1.)

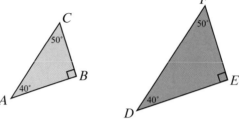

Figure 5.1

We write $\triangle ABC \sim \triangle DEF$. ($\sim$ is read "**is similar to**".) The corresponding sides are proportional. Thus,

$$\frac{AB}{DE} = \frac{BC}{EF} \quad \text{and} \quad \frac{AB}{DE} = \frac{AC}{DF} \quad \text{and} \quad \frac{BC}{EF} = \frac{AC}{DF}.$$

In a pair of similar triangles, we can often find the length of an unknown side by setting up a proportion and solving. Example 3 illustrates such a situation.

Example 3: Similar Triangles

In the figure shown, $\triangle ABC \sim \triangle PQR$. Find the lengths of the sides AB and QR.

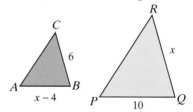

Solution: Set up a proportion involving the corresponding sides and solve for x.

$$\frac{x-4}{10} = \frac{6}{x}$$

$$10x \cdot \frac{x-4}{10} = 10x \cdot \frac{6}{x}$$

$$x(x-4) = 6 \cdot 10$$

$$x^2 - 4x = 60$$

$$x^2 - 4x - 60 = 0$$

$$(x-10)(x+6) = 0$$

$$x - 10 = 0 \quad \text{or} \quad x + 6 = 0$$

$$x = 10 \qquad\qquad x = -6$$

Because the length of a side cannot be negative, the only acceptable solution is $x = 10$. Thus $QR = 10$. Substituting 10 for x gives $AB = 10 - 4 = 6$.

Solving Equations with Rational Expressions

An equation such as

$$\frac{3}{x} + \frac{1}{8} = \frac{13}{4x}$$

that involves the sum of rational expressions is **not a proportion**. However, the method of finding the solution to the equation is similar in that we "clear" the fractions by multiplying both sides of the equation by the LCM of the denominators. In this example the LCM of the denominators is $8x$ and we can proceed as follows:

$$\frac{3}{x} + \frac{1}{8} = \frac{13}{4x} \qquad \text{Note } x \neq 0.$$

$$8x\left(\frac{3}{x} + \frac{1}{8}\right) = 8x \cdot \frac{13}{4x} \qquad \begin{array}{l}\text{Multiply both sides by } 8x, \text{ the}\\ \text{LCM of the denominators.}\end{array}$$

$$8\cancel{x} \cdot \frac{3}{\cancel{x}} + \cancel{8}x \cdot \frac{1}{\cancel{8}} = \overset{2}{\cancel{8}}\cancel{x} \cdot \frac{13}{\cancel{4x}} \qquad \text{Use the distributive property.}$$

$$24 + x = 26 \qquad \text{Simplify.}$$

$$x = 2.$$

As we have seen, rational expressions may contain variables in either the numerator or denominator or both. In any case, a general approach to solving equations that contain rational expressions is as follows.

To Solve an Equation Containing Rational Expressions

1. *Find the LCM of the denominators.*
2. *Multiply both sides of the equation by this LCM and simplify.*
3. *Solve the resulting equation. (This equation will have only polynomials on both sides.)*
4. *Check each solution in the **original equation**. (Remember that no denominator can be 0 and any solution that gives a 0 denominator is to be discarded.)*

Checking is particularly important when equations have rational expressions. Multiplying by the LCM may introduce solutions that are not solutions to the original equation. Such solutions are called **extraneous solutions** or **extraneous roots** and occur because multiplication by a variable expression may in effect be multiplying the original equation by 0.

Example 4: Solving Equations Involving Rational Expressions

State any restrictions on the variable, and then solve the equation.

a. $\dfrac{1}{x-4} = \dfrac{3}{x^2-5x}$

Solution: First find the LCM of the denominators and then multiply both sides of the equation by the LCM.

$$\left.\begin{array}{l} x-4 = x-4 \\ x^2-5x = x(x-5) \end{array}\right\} \quad \text{LCM} = x(x-5)(x-4)$$

$$x(x-5)(x-4)\cdot\dfrac{1}{x-4} = x(x-5)(x-4)\cdot\dfrac{3}{x(x-5)} \quad x \neq 0, 4, 5$$

$$x(x-5) = 3(x-4)$$
$$x^2-5x = 3x-12$$
$$x^2-8x+12 = 0$$
$$(x-6)(x-2) = 0$$
$$x-6=0 \quad \text{or} \quad x-2=0$$
$$x=6 \qquad\qquad x=2$$

Since 6 and 2 are not restrictions, there are two solutions, $x=6$ and $x=2$.

b. $\dfrac{x}{x-2} + \dfrac{x-6}{x(x-2)} = \dfrac{5x}{x-2} - \dfrac{10}{x-2}$

Solution: First find the LCM of the denominators and then multiply each term on both sides of the equation by the LCM.

$$\left.\begin{array}{l} x-2 \\ x(x-2) \end{array}\right\} \quad \text{LCM} = x(x-2)$$

$$x(x-2)\cdot\dfrac{x}{x-2} + x(x-2)\cdot\dfrac{x-6}{x(x-2)} = x(x-2)\cdot\dfrac{5x}{x-2} - x(x-2)\cdot\dfrac{10}{x-2}$$

$$x^2+x-6 = 5x^2-10x \qquad x \neq 0, 2$$
$$0 = 4x^2-11x+6$$
$$0 = (4x-3)(x-2)$$
$$4x-3=0 \quad \text{or} \quad x-2=0$$
$$4x=3 \qquad\qquad \cancel{x=2}$$
$$x=\dfrac{3}{4}$$

The only solution is $x=\dfrac{3}{4}$, since 2 is a restricted value ($x \neq 0, 2$) and thus **not** a solution. No denominator can be 0.

c. $\dfrac{2}{x^2-9}=\dfrac{1}{x^2}+\dfrac{1}{x^2-3x}$

Solution: First find the LCM of the denominators and then multiply each term on both sides of the equation by the LCM.

$$x^2-9=(x+3)(x-3)$$
$$x^2=x^2$$
$$x^2-3x=x(x-3)$$

$$\text{LCM}=x^2(x+3)(x-3)$$

$$x^2\,\cancel{(x+3)}\,\cancel{(x-3)}\cdot\dfrac{2}{\cancel{(x+3)}\,\cancel{(x-3)}}=\cancel{x^2}^{\,}(x+3)(x-3)\cdot\dfrac{1}{\cancel{x^2}}+\cancel{x^2}^{\,x}(x+3)\,\cancel{(x-3)}\cdot\dfrac{1}{\cancel{x}\,\cancel{(x-3)}}$$

$$2x^2=(x+3)(x-3)+x(x+3)\qquad (x\neq 0,-3,3)$$
$$2x^2=x^2-9+x^2+3x$$
$$2x^2=2x^2+3x-9$$
$$9=3x$$
$$\cancel{3=x}\qquad\qquad\text{Note that 3 is one of the restrictions.}$$

There is no solution. The solution set is the empty set, $\varnothing$. The original equation is a contradiction.

Formulas

Many formulas are equations relating more than one variable. The equation is solved for one variable in terms of another variable (maybe more than one). The process of solving equations can be used to manipulate the formula so that it is solved for one of the other variables. Example 5 shows how this can be done.

Example 5: Solving a Formula for a Specified Variable

The formula $S=2\pi r^2+2\pi rh$ is used to find the surface area (S) of a right circular cylinder, where r is the radius of the cylinder and h is the height of the cylinder. Solve the formula for h.

Solution: $\qquad\qquad S=2\pi r^2+2\pi rh \qquad$ Write the formula.

$\qquad\qquad S-2\pi r^2=2\pi rh \qquad$ Add $-2\pi r^2$ to both sides of the equation.

$\qquad\qquad \dfrac{S-2\pi r^2}{2\pi r}=h \qquad$ Divide both sides by $2\pi r$.

$\qquad$ or $\dfrac{S}{2\pi r}-r=h$

Thus the formula solved for h is: $h=\dfrac{S-2\pi r^2}{2\pi r}$.

Practice Problems

Solve each proportion in Exercises 1 and 2. State any restrictions on the variable.

1. $\dfrac{5x}{3} = \dfrac{9x+4}{5}$

2. $\dfrac{16}{x} = \dfrac{4}{x-5}$

3. On a road map, each inch represents 50 miles. What distance is represented by 4.5 inches on the map?

Solve each equation in Exercises 4 and 5. State any restrictions on the variable.

4. $\dfrac{5}{3x+2} = \dfrac{4}{3x+1}$

5. $\dfrac{x}{x-3} - \dfrac{2x+3}{x^2+x-12} = \dfrac{x-1}{x+4}$

6. Solve the formula $A = P + Pr$ for r.

5.4 Exercises

Solve the proportions in Exercises 1 – 8.

1. $\dfrac{4x}{7} = \dfrac{x+5}{3}$

2. $\dfrac{3x+1}{4} = \dfrac{2x+1}{3}$

3. $\dfrac{10}{x} = \dfrac{5}{x-2}$

4. $\dfrac{8}{x-3} = \dfrac{12}{2x-3}$

5. $\dfrac{6}{x-4} = \dfrac{5}{x+7}$

6. $\dfrac{5x+2}{11x} = \dfrac{x-6}{4x}$

7. $\dfrac{x+3}{5x} = \dfrac{x-1}{6x}$

8. $\dfrac{2x}{x-3} = \dfrac{x}{x-4}$

9. Computers: Making a statistical analysis, Ana found 3 defective computers in a sample of 20 computers. If this ratio is consistent, how many defective computers does she expect to find in a batch of 2400 computers?

10. Cartography: On a map, each inch represents 7.5 miles. What is the distance represented by 3.5 inches on the map?

Answers to Practice Problems: **1.** $x = -6$; no restrictions   **2.** $x = \dfrac{20}{3}$; $x \neq 0, 5$ **3.** 225 miles

4. $x = 1$; $x \neq -\dfrac{1}{3}, -\dfrac{2}{3}$ **5.** $x = 1$; $x \neq -4, 3$ **6.** $r = \dfrac{A-P}{P}$

11. **Education:** An elementary school has a ratio of 1 teacher for every 22 children. If the school presently has 23 teachers, how many students are enrolled?

12. **Architecture:** A floor plan is drawn to scale in which 1 inch represents 4 feet. What size will the drawing be for a room that is 30 feet by 40 feet? (Hint: Set up two proportions.)

Exercises 13 – 18 show pairs of similar triangles. Find the lengths of the sides labeled with variables.

13. $\triangle ABC \sim \triangle RST$

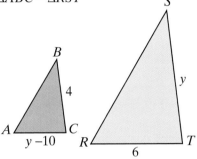

14. $\triangle FED \sim \triangle FGH$

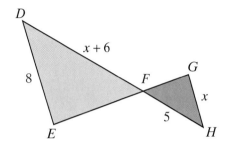

15. $\triangle SUT \sim \triangle PRQ$

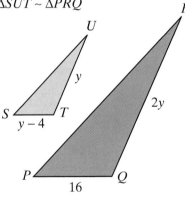

16. $\triangle JKL \sim \triangle JTB$

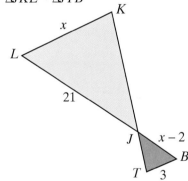

17. $\triangle QRP \sim \triangle TUS$

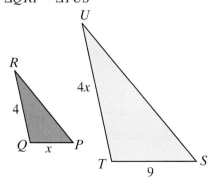

18. $\triangle ABC \sim \triangle DEC$

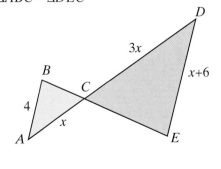

19. In the parallelogram $ABCD$, $AB = CD = 10$ in. Diagonal $AC = 12$ in. The point M on AB is 6 in. from A. Point P is the intersection of DM with AC. The triangles APM and CPD are similar. (Symbolically, $\triangle APM \sim \triangle CPD$.) What are the lengths of AP and PC?

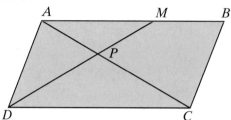

20. If, in the same figure discussed in Exercise 19, the point P is the point of intersection of the two diagonals, AC and DB, what are the lengths of AP and PC?

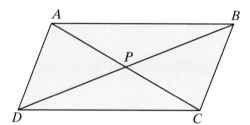

State any restrictions on the variables, and then solve the equations containing rational expressions in Exercises 21 – 50.

21. $\dfrac{5x}{4} - \dfrac{1}{2} = -\dfrac{3}{16}$

22. $\dfrac{x}{6} - \dfrac{1}{42} = \dfrac{1}{7}$

23. $\dfrac{4x}{3} - \dfrac{3}{4} = \dfrac{5x}{6}$

24. $\dfrac{x-2}{3} - \dfrac{x-3}{5} = \dfrac{13}{15}$

25. $\dfrac{2+x}{4} - \dfrac{5x-2}{12} = \dfrac{8-2x}{5}$

26. $\dfrac{8x+10}{5} = 2x+3 - \dfrac{6x+1}{4}$

27. $\dfrac{2}{3x} = \dfrac{1}{4} - \dfrac{1}{6x}$

28. $\dfrac{x-4}{x} + \dfrac{3}{x} = 0$

29. $\dfrac{3}{8x} - \dfrac{7}{10} = \dfrac{1}{5x}$

30. $\dfrac{1}{x} - \dfrac{8}{21} = \dfrac{3}{7x}$

31. $\dfrac{3}{4x} - \dfrac{1}{2} = \dfrac{7}{8x} + \dfrac{1}{6}$

32. $\dfrac{7}{x-3} = \dfrac{6}{x-4}$

33. $\dfrac{2}{3x+2} = \dfrac{4}{5x+1}$

34. $\dfrac{-3}{2x+1} = \dfrac{4}{3x+1}$

35. $\dfrac{9}{5x-3} = \dfrac{5}{3x+7}$

36. $\dfrac{5x+2}{x-6} = \dfrac{11}{4}$

37. $\dfrac{x+9}{3x+2} = \dfrac{5}{8}$

38. $\dfrac{8}{2x+3} = \dfrac{9}{4x-5}$

39. $\dfrac{x}{x-4} - \dfrac{4}{2x-1} = 1$

40. $\dfrac{x}{x+3} + \dfrac{1}{x+2} = 1$

41. $\dfrac{x+2}{x+1} + \dfrac{x+2}{x+4} = 2$

42. $\dfrac{x-2}{x-3} + \dfrac{x-3}{x-2} = \dfrac{2x^2}{x^2-5x+6}$

43. $\dfrac{2}{4x-1} + \dfrac{1}{x+1} = \dfrac{3}{x+1}$

44. $\dfrac{3x+5}{3x+2} + \dfrac{8x+16}{3x^2-4x-4} = \dfrac{x+2}{x-2}$

45. $\dfrac{x}{x-4} - \dfrac{12x}{x^2+x-20} = \dfrac{x-1}{x+5}$

46. $\dfrac{x-2}{x+4} - \dfrac{3}{2x+1} = \dfrac{x-7}{x+4}$

47. $\dfrac{3x+5}{3x+2} - \dfrac{4-2x}{3x^2+8x+4} = \dfrac{x+4}{x+2}$

48. $\dfrac{3}{3x-1} + \dfrac{1}{x+1} = \dfrac{4}{2x-1}$

49. $\dfrac{5}{2x+1} - \dfrac{1}{2x-1} = \dfrac{2}{x-2}$

50. $\dfrac{2}{x+1} + \dfrac{4}{2x-3} = \dfrac{4}{x-5}$

Solve each of the formulas in Exercises 51 – 60 for the specified variables.

51. $S = \dfrac{a}{1-r}$; solve for r (formula for the sum of an infinite geometric sequence)

52. $z = \dfrac{x-\bar{x}}{s}$; solve for x (formula used in statistics)

53. $z = \dfrac{x-\bar{x}}{s}$; solve for s (formula used in statistics)

54. $a_n = a_1 + (n-1)d$; solve for d (formula for the nth term in an arithmetic sequence)

55. $\dfrac{1}{R_{total}} = \dfrac{1}{R_1} + \dfrac{1}{R_2}$; solve for R_{total} (formula used in electronics)

56. $m = \dfrac{y-y_1}{x-x_1}$; solve for y (formula for the slope of a line)

57. $A = P + Pr$; solve for P (formula used for compound interest)

58. $v_{avg} = \dfrac{d_2 - d_1}{t_2 - t_1}$; solve for d_2 (formula for average velocity)

59. $y = \dfrac{ax + b}{cx + d}$; solve for x (formula used in mathematics)

60. $\dfrac{1}{x} = \dfrac{1}{t_1} + \dfrac{1}{t_2}$; solve for x (formula used in mathematics)

Writing and Thinking About Mathematics

In simplifying rational expressions, the result is a rational expression. However, in solving equations with rational expressions, the goal is to find a value (or values) for the variable that will make the equation a true statement. Many students confuse these two ideas. To avoid confusing the techniques for adding and subtracting rational expressions with the techniques for solving equations, simplify the expression in part **a.** and solve the equation in part **b.** Explain, in your own words, the differences in your procedures.

61. a. $\dfrac{10}{x} + \dfrac{31}{x-1} + \dfrac{4x}{x-1}$ **b.** $\dfrac{10}{x} + \dfrac{31}{x-1} = \dfrac{4x}{x-1}$

62. a. $\dfrac{-4}{x^2-16} + \dfrac{x}{2x+8} - \dfrac{1}{4}$ **b.** $\dfrac{-4}{x^2-16} + \dfrac{x}{2x+8} = \dfrac{1}{4}$

63. a. $\dfrac{3x}{x^2-4} + \dfrac{5}{x+2} + \dfrac{2}{x-2}$ **b.** $\dfrac{3x}{x^2-4} + \dfrac{5}{x+2} = \dfrac{2}{x-2}$

64. a. $\dfrac{7}{5x} + \dfrac{2}{x-4} - \dfrac{3}{5x}$ **b.** $\dfrac{7}{5x} + \dfrac{2}{x-4} = \dfrac{3}{5x}$

65. a. $\dfrac{2}{x+9} - \dfrac{2}{x-9} + \dfrac{1}{2}$ **b.** $\dfrac{2}{x+9} - \dfrac{2}{x-9} = \dfrac{1}{2}$

HAWKES LEARNING SYSTEMS: INTERMEDIATE ALGEBRA SOFTWARE

- Solving Equations with Rational Expressions

5.5 Applications

- *Solve applied problems related to fractions.*
- *Solve applied problems related to work.*
- *Solve applications involving distance, rate, and time.*

The following Strategy for Solving Word Problems is valid for all word problems that involve algebraic equations.

Strategy for Solving Word Problems

1. *Read the problem carefully. Read it several times if necessary.*

2. *Decide what is asked for and assign a variable to the unknown quantity.*

3. *Draw a diagram or set up a chart whenever possible as a visual aid.*

4. *Form an equation that relates the information provided.*

5. *Solve the equation.*

6. *Check your solution with the wording of the problem to be sure it makes sense.*

Number Problems Related to Fractions

We now introduce word problems involving rational expressions with problems relating the numerator and denominator of a fraction.

Example 1: Fractions

The denominator of a fraction is 8 more than the numerator. If both the numerator and denominator are increased by 3, the new fraction is equal to $\frac{1}{2}$. Find the original fraction.

Solution: Reread the problem to be sure that you understand all terminology used. Assign variables to the unknown quantities.

$$\text{Let} \quad n = \text{original numerator}$$

$$n + 8 = \text{original denominator}$$

$$\frac{n}{n+8} = \text{original fraction}$$

Continued on the next page...

$$\frac{n+3}{(n+8)+3} = \frac{1}{2}$$

The numerator and the denominator are each increased by 3, making a new

$$\frac{n+3}{n+11} = \frac{1}{2}$$

fraction that is equal to $\frac{1}{2}$.

$$2\cancel{(n+11)} \cdot \frac{n+3}{\cancel{n+11}} = \cancel{2}(n+11) \cdot \frac{1}{\cancel{2}}$$

$$2n+6 = n+11$$

$$n = 5 \qquad \leftarrow \text{Original numerator}$$

$$n+8 = 13 \qquad \leftarrow \text{Original denominator}$$

Check: $\dfrac{(5)+3}{(13)+3} = \dfrac{8}{16} = \dfrac{1}{2}$

The original fraction is $\dfrac{5}{13}$.

Problems Related to Work

Problems involving work usually translate into equations involving rational expressions. The basic idea is to **represent what part of the work is done in one unit of time**. For example, if a man can dig a ditch in 3 hours, what part (of the ditch-digging job) can he do in one hour? The answer is $\frac{1}{3}$ of the work in one hour. If a fence was painted in 2 days, then $\frac{1}{2}$ of the work of painting the fence was done in 1 day. (These ideas assume a steady working pace.) In general, if the total work took x hours, then $\frac{1}{x}$ of the total work would be done in one hour.

Example 2: Work Problems

a. A carpenter can build a certain type of patio cover in 6 hours. His partner takes 8 hours to build the same cover. How long would it take them working together to build this type of patio cover?

Solution: Let x = number of hours to build the cover working together.

Person(s)	Time of Work (in Hours)	Part of Work Done in 1 Hour
Carpenter	6	$\dfrac{1}{6}$
Partner	8	$\dfrac{1}{8}$
Together	x	$\dfrac{1}{x}$

$$\underbrace{\begin{array}{c}\text{Part done in}\\\text{1 hr by carpenter}\end{array}}\quad + \quad \underbrace{\begin{array}{c}\text{Part done in}\\\text{1 hr by partner}\end{array}}\quad = \quad \underbrace{\begin{array}{c}\text{Part done in}\\\text{1 hr together}\end{array}}$$

$$\dfrac{1}{6} \quad + \quad \dfrac{1}{8} \quad = \quad \dfrac{1}{x}$$

$$\dfrac{1}{6}\overset{4}{\left(24x\right)} + \dfrac{1}{8}\overset{3}{\left(24x\right)} = \dfrac{1}{x}\left(24x\right)$$ Multiply each term on both sides by $24x$, the LCM of the denominators.

$$4x + 3x = 24$$
$$7x = 24$$
$$x = \dfrac{24}{7}$$

Together, they can build the patio cover in $\dfrac{24}{7}$ hours, or $3\dfrac{3}{7}$ hours.

(Note that this answer is reasonable because the time is less than either person would take working alone.)

b. A man can wax his car three times faster than his daughter can. Together they can do the job in 4 hours. How long would it take each of them working alone?

Solution: Let t = number of hours for man alone to wax the car and
$3t$ = number of hours for daughter alone to wax the car.

Person(s)	Time of Work (in Hours)	Part of Work Done in 1 Hour
Man	t	$\dfrac{1}{t}$
Daughter	$3t$	$\dfrac{1}{3t}$
Together	4	$\dfrac{1}{4}$

Continued on the next page...

$$\underbrace{\frac{\text{Part done by man}}{\text{alone in 1 hour}}}_{} + \underbrace{\frac{\text{Part done by daughter}}{\text{alone in 1 hour}}}_{} = \underbrace{\frac{\text{Part done working}}{\text{together in 1 hour}}}_{}$$

$$\frac{1}{t} \quad + \quad \frac{1}{3t} \quad = \quad \frac{1}{4}$$

$$\frac{1}{\cancel{t}}\left(12\cancel{t}\right) + \frac{1}{\cancel{3t}}\left(\overset{4}{\cancel{12t}}\right) = \frac{1}{\cancel{4}}\left(\overset{3}{\cancel{12t}}\right) \qquad \text{Multiply each term on both sides by } 12t, \text{ the LCM of the denominators.}$$

$$12 + 4 = 3t$$

$$16 = 3t$$

$$t = \frac{16}{3} \qquad \leftarrow \text{ man's time}$$

$$3t = (3)\left(\frac{16}{3}\right) = 16 \quad \leftarrow \text{ daughter's time}$$

Check: Man's part in 1 hr $= \dfrac{1}{t} = \dfrac{1}{\frac{16}{3}} = \dfrac{3}{16}$

Daughter's part in 1 hr $= \dfrac{1}{3t} = \dfrac{1}{3 \cdot \frac{16}{3}} = \dfrac{1}{16}$

Man's part in 4 hr $= \dfrac{3}{16} \cdot 4 = \dfrac{3}{4}$

Daughter's part in 4 hr $= \dfrac{1}{16} \cdot 4 = \dfrac{1}{4}$

$\dfrac{3}{4} + \dfrac{1}{4} = 1$ car waxed in 4 hours.

Working alone, the man takes $\dfrac{16}{3}$ hours, or 5 hours 20 minutes, and his daughter takes 16 hours.

c. A man was told that his new whirlpool would fill through an inlet valve in 3 hours. He knew something was wrong when the pool took 8 hours to fill. He found he had left the drain valve open. How long will it take to drain the pool once it is completely filled and only the drain valve is open?

Solution: Let t = time to drain pool with only the drain valve open. [**Note:** We use the information gained when the pool was filled with both valves open. In that situation, the inlet and outlet valves worked against each other.]

Valves	Hours to Fill or Drain	Part Filled or Drained in 1 Hour
Inlet	3	$\dfrac{1}{3}$
Outlet	t	$\dfrac{1}{t}$
Together	8	$\dfrac{1}{8}$

Part filled by inlet in 1 hour	−	Part emptied by outlet in 1 hour	=	Part filled together in 1 hour
$\dfrac{1}{3}$	−	$\dfrac{1}{t}$	=	$\dfrac{1}{8}$

$$\frac{1}{3}\overset{8}{(24t)} - \frac{1}{t}(24t) = \frac{1}{8}\overset{3}{(24t)}$$

$$8t - 24 = 3t$$

$$5t = 24$$

$$t = \frac{24}{5}$$

The pool will drain in $\dfrac{24}{5}$ hours, or 4 hours, 48 minutes. (Note that this is more time than the inlet valve would take to fill the pool. If the outlet valve worked faster than the inlet valve, then the pool would never have filled in the first place.)

Problems Related to Distance-Rate-Time: $d = rt$

You may recall that the basic formula involving distance, rate, and time is $d = rt$. This relationship can also be stated in the forms $t = \dfrac{d}{r}$ and $r = \dfrac{d}{t}$.

If distance and rate are known or can be represented, then $t = \dfrac{d}{r}$ is the way to represent time. Similarly, if the distance and time are known or can be represented, then $r = \dfrac{d}{t}$ is the way to represent rate.

> **Example 3: Distance-Rate-Time**

a. On Lake Itasca a man can row his boat 5 miles per hour. On the nearby Mississippi River it takes him the same time to row 5 miles downstream as it does to row 3 miles upstream. What is the speed of the river current in miles per hour?

Solution: Let c = the speed of the current.

Distance and rate are represented first in the table below. Then the time going downstream and coming back upstream is represented in terms of distance and rate. Since the rate is in miles per hour, the distance is in miles and the time is in hours.

	Distance d	Rate r	Time $t = \dfrac{d}{r}$
Downstream	5	$5 + c$	$\dfrac{5}{5+c}$
Upstream	3	$5 - c$	$\dfrac{3}{5-c}$

$$\frac{5}{5+c} = \frac{3}{5-c} \qquad \text{The times are equal.}$$

$$(5+c)(5-c) \cdot \frac{5}{5+c} = (5+c)(5-c) \cdot \frac{3}{5-c}$$

$$25 - 5c = 15 + 3c$$

$$-8c = -10$$

$$c = \frac{5}{4}$$

Check: Time downstream $= \dfrac{5}{5 + \dfrac{5}{4}} = \dfrac{5}{\dfrac{20}{4} + \dfrac{5}{4}} = \dfrac{5}{\dfrac{25}{4}} = 5 \cdot \dfrac{4}{25} = \dfrac{4}{5}$ hr

Time upstream $= \dfrac{3}{5 - \dfrac{5}{4}} = \dfrac{3}{\dfrac{20}{4} - \dfrac{5}{4}} = \dfrac{3}{\dfrac{15}{4}} = 3 \cdot \dfrac{4}{15} = \dfrac{4}{5}$ hr

The times are equal. The rate of the river current is $\dfrac{5}{4}$ mph, or $1\dfrac{1}{4}$ mph.

b. If a passenger train travels three times as fast as a freight train, and the freight train takes 4 hours longer to travel 210 miles, what is the speed of each train?

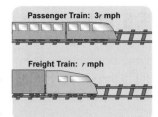

Passenger Train: $3r$ mph

Freight Train: r mph

Solution: Let r = rate of freight train in miles per hour

$3r$ = rate of passenger train in miles per hour

	Distance d	Rate r	Time $t = \dfrac{d}{r}$
Freight	210	r	$\dfrac{210}{r}$
Passenger	210	$3r$	$\dfrac{210}{3r}$

(**Note:** If the rate is faster, then the time is shorter. Thus the fraction $\dfrac{210}{3r}$ is smaller than the fraction $\dfrac{210}{r}$.)

$$\frac{210}{r} - \frac{210}{3r} = 4 \qquad \text{The difference between their times is 4 hours.}$$

$$\frac{210}{r} - \frac{70}{r} = 4$$

$$\frac{210}{\cancel{r}} \cdot \cancel{r} - \frac{70}{\cancel{r}} \cdot \cancel{r} = 4 \cdot r$$

$$210 - 70 = 4r$$

$$140 = 4r$$

$$35 = r$$

$$105 = 3r$$

Check: Time for freight train $= \dfrac{210}{35} = 6\,\text{hr}$

Time for passenger train $= \dfrac{210}{105} = 2\,\text{hr}$

$6 - 2 = 4$ hours difference in time

The freight train travels 35 mph, and the passenger train travels 105 mph.

5.5 Exercises

1. The sum of two numbers is 117 and they are in the ratio of 8 to 5. Find the two numbers.

2. If 4 is subtracted from a certain number and the difference is divided by 2, the result is 1 more than $\frac{1}{5}$ of the original number. Find the original number.

3. What number must be added to both numerator and denominator of $\frac{16}{21}$ to make the resulting fraction equal to $\frac{5}{6}$?

4. Find the number that can be subtracted from both numerator and denominator of the fraction $\frac{69}{102}$ so that the result is $\frac{5}{8}$.

5. The denominator of a fraction exceeds the numerator by 7. If the numerator is increased by 3 and the denominator is increased by 5, the resulting fraction is equal to $\frac{1}{2}$. Find the original fraction.

6. The numerator of a fraction exceeds the denominator by 5. If the numerator is decreased by 4 and the denominator is increased by 3, the resulting fraction is equal to $\frac{4}{5}$. Find the original fraction.

7. One number is $\frac{3}{4}$ of another number. Their sum is 63. Find the numbers.

8. The sum of two numbers is 24. If $\frac{2}{5}$ the larger number is equal to $\frac{2}{3}$ the smaller number, find the numbers.

9. One number exceeds another by 5. The sum of their reciprocals is equal to 19 divided by the product of the two numbers. Find the numbers.

10. One number is 3 less than another. The sum of their reciprocals is equal to 7 divided by the product of the two numbers. Find the numbers.

11. **Shirt sales:** A manufacturer sold a group of shirts for $1026. One-fifth of the shirts were priced at $18 each and the remainder at $24 each. How many shirts were sold?

12. **Paying bills:** Luis spent $\frac{1}{5}$ of his monthly salary for rent and $\frac{1}{6}$ of his monthly salary for his car payment. If $950 was left, what was his monthly salary?

13. **Painting:** Suppose that an artist expects that for every 9 special brushes she orders, 7 will be good and 2 will be defective. If she orders 54 brushes, how many will she expect to be defective?

14. Travel by car: It takes Rosa, traveling at 30 mph, 30 minutes longer to go a certain distance than it takes Melody traveling at 50 mph. Find the distance traveled.

15. Travel by plane: It takes a plane flying at 450 mph 25 minutes longer to travel a certain distance than it takes a second plane to fly the same distance at 500 mph. Find the distance.

16. Landscaping: Toni needs 4 hours to complete the yard work. Her husband, Sonny, needs 6 hours to do the work. How long will the job take if they work together?

17. Cleaning a pool: Manuel can clean his family's pool in 2 hours. His younger sister, Maria, can do it in 3 hours. If they work together, how long will it take them to clean the pool?

18. Mass mailings: Ben's secretary can address the weekly newsletters in $4\frac{1}{2}$ hours. Charlie's secretary needs only 3 hours. How long will it take if they both work on the job?

19. Shoveling snow: Working together, Greg and Cindy can clean the snow from the driveway in 20 minutes. It would have taken Cindy, working alone, 36 minutes. How long would it have taken Greg alone?

20. Carpentry: A carpenter and his partner can put up a patio cover in $3\frac{3}{7}$ hours. If the partner needs 8 hours to complete the patio alone, how long would it take the carpenter working alone?

21. Travel: Beth can travel 208 miles in the same length of time it takes Anna to travel 192 miles. If Beth's speed is 4 mph greater than Anna's, find both rates.

22. Biking: Kirk can bike 32 miles in the same amount of time that his twin brother Karl can bike 24 miles. If Kirk bikes 2 mph faster than Karl, how fast does each man bike?

23. Plane speeds: A commercial airliner can travel 750 miles in the same amount of time that it takes a private plane to travel 300 miles. The speed of the airliner is 60 mph more than twice the speed of the private plane. Find the speed of each aircraft.

24. Car speeds: Gabriela drives her car 350 miles and averages a certain speed. If the average speed had been 9 mph less, she could have traveled only 300 miles in the same length of time. What was her average speed?

25. **Plane speed:** A jet flies twice as fast as a propeller plane. On a 1500 mile trip, the propeller plane took 3 hours longer than the jet. Find the speed of each plane.

26. **Boating:** A family travels 18 miles down river and returns. It takes 8 hours to make the round trip. Their rate in still water is twice the rate of the river's current. How long will the return trip take?

27. **Plane speed:** An airplane can fly 650 mph in still air. If it can travel 2800 miles with the wind in the same time it can travel 2400 miles against the wind, find the wind speed. (**Note:** A tailwind increases the speed of the plane and a headwind decreases the speed of the plane.)

28. **Wind speed:** A one-engine plane can fly 120 mph in still air. If it can fly 490 miles with a tailwind in the same time that it can fly 350 miles against a headwind, what is the speed of the wind?

29. **Filling a pool:** Using a small inlet pipe it takes 9 hours to fill a pool. Using a large inlet pipe it only takes 3 hours. If both are used simultaneously, how long will it take to fill the pool?

30. **Filling a pool:** An inlet pipe on a swimming pool can be used to fill the pool in 36 hours. The drain pipe can be used to empty the pool in 40 hours. If the pool is $\frac{2}{3}$ filled using the inlet pipe and then the drain pipe is accidentally opened, how long from that time will it take to fill the pool?

31. **Clearing land:** A contractor hires two bulldozers to clear the trees from a 20-acre tract of land. One works twice as fast as the other. It takes them 3 days to clear the tract working together. How long would it take each of them alone?

32. **Store maintenance:** John, Ralph, and Denny, working together, can clean their bait and tackle store in 6 hours. Working alone, Ralph takes twice as long to clean the store as does John. Denny needs three times as long as does John. How long would it take each man working alone?

33. **Boating:** Francois rode his jet ski 36 miles downstream and then 36 miles back. The round trip took $5\frac{1}{4}$ hours. Find the speed of the jet ski in still water and the speed of the current if the speed of the current is $\frac{1}{7}$ the speed of the jet ski in still water.

34. Boating: Momence, IL is 12 miles upstream on the same side of the river from Kankakee, IL on the Kankakee River. A motorboat that can travel 8 mph in still water leaves Momence and travels downstream toward Kankakee. At the same time, another boat that can travel 10 mph leaves Kankakee and travels upstream toward Momence. Each boat completes the trip in the same amount of time. Find the rate of the current.

35. Skiing: Samantha rides the ski lift to the top of Blue Mountain, a distance of $1\frac{3}{4}$ kilometers (a little more than 1 mile). She then skis directly down the slope. If she skis five times as fast as the lift travels and the total trip takes 45 minutes, find the rate at which she skis.

Writing and Thinking About Mathematics

36. If n is any integer, then $2n$ is an even integer and $2n + 1$ is an odd integer. Use these ideas to solve the following problems.

  a. Find two consecutive odd integers such that the sum of their reciprocals is $\frac{12}{35}$.

  b. Find two consecutive even integers such that the sum of the first and the reciprocal of the second is $\frac{9}{4}$.

HAWKES LEARNING SYSTEMS: INTERMEDIATE ALGEBRA SOFTWARE

- Applications Involving Rational Expressions

5.6 Variation

- *Solve problems related to direct variation.*
- *Solve problems related to inverse variation.*
- *Solve problems involving combined variation.*

Direct Variation

Suppose that you ride your bicycle at a steady rate of 15 miles per hour (not quite as fast as Lance Armstrong, but you are enjoying yourself). If you ride for 1 hour, the distance you travel would be 15 miles. If you ride for two hours, the distance you travel would be 30 miles. This relationship can be written in the

form of the formula $d = 15t$ (or $\dfrac{d}{t} = 15$) where d is the distance traveled and t is the time in hours. We say that distance and time **vary directly** (or are in **direct variation** or are **directly proportional**). The term proportional implies that the ratio is constant. In this example, 15 is the constant and is called the **constant of variation**. When two variables vary directly, an increase in the value of one variable indicates an increase in the other, and the ratio of the two quantities is constant.

Direct Variation

*A variable quantity y **varies directly as** (or is **directly proportional to**) a variable x if there is a constant k such that*

$$\frac{y}{x} = k \ \text{or} \ y = kx.$$

*The constant k is called the **constant of variation**.*

Example 1: Direct Variation

If y varies directly as x, and $y = 6$ when $x = 2$, find y if $x = 6$.

Solution:

$y = kx$	General formula for direct variation
$6 = 2k$	Substitute the known values and solve for k.
$3 = k$	Use this value for k in the general formula.

So $y = 3x$. Thus, if $x = 6$, then $y = 3 \cdot 6 = 18$.

Example 2: Direct Variation

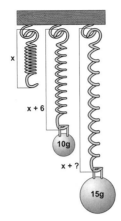

A spring will stretch a greater distance as more weight is placed on the end of the spring. The distance, d, the spring stretches varies directly as the weight, w, placed at the end of the spring. This is a property of springs studied in physics and is known as Hooke's Law. If a weight of 10 g stretches a certain spring 6 cm, how far will the spring stretch with a weight of 15 g? [**Note:** We assume that the weight is not so great as to break the spring.]

Solution: Because the two variables are directly proportional, the relationship can be indicated with the formula

$$d = k \cdot w \quad \text{where } d = \text{distance spring stretches in cm,}$$
$$w = \text{weight in g,}$$
$$\text{and } k = \text{constant of variation.}$$

First, substitute the given information to find the value for k. (The value of k will depend on the particular spring. Springs made of different material or which are wound more tightly, will have different values for k.)

$$d = k \cdot w$$

$$6 = k \cdot 10 \qquad \text{Substitute the known values into the formula.}$$

$$\frac{3}{5} = k \qquad \text{Use this value for } k \text{ in the general formula.}$$

So $\quad d = \frac{3}{5}w.$ $\qquad$ The constant of variation is $\frac{3}{5}$ (or 0.6).

If $w = 15$, we have

$$d = \frac{3}{5} \cdot 15 = 9.$$

The spring will stretch 9 cm if a weight of 15 g is placed at its end.

Listed here are several formulas involving direct variation.

$d = \dfrac{3}{5}w$ $\qquad$ Hooke's Law for a spring where $k = \dfrac{3}{5}$.

$C = 2\pi r$ $\qquad$ The circumference of a circle is directly proportional to the radius.

$A = \pi r^2$ $\qquad$ The area of a circle varies directly as the radius squared.

$P = 625d$ $\qquad$ Water pressure is proportional to the depth of the water.

Inverse Variation

When two variables vary in such a way that their product is constant, we say that the two variables **vary inversely** (or are **inversely proportional**). For example, if a gas is placed in a container (as in an automobile engine) and pressure is increased on the gas, then the product of the pressure and the volume of gas will remain constant. That is, pressure and volume are related by the formula $V \cdot P = k \left(\text{or } V = \dfrac{k}{P} \right)$.

Note that if a product of two variables is to remain constant, then an increase in the value of one variable must be accompanied by a decrease in the other. Or, in the case of a fraction with a constant numerator, if the denominator increases in value, then the fraction decreases in value. For the gas in an engine, an increase in pressure indicates a decrease in the volume of gas.

Inverse Variation

*A variable quantity y **varies inversely as** (or is **inversely proportional to**) a variable x if there is a constant k such that*

$$x \cdot y = k \ \text{or} \ y = \frac{k}{x}.$$

*The constant k is called the **constant of variation**.*

Example 3: Inverse Variation

If y varies inversely as the cube of x, and $y = -1$ when $x = 3$, find y if $x = -3$.

Solution: $y = \dfrac{k}{x^3}$ General formula for inverse variation

$-1 = \dfrac{k}{3^3}$ Substitute the known values and solve for k.

$-1 = \dfrac{k}{27}$

$-27 = k$ Use this value for k in the general formula.

So $y = \dfrac{-27}{x^3}$. Thus, if $x = -3$, then $y = \dfrac{-27}{(-3)^3} = \dfrac{-27}{-27} = 1$.

Example 4: Inverse Variation

The gravitational force, F, between an object and the Earth is inversely proportional to the square of the distance, d, from the object to the center of the Earth. Hence we have the formula

$$F \cdot d^2 = k \text{ or } F = \frac{k}{d^2}, \qquad \text{where } F = \text{force, } d = \text{distance}$$
$$\text{and } k = \text{constant of variation.}$$

(As the distance of an object from the Earth becomes larger, the gravitational force exerted by the Earth on the object becomes smaller.)

If an astronaut weighs 200 pounds on the surface of the Earth, what will he weigh 100 miles above the Earth? Assume that the radius of the Earth is 4000 miles.

Solution: We know $F = 200 = 2 \times 10^2$ pounds   Use scientific notation to make values
when $d = 4000 = 4 \times 10^3$ miles.   simpler to work with in the calculations.

$$2 \times 10^2 = \frac{k}{\left(4 \times 10^3\right)^2} \qquad \text{Substitute and solve for } k.$$

$$k = 2 \times 10^2 \times 16 \times 10^6 = 32 \times 10^8 = 3.2 \times 10^9$$

So, $F = \dfrac{3.2 \times 10^9}{d^2}.$

When the astronaut is 100 miles above the Earth, $d = 4100 = 4.1 \times 10^3$ miles. Then,

$$F = \frac{3.2 \times 10^9}{16.81 \times 10^6} \approx 0.190 \times 10^3 = 190 \text{ pounds.}$$

That is, 100 miles above the Earth the astronaut will weigh 190 pounds.

Combined Variation

If a variable varies either directly or inversely with more than one other variable, the variation is said to be a **combined variation**. If the combined variation is all direct variation (the variables are multiplied), then it is called **joint variation**. For example, the volume of a cylinder varies jointly as its height and the square of its radius.

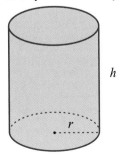

h

$$V = kr^2h$$

where r = radius, h = height,
and k = constant of variation

r

For example, what is the value of k, the constant of variation, if a cylinder has the approximate measurements $V = 198$ cubic feet, $r = 3$ feet, and $h = 7$ feet?

Solution:

$$V = k \cdot r^2 \cdot h \qquad \text{\textbf{V varies jointly} as } r^2 \text{ and } h.$$

$$198 = k \cdot 3^2 \cdot 7 \qquad \text{Substitute the known values.}$$

$$\frac{198}{9 \cdot 7} = k$$

$$k = \frac{22}{7} \approx 3.14$$

We know from experience that $k = \pi$. Since the measurements are only approximate, the estimate for k is only approximate.

Substituting the constant of variation, the formula is $V = \pi r^2 h$.

Example 5: Variation

a. If y is directly proportional to x^2, and $y = 9$ when $x = 2$, what is y when $x = 4$?

Solution:

$$y = k \cdot x^2$$

$$9 = k \cdot 2^2 \qquad \text{Substitute the known values and solve for } k.$$

$$\frac{9}{4} = k$$

So $\quad y = \frac{9}{4}x^2.$ $\quad$ Substitute $\frac{9}{4}$ for k in the general formula.

If $x = 4$, then

$$y = \frac{9}{4} \cdot 4^2$$

$$y = 36.$$

b. The distance an object falls varies directly as the square of the time it falls (until it hits the ground and assuming little or no air resistance). If an object fell 64 feet in two seconds, how far would it have fallen by the end of 3 seconds?

Solution:

$$d = k \cdot t^2 \qquad \text{where } d = \text{distance}, t = \text{time (in seconds)}, \text{ and } k = \text{constant of variation}$$

$$64 = k \cdot 2^2 \qquad \text{Substitute the known values and solve for } k.$$

$$16 = k$$

So $\quad d = 16t^2.$ $\quad$ Substitute 16 for k in the general formula.

If $t = 3$, then

$$d = 16 \cdot 3^2$$

$$d = 144.$$

The object would have fallen 144 feet in 3 seconds.

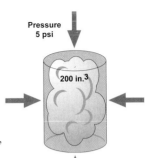

c. The volume of a gas in a container varies inversely as the pressure on the gas. If a gas has a volume of 200 cubic inches under pressure of 5 pounds per square inch, what will be its volume if the pressure is increased to 8 pounds per square inch?

Solution: $V = \dfrac{k}{P}$ where V = volume, P = pressure, and k = constant of variation

$200 = \dfrac{k}{5}$ Substitute the known values and solve for k.

$k = 1000$

So $V = \dfrac{1000}{P}$ Substitute 1000 for k in the general formula.

$V = \dfrac{1000}{8} = 125.$

The volume will be 125 cubic inches.

d. The illumination (in foot-candles, fc) of a light source varies directly as the intensity (in candelas, cd) of the source and inversely as the square of the distance from the source. If a certain light source with intensity of 300 cd provides an illumination of 10 fc at a distance of 20 feet, what is the illumination at a distance of 40 feet? [**Note:** This is an illustration of combined variation.]

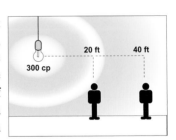

Solution: $I = \dfrac{k \cdot i}{d^2}$ where I = illumination, i = intensity, and d = distance

$10 = \dfrac{k \cdot 300}{(20)^2}$ Substitute the known values and solve for k.

$k = \dfrac{400 \cdot 10}{300}$

$k = \dfrac{40}{3}$

So $I = \dfrac{\dfrac{40}{3} \cdot i}{d^2}$ Substitute $\dfrac{40}{3}$ for k in the general formula.

$I = \dfrac{\dfrac{40}{3} \cdot 300}{(40)^2} = \dfrac{40 \cdot 100}{40 \cdot 40} = \dfrac{5}{2} = 2.5.$

The illumination at 40 feet is 2.5 fc.

Practice Problems

1. *The length that a hanging spring stretches varies directly as the weight placed on the end of the spring. If a weight of 5 mg stretches a certain spring 3 cm, how far will the spring stretch with a weight of 6 mg?*

2. *The volume of propane in a container varies inversely to the pressure on the gas. If the propane has a volume of 200 in.³ under a pressure of 4 lbs per in.², what will be its volume if the pressure is increased to 5 lbs per in.²?*

5.6 Exercises

For Exercises 1 – 20, write an equation or formula that represents the general relationship indicated. Then use the given information to find the unknown value.

1. If y varies directly as x, and $y = 3$ when $x = 9$, find y if $x = 7$.

2. If y is directly proportional to x^2, and $y = 3$ when $x = 2$, what is y when $x = 8$?

3. If y varies inversely as x, and $y = 5$ when $x = 8$, find y if $x = 20$.

4. If y varies inversely as x^2, and $y = -8$ when $x = 2$, find y if $x = 3$.

5. If y is inversely proportional to x, and $y = 5$ when $x = 4$, what is y when $x = 2$?

6. If y is inversely proportional to x^3, and $y = 40$ when $x = \dfrac{1}{2}$, what is y when $x = \dfrac{1}{3}$?

7. If y is proportional to the square root of x, and $y = 6$ when $x = \dfrac{1}{4}$, what is y when $x = 9$?

8. If y is proportional to the square of x, and $y = 80$ when $x = 4$, what is y when $x = 6$?

9. If y varies directly as x^3, and $y = 81$ when $x = 3$, find y if $x = 2$.

10. z varies jointly as x and y, and $z = 60$ when $x = 2$ and $y = 3$. Find z if $x = 3$ and $y = 4$.

11. z varies jointly as x and y, and $z = -6$ when $x = 5$ and $y = 8$. Find z if $x = 12$ and $y = 3$.

12. z varies jointly as x^2 and y, and $z = 20$ when $x = 2$ and $y = 3$. Find z if $x = 4$ and $y = \dfrac{7}{10}$.

13. z varies directly as x and inversely as y^2. If $z = 5$ when $x = 1$ and $y = 2$, find z if $x = 2$ and $y = 1$.

14. z varies directly as x^3 and inversely as y^2. If $z = 24$ when $x = 2$ and $y = 2$, find z if $x = 3$ and $y = 2$.

Answers to Practice Problems: **1.** $\dfrac{18}{5}$ cm   **2.** 160 in.³

15. z varies directly as $\sqrt{x}$ and inversely as y. If $z = 24$ when $x = 4$ and $y = 3$, find z if $x = 9$ and $y = 2$.

16. z is jointly proportional to x^2 and y^3. If $z = 192$ when $x = 4$ and $y = 2$, find z when $x = 2$ and $y = 4$.

17. z varies directly as x^2 and inversely as $\sqrt{y}$. If $z = 108$ when $x = 6$ and $y = 4$, find z if $x = 4$ and $y = 9$.

18. s varies directly as the sum of r and t and inversely as w. If $s = 24$ when $r = 7$ and $t = 8$ and $w = 9$, find s if $r = 9$ and $t = 3$ and $w = 18$.

19. L varies jointly as m and n and inversely as p. If $L = 6$ when $m = 7$ and $n = 8$ and $p = 12$, find L if $m = 15$ and $n = 14$ and $p = 10$.

20. W varies jointly as x and y and inversely as z. If $W = 10$ when $x = 6$ and $y = 5$ and $z = 2$, find W if $x = 12$ and $y = 6$ and $z = 3$.

21. **Weight in space:** If an astronaut weighs 250 pounds on the surface of the earth, what will the astronaut weigh 150 miles above the earth? Assume that the radius of the earth is 4000 miles, and round to the nearest tenth.

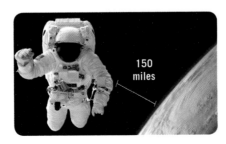

22. **Free falling object:** The distance a free falling object falls is directly proportional to the square of the time it falls (before it hits the ground). If an object fell 144 feet in 3 seconds, how far will it have fallen by the end of 4 seconds?

23. **Stretching a spring:** A hanging spring will stretch 5 in. if a weight of 10 lbs is placed at its end. How far will the spring stretch if the weight is increased to 12 lbs?

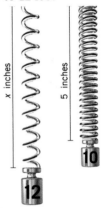

24. **Price of gasoline:** The total price, P, of gasoline purchased varies directly with the number of gallons purchased. If 10 gallons are purchased for $39.80, what will be the price of 15 gallons?

25. **Pizza:** The circumference of a circle varies directly as the diameter. A circular pizza pie with a diameter of 1 foot has a circumference of 3.14 feet. What will be the circumference of a pizza pie with a diameter of 1.5 feet?

26. **Pizza:** The area of a circle varies directly as the square of its radius. A circular pizza pie with a diameter of 12 inches has an area of 113.04 in.^{2} What will be the area of a pizza pie with a diameter of 18 inches?

27. **Elongation of a wire:** The elongation, E, in a wire when a mass, m, is hung at its free end varies jointly as the mass and the length, l, of the wire and inversely as the cross-sectional area, A, of the wire. The elongation is 0.0055 cm when a mass of 120 g is attached to a wire 330 cm long with a cross-sectional area of 0.4 cm^2. Find the elongation if a mass of 160 g is attached to the same wire.

28. **Elongation of a wire:** When a mass of 240 oz is suspended by a wire 49 inches long whose cross-sectional area is 0.035 in.^{2}, the elongation of the wire is 0.016 in. Find the elongation if the same mass is suspended by a 28-in. wire of the same material with a cross-sectional area of 0.04 in.^{2} (See Exercise 27.)

49 inches

240 oz

29. **Triangles:** Several triangles are to have the same area. In this set of triangles the height and base are inversely proportional. In one such triangle the height is 5 m and the base is 12 m. Find the height of the triangle in this set with a base of 10 m.

30. **Safe load of a wooden beam:** The safe load, L, of a wooden beam supported at both ends varies jointly as the width, w, and the square of the depth, d, and inversely as the length, l. A 4 in. × 6 in. beam 12 ft long supports a load of 4800 lb safely. What is the safe load of a beam of the same material that is 6 in. × 10 in. × 15 ft long?

31. **Safe load of a wooden beam:** A wooden beam 2 in. wide, 8 in. deep, and 14 ft long holds up to 2400 lb. What load would a beam 3 in. × 6 in. × 15 ft, of the same material, support? (See Exercise 30.)

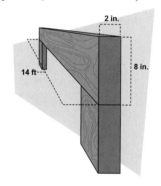

2 in.

8 in.

14 ft

32. **Gravitational force:** The gravitational force of attraction, F, between two bodies varies directly as the product of their masses, m_1 and m_2, and inversely as the square of the distance, d, between them. The gravitational force between a 5-kg mass and a 2-kg mass 1 m apart is 1.5×10^{-10} N. Find the force between a 24-kg mass and a 9-kg mass that are 6 m apart. (N represents a unit of force called a newton.)

33. In Exercise 32, what happens to the force if the distance between the 24-kg mass and the 9-kg mass is cut in half?

Lifting Force

The lifting force (or lift), L, in pounds exerted by the atmosphere on the wings of an airplane is related to the area, A, of the wings in square feet and the speed (or velocity), v, of the plane in miles per hour by the formula

$L = kAv^2$, where k is the constant of variation.

34. If the lift is 9600 lb for a wing area of 120 ft^2 and a speed of 80 mph, find the lift of the same airplane at a speed of 100 mph.

35. The lift for a wing of area 280 ft^2 is 34,300 lb when the plane is traveling at 210 mph. What is the lift if the speed is decreased to 180 mph?

36. The lift for a wing with an area of 144 ft^2 is 10,000 lb when the plane is traveling at 150 mph. What is the lift if the speed is decreased to 120 mph?

37. A plane traveling 140 mph with wing area 195 ft^2 has 12,500 lb of lift exerted on the wings. Find the lift for the same plane traveling at 168 mph.

Pressure

Boyle's Law states that if the temperature of a gas sample remains the same, the pressure, P, of the gas is related to the volume, V, by the formula

$P = \dfrac{k}{V}$, where k is the constant of proportionality.

38. A pressure of 1600 lb per ft^2 is exerted by 2 ft^3 of air in a cylinder. If a piston is pushed into the cylinder until the pressure is 1800 lb per ft^2, what will be the volume of the air? Round to the nearest tenth.

39. The volume of gas in a container is 300 cm^3 when the pressure on the gas is 20 g per cm^2. What will be the volume if the pressure is increased to 30 g per cm^2?

40. The pressure in a canister of gas is 1360 g per in.^{2} when the volume of gas is 5 in.^{3} If the volume is reduced to 4 in.^{3}, what is the pressure?

41. A scuba diver is using a diving tank that can hold 6 liters of air. If the tank has a pressure rating of 220 bar when full, what is the pressure rating when the volume of gas is 4 liters?

Electricity

The resistance, R (in ohms), in a wire is given by the formula

$$R = \frac{kL}{d^2},$$

where k is the constant of variation, L is the length of the wire and d is the diameter.

42. The resistance of a wire 500 ft long with a diameter of 0.01 in. is 20 ohms. What is the resistance of a wire 1500 ft long with a diameter of 0.02 in.?

43. The resistance is 2.6 ohms when the diameter of a wire is 0.02 in. and the wire is 10 ft long. Find the resistance of the same type of wire with a diameter of 0.01 in. and a length of 5 ft.

44. Tristan's car stereo uses a 5 ft audio wire with diameter 0.025 in. and resistance of 1.6 ohms. What is the resistance of 8 ft of the same type of audio wire?

45. Nicole purchased a spool of wire with diameter 0.01 in. for the speakers in her home audio system. If the resistance of 15 ft of this wire is 6 ohms, what is the resistance of 25 ft of the wire?

Levers

If a lever is balanced with weight on opposite sides of its balance point, then the following proportion exists:

$$\frac{W_1}{W_2} = \frac{L_2}{L_1} \quad \text{or} \quad W_1 L_1 = W_2 L_2 \qquad \text{where } L_1 + L_2 = L, \text{ the total length of the lever.}$$

46. How much weight can be raised at one end of a bar 8 ft long by the downward force of 60 lb when the balance point is $\frac{1}{2}$ ft from the unknown weight?

47. Where should the balance point of a bar 12 ft long be located if a 120 lb force is to raise a load weighing 960 lb?

48. Find the location of the balance point of a 25 ft board that can raise a 300 lb package with a downward force of 75 lb.

49. How much weight can be raised on one end of a 17 meter board by 90 kilograms, if the balance point is 5 meters from the unknown weight?

Writing and Thinking About Mathematics

50. Explain in your own words, the meaning of the terms
   a. direct variation,
   b. inverse variation,
   c. joint variation, and
   d. combined variation.

Discuss an example of each type of variation that you have observed in your daily life.

 HAWKES LEARNING SYSTEMS: INTERMEDIATE ALGEBRA SOFTWARE

   ▪ Applications: Variation

Chapter 5 Index of Key Ideas and Terms

Section 5.2 Addition and Subtraction with Rational Expressions (cont.)

To Find the LCM for a Set of Polynomials   page 398
1. Completely factor each polynomial.
2. Form the product of all factors that appear, using each factor the most number of times it appears in any one polynomial.

Procedure for Adding Rational Expressions with Different Denominators   page 398
1. Find the LCD (the LCM of the denominators).
2. Rewrite each fraction in an equivalent form with the LCD as the denominator.
3. Add the numerators and keep the common denominator.
4. Reduce if possible.

Placement of Negative Signs   page 400
If P and Q are polynomials and $Q \neq 0$, then
$$-\frac{P}{Q} = \frac{-P}{Q} = \frac{P}{-Q}.$$

Section 5.3 Complex Fractions

Complex Fractions   page 408
A **complex fraction** is a fraction in which the numerator and/or denominator are themselves fractions or the sum or difference of fractions.

Simplifying Complex Fractions
First Method   page 408
1. Simplify the numerator so that it is a single rational expression.
2. Simplify the denominator so that it is a single rational expression.
3. Divide the numerator by the denominator and reduce to lowest terms.

Second Method   page 410
1. Find the LCM of all the denominators in the numerator and denominator of the complex fraction.
2. Multiply both the numerator and denominator of the complex fraction by this LCM.
3. Simplify both the numerator and denominator and reduce to lowest terms.

Section 5.4 Solving Equations with Rational Expressions

Ratio   page 413
  A **ratio** is a comparison of two numbers by division.

Proportion   pages 413-414
  A **proportion** is an equation stating that two ratios are
  equal. One of the following conditions must be true:
  1. The numerators agree in type and the denominators
     agree in type.
  2. The numerators correspond and the denominators
     correspond.

Similar Figures   pages 415-416
  Similar figures are figures that meet the following two
  conditions:
  1. The corresponding angles are equal.
  2. The corresponding sides are proportional.

To Solve an Equation Containing Rational Expressions   page 417
  1. Find the LCM of the denominators.
  2. Multiply both sides of the equation by this LCM and
     simplify.
  3. Solve the resulting equation.
  4. Check each solution in the original equation.
     (Remember that no denominator can be 0.)

Section 5.5 Applications

Strategy for Solving Word Problems   page 425
  1. Read the problem carefully.
  2. Decide what is asked for and assign a variable to the
     unknown quantity.
  3. Draw a diagram or set up a chart whenever possible.
  4. Form an equation that relates the information provided.
  5. Solve the equation.
  6. Check your solution with the wording of the problem to
     be sure it makes sense.

Applications
  Number problems related to fractions   page 425
  Work problems   page 426
  Distance-Rate-Time problems   page 429

Section 5.6   Variation

Direct Variation   page 436

A variable quantity y **varies directly as** (or is **directly proportional to**) a variable x if there is a constant k such that

$$\frac{y}{x} = k \text{ or } y = kx.$$

The constant k is called the **constant of variation**.

Inverse Variation   page 438

A variable quantity y **varies inversely as** (or is **inversely proportional to**) a variable x if there is a constant k such that

$$x \cdot y = k \text{ or } y = \frac{k}{x}.$$

The constant k is called the **constant of variation**.

Combined Variation   page 439

If a variable varies either directly or inversely with more than one other variable, the variation is said to be **combined variation**.

Joint Variation   page 439

If a combined variation is all direct variation (the variables are multiplied), then it is called **joint variation**.

 HAWKES LEARNING SYSTEMS: INTERMEDIATE ALGEBRA SOFTWARE

For a review of the topics and problems from Chapter 5, look at the following lessons from *Hawkes Learning Systems: Intermediate Algebra.*

- Defining Rational Expressions
- Multiplication and Division with Rational Expressions
- Addition and Subtraction with Rational Expressions
- Complex Fractions
- Solving Equations with Rational Expressions
- Applications Involving Rational Expressions
- Applications: Variation
- Chapter 5 Review and Test

Chapter 5 Review

5.1 Multiplication and Division with Rational Expressions

In Exercises 1 – 10, reduce to lowest terms. State any restrictions on the variable(s).

1. $\dfrac{18xy^3}{96x^2y^3}$

2. $\dfrac{5x+10}{9x^3+15x^2-6x}$

3. $\dfrac{5-2x}{2x-5}$

4. $\dfrac{4x-12}{9-3x}$

5. $\dfrac{2x^3-16}{x^2-4}$

6. $\dfrac{3x^2-75}{x^2-10x+25}$

7. $\dfrac{x^3+64}{x^2-4x+16}$

8. $\dfrac{xy-4y+2x-8}{3x^2-12}$

9. $\dfrac{6x^2+7x-5}{3x^2-4x-15}$

10. $\dfrac{2x^2+x-3}{2x^2+13x+15}$

Perform the indicated operations in Exercises 11 – 20 and reduce to lowest terms. Assume that no denominator has a value of 0.

11. $\dfrac{2x-6}{3x-9} \cdot \dfrac{2x}{2x-4}$

12. $\dfrac{25x^2-9}{15x^2-9x} \cdot \dfrac{6x^2}{5x-3}$

13. $\dfrac{24x^2y}{9xy^6} \div \dfrac{4x^4y}{3x^2y^3}$

14. $\dfrac{x-1}{7x+7} \div \dfrac{3x-3}{x^2+x}$

15. $\dfrac{x^2-8x+15}{x^2+5x-14} \div \dfrac{x^2-9}{x^2-49}$

16. $\dfrac{x-3}{x^2-3x-4} \div \dfrac{x^2-x-2}{3x+3}$

17. $\dfrac{x^2+2x-8}{4x^3} \cdot \dfrac{5x^2}{2x^2-5x+2}$

18. $\dfrac{x-1}{x^2+7x+6} \cdot \dfrac{x^2-4x}{x+4} \cdot \dfrac{x^2-16}{x^2-5x+4}$

19. $\dfrac{x^2+3x}{x^2+8x+12} \cdot \dfrac{2x^2+11x-6}{2x+6} \cdot \dfrac{x^2+4x+4}{2x^2+5x+2}$

20. $\dfrac{x^2+2x+3}{x^2+10x+21} \div \dfrac{2x^2-18}{x^2+14x+49}$

5.2 Addition and Subtraction with Rational Expressions

Perform the indicated operations and reduce if possible in Exercises 21 – 32. Assume that no denominator has a value of 0.

21. $\dfrac{6x}{x+2} + \dfrac{12}{x+2}$

22. $\dfrac{8x}{x+6} + \dfrac{48}{x+6}$

23. $\dfrac{5}{x-2} + \dfrac{x}{x^2-4}$

24. $\dfrac{x+1}{x^2-10x-11} + \dfrac{x}{x-11}$

25. $\dfrac{4}{x+5} - \dfrac{2x}{x^2+3x-10}$

26. $\dfrac{x^2}{x^2+2x+1} - \dfrac{2x-1}{x^2-2x+1}$

27. $\dfrac{x}{3x^2+4x+1} + \dfrac{2x+1}{x^2+5x+6}$

28. $\dfrac{6}{2x-3}-\dfrac{5}{3-2x}$

29. $\dfrac{6x+24}{x^2+8x+16}+\dfrac{3x-12}{x^2-16}$

30. $\dfrac{5x-20}{x^2+x-20}-\dfrac{x^2+5x}{x^2+9x+20}$

31. $\dfrac{x+2}{x^2+x+1}-\dfrac{x-2}{x^3-1}$

32. $\dfrac{3x+2}{x^2+2x-3}-\dfrac{x}{3x^2-2x-1}$

5.3   Complex Fractions

Simplify the complex fractions given in Exercises 33 – 40.

33. $\dfrac{x+y}{\dfrac{2}{x}+\dfrac{2}{y}}$

34. $\dfrac{\dfrac{x^2-36}{x}}{x-6}$

35. $\dfrac{\dfrac{1}{x+2}-\dfrac{1}{x}}{1+\dfrac{2}{x}}$

36. $\dfrac{1-4x^{-2}}{1+2x^{-1}}$

37. $\dfrac{\dfrac{2}{x}}{1+\dfrac{1}{x-1}}$

38. $\dfrac{\dfrac{x^2+5x+4}{x+4}}{x+1}$

39. $\dfrac{2-\dfrac{6}{x}}{\dfrac{x^2-6x-9}{2x}}$

40. $\dfrac{\dfrac{x^2-25}{x^2-10x+25}}{1+\dfrac{5}{x}}$

5.4   Solving Equations with Rational Expressions

Solve the proportions in Exercises 41 – 44.

41. $\dfrac{2}{15}=\dfrac{16}{3x}$

42. $\dfrac{x-3}{x+5}=\dfrac{7}{15}$

43. $\dfrac{4}{x+3}=\dfrac{1}{x-3}$

44. $\dfrac{x-4}{3x}=\dfrac{2}{x-3}$

45. Antifreeze: In a mixture, there are 3 parts antifreeze for each 8 parts water. If you want a total of 220 gallons of mixture, how many gallons of antifreeze are needed?

46. Classroom size: In an English class the ratio of men to women is 4 to 5. If there are 36 students in the class, how many women are in the class?

47. Baseball: Louis is averaging 13 hits for every 50 times at bat. If he maintains this average, how many "at bats" will he need in order to get 156 hits?

48. Shadows: A flagpole casts a 30 foot shadow at the same time as a $5\dfrac{1}{2}$ foot woman casts a $4\dfrac{1}{8}$ foot shadow. How tall is the flagpole?

Exercises 49 and 50 show pairs of similar triangles. Find the lengths of the sides labeled with variables.

49. $\triangle ABC \sim \triangle DEF$

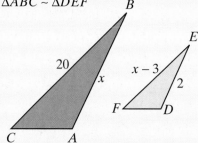

50. $\triangle STU \sim \triangle VWU$

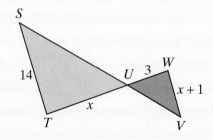

State any restrictions on x and then solve the equations in Exercises 51 – 54.

51. $\dfrac{5}{x} + \dfrac{2}{x-1} = \dfrac{2x}{x-1}$

52. $\dfrac{9}{x^2-9} + \dfrac{x}{3x+9} = \dfrac{1}{3}$

53. $\dfrac{x}{x-4} - \dfrac{6x}{x^2-x-12} = \dfrac{7x}{x+3}$

54. $\dfrac{2x+7}{15} - \dfrac{x+2}{x+6} = \dfrac{2x-6}{5}$

Solve each of the formulas for the specified variables in Exercises 55 and 56.

55. $\dfrac{y-y_1}{x-x_1} = m$; solve for x_1

56. $a_n = a_1 r^{n-1}$; solve for a_1

5.5 Applications

57. The numerator of a fraction is two less than the denominator. If both the numerator and denominator are decreased by 2, the new fraction will reduce to $\dfrac{2}{3}$. What was the original fraction? (Do not reduce answer.)

58. The ratio of two numbers is 7 to 9. If one of the numbers is 6 more than the other, what are the numbers?

59. Preparing a report: Working together, Alice and Judy can prepare an account report in 2.4 hours. Working alone, Alice would need two hours longer to prepare the same report than Judy would. How long would it take each of them working alone?

60. Painting a room: It takes Karl four hours less to paint a room than his son would take to paint the same sized room. If they can paint this size room in $1\dfrac{1}{2}$ hours working together, how long would each of them take to paint such a room alone?

61. Boating: A boat can travel 10 miles downstream in the same time it travels 6 miles upstream. If the rate of the current is 6 mph, find the speed of the boat in still water.

62. Car speed: Mr. Lukin had a golf game scheduled with a friend 80 miles from his home. After traveling 30 miles in heavy traffic, he needed to increase his speed by 25 mph to make the scheduled tee off time. If he traveled the same length of time at each rate, find the rates.

63. Plane speed: After flying 760 miles, the pilot increased the speed by 30 mph and continued on for another 817 miles. If the same amount of time was spent at each rate, find the rates.

5.6 Variation

65. If y is directly proportional to the square of x, and $y = 10$ when x is 3, what is y when x is 7?

66. If y is inversely proportional to x^4, and $y = 50$ when $x = \dfrac{1}{2}$, what is y when $x = \dfrac{1}{10}$?

67. Free falling object: The distance a free falling object falls is directly proportional to the square of the time it falls (before it hits the ground). If an object fell 256 feet in 4 seconds, how far will it have fallen by the end of 6 seconds?

68. Resistance in a wire: The resistance, R (in ohms), in a wire is given by the formula $R = \dfrac{kL}{d^2}$ where k is the constant of variation, L is the length of the wire and d is the diameter of the wire. The resistance of a wire 600 feet long with a diameter of 0.01 in. is 22 ohms. What is the resistance of the same type of wire 1800 feet long with a diameter of 0.02 in.?

64. Research projects: It takes Matilde twice as long to complete a certain type of research project as it takes Raphael to do the same type of project. They found that if they worked together they could finish similar projects in 1 hour. How long would it take each of them working alone to finish this type of project?

69. Volume of a circular cylinder: The volume of a circular cylinder varies jointly as its height and the square of its radius. If a cylinder with $r = 2$ feet and $h = 4$ feet holds 50.24 cubic feet of water, what is the volume of water in the cylinder when $h = 3$ feet?

70. Illumination of a light source: The illumination (in foot-candles, fc) of a light source varies directly as the intensity (in candelas, cd) of the source and inversely as the square of the distance from the source. If a certain light source with intensity of 400 cd provides an illumination of 20 fc at a distance of 25 feet, what is the illumination at a distance of 50 feet?

Chapter 5 Test

In Exercises 1 – 3, reduce to lowest terms. State any restrictions on the variable(s).

1. $\dfrac{x^2+3x}{x^2+7x+12}$

2. $\dfrac{x^3-64}{16-x^2}$

3. $\dfrac{2x+5}{4x^2+20x+25}$

In Exercises 4 – 9, perform the indicated operations and reduce the answers to lowest terms. Assume that no denominator has a value of 0.

4. $\dfrac{x+3}{x^2+3x-4}\cdot\dfrac{x^2+x-2}{x+2}$

5. $\dfrac{6x^2-x-2}{12x^2+5x-2}\div\dfrac{4x^2-1}{8x^2-6x+1}$

6. $\dfrac{x}{x^2+3x-10}+\dfrac{3x}{4-x^2}$

7. $\dfrac{x-4}{3x^2+5x+2}-\dfrac{x-1}{x^2-3x-4}$

8. $\dfrac{x^2-16}{x^2-4x}\cdot\dfrac{x^2}{x+4}\div\dfrac{x-1}{2x^2-2x}$

9. $\dfrac{x}{x+3}-\dfrac{x+1}{x-3}+\dfrac{x^2+4}{x^2-9}$

Simplify the complex fractions given in Exercises 10 and 11.

10. $\dfrac{\dfrac{4}{3x}+\dfrac{1}{6x}}{\dfrac{1}{x^2}-\dfrac{1}{2x}}$

11. $\dfrac{\dfrac{4}{x}+\dfrac{2}{x+3}}{\dfrac{x^2-9}{2x}}$

In Exercises 12 – 15, state any restrictions on the variables and solve the equations.

12. $\dfrac{x-1}{x+4}=\dfrac{4}{5}$

13. $\dfrac{4}{7}-\dfrac{1}{2x}=1+\dfrac{1}{x}$

14. $\dfrac{4}{x+4}+\dfrac{3}{x-1}=\dfrac{1}{x^2+3x-4}$

15. $\dfrac{x}{x+2}=\dfrac{1}{x+1}-\dfrac{x}{x^2+3x+2}$

In Exercises 16 and 17, solve each of the formulas for the specified variables.

16. $S=\dfrac{n\left(a_1+a_n\right)}{2}$; solve for n

17. $y=mx+b$; solve for x

18. The denominator of a fraction is three more than twice the numerator. If eight is added to both the numerator and the denominator, the resulting fraction is equal to $\dfrac{2}{3}$. Find the original fraction.

19. House cleaning: Sonya can clean the apartment in 6 hours. It takes Lucy 12 hours to clean it. If they work together, how long will it take them?

20. Travel: Mario can travel 228 miles in the same time that Carlos travels 168 miles. If Mario's speed is 15 mph faster than Carlos', find their rates.

21. Boating: Bob travels 4 miles upstream. In the same time, he could have traveled 7 miles downstream. If the speed of the current is 3 mph, find the speed of the boat in still water.

22. z varies directly as x^2 and inversely as $\sqrt{y}$. If $z = 24$ when $x = 3$ and $y = 4$, find z if $x = 5$ and $y = 9$.

23. Stretching a spring: Hooke's Law states that the distance a spring will stretch vertically is directly proportional to the weight placed at its end. If a particular spring will stretch 5 cm when a weight of 4 g is placed at its end, how far will the spring stretch if a weight of 6 g is placed at its end?

24. Triangles ABC and DEC are similar. (Symbolically, $\triangle ABC \sim \triangle DEC$.) Find the lengths of sides AC and DC.

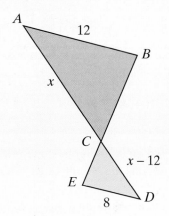

Cumulative Review: Chapters 1–5

1. Given the set of numbers $\left\{-9, -\sqrt{7}, \frac{1}{4}, \frac{\pi}{2}, 3, \sqrt{25}\right\}$, list those numbers in the set that are described below:

    a. $\{x \mid x \text{ is an integer}\}$ **b.** $\{x \mid x \text{ is a rational number}\}$

    c. $\{x \mid x \text{ is an irrational number}\}$ **d.** $\{x \mid x \text{ is a real number}\}$

Find the least common multiple of each set in Exercises 2 – 4.

2. $\{15, 35, 40\}$ **3.** $\{4x^2, 18x^2y, 12xy^3\}$

4. $\{(x+3)^2, 2x+6, 2x^2-18\}$

Solve each of the equations in Exercises 5 – 7.

5. $4(3x-1) = 2(2x-5) - 3$ **6.** $\dfrac{4x-1}{3} + \dfrac{x-5}{2} = 2$ **7.** $\left|\dfrac{3x}{4} - 1\right| = 2$

Solve the inequalities in Exercises 8 – 10 and graph the solutions. Write the solutions in interval notation.

8. $x + 4 - 3x \geq 2x + 5$ **9.** $\dfrac{x}{5} - 3.4 > \dfrac{x}{2} + 1.6$ **10.** $\left|\dfrac{3x}{2} - 1\right| - 2 \leq 3$

11. Find an equation in slope-intercept form for the line that has slope $m = -\dfrac{4}{5}$ and contains the point $(-2, 2)$. Then graph the line.

12. Find an equation in standard form for the line that passes through the point $(3, -4)$ and is parallel to the line $2x - y = 5$. Graph both lines.

13. Find an equation in slope-intercept form for the line passing through the points $(1, -4)$ and $(4, 2)$. Graph the line.

14. Three graphs are shown below. Use the vertical line test to determine whether or not each graph represents a function. If the graph represents a function, state its domain and range.

    a.   **b.**   **c.**

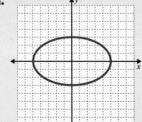
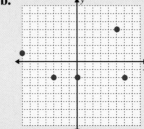
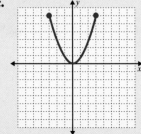

15. Solve the following system of linear equations using Cramer's Rule: $\begin{cases} 3x+8y=-2 \\ -x+2y=-4 \end{cases}$.

16. Solve the following system of equations using Gaussian elimination. You may use your calculator.
$$\begin{cases} x+y+z=-2 \\ 2x-3y-z=15 \\ -x+y-3z=-12 \end{cases}$$

17. Find the value of $\det(A)$ for $A = \begin{bmatrix} 4 & 3 & -1 \\ 2 & -1 & 5 \\ 0 & 6 & -3 \end{bmatrix}$.

18. Solve the following sytem of linear inequalities graphically: $\begin{cases} y>x-2 \\ 3x+y\le 4 \end{cases}$.

Use the rules of exponents to simplify the expressions in Exercises 19 – 21 so that they contain only positive exponents.

19. $\left(3x^2y^{-3}\right)^{-2}$

20. $\left(\dfrac{x^4y}{x^2y^3}\right)^3$

21. $\left(\dfrac{5x^2y^{-1}}{2xy}\right)\left(\dfrac{3x^{-3}y}{5x^2y^0}\right)$

Write each expression in Exercises 22 and 23 in scientific notation and simplify.

22. $(17000)(0.0004)$

23. $\dfrac{6300}{0.006}$

Factor each expression in Exercises 24 – 27 completely.

24. $15x^2+22x+8$

25. $2x^3+8x^2+3x+12$

26. $16x^3-54$

27. $2x^2-5x-3$

28. Given that $P(x)=4x^3-5x^2+2x-1$, find **a.** $P(0)$ **b.** $P(-1)$ **c.** $P(2)$

Solve the equations in Exercises 29 and 30 by factoring.

29. $x^2+22x=-121$

30. $5x^2=14x-x^3$

31. Find an equation with integer coefficients that has $x=-5$ and $x=4$ as roots.

32. Find an equation with integer coefficients that has $x=\dfrac{1}{2}$ and $x=\dfrac{2}{3}$ as roots.

In Exercises 33 and 34, use the division algorithm to divide and write the answer in the form $Q+\dfrac{R}{D}$ where the degree of R < the degree of D.

33. $\dfrac{2x^3+5x^2-x+3}{x+2}$

34. $\dfrac{x^3-4x^2+9}{x^2-6}$

Perform the indicated operations in Exercises 35 and 36.

35. $\dfrac{x+1}{x^2+3x-4} \cdot \dfrac{2x^2+7x-4}{x^2-1} \div \dfrac{x+1}{x-1}$

36. $\dfrac{x}{x^2-2x-8} + \dfrac{x-3}{2x^2-5x-12} - \dfrac{2x-5}{2x^2+7x+6}$

37. State the restrictions on the variables and solve each equation.

a. $\dfrac{x}{x+4} - \dfrac{4}{x^2-16} = \dfrac{-2x}{x-4}$

b. $\dfrac{x-2}{x-2} + \dfrac{x-3}{x-3} = \dfrac{2x^2}{x^2-5x+6}$

Solve each of the formulas in Exercises 38 and 39 for the indicated variables.

38. $PV = nRT$; solve for n

39. $\dfrac{1}{S_1} + \dfrac{1}{S_2} = \dfrac{1}{f}$; solve for f

40. The following table shows the number of feet that Juan drove his car from one minute to another.

a. Plot the points indicated in the table.
b. Calculate the slope of the line segments from point to point.
c. Interpret each slope.

Minutes	Feet
1	1496
2	2024
3	1056
4	440
5	704

41. Determine the values of a and b such that the straight line $ax + by = 14$ passes through the two points $(-1, 19)$ and $(2, 4)$.

42. Falling object: The distance an object falls (in feet) varies directly as the square of the time (in seconds) that it falls. If an object falls 16 feet in 1 s, how far will it fall in 4 s?

43. Ladders: A right triangle is formed when a ladder is leaned against a building. If the ladder is 26 ft long and its bottom is placed 10 ft from the base of the building, how far up the building does the ladder reach?

44. The sum of two numbers is 14. Twice the larger number added to three times the smaller is equal to 31. Find both numbers.

45. The sum of the squares of two consecutive positive even integers is 580. What are the integers?

46. Walking: Emily started walking to a town 10 miles away at a rate of 3 mph. After walking part of the way, she got a ride on a bus. The bus traveled at an average rate of 48 mph, and Emily reached the town 50 minutes after she started walking. How long did she walk before catching the bus?

47. Carpeting a house: LeeAnne is a carpet layer. She has agreed to carpet a house for $18 per square yard. The carpet will cost her $11 per square yard, and she knows that there is approximately 12% waste in cutting and matching. If she plans to make a profit of $580.80, how many square yards of carpet will she buy?

Roots, Radicals, and Complex Numbers

Did You Know?

An important method of reasoning related to mathematical proofs is proof by contradiction. See if you can follow the reasoning in the following proof that $\sqrt{2}$ is an irrational number.

We need the following statement (which can be proven algebraically):

> The square of an integer is even if and only if the integer is even.

Proof: $\sqrt{2}$ is either an irrational number or a rational number.

Suppose that $\sqrt{2}$ is a rational number and

$$\frac{a}{b} = \sqrt{2}$$
    Where a and b are integers and $\frac{a}{b}$ is simplified.

$$\frac{a^2}{b^2} = 2$$
    Square both sides.

$$a^2 = 2b^2$$
    This means a^2 is an even integer.

So $a = 2n$
    Since a^2 is even, a must be even.

$$a^2 = 4n^2$$
    Square both sides.

$$a^2 = 4n^2 = 2b^2$$
    Substitution.

$$2n^2 = b^2$$
    This means b^2 is an even integer.

Therefore, b is an even integer.

But if a and b are both even, 2 is a common factor.

This contradicts the statement that $\frac{a}{b}$ is reduced.

Thus our original supposition that $\sqrt{2}$ is rational is false, making $\sqrt{2}$ an irrational number.

*"The number of grains of sand on the beach at Coney Island is much less than a googol –
10,000,000,000,000,000,000,000,000,000,
000,000,000,000,000,000,000,000,000,000,
000,000,000,000,000,000,000,000,000,000,
000,000,000,000."*

Edward Kasner

In this chapter, we will discuss the meaning of **rational exponents** and their relationship with radical expressions such as square roots $\left(\sqrt{}\right)$ and cube roots $\left(\sqrt[3]{}\right)$. This relationship allows translation from one type of notation to the other with relative ease and a choice for the form of an answer that best suits the purposes of the problem. For example, in higher level mathematics courses, particularly in calculus, an expression with a square root, such as $\sqrt{x^2+1}$, may be changed into an alternate notation using fractional exponents, such as $\left(x^2+1\right)^{\frac{1}{2}}$. Operations learned in calculus can be performed more easily on expressions with exponents in fractional form than on the equivalent expression in radical form. Then, if desired, answers can be changed back into a form with radical notation.

A new category of numbers, called **complex numbers**, is an expansion of the real number system and includes a group of numbers called **imaginary numbers**. The term "imaginary" is somewhat misleading because these numbers have many practical applications and are no more imaginary than any other type of number. They are particularly useful in electrical engineering and hydrodynamics. Complex numbers arise quite naturally as solutions to quadratic equations.

6.1 Roots and Radicals

- *Evaluate square roots and cube roots.*
- *Simplify expressions with square roots and cube roots.*

You are probably familiar with the concept of **square roots** and the square root symbol $\left(\text{or } \textbf{radical sign } \sqrt{}\right)$ from your previous work in algebra and the discussions of real numbers in Chapter 1. For example, $\sqrt{3}$ represents the square root of 3 and is the number whose square is 3.

Perfect Squares and Square Roots

A number is **squared** when it is multiplied by itself. For example,

$$6^2 = 6 \cdot 6 = 36 \quad \text{and} \quad \left(-1.5\right)^2 = \left(-1.5\right)\left(-1.5\right) = 2.25.$$

If an integer is squared, the result is called a **perfect square**. The squares for the integers from 1 to 20 are shown in Table 6.1 for easy reference.

Squares of Integers from 1 to 20 (Perfect Squares)										
Integers (n)	1	2	3	4	5	6	7	8	9	10
Perfect Squares (n^2)	1	4	9	16	25	36	49	64	81	100
Integers (n)	11	12	13	14	15	16	17	18	19	20
Perfect Squares (n^2)	121	144	169	196	225	256	289	324	361	400

Table 6.1

Now, we want to reverse the process of squaring. That is, given a number, we want to find a number that when squared will result in the given number. This is called **finding a square root** of the given number. In general,

if $b^2 = a$, then b is a square root of a.

For example,

- because $5^2 = 25$, 5 is a **square root** of 25 and we write $\sqrt{25} = 5$

- because $9^2 = 81$, 9 is a **square root** of 81 and we write $\sqrt{81} = 9$.

Radical Terminology

*The symbol $\sqrt{}$ is called a **radical sign**.*

*The number under the radical sign is called the **radicand**.*

*The complete expression, such as $\sqrt{64}$, is called a **radical** or **radical expression**.*

Every positive real number has two square roots, one positive and one negative. The positive square root is called the **principal square root**. For example,

- because $(8)^2 = 64$, $\sqrt{64} = 8$ ◀—— the **principal square root**

- because $(-8)^2 = 64$, $-\sqrt{64} = -8$. ◀—— the **negative square root**

The number 0 has only one square root, namely 0.

Square Root

If a is a nonnegative real number and b is a real number such that
$$b^2 = a, \text{ then b is called a } \textbf{square root} \text{ of a.}$$
If b is nonnegative, then we write
$$\sqrt{a} = b \longleftarrow \text{b is called the } \textbf{principal square root},$$
$$\text{and } -\sqrt{a} = -b. \longleftarrow \text{−b is called the } \textbf{negative square root}.$$

Example 1: Evaluating Square Roots

a. 64 has two square roots, one positive and one negative. The $\sqrt{\ }$ sign is understood to represent the **positive square root** or the **principal square root**. Therefore,
$$\sqrt{64} = 8 \qquad \text{and} \qquad -\sqrt{64} = -8.$$

b. Because $11^2 = 121$, we know that $\sqrt{121} = 11$.

c. Because $\left(\dfrac{4}{5}\right)^2 = \dfrac{16}{25}$, we know that $\sqrt{\dfrac{16}{25}} = \dfrac{4}{5}$.

d. $-\sqrt{0.0036} = -0.06$ $\qquad\qquad$ $(0.06)^2 = 0.0036$

e. $\sqrt{16+9} = \sqrt{25} = 5$ $\qquad$ Note that the sum under the radical sign is found first. Then the square root is found. In this situation, the radical sign is similar to parentheses.

f. $\sqrt{16} + \sqrt{9} = 4 + 3 = 7$ $\qquad$ In this case the square roots are evaluated before addition.

[**Note:** As these examples illustrate, the radicand may be a non-negative integer, fraction, or decimal number.]

NOTES   Note that, as illustrated in examples 1e and 1f, in general,
$$\sqrt{a+b} \neq \sqrt{a} + \sqrt{b}.$$

NOTES

The square roots of negative real numbers are **not real numbers**. (There is no real number whose square is negative.) Thus $\sqrt{-4}$ is not a real number and $\sqrt{-25}$ is not a real number. Such numbers are called **imaginary numbers** or **nonreal complex numbers**. These numbers and their properties will be studied later in Sections 6.6 and 6.7.

Using a TI-84 Plus Graphing Calculator to Evaluate Radicals

In Example 1, all the square roots were **integers** or other **rational numbers**. However, most roots written in decimal form are infinite nonrepeating decimals called **irrational numbers**. (Irrational numbers cannot be written in the fraction form $\dfrac{a}{b}$ where a and b are integers.) As discussed briefly in Section 1.2, two examples of irrational numbers are

$$\pi = 3.14159265358979... \quad \text{and} \quad \sqrt{2} = 1.414213562...$$

In the decimal form of an irrational number, there is no repeating pattern in the digits.

Most calculators can be used to find the values of radicals (rational or irrational) accurate to as many as nine decimal places. Example 2 illustrates how to use the TI-84 Plus graphing calculator to find decimal values of square roots.

Example 2: Evaluating Radical Expressions with a Calculator

In each of the following examples the steps (or keys to press) are shown. Although your calculator will probably display the answer to 9 decimal places, your instructor may ask you to round these calculations to fewer than nine places. The answers here are rounded to 4 decimal places.

a. $\sqrt{200}$

Solution: Press the keys in the following sequence.

Note: To get the square root symbol on the TI-84 Plus, press **2ND** then the x^2 key. When the $\sqrt{\ }$ symbol appears, it will appear with a left-hand parenthesis. You should press the right-hand parenthesis key to close the square root operation.

Continued on the next page...

The display will appear as follows.

Thus $\sqrt{200} \approx 14.1421$, rounded to four decimal places.

b. $-\sqrt{35}$

Solution: Press the keys in the following sequence.

The display will appear as follows.

Thus $-\sqrt{35} \approx -5.9161$, rounded to four decimal places.

Simplifying Algebraic Expressions with Square Roots

Various roots can be related to solutions of equations, and we want such numbers to be in a **simplified form** for easier calculations and algebraic manipulations. We need the two properties of radicals stated here for square roots. (Similar properties are true for other roots.)

Properties of Square Roots

If a and b are **positive** real numbers, then

1. $\sqrt{ab} = \sqrt{a}\sqrt{b}$ **2.** $\sqrt{\dfrac{a}{b}} = \dfrac{\sqrt{a}}{\sqrt{b}}$.

As an example, we know that 144 is a perfect square and $\sqrt{144} = 12$. However, in a situation where you may have forgotten this, you can proceed as follows using property 1 of square roots:

$$\sqrt{144} = \sqrt{36} \cdot \sqrt{4} = 6 \cdot 2 = 12.$$

Similarly, using property 2, we can write

$$\sqrt{\frac{49}{36}} = \frac{\sqrt{49}}{\sqrt{36}} = \frac{7}{6}.$$

Simplest Form

A square root is considered to be in **simplest form** when the radicand has no perfect square as a factor.

The number 200 is not a perfect square and in Example 2a we used a calculator to find $\sqrt{200} \approx 14.1421$. Now, to simplify $\sqrt{200}$, we can use property 1 of square roots and any of the following three approaches.

Approach 1: Factor 200 as $4 \cdot 50$ because 4 is a perfect square. This gives,

$$\sqrt{200} = \sqrt{4 \cdot 50} = \sqrt{4} \cdot \sqrt{50} = 2\sqrt{50}.$$

However, $2\sqrt{50}$ is **not in simplest form** because 50 has a perfect square factor, 25. Thus to complete the process, we have

$$\sqrt{200} = 2\sqrt{50} = 2\sqrt{25 \cdot 2} = 2\sqrt{25} \cdot \sqrt{2} = 2 \cdot 5 \cdot \sqrt{2} = 10\sqrt{2}.$$

Approach 2: Note that 100 is a perfect square factor of 200 and $200 = 100 \cdot 2$.

$$\sqrt{200} = \sqrt{100 \cdot 2} = \sqrt{100} \cdot \sqrt{2} = 10\sqrt{2}$$

Approach 3: Use prime factors.

$$\sqrt{200} = \sqrt{10 \cdot 20}$$
$$= \sqrt{2 \cdot 5 \cdot 2 \cdot 2 \cdot 5}$$
$$= \sqrt{2 \cdot 2 \cdot 5 \cdot 5} \cdot \sqrt{2}$$
$$= \sqrt{2 \cdot 2} \cdot \sqrt{5 \cdot 5} \cdot \sqrt{2}$$
$$= 2 \cdot 5 \cdot \sqrt{2}$$
$$= 10\sqrt{2}$$

NOTES

Of these three approaches, the second appears to be the easiest because it has the fewest steps. However, "seeing" the largest perfect square factor may be difficult. If you do not immediately see a perfect square factor, proceed by finding other factors or prime factors as illustrated.

Example 3: Simplifying Numerical Expressions with Square Roots

Simplify each numerical expression so that there are no perfect square numbers in the radicand.

a. $\sqrt{48}$

Solution: $\sqrt{48} = \sqrt{16 \cdot 3} = \sqrt{16} \cdot \sqrt{3} = 4\sqrt{3}$ 16 is the largest perfect square factor.

b. $\sqrt{63}$

Solution: $\sqrt{63} = \sqrt{9 \cdot 7} = \sqrt{9} \cdot \sqrt{7} = 3\sqrt{7}$ 9 is the largest perfect square factor.

c. $\sqrt{\dfrac{75}{16}}$

Solution: $\sqrt{\dfrac{75}{16}} = \dfrac{\sqrt{75}}{\sqrt{16}} = \dfrac{\sqrt{25 \cdot 3}}{\sqrt{16}} = \dfrac{\sqrt{25} \cdot \sqrt{3}}{\sqrt{16}} = \dfrac{5\sqrt{3}}{4}$

If a radical expression is in the form of a fraction with only one term in the denominator, we can simplify the radical and then divide the numerator by the denominator. There are two methods as illustrated in Example 4. Both methods are correct and will give the same answer.

Example 4: Simplifying Radical Expressions

Simplify each radical expression.

a. $\dfrac{12 - \sqrt{20}}{4}$

Solutions:

Method 1: Simplify and factor the numerator.

$$\frac{12 - \sqrt{20}}{4} = \frac{12 - \sqrt{4 \cdot 5}}{4} = \frac{12 - 2\sqrt{5}}{4} = \frac{\overset{}{2}\left(6 - \sqrt{5}\right)}{\underset{2}{\cancel{4}}} = \frac{6 - \sqrt{5}}{2}$$

Method 2: Simplify the numerator and divide each term by the denominator.

$$\frac{12 - \sqrt{20}}{4} = \frac{12 - \sqrt{4 \cdot 5}}{4} = \frac{12 - 2\sqrt{5}}{4} = \frac{12}{4} - \frac{2\sqrt{5}}{4} = 3 - \frac{\sqrt{5}}{2}$$

Note that these answers may look different but they are equivalent. Both are correct, just in different forms.

b. $\dfrac{\sqrt{90} + 6}{3}$

Solutions:

Method 1: Simplify and factor the numerator.

$$\frac{\sqrt{90} + 6}{3} = \frac{\sqrt{9 \cdot 10} + 6}{3} = \frac{3\sqrt{10} + 6}{3} = \frac{\cancel{3}\left(\sqrt{10} + 2\right)}{\cancel{3}} = \sqrt{10} + 2$$

Method 2: Simplify the numerator and divide each term by the denominator.

$$\frac{\sqrt{90} + 6}{3} = \frac{\sqrt{9 \cdot 10} + 6}{3} = \frac{3\sqrt{10} + 6}{3} = \frac{3\sqrt{10}}{3} + \frac{6}{3} = \sqrt{10} + 2$$

To simplify square root expressions that contain variables, such as $\sqrt{x^2}$, we must be aware of whether the variable represents a positive number $(x > 0)$, zero $(x = 0)$, or a negative number $(x < 0)$.

For example,

$$\text{if } x = 0, \text{ then } \sqrt{x^2} = \sqrt{0^2} = \sqrt{0} = 0 = x.$$

$$\text{If } x = 5, \text{ then } \sqrt{x^2} = \sqrt{5^2} = \sqrt{25} = 5 = x.$$

$$\text{But, if } x = -5, \text{ then } \sqrt{x^2} = \sqrt{(-5)^2} = \sqrt{25} = 5 \neq x.$$

$$\text{In fact, if } x = -5, \text{ then } \sqrt{x^2} = \sqrt{(-5)^2} = \sqrt{25} = 5 = |-5| = |x|.$$

Thus, simplifying radical expressions with variables involves more detailed analysis than simplifying radical expressions with only constants. The following definition indicates the correct way to simplify $\sqrt{x^2}$.

Square Root of x^2

If x is a real number, then $\sqrt{x^2} = |x|$.

Note: *If $x \geq 0$ is given, then we can write $\sqrt{x^2} = x$.*

Although using the absolute value when simplifying square roots is correct mathematically, we can avoid some confusion by assuming that the variable under the radical sign represents only positive numbers or 0. This eliminates the need for absolute value signs.

Therefore, for the remainder of this text, we will assume that $x > 0$ and write $\sqrt{x^2} = x$, unless specifically stated otherwise, in all square root expressions such as $\sqrt{x^2}$.

Example 5: Simplifying Square Roots with Variables

Simplify each of the following radical expressions. Assume that all variables represent positive numbers. (Note that, by making this assumption, we need not be concerned about the absolute value sign.)

a. $\sqrt{16y^2}$

Solution: $\sqrt{16y^2} = 4y$

b. $\sqrt{72a^2}$

Solution: $\sqrt{72a^2} = \sqrt{36a^2} \cdot \sqrt{2} = 6a\sqrt{2}$

c. $\sqrt{12x^2y^2}$

Solution: $\sqrt{12x^2y^2} = \sqrt{4x^2y^2} \cdot \sqrt{3} = 2xy\sqrt{3}$

If an exponent is larger than 2, we look for even exponents. The square root of an even power can be found by dividing the exponent by 2. For example,

$$x^2 \cdot x^2 = x^4 \qquad a^3 \cdot a^3 = a^6 \qquad y^5 \cdot y^5 = y^{10}$$

$$\downarrow \qquad\qquad \downarrow \qquad\qquad \downarrow$$

and $\qquad \sqrt{x^4} = x^2 \qquad\quad \sqrt{a^6} = a^3 \qquad\quad \sqrt{y^{10}} = y^5.$

To find the square root of an expression with odd exponents, factor the expression into two terms, one with exponent 1 and the other with an even exponent. For example,

$$x^3 = x^2 \cdot x \qquad \text{and} \qquad y^9 = y^8 \cdot y$$

which means that

$$\sqrt{x^3} = \sqrt{x^2 \cdot x} = \sqrt{x^2} \cdot \sqrt{x} = x \cdot \sqrt{x} \qquad \text{and} \qquad \sqrt{y^9} = \sqrt{y^8 \cdot y} = \sqrt{y^8} \cdot \sqrt{y} = y^4 \sqrt{y}.$$

Square Roots of Expressions with Even and Odd Exponents

For any nonnegative real number x and positive integer m,

$$\sqrt{x^{2m}} = x^m \qquad \text{and} \qquad \sqrt{x^{2m+1}} = x^m \sqrt{x}.$$

Note: *For any integer m, 2m is even and 2m + 1 is odd.*

Example 6: Simplifying Square Roots

Simplify each of the following radical expressions. Look for perfect square factors and even powers of the variables. Assume that all variables represent positive numbers.

a. $\sqrt{81x^4}$

Solution: $\sqrt{81x^4} = 9x^2$ $\qquad$ The exponent 4 is divided by 2.

b. $\sqrt{64x^5y}$

Solution: $\sqrt{64x^5y} = \sqrt{64 \cdot x^4} \cdot \sqrt{xy} = 8x^2\sqrt{xy}$

c. $\sqrt{18a^4b^6}$

Solution: $\sqrt{18a^4b^6} = \sqrt{9a^4b^6} \cdot \sqrt{2} = 3a^2b^3\sqrt{2}$ $\quad$ Each exponent is divided by 2.

d. $\sqrt{\dfrac{9a^{13}}{b^4}}$

Solution: $\sqrt{\dfrac{9a^{13}}{b^4}} = \dfrac{\sqrt{9a^{13}}}{\sqrt{b^4}} = \dfrac{\sqrt{9 \cdot a^{12}} \cdot \sqrt{a}}{\sqrt{b^4}} = \dfrac{3a^6\sqrt{a}}{b^2}$ $\qquad$ Recall $a, b > 0$.

Perfect Cubes and Cube Roots

The cubes of integers are called **perfect cubes**. For easy reference, Table 6.2 lists the cubes of the integers from 1 to 10.

Cubes of Integers from 1 to 10 (Perfect Cubes)										
Integers (n)	1	2	3	4	5	6	7	8	9	10
Perfect Cubes (n^3)	1	8	27	64	125	216	343	512	729	1000

Table 6.2

Cube Root

If a and b are real numbers such that
$b^3 = a$, then b is called the **cube root** of a.
We write $\sqrt[3]{a} = b$. ◄── the **cube root**

NOTES
In the cube root expression $\sqrt[3]{a}$ the number 3 is called the **index**. In a square root expression such as $\sqrt{a}$ the index is understood to be 2 and is not written. That is, $\sqrt{a}$ and $\sqrt[2]{a}$ have the same meaning. Expressions with cube roots (as well as expressions with square roots) are called **radical expressions**.

Example 7: Evaluating Cube Roots

a. $\sqrt[3]{216}$

Solution: Because $6^3 = 216$, $\sqrt[3]{216} = 6$.

b. $\sqrt[3]{-8}$

Solution: Because $(-2)^3 = -8$, $\sqrt[3]{-8} = -2$.

(Note that cube roots of negative numbers are negative.)

c. $\sqrt[3]{500}$

Solution: 500 is not a perfect cube. The value of $\sqrt[3]{500}$ can be estimated using the TI-84 Plus calculator by pressing keys in the following sequence.

[**Note:** The sequence **MATH** **4** selects $\sqrt[3]{(}$ from the math menu.]

The display should appear as follows.

Thus $\sqrt[3]{500} \approx 7.9370,$ accurate to 4 decimal places.

Simplifying Algebraic Expressions with Cube Roots

When simplifying expressions with cube roots, we need to be aware of perfect cube numbers and variables with exponents that are multiples of 3. (Multiples of 3 are 3, 6, 9, 12, 15, and so on.) **Thus exponents are divided by 3 in simplifying cube root expressions.** For example,

$$x^2 \cdot x^2 \cdot x^2 = x^6 \qquad a^3 \cdot a^3 \cdot a^3 = a^9 \qquad y^5 \cdot y^5 \cdot y^5 = y^{15}$$

$$\downarrow \qquad\qquad\qquad \downarrow \qquad\qquad\qquad \downarrow$$

and $\qquad \sqrt[3]{x^6} = x^2 \qquad\qquad \sqrt[3]{a^9} = a^3 \qquad\qquad \sqrt[3]{y^{15}} = y^5$

Simplest Form

*A cube root is considered to be in **simplest form** when the radicand has no perfect cube as a factor.*

Example 8: Simplifying Expressions with Cube Roots

Simplify each of the following radical expressions. Look for perfect cube factors and powers of the variables that are multiples of 3.

a. $\sqrt[3]{54x^6}$

Solution: $\sqrt[3]{54x^6} = \sqrt[3]{27x^6} \cdot \sqrt[3]{2} = 3x^2 \sqrt[3]{2}$

Note that 27 is a perfect cube and the exponent 6 is divisible by 3.

Continued on the next page...

b. $\sqrt[3]{-40x^4y^{13}}$

Solution: $\sqrt[3]{-40x^4y^{13}} = \sqrt[3]{-8x^3y^{12}} \cdot \sqrt[3]{5xy} = -2xy^4\sqrt[3]{5xy}$

Note that −8 is a perfect cube and the exponents on the variables are separated so that one exponent on each variable is divisible by 3.

c. $\sqrt[3]{250a^8b^{11}}$

Solution: $\sqrt[3]{250a^8b^{11}} = \sqrt[3]{125a^6b^9} \cdot \sqrt[3]{2a^2b^2} = 5a^2b^3\sqrt[3]{2a^2b^2}$

Note that 125 is a perfect cube and the exponents on the variables are separated so that one exponent on each variable is divisible by 3.

Practice Problems

Simplify the following radical expressions. Assume that all variables represent positive numbers.

1. $\sqrt{80x^3}$ **2.** $\sqrt{18x^2}$ **3.** $\sqrt[3]{-27y^9}$

4. $\sqrt{128a^2b^5}$ **5.** $\dfrac{4+\sqrt{12}}{2}$

Use a calculator to find the value of each number accurate to four decimal places.

6. $\sqrt{3}$ **7.** $\sqrt{40}$ **8.** $\sqrt[3]{40}$

6.1 Exercises

Use a graphing calculator in Exercises 1 – 10 to find the value of each radical. Round answers to four decimal places, if necessary.

1. $\sqrt{0.0004}$ **2.** $\sqrt{0.0025}$ **3.** $\sqrt{1024}$ **4.** $\sqrt{720}$ **5.** $\sqrt{4500}$

6. $\sqrt[3]{343}$ **7.** $\sqrt[3]{0.000008}$ **8.** $\sqrt[3]{50,000}$ **9.** $\sqrt[3]{12.5}$ **10.** $\sqrt[3]{100}$

Answers to Practice Problems: **1.** $4x\sqrt{5x}$ **2.** $3x\sqrt{2}$ **3.** $-3y^3$ **4.** $8ab^2\sqrt{2b}$ **5.** $2+\sqrt{3}$ **6.** 1.7321   **7.** 6.3246   **8.** 3.4200

In exercises 11 – 60, simplify each radical expression. Assume that the variables represent positive numbers. Say "nonreal" if the expression does not represent a real number.

11. $\sqrt{12}$ **12.** $\sqrt{18}$ **13.** $\sqrt{98}$ **14.** $\sqrt{216}$

15. $-\sqrt{162}$ **16.** $-\sqrt{27}$ **17.** $\sqrt[3]{16}$ **18.** $\sqrt[3]{40}$

19. $\sqrt[3]{108}$ **20.** $\sqrt[3]{-54}$ **21.** $\sqrt{-25}$ **22.** $\sqrt{-100}$

23. $\sqrt{24x^{11}y^2}$ **24.** $\sqrt{20x^{15}y^3}$ **25.** $\sqrt[3]{a^5b^2c^3}$ **26.** $\sqrt[3]{-xy^6}$

27. $\sqrt{-4x^5}$ **28.** $\sqrt{-9a^2}$ **29.** $\sqrt{25y^2}$ **30.** $-\sqrt{81x^2}$

31. $-\sqrt{64a^6}$ **32.** $\sqrt{18x^2y^2}$ **33.** $\sqrt{32x^4y^8}$ **34.** $\sqrt[3]{108ab^9}$

35. $\sqrt[3]{-24x^3y^6}$ **36.** $\sqrt[3]{-64a^{12}}$ **37.** $\sqrt[3]{81x^5y^7}$ **38.** $\sqrt[3]{54a^4b^2}$

39. $\sqrt[3]{8x^9y^{12}}$ **40.** $\sqrt[3]{512a^3b^{27}}$ **41.** $\sqrt[3]{729a^4b^8}$ **42.** $\sqrt[3]{125x^2y^2}$

43. $\sqrt{125x^3y^6}$ **44.** $\sqrt{8x^5y^4}$ **45.** $\sqrt{12ab^2c^3}$ **46.** $\sqrt{45a^2b^3c^4}$

47. $-\sqrt{75x^4y^6z^8}$ **48.** $-\sqrt{200x^2y^2z^2}$ **49.** $\sqrt[3]{24x^5y^7z^9}$ **50.** $\sqrt[3]{250x^6y^9z^{15}}$

51. $\dfrac{6-\sqrt{12}}{2}$ **52.** $\dfrac{4+\sqrt{24}}{2}$ **53.** $\dfrac{7+\sqrt{98}}{14}$ **54.** $\dfrac{12-\sqrt{18}}{3}$

55. $\dfrac{3+\sqrt{54}}{6}$ **56.** $\dfrac{10-\sqrt{108}}{4}$ **57.** $\dfrac{6+\sqrt{12x^2}}{2}$ **58.** $\dfrac{25-\sqrt{75x^2}}{5}$

59. $\dfrac{8+\sqrt{24a^3}}{8}$ **60.** $\dfrac{15+\sqrt{50y^3}}{10}$

61. Puzzle cube: The cubic volume of a puzzle cube is 343 cubic inches. What is the length of one side? [**Hint:** Use the definition of a cube root and determine the values of a and b in this problem. Recall that the volume of a cube is $V = s^3$.]

62. **Building Blocks:** Three perfect cube blocks of different volumes were stacked on top of each other. The top block was 216 cubic centimeters. The middle block was 343 cubic centimeters, and the biggest block was 512 cubic centimeters. How tall was the stack of blocks?

In exercises 63 – 65, use the following two formulas used in electricity to answer the questions below.

$$P = I^2 R$$
$$E^2 = PR$$

$\begin{cases} P = \text{power (in watts)} \\ I = \text{current (in amperes)} \\ E = \text{voltage (in volts)} \\ R = \text{resistance (in ohms, } \Omega) \end{cases}$

63. **Electricity:** What is the current in amperes of a light bulb that produces 150 watts of power and has a 25 Ω resistance?

64. **Electricity:** How many volts of electricity would Marcia need to produce 48 Ω of resistance from a 300 watt lamp?

65. **Electricity:** A 5000 Ω resistor is rated at 2.5 watts. What is the maximum voltage of electricity that should be connected across it?

Writing and Thinking About Mathematics

66. Under what conditions is the expression $\sqrt{a}$ not a real number?

67. Explain why the expression $\sqrt[3]{y}$ is a real number regardless of whether $y > 0$, $y < 0$, or $y = 0$.

HAWKES LEARNING SYSTEMS: INTERMEDIATE ALGEBRA SOFTWARE

- Evaluating Radicals
- Simplifying Radicals

6.2   Rational Exponents

- *Understand the meaning of nth root.*
- *Translate expressions using radicals into expressions using rational exponents and translate expressions using rational exponents into expressions using radicals.*
- *Simplify expressions using the properties of rational exponents.*
- *Evaluate expressions of the form $a^{\frac{m}{n}}$ with a calculator.*

n^{th} Roots: $\sqrt[n]{a} = a^{\frac{1}{n}}$

In Section 6.1 we restricted our discussions to radicals involving square roots and cube roots. In this section we will expand on those ideas by discussing radicals indicating n^{th} roots in general and how to relate radical expressions to expressions with rational (fractional) exponents. For example, the fifth root of x can be written in radical form as $\sqrt[5]{x}$ and with a fractional exponent as $x^{\frac{1}{5}}$.

To understand roots in general, consider the following analysis (assuming that $b > 0$).

Type of Root	Radical Notation and Exponent Notation
For square roots,	if $b^2 = a$, then $b = \sqrt{a}$ $\left(\text{or } b = a^{\frac{1}{2}}\right)$.
For cube roots,	if $b^3 = a$, then $b = \sqrt[3]{a}$ $\left(\text{or } b = a^{\frac{1}{3}}\right)$.
For fourth roots,	if $b^4 = a$, then $b = \sqrt[4]{a}$ $\left(\text{or } b = a^{\frac{1}{4}}\right)$.
For n^{th} roots,	if $b^n = a$, then $b = \sqrt[n]{a}$ $\left(\text{or } b = a^{\frac{1}{n}}\right)$.

Table 6.3

(**Note:** In this discussion, we assume that n is a positive integer and $n \neq 0$.)

For example:

$$\text{Because } 2^4 = 16, \text{ we can say that } \sqrt[4]{16} = 2 \left(\text{or } 2 = 16^{\frac{1}{4}}\right).$$

$$\text{Because } 3^5 = 243, \text{ we can say that } \sqrt[5]{243} = 3 \left(\text{or } 3 = 243^{\frac{1}{5}}\right).$$

The following notation is used for all radical expressions.

Radical Notation

If n is a positive integer and $b^n = a$, then $\boldsymbol{b = \sqrt[n]{a} = a^{\frac{1}{n}}}$ (assuming $\sqrt[n]{a}$ is a real number).

*The expression $\sqrt[n]{a}$ is called a **radical**.*

*The symbol $\sqrt[n]{}$ is called a **radical sign**.*

*n is called the **index**.*

*a is called the **radicand**.*

*(**Note:** If no index is given, it is understood to be 2. For example, $\sqrt{3} = \sqrt[2]{3} = 3^{\frac{1}{2}}$.)*

NOTES

Special Notes about the Index n:

For the expression $\sqrt[n]{a}$ $\left(or\ a^{\frac{1}{n}}\right)$ to be a real number:

1. when a is nonnegative, n can be any index, and

2. when a is negative, n must be odd.

   (If a is negative and n is even, then $\sqrt[n]{a}$ is nonreal.)

Example 1: Evaluating Principal n^{th} Roots

a. $49^{\frac{1}{2}} = \sqrt{49} = 7$, because $7^2 = 49$.

b. $81^{\frac{1}{4}} = \sqrt[4]{81} = 3$, because $3^4 = 81$.

c. $(-8)^{\frac{1}{3}} = \sqrt[3]{-8} = -2$, because $(-2)^3 = -8$.

d. $(0.00001)^{\frac{1}{5}} = \sqrt[5]{0.00001} = 0.1$, because $(0.1)^5 = 0.00001$.

e. $(-16)^{\frac{1}{2}} = \sqrt{-16}$ is not a real number.   (Any even root of a negative number is nonreal.)

Evaluating Roots with a TI-84 Plus Calculator (The $\sqrt[x]{}$ function)

Using a special feature of the TI-84 Plus calculator we can find principal nth roots.

Example 2: Using the TI-84 Plus to find principal nth roots

Use a TI-84 Plus calculator to find the value of each of the following roots accurate to four decimal places.

a. $\sqrt[5]{200}$

Solution: To find $\sqrt[5]{200}$ proceed as follows:

Step 1:   Enter 5. (**Note:** This 5 is the index.)

Step 2:   Press **MATH**.

Step 3:   Choose **5:** $\sqrt[x]{}$.

Step 4:   Enter 200 and press **ENTER**.

The display will appear as follows:

Thus $\sqrt[5]{200} \approx 2.8854$, accurate to four decimal places.

b. $\sqrt[6]{1.25}$

Solution: To find $\sqrt[6]{1.25}$ proceed as follows:

Step 1:   Enter 6. (**Note:** This 6 is the index.)

Step 2:   Press **MATH**.

Step 3:   Choose **5:** $\sqrt[x]{}$.

Step 4:   Enter 1.25 and press **ENTER**.

Continued on the next page...

The display will appear as follows:

Thus $\sqrt[6]{1.25} \approx 1.0379$, accurate to four decimal places.

Rational Exponents of the Form $\dfrac{m}{n}$: $\sqrt[n]{a^m} = a^{\frac{m}{n}}$

In Chapter 4 we discussed the properties of exponents using only integer exponents. These same properties of exponents apply to rational exponents (fractional exponents) as well and are repeated here for easy reference.

Summary of Properties of Exponents

For nonzero real numbers a and b and rational numbers m and n,

The Exponent 1:	$a^1 = a$ (*a is any real number.*)
The Exponent 0:	$a^0 = 1$ $(a \neq 0)$
Product Rule:	$a^m \cdot a^n = a^{m+n}$
Quotient Rule:	$\dfrac{a^m}{a^n} = a^{m-n}$
Negative Exponents:	$a^{-n} = \dfrac{1}{a^n}, \ \dfrac{1}{a^{-n}} = a^n$
Power Rule:	$\left(a^m\right)^n = a^{mn}$
Power Rule for Products:	$(ab)^n = a^n b^n$
Power Rule for Fractions:	$\left(\dfrac{a}{b}\right)^n = \dfrac{a^n}{b^n}$

Now consider the problem of evaluating the expression $8^{\frac{2}{3}}$ where the exponent, $\frac{2}{3}$, is of the form $\frac{m}{n}$. By using the power rule for exponents, we can write

$$8^{\frac{2}{3}} = \left(8^{\frac{1}{3}}\right)^2 = (2)^2 = 4$$

or,

$$8^{\frac{2}{3}} = \left(8^2\right)^{\frac{1}{3}} = (64)^{\frac{1}{3}} = 4.$$

The result is the same with either approach. That is, we can take the cube root first and then square the answer. Or, we can square first and then take the cube root. In general, for an exponent of the form $\frac{m}{n}$, taking the nth root first and then raising this root to the power m is easier because the numbers are smaller.

For example,

$$81^{\frac{3}{4}} = \left(81^{\frac{1}{4}}\right)^3 = (3)^3 = 27$$

is easier to calculate and work with than

$$81^{\frac{3}{4}} = \left(81^3\right)^{\frac{1}{4}} = (531,441)^{\frac{1}{4}} = 27.$$

The fourth root of 81 is more commonly known than the fourth root of 531,441.

The General Form $a^{\frac{m}{n}}$

If n is a positive integer, m is any integer, and $a^{\frac{1}{n}}$ is a real number, then

$$a^{\frac{m}{n}} = \left(a^{\frac{1}{n}}\right)^m = \left(a^m\right)^{\frac{1}{n}}.$$

In radical notation:

$$a^{\frac{m}{n}} = \left(\sqrt[n]{a}\right)^m = \sqrt[n]{a^m}.$$

Example 3: Conversion from Exponential Notation to Radical Notation

Assume that each variable represents a positive real number. Each expression is changed to an equivalent expression in either radical or exponential notation.

a. $x^{\frac{2}{3}} = \sqrt[3]{x^2}$ Note that the index, 3, is the denominator in the rational exponent.

b. $3x^{\frac{4}{5}} = 3\sqrt[5]{x^4}$ Note that the coefficient, 3, is not affected by the exponent.

Continued on the next page...

c. $-a^{\frac{3}{2}} = -\sqrt{a^3}$ Note that -1 is the understood coefficient.

d. $\sqrt[6]{a^5} = a^{\frac{5}{6}}$ Note that the index is the denominator of the rational exponent.

e. $5\sqrt{x} = 5x^{\frac{1}{2}}$ Note that, in a square root, the index is understood to be 2.

f. $-\sqrt[3]{4} = -4^{\frac{1}{3}}$ Note that the coefficient, -1, is not affected by the exponent. Also, we could write $-4^{\frac{1}{3}} = -1 \cdot 4^{\frac{1}{3}}$.

Simplifying Expressions with Rational Exponents

Expressions with rational exponents such as

$$x^{\frac{2}{3}} \cdot x^{\frac{1}{6}}, \quad \frac{x^{\frac{3}{4}}}{x^{\frac{1}{3}}}, \quad \text{and} \quad \left(2a^{\frac{1}{4}}\right)^3$$

can be simplified by using the properties of exponents.

> **NOTES** Unless otherwise stated, we will assume, for the remainder of this chapter, that all variables represent nonnegative real numbers.

Example 4: Simplifying Expressions with Rational Exponents

Each expression is simplified by using one or more of the rules of exponents.

a. $x^{\frac{2}{3}} \cdot x^{\frac{1}{6}} = x^{\frac{2}{3} + \frac{1}{6}}$ Find a common denominator and add the exponents.

$$= x^{\frac{4}{6} + \frac{1}{6}} = x^{\frac{5}{6}}$$

b. $\dfrac{x^{\frac{3}{4}}}{x^{\frac{1}{3}}} = x^{\frac{3}{4} - \frac{1}{3}}$ Find a common denominator and subtract the exponents.

$$= x^{\frac{9}{12} - \frac{4}{12}} = x^{\frac{5}{12}}$$

c. $\left(2a^{\frac{1}{4}}\right)^3 = 2^3 \cdot a^{\frac{1}{4} \cdot 3} = 8a^{\frac{3}{4}}$

d. $\left(27y^{-\frac{9}{10}}\right)^{-\frac{1}{3}} = 27^{-\frac{1}{3}} \cdot y^{-\frac{9}{10}\left(-\frac{1}{3}\right)}$

$\quad = \dfrac{y^{\frac{3}{10}}}{27^{\frac{1}{3}}} = \dfrac{y^{\frac{3}{10}}}{3}$ Multiply the exponents of y and reduce the fraction to $\dfrac{3}{10}$.

e. $(-36)^{-\frac{1}{2}} = \dfrac{1}{(-36)^{\frac{1}{2}}}$ This is not a real number because $(-36)^{\frac{1}{2}} = \sqrt{-36}$ is not real.

f. $9^{\frac{2}{4}} = 9^{\frac{1}{2}} = 3$ The exponent can be reduced as long as the expression is real.

g. $\left(\dfrac{49x^6y^{-2}}{z^{-4}}\right)^{\frac{1}{2}} = \dfrac{\left(49x^6y^{-2}\right)^{\frac{1}{2}}}{\left(z^{-4}\right)^{\frac{1}{2}}}$ **Study this example carefully.**

$\quad = \dfrac{49^{\frac{1}{2}}x^{6 \cdot \frac{1}{2}}y^{-2 \cdot \frac{1}{2}}}{z^{-4 \cdot \frac{1}{2}}}$ Use the power rule three times.

$\quad = \dfrac{7x^3y^{-1}}{z^{-2}}$ Simplify exponents.

$\quad = \dfrac{7x^3z^2}{y}$ Use the properties of negative exponents.

Example 5 shows how to use fractional exponents to simplify rather complicated looking radical expressions. The results may seem surprising at first.

Example 5: Simplifying Radical Notation by Changing to Exponential Notation

Simplify each expression by first changing it into an equivalent expression with rational exponents. Then, rewrite the answer in simplified radical form.

a. $\sqrt[4]{\sqrt[3]{x}} = \left(\sqrt[3]{x}\right)^{\frac{1}{4}} = \left(x^{\frac{1}{3}}\right)^{\frac{1}{4}} = x^{\frac{1}{12}} = \sqrt[12]{x}$ Note that $\dfrac{1}{3} \cdot \dfrac{1}{4} = \dfrac{1}{12}$.

Continued on the next page...

b. $\sqrt[3]{a}\sqrt{a} = a^{\frac{1}{3}} \cdot a^{\frac{1}{2}} = a^{\frac{1}{3}+\frac{1}{2}} = a^{\frac{2}{6}+\frac{3}{6}} = a^{\frac{5}{6}} = \sqrt[6]{a^5}$

c. $\dfrac{\sqrt{x^3}\sqrt[3]{x^2}}{\sqrt[5]{x^2}} = \dfrac{x^{\frac{3}{2}} \cdot x^{\frac{2}{3}}}{x^{\frac{2}{5}}} = \dfrac{x^{\frac{3}{2}+\frac{2}{3}}}{x^{\frac{2}{5}}} = \dfrac{x^{\frac{9}{6}+\frac{4}{6}}}{x^{\frac{2}{5}}} = \dfrac{x^{\frac{13}{6}}}{x^{\frac{2}{5}}}$

$$= x^{\frac{13}{6}-\frac{2}{5}} = x^{\frac{65}{30}-\frac{12}{30}} = x^{\frac{53}{30}}$$

$$= x^{\frac{30}{30}} \cdot x^{\frac{23}{30}} = x\sqrt[30]{x^{23}} \qquad\qquad \text{Note that } \frac{53}{30} = \frac{30}{30} + \frac{23}{30} = 1 + \frac{23}{30}.$$

Evaluating Roots with a TI-84 Plus Calculator (The ⌃ key)

The caret key ⌃ on the TI-84 Plus calculator (and most graphing calculators) is used to indicate exponents. By using this key, roots of real numbers can be calculated with up to nine digit accuracy. To set the number of decimal places you wish in any calculations, press the **MODE** key and highlight the digit opposite the word **FLOAT** that indicates the desired accuracy. If no digit is highlighted, then the accuracy will be to nine decimal places (in some cases ten decimal places).

To Find the Value of $a^{\frac{m}{n}}$ with a Calculator

1. *Enter the value of the base, a.*

2. *Press the caret key* ⌃ .

3. *Enter the fractional exponent enclosed in parentheses. (This exponent may be positive or negative.)*

4. *Press* **ENTER**.

Example 6: Using the TI-84 Plus

a. $125^{\frac{4}{3}}$

Solution: To find $125^{\frac{4}{3}}$ proceed as follows:

Step 1:   Enter the base, 125.

Step 2:   Press the caret key  .

Step 3:   Enter the exponent in parentheses, $\frac{4}{3}$.

Step 4:   Press ENTER .

The display should read as follows:

```
125^(4/3)
                    625
```

b. $36^{\frac{3}{5}}$

Solution: To find $36^{\frac{3}{5}}$ proceed as follows:

Step 1:   Enter the base, 36.

Step 2:   Press the caret key  .

Step 3:   Enter the exponent in parentheses, $\frac{3}{5}$.

Step 4:   Press ENTER .

The display should read as follows:

```
36^(3/5)
            8.585814487
```

Practice Problems

Simplify each of the following expressions. Leave the answers with rational exponents.

1. $64^{\frac{2}{3}}$ **2.** $x^{\frac{3}{4}} \cdot x^{\frac{1}{5}} \cdot x^{\frac{1}{2}}$ **3.** $\dfrac{x^{\frac{1}{6}} \cdot y^{\frac{1}{2}}}{x^{\frac{1}{3}} \cdot y^{\frac{1}{4}}}$ **4.** $\left[16^{\frac{3}{4}} \right]^{-2}$ **5.** $-81^{\frac{1}{4}}$

Simplify each expression by first changing to an equivalent expression with rational exponents. Rewrite the answer in simplified radical form.

6. $\sqrt[4]{x} \cdot \sqrt{x}$ **7.** $\sqrt[5]{\sqrt[3]{a^2}}$ **8.** $\dfrac{\sqrt{36x^5}}{\sqrt[3]{8x^3}}$

Use a graphing calculator to find the following values accurate to 4 decimal places.

9. $128^{\frac{1}{5}}$ **10.** $100^{-\frac{1}{4}}$

6.2 Exercises

Simplify each numerical expression in Exercises 1 – 25.

1. $9^{\frac{1}{2}}$ **2.** $121^{\frac{1}{2}}$ **3.** $100^{-\frac{1}{2}}$ **4.** $25^{-\frac{1}{2}}$ **5.** $-64^{\frac{3}{2}}$

6. $(-64)^{\frac{3}{2}}$ **7.** $\left(-\dfrac{4}{25}\right)^{\frac{1}{2}}$ **8.** $-\left(\dfrac{4}{25}\right)^{\frac{1}{2}}$ **9.** $\left(\dfrac{9}{49}\right)^{\frac{1}{2}}$ **10.** $\left(\dfrac{225}{144}\right)^{\frac{1}{2}}$

11. $(-64)^{\frac{1}{3}}$ **12.** $64^{\frac{2}{3}}$ **13.** $(-125)^{\frac{1}{3}}$ **14.** $(-216)^{-\frac{1}{3}}$ **15.** $8^{-\frac{2}{3}}$

16. $\left(\dfrac{8}{125}\right)^{-\frac{1}{3}}$ **17.** $-\left(\dfrac{16}{81}\right)^{-\frac{3}{4}}$ **18.** $\left(-\dfrac{1}{32}\right)^{\frac{2}{5}}$ **19.** $\left(\dfrac{27}{64}\right)^{\frac{2}{3}}$ **20.** $3 \cdot 16^{-\frac{3}{4}}$

21. $2 \cdot 25^{-\frac{1}{2}}$ **22.** $-100^{-\frac{3}{2}}$ **23.** $\left[(-27)^{\frac{2}{3}} \right]^{-2}$ **24.** $\left[\left(\dfrac{1}{32}\right)^{\frac{2}{5}} \right]^{-3}$ **25.** $-49^{-\frac{5}{2}}$

Answers to Practice Problems: **1.** 16 **2.** $x^{\frac{29}{20}}$ **3.** $\dfrac{y^{\frac{1}{4}}}{x^{\frac{1}{6}}}$ **4.** $\dfrac{1}{64}$ **5.** -3 **6.** $\sqrt[4]{x^3}$ **7.** $\sqrt[15]{a^2}$ **8.** $3x\sqrt{x}$

         9. 2.6390 **10.** 0.3162

In Exercises 26 – 40, use a graphing calculator to find the value of each numerical expression accurate to 4 decimal places, if necessary.

26. $25^{\frac{2}{3}}$ **27.** $81^{\frac{7}{4}}$ **28.** $100^{\frac{7}{2}}$ **29.** $100^{\frac{1}{3}}$ **30.** $250^{\frac{5}{6}}$

31. $18^{-\frac{3}{2}}$ **32.** $24^{-\frac{3}{4}}$ **33.** $2000^{\frac{2}{3}}$ **34.** $\sqrt[4]{0.0025}$ **35.** $\sqrt[4]{3600}$

36. $\sqrt[5]{35.4}$ **37.** $\sqrt[10]{1.8}$ **38.** $\sqrt[6]{4500}$ **39.** $\sqrt[5]{0.00032}$ **40.** $\sqrt[9]{72}$

Simplify each algebraic expression in Exercises 41 – 77. Assume that all variables are positive. Leave the answers in rational exponent form.

41. $\left(2x^{\frac{1}{3}}\right)^{3}$ **42.** $\left(3x^{\frac{1}{2}}\right)^{4}$ **43.** $\left(9a^{4}\right)^{\frac{1}{2}}$ **44.** $\left(16a^{3}\right)^{-\frac{1}{4}}$

45. $8x^{2}\cdot x^{\frac{1}{2}}$ **46.** $3x^{3}\cdot x^{\frac{2}{3}}$ **47.** $5a^{2}\cdot a^{-\frac{1}{3}}\cdot a^{\frac{1}{2}}$ **48.** $a^{\frac{2}{3}}\cdot a^{-\frac{3}{5}}\cdot a^{0}$

49. $\dfrac{a^{2}}{a^{\frac{2}{5}}}$ **50.** $\dfrac{x^{\frac{3}{4}}}{x^{-\frac{1}{6}}}$ **51.** $\dfrac{x^{\frac{2}{5}}}{x^{-\frac{1}{10}}}$ **52.** $\dfrac{a^{\frac{2}{3}}}{a^{\frac{1}{9}}}$

53. $\dfrac{a^{\frac{1}{2}}}{a^{-\frac{2}{3}}}$ **54.** $\dfrac{a^{\frac{3}{4}}\cdot a^{\frac{1}{8}}}{a^{2}}$ **55.** $\dfrac{a^{\frac{1}{2}}\cdot a^{-\frac{3}{4}}}{a^{-\frac{1}{2}}}$ **56.** $\dfrac{x^{\frac{2}{3}}\cdot x^{\frac{4}{3}}}{x^{2}}$

57. $\dfrac{x^{\frac{2}{3}}y}{x^{2}y^{\frac{1}{2}}}$ **58.** $\dfrac{a^{\frac{3}{2}}b^{\frac{4}{5}}}{a^{-\frac{1}{2}}b^{2}}$ **59.** $\dfrac{a^{\frac{3}{4}}b^{-\frac{1}{3}}}{a^{\frac{3}{2}}b^{\frac{1}{6}}}$ **60.** $\left(2x^{\frac{1}{2}}y^{\frac{1}{3}}\right)^{3}$

61. $\left(4x^{-\frac{3}{4}}y^{\frac{1}{5}}\right)^{-2}$ **62.** $\left(a^{\frac{1}{2}}a^{\frac{1}{3}}\right)^{6}$ **63.** $\left(-x^{3}y^{6}z^{-6}\right)^{\frac{2}{3}}$ **64.** $\left(\dfrac{x^{2}y^{-3}}{z^{4}}\right)^{-\frac{1}{2}}$

65. $\left(\dfrac{27a^{3}b^{6}}{c^{9}}\right)^{-\frac{1}{3}}$ **66.** $\left(81a^{-8}b^{2}\right)^{-\frac{1}{4}}$ **67.** $\left(\dfrac{16a^{-4}b^{3}}{c^{4}}\right)^{\frac{3}{4}}$ **68.** $\left(\dfrac{-27a^{2}b^{3}}{c^{-3}}\right)^{\frac{1}{3}}$

69. $\dfrac{\left(x^{\frac{1}{4}}y^{\frac{1}{2}}\right)^{3}}{x^{\frac{1}{2}}y^{\frac{1}{4}}}$ **70.** $\dfrac{\left(x^{\frac{1}{2}}y\right)^{-\frac{1}{3}}}{x^{\frac{2}{3}}y^{-1}}$ **71.** $\dfrac{\left(8x^{2}y\right)^{-\frac{1}{3}}}{\left(5x^{\frac{1}{3}}y^{-\frac{1}{2}}\right)^{2}}$ **72.** $\dfrac{\left(25a^{4}b^{-1}\right)^{\frac{1}{2}}}{\left(2a^{\frac{1}{5}}b^{\frac{3}{5}}\right)^{3}}$

73. $\left(\dfrac{5a^{-3}}{21b^{2}}\right)^{-1}\cdot\left(\dfrac{49a^{4}}{100b^{-8}}\right)^{-\frac{1}{2}}$ **74.** $\left(\dfrac{a^{-3}b^{\frac{1}{3}}}{a^{\frac{1}{2}}b}\right)^{\frac{1}{2}}\cdot\left(\dfrac{ab^{\frac{1}{2}}}{a^{-\frac{2}{3}}b^{-1}}\right)^{\frac{1}{2}}$

75. $\left(\dfrac{x^2 y^{\frac{1}{3}}}{x^{\frac{1}{2}} y^{\frac{3}{2}}}\right)^{\frac{1}{2}} \cdot \left(\dfrac{x^{-\frac{1}{2}} y^{\frac{2}{3}}}{x^{-1} y^{\frac{3}{4}}}\right)^2$ **76.** $\left(\dfrac{a^3 b^{-2}}{ab^4}\right)^{\frac{1}{6}} \cdot \left(\dfrac{a^{\frac{1}{5}} b^{\frac{1}{3}}}{a^{-\frac{1}{2}}}\right)^3$ **77.** $\dfrac{\left(27xy^{\frac{1}{2}}\right)^{\frac{1}{3}}}{\left(25x^{-\frac{1}{2}}y\right)^{\frac{1}{2}}} \cdot \dfrac{\left(x^{\frac{1}{2}}y\right)^{\frac{1}{6}}}{\left(16x^{\frac{1}{3}}y\right)^{\frac{1}{2}}}$

In Exercises 78 – 91, simplify each expression by first changing it into an equivalent expression with rational exponents. Rewrite the answer in simplified radical form.

78. $\sqrt[3]{a} \cdot \sqrt{a}$ **79.** $\sqrt[3]{x^2} \cdot \sqrt[5]{x^3}$ **80.** $\dfrac{\sqrt[4]{y^3}}{\sqrt[6]{y}}$ **81.** $\dfrac{\sqrt[3]{x^4}}{\sqrt[4]{x}}$

82. $\dfrac{\sqrt[3]{x^2}\,\sqrt[5]{x^6}}{\sqrt{x^3}}$ **83.** $\dfrac{a\sqrt[4]{a}}{\sqrt[3]{a}\sqrt{a}}$ **84.** $\sqrt{\sqrt[3]{y}}$ **85.** $\sqrt[5]{\sqrt{x}}$

86. $\sqrt[3]{\sqrt[3]{x}}$ **87.** $\sqrt{\sqrt{a}}$ **88.** $\sqrt[15]{(7a)^5}$ **89.** $\left(\sqrt[4]{a^3 b^6 c}\right)^{12}$

90. $\sqrt[5]{\sqrt[4]{\sqrt[3]{x}}}$ **91.** $\left(\sqrt[3]{a^4 bc^2}\right)^{15}$

Writing and Thinking About Mathematics

92. Is $\sqrt[5]{a} \cdot \sqrt{a}$ the same as $\sqrt[5]{a^2}$? Explain why or why not.

 HAWKES LEARNING SYSTEMS: INTERMEDIATE ALGEBRA SOFTWARE

- Rational Exponents

6.3

Operations with Radicals

- *Perform arithmetic operations with radical expressions.*
- *Rationalize the denominators of radicals.*
- *Use a TI-84 Plus graphing calculator to evaluate radical expressions.*

Addition and Subtraction with Radical Expressions

Recall that to find the sum $2x^2 + 3x^2 - 8x^2$, you can use the distributive property and write

$$2x^2 + 3x^2 - 8x^2 = (2 + 3 - 8)x^2$$
$$= -3x^2.$$

Recall that the terms $2x^2, 3x^2$, and $-8x^2$ are called **like terms** because each term contains the same variable expression, x^2. Similarly,

$$2\sqrt{5} + 3\sqrt{5} - 8\sqrt{5} = (2 + 3 - 8)\sqrt{5}$$
$$= -3\sqrt{5}$$

and $2\sqrt{5}, 3\sqrt{5}$, and $-8\sqrt{5}$ are called **like radicals** because each term contains the same radical expression, $\sqrt{5}$. **Like radicals** have the same index and radicand or they can be simplified so that they have the same index and radicand.

The terms $2\sqrt{3}$ and $2\sqrt{7}$ are **not** like radicals because the radical expressions are not the same, and neither expression can be simplified. Therefore, a sum such as

$$2\sqrt{3} + 2\sqrt{7}$$

cannot be simplified. That is, the terms cannot be combined.

In some cases, radicals that are not like radicals can be simplified, and the results may lead to like radicals. For example, $4\sqrt{12}, \sqrt{75}$, and $-\sqrt{108}$ are not like radicals. However, simplification of each radical allows the sum of these radicals to be found as follows:

$$4\sqrt{12} + \sqrt{75} - \sqrt{108} = 4\sqrt{4 \cdot 3} + \sqrt{25 \cdot 3} - \sqrt{36 \cdot 3}$$
$$= 4 \cdot 2\sqrt{3} + 5\sqrt{3} - 6\sqrt{3}$$
$$= 8\sqrt{3} + 5\sqrt{3} - 6\sqrt{3}$$
$$= (8 + 5 - 6)\sqrt{3}$$
$$= 7\sqrt{3}.$$

Example 1: Addition and Subtraction with Radicals

Perform the indicated operation and simplify, if possible. Assume that all variables are positive.

a. $\sqrt{32x} + \sqrt{18x}$

Solution:
$$\sqrt{32x} + \sqrt{18x} = \sqrt{16 \cdot 2x} + \sqrt{9 \cdot 2x}$$
$$= 4\sqrt{2x} + 3\sqrt{2x}$$
$$= (4+3)\sqrt{2x}$$
$$= 7\sqrt{2x}$$

b. $\sqrt{12} + \sqrt{18} + \sqrt{27}$

Solution:
$$\sqrt{12} + \sqrt{18} + \sqrt{27} = \sqrt{4 \cdot 3} + \sqrt{9 \cdot 2} + \sqrt{9 \cdot 3}$$
$$= 2\sqrt{3} + 3\sqrt{2} + 3\sqrt{3}$$
$$= (2+3)\sqrt{3} + 3\sqrt{2}$$
$$= 5\sqrt{3} + 3\sqrt{2}$$

Note that $\sqrt{3}$ and $\sqrt{2}$ are **not** like radicals. Therefore, the last expression cannot be simplified.

c. $\sqrt[3]{5x} - \sqrt[3]{40x}$

Solution:
$$\sqrt[3]{5x} - \sqrt[3]{40x} = \sqrt[3]{5x} - \sqrt[3]{8 \cdot 5x}$$
$$= \sqrt[3]{5x} - 2\sqrt[3]{5x}$$
$$= (1-2)\sqrt[3]{5x}$$
$$= -\sqrt[3]{5x}$$

d. $x\sqrt{4y^3} - 5\sqrt{x^2 y^3}$

Solution:
$$x\sqrt{4y^3} - 5\sqrt{x^2 y^3} = x\sqrt{4y^2}\sqrt{y} - 5\sqrt{x^2 y^2}\sqrt{y}$$
$$= 2xy\sqrt{y} - 5xy\sqrt{y}$$
$$= -3xy\sqrt{y}$$

Multiplication with Radical Expressions

To find a product such as $\left(\sqrt{3}+5\right)\left(\sqrt{3}-7\right)$ treat the two expressions as two binomials and multiply just as with polynomials. For example, using the FOIL method, we get

$$\left(\sqrt{3}+5\right)\left(\sqrt{3}-7\right) = \left(\sqrt{3}\right)^2 - 7\sqrt{3} + 5\sqrt{3} + 5(-7)$$
$$= 3 - 7\sqrt{3} + 5\sqrt{3} - 35$$
$$= 3 - 35 + (-7+5)\sqrt{3}$$
$$= -32 - 2\sqrt{3}.$$

Example 2: Multiplication of Radicals

Multiply and simplify the following expressions.

a. $\sqrt{5} \cdot \sqrt{15}$

Solution: $\sqrt{5} \cdot \sqrt{15} = \sqrt{5} \cdot \sqrt{5} \cdot \sqrt{3} = \left(\sqrt{5}\right)^2 \cdot \sqrt{3} = 5\sqrt{3}$

b. $\left(3\sqrt{7} - 2\right)\left(\sqrt{7} + 3\right)$

Solution:
$$\left(3\sqrt{7} - 2\right)\left(\sqrt{7} + 3\right) = 3\left(\sqrt{7}\right)^2 + 3 \cdot 3\sqrt{7} - 2\sqrt{7} - 2 \cdot 3$$
$$= 3 \cdot 7 + 9\sqrt{7} - 2\sqrt{7} - 6$$
$$= 21 - 6 + (9 - 2)\sqrt{7}$$
$$= 15 + 7\sqrt{7}$$

c. $\left(\sqrt{6} + \sqrt{2}\right)^2$

Solution:
$$\left(\sqrt{6} + \sqrt{2}\right)^2 = \left(\sqrt{6}\right)^2 + 2\sqrt{6}\sqrt{2} + \left(\sqrt{2}\right)^2 \qquad (a+b)^2 = a^2 + 2ab + b^2$$
$$= 6 + 2\sqrt{12} + 2$$
$$= 8 + 2\sqrt{4}\sqrt{3}$$
$$= 8 + 2 \cdot 2\sqrt{3}$$
$$= 8 + 4\sqrt{3}$$

d. $\left(\sqrt{2x} + 5\right)\left(\sqrt{2x} - 5\right)$

Solution:
$$\left(\sqrt{2x} + 5\right)\left(\sqrt{2x} - 5\right) = \left(\sqrt{2x}\right)^2 - (5)^2 \qquad (a+b)(a-b) = a^2 - b^2$$
$$= 2x - 25$$

Rationalizing Denominators of Rational Expressions

An expression with a radical in the denominator may not be in the most usable form for further algebraic manipulation or operations. In many situations, to make calculations easier, we may want to **rationalize the denominator**. That is, we want to find an equivalent fraction in which the denominator does not have a radical. The numerator may still have a radical in it, but a rational expression with no radicals in the denominator definitely makes arithmetic with radicals much easier. [**Note:** In this section we will deal only with radicals that involve square roots or cube roots, however the ideas discussed here also apply to n^{th} roots in general.]

To Rationalize a Denominator Containing a Square Root or a Cube Root

1. *If the denominator contains a square root, multiply both the numerator and denominator by an expression that will give a denominator with no square roots.*

2. *If the denominator contains a cube root, multiply both the numerator and denominator by an expression that will give a denominator with no cube roots.*

Example 3: Rationalizing the Denominator

Simplify each radical expression so that the denominator contains no radicals. Assume all variables are positive.

a. $\sqrt{\dfrac{5}{12x}}$

Solution: Multiply the numerator and denominator by $\sqrt{3x}$ because $12x \cdot 3x = 36x^2$, a perfect square expression.

$$\sqrt{\frac{5}{12x}} = \frac{\sqrt{5}}{\sqrt{12x}} = \frac{\sqrt{5}}{\sqrt{12x}} \cdot \frac{\sqrt{3x}}{\sqrt{3x}} = \frac{\sqrt{15x}}{\sqrt{36x^2}} = \frac{\sqrt{15x}}{6x}$$

b. $\dfrac{7}{\sqrt[3]{32y}}$

Solution: Multiply the numerator and denominator by $\sqrt[3]{2y^2}$ because $32y \cdot 2y^2 = 64y^3$, a perfect cube expression.

$$\frac{7}{\sqrt[3]{32y}} = \frac{7}{\sqrt[3]{32y}} \cdot \frac{\sqrt[3]{2y^2}}{\sqrt[3]{2y^2}} = \frac{7\sqrt[3]{2y^2}}{\sqrt[3]{64y^3}} = \frac{7\sqrt[3]{2y^2}}{4y}$$

If the denominator has a radical expression with **a sum or difference involving square roots** such as

$$\frac{2}{4-\sqrt{2}} \qquad \text{or} \qquad \frac{12}{3+\sqrt{5}},$$

then a different method is used for rationalizing the denominator. In this method, we think of the denominator in the form of $a - b$ or $a + b$. Thus,

$$\text{if } a-b = 4-\sqrt{2}, \text{ then } a+b = 4+\sqrt{2}$$

and

$$\text{if } a+b = 3+\sqrt{5}, \text{ then } a-b = 3-\sqrt{5}.$$

The two expressions $(a - b)$ and $(a + b)$ are called **conjugates** of each other, and, as we know, their product, $(a - b)(a + b)$, results in the **difference of two squares**:

$$(a-b)(a+b) = a^2 - b^2.$$

Thus, if a or b (or both) contain a square root, the difference of the squares will **not** contain a square root. With this in mind, we can proceed as follows to rationalize a denominator with a sum or difference involving square roots.

To Rationalize a Denominator Containing a Sum or Difference Involving Square Roots

*If the denominator of a fraction contains a sum or difference involving a square root, rationalize the denominator by multiplying both the numerator and denominator by the **conjugate of the denominator**.*

1. *If the denominator is of the form $a - b$, multiply both the numerator and denominator by $a + b$.*

2. *If the denominator is of the form $a + b$, multiply both the numerator and denominator by $a - b$.*

The new denominator will be the difference of two squares and therefore not contain a radical term.

Example 4: Rationalizing a Denominator with a Sum or Difference Involving a Square Root

Simplify each expression by rationalizing the denominator.

a. $\dfrac{2}{4 - \sqrt{2}}$

Solution: Multiply the numerator and denominator by $4 + \sqrt{2}$.

$$\frac{2}{4 - \sqrt{2}} = \frac{2\left(4 + \sqrt{2}\right)}{\left(4 - \sqrt{2}\right)\left(4 + \sqrt{2}\right)}$$ If $a - b = 4 - \sqrt{2}$, then $a + b = 4 + \sqrt{2}$.

$$= \frac{2\left(4 + \sqrt{2}\right)}{4^2 - \left(\sqrt{2}\right)^2}$$ The denominator is the difference of two squares.

$$= \frac{2\left(4 + \sqrt{2}\right)}{16 - 2}$$ The denominator is a rational number. Note that the numerator is now irrational. However, this is generally preferred to having an irrational denominator.

$$= \frac{\overset{1}{\cancel{2}}\left(4 + \sqrt{2}\right)}{\underset{7}{\cancel{14}}}$$

$$= \frac{4 + \sqrt{2}}{7}$$

Continued on the next page...

b. $\dfrac{31}{6+\sqrt{5}}$

Solution: Multiply the numerator and denominator by $6-\sqrt{5}$.

$$\dfrac{31}{6+\sqrt{5}}=\dfrac{31\left(6-\sqrt{5}\right)}{\left(6+\sqrt{5}\right)\left(6-\sqrt{5}\right)}$$

$$=\dfrac{31\left(6-\sqrt{5}\right)}{36-5}$$

$$=\dfrac{\cancel{31}\left(6-\sqrt{5}\right)}{\cancel{31}}$$

$$=6-\sqrt{5}$$

c. $\dfrac{1}{\sqrt{7}-\sqrt{2}}$

Solution: Multiply the numerator and denominator by $\sqrt{7}+\sqrt{2}$.

$$\dfrac{1}{\sqrt{7}-\sqrt{2}}=\dfrac{1\left(\sqrt{7}+\sqrt{2}\right)}{\left(\sqrt{7}-\sqrt{2}\right)\left(\sqrt{7}+\sqrt{2}\right)}$$

$$=\dfrac{\left(\sqrt{7}+\sqrt{2}\right)}{7-2}$$

$$=\dfrac{\sqrt{7}+\sqrt{2}}{5}$$

d. $\dfrac{6}{1+\sqrt{x}}$

Solution: $\dfrac{6}{1+\sqrt{x}}=\dfrac{6\left(1-\sqrt{x}\right)}{\left(1+\sqrt{x}\right)\left(1-\sqrt{x}\right)}$

$$=\dfrac{6\left(1-\sqrt{x}\right)}{1-x}$$

e. $\dfrac{x-y}{\sqrt{x}-\sqrt{y}}$

Solution: $\dfrac{x-y}{\sqrt{x}-\sqrt{y}}=\dfrac{\left(x-y\right)\left(\sqrt{x}+\sqrt{y}\right)}{\left(\sqrt{x}-\sqrt{y}\right)\left(\sqrt{x}+\sqrt{y}\right)}$

$$=\dfrac{\cancel{\left(x-y\right)}\left(\sqrt{x}+\sqrt{y}\right)}{\cancel{\left(x-y\right)}}$$

$$=\sqrt{x}+\sqrt{y}$$

Evaluating Radical Expressions with a TI-84 Plus Graphing Calculator

Techniques for using a TI-84 Plus graphing calculator to evaluate radical expressions and expressions with rational exponents were illustrated in Sections 6.1 and 6.2. These same basic techniques are used to evaluate numerical expressions that contain sums, differences, products, and quotients of radicals. Be careful to use parentheses to ensure that the rules for order of operations are maintained. **In particular, sums and differences in numerators and denominators of fractions must be enclosed in parentheses.** Study the following examples carefully.

Example 5: Using a TI-84 Graphing Calculator to Evaluate Radical Expressions

Use a TI-84 Plus graphing calculator to evaluate each expression. Round answers to 4 decimal places.

a. $3 + 2\sqrt{5}$

Solution: The display should appear as follows:

Thus $3 + 2\sqrt{5} = 7.4721359... \approx 7.4721$ Rounded to 4 decimal places.

b. $\left(\sqrt{2} + 5\right)\left(\sqrt{2} - 5\right)$

Solution: The display should appear as follows:

Note that the right parenthesis on 2 must be included. Otherwise, the calculator will interpret the expression as $\sqrt{(2 + 5)}$ $\left(\text{or } \sqrt{7}\right)$ which is not intended.

Thus $\left(\sqrt{2} + 5\right)\left(\sqrt{2} - 5\right) = -23$.

Continued on the next page...

c. $\dfrac{3}{\sqrt{6}-\sqrt{2}}$

Solution: The display should appear as follows:

Count the parentheses in pairs.

Thus $\dfrac{3}{\sqrt{6}-\sqrt{2}} = 2.8977774... \approx 2.8978.$ Rounded to 4 decimal places.

Practice Problems

Simplify each expression. Assume that all variables are positive.

1. $2\sqrt{10}-6\sqrt{10}$

2. $\sqrt{5}+\sqrt{45}-\sqrt{15}$

3. $\sqrt{8x}-3\sqrt{2x}+\sqrt{18x}$

4. $\sqrt[3]{x^5}+x\sqrt[3]{27x^2}$

5. $\sqrt{\dfrac{3}{8a^2}}$

6. $\dfrac{\sqrt[3]{4ab}}{\sqrt[3]{2a^2b^4}}$

7. $\dfrac{4}{\sqrt{2}+\sqrt{6}}$

8. $\dfrac{x-5}{\sqrt{x}-\sqrt{5}}$

9. $\left(\sqrt{3}+\sqrt{2}\right)^2$

10. $\left(3+\sqrt{2}\right)^2$

11. $\left(\sqrt{3}+\sqrt{8}\right)^2$

12. $\dfrac{\sqrt{5}-3\sqrt{2}}{\sqrt{6}+\sqrt{10}}$

Answers to Practice Problems: **1.** $-4\sqrt{10}$ **2.** $4\sqrt{5}-\sqrt{15}$ **3.** $2\sqrt{2x}$ **4.** $4x\sqrt[3]{x^2}$ **5.** $\dfrac{\sqrt{6}}{4a}$ **6.** $\dfrac{\sqrt[3]{2a^2}}{ab}$

7. $\sqrt{6}-\sqrt{2}$ **8.** $\sqrt{x}+\sqrt{5}$ **9.** $5+2\sqrt{6}$ **10.** $11+6\sqrt{2}$ **11.** $11+4\sqrt{6}$

12. $\dfrac{5\sqrt{2}+6\sqrt{3}-6\sqrt{5}-\sqrt{30}}{4}$

6.3 Exercises

Perform the indicated operations and simplify for Exercises 1 – 45. Assume that all variables are positive.

1. $\sqrt{2} - 7\sqrt{2}$

2. $6\sqrt{11} + 4\sqrt{11} - 3\sqrt{11}$

3. $2\sqrt{x} + 4\sqrt{x} - \sqrt{x}$

4. $8\sqrt[3]{xy} - 3\sqrt[3]{xy} + 4\sqrt[3]{xy}$

5. $9\sqrt[3]{7x^2} - 4\sqrt[3]{7x^2} - 8\sqrt[3]{7x^2}$

6. $12\sqrt[3]{4x} - 10\sqrt[3]{4x} - 6\sqrt[3]{4x}$

7. $2\sqrt{3} + 4\sqrt{12}$

8. $2\sqrt{48} - 3\sqrt{75}$

9. $2\sqrt{18} + \sqrt{8} - 3\sqrt{50}$

10. $2\sqrt{12} + \sqrt{72} - \sqrt{75}$

11. $5\sqrt{48} + 2\sqrt{45} - 3\sqrt{20}$

12. $3\sqrt{28} - \sqrt{63} + 8\sqrt{10}$

13. $2\sqrt{96} + \sqrt{147} - \sqrt{150}$

14. $7\sqrt{12x} - 4\sqrt{27x} + \sqrt{108x}$

15. $6\sqrt{45x^3} + \sqrt{80x^3} - \sqrt{20x^3}$

16. $2\sqrt{18xy^2} + \sqrt{8xy^2} - 3y\sqrt{50x}$

17. $\sqrt{125} - \sqrt{63} + 3\sqrt{45}$

18. $5\sqrt{48} + 2\sqrt{24} - \sqrt{75}$

19. $\sqrt{32x} + 7\sqrt{12x} + \sqrt{98x}$

20. $\sqrt[3]{81x^2} - 5\sqrt[3]{48x^2} - 5\sqrt[3]{24x^2}$

21. $\sqrt[3]{16} - 5\sqrt[3]{54} + 2\sqrt[3]{40}$

22. $x\sqrt{y} + \sqrt{x^2 y} - \sqrt{xy^3}$

23. $x\sqrt{2x^3} - 3\sqrt{8x^5} + x\sqrt{72x^3}$

24. $x\sqrt{y^3} - 2\sqrt{x^2 y^3} - y\sqrt{x^2 y}$

25. $x\sqrt{9x^3 y^2} - 5x^2\sqrt{xy^2} + 6y\sqrt{x^5}$

26. $\left(3 + \sqrt{2}\right)\left(5 - \sqrt{2}\right)$

27. $\left(\sqrt{3x} - 8\right)\left(\sqrt{3x} - 1\right)$

28. $\left(2\sqrt{7} + 4\right)\left(\sqrt{7} - 3\right)$

29. $\left(6 + \sqrt{2x}\right)\left(4 + \sqrt{2x}\right)$

30. $\left(5\sqrt{3} - 2\right)\left(2\sqrt{3} - 7\right)$

31. $\left(\sqrt{6} + 2\right)\left(\sqrt{6} - 2\right)$

32. $\left(3\sqrt{2} + \sqrt{5}\right)\left(\sqrt{2} + \sqrt{5}\right)$

33. $\left(\sqrt{5} + 2\sqrt{2}\right)^2$

34. $\left(2\sqrt{5} + 3\sqrt{2}\right)^2$

35. $\left(3\sqrt{5} + 4\sqrt{3}\right)\left(3\sqrt{5} - 4\sqrt{3}\right)$

36. $\left(\sqrt{2} + \sqrt{3}\right)\left(\sqrt{5} - \sqrt{3}\right)$

37. $\left(\sqrt{6} + \sqrt{5}\right)\left(\sqrt{6} - \sqrt{2}\right)$

38. $\left(3\sqrt{7} + \sqrt{5}\right)\left(3\sqrt{7} - \sqrt{5}\right)$

39. $\left(\sqrt{11} + \sqrt{3}\right)\left(\sqrt{11} - 2\sqrt{3}\right)$

40. $\left(\sqrt{x} + \sqrt{6}\right)\left(\sqrt{x} - 3\sqrt{6}\right)$

41. $\left(7\sqrt{x} + \sqrt{2}\right)\left(7\sqrt{x} - \sqrt{2}\right)$

42. $\left(\sqrt{x} + 5\sqrt{y}\right)^2$

43. $\left(3\sqrt{x} + \sqrt{y}\right)^2$

44. $\left(4\sqrt{x} + 3\sqrt{y}\right)\left(\sqrt{x} - 3\sqrt{y}\right)$

45. $\left(2\sqrt{2x} + \sqrt{y}\right)\left(3\sqrt{2x} + 2\sqrt{y}\right)$

In Exercises 46 – 73, rationalize the denominator and simplify if possible.

46. $\dfrac{\sqrt{14}}{\sqrt{6}}$

47. $\dfrac{\sqrt{12}}{\sqrt{21}}$

48. $\dfrac{\sqrt[3]{35}}{\sqrt[3]{14}}$

49. $\dfrac{\sqrt[3]{10}}{\sqrt[3]{15}}$

50. $-\sqrt{\dfrac{2}{3y}}$

51. $-\sqrt{\dfrac{25}{x^3}}$

52. $\dfrac{\sqrt{5y^2}}{\sqrt{8x}}$

53. $\dfrac{\sqrt{4x}}{\sqrt{3y^2}}$

54. $\dfrac{\sqrt{16y^2}}{\sqrt{2y^3}}$

55. $\dfrac{\sqrt{24a^3b}}{\sqrt{6ab^2}}$

56. $\sqrt[3]{\dfrac{2y^3}{27x^2}}$

57. $\sqrt[3]{\dfrac{7x}{2y^4}}$

58. $\sqrt[3]{\dfrac{6a^2}{25b}}$

59. $\dfrac{\sqrt[3]{x^5}}{\sqrt[3]{9xy}}$

60. $\dfrac{1}{\sqrt{2}+1}$

61. $\dfrac{3}{\sqrt{3}-5}$

62. $\dfrac{\sqrt{3}}{\sqrt{5}-4}$

63. $\dfrac{\sqrt{6}}{\sqrt{7}+3}$

64. $\dfrac{2}{\sqrt{2}+\sqrt{3}}$

65. $\dfrac{8}{\sqrt{5}-\sqrt{3}}$

66. $\dfrac{\sqrt{5}}{\sqrt{7}-\sqrt{3}}$

67. $\dfrac{\sqrt{10}}{\sqrt{5}-2\sqrt{2}}$

68. $\dfrac{2-\sqrt{6}}{\sqrt{6}-3}$

69. $\dfrac{7+2\sqrt{5}}{7-\sqrt{5}}$

70. $\dfrac{\sqrt{3}+\sqrt{7}}{\sqrt{3}-\sqrt{7}}$

71. $\dfrac{2\sqrt{3}+\sqrt{2}}{\sqrt{3}-\sqrt{2}}$

72. $\dfrac{2\sqrt{x}-y}{\sqrt{x}-y}$

73. $\dfrac{x+2\sqrt{y}}{x-2\sqrt{y}}$

*In Exercises 74 – 83, rationalize the numerator (by using the same technique used to rationalize the denominator in Exercises 46 – 73) and simplify if possible. (**Note:** This type of exercise occurs in higher level mathematics courses such as calculus.)*

74. $\dfrac{\sqrt{7}-2}{3}$

75. $\dfrac{\sqrt{5}+1}{2}$

76. $\dfrac{\sqrt{15}+\sqrt{3}}{6}$

77. $\dfrac{\sqrt{10}-\sqrt{2}}{-8}$

78. $\dfrac{\sqrt{x}+\sqrt{5}}{x-5}$

79. $\dfrac{\sqrt{y}-\sqrt{2}}{y-2}$

80. $\dfrac{\sqrt{2y}-\sqrt{x}}{x}$

81. $\dfrac{3\sqrt{x}-y}{3x}$

82. $\dfrac{\sqrt{2+h}-\sqrt{2}}{h}$

83. $\dfrac{\sqrt{5+h}+\sqrt{5}}{h}$

In Exercises 84 – 93, use a graphing calculator to find the value of each expression accurate to 4 decimal places.

84. $13-\sqrt{75}$

85. $5-\sqrt{67}$

86. $\sqrt{900}+\sqrt{2.56}$

87. $\sqrt{1600}-\sqrt{1.69}$

88. $\left(\sqrt{7}+8\right)\left(\sqrt{7}-8\right)$

89. $\left(\sqrt{3}+\sqrt{2}\right)\left(\sqrt{5}-\sqrt{10}\right)$

90. $\left(6\sqrt{8}+5\sqrt{7}\right)\left(3\sqrt{39}-2\sqrt{27}\right)$

91. $\dfrac{19}{35-\sqrt{60}}$

92. $\dfrac{\sqrt{5}}{1+2\sqrt{5}}$

93. $\dfrac{\sqrt{10}-\sqrt{2}}{\sqrt{10}+\sqrt{2}}$

94. Tile patterns: Mary is making a tile decoration for her wall. Using square tiles of different sizes, Mary created one decoration that is five tiles across, with sides touching. The first tile is 10 in.^{2}, the second is 20 in.^{2}, the third is 30 in.^{2}, the fourth is 20 in.^{2}, and the fifth is 10 in.^{2}. What is the length of the decoration?

95. Radio circuits: For a complete radio circuit, $d = \sqrt{2g} + \sqrt{2h}$, where d equals the visual horizon distance and g and h are the heights of the radio antennas at the respective stations. What is d when $g = 75$ ft and $h = 85$ ft?

Writing and Thinking About Mathematics

96. In your own words, explain how to rationalize the denominator of a fraction containing the sum or difference of square roots in the denominator. Why does this work?

 HAWKES LEARNING SYSTEMS: INTERMEDIATE ALGEBRA SOFTWARE

- Addition and Subtraction with Radicals
- Multiplication with Radicals
- Rationalizing Denominators

6.4 Equations with Radicals

- *Solve equations that contain one or more radical expressions.*

Each of the following equations involves at least one radical expression:

$$x + 3 = \sqrt{x + 5} \qquad \sqrt{x} - \sqrt{2x - 14} = 1 \qquad \sqrt[3]{x + 1} = 5.$$

If the radicals are square roots, we solve by squaring both sides of the equations. If the radical is some other root and this root can be isolated on one side of the equation, we solve by raising both sides of the equation to the integer power corresponding to the root. For example, with a cube root both sides are raised to the third power.

Squaring both sides of an equation may introduce new solutions. For example, the first-degree equation $x = -3$ has only one solution, namely, -3. However, squaring both sides gives the quadratic equation

$$x^2 = (-3)^2 \qquad \text{or} \qquad x^2 = 9.$$

The quadratic equation $x^2 = 9$ has two solutions, 3 and -3. Thus a new solution that is not a solution to the original equation has been introduced. Such a solution is called an **extraneous solution**.

When both sides of an equation are raised to a power, an extraneous solution may be introduced. Be sure to check all solutions in the original equation.

The following examples illustrate a variety of situations involving radicals. The steps used are related to the following general method.

Method for Solving Equations with Radicals

1. *Isolate one of the radicals on one side of the equation. (An equation may have more than one radical.)*
2. *Raise both sides of the equation to the power corresponding to the index of the radical.*
3. *If the equation still contains a radical, repeat Steps 1 and 2.*
4. *Solve the equation after all the radicals have been eliminated.*
5. *Be sure to check all possible solutions in the original equation and eliminate any extraneous solutions.*

Example 1: Equations with One Radical

Solve the following equations.

a. $\sqrt{x^2 + 13} = 7$

Solution: The radical is by itself on one side of the equation, so square both sides.

$$\sqrt{x^2 + 13} = 7$$

$$\left(\sqrt{x^2 + 13}\right)^2 = 7^2 \qquad\qquad \text{Square both sides.}$$

$$x^2 + 13 = 49 \qquad\qquad \text{This new equation contains no radical.}$$

$$x^2 - 36 = 0$$

$$(x + 6)(x - 6) = 0 \qquad\qquad \text{Solve by factoring.}$$

$$x = -6 \quad \text{or} \quad x = 6$$

Check both answers in the original equation:

$$\sqrt{(-6)^2 + 13} \overset{?}{=} 7 \qquad \sqrt{(6)^2 + 13} \overset{?}{=} 7$$

$$\sqrt{36 + 13} \overset{?}{=} 7 \qquad \sqrt{36 + 13} \overset{?}{=} 7$$

$$\sqrt{49} \overset{?}{=} 7 \qquad \sqrt{49} \overset{?}{=} 7$$

$$7 = 7 \qquad\qquad 7 = 7$$

Both −6 and 6 are solutions.

b. $\sqrt{y^2 - 10y - 11} = 1 + y$

Solution: Since there is only one radical and it is by itself on one side of the equation, square both sides.

$$\sqrt{y^2 - 10y - 11} = 1 + y$$

$$\left(\sqrt{y^2 - 10y - 11}\right)^2 = (1 + y)^2 \quad \text{Square both sides.}$$

$$y^2 - 10y - 11 = 1 + 2y + y^2$$

$$-12y - 12 = 0 \qquad\qquad \text{Simplifying gives a first-degree equation.}$$

$$-12y = 12$$

$$y = -1$$

Check in the original equation: $\sqrt{(-1)^2 - 10(-1) - 11} \overset{?}{=} 1 + (-1)$

$$\sqrt{1 + 10 - 11} \overset{?}{=} 0$$

$$\sqrt{0} \overset{?}{=} 0$$

$$0 = 0$$

There is one solution, −1.

Continued on the next page...

c. $\sqrt{3x+13}+3=2x$

     Solution: $\sqrt{3x+13}+3=2x$

$$\sqrt{3x+13}=2x-3 \qquad \text{Isolate the radical.}$$

$$\left(\sqrt{3x+13}\right)^2=(2x-3)^2 \qquad \text{Square both sides.}$$

$$3x+13=4x^2-12x+9$$

$$0=4x^2-15x-4$$

$$0=(4x+1)(x-4) \qquad \text{Solve by factoring.}$$

$$x=\frac{-1}{4} \quad \text{or} \quad x=4$$

     Check both answers in the original equation:

$$\sqrt{3\left(\frac{-1}{4}\right)+13}+3\overset{?}{=}2\left(\frac{-1}{4}\right) \qquad \sqrt{3(4)+13}+3\overset{?}{=}2(4)$$

$$\sqrt{\frac{49}{4}}+3\overset{?}{=}\frac{-1}{2} \qquad\qquad\qquad \sqrt{25}+3\overset{?}{=}8$$

$$\frac{7}{2}+3\overset{?}{=}\frac{-1}{2} \qquad\qquad\qquad\qquad 5+3\overset{?}{=}8$$

$$\frac{13}{2}\ne\frac{-1}{2} \qquad\qquad\qquad\qquad\quad 8=8$$

$-\dfrac{1}{4}$ is **not** a solution. The only solution is 4.

Example 2: Equations with Two Radicals

Solve the following equations with two radicals. You may need to rearrange terms and to square twice.

a. $\sqrt{x+4}=\sqrt{3x-2}$

     Solution: There are two radicals on opposite sides of the equation. Squaring both sides will give a new equation with no radicals.

$$\sqrt{x+4}=\sqrt{3x-2}$$

$$\left(\sqrt{x+4}\right)^2=\left(\sqrt{3x-2}\right)^2$$

$$x+4=3x-2$$

$$6=2x \qquad \text{Simplifying gives a first-degree equation.}$$

$$3=x$$

Check in the original equation: $\sqrt{3+4} \overset{?}{=} \sqrt{3 \cdot 3 - 2}$

$$\sqrt{7} = \sqrt{7}$$

There is one solution, 3.

b. $\sqrt{x} - \sqrt{2x - 14} = 1$

Solution: Where there is a sum or difference of radicals, squaring is easier if the radicals are on different sides of the equation. Also, squaring both sides of the equation is easier if one of the radicals is by itself on one side of the equation.

$$\sqrt{x} - \sqrt{2x - 14} = 1$$

$$\sqrt{x} = 1 + \sqrt{2x - 14} \qquad \text{Isolate one of the radicals.}$$

$$\left(\sqrt{x}\right)^2 = \left(1 + \sqrt{2x - 14}\right)^2 \qquad \text{Square both sides.}$$

$$x = 1 + 2\sqrt{2x - 14} + (2x - 14) \qquad \begin{array}{l}\text{Remember, the right-hand}\\ \text{side is the square of a}\\ \text{binomial.}\end{array}$$

$$x = 2\sqrt{2x - 14} + 2x - 13$$

$$-x + 13 = 2\sqrt{2x - 14} \qquad \begin{array}{l}\text{Simplify so that the radical}\\ \text{is on one side by itself.}\end{array}$$

$$\left(-x + 13\right)^2 = \left(2\sqrt{2x - 14}\right)^2 \qquad \text{Square both sides \textbf{again}.}$$

$$x^2 - 26x + 169 = 4(2x - 14)$$

$$x^2 - 26x + 169 = 8x - 56$$

$$x^2 - 34x + 225 = 0$$

$$(x - 9)(x - 25) = 0 \qquad \text{Solve by factoring.}$$

$$x = 9 \quad \text{or} \quad x = 25$$

Check both answers in the original equation:

$$\sqrt{9} - \sqrt{2 \cdot 9 - 14} \overset{?}{=} 1 \qquad\qquad \sqrt{25} - \sqrt{2 \cdot 25 - 14} \overset{?}{=} 1$$

$$3 - 2 \overset{?}{=} 1 \qquad\qquad\qquad 5 - 6 \overset{?}{=} 1$$

$$1 = 1 \qquad\qquad\qquad\qquad -1 \neq 1$$

25 is **not** a solution. The only solution is 9.

NOTES

It is possible that after checking the answers you may find that neither answer is a solution. In this case the answer is **no solution**.

Example 3: Equation Containing a Cube Root

Solve the following equation containing a cube root: $\sqrt[3]{2x+1}+1=3$.

Solution: First, get the radical by itself on one side of the equation. Then, since this radical is a cube root, cube both sides of the equation.

$$\sqrt[3]{2x+1}+1=3$$

$$\sqrt[3]{2x+1}=2 \qquad \text{Add } -1 \text{ to both sides.}$$

$$\left(\sqrt[3]{2x+1}\right)^3=2^3 \qquad \text{Cube both sides.}$$

$$2x+1=8 \qquad \text{Solve the equation.}$$

$$x=\frac{7}{2}$$

Check in the original equation: $\sqrt[3]{2\left(\dfrac{7}{2}\right)+1}+1\overset{?}{=}3$

$$\sqrt[3]{7+1}+1\overset{?}{=}3$$

$$\sqrt[3]{8}+1\overset{?}{=}3$$

$$2+1\overset{?}{=}3$$

$$3=3$$

There is one solution, $\dfrac{7}{2}$.

Practice Problems

Solve the following equations.

1. $2\sqrt{x+4}=x+1$ **2.** $\sqrt{3x+1}+1=\sqrt{x}$ **3.** $\sqrt[3]{2x-9}+4=3$

Answers to Practice Problems: **1.** $x=5$ **2.** No Solution   **3.** $x=4$

6.4 Exercises

In Exercises 1 – 44, solve the following equations. Be sure to check your answers in the original equation.

1. $\sqrt{8x+1} = 5$

2. $\sqrt{7x+1} = 6$

3. $\sqrt{4x-3} = 7$

4. $\sqrt{5x-6} = 8$

5. $\sqrt{2x+5} = \sqrt{4x-1}$

6. $\sqrt{5x-1} = \sqrt{x+7}$

7. $\sqrt{3x+2} = \sqrt{9x-10}$

8. $\sqrt{2-x} = \sqrt{2x-7}$

9. $\sqrt{x(x+3)} = 2$

10. $\sqrt{x(x-5)} = 6$

11. $\sqrt{x(2x+5)} = 5$

12. $\sqrt{x(3x-14)} = 7$

13. $\sqrt{x+6} = x+4$

14. $\sqrt{x+7} = 2x-1$

15. $\sqrt{x^2-16} = 3$

16. $\sqrt{x^2-25} = 12$

17. $5x + \sqrt{x+7} - 13 = 0$

18. $x - 2 - \sqrt{x+4} = 0$

19. $2x = \sqrt{7x-3} + 3$

20. $x - \sqrt{3x-8} = 4$

21. $\sqrt{3x+1} = 1 - \sqrt{x}$

22. $\sqrt{x} = \sqrt{x+16} - 2$

23. $\sqrt{x+4} = \sqrt{x+11} - 1$

24. $\sqrt{1-x} + 2 = \sqrt{13-x}$

25. $\sqrt{x+1} = \sqrt{x+6} + 1$

26. $\sqrt{x+4} = \sqrt{x+20} - 2$

27. $\sqrt{x+5} + \sqrt{x} = 5$

28. $\sqrt{5x-18} - 4 = \sqrt{5x+6}$

29. $\sqrt{2x+3} = 1 + \sqrt{x+1}$

30. $\sqrt{x} + \sqrt{x-3} = 3$

31. $\sqrt{3x+1} - \sqrt{x+4} = 1$

32. $\sqrt{3x+4} - \sqrt{x+5} = 1$

33. $\sqrt{5x-1} = 4 - \sqrt{x-1}$

34. $\sqrt{2x-5} - 2 = \sqrt{x-2}$

35. $\sqrt{2x-3} + \sqrt{x+3} = 6$

36. $\sqrt{2x+3} - \sqrt{x+5} = 1$

37. $\sqrt[3]{4+3x} = -2$

38. $\sqrt[3]{2+9x} = 9$

39. $\sqrt[3]{5x+4} = 4$

40. $\sqrt[3]{7x+1} = -5$

41. $\sqrt{2x+1} = -4$

42. $\sqrt{3x-5} = -2$

43. $\sqrt[4]{2x+1} = 3$

44. $\sqrt[4]{x-6} = 2$

Writing and Thinking About Mathematics

45. Explain, in your own words, why, in general, $(a+b)^2 \neq a^2 + b^2$.

HAWKES LEARNING SYSTEMS: INTERMEDIATE ALGEBRA SOFTWARE

- Solving Radical Equations

6.5 Functions with Radicals

- *Recognize radical functions.*
- *Evaluate radical functions.*
- *Find the domain and range of radical functions.*
- *Graph radical functions.*

Review of Functions and Function Notation

The concept of functions is among the most important and useful ideas in all of mathematics. Functions were introduced in Chapter 2 along with function notation, such as $f(x)$ (read "f of x"). In Chapter 2, we discussed **linear functions** and the use of function notation in evaluating functions. Function notation was used again in Chapter 4 where $P(x)$ was used to represent polynomials. In this section the function concept is expanded to include **radical functions** (functions with radicals). The definitions of relations and functions and the vertical line test are restated here for review and easy reference.

Relation, Domain, and Range

*A **relation** is a set of ordered pairs of real numbers.*

*The **domain**, **D**, of a relation is the set of all first coordinates in the relation.*

*The **range**, **R**, of a relation is the set of all second coordinates in the relation.*

Function

*A **function** is a relation in which each domain element has exactly one corresponding range element.*

Functions have the following two characteristics (these characteristics are simply two ways of saying the same thing):

1. A function is a relation in which each first coordinate appears only once.
2. A function is a relation in which no two ordered pairs have the same first coordinate.

Vertical Line Test

*If **any** vertical line intersects the graph of a relation at more than one point, then the relation graphed is **not** a function.*

We have used the ordered pair notation (x, y) to represent points on the graphs of relations and functions. For example,

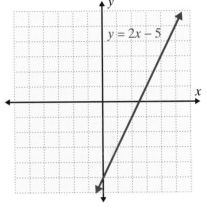

$y = 2x - 5$ represents a linear function and its graph is a straight line (as shown in the figure at the right).

Evaluating Radical Functions

We define **radical functions** (functions with radical expressions) as follows.

Radical Function

*A **radical function** is a function of the form $y = \sqrt[n]{g(x)}$ in which the radicand contains a variable expression.*

*The **domain** of such a function depends on the index, n:*

1. *If n is an even number, the domain is the set of all x such that $g(x) \geq 0$.*

2. *If n is an odd number, the domain is the set of all real numbers, $(-\infty, \infty)$.*

Examples of radical functions are

$$y = 3\sqrt{x}, \quad f(x) = \sqrt{2x + 3}, \quad \text{and} \quad y = \sqrt[3]{x - 7}.$$

Example 1: Finding the Domain of a Radical Function

Determine the domain of each radical function.

a. $y = \sqrt{2x+3}$

Solution: Because the index is 2 (understood), the radicand must be nonnegative. (That is, the expression under the radical sign cannot be negative.) Thus we must have

$$2x + 3 \geq 0$$

$$2x \geq -3$$

$$x \geq -\frac{3}{2}$$

and the domain is the interval of real numbers $\left[-\frac{3}{2}, \infty\right)$.

b. $f(x) = \sqrt[3]{x-5}$

Solution: Because the index is 3, an odd number, the radicand may be any real number.
Thus the domain is the entire interval of real numbers $(-\infty, \infty)$.

As illustrated in Chapters 2 and 4, function notation is particularly useful when evaluating functions for specific values of the variable. For example,

$$\text{if} \quad f(x) = \sqrt{2x+3}, \text{ then } f\left(\frac{1}{2}\right) = \sqrt{2\left(\frac{1}{2}\right)+3} = \sqrt{1+3} = \sqrt{4} = 2$$

$$\text{and} \quad f(0) = \sqrt{2(0)+3} = \sqrt{0+3} = \sqrt{3}.$$

A calculator can be used to find decimal approximations. Such approximations are helpful when estimating the locations of points on a graph. For example,

$$\text{if} \quad f(x) = \sqrt{x-5},$$

$$\text{then} \quad f(8) = \sqrt{(8)-5} = \sqrt{3} \approx 1.7321 \qquad \text{Accurate to 4 decimal places.}$$

$$\text{and} \quad f(25) = \sqrt{(25)-5} = \sqrt{20} \approx 4.4721. \quad \text{Accurate to 4 decimal places.}$$

Example 2: Evaluating Radical Functions

Complete each table by finding the corresponding $f(x)$ values for the given values of x.

a. $f(x) = 3\sqrt{x}$

x	$f(x)$
0	?
4	?
6	?

b. $y = \sqrt[3]{x-7}$

x	$f(x)$
7	?
6	?
−1	?

Solutions:

a.

x	$f(x)$
0	$3\sqrt{0} = 0$
4	$3\sqrt{4} = 3 \cdot 2 = 6$
6	$3\sqrt{6} \approx 7.3485$

b.

x	$f(x)$
7	$\sqrt[3]{7-7} = \sqrt[3]{0} = 0$
6	$\sqrt[3]{6-7} = \sqrt[3]{-1} = -1$
−1	$\sqrt[3]{-1-7} = \sqrt[3]{-8} = -2$

Graphing Radical Functions

To graph a radical function, we need to be aware of its domain and to plot at least a few points to see the nature of the resulting curve. Example 3 shows how to proceed, at least in the beginning, to graph the radical function $y = \sqrt{x+5}$.

Example 3: Graphing a Radical Function

Graph the function $y = \sqrt{x+5}$.

Solution: For the domain we have
$$x + 5 \geq 0$$
$$x \geq -5$$

Continued on the next page...

To see the nature of the graph we select a few values for x in the domain and find the corresponding values of y:

x	y
−5	0
−4	1
−3	$\sqrt{2} \approx 1.41$
0	$\sqrt{5} \approx 2.24$
4	3

Now we plot these points on a graph.

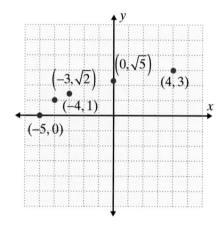

Note that $\sqrt{x+5}$ is the principal square root. This means that $y \geq 0$. Thus, the point $(-5, 0)$ is on the x-axis and the remaining points on the graph are above the x-axis. So, we can complete the graph by drawing a smooth curve that passes through the selected points. The graph of the function is shown here.

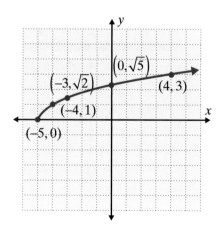

We see that the domain is $[-5, \infty)$ and the range is $[0, \infty)$.

To use a TI-84 Plus graphing calculator to graph this function,

Step 1.   Press  and enter the function as follows:

Step 2.   Press **GRAPH**. (You may need to adjust the **WINDOW**.)

The result will be the graph as shown here:

Example 4 shows how to use a TI-84 Plus graphing calculator to find many points on the graph of a radical function and then how to graph the function.

Example 4: Using a TI-84 Plus to Graph a Radical Function

a.   Use the **TABLE** feature of a TI-84 Plus graphing calculator to locate many points on the graph of the function $y = \sqrt[3]{2x - 3}$.

Solution: Using the **TABLE** feature of a TI-84 Plus:

Step 1:   Press **Y=** and enter the function as follows:

1. Press **MATH**.
2. Choose **4 : $\sqrt[3]{\ }$ (** .
3. Enter $2x - 3$) and press **ENTER**.

Continued on the next page...

Step 2: Press **TBLSET** (which is 2ND WINDOW) and set the display as shown here:

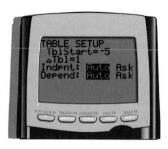

Step 3: Press **TABLE** (which is 2ND GRAPH) and the display will appear as follows:

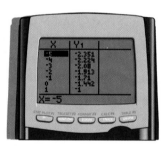

b. Plot several points (approximately) on a graph and then connect them with a smooth curve.

Solution: You may scroll up and down the display to find as many points as you like. A few are shown here to see the nature of the graph.

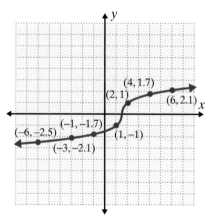

c. Use a TI-84 Plus graphing calculator to graph the function.

Solution: Press ⬤GRAPH and the display will appear with the curve as follows:

Practice Problems

Find the indicated value of the function.

1. For $f(x) = \sqrt{x+7}$, find $f(5)$.

2. For $g(x) = \sqrt{3x-1}$, find $g\left(\dfrac{2}{3}\right)$.

3. For $h(x) = \sqrt[3]{x+5}$, find $h(-13)$.

Use a calculator to estimate the value of the function accurate to 4 decimal places.

4. Estimate $f(2)$ for $f(x) = \sqrt{4x-1}$.

5. Estimate $g(-3)$ for $g(x) = \sqrt[3]{1-3x}$.

6.5 Exercises

In Exercises 1 – 4, evaluate each function as indicated and write the answers in both radical notation and decimal notation (if necessary, round decimal values to 4 decimal places.)

1. Given $f(x) = \sqrt{2x+1}$, find

    a. $f(2)$ **b.** $f(4)$ **c.** $f(24.5)$ **d.** $f(1.5)$

2. Given $f(x) = \sqrt{5-3x}$, find

    a. $f(0)$ **b.** $f(-2)$ **c.** $f\left(-\dfrac{20}{3}\right)$ **d.** $f(-2.4)$

3. Given $g(x) = \sqrt[3]{x+6}$, find

    a. $g(21)$ **b.** $g(-7)$ **c.** $g(-14)$ **d.** $g(18)$

4. Given $h(x) = \sqrt[3]{4-x}$, find

    a. $h(4)$ **b.** $h(-4)$ **c.** $h(3.999)$ **d.** $h(-2.5)$

Answers to Practice Problems: **1.** $2\sqrt{3}$ **2.** 1 **3.** -2 **4.** 2.6458 **5.** 2.1544

In Exercises 5 – 14, use interval notation to indicate the domain of each radical function.

5. $y = \sqrt{x+8}$ **6.** $y = \sqrt{2x-1}$ **7.** $y = \sqrt{2.5-5x}$ **8.** $y = \sqrt{1-3x}$

9. $f(x) = \sqrt[3]{x+4}$ **10.** $f(x) = \sqrt[3]{6x}$ **11.** $g(x) = \sqrt[4]{x}$ **12.** $g(x) = \sqrt[4]{7-x}$

13. $y = \sqrt[5]{4x-1}$ **14.** $y = \sqrt[5]{8+x}$

Match the functions given in Exercises 15 – 20, with the graphs of the functions (A) – (F).

15. $y = \sqrt{x-2}$ **16.** $y = \sqrt{2-x}$ **17.** $y = -\sqrt{x-3}$

18. $y = -\sqrt{3-x}$ **19.** $y = \sqrt{x+4}$ **20.** $y = \sqrt{x-4}$

A.

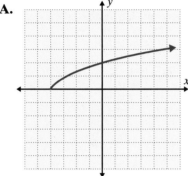

B.

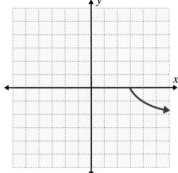

C.

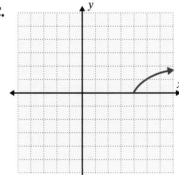

D.

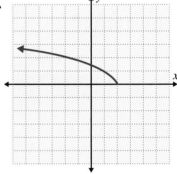

E.

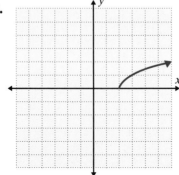

F.

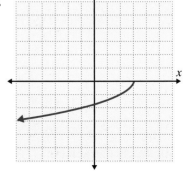

In Exercises 21 – 25, find and label at least 5 points on the graph of the function and then sketch the graph of the function.

21. $y = \sqrt{x-2}$

22. $y = \sqrt{2x+6}$

23. $f(x) = \sqrt[3]{x+2}$

24. $g(x) = \sqrt[3]{x-6}$

25. $y = \sqrt[3]{3x+6}$

Use a graphing calculator to graph each of the functions in Exercises 26 – 35.

26. $y = 3\sqrt{x+2}$

27. $y = 2\sqrt{3-x}$

28. $f(x) = -\sqrt{x+2.5}$

29. $f(x) = -\sqrt{5-x}$

30. $y = -\sqrt[3]{x+2}$

31. $y = \sqrt[3]{3x+4}$

32. $g(x) = -\sqrt{2x}$

33. $g(x) = -\sqrt[4]{x+5}$

34. $y = \sqrt[4]{2x+6}$

35. $y = \sqrt[5]{x+7}$

Writing and Thinking About Mathematics

36. The graph of the radical function $f(x) = \sqrt{x}$ is shown with two values of x on the x-axis, 3 and $3 + h$.

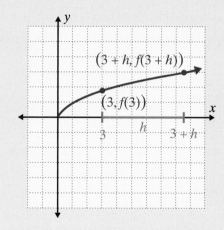

a. Rationalize the numerator and simplify the resulting expression

$$\frac{f(3+h)-f(3)}{h} = \frac{\sqrt{3+h}-\sqrt{3}}{h}$$

by multiplying both the numerator and denominator by the conjugate of the numerator.

b. What do you think this expression represents graphically?
[**Hint:** Two points determine a line.]

Continued on the next page...

Writing and Thinking About Mathematics (cont.)

c. Using your result from part **b.**, what do you see happening on the graph if the value of h shrinks slowly to 0?

d. Using your analysis from part **c.**, what happens to the value of your simplified expression in part **a.** and what do you think this value represents?

 HAWKES LEARNING SYSTEMS: INTERMEDIATE ALGEBRA SOFTWARE

- Functions with Radicals

6.6

Introduction to Complex Numbers

- *Simplify square roots of negative numbers.*
- *Identify the real parts and the imaginary parts of complex numbers.*
- *Solve linear equations with complex numbers by setting the real parts and the imaginary parts equal to each other.*
- *Add and subtract with complex numbers.*

Introduction to Complex Numbers and the Number *i*

One of the properties of real numbers is that the square of any real number is nonnegative. That is, for any real number x, $x^2 \geq 0$. The square roots of negative numbers, such as $\sqrt{-4}$ and $\sqrt{-5}$, are not real numbers. However, they can be defined by expanding the real number system into the system of **complex numbers**.

Complex numbers include all the real numbers and the even roots of negative numbers. In Chapter 7, we will see how these numbers occur as solutions to quadratic equations. At first such numbers seem to be somewhat impractical because they are difficult to picture in any type of geometric setting and they are not solutions to the types of word problems that are familiar. However, complex numbers do occur quite naturally in trigonometry and higher level mathematics and have practical applications in such fields as electrical engineering.

The first step in the development of complex numbers is to define $\sqrt{-1}$.

i and *i*²

$$i = \sqrt{-1} \qquad and \qquad i^2 = \left(\sqrt{-1}\right)^2 = -1$$

Using the definition of $\sqrt{-1}$, the following definition for the square root of a negative number can be made.

$\sqrt{-a} = \sqrt{a}\,i$

If *a* is a positive real number, then

$$\sqrt{-a} = \sqrt{a} \cdot \sqrt{-1} = \sqrt{a}\,i.$$

Note: *The number i is not under the radical sign. To avoid confusion, we sometimes write $i\sqrt{a}$.*

Example 1: $\sqrt{-a}$

Simplify the following radicals.

a. $\sqrt{-25} = \sqrt{-1}\sqrt{25} = i \cdot 5 = 5i$

$(5i)^2 = 5^2 i^2 = 25(-1) = -25$

b. $\sqrt{-36} = \sqrt{-1}\sqrt{36} = i \cdot 6 = 6i$

c. $\sqrt{-24} = \sqrt{-1}\sqrt{4 \cdot 6} = i \cdot 2 \cdot \sqrt{6} = 2\sqrt{6}\,i$ (or $2i\sqrt{6}$)

We can write $2\sqrt{6}\,i$ and $3\sqrt{5}\,i$ as long as we take care not to include

d. $\sqrt{-45} = \sqrt{-1}\sqrt{9 \cdot 5} = i \cdot 3 \cdot \sqrt{5} = 3\sqrt{5}\,i$ (or $3i\sqrt{5}$)

the i under the radical sign.

Complex Numbers

The **standard form** of a **complex number** is **$a + bi$**, where a and b are real numbers. a is called the **real part** and b is called the **imaginary part**.

If $b = 0$, then $a + bi = a + 0i = a$ is a **real number**.

If $a = 0$, then $a + bi = 0 + bi = bi$ is called a **pure imaginary number** (or an **imaginary number**).

Complex Number: $\quad a + bi$

real part ⟶ ⟵ imaginary part

NOTES

The term "imaginary" is somewhat misleading. Complex numbers and imaginary numbers are no more "imaginary" than any other type of number. In fact, all the types of numbers that we have studied (whole numbers, integers, rational numbers, irrational numbers, and real numbers) are products of human imagination.

Example 2: Real and Imaginary Parts

Identify the real and imaginary parts of each complex number.

a. $4 - 2i$

4 is the real part; -2 is the imaginary part.

b. $\dfrac{5 + 2i}{3}$

$\dfrac{5 + 2i}{3} = \dfrac{5}{3} + \dfrac{2}{3}i$ in standard form. Thus $\dfrac{5}{3}$ is the real part; $\dfrac{2}{3}$ is the imaginary part.

c. 7

$7 = 7 + 0i$ in standard form. Thus 7 is the real part; 0 is the imaginary part. (Remember, if $b = 0$, the complex number is a real number.)

d. $-\sqrt{3}\,i$

$-\sqrt{3}\,i = 0 - \sqrt{3}\,i$ in standard form. Thus 0 is the real part; $-\sqrt{3}$ is the imaginary part. (If $a = 0$ and $b \neq 0$, then the complex number is a pure imaginary number.)

In general, if a is a real number, then we can write $a = a + 0i$. This means that a is a complex number. Thus, **every real number is a complex number**. Figure 6.1 illustrates the relationships among the various types of numbers we have studied.

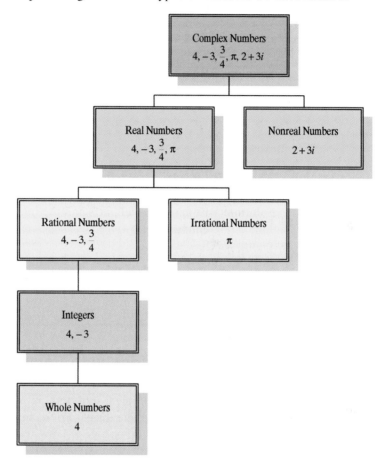

Figure 6.1

If two complex numbers are equal, then the real parts are equal and the imaginary parts are equal. For example, if

$$x + yi = 7 + 2i$$

then

$$x = 7 \quad \text{and} \quad y = 2.$$

This relationship can be used to solve equations involving complex numbers.

Equality of Complex Numbers

For complex numbers $a + bi$ and $c + di$,

if $\boldsymbol{a + bi = c + di}$, then $\boldsymbol{a = c}$ and $\boldsymbol{b = d}$.

Example 3: Solving Equations

Solve each equation for the unknown numbers.

a. $(x + 3) + 2yi = 7 - 6i$

Solution: Equate the real parts and the imaginary parts, and solve the resulting equations.

$$
\begin{array}{lll}
x + 3 = 7 & \text{and} & 2y = -6 \\
x = 4 & & y = -3
\end{array}
$$

b. $2y + 3 - 8i = 9 + 4xi$

Solution: Equate the real parts and the imaginary parts, and solve the resulting equations.

$$
\begin{array}{lll}
2y + 3 = 9 & \text{and} & -8 = 4x \\
2y = 6 & & -2 = x \\
y = 3 &
\end{array}
$$

Addition and Subtraction with Complex Numbers

Adding and subtracting complex numbers is similar to adding and subtracting polynomials. Simply combine like terms. For example,

$$
\begin{aligned}
(2 + 3i) + (9 - 8i) &= 2 + 9 + 3i - 8i \\
&= (2 + 9) + (3 - 8)i \\
&= 11 - 5i.
\end{aligned}
$$

Similarly,

$$
\begin{aligned}
(5 - 2i) - (6 + 7i) &= 5 - 2i - 6 - 7i \\
&= (5 - 6) + (-2 - 7)i \\
&= -1 - 9i.
\end{aligned}
$$

Addition and Subtraction with Complex Numbers

For complex numbers a + bi and c + di,

$$
(a + bi) + (c + di) = (a + c) + (b + d)i
$$

and

$$
(a + bi) - (c + di) = (a - c) + (b - d)i.
$$

Example 4: Addition and Subtraction with Complex Numbers

Find each sum or difference as indicated.

a. $(6 - 2i) + (1 - 2i)$

Solution: $(6 - 2i) + (1 - 2i) = (6 + 1) + (-2 - 2)i$
$$= 7 - 4i$$

b. $\left(-8 - \sqrt{2}\,i\right) - \left(-8 + \sqrt{2}\,i\right)$

Solution: $\left(-8 - \sqrt{2}\,i\right) - \left(-8 + \sqrt{2}\,i\right) = \left(-8 - (-8)\right) + \left(-\sqrt{2} - \sqrt{2}\right)i$
$$= (-8 + 8) + \left(-2\sqrt{2}\right)i$$
$$= 0 - 2\sqrt{2}i$$
$$= -2\sqrt{2}\,i \ (\text{or} - 2i\sqrt{2})$$

c. $\left(\sqrt{3} - 2i\right) + \left(1 + \sqrt{5}\,i\right)$

Solution: $\left(\sqrt{3} - 2i\right) + \left(1 + \sqrt{5}\,i\right) = \left(\sqrt{3} + 1\right) + \left(-2 + \sqrt{5}\right)i$
$$= \left(\sqrt{3} + 1\right) + \left(\sqrt{5} - 2\right)i$$

Note: Here, the coefficients do not simplify. This means that the real part is $\sqrt{3} + 1$ and the imaginary part is $\sqrt{5} - 2$.

Practice Problems

1. Find the imaginary part and the real part of $2 - \sqrt{39}i$.

Add or subtract as indicated. Simplify your answers.

2. $\left(-7 + \sqrt{3}i\right) + (5 - 2i)$ **3.** $(4 + i) - (5 + 2i)$

Solve for x and y.

4. $x + yi = \sqrt{2} - 7i$ **5.** $3y + (x - 7)i = -9 + 2i$

Answers to Practice Problems: **1.** Imaginary part is $-\sqrt{39}$ and real part is 2   **2.** $-2 + \left(-2 + \sqrt{3}\right)i$
3. $-1 - i$ **4.** $x = \sqrt{2}$ and $y = -7$ **5.** $x = 9$ and $y = -3$

6.6 Exercises

Find the real part and the imaginary part of each of the complex numbers in Exercises 1 – 10.

1. $4 - 3i$ **2.** $6 + \sqrt{3}\,i$ **3.** $-11 + \sqrt{2}\,i$ **4.** $\dfrac{3}{4} + i$

5. $\dfrac{2}{3} + \sqrt{17}\,i$ **6.** $\dfrac{4}{7}i$ **7.** $\dfrac{4 + 7i}{5}$ **8.** $\dfrac{2 - i}{4}$

9. $\dfrac{3}{8}$ **10.** $-\sqrt{5} + \dfrac{\sqrt{2}}{2}i$

Simplify the radicals in Exercises 11 – 24.

11. $\sqrt{-49}$ **12.** $\sqrt{-121}$ **13.** $-\sqrt{-64}$ **14.** $\sqrt{-169}$

15. $3\sqrt{147}$ **16.** $\sqrt{128}$ **17.** $2\sqrt{-150}$ **18.** $4\sqrt{-99}$

19. $-2\sqrt{-108}$ **20.** $2\sqrt{175}$ **21.** $\sqrt{242}$ **22.** $\sqrt{-192}$

23. $\sqrt{-1000}$ **24.** $\sqrt{-243}$

Solve the equations in Exercises 25 – 38 for x and y.

25. $x + 3i = 6 - yi$ **26.** $2x - 8yi = -2 + 4yi$

27. $\sqrt{5} - 2i = y + xi$ **28.** $\dfrac{2}{3} - 2yi = 2x + \dfrac{4}{5}$

29. $\sqrt{2} + i - 3 = x + yi$ **30.** $\sqrt{5}\,i - 3 + 4i = x + yi$

31. $2x + 3 + 6i = 7 - (y + 2)i$ **32.** $x + yi + 8 = 2i + 4 - 3yi$

33. $x + 2i = 5 - yi - 3 - 4i$ **34.** $3x + 2 - 7i = i - 2yi + 5$

35. $2 + 3i + x = 5 - 7i + yi$ **36.** $11i - 2x + 4 = 10 - 3i + 2yi$

37. $2x - 2yi + 6 = 6i - x + 2$ **38.** $x + 4 - 3x + i = 8 + yi$

Find each sum or difference as indicated in Exercises 39 – 60.

39. $(2 + 3i) + (4 - i)$ **40.** $(7 - i) + (3 + 6i)$ **41.** $(4 + 5i) - (3 - 2i)$

42. $(-3 + 2i) - (6 + 2i)$ **43.** $-3i + (2 - 3i)$ **44.** $(7 + 5i) + (6 - 2i)$

45. $(8 + 9i) - (8 - 5i)$ **46.** $(-6 + i) - (2 + 3i)$ **47.** $\left(\sqrt{5} - 2i\right) + (3 - 4i)$

48. $\left(4 + 3i\right) - \left(\sqrt{2} + 3i\right)$ **49.** $\left(7 + \sqrt{6}\,i\right) + (-2 + i)$ **50.** $\left(\sqrt{11} + 2i\right) + (5 - 7i)$

51. $\left(\sqrt{3} + \sqrt{2}\,i\right) - \left(5 + \sqrt{2}\,i\right)$ **52.** $\left(\sqrt{5} + \sqrt{3}\,i\right) + (1 - i)$

53. $\left(5 + \sqrt{-25}\right) - \left(7 + \sqrt{-100}\right)$ **54.** $\left(1 + \sqrt{-36}\right) - \left(-4 - \sqrt{-49}\right)$

55. $\left(13 - 3\sqrt{-16}\right) + \left(-2 - 4\sqrt{-1}\right)$ **56.** $\left(7 + \sqrt{-9}\right) - \left(3 - 2\sqrt{-25}\right)$

57. $(4 + i) + (-3 - 2i) - (-1 - i)$ **58.** $(-2 - 3i) + (6 + i) - (2 + 5i)$

59. $(7 + 3i) + (2 - 4i) - (6 - 5i)$ **60.** $(-5 + 7i) + (4 - 2i) - (3 - 5i)$

Writing and Thinking About Mathematics

61. Answer the following questions and give a brief explanation of your answer.
  a. Is every real number a complex number?
  b. Is every complex number a real number?

62. List 5 numbers that do and 5 numbers that do not fit each of the following categories (if possible).
  a. rational number
  b. integer
  c. real number
  d. pure imaginary number
  e. complex number
  f. irrational number

 HAWKES LEARNING SYSTEMS: INTERMEDIATE ALGEBRA SOFTWARE

  ▪ Complex Numbers

6.7   Multiplication and Division with Complex Numbers

- *Multiply with complex numbers.*
- *Divide with complex numbers.*
- *Simplify powers of i.*

Multiplication with Complex Numbers

The product of two complex numbers can be found using the same procedure as when multiplying two binomials. This is similar to multiplying binomial expressions with the sums and differences of radicals as we did in Section 6.3. **Remember that $i^2 = -1$.** For example,

$$
\begin{aligned}
(3 + 5i)(2 + i) &= (3 + 5i)2 + (3 + 5i)i \\
&= 6 + 10i + 3i + 5i^2 \\
&= 6 + 13i - 5 \qquad\qquad 5i^2 = 5(-1) = -5 \\
&= 1 + 13i.
\end{aligned}
$$

Multiplication with Complex Numbers

For complex numbers a + bi and c + di,

$$(a+bi)(c+di) = (ac-bd)+(ad+bc)i.$$

This definition is an application of the FOIL method for multiplying two binomials. **Memorizing the definition is not recommended.** An easier approach is simply to use the FOIL method for each product and do as many steps as you can mentally. **Remember that $i^2 = -1$.**

Example 1: Multiplication with Complex Numbers

Find the following products.

a. $(3i)(2 - 7i)$

Solution: $\begin{aligned} (3i)(2 - 7i) &= 6i - 21i^2 \\ &= 6i + 21 \end{aligned}$

b. $\left(\sqrt{2}-i\right)\left(\sqrt{2}-i\right)$

Solution: $\left(\sqrt{2}-i\right)\left(\sqrt{2}-i\right)=\left(\sqrt{2}\right)^{2}-\sqrt{2}\cdot i-\sqrt{2}\cdot i+i^{2}$

$$= 2-2\sqrt{2}\,i-1 \qquad\qquad \text{Remember } i^{2}=-1.$$

$$= 1-2\sqrt{2}\,i$$

c. $(-1+i)(2-i)$

Solution: $(-1+i)(2-i)=-2+i+2i-i^{2}$

$$=-2+3i+1$$

$$=-1+3i$$

NOTES

COMMON ERROR

Remember that $\sqrt{a}\cdot\sqrt{b}=\sqrt{ab}$ only if a and b are nonnegative real numbers. Applying this rule to negative real numbers can lead to an error. The error can be avoided by first changing the radicals to imaginary form.

INCORRECT

$$\sqrt{-6}\cdot\sqrt{-2}=\sqrt{12}$$

$$=\sqrt{4}\cdot\sqrt{3}$$

$$=2\sqrt{3}$$

CORRECT

$$\sqrt{-6}\cdot\sqrt{-2}=\sqrt{6}\,i\cdot\sqrt{2}\,i$$

$$=\sqrt{12}\,i^{2}$$

$$=2\sqrt{3}(-1)$$

$$=-2\sqrt{3}$$

Division with Complex Numbers

The two complex numbers $a+bi$ and $a-bi$ are called **complex conjugates** or simply **conjugates** of each other. As the following steps show, **the product of two complex conjugates will always be a nonnegative real number**.

$$(a+bi)(a-bi)=a^{2}-abi+abi-b^{2}i^{2}$$

$$=a^{2}-b^{2}i^{2}$$

$$=a^{2}+b^{2}$$

The resulting product, $a^{2}+b^{2}$, is a real number, and it is nonnegative since it is the sum of the squares of real numbers.

Remember that the form $a + bi$ is called the **standard form** of a complex number. The standard form allows for easy identification of the real and imaginary parts. Thus,

$$\frac{1+3i}{5} = \frac{1}{5} + \frac{3}{5}i \text{ in standard form.}$$

The real part is $\frac{1}{5}$ and the imaginary part is $\frac{3}{5}$.

A fraction formed with complex numbers, such as $\frac{1+i}{2-3i}$, indicates division of the numerator by the denominator. However, we do not divide these numbers in the usual sense. The objective is to find an equivalent expression that is in standard form, $a + bi$.

To write the fraction $\frac{1+i}{2-3i}$ in standard form, multiply both the numerator and denominator by $2 + 3i$ and simplify. This will give a positive real number in the denominator.

$$\frac{1+i}{2-3i} = \frac{(1+i)(2+3i)}{(2-3i)(2+3i)} \qquad 2+3i \text{ is the conjugate of the denominator.}$$

$$= \frac{2+3i+2i+3i^2}{2^2+6i-6i-3^2 i^2}$$

$$= \frac{2+5i-3}{4-(9)(-1)} \qquad \text{Reminder: } i^2 = -1.$$

$$= \frac{-1+5i}{13}$$

$$= -\frac{1}{13} + \frac{5}{13}i$$

To Write a Fraction with Complex Numbers in Standard Form

1. *Multiply both the numerator and denominator by the complex conjugate of the denominator.*
2. *Simplify the resulting products in both the numerator and denominator.*
3. *Write the simplified result in standard form.*

Remember the following special product. We restate it here to emphasize its importance.

$$(a + bi)(a - bi) = a^2 + b^2$$

Example 2: Division with Complex Numbers

Write the following fractions in standard form.

a. $\dfrac{4}{-1-5i}$

Solution: $\dfrac{4}{-1-5i} = \dfrac{4(-1+5i)}{(-1-5i)(-1+5i)}$

$$= \dfrac{-4+20i}{(-1)^2 - (5i)^2}$$

$$= \dfrac{-4+20i}{1+25}$$

$$= \dfrac{-4+20i}{26}$$

$$= -\dfrac{4}{26} + \dfrac{20}{26}i$$

$$= -\dfrac{2}{13} + \dfrac{10}{13}i$$

b. $\dfrac{\sqrt{3}+i}{\sqrt{3}-i}$

Solution: $\dfrac{\sqrt{3}+i}{\sqrt{3}-i} = \dfrac{\left(\sqrt{3}+i\right)\left(\sqrt{3}+i\right)}{\left(\sqrt{3}-i\right)\left(\sqrt{3}+i\right)}$

$$= \dfrac{3+2\sqrt{3}\,i+i^2}{\left(\sqrt{3}\right)^2 - i^2} = \dfrac{2+2\sqrt{3}\,i}{3+1}$$

$$= \dfrac{2+2\sqrt{3}\,i}{4} = \dfrac{2}{4} + \dfrac{2\sqrt{3}}{4}i$$

$$= \dfrac{1}{2} + \dfrac{\sqrt{3}\,i}{2}$$

c. $\dfrac{6+i}{i}$

Solution: $\dfrac{6+i}{i} = \dfrac{(6+i)(-i)}{i(-i)}$ Since $i = 0 + i$ and $-i = 0 - i$, the number $-i$ is the conjugate of i.

$$= \dfrac{-6i-i^2}{-i^2} = \dfrac{-6i+1}{1}$$

$$= 1-6i$$

Continued on the next page...

d. $\dfrac{\sqrt{2}+i}{-\sqrt{2}+i}$

Solution: $\dfrac{\sqrt{2}+i}{-\sqrt{2}+i} = \dfrac{\left(\sqrt{2}+i\right)\left(-\sqrt{2}-i\right)}{\left(-\sqrt{2}+i\right)\left(-\sqrt{2}-i\right)}$

$$= \dfrac{-\left(\sqrt{2}\right)^2 - \sqrt{2}\,i - \sqrt{2}\,i - i^2}{\left(-\sqrt{2}\right)^2 - (i)^2}$$

$$= \dfrac{-2 - 2\sqrt{2}\,i + 1}{2+1} = \dfrac{-1 - 2\sqrt{2}\,i}{3}$$

$$= -\dfrac{1}{3} - \dfrac{2\sqrt{2}}{3}\,i$$

Powers of *i*

The powers of *i* form an interesting pattern. Regardless of the particular integer exponent, there are only four possible values for any power of *i*:

$$i, \quad -1, \quad -i, \quad \text{and} \quad 1.$$

The fact that these are the only four possibilities for powers of *i* becomes apparent from studying the following powers.

$$i^1 = i$$
$$i^2 = -1$$
$$i^3 = i^2 \cdot i = -1 \cdot i = -i$$
$$i^4 = i^2 \cdot i^2 = (-1)(-1) = 1$$
$$i^5 = i^4 \cdot i = 1 \cdot i = i$$
$$i^6 = i^4 \cdot i^2 = (1)(-1) = -1$$
$$i^7 = i^4 \cdot i^3 = (1)(-i) = -i$$
$$i^8 = i^4 \cdot i^4 = (1)(1) = 1$$

Higher powers of *i* can be simplified using the fact that when *i* is raised to a power that is a multiple of 4, the result is 1. Thus, if *n* is a positive integer, then

$$i^{4n} = \left(i^4\right)^n = 1^n = 1$$
$$i^{4n+1} = i^{4n} \cdot i = 1 \cdot i = i$$
$$i^{4n+2} = i^{4n} \cdot i^2 = 1 \cdot (-1) = -1$$
$$i^{4n+3} = i^{4n} \cdot i^3 = 1 \cdot (-i) = -i.$$

Example 3: Powers of *i*

a. $i^{45} = i^{44} \cdot i = \left(i^4\right)^{11} \cdot i = 1^{11} \cdot i = i$ 　　　　($i = 0 + i$ in standard form.)

b. $i^{58} = i^{56} \cdot i^2 = \left(i^4\right)^{14} \cdot i^2 = 1^{14} \cdot (-1) = -1$ 　　($-1 = -1 + 0i$ in standard form.)

c. $i^{-7} = \dfrac{1}{i^7} = \dfrac{1}{i^7} \cdot \dfrac{i}{i} = \dfrac{i}{i^8} = \dfrac{i}{1} = i$ 　　　　($i = 0 + i$ in standard form.)

Practice Problems

Write each of the following numbers in standard form.

1. $-2i(3 - i)$ 　　　　　**2.** $(2 + 4i)(1 + i)$ 　　　　**3.** i^{13}

4. i^{-2} 　　　　　　　　**5.** $\dfrac{2}{1 + 5i}$ 　　　　　　**6.** $\dfrac{7 + i}{2 - i}$

6.7 Exercises

In Exercises, 1 – 60, perform the indicated operations and write each result in standard form. Assume k is a positive integer.

1. $8(2 + 3i)$ 　　　**2.** $-3(7 - 4i)$ 　　　**3.** $-7\left(\sqrt{2} - i\right)$ 　　　**4.** $\sqrt{3}\left(\sqrt{3} + 2i\right)$

5. $3i(4 - i)$ 　　　**6.** $-4i(6 - 7i)$ 　　　**7.** $-i\left(\sqrt{3} + i\right)$ 　　　**8.** $2i\left(\sqrt{5} + 2i\right)$

9. $5i\left(2 - \sqrt{2}\,i\right)$ 　　　**10.** $\sqrt{3}\,i\left(2 - \sqrt{3}\,i\right)$ 　　　**11.** $(5 + 3i)(1 + i)$

12. $(2 + 7i)(6 + i)$ 　　　**13.** $(-3 + 5i)(-1 + 2i)$ 　　　**14.** $(6 + 2i)(3 - i)$

15. $(2 - 3i)(2 + 3i)$ 　　　**16.** $(4 + 3i)(7 - 2i)$ 　　　**17.** $(-2 + 5i)(i - 1)$

18. $(5 + 7i)^2$ 　　　**19.** $(3 + 2i)^2$ 　　　**20.** $(4 + 5i)(4 - 5i)$

21. $\left(\sqrt{3} + i\right)\left(\sqrt{3} - 2i\right)$ 　　　**22.** $\left(2\sqrt{5} + 3i\right)\left(\sqrt{5} - i\right)$ 　　　**23.** $\left(4 + \sqrt{5}\,i\right)\left(4 - \sqrt{5}\,i\right)$

24. $\left(\sqrt{7} + 3i\right)\left(\sqrt{7} + i\right)$ 　　　**25.** $\left(5 - \sqrt{2}\,i\right)\left(5 - \sqrt{2}\,i\right)$ 　　　**26.** $\left(7 + 2\sqrt{3}\,i\right)\left(7 - 2\sqrt{3}\,i\right)$

27. $\left(\sqrt{5} + 2i\right)\left(\sqrt{2} - i\right)$ 　　　**28.** $\left(2\sqrt{3} + i\right)\left(4 + 3i\right)$ 　　　**29.** $\left(3 + \sqrt{5}\,i\right)\left(3 + \sqrt{6}\,i\right)$

Answers to Practice Problems: 　**1.** $-2 - 6i$ 　**2.** $-2 + 6i$ 　**3.** $0 + i$ 　**4.** $-1 + 0i$ 　**5.** $\dfrac{1}{13} - \dfrac{5}{13}i$ 　**6.** $\dfrac{13}{5} + \dfrac{9}{5}i$

30. $\left(2-\sqrt{3}\,i\right)\left(3-\sqrt{2}\,i\right)$ **31.** $\dfrac{-3}{i}$ **32.** $\dfrac{7}{i}$ **33.** $\dfrac{5}{4i}$

34. $\dfrac{-3}{2i}$ **35.** $\dfrac{2+i}{-4i}$ **36.** $\dfrac{3-4i}{3i}$ **37.** $\dfrac{-4}{1+2i}$ **38.** $\dfrac{7}{5-2i}$

39. $\dfrac{6}{4-3i}$ **40.** $\dfrac{-8}{6+i}$ **41.** $\dfrac{2i}{5-i}$ **42.** $\dfrac{-4i}{1+3i}$ **43.** $\dfrac{2-i}{2+5i}$

44. $\dfrac{6+i}{3-4i}$ **45.** $\dfrac{2-3i}{-1+5i}$ **46.** $\dfrac{-3+i}{7-2i}$ **47.** $\dfrac{1+4i}{\sqrt{3}+i}$ **48.** $\dfrac{9-2i}{\sqrt{5}+i}$

49. $\dfrac{\sqrt{3}+2i}{\sqrt{3}-2i}$ **50.** $\dfrac{\sqrt{6}-3i}{\sqrt{6}+3i}$ **51.** i^{13} **52.** i^{20} **53.** i^{30} **54.** i^{15}

55. i^{-3} **56.** i^{-5} **57.** i^{4k} **58.** i^{4k+2} **59.** i^{4k+3} **60.** i^{4k+1}

In Exercises 61 – 70, find the indicated products and simplify.

61. $(x+3i)(x-3i)$ **62.** $(y+5i)(y-5i)$ **63.** $\left(x+\sqrt{2}\,i\right)\left(x-\sqrt{2}\,i\right)$

64. $\left(2x+\sqrt{7}\,i\right)\left(2x-\sqrt{7}\,i\right)$ **65.** $\left(\sqrt{5}y+2i\right)\left(\sqrt{5}y-2i\right)$ **66.** $\left(y-\sqrt{3}\,i\right)\left(y+\sqrt{3}\,i\right)$

67. $\left[(x+2)+6i\right]\left[(x+2)-6i\right]$ **68.** $\left[(x+1)-\sqrt{8}\,i\right]\left[(x+1)+\sqrt{8}\,i\right]$

69. $\left[(y-3)+2i\right]\left[(y-3)-2i\right]$ **70.** $\left[(x-1)+5i\right]\left[(x-1)-5i\right]$

Writing and Thinking About Mathematics

71. Explain why the product of every complex number and its conjugate is a positive real number.

72. Explain why $\sqrt{-4}\cdot\sqrt{-4}\neq 4$. What is the correct value of $\sqrt{-4}\cdot\sqrt{-4}$?

73. What condition is necessary for the conjugate of a complex number, $a+bi$, to be equal to the reciprocal of this number?

HAWKES LEARNING SYSTEMS: INTERMEDIATE ALGEBRA SOFTWARE

- Multiplication and Division with Complex Numbers

Chapter 6 Index of Key Ideas and Terms

Section 6.1   Roots and Radicals

Perfect Squares (Table of)   page 463

Radical Terminology   page 463
The symbol $\sqrt{}$ is called a **radical sign**.
The number under the radical sign is called the **radicand**.
The complete expression, such as $\sqrt{64}$, is called a **radical**
or **radical expression**.

Square Root   page 464
If a is a nonnegative real number and b is a real number
such that $b^2 = a$, then b is called a **square root** of a.
If b is nonnegative, then we write
$$\sqrt{a} = b \quad \leftarrow \quad b \text{ is called the \textbf{principal square root},}$$
and $-\sqrt{a} = -b. \leftarrow -b$ is called the **negative square root**.

Properties of Square Roots   page 467
If a and b are positive real numbers, then
1. $\sqrt{ab} = \sqrt{a}\sqrt{b}$

2. $\sqrt{\dfrac{a}{b}} = \dfrac{\sqrt{a}}{\sqrt{b}}.$

Simplest Form of Square Roots   page 467
A square root is considered to be in **simplest form** when the
radicand has no perfect square as a factor.

Square Root of x^2   page 470
If x is a real number, then $\sqrt{x^2} = |x|$. Note: If $x \geq 0$ is given,
then we can write $\sqrt{x^2} = x$.

Square Roots of Expressions with Even and Odd Exponents   page 471
For any nonnegative real number x and positive integer m,
$$\sqrt{x^{2m}} = x^m \qquad \text{and} \qquad \sqrt{x^{2m+1}} = x^m\sqrt{x}.$$
Note: For any integer m, $2m$ is even and $2m + 1$ is odd.

Perfect Cubes (Table of)   page 472

Cube Root   page 472
If a and b are real numbers such that $b^3 = a$, then b is called
the **cube root** of a. We write $\sqrt[3]{a} = b. \leftarrow$ the **cube root**

Section 6.1 Roots and Radicals (cont.)

Simplest Form of Cube Roots   page 473

A cube root is considered to be in **simplest form** when the radicand has no perfect cube as a factor.

Section 6.2 Rational Exponents

Radical Notation   page 478

If n is a positive integer and $b^n = a$, then

$b = \sqrt[n]{a} = a^{\frac{1}{n}}$ (assuming $\sqrt[n]{a}$ is a real number).

The expression $\sqrt[n]{a}$ is called a **radical**.

The symbol $\sqrt[n]{}$ is called a **radical sign**.

n is called the **index**.

a is called the **radicand**.

(Note: If no index is given, it is understood to be 2.

For example, $\sqrt{3} = \sqrt[2]{3} = 3^{\frac{1}{2}}$.)

Special Notes about the Index n   page 478

For the expression $\sqrt[n]{a}$ (or $a^{\frac{1}{n}}$) to be a real number:

1. when a is nonnegative, n can be any index.
2. when a is negative, n must be odd.

(If a is negative and n is even, then $\sqrt[n]{a}$ is nonreal.)

Properties of Exponents   page 480

For nonzero real numbers a and b and rational numbers m and n,

The Exponent 1:	$a^1 = a$ (a is any real number.)
The Exponent 0:	$a^0 = 1$ ($a \neq 0$)
Product Rule:	$a^m \cdot a^n = a^{m+n}$
Quotient Rule:	$\dfrac{a^m}{a^n} = a^{m-n}$
Negative Exponents:	$a^{-n} = \dfrac{1}{a^n}, \ \dfrac{1}{a^{-n}} = a^n$
Power Rule:	$\left(a^m\right)^n = a^{mn}$
Power Rule for Products:	$(ab)^n = a^n b^n$
Power Rule for Fractions:	$\left(\dfrac{a}{b}\right)^n = \dfrac{a^n}{b^n}$

Section 6.2   Rational Exponents (cont.)

The General Form $a^{\frac{m}{n}}$ page 481

If n is a positive integer, m is any integer, and $a^{\frac{1}{n}}$ is a real number, then

$$a^{\frac{m}{n}} = \left(a^{\frac{1}{n}}\right)^m = \left(a^m\right)^{\frac{1}{n}}.$$

In radical notation:

$$a^{\frac{m}{n}} = \left(\sqrt[n]{a}\right)^m = \sqrt[n]{a^m}.$$

To Find the Value of $a^{\frac{m}{n}}$ **with a Calculator**   page 484

1. Enter the value of the base, a.

2. Press the caret key ⌃.

3. Enter the fractional exponent enclosed in parentheses. (This exponent may be positive or negative.)

4. Press ENTER.

Section 6.3   Operations with Radicals

Addition and Subtraction with Radical Expressions   page 489

Like Radicals   page 489

Just as like terms contain the same variable expression, **like radicals** contain the same radical expression.

Multiplication with Radical Expressions   page 490

To Rationalize a Denominator Containing a Square Root or a Cube Root   page 492

1. If the denominator contains a square root, multiply both the numerator and denominator by an expression that will give a denominator with no square roots.

2. If the denominator contains a cube root, multiply both the numerator and denominator by an expression that will give a denominator with no cube roots.

Section 6.3 Operations with Radicals (cont.)

Conjugates page 493

The two expressions $(a - b)$ and $(a + b)$ are called **conjugates** of each other, and their product $(a - b)(a + b)$ results in the difference of two squares.

To Rationalize a Denominator Containing a Sum or Difference Involving Square Roots page 493

If the denominator of a fraction contains a sum or difference involving a square root, rationalize the denominator by multiplying both the numerator and denominator by the conjugate of the denominator.
1. If the denominator is of the form $a - b$, multiply both the numerator and denominator by $a + b$.
2. If the denominator is of the form $a + b$, multiply both the numerator and denominator by $a - b$.

The new denominator will be the difference of two squares and therefore not contain a radical term.

Evaluating Radical Expressions with a Graphing Calculator page 495

Section 6.4 Equations with Radicals

Extraneous Solution page 500

An **extraneous solution** is a solution found when solving an equation that does not satisfy the original equation. It may be introduced by raising both sides of an equation to a power.

Method for Solving Equations with Radicals page 500
1. Isolate one of the radicals on one side of the equation. (An equation may have more than one radical.)
2. Raise both sides of the equation to the power corresponding to the index of the radical.
3. If the equation still contains a radical, repeat Steps 1 and 2.
4. Solve the equation after all the radicals have been eliminated.
5. Be sure to check all possible solutions in the original equation and eliminate any extraneous solutions.

Section 6.5 Functions with Radicals

Evaluating Radical Functions   pages 507-508

Radical Function   page 507
A **radical function** is a function of the form $y = \sqrt[n]{g(x)}$ in
which the radicand contains a variable expression.
The **domain** of such a function depends on the index, n:
  1. If n is an even number, the domain is the set of all
     x such that $g(x) \geq 0$.
  2. If n is an odd number, the domain is the set of all
     real numbers $(-\infty, \infty)$.

Graphing Radical Functions   pages 509-511

Section 6.6 Introduction to Complex Numbers

Complex Numbers   page 517
$$i = \sqrt{-1} \quad \text{and} \quad i^2 = -1$$
If a is a positive real number, then $\sqrt{-a} = \sqrt{a} \cdot \sqrt{-1} = \sqrt{a}\, i$.

Standard Form: $a + bi$ page 518
The **standard form** of a **complex number** is $a + bi$, where
a and b are real numbers. a is called the **real part** and b is
called the **imaginary part**.
If $b = 0$, then $a + bi = a + 0i = a$ is a real number.
If $a = 0$, then $a + bi = 0 + bi = bi$ is called a **pure imaginary
number** (or an **imaginary number**).

Equality of Complex Numbers   page 519
For complex numbers $a + bi$ and $c + di$, if $a + bi = c + di$,
then $a = c$ and $b = d$.

Addition and Subtraction with Complex Numbers   page 520
For complex numbers $a + bi$ and $c + di$,
$$(a + bi) + (c + di) = (a + c) + (b + d)i$$
and $(a + bi) - (c + di) = (a - c) + (b - d)i.$

Section 6.7 Multiplication and Division with Complex Numbers

Multiplication with Complex Numbers   page 524

For complex numbers $a + bi$ and $c + di$,

$$(a + bi)(c + di) = (ac - bd) + (ad + bc)i.$$

Division with Complex Numbers   pages 525-526

Complex Conjugates   page 525

$a + bi$ and $a - bi$ are **complex conjugates**.
The product of two complex conjugates will always be
a nonnegative real number: $(a + bi)(a - bi) = a^2 + b^2$.

To Write a Fraction with Complex Numbers in Standard Form   page 526

1. Multiply both the numerator and denominator by the
   complex conjugate of the denominator.
2. Simplify the resulting products in both the numerator
   and denominator.
3. Write the simplified result in standard form.

Powers of i: i^n   page 528

$$
\begin{aligned}
i^{4n} &= (i^4)^n &&= 1^n &&= 1 \\
i^{4n+1} &= i^{4n} \cdot i &&= 1 \cdot i &&= i \\
i^{4n+2} &= i^{4n} \cdot i^2 &&= 1 \cdot (-1) &&= -1 \\
i^{4n+3} &= i^{4n} \cdot i^3 &&= 1 \cdot (-i) &&= -i
\end{aligned}
$$

🦢 HAWKES LEARNING SYSTEMS: INTERMEDIATE ALGEBRA SOFTWARE

For a review of the topics and problems from Chapter 6, look at the following lessons
from *Hawkes Learning Systems: Intermediate Algebra*

- Evaluating Radicals
- Simplifying Radicals
- Rational Exponents
- Addition and Subtraction with Radicals
- Multiplication with Radicals
- Rationalizing Denominators
- Solving Radical Equations
- Functions with Radicals
- Complex Numbers
- Multiplication and Division with Complex Numbers
- Chapter 6 Review and Test

Chapter 6 Review

6.1 Roots and Radicals

In Exercises 1 – 4, use a graphing calculator to find the value of each radical accurate to 4 decimal places.

1. $\sqrt{33}$ **2.** $\sqrt{72}$ **3.** $\sqrt{500}$ **4.** $-\sqrt{180}$

In Exercises 5 – 20, simplify each radical expression. Assume that the variables represent positive numbers. Say "nonreal" if the expression does not represent a real number.

5. $\sqrt{20}$ **6.** $\sqrt{72}$ **7.** $-\sqrt{98}$ **8.** $-\sqrt{243}$

9. $\sqrt[3]{-27}$ **10.** $\sqrt[3]{24}$ **11.** $\sqrt[3]{-108}$ **12.** $\sqrt[3]{40}$

13. $\sqrt{-16}$ **14.** $\sqrt{-144}$ **15.** $\sqrt{12a^2b^3c}$ **16.** $\sqrt{8x^3y^5}$

17. $\sqrt{128x^5y^6}$ **18.** $\sqrt[3]{216a^6b^{12}}$ **19.** $\sqrt[3]{64x^3y^5}$ **20.** $\sqrt{80x^8y^{10}}$

6.2 Rational Exponents

Simplify each numerical expression in Exercises 21 – 28.

21. $16^{\frac{1}{2}}$ **22.** $81^{\frac{1}{2}}$ **23.** $-36^{-\frac{1}{2}}$ **24.** $225^{-\frac{1}{2}}$

25. $\left(\dfrac{81}{16}\right)^{-\frac{1}{4}}$ **26.** $\left(\dfrac{27}{125}\right)^{\frac{2}{3}}$ **27.** $\left[\left(\dfrac{1}{32}\right)^{\frac{3}{5}}\right]^{-2}$ **28.** $-100^{-\frac{3}{2}}$

In Exercises 29 – 32, use a graphing calculator to find the value of each numerical expression (accurate to 4 decimal places, if necessary).

29. $36^{\frac{2}{3}}$ **30.** $200^{\frac{5}{6}}$ **31.** $\sqrt[4]{2500}$ **32.** $\sqrt[5]{0.00081}$

Simplify each algebraic expression in Exercises 33 – 38. Leave the answers in rational exponent form.

33. $\left(3x^{\frac{2}{3}}\right)^3$ **34.** $\left(16a^2\right)^{\frac{3}{4}}$ **35.** $a^{\frac{1}{4}} \cdot a^{\frac{2}{3}}$ **36.** $5y^3 \cdot y^{\frac{1}{3}}$

37. $\left(\dfrac{27a^6b^3}{c^3}\right)^{\frac{2}{3}}$ **38.** $\dfrac{\left(4x^{-2}y^{\frac{1}{2}}\right)^{-\frac{1}{2}}}{xy^{\frac{1}{4}}}$

In Exercises 39 – 40, simplify each expression by first changing it into an equivalent expression with rational exponents. Rewrite the answer in simplified radical form.

39. $\dfrac{\sqrt[3]{x}\,\sqrt[6]{x^5}}{\sqrt[4]{x^3}}$

40. $\sqrt{\sqrt[3]{64x^2}}$

6.3 Operations with Radicals

Perform the indicated operations and simplify for Exercises 41 – 49. Assume that all variables are positive.

41. $\sqrt{11} - 5\sqrt{11}$

42. $6\sqrt{x} + \sqrt{x} - 2\sqrt{x}$

43. $3\sqrt{28} + 4\sqrt{7} + 2\sqrt{16}$

44. $2\sqrt{12} - 6\sqrt{75} + \sqrt{50}$

45. $2x\sqrt{y} - \sqrt{4x^2 y}$

46. $\left(2\sqrt{6} + 5\right)\left(2\sqrt{6} - 5\right)$

47. $\left(\sqrt{5} + \sqrt{2}\right)^2$

48. $\left(\sqrt{x} - \sqrt{y}\right)\left(\sqrt{y} + \sqrt{x}\right)$

49. $\left(3\sqrt{x} + \sqrt{2}\right)\left(5\sqrt{x} + \sqrt{2}\right)$

In Exercises 50 – 57, rationalize the denominator and simplify if possible.

50. $\sqrt{\dfrac{1}{36x}}$

51. $\sqrt{\dfrac{9}{25y}}$

52. $\dfrac{\sqrt[3]{x}}{\sqrt[3]{16x^2}}$

53. $\dfrac{\sqrt[3]{x^2}}{\sqrt[3]{10xy}}$

54. $\dfrac{20}{3 + \sqrt{5}}$

55. $\dfrac{5}{\sqrt{7} + \sqrt{3}}$

56. $\dfrac{\sqrt{y} - x}{\sqrt{y} + x}$

57. $\dfrac{\sqrt{5} + \sqrt{3}}{2\sqrt{5} - \sqrt{3}}$

In Exercises 58 – 61, use a graphing calculator to find the value of each expression accurate to 4 decimal places.

58. $23 + \sqrt{6}$

59. $\sqrt{70} - 10$

60. $\dfrac{\sqrt{3}}{1 + 4\sqrt{3}}$

61. $\left(\sqrt{8} + 6\right)\left(\sqrt{8} - 6\right)$

6.4 Equations with Radicals

In Exercises 62 – 76, solve the following equations. Be sure to check your answers in the original equation.

62. $\sqrt{3x + 4} = 5$

63. $\sqrt{6x - 20} = 8$

64. $\sqrt{4x + 1} = \sqrt{2x + 5}$

65. $\sqrt{4 - x} = \sqrt{3x - 8}$

66. $\sqrt{x^2 - 23} = 11$

67. $\sqrt{x^2 - 17} = 8$

68. $x - 7 = \sqrt{x - 5}$

69. $\sqrt{x - 3} + \sqrt{x} = 3$

70. $\sqrt{x} + \sqrt{3x + 9} = 9$

71. $\sqrt[3]{x(x - 6)} = -2$

72. $\sqrt{3x + 1} = -4$

73. $\sqrt{4x + 5} = -5$

74. $\sqrt{2x - 5} = \sqrt{x + 9} + 1$

75. $\sqrt{2x + 3} = 9 - \sqrt{x + 5}$

76. $\sqrt[4]{10x + 56} = 4$

6.5 Functions with Radicals

In Exercises 77 – 80, evaluate each function as indicated and write the answers in both radical notation and decimal notation (accurate to 4 decimal places, if necessary).

77. Given $f(x) = \sqrt{3x+1}$, find **a.** $f(5)$ and **b.** $f(0)$.

78. Given $g(x) = \sqrt{1-4x}$, find **a.** $g\left(\dfrac{1}{4}\right)$ and **b.** $g(-6)$.

79. Given $h(x) = \sqrt[3]{6-x}$, find **a.** $h(0)$ and **b.** $h(-10)$.

80. Given $f(x) = \sqrt{2x+5}$, find **a.** $f(-1)$ and **b.** $f(3)$.

In Exercises 81 – 86, use interval notation to indicate the domain of each radical function.

81. $y = \sqrt{x+5}$ **82.** $y = \sqrt{2x+1}$ **83.** $f(x) = \sqrt[3]{4x}$

84. $g(x) = \sqrt{1-5x}$ **85.** $y = \sqrt[4]{3-2x}$ **86.** $f(x) = \sqrt[5]{9-x}$

In Exercises 87 – 90, find and label at least 5 points on the graph of the function and then sketch the graph of the function.

87. $y = \sqrt{x-3}$ **88.** $y = -\sqrt{x+4}$ **89.** $y = \sqrt[3]{x+1}$ **90.** $y = 2\sqrt{x-1}$

Use a graphing calculator to graph each of the functions in Exercises 91 – 96.

91. $y = -2\sqrt{x+3}$ **92.** $y = 3\sqrt{x-1}$ **93.** $f(x) = \sqrt[3]{x-4}$

94. $f(x) = -2\sqrt[3]{x+3}$ **95.** $y = \sqrt[4]{x+1}$ **96.** $y = \sqrt[5]{x+5}$

6.6 Introduction to Complex Numbers

Find the real part and the imaginary part of each of the complex numbers in Exercises 97 – 100.

97. $5 - 2i$ **98.** $-10 + \sqrt{3}i$ **99.** $\dfrac{3}{4}$ **100.** $\dfrac{3+i}{2}$

Simplify the radicals in Exercises 101 – 104.

101. $\sqrt{-36}$ **102.** $\sqrt{144}$ **103.** $-\sqrt{-72}$ **104.** $2\sqrt{725}$

For Exercises 105 – 108, solve each equation for x and for y.

105. $2x + 3i = 18 - (y - 3)i$

106. $\sqrt{3} - 4i = x - 3yi$

107. $10i - x + 4 = 14 + (3 - y)i$

108. $3 + 2i + x = 8 - 6i + yi$

Find each sum or difference as indicated in Exercises 109 – 116.

109. $(3 + 2i) + (2 + 6i)$

110. $(6 + 5i) - (2 - 3i)$

111. $(-7 + i) - (-3 - i)$

112. $(4 - 3i) + (\sqrt{2} - i)$

113. $(\sqrt{5} + \sqrt{3}i) + (2 - \sqrt{3}i)$

114. $(12 - 4\sqrt{-25}) + (3 - 5\sqrt{-1})$

115. $(-3 + 2i) + (7 + i) - (3 - 2i)$

116. $(-6 - 7i) + (5 - 4i) + (6 + 10i)$

6.7 Multiplication and Division with Complex Numbers

In Exercises 117 – 126, perform the indicated operations and write each result in standard form.

117. $4i(3 - \sqrt{2}i)$

118. $(4 + 3i)(1 - i)$

119. $(2 + 3i)^2$

120. $(\sqrt{3} + i)(2\sqrt{3} - i)$

121. $\dfrac{10}{4 - 2i}$

122. $\dfrac{2 - i}{3 + 4i}$

123. $\dfrac{7 - 2i}{\sqrt{5} - i}$

124. i^{15}

125. i^{-5}

126. i^{10}

Chapter 6 Test

Simplify the expressions in Exercises 1 – 4. Assume that all variables are positive.

1. $\sqrt{112}$ **2.** $\sqrt{120xy^4}$ **3.** $\sqrt[3]{48x^2y^5}$ **4.** $\sqrt[3]{81x^4y^3z^3}$

5. Write $(2x)^{\frac{2}{3}}$ in radical notation.

6. Write $\sqrt[6]{8x^2y^4}$ as an equivalent, simplified expression with rational exponents.

Simplify each numerical expression in Exercises 7 – 11. Assume that all variables are positive. Leave the answers in rational exponent form.

7. $(-8)^{\frac{2}{3}}$ **8.** $9^{\frac{-3}{2}}$ **9.** $4x^{\frac{1}{2}}\cdot x^{\frac{2}{3}}$

10. $\left(49x^{\frac{1}{2}}y^{\frac{-2}{3}}\right)^{\frac{1}{2}}$ **11.** $\left(\dfrac{16x^{-4}y}{y^{-1}}\right)^{\frac{3}{4}}$

12. Simplify the expression, $\sqrt[3]{x^2}\cdot\sqrt[4]{x}$, by first changing it into an equivalent expression with rational exponents. Rewrite the answer in simplified radical form.

In Exercises 13 – 20, perform the indicated operations and simplify. Assume that all variables are positive.

13. $2\sqrt{75}+3\sqrt{27}-\sqrt{12}$ **14.** $\sqrt{16x^3}+\sqrt{9x^3}-\sqrt{36x}$

15. $\sqrt[3]{24}+2\sqrt[3]{81}$ **16.** $\left(\sqrt{3}-\sqrt{2}\right)^2$

17. $5x\sqrt{y^3}-2\sqrt{x^2y^3}-4y\sqrt{x^2y}$ **18.** $\left(6+\sqrt{3x}\right)\left(5-2\sqrt{3x}\right)$

19. $\dfrac{14+\sqrt{28}}{2}$ **20.** $\dfrac{15-\sqrt{50}}{5}$

21. Rationalize the denominator and simplify, if possible.

     a. $\sqrt{\dfrac{5y^2}{8x^3}}$ **b.** $\dfrac{1-x}{1-\sqrt{x}}$

In Exercises, 22 – 25, solve the following equations. Be sure to check your answers in the original equation.

22. $\sqrt{5x+1}=1+\sqrt{3x+4}$ **23.** $\sqrt[3]{3x+4}=-2$

24. $\sqrt{x+8}-2=x$ **25.** $\sqrt{2x+9}=\sqrt{x}+3$

26. Find the domain of each of the following radical functions. Write the answer in interval notation.

    a. $y = \sqrt{3x+4}$ **b.** $f(x) = \sqrt[3]{2x+5}$

27. Use a graphing calculator to graph each of the following radical functions. Sketch the graph and label 3 points on the graph.

    a. $f(x) = -\sqrt{x+3}$ **b.** $y = \sqrt[3]{x-4}$

Find the real part and the imaginary part of each of the complex numbers in Exercises 28 and 29.

28. $6 - \dfrac{1}{3}i$ **29.** $\dfrac{-2+3i}{5}$

Simplify the radicals in Exercises 30 and 31.

30. $\sqrt{-8}$ **31.** $3\sqrt{-169}$

In Exercises 32 – 35, perform the indicated operations and write each result in standard form.

32. $(5+8i)+(11-4i)$ **33.** $\left(2+3\sqrt{-4}\right)-\left(7-2\sqrt{-25}\right)$

34. $(4+3i)(2-5i)$ **35.** $\dfrac{2+i}{3+2i}$

36. Solve for x and y: $(2x+3i)-(6+2yi)=5-3i$.

37. Write i^{23} in the standard form, $a + bi$.

38. Find the product and simplify: $(x+2i)(x-2i)$.

39. Find the product and simplify: $\left[(x+3)-\sqrt{3}i\right]\left[(x+3)+\sqrt{3}i\right]$.

In Exercises 40 – 42, use a calculator to find the value of each expression accurate to 4 decimal places.

40. $32^{\frac{-3}{5}}$ **41.** $\left(\sqrt{2}+6\right)\left(\sqrt{2}-1\right)$ **42.** $\dfrac{\sqrt{3}-\sqrt{5}}{\sqrt{7}-\sqrt{10}}$

Cumulative Review: Chapters 1–6

State the property of real numbers that is illustrated in Exercises 1 and 2. All variables represent real numbers.

1. $\sqrt{x} + 0 = \sqrt{x}$

2. $7(4-2) = 7 \cdot 4 - 7 \cdot 2$

Solve the inequalities in Exercises 3 and 4 and graph the solution sets. Write each solution in interval notation. Assume that x is a real number.

3. $6x - 2 < 4x + 10$

4. $2|x-1| + 4 \le 6$

5. Graph the linear equation $4x - 2y = -8$ by locating the *y*-intercept and *x*-intercept.

6. Write the equation $x + 6y = 18$ in slope-intercept form. Then, find the slope and the *y*-intercept, and draw the graph.

7. Find an equation for the vertical line through the point $(5, -7)$.

8. The graphs of three curves are shown. Use the vertical line test to determine whether or not each graph represents a function. Then state the domain and range of each graph.

a.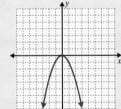

b.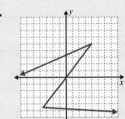

c.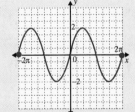

9. Solve the following system of linear equations by graphing both equations and locating the point of intersection.

$$\begin{cases} 3x - 2y = 7 \\ x + 3y = -5 \end{cases}$$

10. Solve the following system of linear equations using Gaussian elimination. You may use your calculator.

$$\begin{cases} x + y - z = 1 \\ 3x - y + z = 3 \\ -x + 2y + 2z = 3 \end{cases}$$

11. Solve the following system of linear equations using Cramer's Rule.

$$\begin{cases} -2x + 5y = 10 \\ 6x - 2y = 30 \end{cases}$$

12. Find the value of $\det(A)$ for the matrix:

$$A = \begin{bmatrix} 5 & -2 & 0 \\ -1 & 1 & 6 \\ 7 & 8 & -3 \end{bmatrix}.$$

13. Solve the following system of inequalities graphically: $\begin{cases} x + y \le 4 \\ 3x - y \le 2 \end{cases}.$

14. Write the expression $\dfrac{0.008 \times 40000}{320 \times 0.001}$ in scientific notation and simplify.

Perform the indicated operations in Exercises 15 – 17.

15. $\left(x^2 + 7x - 5\right) - \left(-2x^3 + 5x^2 - x - 1\right)$ **16.** $(2x + 7)(3x - 1)$ **17.** $(5x + 2)(4 - x)$

18. Use the division algorithm to divide and write the answer in the form $Q + \dfrac{R}{D}$ where the degree of $R <$ the degree of D.

$$\dfrac{x^3 - 7x^2 + 2x - 15}{x^2 + 2x - 1}$$

In Exercises 19 – 22, factor completely.

19. $12x^2 - 7x - 12$ **20.** $28 + x - 2x^2$ **21.** $5x^3 - 320$ **22.** $x^3 + 4x^2 - x - 4$

In Exercises 23 – 25, solve each quadratic equation by factoring.

23. $x^2 - 13x - 48 = 0$ **24.** $x = 2x^2 - 6$ **25.** $0 = 15x^2 - 11x + 2$

26. Given $P(x) = x^3 - 8x^2 + 19x - 12$, find **a.** $P(0)$ **b.** $P(4)$ and **c.** $P(-3)$.

27. a. Find an equation that has $x = 4$ and $x = 7$ as roots.

b. Find an equation that has $x = -2$, $x = 12$, and $x = 1$ as roots.

Perform the indicated operations in Exercises 28 and 29 and reduce if possible.

28. $\dfrac{x+1}{x^2+x-6} + \dfrac{3x-2}{x^2-2x-15}$ **29.** $\dfrac{2x+5}{4x^2-1} - \dfrac{2-x}{2x^2+7x+3}$

Simplify the complex fractions in Exercises 30 and 31.

30. $\dfrac{1 - \dfrac{1}{x^2}}{\dfrac{2}{x} - \dfrac{4}{x^2}}$ **31.** $\dfrac{x + 2 - \dfrac{12}{x+3}}{x - 5 + \dfrac{16}{x+3}}$

State the restrictions on the variables and solve each equation in Exercises 32 and 33.

32. $\dfrac{3}{x}+\dfrac{2}{x+5}=\dfrac{8}{3x}$

33. $\dfrac{9}{x+7}+\dfrac{3x}{x^2+4x-21}=\dfrac{8}{x-3}$

Simplify the expressions in Exercises 34 – 37. Assume that all variables are positive.

34. $8^{\frac{-4}{3}}$

35. $\left(x^{\frac{1}{2}}\cdot x^{\frac{2}{3}}\right)^2$

36. $\sqrt{288}$

37. $\sqrt[3]{16x^6y^{10}}$

In Exercises 38 and 39, rationalize the denominator and simplify if possible.

38. $\dfrac{\sqrt{5}}{\sqrt{2y}}$

39. $\dfrac{\sqrt{5}-\sqrt{6}}{\sqrt{5}+\sqrt{6}}$

Solve the radical equations given in Exercises 40 – 42. Be sure to check your answers in the original equation.

40. $\sqrt{x+10}+1=x+5$

41. $2\sqrt{x}=x-3$

42. $\sqrt[3]{x-7}+3=1$

Perform the indicated operations in Exercises 43 – 45 and simplify. Write your answers in standard form.

43. $(2+2i)(3-4i)$

44. $\dfrac{4-3i}{1+4i}$

45. i^{44}

Solve for the indicated variable in Exercises 46 and 47.

46. Solve $s=a+(n-1)d$ for n

47. Solve $A=p+prt$ for p

48. **Skiing:** The following table shows the number of feet that Linda skied downhill from one second to another. (She never skied uphill, but she did stop occasionally.)

Distance Traveled (in feet)	Time Elapsed (in seconds)
24	3
48	6
96	8
96	15
200	20

a. Plot the points indicated in the table.
b. Calculate the slope of the line segments from point to point.
c. Interpret each slope.

49. The sum of the squares of two consecutive positive odd integers is 514. What are the integers?

50. **Volume of a gas:** V (volume) varies inversely as P (pressure) when a gas is enclosed in a container. In a particular situation, the volume of gas is 25 in^3 when a force of 10 lb per in.^{2} is exerted. What would be the volume of gas if a pressure of 15 lb per in.^{2} were to be used?

51. **Resistance of a wire:** The resistance, R (in ohms), in a wire is directly proportional to the length, L, and inversely proportional to the square of the diameter of the wire. The resistance of a wire 500 ft long with a diameter of 0.01 in. is 20 ohms. What is the resistance of a wire of the same type that is 200 ft long?

52. **Boating:** Susan traveled 25 miles downstream. In the same length of time, she could have traveled 15 miles upstream. If the speed of the current is 2.5 mph, find the speed of Susan's boat in still water.

53. **Investing:** Harold has $50,000 that he wants to invest in two accounts. One pays 6% interest, and the other (at a higher risk) pays 10% interest. If he wants a $3600 annual return on these two investments, how much should he put into each account?

54. **Reporting Sales:** Robin can prepare a monthly sales report in 5 hours. If Mac helps her, together they can prepare the report in 3 hours. How long would it take Mac if he worked alone?

55. **Selling candy:** A grocer plans to make up a special mix of two popular kinds of candy for Halloween. He wants to mix a total of 100 pounds to sell for $1.75 per pound. Individually, the two types sell for $1.25 and $2.50 per pound. How many pounds of each of the two kinds should he put in the mix?

Use a graphing calculator to solve the equations in Exercises 56 and 57. Sketch the graphs that you use and find any approximations accurate to two decimal places.

56. $x^3 = 3x^2 - 3x + 1$

57. $\left| x^2 - 9 \right| = 3$

In Exercises 58 – 60, use a graphing calculator to first graph each function and then use the CALC features to find the zeros of each function.

58. $y = x^2 - 3x - 6$

59. $y = x^3 - 4x + 3$

60. $y = -x^3 + 2x^2 + 19x - 20$

In Exercises 61 – 64, use a calculator to find the value of each expression accurate to 4 decimal places.

61. $36^{\frac{7}{2}}$

62. $125^{\frac{-2}{3}}$

63. $\sqrt{5}\left(\sqrt{21} - 4 \right)$

64. $\dfrac{13 + \sqrt{12}}{5 - \sqrt{6}}$

Quadratic Equations and Quadratic Functions

Did You Know?

Much of classical algebra has focused on the problem of solving equations, such as the second-degree equation $ax^2 + bx + c = 0$, which is studied in this chapter. In Section 7.2, you will be introduced to the quadratic formula, a formula that has been known since approximately 2000 B.C. This formula expresses the roots of the quadratic equation in terms of the coefficients a, b, and c. Italian mathematicians during the 16th century also discovered that general cubic equations, in the form $ax^3 + bx^2 + cx + d = 0$, and general fourth-degree equations, in the form $ax^4 + bx^3 + cx^2 + dx + e = 0$ could be solved by similar types of formulas. Not many people are aware that these general formulas exist. They can be found in a book of mathematical formulas or in a theory of equations text.

After the discovery of general formulas for the first four cases of polynomial equations (linear, quadratic, cubic, and quartic), it was assumed that a fifth-degree equation also could be solved. However, in the early 1800s, two brilliant young mathematicians proved that the fifth-degree equation was not solvable by algebraic formulas involving the coefficients, as the previous four cases had been. The mathematicians were Niels Henrik Abel (1802 – 1829) and Evariste Galois (1811 – 1832).

Galois

Galois was a fascinating person whose life ended in a romantic duel that may have been arranged by right-wing politicians to eliminate the brilliant young radical. He was involved in the French Revolution and was expelled from school because of his political activity. His brilliance in mathematics was unrecognized because of his youth, his bitterness toward organized science, his radical politics, and the sophistication of his work. The night before his duel, anticipating his death, Galois sat down and wrote out all of the creative mathematics that he could–much of it abstract algebra that he had worked out mentally but not recorded. Included in this work was the unsolvability of fifth-degree equations. Galois' turbulent life has been fictionalized in a novel, *Whom the Gods Love*, by Leopold Infeld. Galois' proof that fifth-degree equations could not be solved kept mathematicians from trying to do the impossible and also opened a whole new field in higher mathematics known as **group theory**.

Abel's life ended less dramatically, but also tragically. He died of tuberculosis brought about by poverty, because no one recognized his talent. Ironically, he died two days before he was offered a professorship at the University of Berlin. Abel and Galois both left mathematical legacies that kept future mathematicians engaged in developing their ideas up to the present time.

"Through and through the world is infested with quantity: To talk sense is to talk quantities. It is no use saying the nation is large–How large? It is no use saying that radium is scarce–How scarce? You cannot evade quantity. You may fly to poetry and music, and quantity and number will face you in your rhythms and your octaves."

Abel

Alfred North Whitehead (1861 – 1947)

Quadratic equations appear in one form or another in almost every course in mathematics and in many courses in related fields such as business, biology, engineering, and computer science. In this chapter, we will discuss three techniques for solving quadratic equations: factoring, completing the square, and the quadratic formula. Solving quadratic equations by factoring has already been discussed in Section 4.8 and, when possible, is generally considered the method of first choice because it is easiest to apply and is useful in other mathematical situations. However, the **quadratic formula** is very important and should be memorized as it works in all cases. A part of the formula, called the **discriminant**, gives ready information about the nature of the solutions. Additionally, the formula is easy to use in computer programs.

Relatedly, the concept of a **function** is present in many aspects of our daily lives. The speed at which you drive your car is a **function** of how far you depress the accelerator; your energy level is a **function** of the amount and type of food you eat; your grade in this class is a **function** of the quality time you spend studying. In this text we quantify these ideas by dealing only with functions involving pairs of real numbers. As we have seen in Chapters 2, 4, and 6, this restriction allows us to analyze functions in terms of their graphs in the Cartesian coordinate system. In Section 7.5, we will be particularly interested in a special category of functions called quadratic functions whose graphs are parabolas (curves that, among other things, can be used to describe the paths of projectiles).

7.1 Solving Quadratic Equations

- *Solve quadratic equations by factoring.*
- *Solve quadratic equations by using the definition of square root.*
- *Solve quadratic equations by completing the square.*
- *Find polynomials with given roots.*

Solving Quadratic Equations by Factoring

Not every polynomial can be factored so that the factors have integer coefficients, and not every polynomial equation can be solved by factoring. However, when the solutions of a polynomial equation can be found by factoring, the method depends on the **zero-factor property**, which is restated here for easy reference.

Zero-Factor Property

If the product of two factors is 0, then one or both of the factors must be 0. Symbolically, for factors a and b,

$$\text{\textit{if } } a \cdot b = 0 \text{\textit{, then } } a = 0 \text{\textit{ or } } b = 0 \text{\textit{ or both}}.$$

Also, as discussed in Section 4.8, polynomial equations of second-degree are called **quadratic equations** and, because these are the equations of interest in this chapter, the definition is restated here.

Quadratic Equations

An equation that can be written in the form

$$ax^2 + bx + c = 0 \qquad \text{\textit{where a, b, and c are real numbers and } } a \neq 0$$

*is called a **quadratic equation**.*

The procedure for solving quadratic equations by factoring involves making sure that one side of the equation is 0 and then applying the zero-factor property. The following list of steps outlines the procedure.

To Solve an Equation by Factoring

*1. Add or subtract terms so that **one side of the equation is 0**.*

2. Factor the polynomial expression.

3. Set each factor equal to 0 and solve each of the resulting equations.

*(**Note:** If two of the factors are the same, then the solution is said to be a **double root** or a **root of multiplicity two**.)*

Example 1: Solving Quadratic Equations by Factoring

Solve the following quadratic equations by factoring.

a. $x^2 - 15x = -50$

Solution: $\quad x^2 - 15x = -50$

$x^2 - 15x + 50 = 0$ Add 50 to both sides. **One side must be 0.**

$(x - 5)(x - 10) = 0$ Factor the left-hand side.

$x - 5 = 0 \quad \text{or} \quad x - 10 = 0$ Set each factor equal to 0.

$x = 5 \qquad\qquad x = 10$ Solve each linear equation.

Check: $\quad (5)^2 - 15 \cdot (5) \overset{?}{=} -50 \quad (10)^2 - 15 \cdot (10) \overset{?}{=} -50$

$25 - 75 \overset{?}{=} -50 \qquad 100 - 150 \overset{?}{=} -50$

$-50 = -50 \qquad\qquad -50 = -50$

b. $3x^2 - 24x = -48$

Solution: $\quad 3x^2 - 24x = -48$

$3x^2 - 24x + 48 = 0$ Add 48 to both sides. **One side must be 0.**

$3(x^2 - 8x + 16) = 0$ Factor out the GCF, 3.

$3(x - 4)^2 = 0$ The trinomial is a perfect square.

$x - 4 = 0$ Two factors are the same.

$x = 4$ The solution is a **double root**.

Check: $\quad 3 \cdot (4)^2 - 24 \cdot (4) \overset{?}{=} -48$

$48 - 96 \overset{?}{=} -48$

$-48 = -48$

Now that complex numbers have been introduced and discussed, we will see that quadratic equations may have nonreal complex solutions. **In particular, the sum of two squares (previously declared "not factorable" because real factors were implied) can be factored as the product of complex conjugates.** For example,

$$x^2 + 9 = (x + 3i)(x - 3i).$$

As shown in Example 2, such factors can lead to nonreal solutions to a quadratic equation.

Example 2: Quadratic Equations Involving the Sum of Two Squares

Solve the following quadratic equation by factoring: $x^2 + 4 = 0$.

Solution:
$$x^2 + 4 = 0$$
$$(x + 2i)(x - 2i) = 0$$
$$x + 2i = 0 \quad \text{or} \quad x - 2i = 0$$
$$x = -2i \qquad\qquad x = 2i$$

Note that $x^2 + 4$ is the sum of two squares and can be factored into the product of conjugates of complex numbers.

Check:
$$(-2i)^2 + 4 \stackrel{?}{=} 0 \qquad (2i)^2 + 4 \stackrel{?}{=} 0$$
$$4i^2 + 4 \stackrel{?}{=} 0 \qquad 4i^2 + 4 \stackrel{?}{=} 0$$
$$-4 + 4 \stackrel{?}{=} 0 \qquad -4 + 4 \stackrel{?}{=} 0$$
$$0 = 0 \qquad\qquad 0 = 0$$

Using the Definition of Square Root and the Square Root Property

Consider the equation
$$x^2 = 13.$$

Allowing that the variable, x, might be positive or negative, we use the definition of square root, $\sqrt{x^2} = |x|$.

Taking the square root of both sides of the equation gives:
$$|x| = \sqrt{13}.$$

So, we have two solutions,
$$x = \sqrt{13} \quad \text{and} \quad x = -\sqrt{13}.$$

To indicate both solutions, we write
$$x = \pm\sqrt{13}.$$

Similarly, for the equation
$$(x - 3)^2 = 5$$

the definition of square root gives
$$x - 3 = \pm\sqrt{5}.$$

This leads to the two equations and the two solutions, as follows:

$$x - 3 = \sqrt{5} \quad \text{or} \quad x - 3 = -\sqrt{5}$$
$$x = 3 + \sqrt{5} \qquad x = 3 - \sqrt{5}.$$

We can write the two solutions in the form

$$x = 3 \pm \sqrt{5}.$$

This discussion shows how the definition of a square root can be used to solve quadratic equations in certain forms. In particular, if one side of the equation is a squared expression and the other side is a constant, we can simply take the square root of both sides. If the constant is negative, then the solutions will involve nonreal numbers. We are using the following **square root property**.

Square Root Property

If $x^2 = c$, then $x = \pm\sqrt{c}$.

If $(x - a)^2 = c$, then $x - a = \pm\sqrt{c}$ $\left(or \ x = a \pm \sqrt{c}\right)$.

Note: *If c is negative $(c < 0)$, then the solutions will be nonreal.*

Example 3: Using the Square Root Property to Solve Quadratic Equations

Solve the following quadratic equations using the Square Root Property.

a. $(y + 4)^2 = 8$

Solution: $(y + 4)^2 = 8$ Note that $c > 0$.

$$y + 4 = \pm\sqrt{8}$$
$$y = -4 \pm 2\sqrt{2}$$

b. $x^2 = -25$

Solution: $x^2 = -25$ Note that $c < 0$.

$$x = \pm\sqrt{-25}$$
$$x = \pm 5i$$

Completing the Square

Recall that a perfect square trinomial (Section 4.3) is the result of squaring a binomial. Our objective here is to find the third term of a perfect square trinomial when the first two terms are given. This is called **completing the square**. We will find this procedure useful in solving quadratic equations and in developing the quadratic formula.

Study the following examples to help in your understanding.

Perfect Square Trinomials		Equal Factors		Square of a Binomial
$x^2 - 8x + 16$	=	$(x - 4)(x - 4)$	=	$(x - 4)^2$
$x^2 + 20x + 100$	=	$(x + 10)(x + 10)$	=	$(x + 10)^2$
$x^2 - 9x + \dfrac{81}{4}$	=	$\left(x - \dfrac{9}{2}\right)\left(x - \dfrac{9}{2}\right)$	=	$\left(x - \dfrac{9}{2}\right)^2$
$x^2 - 2ax + a^2$	=	$(x - a)(x - a)$	=	$(x - a)^2$
$x^2 + 2ax + a^2$	=	$(x + a)(x + a)$	=	$(x + a)^2$

Table 7.1

The last two examples are in the form of formulas. We see two things in each case:

1. The leading coefficient (the coefficient of x^2) is 1.

2. The constant term is the square of $\dfrac{1}{2}$ of the coefficient of x.

    For example, $\dfrac{1}{2}(2a) = a$ and the square of this result is the constant a^2.

What constant should be added to $x^2 - 16x$ to get a perfect square trinomial? By following the ideas just discussed, we find that $\dfrac{1}{2}(-16) = -8$ and $(-8)^2 = 64$. Therefore, to complete the square, we add 64. Thus,

$$x^2 - 16x + 64 = (x - 8)^2.$$

By adding 64, we have completed the square for $x^2 - 16x$.

Example 4: Completing the Square

Add the constant that will complete the square for each expression, and write the new expression as the square of a binomial.

a. $x^2 + 10x$

Solution: $x^2 + 10x + \underline{} = (\underline{})^2$

$\frac{1}{2}(10) = 5$ and $(5)^2 = 25$ Find $\frac{1}{2}$ the coefficient of x and square the result.

So, add 25: $x^2 + 10x + \underline{25} = (\underline{x+5})^2$

b. $x^2 - 7x$

Solution: $x^2 - 7x + \underline{} = (\underline{})^2$

$\frac{1}{2}(-7) = -\frac{7}{2}$ and $\left(-\frac{7}{2}\right)^2 = \frac{49}{4}$ Find $\frac{1}{2}$ the coefficient of x and square the result.

So, add $\frac{49}{4}$: $x^2 - 7x + \underline{\dfrac{49}{4}} = \left(\underline{x - \dfrac{7}{2}}\right)^2$

Solving Quadratic Equations by Completing the Square

Now we want to use the process of completing the square to help in solving quadratic equations. This technique involves the following steps.

To Solve a Quadratic Equation by Completing the Square

1. *If necessary, divide or multiply both sides of the equation so that the leading coefficient (the coefficient of x^2) is 1.*

2. *If necessary, isolate the constant term on one side of the equation.*

3. *Find the constant that completes the square of the polynomial and add this constant to both sides. Rewrite the polynomial as the square of a binomial.*

4. *Use the Square Root Property to find the solutions of the equation.*

Example 5: Solving Quadratic Equations by Completing the Square

Solve the following quadratic equations by completing the square.

a. $x^2 - 8x = 25$

Solution: $x^2 - 8x = 25$ The coefficient of x^2 is already 1 and the constant is isolated on one side of the equation.

$$x^2 - 8x + 16 = 25 + 16$$ $\frac{1}{2}(-8) = -4$ and $(-4)^2 = 16$.

Therefore, add 16 to both sides.

$$(x - 4)^2 = 41$$

$$x - 4 = \sqrt{41} \quad \text{or} \quad x - 4 = -\sqrt{41}$$ Use the Square Root Property.

$$x = 4 + \sqrt{41} \qquad x = 4 - \sqrt{41}$$

There are two real solutions: $4 + \sqrt{41}$ and $4 - \sqrt{41}$.

We write $x = 4 \pm \sqrt{41}$.

b. $3x^2 + 6x - 15 = 0$

Solution: $3x^2 + 6x - 15 = 0$

$$\frac{3x^2}{3} + \frac{6x}{3} - \frac{15}{3} = \frac{0}{3}$$ Divide each term by 3. **The leading coefficient must be 1.**

$$x^2 + 2x - 5 = 0$$ Isolate the constant term and complete the

$$x^2 + 2x = 5$$ square: $\frac{1}{2}(2) = 1$ and $1^2 = 1$.

$$x^2 + 2x + 1 = 5 + 1$$ Therefore, add 1 to both sides.

$$(x + 1)^2 = 6$$ Factor the polynomial.

$$x + 1 = \pm\sqrt{6}$$ Use the Square Root Property.

$$x = -1 \pm \sqrt{6}$$

c. $2x^2 + 2x - 7 = 0$

Solution: $2x^2 + 2x - 7 = 0$

$$x^2 + x - \frac{7}{2} = 0$$ Divide each term by 2 so that the leading coefficient will be 1.

$$x^2 + x = \frac{7}{2}$$ Isolate the constant term and complete the

square: $\frac{1}{2}(1) = \frac{1}{2}$ and $\left(\frac{1}{2}\right)^2 = \frac{1}{4}$.

$$x^2 + x + \frac{1}{4} = \frac{7}{2} + \frac{1}{4}$$

Continued on the next page...

$$\left(x+\frac{1}{2}\right)^2 = \frac{15}{4}$$ Factor the polynomial.

$$x+\frac{1}{2} = \pm\sqrt{\frac{15}{4}}$$ Use the Square Root Property.

$$x = -\frac{1}{2} \pm \frac{\sqrt{15}}{2}$$

$$x = \frac{-1 \pm \sqrt{15}}{2}$$

d. $x^2 - 2x + 13 = 0$

Solution: $x^2 - 2x + 13 = 0$

$$x^2 - 2x = -13$$

$$x^2 - 2x + 1 = -13 + 1$$

$$(x-1)^2 = -12$$

$$x - 1 = \pm\sqrt{-12} = \pm i\sqrt{12} = \pm 2i\sqrt{3}$$

$$x = 1 \pm 2i\sqrt{3}$$ The solutions are nonreal complex conjugates.

Writing Equations with Known Roots

In Section 4.8, we found equations with known roots by setting the product of factors equal to 0 and simplifying. The same method is applied here with roots that are nonreal and roots that involve radicals.

Example 6: Equations with Known Roots

Find polynomial equations that have the given roots.

a. $y = 3 + 2i$ and $y = 3 - 2i$

Solution: $y = 3 + 2i$ $y = 3 - 2i$

$y - 3 - 2i = 0$ $y - 3 + 2i = 0$ Get 0 on one side of each equation.

Set the product of the two factors equal to 0 and simplify.

$$[y - 3 - 2i][y - 3 + 2i] = 0$$

$$[(y-3) - 2i][(y-3) + 2i] = 0$$

$$(y-3)^2 - 4i^2 = 0$$

$$y^2 - 6y + 9 + 4 = 0$$

$$y^2 - 6y + 13 = 0$$

Regroup the terms to represent the product of complex conjugates. This makes the multiplication easier.

$i^2 = -1$

This equation has two solutions: $y = 3 + 2i$ and $y = 3 - 2i$.

b. $x = 5 - \sqrt{2}$ and $x = 5 + \sqrt{2}$

Solution: $x = 5 - \sqrt{2}$ $x = 5 + \sqrt{2}$

$x - 5 + \sqrt{2} = 0$ $x - 5 - \sqrt{2} = 0$ Get 0 on one side of each equation.

Set the product of the two factors equal to 0 and simplify.

$$\left[x - 5 + \sqrt{2}\right]\left[x - 5 - \sqrt{2}\right] = 0$$

$$\left[(x - 5) + \sqrt{2}\right]\left[(x - 5) - \sqrt{2}\right] = 0 \qquad \text{Regroup the terms to make the multiplication easier.}$$

$$(x - 5)^2 - \left(\sqrt{2}\right)^2 = 0$$

$$x^2 - 10x + 25 - 2 = 0$$

$$x^2 - 10x + 23 = 0 \qquad \text{This equation has two solutions:} \\ x = 5 - \sqrt{2} \text{ and } x = 5 + \sqrt{2}.$$

c. $x = 3 + i\sqrt{5}$ and $x = 3 - i\sqrt{5}$

Solution: $x = 3 + i\sqrt{5}$ $x = 3 - i\sqrt{5}$

$x - 3 - i\sqrt{5} = 0$ $x - 3 + i\sqrt{5} = 0$ Get 0 on one side of each equation.

Set the product of the two factors equal to 0 and simplify.

$$\left[x - 3 - i\sqrt{5}\right]\left[x - 3 + i\sqrt{5}\right] = 0$$

$$\left[(x - 3) - i\sqrt{5}\right]\left[(x - 3) + i\sqrt{5}\right] = 0 \qquad \text{Regroup the terms to make the multiplication easier.}$$

$$(x - 3)^2 - \left(i\sqrt{5}\right)^2 = 0$$

$$x^2 - 6x + 9 - i^2\left(\sqrt{5}\right)^2 = 0$$

$$x^2 - 6x + 9 - (-1)(5) = 0 \qquad \text{Remember, } i^2 = -1.$$

$$x^2 - 6x + 14 = 0 \qquad \text{This equation has two solutions:} \\ x = 3 + i\sqrt{5} \text{ and } x = 3 - i\sqrt{5}.$$

Practice Problems

Solve each of the following quadratic equations by completing the square.

1. $3x^2 - 6x + 15 = 0$ **2.** $x^2 + 2x + 2 = 0$ **3.** $x^2 - 24x + 72 = 0$

4. $2x^2 + 5x - 3 = 0$ **5.** $x^2 - 3x + 1 = 0$

6. *Find a quadratic equation that has the roots $x = 2 + 3i$ and $2 - 3i$.*

7.1 Exercises

Solve the equations in Exercises 1 – 10 by factoring.

1. $x^2 - 18x + 45 = 0$ **2.** $2x^2 = 34x - 120$ **3.** $3y^2 = 363$

4. $(x+8)(x+2) = -9$ **5.** $(z+3)^2 = 100$ **6.** $12y^2 - 18y - 12 = 0$

7. $x^4 = x^2$ **8.** $(z+4)(z^2 - 3z - 4) = 0$ **9.** $5x^3 - 25x^2 = -30x$

10. $7x^2 = 11x + 6$

Add the correct constant to complete the square in Exercises 11 – 20; then factor the trinomial as indicated.

11. $x^2 - 12x + \underline{\quad} = (\underline{\qquad})^2$ **12.** $y^2 + 14y + \underline{\quad} = (\underline{\qquad})^2$

13. $x^2 + 6x + \underline{\quad} = (\underline{\qquad})^2$ **14.** $x^2 + 8x + \underline{\quad} = (\underline{\qquad})^2$

15. $x^2 - 5x + \underline{\quad} = (\underline{\qquad})^2$ **16.** $x^2 + 7x + \underline{\quad} = (\underline{\qquad})^2$

17. $y^2 + y + \underline{\quad} = (\underline{\qquad})^2$ **18.** $x^2 + \dfrac{1}{2}x + \underline{\quad} = (\underline{\qquad})^2$

19. $x^2 + \dfrac{1}{3}x + \underline{\quad} = (\underline{\qquad})^2$ **20.** $y^2 + \dfrac{3}{4}y + \underline{\quad} = (\underline{\qquad})^2$

Answers to Practice Problems: **1.** $x = 1 \pm 2i$ **2.** $x = -1 \pm i$ **3.** $x = 12 \pm 6\sqrt{2}$ **4.** $x = -3, \ x = \dfrac{1}{2}$

5. $x = \dfrac{3 \pm \sqrt{5}}{2}$ **6.** $x^2 - 4x + 13 = 0$

Solve the equations in Exercises 21 – 35 by using the Square Root Property.

21. $x^2 - 144 = 0$

22. $x^2 - 169 = 0$

23. $x^2 + 25 = 0$

24. $x^2 + 24 = 0$

25. $x^2 + 18 = 0$

26. $(x - 2)^2 = 9$

27. $(x - 4)^2 = 25$

28. $x^2 = 5$

29. $x^2 = 12$

30. $2(x + 3)^2 = 6$

31. $3(x - 1)^2 = 15$

32. $(x - 3)^2 = -4$

33. $(x + 8)^2 = -9$

34. $(x + 2)^2 = -7$

35. $(x - 5)^2 = -10$

Solve the quadratic equations in Exercises 36 – 65 by completing the square.

36. $x^2 + 4x - 5 = 0$

37. $x^2 + 6x - 7 = 0$

38. $y^2 + 2y = 5$

39. $x^2 + 3 = 8x$

40. $x^2 - 10x + 3 = 0$

41. $z^2 + 4z = 2$

42. $x^2 - 6x + 10 = 0$

43. $x^2 + 2x + 6 = 0$

44. $x^2 + 11 = 12x$

45. $y^2 - 10y + 4 = 0$

46. $z^2 + 3z - 5 = 0$

47. $x^2 - 5x + 5 = 0$

48. $x^2 + 5x + 2 = 0$

49. $x^2 + x + 2 = 0$

50. $x^2 - 2x + 5 = 0$

51. $x^2 = 3 - 4x$

52. $3y^2 + 9y + 9 = 0$

53. $4x^2 + 8x + 16 = 0$

54. $x^2 = 6 - x$

55. $3y^2 = 4 - y$

56. $3x^2 - 10x + 5 = 0$

57. $7x + 2 = -4x^2$

58. $3y^2 + 5y - 3 = 0$

59. $4x^2 - 2x + 3 = 0$

60. $2x + 2 = -6x^2$

61. $5y^2 + 15y + 25 = 0$

62. $2x^2 + 9x + 4 = 0$

63. $2x^2 - 8x + 4 = 0$

64. $3 = 3x - 6x^2$

65. $4x^2 + 20x + 32 = 0$

For Exercises 66 – 80, write a quadratic equation with integer coefficients that has the given roots.

66. $x = \sqrt{7}, x = -\sqrt{7}$

67. $x = \sqrt{6}, x = -\sqrt{6}$

68. $x = 1 + \sqrt{3}, x = 1 - \sqrt{3}$

69. $z = 3 + \sqrt{2}, z = 3 - \sqrt{2}$

70. $y = -2 + \sqrt{5}, y = -2 - \sqrt{5}$

71. $x = 1 + 2\sqrt{3}, x = 1 - 2\sqrt{3}$

72. $x = 4i, x = -4i$

73. $x = 7i, x = -7i$

74. $y = i\sqrt{6}, y = -i\sqrt{6}$

75. $y = i\sqrt{5}, y = -i\sqrt{5}$

76. $x = 2 + i, x = 2 - i$

77. $x = -3 + 2i, x = -3 - 2i$

78. $x = 1 + i\sqrt{2}, x = 1 - i\sqrt{2}$

79. $x = 2 + i\sqrt{3}, x = 2 - i\sqrt{3}$

80. $x = -5 + 2i\sqrt{6}, x = -5 - 2i\sqrt{6}$

Writing and Thinking About Mathematics

81. Explain, in your own words, the steps involved in the process of solving a quadratic equation by completing the square.

 HAWKES LEARNING SYSTEMS: INTERMEDIATE ALGEBRA SOFTWARE

- Quadratic Equations: The Square Root Method
- Quadratic Equations: Completing the Square

7.2 The Quadratic Formula: $x = \dfrac{-b \pm \sqrt{b^2 - 4ac}}{2a}$

- *Develop the quadratic formula:* $x = \dfrac{-b \pm \sqrt{b^2 - 4ac}}{2a}$.
- *Use the quadratic formula to solve quadratic equations.*
- *Calculate the discriminant:* $b^2 - 4ac$.
- *Use the discriminant to determine the nature of the solutions (one real, two real, or two nonreal) of a quadratic equation.*

Developing the Quadratic Formula: $x = \dfrac{-b \pm \sqrt{b^2 - 4ac}}{2a}$

The **quadratic formula** gives the roots of any quadratic equation in terms of the coefficients $a, b,$ and c of the **general quadratic equation**

$$ax^2 + bx + c = 0.$$

Therefore, if you have memorized the quadratic formula, you can solve any quadratic equation by simply substituting the coefficients into the formula. To develop the quadratic formula, we solve the general quadratic equation by **completing the square**.

$$ax^2 + bx + c = 0$$
The general quadratic equation

$$x^2 + \frac{b}{a}x + \frac{c}{a} = \frac{0}{a}$$
Divide each term of the equation by a. Since $a \neq 0$, this is permissible.

$$x^2 + \frac{b}{a}x = -\frac{c}{a}$$
Add $-\dfrac{c}{a}$ to both sides of the equation.

$$x^2 + \frac{b}{a}x + \frac{b^2}{4a^2} = \frac{b^2}{4a^2} - \frac{c}{a}$$
$\dfrac{1}{2}\left(\dfrac{b}{a}\right) = \dfrac{b}{2a}$ and $\left(\dfrac{b}{2a}\right)^2 = \dfrac{b^2}{4a^2}$. Add $\dfrac{b^2}{4a^2}$ to both sides of the equation.

$$\left(x + \frac{b}{2a}\right)^2 = \frac{b^2}{4a^2} - \frac{4ac}{4a^2}$$
Factor the left side. $4a^2$ is the common denominator on the right-hand side.

$$\left(x + \frac{b}{2a}\right)^2 = \frac{b^2 - 4ac}{4a^2}$$
Simplify.

$$x + \frac{b}{2a} = \pm\sqrt{\frac{b^2 - 4ac}{4a^2}}$$
Use the Square Root Property.

$$x + \frac{b}{2a} = \pm\frac{\sqrt{b^2 - 4ac}}{2a}$$
Simplify.

$$x = -\frac{b}{2a} \pm \frac{\sqrt{b^2 - 4ac}}{2a}$$
Solve for x.

$$x = \frac{-b \pm \sqrt{b^2 - 4ac}}{2a}$$
This equation is called the Quadratic Formula.

Note about the coefficient a:

For convenience and without loss of generality, in the development of the quadratic formula (and in the examples and exercises) the leading coefficient, a, is positive. If a is a negative number, we can multiply both sides of the equation by -1. This will make the leading coefficient positive without changing any solutions of the original equation.

The Quadratic Formula

For the general quadratic equation

$$ax^2 + bx + c = 0 \qquad \text{where } a \neq 0$$

the solutions are

$$x = \frac{-b \pm \sqrt{b^2 - 4ac}}{2a}.$$

The quadratic formula should be memorized.

Applications of quadratic equations are found in such fields as economics, business, computer science, and chemistry, and in almost all branches of mathematics. Most instructors assume that their students know the quadratic formula and how to apply it. You should recognize that the importance of the quadratic formula lies in the fact that it allows you to solve **any** quadratic equation.

Example 1: The Quadratic Formula

Solve the following quadratic equations by using the quadratic formula.

a. $x^2 - 5x + 3 = 0$

Solution: Substitute $a = 1$, $b = -5$, and $c = 3$ into the formula:

$$x = \frac{-b \pm \sqrt{b^2 - 4ac}}{2a} = \frac{-(-5) \pm \sqrt{(-5)^2 - 4 \cdot 1 \cdot 3}}{2 \cdot 1}$$

$$= \frac{5 \pm \sqrt{25 - 12}}{2}$$

$$= \frac{5 \pm \sqrt{13}}{2}$$

b. $7x^2 - 2x + 1 = 0$

Solution: Substitute $a = 7$, $b = -2$, and $c = 1$ into the formula.

$$x = \frac{-b \pm \sqrt{b^2 - 4ac}}{2a} = \frac{-(-2) \pm \sqrt{(-2)^2 - 4 \cdot 7 \cdot 1}}{2 \cdot 7}$$

$$= \frac{2 \pm \sqrt{4 - 28}}{14}$$

$$= \frac{2 \pm \sqrt{-24}}{14}$$

$$= \frac{2 \pm 2i\sqrt{6}}{14}$$

$$= \frac{\cancel{2}\left(1 \pm i\sqrt{6}\right)}{\cancel{2} \cdot 7} \qquad \text{Factor and reduce.}$$

$$= \frac{1 \pm i\sqrt{6}}{7}$$

The solutions are nonreal complex conjugates. In standard form:

$$\frac{1}{7} + \frac{\sqrt{6}}{7}i \text{ and } \frac{1}{7} - \frac{\sqrt{6}}{7}i$$

c. $\dfrac{3}{4}x^2 - \dfrac{1}{2}x = \dfrac{1}{3}$

Solution: Multiply each term by the LCD, 12, so that the coefficients will be integers. The quadratic formula is easier to use with integer coefficients.

$$12 \cdot \frac{3}{4}x^2 - 12 \cdot \frac{1}{2}x = 12 \cdot \frac{1}{3}$$

$$9x^2 - 6x = 4$$

$$9x^2 - 6x - 4 = 0 \qquad \text{To apply the formula, one side must be 0.}$$

$$x = \frac{-(-6) \pm \sqrt{(-6)^2 - 4(9)(-4)}}{2 \cdot 9}$$

$$= \frac{6 \pm \sqrt{36 + 144}}{18}$$

$$= \frac{6 \pm \sqrt{180}}{18} = \frac{6 \pm 6\sqrt{5}}{18}$$

$$= \frac{\cancel{6}\left(1 \pm \sqrt{5}\right)}{\cancel{6} \cdot 3} = \frac{1 \pm \sqrt{5}}{3} \qquad \text{Factor out 6 and reduce.}$$

NOTES

COMMON ERROR

Many students make a mistake when simplifying fractions by dividing the denominator into only one of the terms in the numerator.

INCORRECT $\dfrac{4 + \cancel{2}\sqrt{3}}{\cancel{2}} = 4 + \sqrt{3}$

The correct method is to divide both terms by the denominator or to factor out a common factor in the numerator and then reduce.

CORRECT $\dfrac{4 + 2\sqrt{3}}{2} = \dfrac{4}{2} + \dfrac{2\sqrt{3}}{2} = 2 + \sqrt{3}$

CORRECT $\dfrac{4 + 2\sqrt{3}}{2} = \dfrac{\cancel{2}\left(2 + \sqrt{3}\right)}{\cancel{2}} = 2 + \sqrt{3}$

In Example 2, the equation is third-degree (a cubic equation) and one of the factors is quadratic. The quadratic formula can be applied to this factor.

Example 2: Cubic Equation

Solve the following cubic equation by factoring and by using the quadratic formula.
$$2x^3 - 10x^2 + 6x = 0$$

Solution: $2x^3 - 10x^2 + 6x = 0$

$2x\left(x^2 - 5x + 3\right) = 0$ Factor out $2x$.

$2x = 0$ or $x^2 - 5x + 3 = 0$ Set each factor equal to 0.

$x = 0$ $x = \dfrac{5 \pm \sqrt{13}}{2}$ Solve each equation. (The quadratic equation was solved in Example 1a by using the quadratic formula.)

The Discriminant: $b^2 - 4ac$

The expression $b^2 - 4ac$, the part of the quadratic formula that lies under the radical sign, is called the **discriminant**. The discriminant identifies the number and type of solutions to a quadratic equation. Assuming a, b, and c are all real numbers, there are three possibilities: the discriminant is either positive, negative, or zero.

In Example 1a, the discriminant was positive, $b^2 - 4ac = (-5)^2 - 4(1)(3) = 13$, and there were two real solutions: $x = \dfrac{5 \pm \sqrt{13}}{2}$. In Example 1b, the discriminant was negative, $b^2 - 4ac = (-2)^2 - 4(7)(1) = -24$, and there were two nonreal solutions: $x = \dfrac{1 \pm i\sqrt{6}}{7}$.

The discriminant gives the following information:

Discriminant	Nature of Solutions
$b^2 - 4ac > 0$	Two real solutions
$b^2 - 4ac = 0$	One real solution, $x = \dfrac{-b \pm 0}{2a} = -\dfrac{b}{2a}$
$b^2 - 4ac < 0$	Two nonreal solutions

Table 7.2

In the case where $b^2 - 4ac = 0$, we say $x = -\dfrac{b}{2a}$ is a **double root**. Additionally, if the discriminant is a perfect square, the equation is factorable.

Example 3: Finding the Discriminant

Find the discriminant and determine the nature of the solutions to each of the following quadratic equations.

a. $3x^2 + 11x - 7 = 0$

Solution: $b^2 - 4ac = 11^2 - 4(3)(-7)$

$$= 121 + 84$$

$$= 205 > 0 \qquad \text{There are two real solutions.}$$

b. $x^2 + 6x + 9 = 0$

Solution: $b^2 - 4ac = 6^2 - 4(1)(9)$

$$= 36 - 36$$

$$= 0 \qquad \text{There is one real solution.}$$

c. $x^2 + 1 = 0$

Solution: Here $b = 0$. We could write $x^2 + 0x + 1 = 0$.

$$b^2 - 4ac = 0^2 - 4(1)(1)$$

$$= 0 - 4$$

$$= -4 \qquad \text{There are two nonreal solutions.}$$

Example 4: Using the Discriminant

a. Determine the values for k such that $x^2 + 8x - k = 0$ will have one real solution. (**Hint:** Set the discriminant equal to 0 and solve the equation for k.)

Solution:
$$b^2 - 4ac = 0$$
$$8^2 - 4(1)(-k) = 0$$
$$64 + 4k = 0$$
$$4k = -64$$
$$k = -16$$

Check:
$$x^2 + 8x - (-16) \overset{?}{=} 0$$
$$x^2 + 8x + 16 \overset{?}{=} 0$$
$$(x+4)^2 \overset{?}{=} 0$$
$$x = -4$$

There is only one real solution. Thus, -4 is a double root.

b. Determine the values for k such that $kx^2 - 8x + 4 = 0$ will have two nonreal solutions. (**Hint:** Set the discriminant less than 0 and solve for k.)

Solution:
$$b^2 - 4ac < 0$$
$$(-8)^2 - 4(k)(4) < 0$$
$$64 - 16k < 0$$
$$-16k < -64$$
$$k > 4$$

Thus, if k is any real number greater than 4, the discriminant will be negative and the equation will have two nonreal soutions.

Practice Problems

Solve each of the following quadratic equations using the quadratic formula.

1. $x^2 + 2x - 4 = 0$

2. $2x^2 - 3x + 4 = 0$

3. $5x^2 - x - 4 = 0$

4. $\dfrac{1}{4}x^2 - \dfrac{1}{2}x = -\dfrac{1}{4}$

5. $3x^2 + 5 = 0$

6. Determine the values for k such that $x^2 - 6x + k = 0$ will have two real solutions.

7.2 Exercises

Find the discriminant and determine the nature of the solutions of each quadratic equation in Exercises 1 – 12.

1. $x^2 + 6x - 8 = 0$

2. $x^2 + 3x + 1 = 0$

3. $x^2 - 8x + 16 = 0$

4. $x^2 + 3x + 5 = 0$

5. $4x^2 + 2x + 3 = 0$

6. $3x^2 - x + 2 = 0$

7. $5x^2 + 8x + 3 = 0$

8. $4x^2 + 12x + 9 = 0$

9. $100x^2 - 49 = 0$

10. $9x^2 + 121 = 0$

11. $3x^2 + x + 1 = 0$

12. $5x^2 - 3x - 2 = 0$

Solve each of the quadratic equations in Exercises 13 – 24 using the quadratic formula.

13. $x^2 + 4x - 4 = 0$

14. $x^2 - 6x - 1 = 0$

15. $9x^2 + 12x + 4 = 0$

16. $4x^2 - 20x + 25 = 0$

17. $x^2 - 2x + 7 = 0$

18. $x^2 - 2x + 3 = 0$

19. $2x^2 + 5x - 3 = 0$

20. $3x^2 - 7x + 4 = 0$

21. $4x^2 + 6x + 1 = 0$

22. $2x^2 - 3x - 1 = 0$

23. $4x^2 + 6x + 3 = 0$

24. $x^2 - 5x + 7 = 0$

Solve the equations in Exercises 25 – 52 by using any of the techniques discussed for solving quadratic equations: factoring, completing the square, or the quadratic formula.

25. $x^2 + 3x - 5 = 0$

26. $x^2 = 7x + 3$

27. $x^2 - 5x + 2 = 0$

28. $x^2 + 4x + 3 = 0$

29. $16x^2 + 8x = -1$

30. $3x^2 + 2x - 2 = 0$

Answers to Practice Problems: **1.** $x = -1 \pm \sqrt{5}$ **2.** $x = \dfrac{3 \pm i\sqrt{23}}{4}$ **3.** $x = -\dfrac{4}{5}, x = 1$ **4.** $x = 1$

5. $x = \dfrac{\pm i\sqrt{15}}{3}$ **6.** $k < 9$

31. $6x^2 = 5x + 1$

32. $3x^2 = 18x - 33$

33. $3x^2 - 4 = 0$

34. $42x + 147 = -3x^2$

35. $x^3 - 9x^2 + 4x = 0$

36. $x^3 - 8x^2 = 3x^2 + 3x$

37. $x^3 + 3x^2 + x = 0$

38. $4x^3 + 10x^2 - 3x = 0$

39. $9x^2 - 12x + 4 = 0$

40. $9x^2 - 6x + 1 = 0$

41. $2x^2 = -8x - 9$

42. $3x^2 + 7x - 4 = 0$

43. $x^2 - 7 = 0$

44. $3x^2 = 6x - 4$

45. $x^2 + 4x = x - 2x^2$

46. $3x^2 + 4x = 0$

47. $5x^2 + 5 = 7x$

48. $4x^2 - 5x + 3 = 0$

49. $6x^2 + 2x = 20$

50. $10x^2 + 30 = -35x$

51. $4x^2 + 9 = 0$

52. $3x^2 - 8x + 6 = 0$

In Exercises 53 – 60, first multiply each side of the equation by the LCM of the denominator to get integer coefficients and then solve the resulting equation.

53. $3x^2 - 4x + \dfrac{1}{3} = 0$

54. $\dfrac{3}{4}x^2 - 2x + \dfrac{1}{8} = 0$

55. $\dfrac{3}{7}x^2 - \dfrac{1}{2}x + 1 = 0$

56. $2x^2 + 3x + \dfrac{5}{4} = 0$

57. $\dfrac{1}{2}x^2 - x + \dfrac{1}{4} = 0$

58. $\dfrac{2}{3}x^2 - \dfrac{1}{3}x + \dfrac{1}{2} = 0$

59. $\dfrac{1}{4}x^2 + \dfrac{7}{8}x + \dfrac{1}{2} = 0$

60. $\dfrac{5}{12}x^2 - \dfrac{1}{2}x - \dfrac{1}{4} = 0$

In Exercises 61 – 72, find the indicated value for k.

61. Determine the values for k such that $x^2 - 8x + k = 0$ will have two real solutions.

62. Determine the values for k such that $x^2 + 5x + k = 0$ will have two real solutions.

63. Determine the values for k such that $x^2 + 9x + k = 0$ will have one real solution.

64. Determine the values for k such that $x^2 - 7x + k = 0$ will have one real solution.

65. Determine the values for k such that $kx^2 - 6x + 3 = 0$ will have two nonreal solutions.

66. Determine the values for k such that $kx^2 + 4x - 2 = 0$ will have two nonreal solutions.

67. Determine the values for k such that $kx^2 + x - 9 = 0$ will have two real solutions.

68. Determine the values for k such that $kx^2 + 6x + 3 = 0$ will have two real solutions.

69. Determine the values for k such that $kx^2 + 7x + 12 = 0$ will have one real solution.

71. Determine the values for k such that $3x^2 + 4x + k = 0$ will have two nonreal solutions.

70. Determine the values for k such that $kx^2 - 2x + 8 = 0$ will have one real solution.

72. Determine the values for k such that $2x^2 + 3x + k = 0$ will have two nonreal solutions.

In Exercises 73 – 80, solve the quadratic equations by using the quadratic formula and your calculator. Write the solutions accurate to 4 decimal places.

73. $0.02x^2 - 1.26x + 3.14 = 0$

74. $0.5x^2 + 0.07x - 5.6 = 0$

75. $\sqrt{2}x^2 - \sqrt{3}x - \sqrt{5} = 0$

76. $x^2 - 2\sqrt{10}x + 10 = 0$

77. $0.3x^2 + \sqrt{2}x + 0.72 = 0$

78. $\sqrt[3]{4}x^2 - \sqrt[4]{2}x - \sqrt{11} = 0$

79. $x^2 + 2\sqrt{15} - 15 = 0$

80. $0.05x^2 - \sqrt{30} = 0$

Writing and Thinking About Mathematics

81. Find an equation of the form $Ax^4 + Bx^2 + C = 0$ that has the four roots ± 2 and ± 3. Explain how you arrived at this equation.

82. The surface area of a right circular cylinder can be found with the following formula:

$S = 2\pi r^2 + 2\pi rh$, where r is the radius of the cylinder and h is the height.

Estimate the radius of a circular cylinder of height 30 cm and surface area 300 cm^2. Explain how you used your knowledge of quadratic equations.

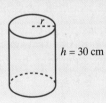

$h = 30$ cm

 HAWKES LEARNING SYSTEMS: INTERMEDIATE ALGEBRA SOFTWARE

- Quadratic Equations: The Quadratic Formula

7.3 Applications

- *Use quadratic equations to solve applied problems.*

The following Strategy for Solving Word Problems, given in Section 1.6 and summarized in Section 5.5, is a valid approach to solving word problems at all levels.

Strategy for Solving Word Problems

1. *Understand the problem.*

  a. *Read the problem carefully. (Read it several times if necessary.)*

  b. *If it helps, restate the problem in your own words.*

2. *Devise a plan.*

  a. *Decide what is asked for; assign a variable to the unknown quantity. Label this variable so you know exactly what it represents.*

  b. *Draw a diagram or set up a chart whenever possible.*

  c. *Write an equation that relates the information provided.*

3. *Carry out the plan.*

  a. *Study your picture or diagram for insight into the solution.*

  b. *Solve the equation.*

4. *Look back over the results.*

  a. *Does your solution make sense in terms of the wording of the problem?*

  b. *Check your solution in the equation.*

The problems in this section can be solved by setting up quadratic equations and then solving these equations by factoring, completing the square, or using the quadratic formula.

The Pythagorean Theorem

The **Pythagorean Theorem** is one of the most interesting and useful ideas in mathematics. We discussed the Pythagorean Theorem in Section 4.8 and do so again here because problems with right triangles often generate quadratic equations.

In a **right triangle**, one of the angles is a right angle (measures 90°), and the side opposite this angle (the longest side) is called the **hypotenuse**. The other two sides are called **legs**.

The Pythagorean Theorem

In a right triangle, the square of the hypotenuse is equal to the sum of the squares of the legs.

$$c^2 = a^2 + b^2$$

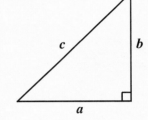

Example 1: The Pythagorean Theorem

The length of a rectangular field is 6 meters longer than its width. If a diagonal foot path stretching from one corner of the field to the opposite corner is 30 meters long, what are the dimensions of the field?

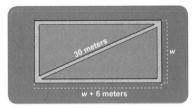

Solution: Let w = width,
then $w + 6$ = length.

$$(w+6)^2 + w^2 = 30^2 \qquad \text{Use the Pythagorean Theorem.}$$

$$w^2 + 12w + 36 + w^2 = 900$$

$$2w^2 + 12w - 864 = 0$$

$$w^2 + 6w - 432 = 0$$

$$(w + 24)(w - 18) = 0$$

$$\cancel{w = -24} \quad \text{or} \quad w = 18 \qquad \text{A negative number does not fit}$$
the conditions of the problem.

$$w = 18 \, \text{meters}$$
$$w + 6 = 24 \, \text{meters}$$

The length is 24 meters and the width is 18 meters.

Projectiles

The formula $h = -16t^2 + v_0 t + h_0$ is used in physics and relates to the height of a projectile such as a thrown ball, a bullet, or a rocket.

$$h = \text{height of object, in feet}$$
$$t = \text{time object is in the air, in seconds}$$
$$v_0 = \text{beginning velocity, in feet per second}$$
$$h_0 = \text{beginning height} \quad (h_0 = 0 \text{ if the object is initially at ground level.})$$

Example 2: Projectiles

A bullet is fired straight up from ground level with a muzzle velocity of 320 feet per second.
a. When will the bullet hit the ground?
b. When will the bullet be 1200 feet above the ground?

Solution: In this problem, $v_0 = 320$ feet per second
and $h_0 = 0$ feet.

a. The bullet hits the ground when $h = 0$.

$$h = -16t^2 + v_0 t + h_0$$

$$0 = -16t^2 + 320t + 0$$

$$0 = t^2 - 20t \qquad\qquad \text{Divide both sides by } -16.$$

$$0 = t(t - 20) \qquad\qquad \text{Factor.}$$
$$t = 0 \quad \text{or} \quad t = 20$$

The bullet hits the ground in 20 seconds. The solution $t = 0$ confirms the fact that the bullet was fired from the ground.

b. Let $h = 1200$.

$$1200 = -16t^2 + 320t$$

$$0 = -16t^2 + 320t - 1200$$

$$0 = t^2 - 20t + 75$$

$$0 = (t - 5)(t - 15)$$
$$t = 5 \quad \text{or} \quad t = 15$$

Both solutions are meaningful. The bullet is at a height of 1200 feet twice; once at 5 seconds while going up and once at 15 seconds while coming down.

Geometry

Example 3: Geometry

A rectangular sheet of copper was 6 in. longer than it was wide. Three inch by three inch squares were cut from each corner and the sides were folded up to form an open box. If the box has a volume of 336 in.^{3}, what were the dimensions of the original sheet of copper?

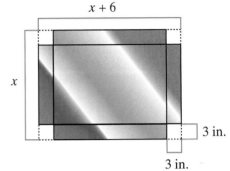

Solution:

Let x = width of copper sheet (before the squares are cut out),
then $x + 6$ = length of copper sheet (before the squares are cut out).

Now,
$$3 = \text{height of box}$$
$$(x+6)-3-3 = x = \text{length of box} \quad \text{3 is subtracted twice.}$$
$$x-3-3 = x-6 = \text{width of box.}$$

The volume of the box is length $\times$ width $\times$ height: $V = lwh$.

Thus,

length $\times$ width $\times$ height = volume

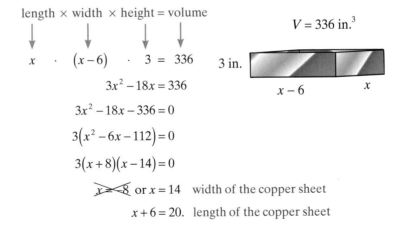

$V = 336$ in.^{3}

$$x \cdot (x-6) \cdot 3 = 336$$

$$3x^2 - 18x = 336$$

$$3x^2 - 18x - 336 = 0$$

$$3(x^2 - 6x - 112) = 0$$

$$3(x+8)(x-14) = 0$$

$$\cancel{x = -8} \text{ or } x = 14 \quad \text{width of the copper sheet}$$

$$x + 6 = 20. \quad \text{length of the copper sheet}$$

The length of the sheet was 20 in. and the width was 14 in.

Cost Per Person

In the following example, note that the **cost per person** is found by dividing the total cost by the number of people going to the tournament. The cost per person changes because the total cost remains fixed while the number of people changes.

Example 4: Cost Per Person

The members of a bowling club were going to fly commercially to a tournament at a total cost of $2420, which was to be divided equally among the members. At the last minute, two of the members decided to fly their own private planes. The cost to the remaining members increased $11 each. How many members flew commercially?

Solution: Let $\quad x =$ number of club members,

then $\quad x - 2 =$ number of club members that flew commercially.

$$\underbrace{\frac{\text{Final cost}}{\text{per member}}} \quad - \quad \underbrace{\frac{\text{Initial cost}}{\text{per member}}} \quad = \quad \underbrace{\frac{\text{Difference in cost}}{\text{per member}}}$$

$$\frac{2420}{x-2} \quad - \quad \frac{2420}{x} \quad = \quad 11$$

$$x(x-2)\frac{2420}{x-2} - x(x-2)\frac{2420}{x} = x(x-2) \cdot 11$$

$$2420x - 2420(x-2) = 11x(x-2)$$

$$2420x - 2420x + 4840 = 11x^2 - 22x$$

$$0 = 11x^2 - 22x - 4840$$

$$0 = 11(x^2 - 2x - 440)$$

$$0 = 11(x-22)(x+20)$$

$$x = 22 \text{ or } x = -20$$

$$x - 2 = 20 \qquad \begin{array}{l} -20 \text{ does not fit the conditions.} \\ \text{That is, the number of people in} \\ \text{a club is a positive number.} \end{array}$$

Check: Final cost per member $= \dfrac{2420}{20} = \$121$

Initial cost per member $= \dfrac{2420}{22} = \$110$

$\$121 - \$110 = \$11 \qquad$ Difference in cost per member

Twenty members flew commercially.

7.3 Exercises

1. A positive integer is one more than twice another. Their product is 78. Find the two integers.

2. One number is equal to the square of another. Find the numbers if their sum is 72.

3. Find two positive numbers whose difference is 10 and whose product is 56. (**Hint:** If x is one number, then the other is either $x + 10$ or $x - 10$. Try both and analyze your answers.)

4. One number is three more than twice a second number. Their product is 119. Find the numbers.

5. The sum of two numbers is -17. Their product is 66. Find the numbers. [**Hint:** If x is one number, then the other is $-17 - x$.]

6. Find a positive real number such that its square is equal to twice the number increased by 2.

7. Find a negative real number such that the square of the sum of the number and 5 is equal to 48.

8. The square of a negative real number is decreased by 2.25 and the result is equal to 3 times the number. What is the number?

9. Twice the square of a positive real number is equal to 4 more than the number. What is the number?

10. The sum of three positive integers is 68. The second is one more than twice the first. The third is three less than the square of the first. Find the integers.

11. Right triangles: A right triangle has two equal sides. The hypotenuse is 14 centimeters. Find the length of the sides.

12. Right triangles: The length of one leg of a right triangle is twice the length of the second leg. The hypotenuse is 15 meters. Find the lengths of the two legs.

13. Average speed: Mel and John leave Desert Point at the same time. Mel drives north and John drives east. Mel's average speed is 10 mph slower than John's. At the end of one hour they are 50 miles apart. Find the average speed of each driver.

14. Height of a telephone pole: A telephone pole was bent over at a point $\frac{4}{9}$ of the distance from its base to the top. The top of the pole reached a point on the ground 9 meters from the base of the pole (thus forming a triangle with the ground). What was the original height of the pole?

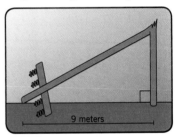

15. Dimensions of a rectangle: The length of a rectangle is 2 feet less than three times the width. If the area of the rectangle is 40 square feet, find the dimensions.

16. Dimensions of a rectangle: A rectangle is 3 meters longer than it is wide. If the width is doubled and the length is decreased by 4 meters, the area is unchanged. Find the original dimensions.

17. Dimensions of a rectangle: The area of a rectangle is 102 square inches and the perimeter of the rectangle is 46 inches. Find the length and width.

18. **Dimensions of a box:** A rectangular piece of cardboard that is twice as long as it is wide has a small square 4 cm by 4 cm cut from each corner. The edges are then folded up to form an open box with a volume of 1536 cubic cm. What are the dimensions of the box? (See example 3.)

19. **Size of an orchard:** An orchard has 2030 trees. The number of trees in each row exceeds twice the number of rows by 12. How many trees are in each row?

20. **Capacity of an auditorium:** A rectangular auditorium seats 960 people. The number of seats in each row exceeds the number of rows by 16. Find the number of seats in each row.

21. **Size of an apartment complex:** An apartment building has the same number of units on each floor. The building has five times as many units per floor as number of floors, and there are 405 units total. How many floors does the building have?

22. **Dimensions of a rectangle:** The length of a rectangle is 2 cm greater than its width. If the length and the width are each increased by 3 cm, the area is increased by 57 cm^2. Find the dimensions of the original rectangle.

23. **Dimensions of a frame:** A watercolor 9 in. wide and 12 in. long is surrounded by a frame of uniform thickness. The area of the frame itself, not including the center, is 162 sq. in. Find the thickness of the frame.

Area of Frame = 162 sq. in.

9 in.

12 in.

24. **Power of a generator:** A 40-volt generator with a resistance of 4 ohms delivers power externally of $40I - 4I^2$ watts, where I is the current measured in amperes. Find the current needed for the generator to deliver 100 watts of power.

25. **Power of a generator:** Find the current needed for the 40-volt generator in Exercise 24 to deliver 64 watts of power.

26. **Selling signs:** Ray operates a small sign-making business. He finds that if he charges x dollars for each sign, he sells $40 - x$ signs per week. What is the least number of signs he can sell to have an income of $336 in one week?

27. **Shopping:** J.B. bought some shirts and pants. He bought two more pairs of pants than shirts. He spent $154 on pants and $65 on shirts. Find the price of each type of clothing if the price of a pair of pants exceeds the price of a shirt by $9.

28. **Selling picture frames:** It costs Mrs. Snow $3 to build a picture frame. She estimates that if she charges x dollars each, she can sell $60 - x$ frames per week. What is the lowest price necessary to make a profit of $432 each week?

29. **Selling peanuts:** Samuel operates a small peanut stand. He estimates that he can sell 600 bags of peanuts per day if he charges 50¢ for each bag. He determines that he can sell 20 more bags for each 1¢ reduction in price.

   a. What would his revenue be if he charged 48¢ per bag?
   b. What should he charge in order to have receipts of $315?

30. **Rock climbing:** The Piton Rock Climbing Club planned a climbing expedition. The total cost was $900, which was to be divided equally among the members going. While practicing, three members fell and were injured so they were unable to go. If the cost per person increased by $15, how many people went on the expedition?

31. **Manufacturing:** A manufacturing crew needs to assemble 1000 boxes per day, divided equally among the workers. One day, three workers call in sick, and the remaining members each need to assemble 75 more boxes than usual. How many workers are on the manufacturing crew?

32. **Traveling by car:** Jim traveled 240 miles to a convention. Later, his wife Ann drove up to meet him. Ann's average speed exceeded Jim's by 4 mph, and the trip took her 15 minutes less time. Find Ann's speed.

33. **Boating:** In two hours, a motorboat can travel 8 miles down a river and return 4 miles back. If the river flows at a rate of 2 miles per hour, how fast can the boat travel in still water?

34. **Product assembly:** Two employees together can prepare a large order in 2 hrs. Working alone, one employee takes three hours longer than the other. How long does it take each person working alone?

35. **Library processing:** Two librarians working together can catalog a new shipment of books in 3 hours. Working alone, one librarian takes 8 hours longer than the other. How long does it take each librarian working alone?

36. **Traveling by plane:** Fern can fly her plane 240 miles against the wind in the same time it takes her to fly 360 miles with the wind. The speed of the plane in still air is 30 mph more than four times the speed of the wind. Find the speed of the plane in still air.

37. **Selling coffee:** A grocer mixes $9.00 worth of Grade A coffee with $12.00 worth of Grade B coffee to obtain 30 pounds of a blend. If Grade A costs 30¢ a pound more than Grade B, how many pounds of each were used?

38. **Theater:** The Andersonville Little Theater Group sold 340 tickets to their spring production. Receipts from the sale of reserved tickets were $855. Receipts from general admission tickets were $375. How many of each type ticket were sold if the cost of a reserved ticket is $2 more than a general admission ticket?

In Exercises 39 – 42 use the formula for a projectile, $h = -16t^2 + v_0t + h_0$, where t is in seconds, v_0 is in ft/s and h_0 is in feet.

39. Throwing a ball: A ball is thrown vertically from ground level with an initial speed of 108 ft/s.
  a. When will the ball hit the ground?
  b. When will the ball be 180 ft above the ground?

40. Throwing a ball: A ball is thrown vertically from the ground with an initial speed of 160 ft/s.
  a. When will the ball strike the ground?
  b. When will the ball be 400 ft above the ground?

41. Shooting an arrow: An arrow is shot vertically upward from a platform 40 ft high at a rate of 224 ft/s.
  a. When will the arrow be 824 ft above the ground?
  b. When will it be 424 ft above the ground?

42. Dropping a stone: A stone is dropped from a platform 196 ft high.
  Hint: Since the stone is dropped, $v_0 = 0$.
  a. When will it hit the ground?
  b. How far has it fallen after 3 seconds?

In Exercises 43 – 50, use your calculator to find the answers accurate to two decimal places.

43. Geometry: If a triangle is inscribed in a circle so that one side of the triangle is a diameter of the circle, the triangle will be a right triangle (every time). If an isosceles triangle (two sides equal) is inscribed in this manner in a circle with diameter 20 cm, find the length of the two equal sides.

44. Geometry: If a triangle is inscribed in a semicircle such that one side of the triangle is the diameter of a circle with radius 6 in., and one side of the triangle is 5 in., what is the length of the third side? (See Exercise 43.)

45. Geometry: A square is said to be inscribed in a circle if each corner of the square lies on the circle. (Use $\pi = 3.14$.)
  a. Find the circumference and area of a circle with diameter 30 feet.
  b. Find the perimeter and area of a square inscribed in the circle.

46. Baseball: The shape of a baseball infield is a square with sides 90 feet long.
   a. Find the distance from home plate to second base.
   b. Find the distance from first base to third base.

47. Baseball: The distance from home plate to the pitcher's mound for the field in Exercise 46 is 60.5 feet.
   a. Is the pitcher's mound exactly half way between home plate and second base?
   b. If not, which base is it closer to, home plate or second base?
   c. Do the two diagonals of the square intersect at the pitcher's mound?

48. Flying an airplane: If an airplane passes directly over your head at an altitude of 1 mile, how far is the airplane from your position after it has flown 2 miles farther at the same altitude?

49. Length of a shadow: The GE Building in New York is 850 feet tall (70 stories). At a certain time of day, the building casts a shadow 100 feet long. Find the distance from the top of the building to the tip of the shadow (to the nearest hundredth of a foot).

50. Quilting: To create a square inside a square, a quilting pattern requires four triangular pieces like the one shaded in the figure shown here. If the square in the center measures 12 centimeters on a side, and the two legs of each triangle are of equal length, how long are the legs of each triangle?

Writing and Thinking About Mathematics

51. Develop the Quadratic Formula by using the technique of completing the square with the general quadratic equation $ax^2 + bx + c = 0$.

 HAWKES LEARNING SYSTEMS: INTERMEDIATE ALGEBRA SOFTWARE

- Applications: Quadratic Equations

Equations in Quadratic Form

7.4

- *Make substitutions that allow equations to be written in quadratic form.*
- *Solve equations that can be written in quadratic form.*
- *Solve equations that contain rational expressions.*

Solving Equations in Quadratic Form

The general quadratic equation is $ax^2 + bx + c = 0$, where $a \neq 0$.

The equations

$$x^4 - 7x^2 + 12 = 0 \qquad \text{and} \qquad x^{\frac{2}{3}} - 4x^{\frac{1}{3}} - 21 = 0$$

are **not** quadratic equations, but they are in **quadratic form** because the degree of the middle term is one-half the degree of the first term. Specifically,

$$\frac{1}{2}(4) = 2 \qquad \text{and} \qquad \frac{1}{2}\left(\frac{2}{3}\right) = \frac{1}{3}.$$

first term   middle term   first term   middle term
exponent   exponent   exponent   exponent

Equations in quadratic form can be solved by using the quadratic formula or by factoring just as if they were quadratic equations. In each case, a substitution can be made to clarify the problem. Follow these suggestions:

Solving Equations in Quadratic Form by Substitution

1. *Look at the middle term.*

2. *Substitute a first-degree variable, such as u, for the variable expression in the middle term.*

3. *Substitute the square of this variable, u^2, for the variable expression in the first term.*

4. *Solve the resulting quadratic equation for u.*

5. *Substitute the results "back" for u in the beginning substitution and solve for the original variable.*

The following examples illustrate how such a substitution may help. Study these examples carefully and note the variety of algebraic manipulations used.

Example 1: Using Substitution to Solve Equations in Quadratic Form

Solve the following equations. These equations are in quadratic form and a substitution will reveal the quadratic expression.

a. $x^4 - 7x^2 + 12 = 0$

    Solution:

$$x^4 - 7x^2 + 12 = 0$$

$$(u)^2 - 7(u) + 12 = 0 \qquad \text{Substitute } u = x^2 \text{ and } (u)^2 = x^4.$$

$$(u - 3)(u - 4) = 0 \qquad \text{Solve for } u \text{ by factoring.}$$

$$u = 3 \quad \text{or} \quad u = 4$$

$$x^2 = 3 \quad \text{or} \quad x^2 = 4 \qquad \text{Now substitute back } x^2 \text{ for } u.$$

$$x = \pm\sqrt{3} \qquad x = \pm 2 \qquad \text{Solve quadratic equations for } x.$$

There are four solutions: $-\sqrt{3}, \sqrt{3}, -2,$ and 2.

b. $x^{\frac{2}{3}} - 4x^{\frac{1}{3}} - 21 = 0$

    Solution:

$$x^{\frac{2}{3}} - 4x^{\frac{1}{3}} - 21 = 0$$

$$(u)^2 - 4(u) - 21 = 0 \qquad \text{Let } u = x^{\frac{1}{3}} \text{ and } (u)^2 = x^{\frac{2}{3}}.$$

$$(u - 7)(u + 3) = 0 \qquad \text{Solve for } u \text{ by factoring.}$$

$$u = 7 \qquad \text{or} \qquad u = -3$$

$$x^{\frac{1}{3}} = 7 \qquad \text{or} \qquad x^{\frac{1}{3}} = -3 \qquad \text{Substitute back } x^{\frac{1}{3}} \text{ for } u.$$

$$\left(x^{\frac{1}{3}}\right)^3 = 7^3 \qquad \text{or} \qquad \left(x^{\frac{1}{3}}\right)^3 = (-3)^3 \qquad \text{Cube both sides.}$$

$$x = 343 \qquad\qquad x = -27$$

There are two solutions: -27 and 343.

c. $x^{-4} - 7x^{-2} + 10 = 0$

    Solution:

$$x^{-4} - 7x^{-2} + 10 = 0$$

$$(u)^2 - 7(u) + 10 = 0 \qquad \text{Let } u = x^{-2} \text{ and } (u)^2 = x^{-4}.$$

$$(u - 2)(u - 5) = 0 \qquad \text{Solve for } u \text{ by factoring.}$$

$$u = 2 \quad \text{or} \quad u = 5$$

$$x^{-2} = 2 \qquad x^{-2} = 5 \qquad \text{Substitute back } x^{-2} \text{ for } u.$$

Continued on the next page...

$$\frac{1}{x^2} = 2 \qquad \frac{1}{x^2} = 5 \qquad \text{Remember, } x^{-2} = \frac{1}{x^2}.$$

$$x^2 = \frac{1}{2} \qquad x^2 = \frac{1}{5} \qquad \text{Reciprocals}$$

$$x = \pm\sqrt{\frac{1}{2}} \qquad x = \pm\sqrt{\frac{1}{5}}$$

$$x = \pm\frac{1}{\sqrt{2}} \qquad x = \pm\frac{1}{\sqrt{5}}$$

Rationalizing denominators, we have $x = \pm\dfrac{\sqrt{2}}{2}, x = \pm\dfrac{\sqrt{5}}{5}$.

There are four solutions: $\dfrac{-\sqrt{2}}{2}, \dfrac{\sqrt{2}}{2}, \dfrac{-\sqrt{5}}{5}$, and $\dfrac{\sqrt{5}}{5}$.

d. $(x + 2)^2 - (x + 2) - 12 = 0$

Solution: $(x+2)^2 - (x+2) - 12 = 0$

$$(u)^2 - (u) - 12 = 0 \qquad \text{Let } u = x + 2.$$

$$(u - 4)(u + 3) = 0 \qquad \text{Solve for } u \text{ by factoring.}$$

$$u = 4 \quad \text{or} \quad u = -3$$
$$x + 2 = 4 \qquad x + 2 = -3 \qquad \text{Substitute back } x + 2 \text{ for } u.$$
$$x = 2 \qquad \quad x = -5$$

There are two solutions: -5 and 2.

Solving Equations with Rational Expressions

Equations with rational expressions were discussed in Section 5.4. In solving equations with rational expressions, first multiply every term on both sides of the equation by the least common multiple (LCM) of the denominators. This will "clear" the equation of fractions. Remember to check the restrictions on the variables. That is, no denominator can have a value of 0.

Example 2: Solving Equations with Rational Expressions

Solve the following equation containing rational expressions: $\dfrac{2}{3x-1} + \dfrac{1}{x+1} = \dfrac{x}{x+1}$.

Solution: This equation is not in quadratic form. However, multiplying both sides of the equation by the LCM, $(x+1)(3x-1)$, of the denominators does give a quadratic equation. The restrictions on x are: $x \neq \dfrac{1}{3}, -1$.

In this case, use the quadratic formula to solve the resulting quadratic equation.

$$(x+1)\,\cancel{(3x-1)} \cdot \frac{2}{\cancel{3x-1}} + \cancel{(x+1)}\,(3x-1) \cdot \frac{1}{\cancel{x+1}} = \cancel{(x+1)}\,(3x-1) \cdot \frac{x}{\cancel{x+1}}$$

$$2(x+1)+3x-1=(3x-1)x$$

$$2x+2+3x-1=3x^2-x$$

$$0=3x^2-6x-1$$

$$x=\frac{6\pm\sqrt{(-6)^2-4\cdot3(-1)}}{6}$$

$$=\frac{6\pm\sqrt{48}}{6}$$

$$=\frac{6\pm4\sqrt{3}}{6}$$

$$=\frac{3\pm2\sqrt{3}}{3}$$

Solving Higher-Degree Equations

Example 3: Solving Higher-Degree Equations

Solve the following higher-degree equations.

a. $x^5-16x=0$

> **Solution:** This equation can be solved by factoring and using the square root property.
>
> $$x^5-16x=0$$
> $$x(x^4-16)=0 \qquad \text{Factor out the common term } x.$$
> $$x(x^2+4)(x^2-4)=0 \qquad \text{Factor the difference of two squares.}$$
> $$x=0 \quad \text{or} \quad x^2=-4 \quad \text{or} \quad x^2=4$$
> $$x=\pm2i \qquad\qquad x=\pm2$$

There are five solutions: $0, -2i, 2i, -2, 2$.

Continued on the next page...

b. $x^3 - 27 = 0$

Solution: The polynomial is the difference of two cubes and can be factored. In this case, complex solutions can be found by using the quadratic formula.

$$x^3 - 27 = 0$$

$$(x-3)(x^2 + 3x + 9) = 0$$

$x - 3 = 0$	$x^2 + 3x + 9 = 0$
$x = 3$	Using the quadratic formula:

$$x = \frac{-3 \pm \sqrt{3^2 - 4 \cdot 9}}{2} = \frac{-3 \pm \sqrt{-27}}{2} = \frac{-3 \pm 3i\sqrt{3}}{2}$$

There are three solutions: $3, \dfrac{-3 + 3i\sqrt{3}}{2}, \dfrac{-3 - 3i\sqrt{3}}{2}$.

Practice Problems

Solve the following equations.

1. $x - x^{\frac{1}{2}} - 2 = 0$

　　$(Let\ u = x^{\frac{1}{2}}\ and\ u^2 = x.)$

2. $x^4 + 16x^2 = -48$

　　$(Let\ u = x^2\ and\ u^2 = x^4.)$

3. $\dfrac{3(x-2)}{x-1} = \dfrac{2(x+1)}{x-2} + 2$

4. $x^3 + 8 = 0$

7.4 Exercises

Solve the equations in Exercises 1 – 56.

1. $x^4 - 13x^2 + 36 = 0$ **2.** $x^4 - 29x^2 + 100 = 0$ **3.** $x^4 - 9x^2 + 20 = 0$

4. $y^4 - 11y^2 + 18 = 0$ **5.** $y^4 - 3y^2 - 28 = 0$ **6.** $y^4 + y^2 - 12 = 0$

7. $y^4 - 25 = 0$ **8.** $x^{-2} - 12x^{-1} + 35 = 0$ **9.** $z^{-2} - 2z^{-1} - 24 = 0$

10. $16x^3 + 100x = 0$ **11.** $2x - 9x^{\frac{1}{2}} + 10 = 0$ **12.** $2x - 3x^{\frac{1}{2}} + 1 = 0$

13. $x^3 - 9x^{\frac{3}{2}} + 8 = 0$ **14.** $y^3 - 28y^{\frac{3}{2}} + 27 = 0$ **15.** $2x^{\frac{2}{3}} + 3x^{\frac{1}{3}} - 2 = 0$

Answers to Practice Problems: **1.** $x = 4$ **2.** $x = \pm 2i, x = \pm 2i\sqrt{3}$ **3.** $x = -3 \pm \sqrt{19}$ **4.** $x = -2, x = 1 \pm i\sqrt{3}$

16. $2x^{-\frac{2}{3}} + x^{-\frac{1}{3}} - 6 = 0$ **17.** $x^{-1} + 5x^{-\frac{1}{2}} - 50 = 0$ **18.** $2x^{-2} - 7x^{-1} + 6 = 0$

19. $3x^{-2} + x^{-1} - 24 = 0$ **20.** $3y^{-1} - 7y^{-\frac{1}{2}} + 2 = 0$ **21.** $3x^{\frac{5}{3}} + 15x^{\frac{4}{3}} + 18x = 0$

22. $2x^2 - 30x^{\frac{3}{2}} + 112x = 0$ **23.** $(3x - 5)^2 + (3x - 5) - 2 = 0$

24. $(x - 1)^2 + (x - 1) - 6 = 0$ **25.** $(2x + 3)^2 + 7(2x + 3) + 12 = 0$

26. $(5x - 4)^2 + 2(5x - 4) - 8 = 0$ **27.** $(x - 3)^2 - 2(x - 3) - 15 = 0$

28. $(x + 4)^2 - 2(x + 4) = 3$ **29.** $(2x + 1)^2 + (2x + 1) = 0$

30. $(x + 7)^2 + 5(x + 7) = 50$ **31.** $x^4 - 2x^2 + 2 = 0$

32. $x^4 - 4x^2 + 5 = 0$ **33.** $x^4 - 2x^2 + 10 = 0$

34. $x^4 + 16 = 0$
   [**Hint:** $x^4 + 16 = x^4 + 0x^2 + 16$]   **35.** $x^4 - 4x^2 + 7 = 0$

36. $x^4 - 6x^2 + 11 = 0$ **37.** $x^{-4} - 6x^{-2} + 5 = 0$

38. $3x^{-4} - 5x^{-2} + 2 = 0$ **39.** $3x^{-4} + 25x^{-2} - 18 = 0$

40. $2x^{-4} + 3x^{-2} - 20 = 0$ **41.** $\dfrac{2}{4x - 1} + \dfrac{1}{x + 1} = \dfrac{-x}{x + 1}$

42. $\dfrac{3x - 2}{15} - \dfrac{16 - 3x}{x + 6} = \dfrac{x + 3}{5}$ **43.** $\dfrac{2x}{x - 4} - \dfrac{12x}{x^2 + x - 20} = \dfrac{x - 1}{x + 5}$

44. $\dfrac{x + 1}{x + 3} + \dfrac{2x - 1}{x - 2} = \dfrac{12x - 2}{x^2 + x - 6}$ **45.** $\dfrac{x + 5}{3x + 2} - \dfrac{4 - 2x}{3x^2 + 8x + 4} = \dfrac{x + 4}{x + 2}$

46. $\dfrac{x + 5}{3x + 4} + \dfrac{16x^2 + 5x + 6}{3x^2 - 2x - 8} = \dfrac{4x}{x - 2}$ **47.** $\dfrac{4x + 1}{x - 6} - \dfrac{3x^2 - 8x + 20}{2x^2 - 13x + 6} = \dfrac{3x + 7}{2x - 1}$

48. $\dfrac{3x + 2}{x + 3} + \dfrac{22x - 31}{x^2 - x - 12} = \dfrac{3(x + 4)}{x + 3}$ **49.** $\dfrac{5(x - 10)}{x - 7} = \dfrac{2(x + 1)}{x - 4} + 3$

50. $2 + \dfrac{2 - x}{x + 2} = \dfrac{x - 3}{x + 5}$ **51.** $x^5 - 64x = 0$ **52.** $x^5 = 36x$

53. $8x^3 = 64$ **54.** $x^3 - 125 = 0$ **55.** $x^3 + 1000 = 0$

56. $2x^5 + 54x^2 = 0$

Writing and Thinking About Mathematics

57. One of the most studied and interesting visual and numerical concepts in algebra is the **Golden Ratio**. Ancient Greeks thought (and many people still do) that a rectangle was most aesthetically pleasing to the eye if the ratio of its length to its width is the Golden Ratio (about 1.618). In fact, the Parthenon, built by Greeks in the fifth century B.C., utilizes the Golden Ratio. A rectangle is "golden" if its length, l, and width, w, satisfy the equation $\dfrac{l}{w} = \dfrac{w}{l-w}$.

a. By letting $w = 1$ unit in the equation above, we get the equation $\dfrac{l}{1} = \dfrac{1}{l-1}$.

Solve this equation for the positive value of l (which is the algebraic expression for the golden ratio).

b. Suppose that an architect is constructing a building with a rectangular front that is to be 60 feet high. About how long should the front be if he wants the appearance of a golden rectangle? (Assume $w = 60$ ft and that you are looking for l.) Round to 2 decimal places.

c. Look at the two rectangles shown here. Which seems most pleasing to your eye? Measure the length and width of each rectangle and see if you chose the golden rectangle.

58. Consider the following equation: $x - x^{\frac{1}{2}} - 6 = 0$.

In your own words, explain why, even though it is in quadratic form, this equation has only one solution.

HAWKES LEARNING SYSTEMS: INTERMEDIATE ALGEBRA SOFTWARE

▪ Equations in Quadratic Form

Graphing Quadratic Functions: Parabolas

7.5

- *Graph a parabola (a quadratic function) and determine its vertex, domain, range, line of symmetry, and zeros.*
- *Solve applied problems by using quadratic functions and the concepts of maximum and minimum.*

Introduction to Quadratic Functions: $y = ax^2 + bx + c$ ($a \neq 0$)

We have studied various types of **functions** and their corresponding graphs: linear functions, polynomial functions, and functions with radicals. Related concepts include **domain**, **range**, and **zeros** (points where the y-value is 0 and the graph crosses the x-axis). The vertical line test (restated here) can be used to tell whether or not a graph represents a function.

Vertical Line Test

If **any** vertical line intersects a graph in more than one point, then the relation graphed is **not** a function.

In this section, we expand our interest in functions to include a detailed analysis of **quadratic functions**, functions that are represented by quadratic expressions. For example, consider the function

$$y = x^2 - 4x + 3.$$

What is the graph of this function? Since the equation is not linear, the graph will not be a straight line. The nature of the graph can be investigated by plotting several points (See Figure 7.1).

x	$x^2 - 4x + 3 = y$
-1	$(-1)^2 - 4(-1) + 3 = 8$
0	$0^2 - 4(0) + 3 = 3$
$\dfrac{1}{2}$	$\left(\dfrac{1}{2}\right)^2 - 4\left(\dfrac{1}{2}\right) + 3 = \dfrac{5}{4}$
1	$1^2 - 4(1) + 3 = 0$
2	$2^2 - 4(2) + 3 = -1$
3	$3^2 - 4(3) + 3 = 0$
$\dfrac{7}{2}$	$\left(\dfrac{7}{2}\right)^2 - 4\left(\dfrac{7}{2}\right) + 3 = \dfrac{5}{4}$
4	$4^2 - 4(4) + 3 = 3$
5	$5^2 - 4(5) + 3 = 8$

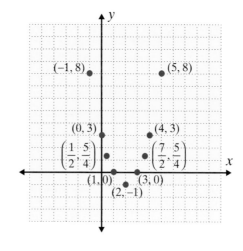

Figure 7.1

The complete graph of $y = x^2 - 4x + 3$ is shown in Figure 7.2. The curve is called a **parabola**. The point $(2, -1)$ is the "turning point" of the parabola and is called the **vertex** of the parabola. The line $x = 2$ is the **line of symmetry** or **axis of symmetry** for the parabola. That is, the curve is a "mirror image" of itself with respect to the line $x = 2$.

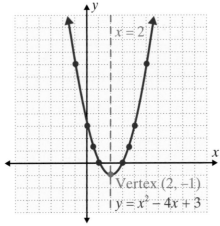

Vertex is $(2, -1)$.
$y = x^2 - 4x + 3$ is a parabola.
$x = 2$ is the line of symmetry.

Figure 7.2

Quadratic Functions

Any function that can be written in the form

$$y = ax^2 + bx + c$$

*where a, b, and c are real numbers and $a \neq 0$ is a **quadratic function**.*

The graph of every quadratic function is a parabola. The position of the parabola, its shape, and whether it "opens upward" or "opens downward" can be determined by investigating the function itself. For convenience, we will refer to parabolas that open up or down as **vertical parabolas**. Parabolas that open left or right will be called **horizontal parabolas**. As we will see in Section 9.2, **horizontal parabolas do not represent functions**.

We will discuss quadratic functions in each of the following five forms where a, b, c, h, and k are constants:

$$y = ax^2$$
$$y = ax^2 + k$$
$$y = a(x - h)^2$$
$$y = a(x - h)^2 + k$$
$$y = ax^2 + bx + c.$$

Functions of the Form $y = ax^2$

For any real number x, $x^2 \geq 0$. So, $ax^2 \geq 0$ if $a > 0$ and $ax^2 \leq 0$ if $a < 0$. This means that the graph of $y = ax^2$ is "above" the x-axis if $a > 0$ and "below" the x-axis if $a < 0$. The **vertex** is at the origin $(0, 0)$ in either of these cases and is the one point where each graph touches (or is tangent to) the x-axis.

For all quadratic functions, the **domain** is the set of all real numbers. That is, x can be replaced by any real number and there will be one corresponding y-value. The **range** of the function $y = ax^2$ depends on the value of a. If $a > 0$, then $y \geq 0$. If $a < 0$, then $y \leq 0$.

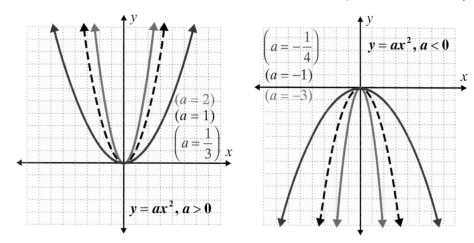

Domain: $\{x \,|\, x$ is any real number$\}$

Range: $\{y \,|\, y \geq 0\}$

In interval notation:

Domain: $(-\infty, \infty)$, Range: $[0, \infty)$

Domain: $\{x \,|\, x$ is any real number$\}$

Range: $\{y \,|\, y \leq 0\}$

In interval notation:

Domain: $(-\infty, \infty)$, Range: $(-\infty, 0]$

Figure 7.3

Figure 7.3 illustrates the following characteristics of quadratic functions of the form $y = ax^2$:

    a. If $a > 0$, the parabola "opens upward."

    b. If $a < 0$, the parabola "opens downward."

    c. The bigger $|a|$ is, the narrower the opening.

    d. The smaller $|a|$ is, the wider the opening.

    e. The line $x = 0$ (the y-axis) is the line of symmetry.

Functions of the Form $y = ax^2 + k$

Adding k to ax^2 simply changes each y-value of $y = ax^2$ by k units (increase if k is positive, decrease if k is negative). That is, the graph of $y = ax^2 + k$ can be found by "sliding" or "shifting" the graph of $y = ax^2$ up k units if $k > 0$ or down $|k|$ units if $k < 0$ (Figure 7.4).

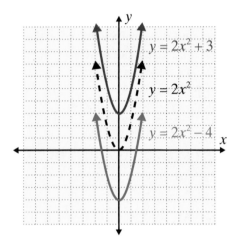

Domain: $\{x \,|\, x \text{ is any real number}\}$

Range: $\{y \,|\, y \geq k\}$

In interval notation:

Domain: $(-\infty, \infty)$, Range: $[k, \infty)$

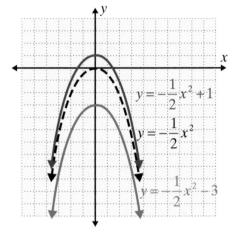

Domain: $\{x \,|\, x \text{ is any real number}\}$

Range: $\{y \,|\, y \leq k\}$

In interval notation:

Domain: $(-\infty, \infty)$, Range: $(-\infty, k]$

Figure 7.4

The vertex of $y = ax^2 + k$ is at the point $(0, k)$. The graph of $y = ax^2 + k$ is a **vertical shift** (or **vertical translation**) of the graph of $y = ax^2$. The line $x = 0$ (the y-axis) is the line of symmetry just as with equations of the form $y = ax^2$.

Functions of the Form $y = a(x - h)^2$

We know that the square of a real number is nonnegative. Therefore, $(x - h)^2 \geq 0$. So, if $a > 0$, then $y = a(x - h)^2 \geq 0$. If $a < 0$, then $y = a(x - h)^2 \leq 0$. Also notice that $ax^2 = 0$ when $x = 0$, and $a(x - h)^2 = 0$ when $x = h$. Thus the vertex is at $(h, 0)$, and the parabola "opens upward" if $a > 0$ and "opens downward" if $a < 0$ (Figure 7.5).

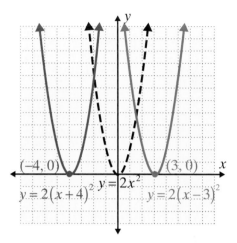 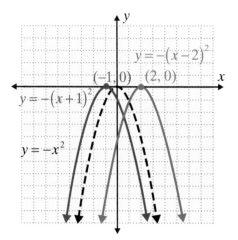

Domain: $\{x \mid x$ is any real number$\}$ Domain: $\{x \mid x$ is any real number$\}$

Range: $\{y \mid y \geq 0\}$ Range: $\{y \mid y \leq 0\}$

In interval notation: In interval notation:

Domain: $(-\infty, \infty)$, Range: $[0, \infty)$ Domain: $(-\infty, \infty)$, Range: $(-\infty, 0]$

Figure 7.5

The graph of $y = a(x - h)^2$ is a **horizontal shift** (or **horizontal translation**) of the graph of $y = ax^2$. The shift is to the right if $h > 0$ and to the left if $h < 0$. As a special comment, note that if $h = -3$, then

$$y = a(x - h)^2$$

gives $y = a\left(x - (-3)\right)^2$

or $y = a(x + 3)^2$.   (A shift of 3 units to the left)

Thus, if h is negative, the expression $(x - h)^2$ appears with a plus sign. If h is positive, the expression $(x - h)^2$ appears with a minus sign. In either case, the line $x = h$ is the **line of symmetry**.

Example 1: Quadratic Functions

Graph the following quadratic functions. Set up a table of values for x and y as an aid, and choose values of x on each side of the line of symmetry. Find the line of symmetry and the vertex, and state the domain and range of each function.

a. $y = 2x^2 - 3$

   Solution: Line of symmetry is $x = 0$. (The parabola opens upward since a is positive.)

Continued on the next page...

Vertex: $(0,-3)$; Domain: $\{x \mid x \text{ is any real number}\}$ or $(-\infty,\infty)$;

Range: $\{y \mid y \ge -3\}$ or $[-3,\infty)$

x	y
0	−3
$\dfrac{1}{2}$	$-\dfrac{5}{2}$
$-\dfrac{1}{2}$	$-\dfrac{5}{2}$
1	−1
−1	−1
2	5
−2	5

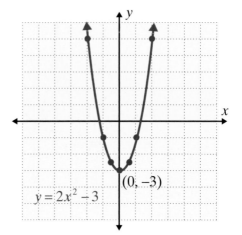

b. $y = -\left(x - \dfrac{5}{2}\right)^2$

Solution: Line of symmetry is $x = \dfrac{5}{2}$. (The parabola opens down since a is negative.)

Vertex: $\left(\dfrac{5}{2},0\right)$; Domain: $\{x \mid x \text{ is any real number}\}$ or $(-\infty,\infty)$;

Range: $\{y \mid y \le 0\}$ or $(-\infty,0]$

x	y
$\dfrac{5}{2}$	0
2	$-\dfrac{1}{4}$
3	$-\dfrac{1}{4}$
1	$-\dfrac{9}{4}$
4	$-\dfrac{9}{4}$
0	$-\dfrac{25}{4}$
5	$-\dfrac{25}{4}$

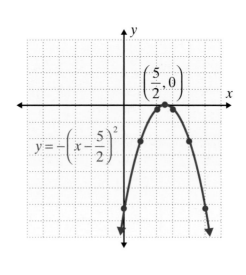

Functions of the Form $y = a(x - h)^2 + k$ and $y = ax^2 + bx + c$

The graphs of equations of the form

$$y = a(x - h)^2 + k$$

combine both the vertical shift of k units and the horizontal shift of h units. The **vertex is at (h, k)**. For example, the graph of the function $y = -2(x - 3)^2 + 5$ is a shift of the graph of $y = -2x^2$ up 5 units and to the right 3 units and has its vertex at $(3, 5)$.

The graph of $y = \left(x + \dfrac{1}{2}\right)^2 - 2$ is the same as the graph of $y = x^2$ but is shifted left $\dfrac{1}{2}$ unit and down 2 units. The vertex is at $\left(-\dfrac{1}{2}, -2\right)$ (Figure 7.6).

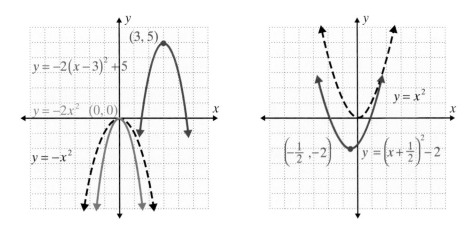

Figure 7.6

The general form of a quadratic function is $y = ax^2 + bx + c$. However, this form does not give as much information about the graph of the corresponding parabola as the form $y = a(x - h)^2 + k$. **Therefore, to easily find the vertex, line of symmetry, and range, and to graph the parabola, we want to change the general form $y = ax^2 + bx + c$ into the form $y = a(x - h)^2 + k$.** This can be accomplished by completing the square using the following technique. This technique is also useful in other courses in mathematics and should be studied carefully. (Be aware that you are not solving an equation. You do not "do something" to both sides. You are **changing the form** of a function.) [**Note:** A graphing calculator will give the same graph regardless of the form of the function.]

$$y = ax^2 + bx + c \qquad \text{Write the function.}$$

$$= a\left(x^2 + \frac{b}{a}x\right) + c \qquad \text{Factor } a \text{ from just the first two terms.}$$

$$= a\left(x^2 + \frac{b}{a}x + \frac{b^2}{4a^2} - \frac{b^2}{4a^2}\right) + c \quad \text{Complete the square of } x^2 + \frac{b}{a}x.$$

$$\frac{1}{2} \cdot \frac{b}{a} = \frac{b}{2a} \text{ and } \left(\frac{b}{2a}\right)^2 = \frac{b^2}{4a^2}.$$

Add and subtract $\dfrac{b^2}{4a^2}$ inside the parentheses.

$$= a\left(x^2 + \frac{b}{a}x + \frac{b^2}{4a^2}\right) - \frac{b^2}{4a} + c \quad \text{Multiply } a \cdot \left(\frac{-b^2}{4a^2}\right) \text{ and write this term outside}$$

the parentheses.

$$= a\left(x + \frac{b}{2a}\right)^2 + \frac{4ac - b^2}{4a} \qquad \text{Write the square of the binomial and simplify}$$

the fraction to get the form $y = a(x - h)^2 + k$.

In terms of the coefficients $a, b,$ and c,

$$x = -\frac{b}{2a} \quad \text{is the \textbf{line of symmetry}}$$

and

$$(h, k) = \left(-\frac{b}{2a}, \frac{4ac - b^2}{4a}\right) \quad \text{is the \textbf{vertex}.} \quad \left(\text{In function notation, } \left[-\frac{b}{2a}, f\left(-\frac{b}{2a}\right)\right].\right)$$

 NOTES Rather than memorize the formula for the coordinates of the vertex, you should just remember that the x-coordinate of the vertex is $x = -\dfrac{b}{2a}$. Substituting this value for x in the function will give the y-value for the vertex.

Zeros of a Quadratic Function

The points where a parabola crosses the x-axis, if any, are the x-intercepts. This is where $y = 0$. These points are called the **zeros of the function**. We find these points by substituting 0 for y and solving the resulting quadratic equation.

$$y = ax^2 + bx + c \quad \textbf{quadratic function}$$

$$0 = ax^2 + bx + c \quad \textbf{quadratic equation}$$

If the solutions are nonreal complex numbers, then the graph does not cross the x-axis. It is either entirely above the x-axis or entirely below the x-axis.

The following examples illustrate how to apply all our knowledge about quadratic functions.

Example 2: Graphing Quadratic Equations

a. For $y = x^2 - 6x + 1$, find the zeros of the function, the line of symmetry, the vertex, the domain, the range, and graph the parabola.

Solution: $x^2 - 6x + 1 = 0$

$$x = \frac{6 \pm \sqrt{(-6)^2 - 4}}{2}$$ Quadratic Formula

$$= \frac{6 \pm \sqrt{32}}{2}$$

$$= \frac{6 \pm 4\sqrt{2}}{2}$$

$$= 3 \pm 2\sqrt{2}$$

The zeros are $3 \pm 2\sqrt{2}$.

Change the form of the function for easier graphing.

$$y = x^2 - 6x + 1$$

$$= \left(x^2 - 6x + 9 - 9\right) + 1$$ Add $0 = 9 - 9$ inside the parentheses.

$$= \left(x^2 - 6x + 9\right) - 9 + 1$$

$$= (x - 3)^2 - 8$$

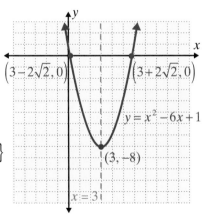

Summary:

Zeros: $3 \pm 2\sqrt{2}$

Line of symmetry is $x = 3$.

Vertex: $(3, -8)$

Domain: $(-\infty, \infty)$ or
$\{x \mid x \text{ is any real number}\}$

Range: $[-8, \infty)$ or $\{y \mid y \geq -8\}$

Continued on the next page...

b. For $y = -x^2 - 4x + 2$, find the zeros of the function, the line of symmetry, the vertex, the domain, the range, and graph the parabola.

Solution: $-x^2 - 4x + 2 = 0$

$$x = \frac{4 \pm \sqrt{(-4)^2 - 4(-1)(2)}}{2(-1)} \qquad \text{Quadratic Formula}$$

$$= \frac{4 \pm \sqrt{24}}{-2}$$

$$= \frac{4 \pm 2\sqrt{6}}{-2}$$

$$= -2 \pm \sqrt{6}$$

The zeros are $-2 \pm \sqrt{6}$.

Change the form of the function for easier graphing.

$y = -x^2 - 4x + 2$

$\quad = -\left(x^2 + 4x\right) + 2$ Factor -1 from the first two terms only.

$\quad = -\left(x^2 + 4x + 4 - 4\right) + 2$ Add $0 = 4 - 4$ inside the parentheses.

$\quad = -\left(x^2 + 4x + 4\right) + 4 + 2$ Multiply $-1(-4)$ and put this outside the parentheses.

$\quad = -(x+2)^2 + 6$

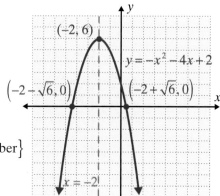

Summary:

Zeros: $-2 \pm \sqrt{6}$

Line of symmetry is $x = -2$.

Vertex: $(-2, 6)$

Domain: $(-\infty, \infty)$ or
$\qquad \left\{ x \mid x \text{ is any real number} \right\}$

Range: $(-\infty, 6]$ or $\left\{ y \mid y \le 6 \right\}$

c.   For $y = 2x^2 - 6x + 5$,   find the zeros of the function, the line of symmetry, the vertex, the domain, the range, and graph the parabola.

Solution: $2x^2 - 6x + 5 = 0$

$$x = \frac{6 \pm \sqrt{(-6)^2 - 4(2)(5)}}{2(2)}$$ Quadratic Formula

$$= \frac{6 \pm \sqrt{-4}}{4}$$

There are **no real zeros** because the discriminant is negative. The graph will not cross the x–axis. Now, using another approach, the vertex is at

$$x = -\frac{b}{2a} = -\frac{-6}{2 \cdot 2} = \frac{3}{2}$$

and

$$f\left(\frac{3}{2}\right) = 2\left(\frac{3}{2}\right)^2 - 6\left(\frac{3}{2}\right) + 5$$

$$= \frac{9}{2} - 9 + 5$$

$$= \frac{1}{2}.$$

So we have the vertex at $\left(\frac{3}{2}, \frac{1}{2}\right)$ and the following results:

Summary:

Zeros: No real zeros

Line of symmetry is $x = \frac{3}{2}$.

Vertex: $\left(\frac{3}{2}, \frac{1}{2}\right)$

Domain: $(-\infty, \infty)$ or
$\{x \,|\, x \text{ is any real number}\}$

Range: $\left[\frac{1}{2}, \infty\right)$ or $\left\{y \,|\, y \geq \frac{1}{2}\right\}$

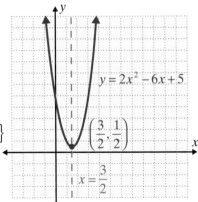

Applications with Maximum and Minimum Values

The vertex of a vertical parabola is either the lowest point or the highest point on the graph of the parabola.

Minimum and Maximum Values

For a parabola with its equation in the form $y = a(x - h)^2 + k,$

1. *If $a > 0$, then the parabola opens upward and (h, k) is the lowest point and $y = k$ is called the **minimum value** of the function.*

2. *If $a < 0$, then the parabola opens downward and (h, k) is the highest point and $y = k$ is called the **maximum value** of the function.*

If the function is in the general quadratic form $y = ax^2 + bx + c$, then the maximum or minimum value can be found by letting $x = -\dfrac{b}{2a}$ and solving for y.

The concepts of maximum and minimum values of a function help not only in graphing but also in solving many types of applications. Applications involving quadratic functions are discussed here. Other types of applications are discussed in Chapter 9 and in more advanced courses in mathematics.

Example 3: Minimum and Maximum Values

a. A sandwich company sells hot dogs at the local baseball stadium for $3.00 each. On average they sell 2000 hot dogs per game. The company estimates that each time the price is raised by 25¢, they will sell 100 fewer hot dogs. What price should they charge to maximize their revenue (income) per game? What will be the maximum revenue?

Solution: Let x = number of 25¢ increases in price.

Then $3.00 + 0.25x$ = price per hot dog

and $2000 - 100x$ = number of hot dogs sold.

Revenue = (price per unit)·(number of units sold)

So, $R = (3.00 + 0.25x)(2000 - 100x)$

$$= 6000 - 300x + 500x - 25x^2$$

$$= -25x^2 + 200x + 6000$$

The revenue is represented by a quadratic function and the maximum revenue occurs at the point where

$$x = -\frac{b}{2a} = -\frac{200}{-50} = 4.$$

For $x = 4$,

$$\text{price per hot dog} = 3.00 + 0.25(4) = \$4.00$$

and $$\text{Revenue} = R = (4)(2000 - 100 \cdot 4) = \$6400.$$

Thus the company will make its maximum revenue of $6400 by charging $4 per hot dog.

b. A rancher is going to build three sides of a rectangular corral next to a river. He has 240 feet of fencing and wants to enclose the maximum area possible inside the corral. What are the dimensions of the corral with the maximum area and what is this area?

Solution: Let x = length of one of the two equal sides of the rectangle.
Then $240 - 2x$ = length of third side of the rectangle.

Since area equals length times width, the area, A, of the corral is represented by the quadratic function $A = x(240 - 2x) = 240x - 2x^2$, the maximum area occurs at the point where $x = -\dfrac{b}{2a} = -\dfrac{240}{-4} = 60$.

Two sides of the rectangle are 60 feet and the third side is $240 - 2(60) = 120$ feet.

The maximum area possible is $60(120) = 7200 \text{ ft}^2$.

Practice Problems

1. Write the function $y = 2x^2 - 4x + 3$ in the form $y = a(x - h)^2 + k$.

2. Find the zeros of the function $y = x^2 - 7x + 10$.

3. Find the vertex and the range of the function $y = -x^2 + 4x - 5$.

Answers to Practice Problems: **1.** $y = 2(x - 1)^2 + 1$ **2.** $x = 2$, $x = 5$ **3.** Vertex: $(2, -1)$, Range: $(-\infty, -1]$

7.5 Exercises

For each of the quadratic functions in Exercises 1 – 20, determine the line of symmetry, the vertex, the domain, and the range.

1. $y = 3x^2 - 4$

2. $y = \dfrac{2}{3}x^2 + 7$

3. $y = 7x^2 + 9$

4. $y = 5x^2 - 1$

5. $y = -4x^2 + 1$

6. $y = -2x^2 - 6$

7. $y = -\dfrac{3}{4}x^2 + \dfrac{1}{5}$

8. $y = \dfrac{5}{3}x^2 + \dfrac{7}{8}$

9. $y = (x+1)^2$

10. $y = (x-1)^2$

11. $y = -\dfrac{2}{3}(x-4)^2$

12. $y = -5(x+2)^2$

13. $y = 2(x+3)^2 - 2$

14. $y = 4(x-5)^2 + 1$

15. $y = \dfrac{3}{4}(x+2)^2 - 6$

16. $y = -2(x+1)^2 - 4$

17. $y = -\dfrac{1}{2}\left(x - \dfrac{3}{2}\right)^2 + \dfrac{7}{2}$

18. $y = -\dfrac{5}{3}\left(x - \dfrac{9}{2}\right)^2 + \dfrac{3}{4}$

19. $y = \dfrac{1}{4}\left(x - \dfrac{4}{5}\right)^2 - \dfrac{11}{5}$

20. $y = \dfrac{10}{3}\left(x + \dfrac{7}{8}\right)^2 - \dfrac{9}{16}$

21. Graph the function $y = x^2$. Then, without additional computation, graph the following translations.

    a. $y = x^2 - 2$

    b. $y = (x-3)^2$

    c. $y = -(x-1)^2$

    d. $y = 5 - (x+1)^2$

22. Graph the function $y = 2x^2$. Then, without additional computation, graph the following translations.

    a. $y = 2x^2 - 3$

    b. $y = 2(x-4)^2$

    c. $y = -2(x+1)^2$

    d. $y = -2(x+2)^2 - 4$

23. Graph the function $y = \dfrac{1}{2}x^2$. Then, without additional computation, graph the following translations.

a. $y = \dfrac{1}{2}x^2 + 3$ **b.** $y = \dfrac{1}{2}(x+2)^2$

c. $y = -\dfrac{1}{2}x^2$ **d.** $y = \dfrac{1}{2}(x-1)^2 - 4$

24. Graph the function $y = \dfrac{1}{4}x^2$. Then, without additional computation, graph the following translations.

a. $y = -\dfrac{1}{4}x^2$ **b.** $y = \dfrac{1}{4}x^2 - 5$

c. $y = \dfrac{1}{4}(x+4)^2$ **d.** $y = 2 - \dfrac{1}{4}(x+2)^2$

Rewrite each of the quadratic functions in Exercises 25 – 40 in the form $y = a(x-h)^2 + k$. Find the vertex, range, and zero(s), if any, of each function. Graph the function.

25. $y = 2x^2 - 4x + 2$ **26.** $y = -3x^2 + 12x - 12$ **27.** $y = x^2 - 2x - 3$

28. $y = x^2 - 4x + 5$ **29.** $y = x^2 + 6x + 5$ **30.** $y = x^2 - 8x + 12$

31. $y = 2x^2 - 8x + 5$ **32.** $y = 2x^2 - 12x + 16$ **33.** $y = -3x^2 - 12x - 9$

34. $y = 3x^2 - 6x - 1$ **35.** $y = 5x^2 - 10x + 8$ **36.** $y = -4x^2 + 16x - 11$

37. $y = -x^2 - 5x - 2$ **38.** $y = x^2 + 3x - 1$ **39.** $y = 2x^2 + 7x + 5$

40. $y = 2x^2 + x - 3$

In Exercises 41 – 44, graph the two given functions and answer the following questions:

   a. Are the graphs the same?

   b. Do the functions have the same zeros?

   c. Briefly, discuss your interpretation of the results in parts **a** and **b**.

41. $\begin{cases} y = x^2 - 3x - 10 \\ y = -x^2 + 3x + 10 \end{cases}$

42. $\begin{cases} y = x^2 - 5x + 6 \\ y = -x^2 + 5x - 6 \end{cases}$

43. $\begin{cases} y = 2x^2 - 5x - 3 \\ y = -2x^2 + 5x + 3 \end{cases}$

44. $\begin{cases} y = -4x^2 - 15x + 4 \\ y = 4x^2 + 15x - 4 \end{cases}$

In Exercises 45 – 48, use the function $h = -16t^2 + v_0 t + h_0$, where h is the height of the object after time t, v_0 is the initial velocity, and h_0 is the initial height.

45. Throwing a ball: A ball is thrown vertically upward from the ground with an initial velocity of 112 ft/s.

   a. When will the ball reach its maximum height?

   b. What will be the maximum height?

46. Throwing a ball: A ball is thrown vertically upward from the ground with an initial velocity of 104 ft/s.

   a. When will the ball reach its maximum height?

   b. What will be the maximum height?

47. Throwing a stone: A stone is projected vertically upward from a platform that is 20 feet high, at a rate of 160 feet per second.

   a. When will the stone reach its maximum height?

   b. What will be the maximum height?

48. Throwing a stone: A stone is projected vertically upward from a platform that is 32 feet high, at a rate of 128 feet per second.

   a. When will the stone reach its maximum height?

   b. What will be the maximum height?

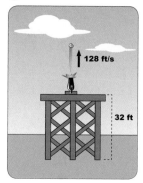

49. Selling lamps: A store owner estimates that by charging x dollars each for a certain lamp, he can sell $40 - x$ lamps each week. What price will give him maximum receipts?

50. Selling radios: A retailer sells radios. He estimates that by selling them for x dollars each, he will be able to sell $100 - x$ radios each month.

   a. What price will yield maximum revenue?

   b. What will be the maximum revenue?

51. Selling picture frames: Mrs. Richey can sell 72 picture frames each month if she charges $24 each. She estimates that for each $1 increase in price, she will sell 2 fewer frames.

   a. Find the price that will yield maximum revenue.

   b. What will be the maximum revenue?

52. Construction: A contractor is to build a brick wall 6 feet high to enclose a rectangular garden. The wall will be on three sides of the rectangle because the fourth side is a building. The owner wants to enclose the maximum area but wants to pay for only 150 feet of wall. What dimensions should the contractor make the garden?

In Exercises 53 – 58, use a graphing calculator to graph each function by pressing [Y=] *and entering the function. Find the zeros (or estimate the zeros) as follows. Round answers to four decimal places.*

Step 1: Press **CALC** ([2ND] [TRACE]).

Step 2: Press or choose **2: zero**.

Step 3: Follow the directions for moving the cursor to **Left Bound?**, **Right Bound?**, and **Guess?**. (Press [ENTER] each time.)

53. $y = x^2 - 2x - 2$

54. $y = 3x^2 + x - 1$

55. $y = -2x^2 + 2x + 5$

56. $y = -x^2 - 2x + 7$

57. $y = x^2 + 3x + 3$

58. $y = -4x^2 - x - 6$

Writing and Thinking About Mathematics

59. Discuss the following features of the general quadratic function $y = ax^2 + bx + c$.

   a. What type of curve is its graph?

   b. What is the value of x at the vertex of the parabola?

   c. What is the equation of the line of symmetry?

   d. Does the graph always cross the x-axis? Explain.

60. Discuss the discriminant of the general quadratic equation $y = a(x - h)^2 + k$ and how the value of the discriminant is related to the graph of the corresponding quadratic function $y = ax^2 + bx + c$.

Writing and Thinking About Mathematics (cont.)

61. Discuss the domain and range of a quadratic function in the form $y = a(x-h)^2 + k$.

 HAWKES LEARNING SYSTEMS: INTERMEDIATE ALGEBRA SOFTWARE

- Graphing Parabolas

Solving Quadratic and Rational Inequalities

- *Solve quadratic inequalities.*
- *Solve higher-degree inequalities.*
- *Solve rational inequalities.*
- *Graph the solutions of inequalities on real number lines.*
- *Use a graphing calculator as an aid in solving quadratic inequalities.*

In this section we will develop algebraic and graphical techniques for solving polynomial inequalities **with emphasis on quadratic inequalities** and **rational inequalities**. For example, we will see that inequalities such as

$$x^2 + 3x + 2 \geq 0, \quad x^2 - 2x > 8, \quad x^3 + 4x^2 - 5x < 0 \quad \text{and} \quad \frac{x+2}{x-2} > 0$$

can be solved by factoring or by using the quadratic formula and then analyzing the graphs of the corresponding functions. The solutions to these inequalities consist of intervals on the real number line.

Quadratic Inequalities: Solved Algebraically

The technique of factoring to solve inequalities is based on the simple idea that for values of x on either side of a number a, the sign for an expression of the form $(x - a)$ changes. More specifically:

If $x > a$, then $(x - a)$ is positive.

If $x < a$, then $(x - a)$ is negative.

Graphically,

$$x - a < 0 \qquad\qquad x - a > 0$$
$$\text{any } x < a \qquad a \qquad \text{any } x > a$$

To solve a polynomial inequality algebraically, the idea is to get 0 on one side of the inequality and then factor the polynomial on the other side. Locate the points where each factor is 0, and then analyze the sign of the polynomial in the intervals on each side of these points.

To Solve a Polynomial Inequality Algebraically

1. *Arrange the terms so that one side of the inequality is 0.*

2. *Factor the algebraic expression, if possible, and find the points where each factor is 0. (Use the quadratic formula, if necessary.)*

Continued on the next page...

To Solve a Polynomial Inequality Algebraically (cont.)

3. *Mark each of these points on a number line. These are the interval endpoints.*

4. *Test one point from each interval to determine the sign of the polynomial expression for all points in that interval.*

5. *The solution consists of those intervals where the test points satisfy the original inequality.*

6. *Mark a bracket for an endpoint that is included and a parenthesis for an endpoint that is not included.*

The following examples illustrate this technique.

Example 1: Solving Polynomial Inequalities by Factoring

Solve the following inequalities by factoring and by using a number line. Graph the solution set on a number line.

a. $x^2 - 2x > 8$

Solution:

$$x^2 - 2x > 8$$

$$x^2 - 2x - 8 > 0 \quad \text{Add } -8 \text{ to both sides so that one side is 0.}$$

$$(x+2)(x-4) > 0 \quad \text{Factor.}$$

Set each factor equal to 0 to locate the interval endpoints.

$$x + 2 = 0 \qquad x - 4 = 0$$

$$x = -2 \qquad x = 4$$

Mark these points on a number line and test one point from each of the intervals formed.

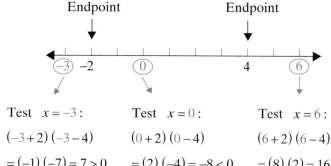

Test $x = -3$:	Test $x = 0$:	Test $x = 6$:
$(-3+2)(-3-4)$	$(0+2)(0-4)$	$(6+2)(6-2)$
$= (-1)(-7) = 7 > 0$	$= (2)(-4) = -8 < 0$	$= (8)(2) = 16 > 0$

This means that $(x+2)(x-4) > 0$ if $x < -2$.

This means that $(x+2)(x-4) < 0$ if $-2 < x < 4$.

This means that $(x+2)(x-4) > 0$ if $x > 4$.

Graphically,

The solution is:

algebraic notation   or   interval notation

$x < -2$ or $x > 4$ $(-\infty, -2) \cup (4, \infty)$

b. $2x^2 + 15 \le 13x$

Solution: $2x^2 + 15 \le 13x$

$2x^2 - 13x + 15 \le 0$ Add $-13x$ to both sides so that one side is 0.

$(2x - 3)(x - 5) \le 0$ Factor.

Set each factor equal to 0 to locate the interval endpoints.

$2x - 3 = 0$ $x - 5 = 0$

$x = \dfrac{3}{2}$ $x = 5$

Test one point from each interval formed.

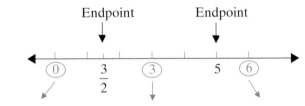

Test $x = 0$:	Test $x = 3$:	Test $x = 6$:
$(2 \cdot 0 - 3)(0 - 5)$	$(2 \cdot 3 - 3)(3 - 5)$	$(2 \cdot 6 - 3)(6 - 5)$
$= (-3)(-5) = 15 > 0$	$= (3)(-2) = -6 < 0$	$= (9)(1) = 9 > 0$
This means that	This means that	This means that
$(2x - 3)(x - 5) > 0$	$(2x - 3)(x - 5) < 0$	$(2x - 3)(x - 5) > 0$
if $x < \dfrac{3}{2}$.	if $\dfrac{3}{2} < x < 5$.	if $x > 5$.

The solution includes both endpoints since the inequality ($\le$) includes 0:

Continued on the next page...

The solution is:

algebraic notation   or   interval notation

$$\frac{3}{2} \le x \le 5 \qquad\qquad \left[\frac{3}{2}, 5\right]$$

c. $x^3 + 4x^2 - 5x < 0$

Solution: $x^3 + 4x^2 - 5x < 0$

$x\left(x^2 + 4x - 5\right) < 0$

$x(x + 5)(x - 1) < 0$

Set each factor equal to 0 to locate the interval endpoints.

$$x = 0 \qquad x + 5 = 0 \qquad\quad x - 1 = 0$$
$$x = -5 \qquad\qquad x = 1$$

Test one point from each interval formed.

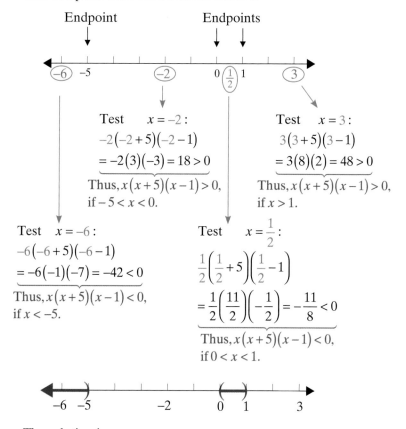

Endpoint   Endpoints

Test $x = -2$:
$-2(-2 + 5)(-2 - 1)$
$= -2(3)(-3) = 18 > 0$
Thus, $x(x + 5)(x - 1) > 0$,
if $-5 < x < 0$.

Test $x = 3$:
$3(3 + 5)(3 - 1)$
$= 3(8)(2) = 48 > 0$
Thus, $x(x + 5)(x - 1) > 0$,
if $x > 1$.

Test $x = -6$:
$-6(-6 + 5)(-6 - 1)$
$= -6(-1)(-7) = -42 < 0$
Thus, $x(x + 5)(x - 1) < 0$,
if $x < -5$.

Test $x = \dfrac{1}{2}$:
$\dfrac{1}{2}\left(\dfrac{1}{2} + 5\right)\left(\dfrac{1}{2} - 1\right)$
$= \dfrac{1}{2}\left(\dfrac{11}{2}\right)\left(-\dfrac{1}{2}\right) = -\dfrac{11}{8} < 0$
Thus, $x(x + 5)(x - 1) < 0$,
if $0 < x < 1$.

The solution is:

algebraic notation   or   interval notation

$x < -5$ or $0 < x < 1$ $(-\infty, -5) \cup (0, 1)$

Example 2: Solving Polynomial Inequalities Using the Quadratic Formula

Solve the following inequalities using the quadratic formula and a number line. Graph the solution set on a number line.

a. $x^2 - 2x - 1 > 0$

Solution: The quadratic expression $x^2 - 2x - 1$ will not factor with integer coefficients. Use the quadratic formula and find the roots of the equation $x^2 - 2x - 1 = 0$. Then use these roots as endpoints for the intervals. The test points can themselves be integers.

$x^2 - 2x - 1 = 0$

$$x = \frac{2 \pm \sqrt{(-2)^2 - 4(1)(-1)}}{2(1)} \quad \text{Use the quadratic formula.}$$

$$x = \frac{2 \pm \sqrt{4 + 4}}{2}$$

$$x = \frac{2 \pm 2\sqrt{2}}{2} = 1 \pm \sqrt{2}$$

The endpoints are $x = 1 - \sqrt{2}$ and $x = 1 + \sqrt{2}$.

Test one point from each interval formed.
Note: With a calculator, you can determine that $1 - \sqrt{2} \approx -0.414$ and $1 + \sqrt{2} \approx 2.414$.

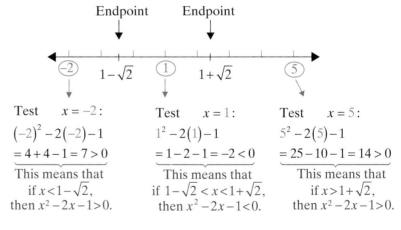

The solution is:

    algebraic notation   or   interval notation

$x < 1 - \sqrt{2}$ or $x > 1 + \sqrt{2}$ $\left(-\infty, 1 - \sqrt{2}\right) \cup \left(1 + \sqrt{2}, \infty\right)$

Continued on the next page...

b. $x^2 - 2x + 13 > 0$

Solution: To find where $x^2 - 2x + 13 > 0$, use the quadratic formula:

$$x = \frac{2 \pm \sqrt{(-2)^2 - 4(1)(13)}}{2} = \frac{2 \pm \sqrt{-48}}{2}$$

$$= \frac{2 \pm 4i\sqrt{3}}{2} = 1 \pm 2i\sqrt{3}.$$

Since these values are nonreal, the polynomial is either always positive or always negative for real values of x. Therefore, we only need to test one point. If that point satisfies the inequality, then the solution is all real numbers. If it does not, then there is no solution. In this example, we test $x = 0$ because the polynomial is easy to evaluate for $x = 0$.

$$0^2 - 2(0) + 13 = 13 > 0$$

Since the real number 0 satisfies the inequality, the solution is all real numbers. In interval notation we write $(-\infty, \infty)$, and in algebraic notation we write $\mathbb{R}$. Graphically,

Solving Rational Inequalities

A rational inequality may involve the product or quotient of several first-degree expressions. For example, the inequality

$$\frac{x+3}{x-2} > 0$$

involves the two first-degree expressions $x + 3$ and $x - 2$.

Inequalities of this form can be solved using a similar procedure to that used for quadratic inequalities.

To Solve a Polynomial Inequality with Rational Expressions

1.   *Simplify the inequality so that one side is 0 and the other side has a single fraction with both the numerator and denominator in factored form.*

2.   *Find the points that cause the factors in the numerator or in the denominator to be 0.*

Continued on the next page...

To Solve a Polynomial Inequality with Rational Expressions (cont.)

3. *Mark each of these points on a number line. These are the interval endpoints.*

4. *Test one point from each interval to determine the sign of the polynomial expression for all points in that interval.*

5. *The solution consists of those intervals where the test points satisfy the original inequality.*

6. *Mark a bracket for an endpoint that is included and a parenthesis for an endpoint that is not included. Remember that no denominator can be 0.*

Example 3: Solving Rational Inequalities

Solve and graph the following rational inequalities.

a. $\dfrac{x+3}{x-2} > 0$

Solution: Set each factor in the numerator and the denominator equal to 0 to locate the interval endpoints.

$$x+3=0 \qquad x-2=0$$
$$x=-3 \qquad x=2$$

Mark each of these points on a number line and test one point from each of the intervals formed. (Consider these points as endpoints of intervals.)

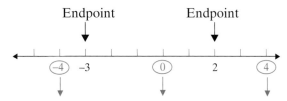

Test $x=-4$:

$$\frac{-4+3}{-4-2}=\frac{-1}{-6}=\frac{1}{6}>0$$

This means that

$$\frac{x+3}{x-2}>0 \text{ if } x<-3.$$

Test $x=0$:

$$\frac{0+3}{0-2}=-\frac{3}{2}<0$$

This means

that $\dfrac{x+3}{x-2}<0$

if $-3<x<2.$

Test $x=4$:

$$\frac{4+3}{4-2}=\frac{7}{2}>0$$

This means that

$$\frac{x+3}{x-2}>0 \text{ if } x>2.$$

Continued on the next page...

Graphically,

The solution is:

algebraic notation	or	interval notation
$x < -3$ or $x > 2$		$(-\infty, -3) \cup (2, \infty)$

b. $\dfrac{x+3}{x-2} < 0$

Solution: Using the graph and test points from Example 2a, we know that $\dfrac{x+3}{x-2} < 0$ if $-3 < x < 2$.

Graphically,

The solution is:

algebraic notation	or	interval notation
$-3 < x < 2$		$(-3, 2)$

Example 4: Solving Rational Inequalities

Solve the following rational inequality.

$$\frac{x+5}{x-4} \geq -1$$

Solution:

$\dfrac{x+5}{x-4} + 1 \geq 0$　　　One side must be 0.

$\dfrac{x+5}{x-4} + \dfrac{x-4}{x-4} \geq 0$　　　Rewrite 1 as a fraction using the LCD, $(x-4)$, as the denominator.

$\dfrac{2x+1}{x-4} \geq 0$　　　Simplify to get one fraction. $(x \neq 4)$

Set each linear expression equal to 0 to find the interval endpoints.

$$2x + 1 = 0 \qquad x - 4 = 0$$

$$x = -\frac{1}{2} \qquad x = 4$$

Test a value from each of the intervals:

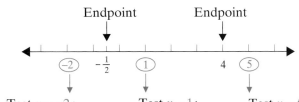

Test $x = -2$:

$$\underbrace{\frac{2(-2)+1}{-2-4} = \frac{-3}{-6} = \frac{1}{2} > 0}$$

This means that
$$\frac{2x+1}{x-4} > 0 \text{ if } x < -\frac{1}{2}.$$

Test $x = 1$:

$$\underbrace{\frac{2(1)+1}{1-4} = \frac{3}{-3} < 0}$$

This means
that $\frac{2x+1}{x-4} < 0$
if $-\frac{1}{2} < x < 4.$

Test $x = 5$:

$$\underbrace{\frac{2(5)+1}{5-4} = \frac{11}{1} > 0}$$

This means that
$$\frac{2x+1}{x-4} > 0 \text{ if } x > 4.$$

The solution includes the endpoint $-\frac{1}{2}$ since the inequality ($\leq$) includes 0. However, the endpoint 4 is not included in the solution because the rational inquality is undefined (the denominator equals 0) when $x = 4$.

Graphically,

The solution is:

algebraic notation   or   interval notation

$x \leq -\frac{1}{2}$ or $x > 4$ $\left(-\infty, -\frac{1}{2}\right] \cup (4, \infty)$

NOTES   Notice that in the first step, we do **not** multiply by the denominator $x - 4$. The reason is that the variable expression is positive for some values of x and negative for other values of x. Therefore, if we did multiply by $x - 4$, we would not be able to determine whether the inequality should stay as $\geq$ or be reversed to $\leq$.

Solving Quadratic Inequalities Using a Graphing Calculator

To solve quadratic (or other polynomial) inequalities with a graphing calculator by graphing the related polynomial function, look for the intervals of x for which the graph of the function is above the x-axis (the y-values will be positive) and where it is below the x-axis (the y-values will be negative). The y-values represent the values of the related polynomial function.

To Solve a Polynomial Inequality Using a Graphing Calculator

1. *Arrange the terms so that one side of the inequality is 0.*

2. *Form a function by letting y = [the polynomial], and graph the function. Be sure to set the* **WINDOW** *so that all of the zeros are easily seen. This may be difficult if large numbers are involved.*

3. *Use the* **CALC** *key (* 2ND   TRACE *) and select* **2:(zero)** *to find (or approximate) the real zeros of the function, if there are any.*

   a. *The values of x for which the y-values are above the x-axis satisfy* $y > 0$.

   b. *The values of x for which the y-values are below the x-axis satisfy* $y < 0$.

4. *Endpoints of intervals are included if the inequality includes 0, as in* $y \geq 0$ *or* $y \leq 0$.

Example 5: Solving a Quadratic Inequality Using a Graphing Calculator

Solve the following quadratic inequalities using a graphing calculator. Graph the solution set on a real number line.

a. $x^2 - 2x > 8$

Solution: This inequality was solved algebraically in Example 1a. We repeat the solution here using a graphing calculator to show how the two methods are related.

Manipulate the inequality so that one side is 0.

$$x^2 - 2x > 8$$
$$x^2 - 2x - 8 > 0$$

Press [Y=] and enter the function: $y = x^2 - 2x - 8$.

Press [GRAPH] to graph the function.

The graph will appear as illustrated here. (In this case the graph is a parabola.)

Find the zeros (or estimate the zeros) as follows:

Step 1: Press **CALC** ().

Step 2: Press or choose **2: zero**.

Step 3: Follow the directions for moving the cursor to **Left Bound?**, **Right Bound?**, and **Guess?** for each point. (Press **ENTER** each time.)

[**Note:** If there are no real zeros, then the graph will be entirely above or entirely below the *x*-axis. The solution is then dependent on the nature of the inequality. It will be either the empty set (no solution) or the entire set of real numbers $(-\infty, \infty)$.]

In this case, the zeros are $x = -2$ and $x = 4$ (just as we found in Example 1a). Because we want to know where $y > 0$, we look at the graph and choose the intervals for *x* where the curve is above the *x*-axis. [**Note:** Endpoints are not included because the inequality does not include 0.]

Thus the solution consists of the union of two intervals: $(-\infty, -2) \cup (4, \infty)$.

b. $2x^2 + 3x - 10 \le 0$

Solution: Press  and enter the function: $y = 2x^2 + 3x - 10$.

Press **GRAPH** to graph the function.

The graph will appear as illustrated here. (In this case the graph is a parabola.)

Find the zeros (or estimate the zeros) as follows:

Step 1: Press **CALC** (**2ND** **TRACE**).

Step 2: Press or choose **2: zero**.

Step 3: Follow the directions for moving the cursor to **Left Bound?**, **Right Bound?**, and **Guess?** for each point. (Press **ENTER** each time.)

Continued on the next page...

In this case the zeros are estimates: $x \approx -3.1085$ and $x \approx 1.6085$.

Because we want to know where $y \leq 0$, we look at the graph and choose the intervals for x where the curve is below the x-axis and include the endpoints because 0 is included in the inequality.

Thus the solution is the closed interval $[-3.1085, 1.6085]$.

-3.1085 1.6085

Example 6: Solving a Higher-Degree Polynomial Inequality Using a Graphing Calculator

Solve the following 3rd-degree polynomial inequality using a graphing calculator: $x^3 + 2x^2 - 11x - 12 > 0$.

Solution: Press ⬜ Y= and enter the function: $y = x^3 + 2x^2 - 11x - 12$.
Press ⬜ GRAPH to graph the function.

The graph will appear as illustrated here. (In this case the graph is not a parabola.)

Find the zeros (or estimate the zeros) as follows:
    Step 1: Press **CALC** (**2ND** **TRACE**).
    Step 2: Press or choose **2: zero**.
    Step 3: Follow the directions for moving the cursor to **Left Bound?**, **Right Bound?**, and **Guess?** for each point. (Press **ENTER** each time.)

In this case there are three zeros: $x = -4$, $x = -1$, and $x = 3$.

Because we want to know where $y > 0$, we look at the graph and choose the intervals for x where the curve is above the x-axis. Do not include the endpoints because 0 is not included in the inequality.

Thus the solution is the union of two intervals: $(-4, -1) \cup (3, \infty)$.

Practice Problems

Solve each of the following inequalities algebraically. Graph each solution set on a real number line. Write the answers in interval notation.

1. $(x-5)(x-7) > 0$ **2.** $x^2 - 5x > -4$ **3.** $\dfrac{4+x}{x-2} \le 0$

Use a graphing calculator to solve the following inequalities. Write the answers in interval notation.

4. $2x^2 - 2x - 3 \ge 0$ **5.** $x^3 - 4x + 6 > 0$

Answers to Practice Problems: **1.** $(-\infty, 5) \cup (7, \infty)$ **2.**
$(-\infty, 1) \cup (4, \infty)$ **3.** $[-4, 2)$
4. $(-\infty, -0.8229) \cup (1.8229, \infty)$ **5.** $(-2.5251, \infty)$

7.6 Exercises

*In Exercises 1 – 50, solve the quadratic (and higher degree) inequalities algebraically, write the answers in interval notation, and then graph each solution set on a number line. (**Note**: You may need to use the quadratic formula to find endpoints of intervals.)*

1. $(x-6)(x+2)<0$ **2.** $(x+4)(x-2)>0$ **3.** $(3x-2)(x-5)>0$

4. $(4x+1)(x+1)\leq 0$ **5.** $(x+7)(2x-5)\geq 0$ **6.** $(x-3)(5x-3)\leq 0$

7. $(3x+1)(x+2)\leq 0$ **8.** $(x-4)(3x-8)>0$

9. $x(3x+4)(x-5)<0$ **10.** $(x-1)(x+4)(2x+5)<0$

11. $x^2+4x+4\leq 0$ **12.** $5x^2+4x-12>0$

13. $2x^2>x+15$ **14.** $6x^2+x>2$ **15.** $8x^2<10x+3$

16. $2x^2<x+10$ **17.** $2x^2-5x+2\geq 0$ **18.** $15y^2-21y-18<0$

19. $6y^2+7y<-2$ **20.** $3x^2+3\geq 10x$ **21.** $4z^2-20z+25>0$

22. $14+11x\geq 15x^2$ **23.** $8x^2+6x\leq 35$ **24.** $7x<6x^2+x^3$

25. $x^3>2x^2+3x$ **26.** $x^3<6x^2-9x$ **27.** $x^3>5x^2-4x$

28. $4x^2\leq x^3+3x$ **29.** $(x+2)(x-2)>3x$ **30.** $(x+4)(x-1)<2x+2$

31. $x^4-5x^2+4>0$ **32.** $x^4-25x^2+144<0$ **33.** $y^4-13y^2+36\leq 0$

34. $y^4-13y^2-48\geq 0$ **35.** $(x+1)^2-9\geq 0$ **36.** $(3x-1)^2-16<0$

37. $(2x-3)(3x+2)-(3x+2)<0$ **38.** $2(x-1)(x-3)>(x-1)(x-6)$

39. $x^2+2x-4>0$ **40.** $x^2-8x+14<0$ **41.** $x^2+6x+7\geq 0$

42. $2x^2+4x-3<0$ **43.** $3x^2+5x+1<0$ **44.** $3x^2+8x+5\geq 0$

45. $2x^3 \le 7x^2 + 4x$ **46.** $2x^2 > 9x - 8$ **47.** $x^2 - 2x + 2 > 0$

48. $x^2 + 3x + 3 < 0$ **49.** $2x - 1 > 3x^2$ **50.** $6x - 10 < x^2$

In Exercises 51 – 54, the graph of a quadratic function is given. Use the information in the graph to solve the related equations and inequalities in parts a – c.

51. $y = x^2 - 7x - 10$

52. $y = x^2 + 5x - 6$

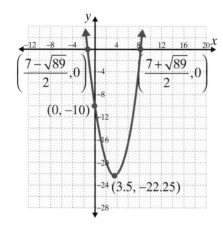

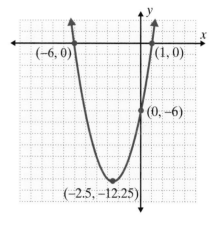

a. $x^2 - 7x - 10 = 0$
b. $x^2 - 7x - 10 > 0$
c. $x^2 - 7x - 10 < 0$

a. $x^2 + 5x - 6 = 0$
b. $x^2 + 5x - 6 > 0$
c. $x^2 + 5x - 6 < 0$

53. $y = -x^2 - 4x + 5$

54. $y = -3x^2 - 6x + 15$

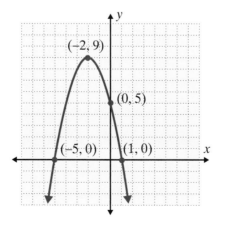

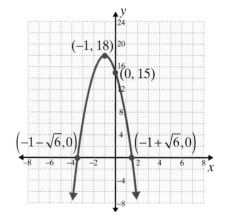

a. $-x^2 - 4x + 5 = 0$
b. $-x^2 - 4x + 5 > 0$
c. $-x^2 - 4x + 5 < 0$

a. $-3x^2 - 6x + 15 = 0$
b. $-3x^2 - 6x + 15 > 0$
c. $-3x^2 - 6x + 15 < 0$

In Exercises 55 – 72, solve the rational inequalities algebraically, write the answers in interval notation, and then graph each solution set on a number line.

55. $\dfrac{x+4}{2x} \geq 0$

56. $\dfrac{x}{x-4} \geq 0$

57. $\dfrac{x+6}{x^2} < 0$

58. $\dfrac{3-x}{x+1} < 0$

59. $\dfrac{x+3}{x+9} > 0$

60. $\dfrac{2x+3}{x-4} < 0$

61. $\dfrac{3x-6}{2x-5} < 0$

62. $\dfrac{4-3x}{2x+4} \leq 0$

63. $\dfrac{x+5}{x-7} \geq 1$

64. $\dfrac{x+4}{2x-1} > 2$

65. $\dfrac{2x+5}{x-4} \leq -3$

66. $\dfrac{3x+2}{4x-1} < 3$

67. $\dfrac{5-2x}{3x+4} < -1$

68. $\dfrac{8-x}{x+5} < -4$

69. $\dfrac{x(x+4)}{x-3} \leq 0$

70. $\dfrac{(x+3)(x-2)}{x+1} > 0$

71. $\dfrac{(x-5)}{x(x+2)} \geq 0$

72. $\dfrac{-(x-3)^2}{(x-1)(x-4)} < 0$

Use a graphing calculator to solve the inequalities in Exercises 73 – 82. Write the answers in interval notation, and then graph each solution set on a number line. (Estimate endpoints, when necessary, to 4 decimal places.)

73. $x^2 > 10$

74. $20 \geq x^2$

75. $x^2 - 2.5x + 6.25 < 0$

76. $x^2 + 2x \geq -1$

77. $x^3 - 9x < 0$

78. $x^3 - 4x^2 + 4x \leq 0$

79. $2x^3 - 5x + 4 \geq 0$

80. $x^3 - 4x^2 + 3 < 0$

81. $-x^4 + 6x^2 - 3 > 0$

82. $x^4 - 2x^3 - x^2 - 1 < 0$

Writing and Thinking About Mathematics

83. Use a graphing calculator to graph the rational function $y = \dfrac{x^2 + 3x - 4}{x}$.

  a. Use the graph to find the solution set for $y > 0$.
  b. Use the graph to find the solution set for $y < 0$.
  c. Explain the effect of $x = 0$ on the graph and why 0 is not included in either
     a. or **b.**

Writing and Thinking About Mathematics (cont.)

84. In your own words, explain why (as in Example 2b), when the quadratic formula gives nonreal values, the quadratic polynomial is either always positive or always negative.

 HAWKES LEARNING SYSTEMS: INTERMEDIATE ALGEBRA SOFTWARE

- Solving Quadratic Inequalities
- Solving Inequalities with Rational Expressions

Chapter 7 Index of Key Ideas and Terms

Section 7.2 The Quadratic Formula: $x = \dfrac{-b \pm \sqrt{b^2 - 4ac}}{2a}$

Quadratic Formula   page 562

For the general quadratic equation $ax^2 + bx + c = 0$

where $a \neq 0$, the solutions are $x = \dfrac{-b \pm \sqrt{b^2 - 4ac}}{2a}$.

Discriminant   pages 564-565

The expression $b^2 - 4ac$, the part of the quadratic formula
that lies under the radical sign, is called the **discriminant**.

If $b^2 - 4ac > 0$, there are two real solutions.

If $b^2 - 4ac = 0$, there is one real solution.

If $b^2 - 4ac < 0$, there are two nonreal solutions.

Section 7.3 Applications

The Pythagorean Theorem   page 571

In a right triangle, the square of the hypotenuse is equal to
the sum of the squares of the legs: $c^2 = a^2 + b^2$.

Applications

The Pythagorean Theorem   page 571

Projectiles   page 572

Geometry   page 573

Cost Per Person   pages 573-574

Section 7.4 Equations in Quadratic Form

Solving Equations in Quadratic Form by Substitution   page 580

1. Look at the middle term.
2. Substitute a first-degree variable, such as u, for the
    variable expression in the middle term.
3. Substitute the square of this variable, u^2, for the
    variable expression in the first term.
4. Solve the resulting quadratic equation for u.
5. Substitute the results "back" for u in the beginning
    substitution and solve for the original variable.

Solving Equations with Rational Expressions   page 582

Solving Higher-Degree Equations   page 583

Section 7.5 Graphing Quadratic Functions: Parabolas

Zeros of a Function   page 587

The **zeros of a function** are the points where the y-value is 0 and the graph crosses the x-axis.

Vertical Line Test   page 587

If **any** vertical line intersects a graph in more than one point, then the relation graphed is **not** a function.

Quadratic Function   page 588

Any function that can be written in the form $y = ax^2 + bx + c$, where a, b, and c are real numbers and $a \neq 0$ is a **quadratic function**.

Parabola   page 588

The graph of every quadratic function is a **parabola**.

Forms of Quadratic Functions

$y = ax^2$ page 589

$y = ax^2 + k$ page 590

$y = a(x - h)^2$ pages 590-591

$y = a(x - h)^2 + k$ page 593

$y = ax^2 + bx + c$ page 593

Line of Symmetry   page 594

The line $x = -\dfrac{b}{2a}$ is the **line of symmetry** of the graph of the quadratic function $y = ax^2 + bx + c$.

Vertex   page 594

The **vertex** of the graph of $y = ax^2 + bx + c$ is at the point $\left(-\dfrac{b}{2a}, \dfrac{4ac - b^2}{4a}\right)$.

Zeros of a Quadratic Function   page 594

Minimum and Maximum Values   page 598

For a parabola with equation of the form $y = a(x - h)^2 + k$,

1. If $a > 0$, then (h, k) is the lowest point and $y = k$ is called the **minimum value** of the function.
2. If $a < 0$, then (h, k) is the highest point and $y = k$ is called the **maximum value** of the function.

Section 7.6   Solving Quadratic and Rational Inequalities

To Solve a Polynomial Inequality Algebraically pages 605-606

1. Arrange the terms so that one side of the inequality is 0.
2. Factor the algebraic expression, if possible, and find the points where each factor is 0. (Use the quadratic formula, if necessary.)
3. Mark each of these points on a number line. These are the interval endpoints.
4. Test one point from each interval to determine the sign of the polynomial expression for all points in that interval.
5. The solution consists of those intervals where the test points satisfy the original inequality.
6. Mark a bracket for an endpoint that is included and a parenthesis for an endpoint that is not included.

To Solve a Polynomial Inequality with Rational Expressions pages 610-611

1. Simplify the inequality so that one side is 0 and the other side has a single fraction with both the numerator and the denominator in factored form.
2. Find the points that cause the factors in the numerator or in the denominator to be 0.
3. Mark each of these points on a number line. These are the interval endpoints.
4. Test one point from each interval to determine the sign of the polynomial expression for all points in that interval.
5. The solution consists of those intervals where the test points satisfy the original inequality.
6. Mark a bracket for an endpoint that is included and a parenthesis for an endpoint that is not included. Remember that no denominator can be 0.

To Solve a Polynomial Inequality Using a Graphing Calculator page 614

1. Arrange the terms so that one side of the inequality is 0.
2. Form a function by letting $y = $ [the polynomial], and graph the function. Be sure to set the WINDOW so that all of the zeros are easily seen. This may be difficult if large numbers are involved.

Continued on next page...

Section 7.6 Solving Quadratic and Rational Inequalities (cont.)

3. Use the CALC key (2nd TRACE) and select 2:(zero) to find (or approximate) the real zeros of the function, if there are any.

   a. The values of x for which the y-values are above the x-axis satisfy $y > 0$.

   b. The values of x for which the y-values are below the x-axis satisfy $y < 0$.

4. Endpoints of the intervals are included if the inequality includes 0, as in $y \geq 0$ or $y \leq 0$.

HAWKES LEARNING SYSTEMS: INTERMEDIATE ALGEBRA SOFTWARE

For a review of the topics and problems from Chapter 7, look at the following lessons from *Hawkes Learning Systems: Intermediate Algebra*

- Quadratic Equations: The Square Root Method
- Quadratic Equations: Completing the Square
- Quadratic Equations: The Quadratic Formula
- Applications: Quadratic Equations
- Equations in Quadratic Form
- Graphing Parabolas
- Solving Quadratic Inequalities
- Solving Inequalities with Rational Expressions
- Chapter 7 Review and Test

Chapter 7 Review

7.1 Solving Quadratic Equations

Solve the equations in Exercises 1 – 4 by factoring.

1. $x^2 + 5x - 36 = 0$ **2.** $x^2 - 6x = 14 - x$ **3.** $2x^2 = 24x - 22$ **4.** $-13x = x^2$

Add the correct constant to complete the square in Exercises 5 – 8, then factor the trinomial as indicated.

5. $x^2 - 6x + \underline{\hspace{0.5cm}} = \left(\underline{\hspace{0.8cm}}\right)^2$ **6.** $x^2 + 10x + \underline{\hspace{0.5cm}} = \left(\underline{\hspace{0.8cm}}\right)^2$

7. $y^2 - \dfrac{1}{2}y + \underline{\hspace{0.5cm}} = \left(\underline{\hspace{0.8cm}}\right)^2$ **8.** $x^2 - x + \underline{\hspace{0.5cm}} = \left(\underline{\hspace{0.8cm}}\right)^2$

Solve the equations given in Exercises 9 – 12 by using the square root property.

9. $x^2 = -36$ **10.** $x^2 - 100 = 0$ **11.** $x^2 = 24$ **12.** $2(x+4)^2 = 20$

Solve the quadratic equations in Exercises 13 – 18 by completing the square.

13. $x^2 - 6x + 4 = 0$ **14.** $x^2 + 2x = 7$ **15.** $3x^2 + 6x + 12 = 0$

16. $4x^2 - 6x + 2 = 0$ **17.** $6x^2 = 3 - 3x$ **18.** $5x^2 - 15x + 25 = 0$

For Exercises 19 – 22, write a quadratic equation with integer coefficients that has the given roots.

19. $x = \sqrt{3}$, $x = -\sqrt{3}$ **20.** $x = 1 + \sqrt{2}$, $x = 1 - \sqrt{2}$

21. $x = i\sqrt{6}$, $x = -i\sqrt{6}$ **22.** $x = -3 + i\sqrt{5}$, $x = -3 - i\sqrt{5}$

7.2 The Quadratic Formula: $x = \dfrac{-b \pm \sqrt{b^2 - 4ac}}{2a}$

Find the discriminant and determine the nature of the solutions of each quadratic equation in Exercises 23 – 26.

23. $2x^2 - 3x + 4 = 0$ **24.** $3x^2 + x - 2 = 0$

25. $5x^2 = 3x$ **26.** $x^2 - 20x + 100 = 0$

Solve the equations in Exercises 27 – 32 by using the quadratic formula.

27. $5x^2 + 2x - 2 = 0$ **28.** $x^2 - 2x - 4 = 0$ **29.** $2x^2 - 4x + 3 = 0$

30. $4x^2 + 169 = 0$ **31.** $3x^2 + 1 = -x$ **32.** $x^2 + 10x = -25$

Solve the equations in Exercises 33 – 42 by using any of the techniques discussed for solving quadratic equations: factoring, completing the square, or the quadratic formula.

33. $2x^2 - 5x - 3 = 0$

34. $x^2 - 6x - 7 = 0$

35. $3x^2 + 2x = 5$

36. $x^2 = 11x - 28$

37. $x^2 + 6x + 7 = 0$

38. $2x^2 + 7x + 9 = 0$

39. $4x^2 - 6x + 3 = 0$

40. $x^3 - 5x^2 = x^2 + 3x$

41. $6x^2 = 10x$

42. $9x^2 + 4 = 0$

In Exercises 43 and 44, find the indicated values for k.

43. Determine the values for k such that $x^2 + 6x - k = 0$ will have two real solutions.

44. Determine the values for k such that $kx^2 + 4x + k = 0$ will have two nonreal solutions.

7.3 Applications

45. Right triangles: The length of one leg of a right triangle is 10 feet longer than the length of the other leg. If the hypotenuse is 50 feet long, what are the lengths of the two legs?

46. Find a positive real number such that its square is equal to 144 more than 10 times the number.

47. One integer is five more than twice a second integer. Their product is 168. What are the integers?

48. Football: The seating on one side of a high school football field is in the shape of a rectangle. The capacity of the seating is 1920 fans. The number of seats in each row is 40 less than ten times the number of rows. How many seats are in each row?

49. Area of a rectangle: The area of a rectangle is 72 square meters, and the perimeter of the rectangle is 34 meters. Find the length and the width of the rectangle.

50. Gardening: A garden measures 20 meters by 18 meters. The gardener is going to till strips of equal width along the four sides of the garden to plant a grass border. How wide will the strips be if the non-grassy area of the garden is now 288 square meters? [**Hint:** Draw a picture.]

51. Bridge club: The women in a bridge club are to travel to a bridge tournament by bus for a total cost of $1980 which is to be divided equally among the women. However, two women became ill and could not make the trip. The cost to the remaining members increased $11 each. How many members made the trip? What was the final cost per member?

52. Biking: Martin and Milton leave Denver at the same time. Martin rides his bicycle south and Milton rides his bicycle east. Milton's average speed is 5 mph faster than Martin's. At the end of two hours they are 50 miles apart. Find the average speed of each rider.

53. Launching a rocket: A rocket is fired straight up from the ground with an initial velocity of 640 ft/s. Using the formula $h = -16t^2 + v_0 t + h_0$ determine the following:

a. When will the projectile hit the ground?

b. When will the projectile be 1200 ft above the ground? (Round answers to nearest hundredth.)

54. Throwing a ball: A ball is tossed straight up with an initial velocity of 32 ft/s from a height of 20 ft above the beach at the edge of a cliff. Using the formula $h = -16t^2 + v_0 t + h_0$ determine the following:

a. Will the ball reach a height of 60 ft above the beach?

b. When will the ball hit the beach?

7.4 Equations in Quadratic Form

Solve the equations in Exercises 55 – 64.

55. $x^4 - 13x^2 + 36 = 0$

56. $x^4 + 4x^2 - 12 = 0$

57. $y^{-2} + y^{-1} - 2 = 0$

58. $2x^{\frac{2}{3}} - 7x^{\frac{1}{3}} - 4 = 0$

59. $y^{-1} + 6y^{-\frac{1}{2}} - 7 = 0$

60. $2 + \dfrac{1-x}{x+1} = \dfrac{x-2}{x+6}$

61. $\dfrac{y+2}{y-6} + \dfrac{2y-6}{y+1} = \dfrac{11y+3}{y+1}$

62. $x^4 = -81$

63. $(x+1)^2 + 8(x+1) + 7 = 0$

64. $3(x+5)^2 + 14(x+5) = 5$

7.5 Graphing Quadratic Functions: Parabolas

For each of the quadratic functions in Exercises 65 – 70, determine the line of symmetry, the vertex, the domain, and the range. Graph the function.

65. $y = 2x^2 - 3$

66. $y = -3x^2 + 4$

67. $y = (x+4)^2$

68. $y = -2x^2 + 8$

69. $y = (x+1)^2 - 4$

70. $y = -2\left(x - \dfrac{3}{2}\right)^2 - \dfrac{9}{2}$

Rewrite each of the quadratic functions in Exercises 71 – 76 in the form $y = a(x-h)^2 + k$. Find the vertex, range, and zero(s), if any, of each function. Graph the function.

71. $y = 2x^2 - 12x + 18$

72. $y = x^2 - 2x - 4$

73. $y = x^2 - 4x + 1$

74. $y = -x^2 - 6x - 2$

75. $y = 2x^2 + 9x + 6$

76. $y = x^2 + 2x + 4$

77. Selling computers: An electronics store sells computers. The owner estimates that by selling them for x dollars each, the sales will be $1600 - x$ each month.

  a. What price will yield the maximum monthly revenue?

  b. What will be the maximum monthly revenue?

78. Publishing: The profit function (in dollars) for a publisher selling software to accompany an English textbook is given by the function $P(x) = -0.2x^2 + 120x - 3000$ where x is the number of copies of the software sold in one semester.

  a. What number of copies of the software should be sold to give the maximum profit?

  b. What will be the maximum profit?

79. Rockets: A toy rocket is projected vertically upward from the ground with an initial velocity of 208 ft/s. Using the formula $h = -16t^2 + v_0t + h_0$, answer the following questions.

  a. When will the rocket reach its maximum height?

  b. What will be the maximum height?

  c. When will the rocket be on the ground?

80. Building a fence: Manuel decides to build a fence to enclose an area for his dogs to play. The area is to be rectangular in shape. If he has 100 feet of fencing, find the dimensions of the rectangle that will give the dogs the maximum area in which to play. Only 3 sides need to be fenced because his house is to be one border of the rectangle.

7.6 Solving Quadratic and Rational Inequalities

In Exercises 81 – 92, solve the inequalities algebraically, write the answers in interval notation, and then graph each solution set on a real number line.

81. $(x-5)(x+1) < 0$

82. $(x+3)(2x-1) > 0$

83. $2x^2 + x \geq 3$

84. $x^2 + 6x + 9 \leq 0$

85. $x(3x+5)(x-4) < 0$

86. $4x^2 - 12x + 9 > 0$

87. $3x^2 + 1 \geq 4x$

88. $x^2 + 2x - 5 > 0$

89. $\dfrac{x-6}{2x+1} \leq 0$

90. $\dfrac{3x-3}{x-4} > 0$

91. $\dfrac{2x-5}{3-x} < -1$

92. $\dfrac{4x+1}{x+2} \geq \dfrac{5}{2}$

Use a graphing calculator to solve the inequalities in Exercises 93 – 98. Write the answers in interval notation and graph each solution set on a real number line. (Estimate endpoints when necessary to 4 decimal places.)

93. $x^2 > 8$

94. $x^2 - 6x > 0$

95. $x^2 - 5x - 6 \leq 0$

96. $x^3 - 2x^2 + 4 < 0$

97. $-x^4 + 4x^2 - 2 \leq 0$

98. $2x^3 - 6x + 3 \geq 0$

Chapter 7 Test

Solve the equations in Exercises 1 and 2 by factoring.

1. a. $x^2 - 16 = 0$
   b. $x^2 + 16 = 0$

2. $4x^3 = -4x^2 - x$

3. Add the correct constant to complete the square in each expression and then factor the trinomials as indicated.

   a. $x^2 - 30x + \underline{\quad} = (\underline{\quad\quad})^2$ **b.** $x^2 + 5x + \underline{\quad} = (\underline{\quad\quad})^2$

4. Write a quadratic equation with integer coefficients that has
   a. the two numbers $\pm 2i\sqrt{2}$ as roots.
   b. the two numbers $1 \pm \sqrt{5}$ as roots.

5. Solve the following equation by using the square root property: $(x+1)^2 = 9$.

6. Solve the following equation by completing the square: $x^2 + 4x + 1 = 0$.

7. Find the discriminant of the quadratic equation $4x^2 + 5x - 3 = 0$ and determine the nature of its solutions.

8. By using the discriminant, determine the value for k such that the equation $2x^2 - kx + 3 = 0$ will have exactly one real root.

Solve the equations in Exercises 9 – 14 by using any of the techniques discussed for solving quadratic equations: factoring, completing the square, or the quadratic formula.

9. $2x^2 + x + 1 = 0$

10. $2x^2 - 3x - 4 = 0$

11. $2x^2 + 3 = 4x$

12. $x^4 = 10x^2 - 9$

13. $3x^{-2} + x^{-1} - 2 = 0$

14. $\dfrac{2x}{x-3} - \dfrac{2}{x-2} = 1$

For each of the quadratic functions in Exercises 15 – 17, determine the line of symmetry, the vertex, the domain, and the range. Graph each function.

15. $y = (x-3)^2 - 1$

16. $y = -2\left(x - \dfrac{3}{2}\right)^2 + \dfrac{15}{2}$

17. $y = 2(x-3)^2 - 9$

Rewrite each of the quadratic functions in Exercises 18 and 19 in the form $y = a(x - h)^2 + k$. Find the vertex, range, and zero(s) of each function. Graph the function.

18. $y = x^2 + 4x - 6$

19. $y = x^2 - 6x + 8$

20. Solve the inequalities algebraically. Write the answers in interval notation and then graph each solution set on a number line.

  a. $x^2 + 5x + 6 > 0$ **b.** $x^2 + 5x + 6 < 0$ **c.** $x(x+7)(x-3) \geq 0$

21. Solve the rational inequalities algebraically. Write the answers in interval notation and then graph each solution set on a number line.

  a. $\dfrac{2x+5}{x-3} \geq 0$ **b.** $\dfrac{x-3}{2x+1} < 2$

22. Use a graphing calculator to solve the following inequality: $y = x^3 + 3x^2 - 10x > 0$. Write the solution in interval notation and then graph the solution set on a number line.

23. Right triangles: The length of one leg of a right triangle is one meter less than twice the length of the second leg. The hypotenuse is 17 meters. Find the lengths of the two legs.

24. Throwing a ball: A man standing at the edge of a cliff 112 ft above the beach throws a ball into the air with a velocity of 96 ft/s. Use the formula $h = -16t^2 + v_0 t + h_0$. (Round answers to the nearest hundredth.)
  a. When will the ball hit the beach?
  b. When will the ball be 64 ft above the beach?

25. Dimensions of a rectangle: The length of a rectangle is 4 inches longer than the width. If the diagonal is 20 inches long, what are the dimensions of the rectangle?

26. Traveling by car: Sandy made a business trip to a city 200 miles away and then returned home. Her average speed on the return trip was 10 mph less than her average speed going. If her total travel time was 9 hours, what was her average rate in each direction?

27. One number exceeds another by 10. Form a quadratic function that will allow you to find the minimum product of the two numbers. Find the numbers and the minimum product.

28. Maximize the area of a rectangle: The perimeter of a rectangle is 22 inches. Find the dimensions that will maximize the area.

Cumulative Review: Chapters 1–7

Simplify the expressions in Exercises 1 – 8. Assume that all variables are positive.

1. $\left(4x^{-3}\right)\left(2x\right)^{-2}$

2. $\dfrac{x^{-2}y^4}{x^{-5}y^{-2}}$

3. $\left(27x^{-3}y^{\frac{3}{4}}\right)^{\frac{2}{3}}$

4. $\left(\dfrac{9x^{\frac{4}{3}}}{4y^{\frac{2}{3}}}\right)^{\frac{3}{2}}$

5. $\sqrt[3]{-27x^6y^8}$

6. $\sqrt[4]{32x^9y^{15}}$

7. $\dfrac{\sqrt{72}}{3}+5\sqrt{\dfrac{1}{2}}$

8. $\dfrac{1}{2}\sqrt{\dfrac{4}{3}}+3\sqrt{\dfrac{1}{3}}$

In Exercises 9 – 11, completely factor each polynomial.

9. $xy+2x-5y-10$

10. $64x^2-81$

11. $2x^6-432y^3$

Rationalize the denominator and simplify each expression in Exercises 12 and 13.

12. $\dfrac{5}{\sqrt{2}+\sqrt{3}}$

13. $\dfrac{x^2-81}{3+\sqrt{x}}$

In Exercises 14 – 16, perform the indicated operations and write the results in the standard form a + bi.

14. $(2+5i)+(2-3i)$

15. $(2+5i)(2-3i)$

16. $\dfrac{2+5i}{2-3i}$

Perform the indicated operations in Exercises 17 and 18 and simplify the results.

17. $\dfrac{x^3+3x^2}{6x^2-36x+30}\cdot\dfrac{2x^2+2x-4}{x^3+2x^2}$

18. $\dfrac{2x^2+x}{x^2-2x+1}\div\dfrac{4x^2-6x}{x^2-1}$

Solve the equations in Exercises 19 – 28.

19. $7(2x-5)=5(x+3)+4$

20. $(2x+1)(x-4)=(2x-3)(x+6)$

21. $|3x-2|=|x+4|$

22. $10x^2+11x-6=0$

23. $4x^2+7x+1=0$

24. $\sqrt{x+5}-2=x+1$

25. $8x^{-2}-2x^{-1}-3=0$

26. $x^4-34x^2+225=0$

27. $3x^{\frac{2}{3}}+5x^{\frac{1}{3}}+2=0$

28. $3+\dfrac{1-x}{x+2}=\dfrac{4x+5}{2(x-2)}$

29. Given $P(x)=x^4-10x^3+20x^2-8x-2$, find **a.** $P(2)$ and **b.** $P(-1)$.

In Exercises 30 and 31, find an equation with integer coefficients that has the indicated roots.

30. $x = \dfrac{3}{4}, x = -5$

31. $x = 1 - 2\sqrt{5}, x = 1 + 2\sqrt{5}$

Solve the inequalities in Exercises 32 – 37 algebraically, write the answers in interval notation, and then graph each solution set on a real number line.

32. $4(x+3) - 1 \geq 2(x-4)$ **33.** $\dfrac{7}{2}x + 3 \leq x + \dfrac{13}{2}$ **34.** $(x+2)(x-3) < 0$

35. $x^2 - 8x + 16 \leq 0$ **36.** $x(x+4)(3x-5) < 0$ **37.** $\dfrac{3x-10}{x+2} \leq 1$

38. Find an equation in slope-intercept form for the line passing through the points $(-1, -8)$ and $(4, -3)$. Graph the line.

39. Write an equation in standard form for the line parallel to $y = 3x - 4$ that passes through the point $(4, 0)$. Graph both lines.

40. What is the vertical line test for functions? Why does it work?

41. Solve the following system of two linear equations graphically: $\begin{cases} -2x + 3y = -4 \\ x - 2y = 3 \end{cases}$.

42. Solve the following system of two linear equations algebraically: $\begin{cases} 3x - 2y = 7 \\ -2x + y = -6 \end{cases}$.

43. Use Gaussian elimination to solve the following system: $\begin{cases} x - 3y - z = 1 \\ 2x + y - 2z = -5 \\ 3x - y + 2z = 10 \end{cases}$.

44. If $A = \begin{bmatrix} 1 & 5 & 2 \\ -3 & 4 & 6 \\ -2 & -5 & 3 \end{bmatrix}$ find the value of det (A).

45. Solve the following system of linear inequalities graphically: $\begin{cases} x < 4 \\ x - y \geq 3 \end{cases}$.

For each of the quadratic functions in Exercises 46 – 48, determine the line of symmetry, the vertex, the domain, the range, and the zeros. Graph the function.

46. $y = x^2 + x - 6$ **47.** $y = -x^2 + x + 12$ **48.** $y = 2x^2 - 9x - 5$

Solve for the indicated variable in Exercises 49 and 50.

49. $3x + 2y = 6$ for y

50. $\frac{3}{4}x + \frac{1}{2}y = 5$ for y

51. z varies directly as x^2 and inversely as $\sqrt{y}$. If $z = 50$ when $x = 2$ and $y = 25$, find z if $x = 4$ and $y = 100$.

52. Medicine: A doctor needs to administer two drugs to a patient. The amount of Drug A administered must be three times the amount of Drug B, and the total amount of medication must equal 20 mg. How much of each drug should be given to the patient?

53. Renting cars: A car rental agency rents 200 cars per day at a rate of $30 per day for each car. For each $1 increase in the daily rate, the owners have found that they will rent 5 fewer cars per day. What daily rate would give total receipts of $6125? [**Hint:** Let x = the number of $1 increases.]

54. Find the dimensions of a rectangle that has an area of 520 m^2 and a perimeter of 92 m.

55. Find three consecutive even integers such that the square of the first added to the product of the second and third gives a result of 368.

56. The base of a triangle is 3 cm more than twice its height. If the area of the triangle is 76 cm^2, find the length of the base and the height of the triangle.

57. Throwing a ball: A ball is thrown vertically upward from the ground with an initial velocity of 144 ft/s. $\left[h = -16t^2 + v_0 t + h_0\right]$
   a. When will the ball reach its maximum height?
   b. What will be the maximum height?
   c. When will the ball be on the ground?

58. Selling golf balls: The profit function (in dollars) for a golf ball manufacturer is given by the function $P(x) = -0.01x^2 + 200x - 5000$ where x is the number of dozens of ball sold in one week.
   a. How many dozens of golf balls need to be sold to give the maximum profit in a week?
   b. What will be the maximum profit?
   c. How many golf balls is this?

59. Use a calculator to estimate the value of each number accurate to 4 decimal places.

   a. $\sqrt{6} + 2\sqrt{3}$ **b.** $\sqrt{2}\left(1 + 3\sqrt{10}\right)$ **c.** $\frac{\sqrt{2} + 2}{\sqrt{2} - 2}$

Use a graphing calculator and the CALC and zero features to graph the functions in Exercises 60 – 63 and estimate the x-intercepts to two decimal places.

60. $f(x) = 2x^2 - 5$

61. $y = -x^3 + 2x^2 - 1$

62. $g(x) = x^4 - x^2 + 8$

63. $h(x) = x^3 + 3x + 2$

Use a graphing calculator to graph each of the functions in Exercises 64 – 66. State the domain and range of each function.

64. $f(x) = \sqrt{x - 3}$

65. $g(x) = \sqrt{1 - x}$

66. $y = -\sqrt{x + 1}$

Exponential and Logarithmic Functions

Did You Know?

In this chapter, you will study exponential functions and their related inverses, logarithmic functions. You will also see how the use of logarithms can simplify calculations involving multiplication and division. Although electronic calculators have made calculation with logarithms obsolete, it is still important to study the logarithmic functions because they have many applications other than computing.

Napier

The inventor of logarithms was John Napier (1550 – 1617). Napier was a Scottish nobleman, Laird of Merchiston Castle, a stronghold on the outskirts of the town of Edinburgh. An eccentric, Napier was intensely involved in the political and religious struggles of his day. He had interests in many areas, including mathematics. In 1614, Napier published his "Description of the Laws of Logarithms," and thus he is given credit for first publishing and popularizing the idea of logarithms. Napier used a base close to the number e for his system, and natural logarithms (base e) are often called **Napierian logarithms**. Napier soon saw that a base of 10 would be more appropriate for calculations since our decimal number system uses base 10. Napier began work on a base-10 system but was unable to complete it before his death. Henry Briggs (1560 – 1630) completed Napier's work, and base-10 logarithms are often called **Briggsian logarithms** in his honor.

Napier's interest in simplifying calculations was based on the need at that time to do many calculations by hand for astronomical and scientific research. He also invented the forerunner of the slide rule and predicted tanks, submarines, and other advanced war technology. Napier's remarkable ingenuity led the local people to consider him either crazy or a dealer in the black art of magic.

A particularly amusing story is told of Napier's method of identifying which of his servants was stealing from him. He told his servants that his black rooster would identify the thief. Each servant was sent alone into a darkened room to pet the rooster on the back. Napier had coated the back of the rooster with soot, and the guilty servant came out of the room with clean hands.

Napier was a staunch Presbyterian, and he felt that his claim to immortality would be an attack that he had written on the Catholic Church. The scientific community more correctly judged that logarithms would be his greatest contribution.

The invention of logarithms: "by shortening the labors doubled the life of the astronomer."

Pierre de Laplace (1749 – 1827)

A s discussed throughout this text, functions are an important topic in mathematics. Function notation, $f(x)$, is particularly helpful in evaluating functions and indicating graphical relationships. Operating algebraically with functions as well as understanding and finding the **composition** and **inverses** of functions rely heavily on function notation. The concepts of composite and inverse functions form the basis of the relationship between logarithmic and exponential functions.

Logarithms are exponents. Traditionally, logarithmic and exponential values were calculated with the extensive use of printed tables and techniques for estimating values not found in the tables. As some of your "older" teachers will tell you, this was a long and detailed process. These tables are no longer printed in textbooks. Now hand-held calculators have programs stored in their electronic memories that calculate the values in these tables with even greater accuracy, and complicated expressions can be evaluated by pressing a few keys.

Of all the topics discussed in algebra, logarithmic functions and exponential functions probably have the most value in terms of applied problems. Learning curves, important in business and education, can be described with logarithmic and exponential functions. Exponential growth and decay are basic concepts in biology and medicine. (Cancer cells grow exponentially and radium decays exponentially.) Computers use logarithmic and exponential concepts in their design and implementation. You are likely to encounter these concepts at some point in almost any field you choose to study.

8.1   Algebra of Functions

- *Find the sum, difference, product, and quotient of two functions.*
- *Graph the sum of two functions.*
- *Use a graphing calculator to graph the sum of two functions.*

If two (or more) functions have the **same domain**, then we can perform the operations of addition, subtraction, multiplication, and division with these functions. For example, consider the following quadratic functions:

$$f(x) = 2x^2 - 1 \quad \text{and} \quad g(x) = x^2 + 2x - 5.$$

Both have the same domain: $\mathbb{R}$ = all real numbers = $(-\infty, \infty)$. This means that we can
     a. choose any value for x from the common domain,
     b. evaluate each function for that value of x, and
     c. perform operations with those functional values.

The functional values are the y-values. For example, if we choose $x = 3$, then

$$f(3) = 2 \cdot (3)^2 - 1 = 17 \quad \text{and} \quad g(3) = (3)^2 + 2 \cdot (3) - 5 = 10.$$

Now we can easily find the sum and difference

$$f(3)+g(3) = 17+10 = 27 \qquad \text{and} \qquad f(3)-g(3) = 17-10 = 7.$$

However, if we want to find, say $f(5)+g(5)$ and $f(5)-g(5)$, we would need to again evaluate both functions, this time at $x = 5$. Another way to find sums and differences of functions is to find the algebraic sum (or difference) of the two expressions. Then, these new expressions will allow us to find the sum (or difference) directly for any value of x that is in the original domain of both functions. For example, we find the sum $f + g$

$$(f+g)(x) = f(x)+g(x)$$
$$= \left(2x^2 - 1\right) + \left(x^2 + 2x - 5\right)$$
$$= 3x^2 + 2x - 6.$$

With this new function, we find directly that

$$(f+g)(3) = 3\cdot(3)^2 + 2\cdot(3) - 6 = 27.$$

Similarly,

$$(f-g)(x) = f(x)-g(x)$$
$$= \left(2x^2 - 1\right) - \left(x^2 + 2x - 5\right)$$
$$= x^2 - 2x + 4,$$

and we have

$$(f-g)(3) = (3)^2 - 2\cdot(3) + 4 = 7.$$

Similar notation is used for the product and quotient of two functions. One important condition is that both functions **must have the same domain**. If not, then the algebraic sums, differences, products, and quotients are **restricted to portions of the domains that are in common**. **Also, in the case of quotients, no denominator can be 0.**

Algebraic Operations with Functions

*If $f(x)$ and $g(x)$ represent two functions and x is a value in the **domain of both functions**, then:*

1. *Sum of Two Functions:* $(f+g)(x) = f(x)+g(x)$

2. *Difference of Two Functions:* $(f-g)(x) = f(x)-g(x)$

3. *Product of Two Functions:* $(f \cdot g)(x) = f(x) \cdot g(x)$

4. *Quotient of Two Functions:* $\left(\dfrac{f}{g}\right)(x) = \dfrac{f(x)}{g(x)}$ *where $g(x) \neq 0$.*

Example 1: Algebraic Operations with Functions

Let $f(x) = 3x^2 + x - 4$ and $g(x) = x - 6$. Find the following functions.

a. $(f+g)(x)$ **b.** $(f-g)(x)$ **c.** $(f \cdot g)(x)$

d. Evaluate each of the functions found in parts **a** – **c** at $x = 2$.

Solutions: **a.** $(f+g)(x) = \left(3x^2 + x - 4\right) + (x - 6) = 3x^2 + 2x - 10$

b. $(f-g)(x) = \left(3x^2 + x - 4\right) - (x - 6) = 3x^2 + 2$

c. $(f \cdot g)(x) = \left(3x^2 + x - 4\right)(x - 6) = 3x^3 - 17x^2 - 10x + 24$

d. Evaluating each of these functions at $x = 2$ gives the following results.

$$(f+g)(2) = 3 \cdot (2)^2 + 2 \cdot (2) - 10 = 6$$

$$(f-g)(2) = 3 \cdot (2)^2 + (2) = 14$$

$$(f \cdot g)(2) = 3 \cdot (2)^3 - 17 \cdot (2)^2 - 10 \cdot (2) + 24 = -40$$

Example 2: Algebraic Operations with Functions

Let $f(x) = x^2 - x$ and $g(x) = 2x + 1$. Find the following functions.

a. $(f+g)(x)$ **b.** $(g-f)(x)$ **c.** $\left(\dfrac{f}{g}\right)(x)$

d. Evaluate each of the functions found in parts **a** – **c** at $x = 3$.

Solutions: **a.** $(f+g)(x) = \left(x^2 - x\right) + (2x + 1) = x^2 + x + 1$

b. $(g-f)(x) = (2x + 1) - \left(x^2 - x\right) = -x^2 + 3x + 1$

c. $\left(\dfrac{f}{g}\right)(x) = \dfrac{x^2 - x}{2x + 1}$ where $2x + 1 \neq 0$ $\left(\text{or } x \neq -\dfrac{1}{2}\right)$

d. Evaluating each of these functions at $x = 3$ gives the following results.

$$(f+g)(3) = (3)^2 + (3) + 1 = 13$$

$$(g-f)(3) = -(3)^2 + 3 \cdot (3) + 1 = 1$$

$$\left(\frac{f}{g}\right)(3) = \frac{(3)^2 - (3)}{2 \cdot (3) + 1} = \frac{6}{7}$$

Note that, except for Example 2c, the domain for all of the functions discussed in Examples 1 and 2 is the set of all real numbers, $(-\infty, \infty)$. In Example 2c we noted that the denominator cannot equal 0. In Example 3 we show how the domain may need to be limited before performing algebra with functions that contain radical expressions.

Example 3: Algebraic Operations with Functions with Limited Domains

Let $f(x) = x + 5$ and $g(x) = \sqrt{x-2}$. Find the following functions and state the domain of each function.

a. $(f+g)(x)$ **b.** $\left(\dfrac{f}{g}\right)(x)$

Solutions: **a.** $(f+g)(x) = x + 5 + \sqrt{x-2}$

The domain of f is the set of all real numbers. However, the domain of the sum is restricted to the domain of g, the radical function. In this case we must have $x - 2 \geq 0$. Thus, in interval notation, the domain is $[2, \infty)$.

b. $\left(\dfrac{f}{g}\right)(x) = \dfrac{x+5}{\sqrt{x-2}}$

For this function, the denominator cannot be 0, so $x \neq 2$. Therefore, we must have $x - 2 > 0$ and the domain, in interval notation, is $(2, \infty)$. (Note that the domain can become smaller than, but never larger than the domain of the two original functions.)

Graphing the Sum of Two Functions

In this section we discuss how to graph the sum of two functions. (Graphing the difference, product, and quotient can be accomplished in a similar manner.) Remember that, in any case, algebraic operations with functions can be performed only over a common domain.

We begin with two functions f and g that have the same domain and only a finite number of ordered pairs:

$$f = \{(-2,1),(0,4),(3,5)\}$$
$$g = \{(-2,3),(0,1),(3,2)\}$$

Note that the domain of both functions is $\{-2,0,3\}$.

Both functions have only three points and are graphed in Figure 8.1.

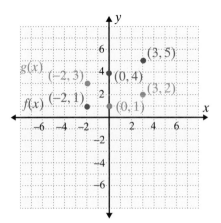

Figure 8.1

To add the two functions graphically, look at each of the x-values on the x-axis and the points directly above (or below) them and add the corresponding y-values. This process will give new points, and these new points represent a function that is the sum of the two original functions.

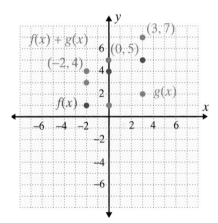

Figure 8.2

From the points in the figure we see that

$$(f+g)(x) = f(x) + g(x)$$
$$= \{(-2, 1+3), (0, 4+1), (3, 5+2)\}$$
$$= \{(-2, 4), (0, 5), (3, 7)\}.$$

In general, graphing the sum (difference, product, or quotient) of two functions will involve an infinite number of points. Certainly not all of these points can be plotted one at a time. However, by making a table of a few key points and joining these points with smooth curves or line segments, the general nature of the result can be found. In fact, if the two functions consist of line segments, then the sum will also consist of line segments. Table 8.1 and Figures 8.3 and 8.4 illustrate just such a case.

Remember that when operating with functions, the operations are performed with the *y*-values for each value of *x* in the common domain. (Don't mess with the *x*-values!)

Points Illustrated in Figure 8.3 and Figure 8.4

x	f(x)	g(x)	f(x) + g(x)
−3	1	−3	$1 + (−3) = −2$
−2	1	0	$1 + 0 = 1$
−1	1	3	$1 + 3 = 4$
0	−2	3	$(−2) + 3 = 1$
1	1	3	$1 + 3 = 4$
2	1	4	$1 + 4 = 5$
3	0	5	$0 + 5 = 5$

Table 8.1

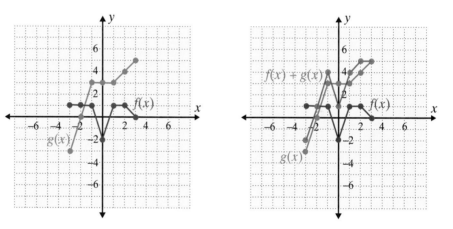

Figure 8.3   Figure 8.4

Using a TI-84 Plus Graphing Calculator to Graph the Sum of Two Functions

A graphing calculator can be used to graph the sum (difference, product, or quotient) of two functions. An interesting way to do this is to first press ▬▬▬ and enter the functions as **Y1** and **Y2**; then assign **Y3** to be the sum of the first two. Figures 8.5a and 8.5b show the displays for entering two functions and their graphs.

Using the TI-84 Plus graphing calculator, we enter:

$$\texttt{Y1} = x^2 - 3 \quad \text{(a parabola)}$$

and

$$\texttt{Y2} = 5x - 1 \quad \text{(a straight line)}.$$

Figure 8.5a: The two functions entered   **Figure 8.5b:** The two functions graphed

Now we can graph the sum of the two functions, $f(x) + g(x)$, (or $\texttt{Y1} + \texttt{Y2}$ on the calculator) by assigning $\texttt{Y3} = \texttt{Y1} + \texttt{Y2}$ as follows. (This saves performing the actual algebraic operations. The calculator does it for us.)

Step 1: Press ⬤ Y= and select $\texttt{Y3}$.

Step 2: Press the **VARS** key.

Step 3: Move the cursor to the $\texttt{Y-VARS}$ at the top of the display.

Step 4: Select $\texttt{1:FUNCTION...}$ and press **ENTER**.

Step 5: Select $\texttt{Y1}$ and press **ENTER**.

Step 6: Press the ➕ key.

Step 7: Go to $\texttt{Y-VARS}$ again, select $\texttt{1:FUNCTION...}$ and select $\texttt{Y2}$.

The display will now appear as follows.

Figure 8.6

Now press ⬤ GRAPH and the display will show all three curves, $\texttt{Y1}$, $\texttt{Y2}$, and the sum $\texttt{Y3} = \texttt{Y1} + \texttt{Y2}$. You may need to reset the ⬤ WINDOW to allow a more complete display of all of the graphs.

Figure 8.7

If this graph is somewhat confusing (because three graphs are shown), you can turn off the first two graphs and show just the sum. To turn off a graph, go to the highlighted equal sign, **=**, next to the equation and hit **ENTER**. You should see the following screens.

Figure 8.8a   Figure 8.8b

Practice Problems

1. For $f(x) = x^2 - 2x - 3$ and $g(x) = x - 3$ find:

   a. $(f+g)(x)$ **b.** $(f-g)(x)$ **c.** $(f \cdot g)(x)$ **d.** $\left(\dfrac{f}{g}\right)(x)$

2. Evaluate each of the functions **a** – **d** in Exercise 1 for $x = 4$.

3. For $f(x) = 2x - 5$ and $h(x) = \sqrt{x+6}$,

   a. find $f(x) + h(x)$ and state the domain of the sum function.

   b. find $f(x) \cdot h(x)$ and state the domain of the product function.

4. Given $f(x) = \{(-3,2),(-1,1),(0,-1),(2,3)\}$ and $g(x) = \{(-3,4),(-1,2),(0,1),(2,1)\}$ graph the sum of the two functions.

Answers to Practice Problems:

  1. a. $x^2 - x - 6$ **b.** $x^2 - 3x$ **c.** $x^3 - 5x^2 + 3x + 9$

     d. $x + 1, x \neq 3$ (Note the restriction holds even after simplifying.)

  2. a. 6 **b.** 4 **c.** 5 **d.** 5

  3. a. $2x - 5 + \sqrt{x+6}$ Domain: $[-6, \infty)$

     b. $2x\sqrt{x+6} - 5\sqrt{x+6}$ Domain: $[-6, \infty)$

4.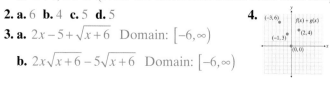

8.1 Exercises

*In Exercises 1 – 10, graph **a.** the sum (f + g) and **b.** the difference (f – g) on two different graphs.*

1. $f = \{(-2,5),(0,-3),(2,1)\}$

$g = \{(-2,-2),(0,-1),(2,-3)\}$

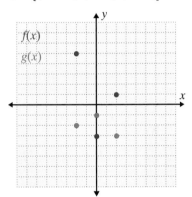

2. $f = \{(0,2),(1,1),(2,-1)\}$

$g = \{(0,4),(1,2),(2,-2)\}$

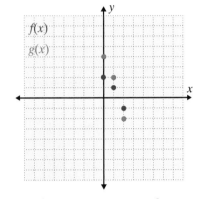

3. $f = \{(-3,-1),(-1,0),(2,-4)\}$

$g = \{(-3,-3),(-1,5),(2,1)\}$

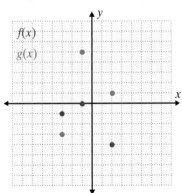

4. $f = \{(-4,4),(-2,-5),(1,3)\}$

$g = \{(-4,1),(-2,0),(1,-2)\}$

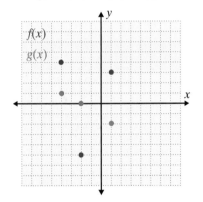

5. $f = \{(-2,-5),(-1,4),(0,-1),(1,5)\}$

$g = \{(-2,-1),(-1,0),(0,-4),(1,1)\}$

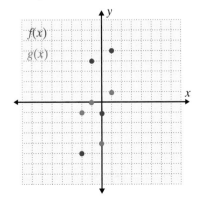

6. $f = \{(-2,3),(1,2),(2,1),(3,0)\}$

$g = \{(-2,-4),(1,-3),(2,-2),(3,-1)\}$

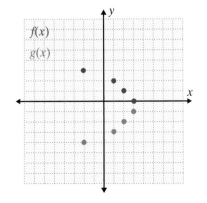

7. $f = \{(-3,1),(-1,2),(1,1),(3,2)\}$

$g = \{(-3,-3),(-1,-4),(1,3),(3,4)\}$

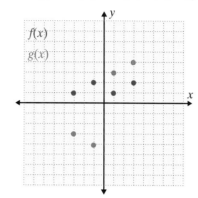

8. $f = \{(-4,6),(-2,0),(0,-2),(3,-3)\}$

$g = \{(-4,0),(-2,-4),(0,1),(3,3)\}$

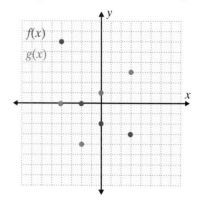

9. $f = \{(-5,1),(-1,2),(2,3),(3,2)\}$

$g = \{(-5,2),(-1,1),(2,0),(3,-1)\}$

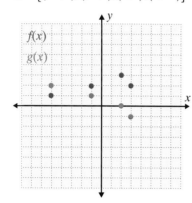

10. $f = \{(-3,2),(0,0),(3,-1),(4,1)\}$

$g = \{(-3,-4),(0,-3),(3,2),(4,-1)\}$

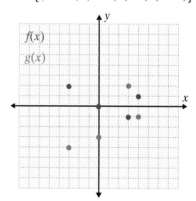

In Exercises 11 – 20, find **a.** $(f+g)(x)$, **b.** $(f-g)(x)$, **c.** $(f \cdot g)(x)$, *and* **d.** $\left(\dfrac{f}{g}\right)(x)$.

11. $f(x) = x+2,\ g(x) = x-5$

12. $f(x) = 2x,\ g(x) = x+4$

13. $f(x) = x^2,\ g(x) = 3x-4$

14. $f(x) = x-3,\ g(x) = x^2+1$

15. $f(x) = x^2-9,\ g(x) = x-3$

16. $f(x) = x^2-25,\ g(x) = x+5$

17. $f(x) = 2x^2+x,\ g(x) = x^2+2$

18. $f(x) = x^3+6x,\ g(x) = x^2+6$

19. $f(x) = x^2+4x+1,\ g(x) = x^2-4x+1$

20. $f(x) = x^3-x^2,\ g(x) = 6-x^2$

For Exercises 21 – 30, let $f(x) = x^2 + 4$ and $g(x) = -x + 3$. Find the value of the indicated expression.

21. $f(2) + g(2)$ **22.** $f(2) \cdot g(2)$ **23.** $g(a) - f(a)$ **24.** $\dfrac{g(a)}{f(a)}$

25. $(f + g)(-4)$ **26.** $(f - g)(0.5)$ **27.** $\left(\dfrac{f}{g}\right)(-2)$ **28.** $(f \cdot g)(-3)$

29. $(g - f)(-6)$ **30.** $\left(\dfrac{g}{f}\right)(-1)$

In Exercises 31 – 40, find the indicated function and state its domain in interval notation.

31. If $f(x) = \sqrt{2x - 6}$ and $g(x) = x + 4$, find $(f + g)(x)$.

32. If $f(x) = x^2 - 2x + 1$ and $g(x) = x - 1$, find $\left(\dfrac{f}{g}\right)(x)$.

33. Find $f(x) \cdot g(x)$ given that $f(x) = 3x + 2$ and $g(x) = x - 7$.

34. Find $f(x) - g(x)$ given that $f(x) = x^2$ and $g(x) = x^2 - 2$.

35. For $f(x) = x - 5$ and $g(x) = \sqrt{x + 3}$ find $\dfrac{f(x)}{g(x)}$.

36. For $f(x) = 2x - 8$ and $g(x) = \sqrt{2 - x}$ find $f(x) \cdot g(x)$.

37. If $f(x) = -\sqrt{x - 3}$ and $g(x) = 3x$, find $(f \cdot g)(x)$.

38. If $f(x) = -\sqrt{4 - x}$ and $g(x) = 5 - x$, find $(g - f)(x)$.

39. If $f(x) = \sqrt[3]{x + 3}$ and $g(x) = \sqrt{5 + x}$, find $f(x) + g(x)$.

40. If $f(x) = \sqrt{x - 1}$ and $g(x) = \sqrt[3]{2x + 1}$, find $f(x) - g(x)$.

In Exercises 41 – 50, graph each pair of functions and the sum of these functions on the same set of axes.

41. $y = x^2$ and $y = -1$ **42.** $y = x^2$ and $y = 2$

43. $y = x + 1$ and $y = 2x$ **44.** $y = 2 - x$ and $y = x$

45. $y = x + 4$ and $y = -x$ **46.** $y = x + 5$ and $y = x - 5$

47. $f(x) = x + 1$ and $g(x) = x^2 - 1$ **48.** $f(x) = x^2 + 2$ and $g(x) = x^2 - 2$

49. $f(x) = \sqrt{x - 6}$ and $g(x) = 2$ **50.** $f(x) = \sqrt{3 - x}$ and $g(x) = -1$

Use the graph shown here to find the values indicated in Exercises 51 – 56.

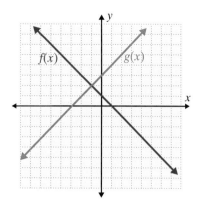

51. $(f+g)(-2)$ **52.** $(f-g)(2)$ **53.** $(f \cdot g)(3)$

54. $(g-f)(0)$ **55.** $\left(\dfrac{f}{g}\right)(4)$ **56.** $(g \cdot f)(4)$

In Exercises 57 – 62, the graphs of two functions are given. Graph the sum of these two functions.

57.
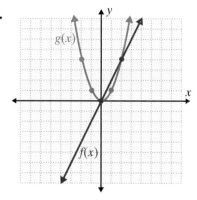

58.
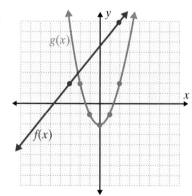

59.
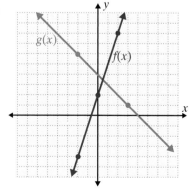

60.
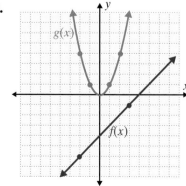

61.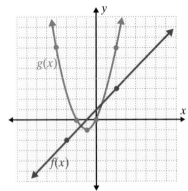

62.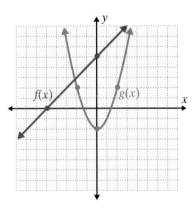

In Exercises 63 – 70, use a graphing calculator to graph each pair of functions and the sum of these functions on the same set of axes.

63. $y = x^2$ and $y = 2x + 1$

64. $y = x^2 + x$ and $y = 3x + 4$

65. $y = \sqrt{x + 4}$ and $y = -2$

66. $y = -\sqrt{x - 1}$ and $y = 3$

67. $f(x) = \sqrt[3]{x + 5}$ and $h(x) = 2x$

68. $h(x) = \sqrt[3]{x - 1}$ and $g(x) = x - 1$

69. $g(x) = -\sqrt{x - 2}$ and $h(x) = 1$

70. $f(x) = x^2 + 5$ and $g(x) = 4 - x^2$

Writing and Thinking About Mathematics

71. Explain why, in general, $(f - g)(x) \neq (g - f)(x)$.

72. Given the two functions f and g,

$$f = \{(-2, 0), (-1, 1), (0, 4), (2, 4), (3, 5), (4, 1)\}$$
$$g = \{(-2, 3), (-1, 4), (0, 1), (2, -1), (3, 2), (4, 6)\}$$

find and graph the following.

a. $f - g$ **b.** $f \cdot g$ **c.** $\dfrac{f}{g}$

73. Use the graphs of the two functions f and g shown in Figure 8.3 on page 643.

a. Sketch the graph of $f - g$.

b. Sketch the graph of $f \cdot g$.

c. Is $\dfrac{f}{g}$ defined on the entire interval $[-3, 3]$? Briefly discuss your reasoning.

HAWKES LEARNING SYSTEMS: INTERMEDIATE ALGEBRA SOFTWARE

- Algebra of Functions

8.2 Composition of Functions and Inverse Functions

- *Form the **composition** of two functions.*
- *Determine if a function is one-to-one by using the horizontal line test.*
- *Show that two functions are **inverses** by verifying that $f(g(x)) = x$ and $g(f(x)) = x$.*
- *Find the **inverse** of a one-to-one function.*
- *Graph the inverses of functions by **reflecting** the graphs of the functions across the line $y = x$.*

Composition of Functions

Previously, function notation, $f(x)$, has proven useful in evaluating functions and in indicating arithmetic operations with functions. This same notation is needed in developing the concept of **a function of a function**, called the **composition** of two functions. For example, suppose that

$$g(x) = 2x - 4 \quad \text{and} \quad f(x) = x^2 - 4x + 1.$$

Now, for $x = 3$, we have

$$g(3) = 2 \cdot (3) - 4 = 2 \quad \text{and} \quad f(2) = (2)^2 - 4 \cdot (2) + 1 = -3.$$

For the **composition** $f(g(3))$ (read "f of g of 3"), we have

$$f(g(3)) = f(2) = -3. \qquad \text{We see that } g(3), \text{which equals } 2, \text{replaces } x \text{ in the function } f(x).$$

More generally, given two functions $f(x)$ and $g(x)$, a new function $f(g(x))$ called the **composition** (or **composite**) of f and g, is found by substituting the expression for $g(x)$ in place of x in the function f. In this case the value of $g(x)$ must be in the domain of f. Thus, for

$$g(x) = 2x - 4 \quad \text{and} \quad f(x) = x^2 - 4x + 1,$$

the composition

$$f(g(x)) \qquad \text{(read "f of g of x")}$$

is found as follows:

$$f(x) = x^2 - 4x + 1$$

$$f\big(g(x)\big) = \big(g(x)\big)^2 - 4\big(g(x)\big) + 1 \qquad \text{Replace the } x \text{ in } f(x) \text{ with } g(x).$$

$$= (2x - 4)^2 - 4(2x - 4) + 1 \qquad \text{Replace } g(x) \text{ with } 2x - 4.$$

$$= 4x^2 - 16x + 16 - 8x + 16 + 1 \qquad \text{Simplify.}$$

$$= 4x^2 - 24x + 33.$$

The composition of g and f (reversing the order of f and g) is indicated by

$$g\big(f(x)\big) \qquad \text{read} \qquad \text{"}g \text{ of } f \text{ of } x\text{"}$$

and is found by substituting the expression for $f(x)$ in place of x in the function g. Thus,

$$g(x) = 2x - 4$$

$$g\big(f(x)\big) = 2\big(f(x)\big) - 4 \qquad \text{Replace the } x \text{ in } g(x) \text{ with } f(x).$$

$$= 2(x^2 - 4x + 1) - 4 \qquad \text{Replace } f(x) \text{ with } x^2 - 4x + 1.$$

$$= 2x^2 - 8x + 2 - 4 \qquad \text{Simplify.}$$

$$= 2x^2 - 8x - 2.$$

As we can see with these examples, in general, $f(g(x)) \neq g(f(x))$ and substitutions must be done carefully and accurately. The following definition shows another notation (a small raised circle) often used to indicate the composition of functions.

Composite Function

*For two functions f and g, the **composite function** $f \circ g$ is defined as follows:*

$$(\boldsymbol{f \circ g})(\boldsymbol{x}) = \boldsymbol{f\big(g(x)\big)}.$$

Domain of $f \circ g$: *The domain of $f \circ g$ consists of those values of x in the domain of g for which $g(x)$ is in the domain of f.*

Example 1: Compositions

a. Form the compositions $(f \circ g)(x)$ and $(g \circ f)(x)$ if $f(x) = 5x + 2$ and $g(x) = 3x - 7$.

Solution: $(f \circ g)(x) = f\big(g(x)\big) = 5 \cdot g(x) + 2 = 5(3x - 7) + 2 = 15x - 33$

$(g \circ f)(x) = g\big(f(x)\big) = 3 \cdot f(x) - 7 = 3(5x + 2) - 7 = 15x - 1$

Note: Both $(f \circ g)(x)$ and $(g \circ f)(x)$ are **defined for all real numbers**.

Continued on the next page...

b. Form the composite functions $(f \circ g)(x)$ and $(g \circ f)(x)$ if $f(x) = \sqrt{x-3}$ and $g(x) = x^2 + 4$.

Solution: $(f \circ g)(x) = \sqrt{g(x) - 3}$

$$= \sqrt{(x^2 + 4) - 3} = \sqrt{x^2 + 1}$$

Note: For the expression under the radical to be defined, we must have $g(x) \geq 3$. Because $x^2 + 4 \geq 3$ for all real numbers, the domain of $(f \circ g)(x)$ is all real numbers.

$$(g \circ f)(x) = (f(x))^2 + 4$$

$$= (\sqrt{x-3})^2 + 4 = x - 3 + 4 = x + 1$$

Note: $(g \circ f)(x)$ is defined only for $x \geq 3$. (Here the domain comes from f before the simplification.)

c. Find $f(g(x))$ and $g(f(x))$ if $f(x) = \sqrt{x+3}$ and $g(x) = 2x - 5$.

Solution: $f(g(x)) = \sqrt{g(x) + 3}$

$$= \sqrt{(2x - 5) + 3} = \sqrt{2x - 2}$$

Note: $f(g(x))$ is defined only for $2x - 2 \geq 0$ or $x \geq 1$.

$$g(f(x)) = 2(f(x)) - 5$$

$$= 2(\sqrt{x+3}) - 5 = 2\sqrt{x+3} - 5$$

Note: $g(f(x))$ is defined only for $x \geq -3$.

One-to-One Functions

By the definition of a function, there is only one corresponding y-value for each x-value in a function's domain. Graphically, the **vertical line test** can be used to help determine whether or not a graph represents a function. Now, in order to develop the concept of **inverse functions**, we need to study functions that have only one x-value for each y-value in the range. Such functions are said to be **one-to-one functions** (or **1–1 functions**).

Consider the following **functions**:

$$f = \{(1, 2), (2, 4), (3, 6), (4, 8), (5, 10)\}$$
$$g = \{(-2, 6), (0, 6), (1, 5), (2, 4), (4, 1)\}.$$

Both sets of ordered pairs are functions because each value of x appears only once. In the function f, each y-value appears only once. But, in function g, the y-value 6 appears twice: in $(-2, 6)$ and $(0, 6)$. Figure 8.9 illustrates both functions.

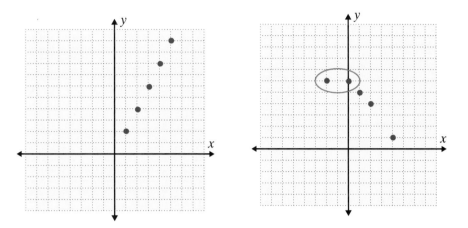

a. Function f is a 1–1 function.   **b.** Function g is not a 1–1 function.

Figure 8.9

One-to-One Functions

*A function is a **one-to-one function** (or **1–1 function**) if for each value of y in the range there is only one corresponding value of x in the domain.*

Graphically, as illustrated in Figure 8.9(a), if a horizontal line intersects the graph of a function in more than one point then it is **not** one-to-one. This is, in effect, the **horizontal line test**.

Horizontal Line Test

*A function is one-to-one if no **horizontal line** intersects the graph of the function at more than one point.*

The graphs in Figure 8.10 (on the next page) illustrate the concept of one-to-one functions. (Note that each graph is indeed a function and passes the vertical line test.)

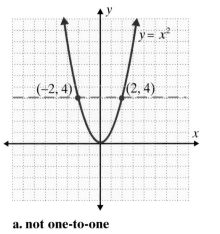

a. not one-to-one

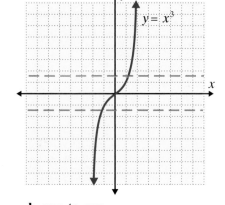

b. one-to-one

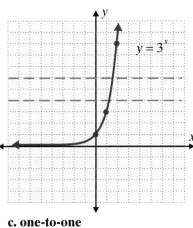

c. one-to-one

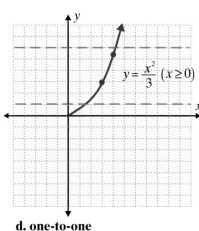

d. one-to-one

Figure 8.10

Example 2: One-to-One Functions

Determine whether each function **is** or **is not** 1-1.

a. $y = \sqrt{x+5}$

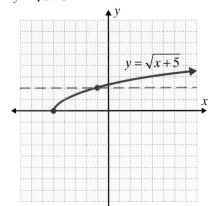

Solution:

The horizontal line test shows that this function is 1–1.

b. $f = \{(-3,4),(-2,1),(0,4),(3,1)\}$

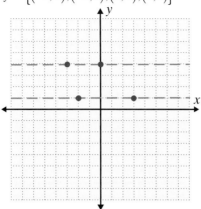

Solution:

This function is not 1–1. Both y-values, 4 and 1, have more than one corresponding x-value.

c. $y = 2x - 1$

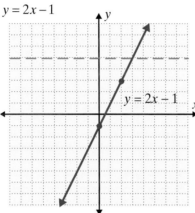

Solution:

The graph of the function $y = 2x - 1$ is a straight line. Straight lines that are not vertical and not horizontal represent 1–1 functions. (Vertical lines are not functions in the first place and horizontal lines fail the horizontal line test.)

d. $y = -x^2 + 1$

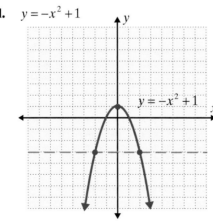

Solution:

The graph of the function $y = -x^2 + 1$ is a parabola and the horizontal line test shows that the function is not 1–1.

Inverse Functions

Now that we have discussed 1–1 functions, we can develop the concept of **inverse functions**. We will find that **only 1–1 functions have inverse functions**. To find the inverse of a 1–1 function represented by a set of ordered pairs, exchange x and y in each ordered pair. That is, if (x, y) is in the original 1–1 function, then (y, x) is in the inverse function.

For example,

$$\text{if } f = \{(-1, 1), (0, 2), (1, 4)\},$$

then interchanging the coordinates in each ordered pair gives

$$g = \{(1, -1), (2, 0), (4, 1)\}.$$

The functions f and g are called **inverses** of each other. If g is the inverse of f we write

$$f^{-1} \text{ (read "} f \text{ inverse") rather than use } g.$$

Thus, in this example, we can write $f^{-1} = \{(1, -1), (2, 0), (4, 1)\}$.

Inverse Functions

*If f is a 1–1 function with ordered pairs of the form (x, y), then its **inverse function**, denoted as f^{-1}, is also a 1–1 function with ordered pairs of the form (y, x).*

Note the importance of the original function being 1–1. If it was not 1–1, then interchanging the x- and y-values would yield a relation that is not a function. Graphically, the function must satisfy the horizontal line test. If it did not, then the inverse would not pass the vertical line test and would not be a function.

NOTES   The notation $f^{-1}(x)$ represents the inverse of a 1–1 function. This inverse is a new function in which the x- and y-values have been interchanged. $f^{-1}(x)$ does NOT mean $\dfrac{1}{f(x)}$ because the –1 is NOT an exponent.

The graph of any point (b, a) is the reflection of the point (a, b) across the line $y = x$. Thus the points of the inverse function f^{-1} are reflections of the points of the function f across the line $y = x$. We say that the graphs are **symmetric about the line $y = x$**. Figure 8.11 illustrates these reflections and the symmetry. **We see that the domain, denoted D_f, and range, denoted R_f, of the two functions are interchanged.**

That is,

$$D_f = R_{f^{-1}} = \{-1, 0, 1\}$$
$$R_f = D_{f^{-1}} = \{1, 2, 4\}.$$

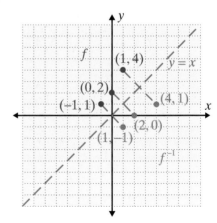

Figure 8.11

In general, every 1–1 function has an inverse function (or just inverse), and the graph of the inverse function of any 1–1 function f can be found by reflecting the graph of f across the line $y = x$. Figure 8.12 shows two more illustrations of this concept.

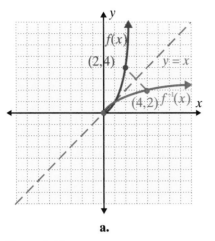

a.

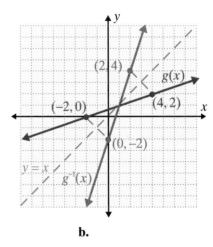

b.

Figure 8.12

The following definition of inverse functions helps to determine whether or not two functions are inverses of each other.

Inverse Functions

If f and g are one-to-one functions and

$$f(g(x)) = x \qquad \text{for all } x \text{ in } D_g, \text{ and}$$
$$g(f(x)) = x \qquad \text{for all } x \text{ in } D_f$$

*then f and g are **inverse functions**.*
That is, $g = f^{-1}$ and $f = g^{-1}$.

Example 3: Inverse Functions

Use the definition of inverse functions to show that f and g are inverse functions.

a. $f(x) = 2x + 6$ and $g(x) = \dfrac{x-6}{2}$.

Solution: $f(x) = 2x + 6$ and $g(x) = \dfrac{x-6}{2}$. The domain of both functions is the set of all real numbers.

Continued on the next page...

We have

$$f\big(g(x)\big) = 2 \cdot g(x) + 6 \qquad$$ Replace the x in $f(x)$ with $g(x)$ and simplify.

$$= 2\left(\frac{x-6}{2}\right) + 6$$

$$= (x-6) + 6$$

$$= x.$$

Also,

$$g\big(f(x)\big) = \frac{f(x) - 6}{2} \qquad$$ Replace the x in $g(x)$ with $f(x)$ and simplify.

$$= \frac{(2x+6) - 6}{2}$$

$$= \frac{2x}{2}$$

$$= x.$$

Therefore, $g = f^{-1}$ and $f = g^{-1}$.

The graph shows that the line $y = x$ is a line of symmetry for the graphs of the inverse functions.

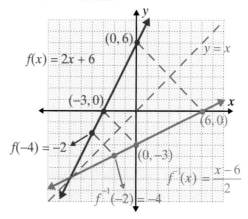

b. $f(x) = \sqrt{x-3}$ and $g(x) = x^2 + 3$ for $x \geq 0$.

Solution: $f(x) = \sqrt{x-3}$ and $g(x) = x^2 + 3$ for $x \geq 0$. The domain of f is the interval $[3, \infty)$ and the domain of g is the interval $[0, \infty)$.

We have

$$f\big(g(x)\big) = \sqrt{g(x) - 3} \qquad$$ Replace the x in $f(x)$ with $g(x)$ and simplify.

$$= \sqrt{(x^2 + 3) - 3}$$

$$= \sqrt{x^2}$$

$$= x \qquad \text{for } x \geq 0.$$

Also,

$$g\big(f(x)\big) = \big(f(x)\big)^2 + 3$$

Replace the x in $g(x)$ with $f(x)$ and simplify.

$$= \Big(\sqrt{x-3}\Big)^2 + 3$$

$$= x - 3 + 3$$

$$= x \qquad \text{for } x \geq 3.$$

Therefore, $g = f^{-1}$ and $f = g^{-1}$.

The graph shows that the line $y = x$ is a line of symmetry for the graphs of the inverse functions. Note that these graphs are only parts of parabolas and the domains have been restricted accordingly.

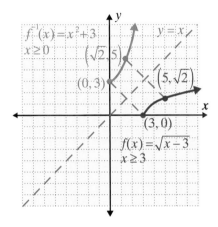

Example 4: Evaluating Compositions of Inverses

Given $f(x) = \sqrt{x+4}$ for $x \geq -4$ and $f^{-1}(x) = x^2 - 4$ for $x \geq 0$, evaluate the indicated compositions.

a. $f\big(f^{-1}(2)\big)$

   Solution: $f^{-1}(2) = (2)^2 - 4 = 0$ so $f\big(f^{-1}(2)\big) = f(0) = \sqrt{(0)+4} = \sqrt{4} = 2$

b. $f\big(f^{-1}(5)\big)$

   Solution: $f^{-1}(5) = (5)^2 - 4 = 21$ so $f\big(f^{-1}(5)\big) = f(21) = \sqrt{(21)+4} = \sqrt{25} = 5$

c. $f^{-1}\big(f(-2)\big)$

   Solution: $f(-2) = \sqrt{(-2)+4} = \sqrt{2}$ so $f^{-1}\big(f(-2)\big) = f^{-1}\big(\sqrt{2}\big) = \big(\sqrt{2}\big)^2 - 4 = 2 - 4 = -2$

Continued on the next page...

d. $f^{-1}\left(f\left(-5\right)\right)$

Solution: $f(-5)$ does not exist because -5 is not in the domain of f. Therefore $f^{-1}\left(f\left(-5\right)\right)$ does not exist.

Finding the Inverse of a 1–1 Function

In Examples 3 and 4 the given functions were inverses of each other. The next question is how to find the inverse of a given 1–1 function. The following procedure shows one method for finding the inverse by using the fact that if an ordered pair (x, y) belongs to the function f, then (y, x) belongs to f^{-1}.

To Find the Inverse of a 1–1 Function

1. Let $y = f(x)$. *(In effect, substitute y for f(x).)*

2. Interchange *x* and *y*.

3. In the new equation, solve for *y* in terms of *x*.

4. Substitute $f^{-1}(x)$ for *y*. *(This new function is the inverse of f.)*

Example 5: Finding the Inverse

a. Find $f^{-1}(x)$ if $f(x) = 5x - 7$.

Solution:

$$f(x) = 5x - 7$$

$$y = 5x - 7 \qquad \text{Substitute } y \text{ for } f(x).$$

$$x = 5y - 7 \qquad \text{Interchange } x \text{ and } y.$$

$$x + 7 = 5y \qquad \text{Solve for } y \text{ in terms of } x.$$

$$\frac{x + 7}{5} = y$$

$$f^{-1}(x) = \frac{x + 7}{5} \qquad \text{Substitute } f^{-1}(x) \text{ for } y.$$

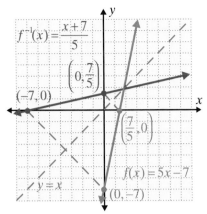

b. Find $g^{-1}(x)$ if $g(x) = x^2 - 2$ for $x \geq 0$.

Solution:

$g(x) = x^2 - 2$	For $x \geq 0$. (Note that g is 1–1 for $x \geq 0$.)
$y = x^2 - 2$	Substitute y for $g(x)$.
$x = y^2 - 2$	Interchange y and x.
$\pm\sqrt{x+2} = y$	Solve for y in terms of x.
$g^{-1}(x) = \sqrt{x+2}$	Take the positive square root because we must have $y \geq 0$. (The domain of g is $x \geq 0$, so the range of g^{-1} is $y \geq 0$.)

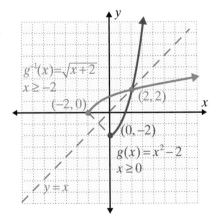

Practice Problems

1. For $f(x) = 2x - 1$ and $g(x) = x^2$, find:

   a. $f(g(x))$ **b.** $g(f(x))$

2. Given $f(x) = 3x + 4$ and $f^{-1}(x) = \dfrac{x - 4}{3}$, find the following:

   a. $f(f^{-1}(2))$ **b.** $f^{-1}(f(1))$

3. Given $g(x) = \sqrt{x - 2}$ and $f(x) = \dfrac{1}{x - 1}$, find $f(g(3))$.

4. Find the inverse of the function $f(x) = 3x - 1$.

5. Find $f^{-1}(x)$ if $f(x) = x^2 - 4$, $x \geq 0$. (Note that f is 1–1 for $x \geq 0$.)

8.2 Exercises

1. $f(x) = 3x + 5$, $g(x) = \dfrac{x + 4}{2}$

   Find **a.** $f(g(2))$ and **b.** $g(f(2))$.

2. $f(x) = \dfrac{1}{4}x + 1$, $g(x) = 6x - 7$

   Find **a.** $f(g(4))$ and **b.** $g(f(x))$.

3. $f(x) = x^2$, $g(x) = 2x + 3$

   Find **a.** $(f \circ g)(-5)$ and **b.** $(g \circ f)(-1)$.

4. $f(x) = x^2 + 1$, $g(x) = x - 6$

   Find **a.** $(f \circ g)\left(\dfrac{1}{2}\right)$ and **b.** $(g \circ f)(-2)$.

Form the compositions $f(g(x))$ and $g(f(x))$ for Exercises 5 – 20.

5. $f(x) = \dfrac{1}{x}$, $g(x) = 5x - 8$

6. $f(x) = \dfrac{1}{x + 1}$, $g(x) = x^2 + x - 3$

7. $f(x) = x - 1$, $g(x) = \dfrac{1}{x^2}$

8. $f(x) = \dfrac{1}{x^2}$, $g(x) = x^2 + 1$

9. $f(x) = x^3 + x + 1$, $g(x) = x + 1$

10. $f(x) = x^3$, $g(x) = 2x - 1$

11. $f(x) = \sqrt{x}$, $g(x) = x - 2$

12. $f(x) = \sqrt{x}$, $g(x) = x^2 - 9$

13. $f(x) = \sqrt{x}$, $g(x) = x^2$

14. $f(x) = \dfrac{1}{\sqrt{x}}$, $g(x) = x^2$

Answers to Practice Problems: **1. a.** $f(g(x)) = 2x^2 - 1$ **b.** $g(f(x)) = (2x - 1)^2$ **2. a.** 2 **b.** 1

            3. Undefined **4.** $f^{-1}(x) = \dfrac{x + 1}{3}$ **5.** $f^{-1}(x) = \sqrt{x + 4}$, $x \geq -4$

15. $f(x) = \dfrac{1}{\sqrt{x}}$, $g(x) = x^2 - 4$

16. $f(x) = x^{3n}$, $g(x) = 2x - 6$

17. $f(x) = \dfrac{1}{x}$, $g(x) = \dfrac{1}{x}$

18. $f(x) = x^{\frac{1}{3}}$, $g(x) = 4x + 7$

19. $f(x) = x^3$, $g(x) = \sqrt{x - 8}$

20. $f(x) = x^3 + 1$, $g(x) = \dfrac{1}{x}$

21. For the functions $f(x) = 6x - 3$ and $g(x) = \dfrac{1}{3}x + 3$, find:

    a. $f\big(g(3)\big)$ **b.** $g\big(f(0)\big)$

    c. Does it appear that f and g are inverses of each other? Explain.

22. For the functions $h(x) = -2x + 4$ and $g(x) = \dfrac{4 - x}{2}$, find:

    a. $h\big(g(6)\big)$ **b.** $g\big(h(-4)\big)$

    c. Does it appear that h and g are inverses of each other? Explain.

23. Given $f(x) = \dfrac{1}{x}$ and $h(x) = -\dfrac{1}{x}$, find:

    a. $f\big(h(-2)\big)$ **b.** $h\big(f(5)\big)$

    c. Does it appear that f and h are inverses of each other? Explain.

24. Given $f(x) = \sqrt{x - 9}$ and $g(x) = x - 9$, find:

    a. $g\big(f(109)\big)$ **b.** $f\big(g(9)\big)$ **c.** Explain the different result in parts **a** and **b**.

In Exercises 25 – 34, show that the given 1–1 functions are inverses of each other. Graph both functions on the same set of axes and show the line y = x as a dotted line on each graph. (You may use a calculator as an aid in finding the graphs.)

25. $f(x) = 3x + 1$ and $g(x) = \dfrac{x - 1}{3}$ **26.** $f(x) = -2x + 3$ and $g(x) = \dfrac{3 - x}{2}$

27. $f(x) = \sqrt[3]{x - 1}$ and $g(x) = x^3 + 1$ **28.** $f(x) = x^3 - 4$ and $g(x) = \sqrt[3]{x + 4}$

29. $f(x) = x^2$ for $x \ge 0$ and $g(x) = \sqrt{x}$

30. $f(x) = \sqrt{x + 3}$ and $g(x) = x^2 - 3$ for $x \ge 0$

31. $f(x) = x^3 + 2$ and $g(x) = \sqrt[3]{x - 2}$

32. $f(x) = \sqrt[5]{x + 6}$ and $g(x) = x^5 - 6$

33. $f(x) = x^2 + 4$ for $x \ge 0$ and $g(x) = \sqrt{x - 4}$

34. $f(x) = \dfrac{3}{x}$ and $g(x) = \dfrac{3}{x}$

In Exercises 35 – 48, find the inverse of the given function. Graph both functions on the same set of axes and show the line y = x as a dotted line in the graph.

35. $f(x) = 2x - 3$

36. $f(x) = 2x - 5$

37. $g(x) = x$

38. $g(x) = 1 - 4x$

39. $f(x) = 5x + 1$

40. $g(x) = \dfrac{2}{3}x + 2$

41. $g(x) = -3x + 1$

42. $f(x) = -\dfrac{1}{2}x - 3$

43. $f(x) = -\sqrt{x},\ x \geq 0$

44. $f(x) = -2x + 4$

45. $f(x) = x^2 + 1, x \geq 0$

46. $f(x) = x^2 - 1, x \geq 0$

47. $f(x) = -x - 2$

48. $f(x) = -\sqrt{x - 2},\ x \geq 2$

*Using the horizontal line test, determine which of the graphs in Exercises 49 – 58 are graphs of 1–1 functions. If the graph represents a 1–1 function, graph its inverse by reflecting the graph of the function across the line y = x. [**Hint:** If a function is 1–1, label a few points on the graph and use the fact that the x- and y-coordinates are interchanged on the graph of the inverse.]*

49.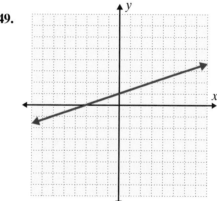

50.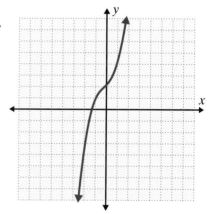

51.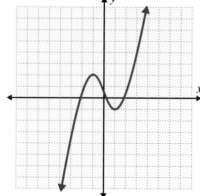

52.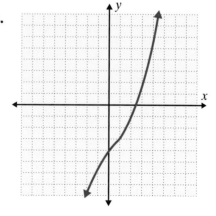

53.

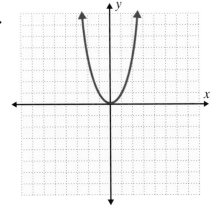

54.

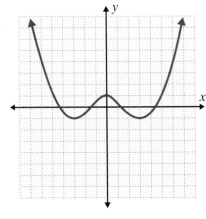

55.

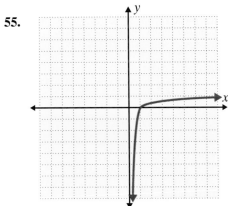

56.

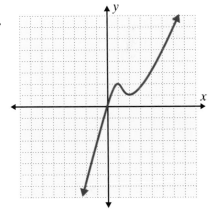

57.

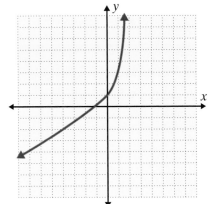

58.

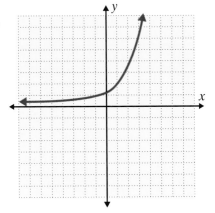

Use a graphing calculator to graph each of the functions in Exercises 59 – 68, and determine which of the functions are 1–1 by inspecting the graph and using the horizontal line test.

59. $f(x) = 2x + 3$ **60.** $f(x) = 7 - 4x$ **61.** $g(x) = x^2 - 2$ **62.** $g(x) = 9 - x^2$

63. $f(x) = x^3 + 2$ **64.** $g(x) = \dfrac{1}{x}$ **65.** $g(x) = \sqrt{x - 3}$ **66.** $f(x) = \sqrt{x + 5}$

67. $f(x) = |x + 1|$ **68.** $f(x) = |x - 5|$

*In Exercises 69 – 80, find the inverse of the given function. Then use a graphing calculator to graph both the function and its inverse. Set the **WINDOW** so that it is "square."*

69. $f(x) = x^3$ **70.** $f(x) = (x + 1)^3$ **71.** $f(x) = \dfrac{1}{x - 3}$

72. $f(x) = \dfrac{1}{x}$ **73.** $g(x) = x^3 + 2$ **74.** $f(x) = x^2, \ x \geq 0$

75. $f(x) = x^2 + 2, \ x \geq 0$ **76.** $g(x) = \sqrt{x + 7}, \ x \geq -7$ **77.** $f(x) = \sqrt{x + 5}, \ x \geq -5$

78. $g(x) = \sqrt{x - 3}, \ x \geq 3$ **79.** $f(x) = -x^2 + 1, \ x \geq 0$ **80.** $g(x) = -x^2 - 2, \ x \geq 0$

Writing and Thinking About Mathematics

81. Explain, in your own words, why the domains of the two composite functions $f(g(x))$ and $g(f(x))$ might not be the same. Give an example of two functions that illustrate this possibility.

82. Explain briefly why a function must be 1–1 to have an inverse.

 HAWKES LEARNING SYSTEMS: INTERMEDIATE ALGEBRA SOFTWARE

- Composition of Functions and Inverse Functions

8.3

Exponential Functions

- *Graph exponential functions.*
- *Solve applied problems involving exponential functions: exponential growth, exponential decay, and compound interest.*

Exponential Functions

You may have read that the population of the world is growing exponentially or studied the exponential growth of bacteria in a biology class. Radioactive materials decay exponentially and never actually disappear. The graph in Figure 8.13 illustrates that exponential growth has a relatively slow beginning and then builds at an exceedingly rapid rate. This can be extremely important if you are a doctor trying to curb the growth of "bad" bacteria in a patient.

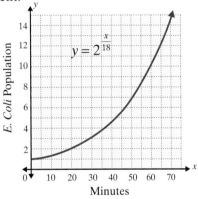

The population of certain strains of *E. Coli* doubles every 18 minutes under optimal conditions.

Figure 8.13

Quadratic functions have a variable base and a constant exponent, as in $f(x) = x^2$. However, in **exponential functions**, the base is constant and the exponent is a variable, as in $f(x) = 2^x$. As we will see, these two types of functions have major differences in their characteristics. Exponential functions are defined as follows.

Exponential Functions

> *An **exponential function** is a function of the form*
> $$f(x) = b^x$$
> *where $b > 0$, $b \neq 1$, and x is any real number.*

Examples of exponential functions are:

$$f(x) = 2^x, \quad f(x) = 3^x, \quad \text{and} \quad y = \left(\frac{1}{3}\right)^x.$$

NOTES

The two conditions $b > 0$ and $b \neq 1$ in the definition are important. We must have $b > 0$ so that b^x is defined for all real x. For example, we do not consider $y = (-2)^x$ to be an exponential function because $(-2)^{\frac{1}{2}}$ is not a real number. Also, $b \neq 1$. Because the function $y = 1^x = 1$ for all real x, this function is not considered to be an exponential function.

Exponential Growth

The following table of values and the graphs of the corresponding points give a very good idea of what the graph of the **exponential growth** function $y = 2^x$ looks like (see Figure 8.14a.). Because we know that 2^x is defined for all real exponents, points such as $\left(\sqrt{2}, 2^{\sqrt{2}}\right)$, $\left(\pi, 2^\pi\right)$, and $\left(\sqrt{5}, 2^{\sqrt{5}}\right)$ are on the graph, and the graph for $f(x) = 2^x$ is a smooth curve as shown in Figure 8.14b.

x	$y = 2^x$
3	$2^3 = 8$
2	$2^2 = 4$
1	$2^1 = 2$
$\dfrac{1}{2}$	$2^{\frac{1}{2}} = \sqrt{2} \approx 1.41$
0	$2^0 = 1$
$-\dfrac{1}{2}$	$2^{-\frac{1}{2}} = \dfrac{1}{\sqrt{2}} \approx 0.707$
-1	$2^{-1} = \dfrac{1}{2}$
-2	$2^{-2} = \dfrac{1}{2^2} = \dfrac{1}{4}$
-3	$2^{-3} = \dfrac{1}{2^3} = \dfrac{1}{8}$
-4	$2^{-4} = \dfrac{1}{2^4} = \dfrac{1}{16}$

Domain $= (-\infty, \infty)$
Range $= (0, \infty)$

Figure 8.14

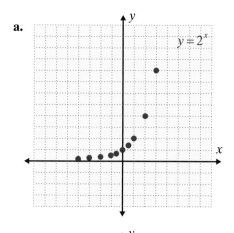

a.

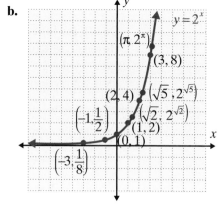

b.

Figure 8.15 shows a table of values and the graph of the function $y = 3^x$. Note that the graphs of $y = 2^x$ and $y = 3^x$ are quite similar, but that the graph of $y = 3^x$ rises faster. That is, the **exponential growth is faster if the base is larger**.

x	$y = 3^x$
2	9
1	3
0	1
−1	$\frac{1}{3} \approx 0.3333$
−3	$\frac{1}{27} \approx 0.037$

Domain = $(-\infty, \infty)$
Range = $(0, \infty)$

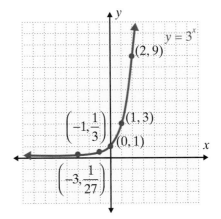

Figure 8.15

Notice that in both graphs the curves tend to get **very close** to the line $y = 0$ (the x-axis) without ever touching the x-axis. When this happens, the line is called an **asymptote**. If the line is horizontal, as in the cases of exponential growth and (as we will see) exponential decay, the line is called a **horizontal asymptote**. We say that the curve (or function) approaches the line **asymptotically**. As you study more mathematics, you will see that this phenomenon happens frequently.

Exponential Decay

Now consider the **exponential decay** function $f(x) = \left(\frac{1}{2}\right)^x$. The table and the graph of the corresponding points shown in Figure 8.16 indicate the nature of the graph of this function.

x	$y = \left(\frac{1}{2}\right)^x = 2^{-x}$
−3	$2^{-(-3)} = 2^3 = 8$
−2	$2^{-(-2)} = 2^2 = 4$
−1	$2^{-(-1)} = 2^1 = 2$
$-\frac{1}{2}$	$2^{-\left(\frac{1}{2}\right)} = 2^{\frac{1}{2}} = \sqrt{2} \approx 1.41$
0	$2^{-0} = 2^0 = 1$
$\frac{1}{2}$	$2^{-\frac{1}{2}} = \frac{1}{\sqrt{2}} \approx 0.707$
1	$2^{-1} = \frac{1}{2}$
2	$2^{-2} = \frac{1}{2^2} = \frac{1}{4}$
3	$2^{-3} = \frac{1}{2^3} = \frac{1}{8}$

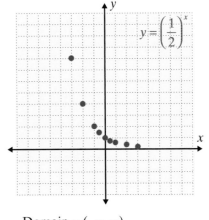

Domain = $(-\infty, \infty)$
Range = $(0, \infty)$

Figure 8.16

NOTES Because $\dfrac{1}{2} = 2^{-1}$, $\dfrac{1}{3} = 3^{-1}$, $\dfrac{1}{4} = 4^{-1}$, and so on, for fractions between 0 and 1, we can write an exponential function with a fractional base between 0 and 1 (these are exponential decay functions) in the form of an exponential function with a base greater than 1 and a negative exponent. Thus we write

$$y = \left(\frac{1}{2}\right)^x = \left(2^{-1}\right)^x = 2^{-x} \quad \text{and} \quad y = \left(\frac{1}{3}\right)^x = \left(3^{-1}\right)^x = 3^{-x}.$$

Figure 8.17 shows the complete graphs of the two exponential decay functions

$$y = \left(\frac{1}{2}\right)^x = 2^{-x} \quad \text{and} \quad y = \left(\frac{1}{3}\right)^x = 3^{-x}.$$

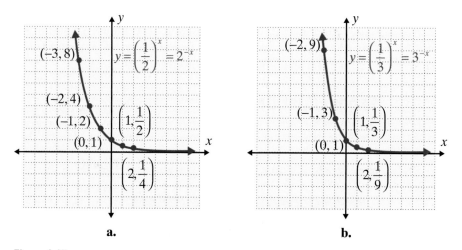

a.   **b.**

Figure 8.17

Note again, in Figure 8.17, that the line $y = 0$ (the x-axis) is a **horizontal asymptote** for both curves. For exponential growth the curves approach the asymptote as x moves further in the negative direction as in Figures 8.14b and 8.15. For exponential decay, the graph approaches the asymptote as x moves further in the positive direction as in Figure 8.17.

The following general concepts are helpful in understanding the graphs and the nature of exponential functions, both exponential growth and exponential decay.

General Concepts of Exponential Functions

For $b > 1$:

1. $b^x > 0$

2. b^x increases to the right and is called an **exponential growth function**.

3. $b^0 = 1$, so $(0, 1)$ is on the graph.

4. b^x approaches the x-axis for negative values of x. (The x-axis is a horizontal asymptote. **See Figure 8.15 on page 671.**)

For $0 < b < 1$:

1. $b^x > 0$

2. b^x decreases to the right and is called an **exponential decay function**.

3. $b^0 = 1$, so $(0, 1)$ is on the graph.

4. b^x approaches the x-axis for positive values of x. (The x-axis is a horizontal asymptote. **See Figure 8.17 on page 672.**)

As with all functions, exponential functions can be multiplied by constants, shifted horizontally, and shifted vertically. [**Note:** We did this with parabolas in Chapter 7 and will be more detailed on this topic in Chapter 9.] Thus

$$y = a \cdot b^{r \cdot x}, \qquad y = b^{x-h}, \qquad \text{and} \qquad y = b^x + k$$

and various combinations of these expressions are all exponential functions.

Bacterial Growth

Exponential functions are related to many practical applications, among which are bacterial growth, radioactive decay, compound interest, and light absorption. For example, a bacteria culture kept at a certain temperature may grow according to the exponential function

$$y = y_0 \cdot 2^{0.3t}$$

where t = time in hours and
y_0 = amount of bacteria present when $t = 0$.
(y_0 is called the **initial value** of y.)

Example 1: Exponential Growth

a. A scientist has 10,000 bacteria present when $t = 0$. She knows the bacteria grow according to the function $y = y_0 \cdot 2^{0.5t}$ where t is measured in hours. How many bacteria will be present at the end of one day?

Solution: Substitute $t = 24$ hours and $y_0 = 10,000$ into the function.

$$y = 10,000 \cdot 2^{0.5(24)} = 10,000 \cdot 2^{12}$$

$$= 10,000(4096)$$

$$= 40,960,000$$

$$= 4.096 \times 10^7 \text{ bacteria}$$

Continued on the next page...

You could also use your TI-84 Plus graphing calculator by entering the numbers as shown in the following display. Press **ENTER** to get the result.

b. Use the formula for exponential growth, $y = y_0 b^t$, to determine the exponential function that fits the following information: $y_0 = 5000$ bacteria, and there are 135,000 bacteria present after 3 days.

Solution: Use $y = y_0 b^t$ where t is measured in days. Substitute 135,000 for y, 3 for t, and 5000 for y_0, then solve for b.

$$135{,}000 = 5000 b^3$$

$$27 = b^3$$

$$\sqrt[3]{(27)} = \sqrt[3]{\left(b^3\right)}$$

$$3 = b$$

The function is $y = 5000 \cdot 3^t$.

Compound Interest

The topic of compound interest (interest paid on interest) leads to a particularly interesting (and useful) exponential function. The formula $A = P(1+r)^t$ can be used for finding the value (amount) accumulated when a principal, P, is invested and interest is compounded once a year. If compounding is performed more than once a year, we use the following formula to find A.

Compound Interest

Compound interest on a principal P invested at an annual interest rate r (in decimal form) for t years that is compounded n times per year can be calculated using the following formula:

$$A = P\left(1 + \frac{r}{n}\right)^{nt}$$

where A is the amount accumulated.

Example 2: Compound Interest

a. If P dollars are invested at a rate of interest r (in decimal form) compounded annually (once a year, $n = 1$) for t years, the formula for the amount A becomes

$A = P\left(1 + \dfrac{r}{1}\right)^{1 \cdot t} = P(1 + r)^t$. Find the value of $1000 invested at $r = 6\% = 0.06$ for 3 years.

Solution: We have $P = 1000$, $r = 0.06$, and $t = 3$.

$$A = 1000(1 + 0.06)^3$$

$$= 1000(1.06)^3$$

$$= 1000(1.191016) \approx \$1191.02$$

To use your calculator to evaluate $(1.06)^3$, enter 1.06, press the

key, enter 3, then press the **ENTER** key.

b. What will be the value of a principal investment of $1000 invested at 6% for 3 years if interest is compounded monthly (12 times per year)?

Solution: Use the formula for compound interest.
We have $P = 1000$, $r = 0.06$, $n = 12$, and $t = 3$.

$$A = 1000\left(1 + \frac{0.06}{12}\right)^{12(3)}$$

$$= 1000(1 + 0.005)^{36}$$

$$= 1000(1.005)^{36}$$

$$= 1000(1.196680524\ldots) \qquad \text{Using a calculator}$$

$$\approx \$1196.68$$

c. Find the value of A if $1000 is invested at 6% for 3 years and interest is compounded daily. [**Note:** Banks and savings institutions often use 360 days per year.]

Solution: Use the formula for compound interest.
We have $P = 1000$, $r = 0.06$, $n = 360$, and $t = 3$.

$$A = 1000\left(1 + \frac{0.06}{360}\right)^{360(3)}$$

$$= 1000(1.000166666\ldots)^{1080}$$

$$= 1000(1.197199407\ldots) \qquad \text{Using a calculator}$$

$$\approx \$1197.20$$

Examples 2a, 2b, and 2c illustrate the effects of compounding interest more frequently over 3 years. The formula gives the results

$$A = \$1191.02 \quad \text{for} \quad n = 1,$$
$$A = \$1196.68 \quad \text{for} \quad n = 12,$$
$$A = \$1197.20 \quad \text{for} \quad n = 360.$$

These numbers might not seem very dramatic, only a difference of $6.18 for 3 years; but, if you use your calculator and 20 years you will see a difference of $112.65 for a $1000 investment. An investment of $10,000 for 20 years at 9% will show a difference of $4438.76. The results show that more frequent compounding will result in higher income.

Interest Compounded Continuously

If interest is **compounded continuously** (even faster than every second), then the irrational number e $(e \approx 2.718)$ can be shown to be the base of the corresponding exponential function for calculating interest. Table 8.2 shows how the expression $\left(1+\dfrac{1}{n}\right)^n$ changes as n **takes on larger and larger values**. We say that the number e is the **limit (or limiting value)** of the expression "as n approaches infinity" $(n \to \infty)$ and we write $e = \lim\limits_{n \to \infty}\left(1+\dfrac{1}{n}\right)^n$. Study the following table to help understand the ideas.

Values of the expression $\left(1 + \dfrac{1}{n}\right)^n$ as n approaches infinity $(n \to \infty)$

n	$\left(1+\dfrac{1}{n}\right)$	$\left(1+\dfrac{1}{n}\right)^n$
1	$\left(1+\dfrac{1}{1}\right)=2$	$2^1 = 2$
2	$\left(1+\dfrac{1}{2}\right)=1.5$	$(1.5)^2 = 2.25$
5	$\left(1+\dfrac{1}{5}\right)=1.2$	$(1.2)^5 = 2.48832$
10	$\left(1+\dfrac{1}{10}\right)=1.1$	$(1.1)^{10} \approx 2.59374246$
100	$\left(1+\dfrac{1}{100}\right)=1.01$	$(1.01)^{100} \approx 2.704813829$
1000	$\left(1+\dfrac{1}{1000}\right)=1.001$	$(1.001)^{1000} \approx 2.716923932$

10,000	$\left(1+\dfrac{1}{10,000}\right)=1.0001$	$(1.0001)^{10,000}\approx 2.718145927$
100,000	$\left(1+\dfrac{1}{100,000}\right)=1.00001$	$(1.00001)^{100,000}\approx 2.718268237$
$\downarrow$		$\downarrow$
∞		$e=2.718281828459...$

Table 8.2

This gives us the following definition of e. (Be aware that these are very sophisticated mathematical concepts and it will take some careful reading to understand how to arrive at e and the formula $A=Pe^{rt}$.)

The Number e

The Number e

The number e is defined to be
$$e = 2.718281828459\ldots$$

As we know, the formula for compound interest is

$$A = P\left(1+\frac{r}{n}\right)^{nt}.$$

This formula for compound interest takes a different form when interest is **compounded continuously**. The new form involves the number e and is stated below.

Continuously Compounded Interest

Continuously compounded interest *on a principal P invested at an annual interest rate r for t years, can be calculated using the following formula:*

$$A = Pe^{rt}$$

where A is the amount accumulated.

As illustrated in Example 3, a calculator is needed to use the formula for continuously compounded interest.

Example 3: Calculating Continuously Compounded Interest Using a TI-84 Plus Calculator

Find the value of $1000 invested at 6% for 3 years if interest is compounded continuously. (In this case, $P = \$1000$, $r = 6\% = 0.06$, and $t = 3$.)

Solution: (The number e is on the TI-84 Plus calculator above the divide key.

Press **2ND** and **LN** and $e^{\wedge}($ will appear on the display.)

To find the value of

$$A = Pe^{rt} = 1000e^{0.06 \cdot 3}$$

enter the numbers as shown and press **ENTER** to get the result.

The entire exponent must be in parentheses.

Thus the value of $1000 compounded continuously at 6% for 3 years will be $1197.22. (Note that from Example 2c there is only a 2¢ gain in A when $1000 is compounded continuously instead of daily at 6% for 3 years.)

Practice Problems

1. Sketch the graph of the exponential function $f(x) = 2 \cdot 3^x$ and label 3 points on the graph.

2. Sketch the graph of the exponential decay function $y = 0.5 \cdot 2^{-x}$ and label 3 points on the graph.

3. Find the value of $5000 invested at 8% for 10 years if interest is (a) compounded monthly, (b) compounded continuously.

Answers to Practice Problems: **1.**   **2.**   **3. a.** $11,098.20   **b.** $11,127.70

8.3 Exercises

In Exercises 1 – 20, sketch the graph of each of the exponential functions and label three points on each graph. (Note that some of the graphs are shifts, horizontal or vertical, of the basic exponential functions. These are similar to the shifts performed on parabolas in Chapter 7.)

1. $y = 4^x$

2. $y = \left(\dfrac{1}{3}\right)^x$

3. $y = \left(\dfrac{1}{5}\right)^x$

4. $y = 5^x$

5. $y = 10^x$

6. $y = \left(\dfrac{2}{3}\right)^x$

7. $y = \left(\dfrac{5}{2}\right)^x$

8. $y = \left(\dfrac{1}{2}\right)^{-x}$

9. $y = 2^{x-1}$

10. $y = 3^{x+1}$

11. $f(x) = 2^x + 1$

12. $f(x) = 2^{x+1}$

13. $f(x) = 3^{2x}$

14. $f(x) = 2^{0.5x}$

15. $g(x) = 0.5 \cdot 3^x - 1$

16. $g(x) = 10^{-x} - 3$

17. $g(x) = -2^{-x}$

18. $g(x) = 10^{0.5x}$

19. $y = 3 \cdot \left(\dfrac{1}{2}\right)^{0.2x}$

20. $y = -4 \cdot \left(\dfrac{1}{3}\right)^{x-1}$

21. If $f(t) = 3 \cdot 4^t$ what is the value of $f(2)$?

22. Use your calculator to find the value (to the nearest hundredth) of $f(2)$ if $f(x) = 27.3 \cdot e^{-0.4x}$.

23. For $f(x) = 3 \cdot 10^{2x}$, find the value of $f(0.5)$.

24. Use your calculator to find the value (to the nearest hundredth) of $f(9)$ if $f(t) = 2000 \cdot e^{0.08t}$. What does this value indicate to you about investing money?

25. Use your calculator to find the value (to the nearest hundredth) of $f(22)$ if $f(t) = 2000 \cdot e^{0.05t}$. What does this value indicate to you about investing money?

26. Bacteria growth: A biologist knows that in the laboratory, bacteria in a culture grow according to the function $y = y_0 \cdot 5^{0.2t}$, where y_0 is the initial number of bacteria present and t is time measured in hours. How many bacteria will be present in a culture at the end of 5 hours if there were 5000 present initially?

27. Bacteria growth: Referring to Exercise 26, how many bacteria were present initially if at the end of 15 hours, there were 2,500,000 bacteria present?

28. Savings accounts: Four thousand dollars is deposited into a savings account with a rate of 8% per year. Find the total amount, A, on deposit at the end of 5 years if the interest is compounded:
a. annually
b. semiannually
c. quarterly
d. daily (use 360 days)
e. continuously

29. Savings accounts: Find the amount, A, in a savings account if $2000 is invested at 7% for 4 years and the interest is compounded:
a. annually
b. semiannually
c. quarterly
d. daily (use 360 days)
e. continuously

30. Investing: Find the value of $1800 invested at 6% for 3 years if the interest is compounded continuously.

31. Investing: Find the value of $2500 invested at 5% for 5 years if the interest is compounded continuously.

32. Sales revenue: The revenue function is given by $R(x) = x \cdot p(x)$ dollars, where x is the number of units sold and $p(x)$ is the unit price. If $p(x) = 25(2)^{\frac{-x}{5}}$, find the revenue if 15 units are sold.

33. Sales revenue: Referring to Exercise 32, if $p(x) = 40(3)^{\frac{-x}{6}}$, find the revenue if 12 units are sold.

34. Broadcasting: A radio station knows that during an intense advertising campaign, the number of people, N, who will hear a commercial is given by $N = A\left(1 - 2^{-0.05t}\right)$, where A is the number of people in the broadcasting area and t is the number of hours the commercial has been run. If there are 500,000 people in the area, how many will hear a commercial during the first 20 hours?

35. Battery reliability: Statistics show that the fractional part of flashlight batteries, f, that are still good after t hours of use is given by $f = 4^{-0.02t}$. What fractional part of the batteries are still operating after 150 hours of use?

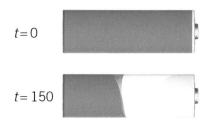

$t = 0$

$t = 150$

36. Investing: If a principal, P, is invested at a rate, r (expressed as a decimal), compounded continuously, the interest earned is given by $I = A - P$. How much interest will be earned in 20 years on an investment of $10,000 invested at 10% and compounded continuously?

37. Investing: Referring to Exercise 36, find the interest earned in 20 years on $10,000 invested at 5% and compounded continuously. Explain why the interest earned at 5% is not just one-half of the interest earned at 10% in Exercise 36.

38. Machine value: The value, V, of a machine at the end of t years is given by $V = C(1-r)^t$, where C is the original cost and r is the rate of depreciation. Find the value of a machine at the end of 4 years if the original cost was $1200 and $r = 0.20$.

39. Machine value: Referring to Exercise 38, find the value of a machine at the end of 3 years if the original cost was $2000 and $r = 0.15$.

40. Drug absorption: A cancer patient is given a dose of 50 mg of a particular drug. In five days the amount of the drug in her system is reduced to 1.5625 mg. If the drug decays (or is absorbed) at an exponential rate, find the function that represents the amount of the drug. [**Hint:** Use the formula $y = y_0 b^{-t}$ and solve for b.]

41. Growth of cancer cells: Determine the exponential function that fits the following information concerning exponential growth of cancer cells: $y_0 = 10,000$ cancer cells, and there are 160,000 cancer cells present after 4 days. [**Hint:** Use the formula $y = y_0 b^t$ and solve for b.]

42. Use a graphing calculator to graph each of the following functions. In each case the x-axis is a horizontal asymptote. Explain why the graphing calculator does not seem to indicate this fact.

  a. $y = e^x$

  b. $y = e^{-x}$

  c. $y = e^{-x^2}$

Writing and Thinking About Mathematics

43. Discuss, in your own words, how the graph of each of the following functions is related to the graph of the exponential function $y = b^x$.

  a. $y = a \cdot b^x$ **b.** $y = b^{x-h}$ **c.** $y = b^x + k$

44. Discuss, in your own words, the symmetrical relationship of the graphs of the two exponential functions $y = 10^x$ and $y = 10^{-x}$.

45. Discuss, in your own words, the symmetrical relationship of the graphs of the two exponential functions $y = 10^x$ and $y = -10^x$.

Writing and Thinking About Mathematics (cont.)

46. The following formula can be used to calculate monthly mortgage payments:

$$A = \frac{P\left(1+\dfrac{r}{12}\right)^{n} \cdot \dfrac{r}{12}}{\left(1+\dfrac{r}{12}\right)^{n} - 1}$$

where A = the monthly payment,
  P = amount initially borrowed (the mortgage),
  r = the annual interest rate (in decimal form), and
  n = the total number of monthly payments (12 times the number of years).

With the class divided into teams of 3 or 4 students, have each team complete one table (using different values for r and for P) and then have a class discussion about the results and what this might mean to them personally.

For annual rate $r =$ _____ and initial mortgage $P =$ _____

Length of Mortgage (in years)	Monthly Payment A	Total Cost of Mortgage n times A
15		
20		
25		
30		

 HAWKES LEARNING SYSTEMS: INTERMEDIATE ALGEBRA SOFTWARE

- Exponential Functions and the Number e

8.4

Logarithmic Functions

- *Write exponential expressions in logarithmic form.*
- *Write logarithmic expressions in exponential form.*
- *Use the basic properties of logarithms to evaluate logarithms.*
- *Solve equations by using the definitions of exponential and logarithmic functions.*
- *Graph exponential functions and logarithmic functions on the same set of axes.*

Logarithms

Exponential functions of the form $y = b^x$ are 1–1 functions and, therefore, have inverses. To find the inverse of a function, we interchange x and y in the equation and solve for y. Thus, for the function

$$y = b^x,$$

interchanging x and y gives the inverse function

$$x = b^y.$$

Figure 8.18 shows the graphs of these two functions with $b > 1$. Each function is a reflection of the other across the line $y = x$. Note that the exponential function $y = b^x$ has **the line $y = 0$ (the x-axis) as a horizontal asymptote**, and the inverse function $x = b^y$ has **the line $x = 0$ (the y-axis) as a vertical asymptote**.

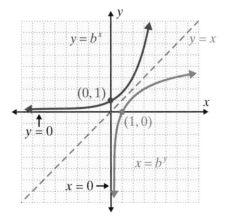

Figure 8.18

To solve the inverse equation $x = b^y$ for y, mathematicians have simply created a name for y. This name is **logarithm** (abbreviated **log**). This means that **the inverse of an exponential function is a logarithmic function**.

Definition of Logarithm (base *b*)

For $b > 0$ and $b \neq 1$,

$$x = b^y \text{ is equivalent to } y = \log_b x.$$

$y = \log_b x$ is read "y is the logarithm (base b) of x."

A logarithm is an exponent. Example 1 shows how exponential forms and logarithmic forms of equations are related.

Example 1: Translations Between Exponential Form and Logarithmic Form

	Exponential Form		Logarithmic Form	
a.	$2^3 = 8$	$\Leftrightarrow$	$\log_2 8 = 3$	The base is 2. The logarithm is 3.
b.	$2^4 = 16$	$\Leftrightarrow$	$\log_2 16 = 4$	The base is 2. The logarithm is 4.
c.	$10^3 = 1000$	$\Leftrightarrow$	$\log_{10} 1000 = 3$	The base is 10. The logarithm is 3.
d.	$3^0 = 1$	$\Leftrightarrow$	$\log_3 1 = 0$	The base is 3. The logarithm is 0.
e.	$5^{-1} = \dfrac{1}{5}$	$\Leftrightarrow$	$\log_5 \dfrac{1}{5} = -1$	The base is 5. The logarithm is -1.

Note that in each case the base of the exponent is the base of the logarithm.

REMEMBER, a logarithm is an exponent. For example,

logarithm   logarithm   logarithm

$$100 = 10^2 \quad \text{and} \quad 2 = \log_{10} 100 \quad \text{and} \quad 100 = 10^{\log_{10} 100}$$

are all equivalent. In words,

2 is the **exponent** of the base 10 to get 100 $\left(10^2 = 100\right)$; and

2 is the **logarithm** base 10 of 100 $\left(2 = \log_{10} 100\right)$.

Basic Properties of Logarithms

We know that **exponents are logarithms** and from our previous knowledge of exponents, we can make the following equivalent statements for logarithms, base b.

$$b^0 = 1 \qquad \Leftrightarrow \qquad \log_b 1 = 0$$

$$b^1 = b \qquad \Leftrightarrow \qquad \log_b b = 1$$

Also, directly from the definition of $y = \log_b x$, we can make two more general statements:

$$x = b^{\log_b x} \qquad \text{and} \qquad \log_b b^x = x.$$

In summary, we have the following four basic properties of logarithms.

Basic Properties of Logarithms

For $b > 0$ and $b \neq 1$,

   1. $\log_b 1 = 0$ *Regardless of the base, the logarithm of 1 is 0.*

   2. $\log_b b = 1$ *The logarithm of the base is always 1.*

   3. $x = b^{\log_b x}$ *For $x > 0$*

   4. $\log_b b^x = x$

REMEMBER, A logarithm is an exponent.

Example 2: Evaluating Logarithms

Use the four basic properties of logarithms to evaluate the following logarithms.

a. $\log_3 1$

   Solution: $\log_3 1 = 0$ By property 1

b. $\log_8 8$

   Solution: $\log_8 8 = 1$ By property 2

c. $10^{\log_{10} 20}$

   Solution: $10^{\log_{10} 20} = 20$ By property 3

Continued on the next page...

d. $\log_2 32$

Solution: $\log_2 32 = \log_2 2^5$ Write 32 as 2^5 so the base is 2.

$= 5$ By property 4

e. $\log_{10}(0.01)$

Solution: $\log_{10}(0.01) = \log_{10}\dfrac{1}{100}$

$= \log_{10}\dfrac{1}{10^2}$

$= \log_{10} 10^{-2}$ Write $\dfrac{1}{10^2}$ as 10^{-2} so the base is 10.

$= -2$ By property 4

Example 3: Using the Exponential Form to Solve Logarithmic Equations

a. Find the value of x if $\log_{16} x = \dfrac{3}{4}$.

Solution: $\log_{16} x = \dfrac{3}{4}$

$x = 16^{\frac{3}{4}}$ Write the equation in exponential form and solve for x.

$x = \left(16^{\frac{1}{4}}\right)^3 = 2^3 = 8$

Thus $\log_{16} 8 = \dfrac{3}{4}$.

b. Find the value of x if $\log_4 8 = x$.

Solution: $\log_4 8 = x$

$4^x = 8$ Write the equation in exponential form and solve for x.

$\left(2^2\right)^x = 2^3$ Use the common base, 2.

$2^{2x} = 2^3$

$2x = 3$ The exponents are equal because the bases are the same.

$x = \dfrac{3}{2}$

Thus $\log_4 8 = \dfrac{3}{2}$.

Graphs of Logarithmic Functions

As illustrated in Figure 8.19, because logarithmic functions are the inverses of exponential functions, the graphs of logarithmic functions can be found by reflecting the corresponding exponential functions across the line $y = x$. Figure 8.19a shows how the graphs of $y = 2^x$ and $y = \log_2 x$ are related. Figure 8.19b shows how the graphs of $y = 10^x$ and $y = \log_{10} x$ are related. Note that in the graphs of both logarithmic functions the values of y are negative when x is between 0 and 1 $(0 < x < 1)$.

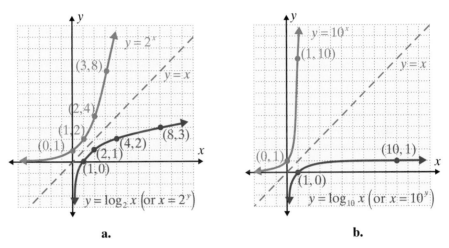

a.   **b.**

Figure 8.19

Notice that points on the graphs of inverse functions can be found by reversing the coordinates of ordered pairs.

Recall that the domain and range of a function and its inverse are interchanged. Thus, for exponential functions and logarithmic functions, we have the following.

- For the exponential function $y = b^x$,

    the domain is all real x, and
    the range is all $y > 0$.   (The graph is above the x-axis.)
    There is a horizontal asymptote at $y = 0$.

- For the logarithmic function $y = \log_b x$ $\left(\text{or } x = b^y\right)$,

    the domain is all $x > 0$, and   (The graph is to the right of the y-axis.)
    the range is all real y.
    There is a vertical asymptote at $x = 0$.

Practice Problems

Express the given equation in logarithmic form.

1. $4^2 = 16$

2. $5^3 = 125$

3. $10^{-1} = \dfrac{1}{10}$

Express the given equation in exponential form.

4. $\log_2 x = -1$

5. $\log_5 x = 2$

6. $\log_3 x = 0$

Find the value of each expression.

7. $\log_2 64$

8. $\log_3 27$

9. $\log_7 7$

8.4 Exercises

In Exercises 1 – 12, express each equation in logarithmic form.

1. $7^2 = 49$

2. $3^3 = 27$

3. $5^{-2} = \dfrac{1}{25}$

4. $10^2 = 100$

5. $2^{-5} = \dfrac{1}{32}$

6. $1 = \pi^0$

7. $\left(\dfrac{2}{3}\right)^2 = \dfrac{4}{9}$

8. $10^k = 23$

9. $17 = 2^x$

10. $4^k = 11.6$

11. $10^1 = 10$

12. $6^0 = 1$

In Exercises 13 – 24, express each equation in exponential form.

13. $\log_3 9 = 2$

14. $\log_5 125 = 3$

15. $\log_9 3 = \dfrac{1}{2}$

16. $\log_b 4 = \dfrac{2}{3}$

17. $\log_7 \dfrac{1}{7} = -1$

18. $\log_{\frac{1}{2}} 8 = -3$

19. $\log_{10} N = 1.74$

20. $\log_2 42.3 = x$

21. $\log_b 18 = 4$

22. $\log_b 39 = 10$

23. $\log_n y^2 = x$

24. $\log_b a = x$

Solve Exercises 25 – 40 by first changing each equation to exponential form.

25. $\log_4 x = 2$

26. $\log_3 x = 4$

27. $\log_{14} 196 = x$

28. $\log_5 \dfrac{1}{125} = x$

29. $\log_{36} x = -\dfrac{1}{2}$

30. $\log_x 32 = 5$

31. $\log_x 121 = 2$

32. $\log_{81} x = -\dfrac{3}{4}$

Answers to Practice Problems: **1.** $\log_4 16 = 2$ **2.** $\log_5 125 = 3$ **3.** $\log_{10} \dfrac{1}{10} = -1$ **4.** $x = 2^{-1}$ **5.** $x = 5^2$

6. $x = 3^0$ **7.** 6   **8.** 3   **9.** 1

33. $\log_8 x = \dfrac{5}{3}$ **34.** $\log_{25} 125 = x$ **35.** $\log_3 \dfrac{1}{9} = x$ **36.** $\log_8 8^{3.7} = x$

37. $\log_{10} 10^{1.52} = x$ **38.** $\log_5 5^{\log_5 25} = x$ **39.** $\log_4 4^{\log_2 8} = x$ **40.** $\log_p p^{\log_3 81} = x$

In Exercises 41 – 50, graph each function and its inverse on the same set of axes. Label two points on each graph.

41. $f(x) = 6^x$ **42.** $f(x) = 2^x$ **43.** $y = \left(\dfrac{2}{3}\right)^x$ **44.** $y = \left(\dfrac{1}{4}\right)^x$

45. $f(x) = \log_4 x$ **46.** $f(x) = \log_5 x$ **47.** $y = \log_{\frac{1}{2}} x$ **48.** $y = \log_{\frac{1}{3}} x$

49. $y = \log_8 x$ **50.** $y = \log_7 x$

51. Consider the function $y = c\left(3^x\right)$ where c is a constant greater than zero. Please list the following:
   a. The domain of the function.
   b. The range of the function.
   c. Any asymptotes of the graph of the function.
   d. Give c two different values and sketch the graphs of both functions.

52. Consider the function $y = c\left(3^{-x}\right)$ where c is a constant greater than zero. Please list the following:
   a. The domain of the function.
   b. The range of the function.
   c. Any asymptotes of the graph of the function.
   d. Give c two different values and sketch the graphs of both functions.

Writing and Thinking About Mathematics

53. Discuss, in your own words, the symmetrical relationship of the graphs of the two functions $y = 10^x$ and $y = \log_{10} x$.

54. Discuss, in your own words, the symmetrical relationship of the graphs of the two logarithmic functions $y = \log_{10} x$ and $y = -\log_{10} x$.

HAWKES LEARNING SYSTEMS: INTERMEDIATE ALGEBRA SOFTWARE

- Logarithmic Functions

Properties of Logarithms

- *Understand the rules for changing the form of logarithmic expressions.*
- *Use the rules for changing the form of logarithmic expressions to evaluate logarithms.*
- *Use the rules to expand logarithmic expressions and to write expanded forms as single logarithms.*

We know that calculators are effective in giving numerical evaluations. However, they do not, in general, simplify or solve equations. (You might be aware that there are calculators with computer algebra systems, CAS, that do simplify expressions and solve equations.) In this section, we will discuss several properties (or rules) of logarithms that are helpful in solving equations and simplifying expressions that involve logarithms or exponential functions.

The following four basic properties have been discussed.

> For $b > 0$ and $b \neq 1$,
>
> **1.** $\log_b 1 = 0$
>
> **2.** $\log_b b = 1$
>
> **3.** $x = b^{\log_b x}$ for $x > 0$
>
> **4.** $\log_b b^x = x$.

With these four properties as a basis, we now develop three more properties (or rules) for logarithms: the Product Rule, the Quotient Rule, and the Power Rule.

The Product Rule

Because logarithms are exponents, their properties are similar to those of exponents. In fact, the properties of exponents are used to prove the rules of logarithms. Study the developments carefully and you will see that logarithms are handled just as exponents.

NOTES

We will discuss using a calculator to find the values of logarithms in the next section. To illustrate some of the properties of logarithms in this section, we will use the following base 10 logarithmic values, accurate to 4 decimal places.

$$\log_{10} 2 \approx 0.3010 \quad \log_{10} 3 \approx 0.4771$$
$$\log_{10} 5 \approx 0.6990 \quad \log_{10} 6 \approx 0.7782$$

Also, because we are using logarithms rounded to 4 decimal places, there may be a slight difference in the answers if a calculator is used to find the logarithms in some of the illustrations.

Consider the following analysis.

We know that $$6 = 2 \cdot 3.$$

Using property 3, we have

$$10^{\log_{10} 6} = 6, \qquad 10^{\log_{10} 2} = 2, \qquad \text{and} \qquad 10^{\log_{10} 3} = 3.$$

Now, by using the properties of exponents, we can write

$$
\begin{array}{ccccc}
6 & = & 2 & \cdot & 3 \\
\downarrow & & \downarrow & & \downarrow \\
10^{\log_{10} 6} & = & 10^{\log_{10} 2} & \cdot & 10^{\log_{10} 3} = 10^{\log_{10} 2 + \log_{10} 3}.
\end{array}
$$

Equating exponents (with the same base) gives the result:

$$\log_{10} 6 = \log_{10} 2 + \log_{10} 3$$

$$\downarrow \qquad \downarrow \qquad \downarrow$$

$$0.7782 \approx 0.3010 + 0.4771.$$ Note that the fourth digit is off because of rounding.

This technique can be used to prove the following **Product Rule of Logarithms**.

Product Rule of Logarithms

For $b > 0$, $b \neq 1$, and $x, y > 0$,

$$\log_b (xy) = \log_b x + \log_b y.$$

*In words, **the logarithm of a product is equal to the sum of the logarithms of the factors.***

Proof of the Product Rule

$$
\overbrace{b^{\log_b(xy)}}^{\text{Property 3}} = xy = x \cdot y = \overbrace{b^{\log_b x} \cdot b^{\log_b y}}^{\text{Property 3 again}} = \overbrace{b^{\log_b x + \log_b y}}^{\text{Add exponents}}
$$

Thus

$$b^{\log_b(xy)} = b^{\log_b x + \log_b y}.$$

Equating the exponents gives the Product Rule of Logarithms:

$$\log_b (xy) = \log_b x + \log_b y.$$

Example 1: Using the Product Rule of Logarithms

Simplify each of the following expressions by using the Product Rule.

a. $\log_{10} 1000$

> **Solution:** $\log_{10} 1000 = \log_{10}(10 \cdot 10 \cdot 10)$
>
> $$= \log_{10} 10 + \log_{10} 10 + \log_{10} 10$$
>
> $$= 1 + 1 + 1$$
>
> $$= 3$$
>
> (We knew this result from another approach: $\log_{10} 1000 = \log_{10} 10^3 = 3$. But the approach shown helps to confirm that the Product Rule works.)

b. $\log_{10} 30$

> **Solution:** $\log_{10} 30 = \log_{10} 10 \cdot 3$
>
> $$= \log_{10} 10 + \log_{10} 3$$
>
> $$\approx 1 + 0.4771 = 1.4771$$

c. $\log_{10} 30$

> **Solution:** $\log_{10} 30 = \log_{10}(2 \cdot 3 \cdot 5)$
>
> $$= \log_{10} 2 + \log_{10} 3 + \log_{10} 5$$
>
> $$\approx 0.3010 + 0.4771 + 0.6990 = 1.4771$$

Note that in **b.** and **c.** using different factors of 30 gives the same result.

The Quotient Rule

Now consider the problem of finding the logarithm of a quotient. For example, to find $\log_{10} \dfrac{3}{2}$, we can proceed as follows:

$$10^{\log_{10} \frac{3}{2}} = \frac{3}{2}$$

$$= \frac{10^{\log_{10} 3}}{10^{\log_{10} 2}}$$

$$= 10^{\log_{10} 3 - \log_{10} 2}.$$

Equating exponents gives:

$$\log_{10} \frac{3}{2} = \log_{10} 3 - \log_{10} 2$$

$$\approx 0.4771 - 0.3010$$

$$= 0.1761.$$

These ideas lead to the following **Quotient Rule of Logarithms**. The proof is left as an exercise for the student.

Quotient Rule of Logarithms

For b > 0, b ≠ 1, and x, y > 0,

$$log_b \frac{x}{y} = log_b x - log_b y.$$

*In words, **the logarithm of a quotient is equal to the difference between the logarithm of the numerator and the logarithm of the denominator.***

Example 2: Using the Quotient Rule of Logarithms

Simplify each of the following expressions by using the Quotient Rule.

a. $\log_{10} \frac{1}{2}$

   Solution: $\log_{10} \frac{1}{2} = \log_{10} 1 - \log_{10} 2 \approx 0 - \log_{10} 2 \approx -0.3010$

b. $\log_2 \frac{1}{8}$

   Solution: $\log_2 \frac{1}{8} = \log_2 1 - \log_2 8 = \log_2 1 - \log_2 2^3 = 0 - 3 = -3$

   (Note that if we write $\frac{1}{8}$ as 2^{-3} we get the same result.)

c. $\log_{10} \frac{15}{2}$

   Solution: $\log_{10} \frac{15}{2} = \log_{10} 15 - \log_{10} 2 = \log_{10} 3 + \log_{10} 5 - \log_{10} 2$

   $$\approx 0.4771 + 0.6990 - 0.3010$$

   $$= 0.8751$$

The Power Rule

The next property of logarithms involves a number raised to a power and the multiplication property of exponents. For example,

$$10^{\log_{10}\left(2^3\right)} = 2^3$$
$$= \left(10^{\log_{10} 2}\right)^3$$
$$= 10^{3 \cdot \log_{10} 2}. \quad \text{By the Power Rule of Exponents}$$

Now equating exponents in the first and last expressions gives

$$\log_{10}\left(2^3\right) = 3 \cdot \log_{10} 2.$$

These ideas lead to the following **Power Rule of Logarithms**.

Power Rule of Logarithms

For $b > 0$, $b \neq 1$, $x > 0$, and any real number r,

$$log_b x^r = r \cdot log_b x.$$

*In words, **the logarithm of a number raised to a power is equal to the product of the exponent and the logarithm of the number.***

Proof of the Power Rule of Logarithms

$$\overbrace{b^{\log_b\left(x^r\right)} = x^r = \left(b^{\log_b x}\right)^r}^{\text{Property 3 twice}} = \overbrace{b^{r \cdot \log_b x}}^{\text{Multiply exponents}}$$

Equating the exponents gives the result called the Power Rule of Logarithms:

$$\log_b x^r = r \cdot \log_b x.$$

Example 3: Using the Power Rule

a. $\log_{10}\left(\sqrt{2}\right)$

Solution: $\log_{10}\left(\sqrt{2}\right) = \log_{10}\left(2^{\frac{1}{2}}\right) = \frac{1}{2} \cdot \log_{10} 2 \approx \frac{1}{2}(0.3010) = 0.1505$

b. $\log_{10} 8$

Solution: $\log_{10} 8 = \log_{10}\left(2^3\right) = 3 \cdot \log_{10} 2 \approx 3(0.3010) = 0.9030$

c. $\log_{10} 25$

Solution: $\log_{10} 25 = \log_{10}\left(5^2\right) = 2 \cdot \log_{10} 5 \approx 2(0.6990) = 1.3980$

Summary of Properties of Logarithms

For $b > 0$, $b \neq 1$, $x, y > 0$, and any real number r,

1. $\log_b 1 = 0$

2. $\log_b b = 1$

3. $x = b^{\log_b x}$

4. $\log_b b^x = x$

5. $\log_b (xy) = \log_b x + \log_b y$ The Product Rule

6. $\log_b \dfrac{x}{y} = \log_b x - \log_b y$ The Quotient Rule

7. $\log_b x^r = r \cdot \log_b x$ The Power Rule

Example 4: Using the Properties of Logarithms to Expand Expressions

Use the properties of logarithms to expand each expression as much as possible.

a. $\log_b (2x^3)$

Solution: $\log_b (2x^3) = \log_b 2 + \log_b x^3$ Product Rule

$= \log_b 2 + 3\log_b x$ Power Rule

b. $\log_b \left(\dfrac{xy^2}{z} \right)$

Solution: $\log_b \left(\dfrac{xy^2}{z} \right) = \log_b (xy^2) - \log_b z$ Quotient Rule

$= \log_b x + \log_b y^2 - \log_b z$ Product Rule

$= \log_b x + 2\log_b y - \log_b z$ Power Rule

c. $\log_b (xy)^{-3}$

Solution: $\log_b (xy)^{-3} = -3\log_b (xy)$ Power Rule

$= -3 \left(\log_b x + \log_b y \right)$ Product Rule

$= -3\log_b x - 3\log_b y$

d. $\log_b \left(\sqrt{3x} \right)$

Solution: $\log_b \left(\sqrt{3x} \right) = \log_b (3x)^{\frac{1}{2}} = \dfrac{1}{2}\log_b (3x)$ Power Rule

$= \dfrac{1}{2} \left(\log_b 3 + \log_b x \right)$ Product Rule

$= \dfrac{1}{2}\log_b 3 + \dfrac{1}{2}\log_b x$

Example 5: Using the Properties of Logarithms to Write Single Logarithm Expressions

Use the properties of logarithms to write each expression as a single logarithm. Note that the bases must be the same when simplifying sums and differences of logarithms.

a. $2\log_b x - 3\log_b y$

Solution: $2\log_b x - 3\log_b y = \log_b x^2 - \log_b y^3$ Power Rule

$$= \log_b\left(\frac{x^2}{y^3}\right)$$ Quotient Rule

b. $\dfrac{1}{2}\log_a 4 - \log_a 5 - \log_a y$

Solution:

$$\frac{1}{2}\log_a 4 - \log_a 5 - \log_a y = \log_a 4^{\frac{1}{2}} - \log_a 5 - \log_a y$$ Power Rule

$$= \log_a 2 - \left(\log_a 5 + \log_a y\right) \quad \left(4^{\frac{1}{2}} = 2\right) \text{ Factor out } -1.$$

$$= \log_a 2 - \log_a(5y)$$ Product Rule

$$= \log_a\left(\frac{2}{5y}\right)$$ Quotient Rule

c. $\log_a(x+1) + \log_a(x-1)$

Solution: $\log_a(x+1) + \log_a(x-1) = \log_a\left[(x+1)(x-1)\right]$ Product Rule

$$= \log_a\left(x^2 - 1\right)$$

d. $\log_b \sqrt{x} + \log_b \sqrt[3]{x}$

Solution: $\log_b \sqrt{x} + \log_b \sqrt[3]{x} = \log_b x^{\frac{1}{2}} + \log_b x^{\frac{1}{3}}$

$$= \frac{1}{2}\log_b x + \frac{1}{3}\log_b x$$ Power Rule

$$= \left(\frac{1}{2} + \frac{1}{3}\right)\log_b x$$ Distributive Property

$$= \frac{5}{6}\log_b x$$ $\dfrac{1}{2} + \dfrac{1}{3} = \dfrac{3}{6} + \dfrac{2}{6} = \dfrac{5}{6}$

OR

$$\log_b \sqrt{x} + \log_b \sqrt[3]{x} = \log_b x^{\frac{1}{2}} + \log_b x^{\frac{1}{3}}$$

$$= \log_b \left(x^{\frac{1}{2}} \cdot x^{\frac{1}{3}} \right) \qquad \text{Product Rule}$$

$$= \log_b x^{\left(\frac{1}{2} + \frac{1}{3} \right)}$$

$$= \log_b x^{\frac{5}{6}}$$

$$= \frac{5}{6} \log_b x \qquad \text{Power Rule}$$

NOTES

Common Misunderstandings about Logarithms

There is no logarithmic property for the logarithm of a sum or a difference.

$\log_b (x + y)$ **Cannot be simplified**

$\log_b (x - y)$ **Cannot be simplified**

Also,

$\log_b (xy) \neq \log_b x \cdot \log_b y$ **The log of a product does not equal the product of the logs.**

$\log_b \dfrac{x}{y} \neq \dfrac{\log_b x}{\log_b y}$ **The log of a quotient does not equal the quotient of the logs.**

Practice Problems

1. Write $\log_b \left(x^3 y \right)$ *as a sum of logarithmic expressions.*

2. Write the expression $2\log_b 5 + \log_b x - \log_b 3$ *as a single logarithm.*

3. Write the expression $2\log_b x - \log_b (x + 1)$ *as a single logarithm.*

Answers to Practice Problems: **1.** $3\log_b x + \log_b y$ **2.** $\log_b \left(\dfrac{25x}{3} \right)$ **3.** $\log_b \left(\dfrac{x^2}{x+1} \right)$

8.5 Exercises

The following logarithms (accurate to 4 decimal places) are used in exercises 1 and 2.

$$\log_{10} 2 \approx 0.3010 \qquad \log_{10} 3 \approx 0.4771$$
$$\log_{10} 5 \approx 0.6990 \qquad \log_{10} 6 \approx 0.7782$$

1. Find the value of the following expressions.

    a. $10^{0.3010}$ **b.** $10^{0.4771}$ **c.** $10^{0.6990}$ **d.** $10^{0.7782}$

2. Find the value of the following expressions.

    a. $10^{0.3010+0.7782}$ **b.** $10^{0.4771+0.7782}$ **c.** $10^{0.4771+0.6990}$ **d.** $10^{0.6990-0.3010}$

In Exercises 3 – 12, use your knowledge of logarithms and exponents to find the value of each expression.

3. $\log_2 32$ **4.** $\log_3 9$ **5.** $\log_4 \dfrac{1}{16}$ **6.** $\log_5 \dfrac{1}{125}$

7. $\log_3 \sqrt{3}$ **8.** $\log_2 \sqrt{8}$ **9.** $5^{\log_5 10}$ **10.** $3^{\log_3 17}$

11. $6^{\log_6 \sqrt{3}}$ **12.** $5^{\log_5 5}$

In Exercises 13 – 32, use the properties of logarithms to expand each expression as much as possible.

13. $\log_b \left(5x^4\right)$ **14.** $\log_b \left(3x^2 y\right)$ **15.** $\log_b \left(2x^{-3} y\right)$ **16.** $\log_5 \left(xy^2 z^{-1}\right)$

17. $\log_6 \left(\dfrac{2x}{y^3}\right)$ **18.** $\log_3 \left(\dfrac{xy}{4z}\right)$ **19.** $\log_b \left(\dfrac{x^2}{yz}\right)$ **20.** $\log_3 \left(\dfrac{xy^2}{z^2}\right)$

21. $\log_5 \left(xy\right)^{-2}$ **22.** $\log_b \left(x^2 y\right)^4$ **23.** $\log_6 \sqrt[3]{xy^2}$ **24.** $\log_5 \sqrt{2x^3 y}$

25. $\log_3 \sqrt{\dfrac{xy}{z}}$ **26.** $\log_6 \sqrt[3]{\dfrac{x^2}{y}}$ **27.** $\log_5 \left(21x^2 y^{\frac{2}{3}}\right)$ **28.** $\log_b \left(15x^{-\frac{1}{2}} y^{\frac{1}{3}}\right)$

29. $\log_6 \dfrac{x}{\sqrt{x^3 y^5}}$ **30.** $\log_3 \dfrac{1}{\sqrt{x^4 y}}$ **31.** $\log_b \left(\dfrac{x^3 y^2}{z}\right)^{-3}$ **32.** $\log_4 \left(\dfrac{x^{-\frac{1}{2}} y}{z^2}\right)^{-2}$

Use the properties of logarithms to write each expression in Exercises 33 – 54 as a single logarithm of a single expression.

33. $2\log_b 3 + \log_b x - \log_b 5$ **34.** $\dfrac{1}{2}\log_b 25 + \log_b 3 - \log_b x$

35. $\log_2 7 - \log_2 9 + 2\log_2 x$ **36.** $\log_5 4 + \log_5 6 + \log_5 y$

37. $2 \log_b x + \log_b y$

38. $\log_2 x + 3 \log_2 y$

39. $3 \log_{10} x - 2 \log_{10} y$

40. $3 \log_5 y - \dfrac{1}{2} \log_5 x$

41. $\dfrac{1}{2} \left(\log_5 x - \log_5 y \right)$

42. $\dfrac{1}{3} \left(\log_{10} x - 2 \log_{10} y \right)$

43. $\log_2 x - \log_2 y + \log_2 z$

44. $\log_b x + 2 \log_b y - \dfrac{1}{2} \log_b z$

45. $\log_b x - 2 \log_b y - 2 \log_b z$

46. $-\dfrac{2}{3} \log_2 x - \dfrac{1}{3} \log_2 y + \dfrac{2}{3} \log_2 z$

47. $\log_5 x + \log_5 (2x + 1)$

48. $\log_{10} (x + 3) + \log_{10} (x - 3)$

49. $\log_2 (x - 1) + \log_2 (x + 3)$

50. $\log_b (3x + 1) + 2 \log_b x$

51. $\log_b \left(x^2 - 2x - 3 \right) - \log_b (x - 3)$

52. $\log_2 (x - 4) - \log_2 \left(x^2 - 2x - 8 \right)$

53. $\log_{10} (x + 6) - \log_{10} \left(2x^2 + 9x - 18 \right)$

54. $\log_5 \left(3x^2 + 5x - 2 \right) - \log_5 (3x - 1)$

Writing and Thinking About Mathematics

55. Prove the Quotient Rule of Logarithms: For $b > 0, b \neq 1$, and $x, y > 0$,

$\log_b \dfrac{x}{y} = \log_b x - \log_b y.$

56. Prove the following property of logarithms: For $b > 0, b \neq 1$, and $x > 0$,

$\log_b b^x = x.$

HAWKES LEARNING SYSTEMS: INTERMEDIATE ALGEBRA SOFTWARE

- Properties of Logarithmic Functions

8.6

Common Logarithms and Natural Logarithms

- *Understand that base 10 logarithms are called **common logarithms**.*
- *Use a calculator to find the values of expressions with common logarithms.*
- *Understand that base e logarithms are called **natural logarithms**.*
- *Use a calculator to find the values of expressions with natural logarithms.*
- *Use a calculator to find inverse logarithms (called antilogs) for common and for natural logarithms.*

Base 10 logarithms (called **common logarithms**) and base e logarithms (called **natural logarithms**) are the two most commonly used logarithms. Common logarithms occur frequently because of their close relationship with the decimal system and are used in chemistry, astronomy, and physical science. Natural logarithms are an important tool in calculus and appear "naturally" in growth and decay problems. Both common and natural logarithms have designated keys on calculators.

Common Logarithms (Base 10 Logarithms)

The abbreviated notation $\log x$ is used to indicate common logarithms (base 10 logarithms). That is, whenever the base notation is omitted, the base is understood to be 10:

$$\log_{10} x = \log x.$$

Example 1: Translations Between Exponential Form and Logarithmic Form Using Common Logarithms

	Exponential Form		Logarithmic Form	
a.	$10^4 = 10,000$	$\Leftrightarrow$	$\log 10,000 = 4$	This is a common logarithm that equals 4.
b.	$10^{-2} = 0.01$	$\Leftrightarrow$	$\log 0.01 = -2$	This is a common logarithm that equals -2.
c.	$10^x = 6.3$	$\Leftrightarrow$	$\log 6.3 = x$	This is a common logarithm that equals x.

The TI-84 Plus graphing calculator has the designated key $\boxed{\text{LOG}}$ that can be used in the following 3-step process to find the values of common logarithms.

Using a Calculator to Evaluate Common Logarithms (Base 10)

To evaluate a common logarithm on a TI-84 Plus graphing calculator:

1. *Press the* $\boxed{\text{LOG}}$ *key.* **log(** *will appear on the display.*

2. *Enter the number and a right-hand parenthesis,* $\boxed{)}$.

3. *Press* $\boxed{\text{ENTER}}$.

Example 2: Evaluating Common Logarithms Using a TI-84 Plus Graphing Calculator

Use a TI-84 Plus graphing calculator to find the values of the following common logarithms.

a. $\log 200$ **b.** $\log 50{,}000$ **c.** $\log 0.0006$

Solutions:

From the display we see the results (accurate to 8 or 9 decimal places).

a. $\log 200 \approx 2.301029996$

b. $\log 50{,}000 \approx 4.698970004$

c. $\log 0.0006 \approx -3.22184875$

Example 2c shows a negative value for the logarithm. This is because 0.0006 is between 0 and 1. **In fact, the logarithm (with base greater than 1) of any number between 0 and 1 will always be negative.** See Figure 8.20 on the following page.

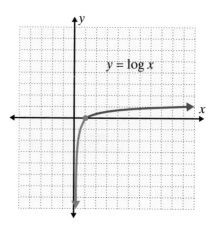

Figure 8.20

Logarithms are exponents and they may be positive, negative, or 0. For example,

$$10^{-2} = \frac{1}{100} = 0.01 \qquad \text{and} \qquad \log 0.01 = -2 \ .$$

However, there are no logarithms of negative numbers or 0. The domains of logarithmic functions are nonnegative real numbers. The logarithm of a negative number is **undefined**. For example, $\log(-2)$ is undefined. That is, there is no real number x such that $10^x = -2$. If you enter $\log(-2)$ in a calculator, an error message will appear as shown here.

Figure 8.21

Finding the Inverse of a Logarithm (Base 10)

Because logarithms are exponents, if we know the value of a common logarithm, we know an exponent. Finding the value of the related exponential expression is called finding the **inverse of the logarithm** (or finding the **antilog**). For example,

if $\log x = N$, then $x = 10^N$ and the number x is called the **inverse log of N**.

Using a TI-84 Plus Graphing Calculator to find the Inverse Log of *N*

Finding the inverse log of *N* is a three-step process.

Using a Calculator to Find the Inverse Log of *N*

*To evaluate the **inverse log of N** on a TI-84 Plus graphing calculator:*

1. Press **2ND** *and* **LOG** *. The expression* **10^(** *will appear on the display.*

2. Enter the value of N and a right-hand parenthesis **)** *.*

3. Press **ENTER** *.*

Example 3: Using a TI-84 Plus Graphing Calculator to find the Inverse Log of *N*

Use a TI – 84 Plus graphing calculator to find the **inverse log of *N*** for each expression. (That is, find the value of *x*.)

a. $\log x = 5$ **b.** $\log x = -3$ **c.** $\log x = 2.4142$ **d.** $\log x = 16.5$

Solutions:

Thus the calculator gives,

a. $x = 10^5 = 100,000$

b. $x = 10^{-3} = 0.001$

c. $x = 10^{2.4142} \approx 259.5374301$

d. $x = 10^{16.5} \approx 3.16227766\text{E}16$

The letter E in the solution is the calculator version of scientific notation. Thus, $3.16227766\text{E}16 = 3.16227766 \times 10^{16}$.

Natural Logarithms (Base *e* Logarithms)

The number $e \approx 2.718281828$ is an irrational number, and we have seen this number appear quite naturally in continuously compounded interest. (We will see more applications involving e in Section 8.8.) Base e logarithms are called natural logarithms. **The notation for natural logarithms is shortened to ln x** (read "L N of x" or "natural log of x"). That is,

$$\log_e x = \ln x.$$

Figure 8.22 shows the graphs of the two functions $y = e^x$ and its inverse $y = \ln x$.

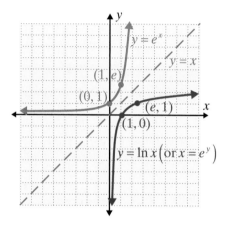

Figure 8.22

Example 4: Translations Between Exponential Form and Logarithmic Form Using Natural Logarithms				
	Exponential Form		**Logarithmic Form**	
a.	$e^t = 3.21$	$\Leftrightarrow$	$\ln 3.21 = t$	This is a natural logarithm that equals t.
b.	$e^0 = 1$	$\Leftrightarrow$	$\ln 1 = 0$	This is a natural logarithm that equals 0.
c.	$e^2 = n$	$\Leftrightarrow$	$\ln n = 2$	This is a natural logarithm that equals 2.

To evaluate a natural logarithm on a TI-84 Plus graphing calculator:

1. *Press the* **LN** *key.* `ln (` *will appear on the display.*

2. *Enter the number and a right-hand parenthesis,* **)** *.*

3. *Press* **ENTER** *.*

Example 5: Evaluating Natural Logarithms by Using a TI-84 Plus Graphing Calculator

Use a TI-84 Plus graphing calculator to find the values of the following natural logarithms.

a. $\ln 1$ **b.** $\ln 3$ **c.** $\ln 0.02$

Solutions:

From the display we see the results (accurate to 9 decimal places).

a. $\ln 1 = 0$

b. $\ln 3 \approx 1.098612289$

c. $\ln 0.02 \approx -3.912023005$

Finding the Inverse of a Natural Logarithm (Base *e*)

Again, because logarithms are exponents, if we know the value of a natural logarithm, we know an exponent. Finding the value of the related exponential expression is called finding the **inverse of the logarithm**. For example,

if $\ln x = N$, then $x = e^{N}$ and the number x is called the **inverse ln of *N***.

Using a Calculator to Find the Inverse ln of N

To evaluate the **inverse ln of N** on a TI-84 Plus graphing calculator:

1. Press [2ND] and [LN]. The expression **e^(** will appear on the display.

2. Enter the value of N and a right-hand parenthesis [)].

3. Press [ENTER].

Example 6: Using a TI-84 Plus Graphing Calculator to find the Inverse ln of N

Use a TI-84 Plus graphing calculator to find the inverse ln of N for each expression. (That is, find the value of x.)

    a. $\ln x = 3$ **b.** $\ln x = -1$ **c.** $\ln x = -0.1$ **d.** $\ln x = 50$

Solutions:

Thus the calculator gives,

    a. $x = e^{3} \approx 20.08553692$

    b. $x = e^{-1} \approx 0.3678794412$

    c. $x = e^{-0.1} \approx 0.904837418$

    d. $x = e^{50} \approx 5.184705529\text{E}21$

Practice Problems

Express each equation in logarithmic form.

1. $10^1 = 10$ **2.** $e^3 = x$ **3.** $10^{-4} = 0.0001$

Express each equation in exponential form.

4. $log(0.1) = -1$ **5.** $log\dfrac{1}{100} = -2$ **6.** $ln\,x = 6$

Use a calculator to evaluate each logarithm (accurate to 5 decimal places).

7. $log\,175$ **8.** $ln\,52$ **9.** $log(-6.1)$

Use a calculator to find the value of x in each equation (accurate to 5 decimal places).

10. $ln\,x = -2$ **11.** $log\,x = 0.5$ **12.** $ln\,x = 30$

8.6 Exercises

Express each equation in logarithmic form in Exercises 1 – 10.

1. $10^{-3} = \dfrac{1}{1000}$ **2.** $10^{-4} = 0.0001$ **3.** $10^{1.5} = x$ **4.** $10^k = 23$

5. $e^x = 27$ **6.** $e^k = 12.4$ **7.** $e^0 = 1$ **8.** $e^4 = x$

9. $10^x = 3.2$ **10.** $10^y = x$

Express each equation in exponential form in Exercises 11 – 20.

11. $log\,1 = 0$ **12.** $log\,100 = 2$ **13.** $ln\,5.4 = y$ **14.** $ln\,40.1 = x$

15. $ln\,x = 1.54$ **16.** $log\,x = 25.3$ **17.** $log\,25.3 = x$ **18.** $ln\,e = 1$

19. $log\,10 = 1$ **20.** $ln\,x = a$

Answers to Practice Problems: **1.** $log\,10 = 1$ **2.** $ln\,x = 3$ **3.** $log\,0.0001 = -4$ **4.** $10^{-1} = 0.1$
5. $10^{-2} = \dfrac{1}{100}$ **6.** $e^6 = x$ **7.** 2.24304 **8.** 3.95124 **9.** Error (undefined)
10. 0.13534 **11.** 3.16228 **12.** 1.06865×10^{13}

Use a calculator to evaluate the logarithms in Exercises 21 – 32 accurate to 5 decimal places.

21. log 173   **22.** log 396   **23.** log 88.4   **24.** log 0.0061

25. log 0.0573   **26.** log (–8.47)   **27.** ln 37.5   **28.** ln 96

29. ln (–14.9)   **30.** ln 157.6   **31.** ln 0.00461   **32.** ln 0.0139

Use a calculator to find the value of x in each equation in Exercises 33 – 44 accurate to 5 decimal places.

33. $\log x = 2.31$ **34.** $\log x = -3$ **35.** $\log x = -1.7$

36. $\log x = 4.1$ **37.** $2 \log x = -0.038$ **38.** $5 \log x = 9.4$

39. $\ln x = 5.17$ **40.** $\ln x = 4.9$ **41.** $\ln x = -8.3$

42. $\ln x = 6.74$ **43.** $0.2 \ln x = 0.0079$ **44.** $3 \ln x = -0.066$

Writing and Thinking About Mathematics

45. Explain the difference in the meaning of the expressions $\log x$ and $\ln x$.

46. The function $y = \log x$ is defined only for $x > 0$. Discuss the function $y = \log (-x)$. That is, does this function even exist? If it does exist, what is its domain? Sketch its graph and the graph of the function $y = \log x$.

47. What is the domain of the function $y = \ln |x|$? Graph the function.

 HAWKES LEARNING SYSTEMS: INTERMEDIATE ALGEBRA SOFTWARE

▪ Common Logarithms and Natural Logarithms

Logarithmic and Exponential Equations and Change-of-Base

8.7

- *Solve exponential equations in which the bases are the same.*
- *Solve exponential equations in which the bases are not the same.*
- *Solve equations with logarithms.*
- *Use the change-of-base formula and a calculator to evaluate logarithmic expressions.*

Solving Exponential Equations with the Same Base

All the properties of exponents that have been discussed are valid for real exponents, not just integers or fractions. For example, $2^{\sqrt{2}} \cdot 2^3 = 2^{\sqrt{2}+3}$ is a valid operation and gives a real number. To be complete in our development of the properties of exponents, the properties for real exponents and positive real bases are stated here.

Properties of Real Exponents

If a and b are positive real numbers and x and y are any real numbers, then:

1. $b^0 = 1$
2. $b^{-x} = \dfrac{1}{b^x}$
3. $b^x \cdot b^y = b^{x+y}$
4. $\dfrac{b^x}{b^y} = b^{x-y}$
5. $\left(b^x\right)^y = b^{xy}$
6. $(ab)^x = a^x b^x$
7. $\left(\dfrac{a}{b}\right)^x = \dfrac{a^x}{b^x}$

The following properties are used in solving equations containing exponents and logarithms.

Properties of Equations with Exponents and Logarithms

For b > 0 and b ≠ 1,

1. *If $b^x = b^y$, then x = y.*
2. *If x = y, then $b^x = b^y$.*
3. *If $\log_b x = \log_b y$, then x = y (x > 0 and y > 0).*
4. *If x = y, then $\log_b x = \log_b y$ (x > 0 and y > 0).*

Example 1: Solving Exponential Equations with the Same Base

Solve each equation for x.

a. $2^{x^2-7} = 2^{6x}$

Solution:

$$2^{x^2-7} = 2^{6x}$$ Both bases are 2.

$$x^2 - 7 = 6x$$ The exponents are equal because the bases are the same.

$$x^2 - 6x - 7 = 0$$ Solve for x.

$$(x-7)(x+1) = 0$$

$$x = 7 \quad \text{or} \quad x = -1$$

b. $8^{4-2x} = 4^{x+2}$

Solution:

$$8^{4-2x} = 4^{x+2}$$ Here the bases are different.

$$\left(2^3\right)^{4-2x} = \left(2^2\right)^{x+2}$$ Rewrite both sides so that the bases are the same.

$$2^{12-6x} = 2^{2x+4}$$ Use the property $\left(b^x\right)^y = b^{xy}$.

$$12 - 6x = 2x + 4$$ The exponents are equal because the bases are the same.

$$8 = 8x$$ Solve for x.

$$1 = x$$

Solving Exponential Equations with Different Bases

As we have seen in Example 1, if the bases are the same in an exponential equation, there is no need to involve logarithms. However, as illustrated in Example 2, **when the bases are not the same, we involve logarithms by taking the logarithm of both sides**. (See Property 4.)

Example 2: Solving Exponential Equations with Different Bases

Solve each of the following exponential equations by taking the log (or ln) of both sides or by using the definition of logarithm as an exponent.

a. $10^{3x} = 2.1$

Solution: First, take the log of both sides, and then use the definition of logarithm as an exponent as follows.

$$10^{3x} = 2.1$$

$$\log\left(10^{3x}\right) = \log 2.1 \quad \text{Take the log of both sides.}$$

$$3x = \log 2.1 \quad \text{By the definition of a common logarithm}$$

$$x = \frac{\log 2.1}{3}$$

Use a calculator to find a decimal approximation:

$$x = \frac{\log(2.1)}{3} \approx \frac{0.3222}{3} = 0.1074.$$

b. $6^x = 18$

   Solution: The base is 6, not 10 or e, but we can solve by taking the **log** of both sides or by taking the **ln** of both sides. The result is the same.

Taking the log of both sides:	**Taking the ln of both sides:**
$6^x = 18$	$6^x = 18$
$\log 6^x = \log 18$	$\ln 6^x = \ln 18$
$x \cdot \log 6 = \log 18$	$x \cdot \ln 6 = \ln 18$
$x = \dfrac{\log 18}{\log 6}$	$x = \dfrac{\ln 18}{\ln 6}$
Using a calculator,	Using a calculator,
$x = \dfrac{\log 18}{\log 6} \approx \dfrac{1.2553}{0.7782}$	$x = \dfrac{\ln 18}{\ln 6} \approx \dfrac{2.8904}{1.7918}$
$\approx 1.6131.$	$\approx 1.6131.$

c. $5^{2x-1} = 10^x$

   Solution:

$$5^{2x-1} = 10^x$$

$$\log 5^{2x-1} = \log 10^x \qquad \text{Take the log of both sides.}$$

$$(2x-1)\log 5 = x \log 10 \qquad \text{Power Rule}$$

$$2x \cdot \log 5 - 1 \cdot \log 5 = x \cdot 1 \qquad \log 10 = 1$$

$$2x \log 5 - x = \log 5 \qquad \text{Arrange } x\text{-terms on one side.}$$

$$x(2\log 5 - 1) = \log 5 \qquad \text{Factor out the } x.$$

$$x = \frac{\log 5}{2\log 5 - 1}$$

As a decimal approximation,

$$x = \frac{\log 5}{2\log 5 - 1} \approx \frac{0.6990}{2(0.6990) - 1} \approx 1.7563. \qquad \text{Using rounded values}$$

Solving Equations with Logarithms

All the various properties of logarithms can be used to solve equations that involve logarithms. **Remember that logarithms are defined only for positive real numbers, so each answer should be checked in the original equation.**

Example 3: Solving Equations with Logarithms

Use the properties of logarithms to solve the following equations.

a. $\log(5x) = 3$

> **Solution:** $\log(5x) = 3$
>
> $\qquad 5x = 10^3$ Definition of a common logarithm
>
> $\qquad 5x = 1000$
>
> $\qquad x = 200$

b. $\log(x-1) + \log(x-4) = 1$

> **Solution:** $\log(x-1) + \log(x-4) = 1$
>
> $\qquad \log\big[(x-1)(x-4)\big] = 1$ Product Rule
>
> $\qquad (x-1)(x-4) = 10^1$ Definition of a common logarithm
>
> $\qquad x^2 - 5x + 4 = 10$
>
> $\qquad x^2 - 5x - 6 = 0$
>
> $\qquad (x-6)(x+1) = 0$ Solve by factoring.
>
> $\qquad x = 6 \quad \text{or} \quad \cancel{x = -1}$ Checking $x = -1$ yields $\log(-1-1) = \log(-2)$, which is undefined.

c. $\log x - \log(x-1) = \log 3$

> **Solution:** $\log x - \log(x-1) = \log 3$
>
> $\qquad \log\left(\dfrac{x}{x-1}\right) = \log 3$ Quotient Rule
>
> $\qquad \dfrac{x}{x-1} = 3$ If $\log_b x = \log_b y$, then $x = y$.

$$x = 3(x-1) \qquad \text{Solve for } x.$$
$$x = 3x - 3$$
$$3 = 2x$$
$$\frac{3}{2} = x$$

d. $\ln(x^2 - x - 6) - \ln(x + 2) = 2$

Solution: $\ln(x^2 - x - 6) - \ln(x + 2) = 2$

$$\ln\left(\frac{x^2 - x - 6}{x + 2}\right) = 2 \qquad \text{Quotient Rule}$$

$$\ln\left(\frac{(x+2)(x-3)}{(x+2)}\right) = 2 \qquad \text{Factor the numerator.}$$

$$\ln(x - 3) = 2 \qquad \text{Simplify.}$$

$$x - 3 = e^2 \qquad \text{Change to exponential form with base } e.$$

$$x = 3 + e^2$$

Or, using a calculator, $x = 3 + e^2 \approx 3 + 7.3891 = 10.3891$.

Change-of-Base

Because a calculator can be used to evaluate common logarithms and natural logarithms, we have restricted most of the examples to base 10 or base e expressions. If an equation involves logarithms of other bases, the following discussion shows how to rewrite each logarithm using any base you choose.

Change-of-Base Formula

For a, b, x > 0 and a, b ≠ 1,

$$\log_b x = \frac{\log_a x}{\log_a b}.$$

The change-of-base formula can be derived by using properties of logarithms as follows.

$$b^{\log_b x} = x \qquad \text{Property 3 in Section 8.5}$$

$$\log_a\left(b^{\log_b x}\right) = \log_a x \qquad \text{Take the } \log_a \text{ of both sides.}$$

$$\log_b x\left(\log_a b\right) = \log_a x \qquad \text{By the Power Rule using } \log_b x \text{ as the exponent } r.$$

$$\log_b x = \frac{\log_a x}{\log_a b}. \qquad \text{Divide both sides by } \log_a b \text{ to arrive at the change-of-base formula.}$$

Example 4: Change-of-Base

Use the change-of-base formula to evaluate the expressions in **a.** and **b.** and to solve the equation in **c.**

a. $\log_2 3.42$

Solution: This expression can be evaluated by using either base 10 or base e since both are easily available on a calculator.

$$\log_2 3.42 = \frac{\ln 3.42}{\ln 2} \approx \frac{1.2296}{0.6931} \approx 1.7741 \qquad \text{Using rounded values}$$

(The student can show that $\dfrac{\log 3.42}{\log 2}$ gives the same result.)

In exponential form: $2^{1.7741} \approx 3.42$

b. $\log_3 0.3333$

Solution: $\log_3 0.3333 = \dfrac{\log 0.3333}{\log 3} \approx \dfrac{-0.4772}{0.4771} \approx -1.0002 \qquad \text{Using rounded values}$

c.   Use the change-of-base formula to find the value of x (accurate to 4 decimal places) in the equation $5^x = 16$.

Solution: Because the base is 5, we can take $\log_5$ of both sides. (This method is not necessary, but it does show how the change-of-base formula can be used.)

$$5^x = 16$$

$$\log_5\left(5^x\right) = \log_5 16$$

$$x = \log_5 16$$

$$x = \frac{\ln 16}{\ln 5} \approx 1.7227$$

Practice Problems

Solve each of the following equations.

1. $4^x = 64$

2. $10^x = 64$

3. $2^{3x-1} = 0.1$

4. $15 \log x = 45.15$

5. $\ln(x^2 + 5x + 6) - \ln(x + 3) = 1$

8.7 Exercises

Use the properties of exponents and logarithms to solve each of the equations in Exercises 1 – 84. If necessary, round answers to four decimal places.

1. $2^4 \cdot 2^7 = 2^x$

2. $3^7 \cdot 3^{-2} = 3^x$

3. $\left(3^5\right)^2 = 3^{x+1}$

4. $\left(5^x\right)^2 = 5^6$

5. $\left(2^x\right)^3 = \sqrt{2}$

6. $\dfrac{10^4 \cdot 10^{\frac{1}{2}}}{10^x} = 10$

7. $\left(10^2\right)^x = \dfrac{10 \cdot 10^{\frac{2}{3}}}{10^{\frac{1}{2}}}$

8. $2^{5x} = 4^3$

9. $(25)^x = 5^3 \cdot 5^4$

10. $7^{3x} = 49^4$

11. $10^x \cdot 10^8 = 100^3$

12. $8^{x+3} = 2^{x-1}$

13. $27^x = 3 \cdot 9^{x-2}$

14. $100^{2x+1} = 1000^{x-2}$

15. $2^{3x+5} = 2^{x^2+1}$

16. $10^{x^2+x} = 10^{x+9}$

17. $10^{2x^2+3} = 10^{x+6}$

18. $3^{x^2+5x} = 3^{2x-2}$

19. $\left(3^{x+1}\right)^x = \left(3^{x+3}\right)^2$

20. $\left(10^x\right)^{x+3} = \left(10^{x+2}\right)^{-2}$

21. $3^x = 9$

22. $2^{5x-8} = 4$

23. $4^{x^2} = \left(\dfrac{1}{2}\right)^{3x}$

24. $25^{x^2+2x} = 5^{-x}$

25. $5^{2x-x^2} = \dfrac{1}{125}$

26. $10^{x^2-2x} = 1000$

27. $10^{3x} = 140$

28. $10^{2x} = 97$

29. $10^{0.32x} = 253$

30. $10^{-0.48x} = 88.6$

31. $4.10^{-0.94x} = 126.2$

32. $3 \cdot 10^{-2.1x} = 83.5$

33. $e^{0.03x} = 2.1$

34. $e^{-0.5x} = 47$

35. $e^{-0.006t} = 50.3$

36. $e^{4t} = 184$

37. $3e^{-0.12t} = 3.6$

38. $5e^{2.4t} = 44$

39. $2^x = 10$

Answers to Practice Problems: **1.** $x = 3$ **2.** $x \approx 1.8062$ **3.** $x \approx -0.7740$ **4.** $x = 10^{3.01} \approx 1023.2930$
 5. $x = e - 2 \approx 0.7183$

40. $3^{x-2} = 100$

41. $5^{2x} = \dfrac{1}{100}$

42. $7^{2x-3} = 10$

43. $5^{1-x} = 1$

44. $4^{2x+5} = 0.01$

45. $4^{2-3x} = 0.1$

46. $14^{3x-1} = 10^3$

47. $12^{2x+7} = 10^4$

48. $12^{5x+2} = 1$

49. $7^x = 9$

50. $2^x = 20$

51. $3^{x-2} = 23$

52. $5^{2x} = 23$

53. $6^{2x-1} = 14.8$

54. $4^{7-3x} = 26.3$

55. $5\log x = 7$

56. $3\log x = 13.2$

57. $4\log x - 6 = 0$

58. $2\log x - 15 = 0$

59. $4\log x^{\frac{1}{2}} + 8 = 0$

60. $\dfrac{2}{3}\log x^{\frac{2}{3}} + 9 = 0$

61. $5\ln x - 8 = 0$

62. $2\ln x + 3 = 0$

63. $\ln x^2 + 2.2 = 0$

64. $\ln x^2 - 41.6 = 0$

65. $\log x + \log 2x = \log 18$

66. $\log(x+4) + \log(x-4) = \log 9$

67. $\log x^2 - \log x = 2$

68. $\log x + \log x^2 = 3$

69. $\ln(x-3) + \ln x = \ln 18$

70. $\ln(x^2 - 3x + 2) - \ln(x-1) = \ln 4$

71. $\log(x-15) + \log x = 2$

72. $\log(3x-5) + \log(x-1) = 1$

73. $\log(2x-17) = 2 - \log x$

74. $\log(x-3) - 1 = \log(x+1)$

75. $\log(x^2 + 2x - 3) = 3 + \log(x+3)$

76. $\log(x^2 - 9) - \log(x-3) = -2$

77. $\log(x^2 - x - 12) + 2 = \log(x-4)$

78. $\log(x^2 - 4x - 5) - \log(x+1) = 2$

79. $\ln(x^2 + 4x - 5) - \ln(x+5) = -2$

80. $\ln(x+1) + \ln(x-1) = 0$

81. $\ln(x^2 - 4) - \ln(x+2) = 3$

82. $\ln(x^2 + 2x - 3) = 1 + \ln(x-1)$

83. $\log \sqrt[3]{x^2 + 2x + 20} = \dfrac{2}{3}$

84. $\log \sqrt{x^2 - 24} = \dfrac{3}{2}$

Use the change-of-base formula to evaluate each of the expressions or solve the equations in Exercises 85 – 100.

85. $\log_3 12$

86. $\log_4 36$

87. $\log_5 1.68$

88. $\log_{11} 39.6$

89. $\log_8 0.271$

90. $\log_7 0.849$

91. $\log_{15} 739$

92. $\log_2 14.2$

93. $\log_{20} 0.0257$

94. $\log_9 2.384$

95. $2^x = 5$

96. $3^{2x} = 10$

97. $5^{x-1} = 30$

98. $9^{2x-1} = 100$

99. $4^{3-x} = 20$

100. $6^{3x-4} = 25$

Writing and Thinking About Mathematics

101. Solve the following equation for x two different ways: $a^{2x-1} = 1$.

102. Rewrite each of the following expressions as products.

a. 5^{x+2} **b.** 3^{x-2}

103. Explain, in your own words, why $7 \cdot 7^x \neq 49^x$. Show each of the expressions $7 \cdot 7^x$ and 49^x as a single exponential expression with base 7.

 HAWKES LEARNING SYSTEMS: INTERMEDIATE ALGEBRA SOFTWARE

- Exponential and Logarithmic Equations

Applications

8.8

> ▪ *Solve applied problems by using logarithmic and exponential equations.*

In Section 8.3 we found that the number e appears in a surprisingly natural way in the formula for continuously compounding interest,

$$A = Pe^{rt},$$

which was developed from the formula for compounding n times per year:

$$A = P\left(1+\frac{r}{n}\right)^{nt}.$$

There are many formulas that involve exponential functions. A few are shown here and in the exercises.

$A = A_0 e^{-0.04t}$ This is a law for decomposition of radium where t is in centuries.

$A = A_0 e^{-0.1t}$ This is one law for skin healing where t is measured in days.

$A = A_0 2^{-t/5600}$ This law is used for carbon-14 dating to determine the age of fossils where t is measured in years.

$T = Ae^{-kt} + C$ This is Newton's law of cooling where C is the constant temperature of the surrounding medium. The values of A and k depend on the particular object that is cooling.

Example 1: Exponential Growth

Suppose that the formula $y = y_0 e^{0.4t}$ represents the number of bacteria present after t days, where y_0 is the initial number of bacteria. In how many days will the bacteria double in number?

Solution: $y = y_0 e^{0.4t}$

$2y_0 = y_0 e^{0.4t}$ $2y_0$ is double the initial number present.

$2 = e^{0.4t}$ Divide both sides by y_0.

$\ln 2 = \ln\left(e^{0.4t}\right)$ Take the natural log of both sides.

$\ln 2 = 0.4t \cdot (1)$ Power rule of logarithms; $\ln(e) = 1$

$t = \dfrac{\ln 2}{0.4} \approx \dfrac{0.6931}{0.4}$

$t \approx 1.73$ days

The number of bacteria will double in approximately 1.73 days. Note that this number is completely independent of the number of bacteria initially present. That is, if $y_0 = 10$ or $y_0 = 1000$, the doubling time is the same, namely 1.73 days.

Example 2: Continuously Compounded Interest

If $1000 is invested at a rate of 6% compounded continuously, in how many years will it grow to $5000?

Solution:

$$A = Pe^{rt} \qquad \text{Formula for continuously compounded interest}$$

$$5000 = 1000e^{0.06t}$$

$$5 = e^{0.06t}$$

$$\ln 5 = \ln\left(e^{0.06t}\right) \qquad \text{Take the natural log of both sides.}$$

$$\ln 5 = 0.06t \cdot (1) \qquad \text{Power rule of logarithms; } \ln(e) = 1$$

$$t = \frac{\ln 5}{0.06} \approx \frac{1.6094}{0.06}$$

$$t \approx 26.82 \text{ years}$$

$1000 will grow to $5000 in approximately 26.82 years.

Example 3: The Richter Scale

a. The magnitude of an earthquake is measured on the **Richter scale** as a logarithm of the intensity of the shock wave. For magnitude R and intensity I, the formula is $R = \log I$. The 1994 earthquake in Northridge, California measured 6.7 on the Richter scale. What was the intensity of this earthquake?

Solution: Substitute 6.7 for R in the formula and solve for I: $6.7 = \log I$

$$I = 10^{6.7}$$

b. The Long Beach earthquake in 1933 measured 6.2 on the Richter scale. How much stronger was the Northridge earthquake than the 1933 Long Beach earthquake?

Solution: The comparative sizes of the quakes can be found by finding the ratio of the intensities. For the Long Beach quake, $6.2 = \log I$ and $I = 10^{6.2}$. Therefore, the ratio of the two intensities is

$$\frac{I \text{ for Northridge}}{I \text{ for Long Beach}} = \frac{10^{6.7}}{10^{6.2}} = 10^{0.5} \approx 3.16.$$

Thus the Northridge earthquake had an intensity about 3.16 times the intensity of the Long Beach earthquake.

Example 4: Half-life of Radium

The **half-life** of a substance is the time needed for the substance to decay to one-half of its original amount. The half-life of radium-226, a common isotope of radium, is 1620 years. If 10 grams are present today, how many grams will remain in 500 years?

Solution: The model for radioactive decay is $y = y_0 e^{-kt}$. Since the half-life is 1620 years, if we assume $y_0 = 10$ g, then y would be 5 g after 1620 years. We solve for k as follows.

$$5 = 10e^{-k(1620)} \qquad \text{Substitute } y = 5, y_0 = 10, \text{ and } t = 1620.$$

$$\frac{5}{10} = e^{-1620k} \qquad \text{Solve for } k.$$

$$\ln(0.5) = \ln\left(e^{-1620k}\right) \qquad \text{Take the natural log of both sides.}$$

$$\ln(0.5) = -1620k \cdot (1) \qquad \text{Power rule for logarithms; } \ln(e) = 1$$

$$k = \frac{\ln(0.5)}{-1620} \approx \frac{-0.6931}{-1620} \approx 0.0004278$$

The model is $y = 10e^{-0.0004278t}$.

Substituting $t = 500$ gives

$$y = 10e^{(-0.0004278)(500)} = 10e^{-0.2139} \approx 10(0.8074) \approx 8.07.$$

Thus there will still be about 8.07 g of the radium-226 remaining after 500 years.

Example 5: Newton's Law of Cooling

Suppose that the room temperature is 70°, and the temperature of a cup of tea is 150° when it is placed on the table. In 5 minutes, the tea cools to 120°. How long will it take for the tea to cool to 100° ?

Solution: Using the formula $T = Ae^{-kt} + C$ (Newton's law of cooling), first find A and then k. We know that $C = 70°$ and that $T = 150°$ when $t = 0$.

Find A by substituting these values.

$$150 = Ae^{-k(0)} + 70$$

$$150 = A \cdot 1 + 70 \qquad e^{-k(0)} = e^0 = 1$$

$$80 = A$$

Therefore, the formula can be written as $T = 80e^{-kt} + 70$.

Since $T = 120°$ when $t = 5$, substituting these values allows us to find k.

$$120 = 80e^{-k(5)} + 70$$

$$50 = 80e^{-5k}$$

$$\frac{50}{80} = e^{-5k}$$

$$\ln\frac{5}{8} = \ln e^{-5k} \qquad \text{Take the natural log of both sides.}$$

$$\ln 0.625 = -5k$$

$$k = \frac{\ln 0.625}{-5} \approx \frac{-0.4700}{-5} \approx 0.0940$$

The formula can now be written as $T = 80e^{-0.0940t} + 70$.

With all the constants in the formula known, we can find t when $T = 100°$.

$$100 = 80e^{-0.0940t} + 70$$

$$30 = 80e^{-0.0940t}$$

$$\frac{30}{80} = e^{-0.0940t}$$

$$\ln\frac{3}{8} = \ln e^{-0.0940t} \qquad \text{Take the natural log of both sides.}$$

$$\ln 0.375 = -0.0940t$$

$$t = \frac{\ln 0.375}{-0.0940}$$

$$\approx \frac{-0.9808}{-0.0940} \approx 10.43 \text{ minutes}$$

The tea will cool to $100°$ in about 10.43 minutes.

8.8 Exercises

Solve the following application problems. If necessary, round answers to two decimal places.

1. **Investing:** If Kim invests $2000 at a rate of 7% compounded continuously, what will be her balance after 10 years?

2. **Savings accounts:** Find the amount of money that will be accumulated in a savings account if $3200 is invested at 6.5% for 6 years and the interest is compounded continuously.

3. **Investing:** Four thousand dollars is invested at 6% compounded continuously. How long will it take for the balance to be $8000?

4. **Investing:** How long does it take $1000 to double if it is invested at 5% compounded continuously?

5. **Battery reliability:** The reliability of a certain type of flashlight battery is given by $f = e^{-0.03x}$, where f is the fractional part of the batteries produced that last x hours. What fraction of the batteries produced are good after 40 hours of use?

6. **Battery reliability:** From Exercise 5, how long will at least one-half of the batteries last?

7. **Concentration of a drug:** The concentration of a drug in the blood stream is given by $C = C_0 e^{-0.8t}$, where C_0 is the initial dosage and t is the time in hours elapsed after administering the dose. If 20 mg of the drug is given, how much time elapses until 5 mg of the drug remains?

8. **Concentration of a drug:** Using the formula in Exercise 7, determine the amount of the drug present after 3 hours if 0.60 mg is given.

9. **Healing of the skin:** One law for skin healing is $A = A_0 e^{-0.1t}$, where A is the number of cm^2 of unhealed area after t days and A_0 is the number of cm^2 of the original wound. Find the number of days needed to reduce the wound to one-third the original size.

10. **Beekeeping:** A swarm of bees grows according to the formula $P = P_0 e^{0.35t}$, where P_0 is the number present initially and t is the time in days. How many bees will be present in 6 days if there were 1000 present initially?

11. **Inversion of raw sugar:** If inversion of raw sugar is given by $A = A_0 e^{-0.03t}$, where A_0 is the initial amount and t is the time in hours, how long will it take for 1000 lb of raw sugar to be reduced to 800 lb?

12. **Atmospheric pressure:** Atmospheric pressure, P, is related to the altitude, h, by the formula $P = P_0 e^{-0.00004h}$, where P_0, the pressure at sea level, is approximately 15 lb per in.^{2} Determine the pressure at 5000 in.

13. **Radioactive decay:** A radioactive substance decays according to $A = A_0 e^{-0.0002t}$, where A_0 is the initial amount and t is the time in years. If $A_0 = 640$ grams, find the time for A to decay to 400 grams.

14. **Substance decay:** A substance decays according to $A = A_0 e^{-0.045t}$, where t is in hours and A_0 is the initial amount. Determine the half-life of the substance.

15. **Employee training:** An employee is learning to assemble remote-control units. The number of units per day he can assemble after t days of intensive training is given by $N = 80\left(1 - e^{-0.3t}\right)$. How many days of training will be needed before the employee is able to assemble 40 units per day?

16. **Lava analysis:** A scientist collects a lava sample and measures that its temperature is 1650°. To safely analyze the sample, it must be no warmer than 500°. The scientist stores the sample in a cooling chamber with an ambient temperature of 50° and finds that in 2 hours, the lava has cooled to 1000°. When will the lava sample be safe to analyze?

17. **Baking:** The temperature of a carrot cake is 350° when it is removed from the oven. The temperature in the room is 72°. In 10 minutes, the cake cools to 280°. How long will it take for the cake to cool to 160°?

18. **Investing:** How long does it take $10,000 to double if it is invested at 8% compounded quarterly?

19. **Investing:** If $1000 is deposited at 6% compounded monthly, how long before the balance is $1520?

20. **Value of a machine:** The value, V, of a machine at the end of t years is given by $V = C\left(1 - r\right)^{t}$, where C is the original cost of the machine and r is the rate of depreciation. A machine that originally cost $12,000 is now valued at $3800. How old is the machine if $r = 0.12$?

21. **Carbon-14 dating:** The formula $A = A_0 2^{-t/5600}$ is used for carbon-14 dating to determine the age of fossils where t is measured in years. Determine the half-life of carbon-14.

22. **Half-life of radioactive iodine:** Radioactive iodine has a half-life of 60 days. If an accident occurs at a nuclear plant and 30 grams of radioactive iodine are present, in how many days will 1 gram be present? (Round k to at least 7 decimal places.)

23. **Investing:** If a principal, P, is doubled, then $A = 2P$. Use the formula for continuously compounded interest to find the time it takes the principal to double in value if the rate of interest is:
    a. 5%   **b.** 10%
    (Note that the time for doubling the principal is completely independent of the principal itself.)

24. **Investing:** If a principal, P, is tripled, then $A = 3P$. Use the formula for continuously compounded interest to find the time it takes the principal to triple in value if the interest rate is
    a. 4%   **b.** 8%
    (Note that the time for tripling the principal is completely independent of the principal itself.)

25. **The Richter scale:** The 1906 earthquake in San Francisco measured 8.6 on the Richter scale. In 1971, an earthquake in the San Fernando Valley measured 6.6 on the Richter scale. How many times greater was the 1906 earthquake than the 1971 earthquake? (See Example 3.)

26. The Richter scale: In 1985, an earthquake in Mexico measured 8.1 on the Richter scale. How many times greater was this earthquake than the one in Landers, California in 1992 that measured 7.3 on the Richter scale? (See Example 3.)

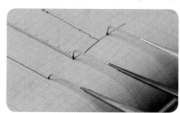

27. Population growth: Population does not generally grow in a linear fashion. In fact, the population of many species grows exponentially, at least for a limited time. Using the exponential model $y = y_0 e^{kt}$ for population growth, estimate the population of a state in 2020 if the population was 5 million in 1990 and 6 million in 2000. (Assume that t is measured in years and $t = 0$ corresponds to 1990.)

28. Fish population growth: Suppose that a lake is stocked with 500 fish, and biologists predict that the population of these fish will be approximated by the function $P(t) = 500 \ln(2t + e)$ where t is measured in years. What will be the fish population in 3 years? in 5 years? in 10 years?

29. Sales revenue: Sales representatives of a new type of computer predict that sales can be approximated by the function $S(t) = 1000 + 500 \ln(3t + e)$ where t is measured in years. What are the predicted sales in 2 years? in 5 years? in 10 years?

30. pH of a solution: In chemistry, the pH of a solution is a measure of the acidity or alkalinity of a solution. Water has a pH of 7 and, in general, acids have a pH less than 7 and alkaline solutions have a pH greater than 7. The model for pH is $pH = -\log[H^+]$ where $[H^+]$ is the hydrogen ion concentration in moles per liter of a solution.

a. Find the pH of a solution with a hydrogen ion concentration of 8.6×10^{-7}.

b. Find the hydrogen ion concentration $[H^+]$ of a solution if the pH of the solution is 4.5. Write the answer in scientific notation.

31. Sound levels: A decibel (abbreviated dB) is a unit used to measure the loudness of sound. The decibel level D of a sound of intensity I is measured by comparing it to a barely audible sound of intensity I_0 with the following formula:

$$D = 10 \log\left(\frac{I}{I_0}\right).$$

Sounds measuring over 85 dB are not considered safe.

a. Find the decibel level of a rock concert with an intensity of $6.24 \times 10^{11} I_0$.

b. What is the intensity level of 85 dB?

c. What is the intensity level of 60 dB (normal conversation)?

 HAWKES LEARNING SYSTEMS: INTERMEDIATE ALGEBRA SOFTWARE

▪ Applications: Exponential and Logarithmic Functions

Chapter 8 Index of Key Ideas and Terms

Section 8.1 Algebra of Functions

Algebraic Operations with Functions   page 639

If $f(x)$ and $g(x)$ represent two functions and x is a value
in the domain of both functions, then:

1. Sum of Two Functions: $(f+g)(x) = f(x) + g(x)$

2. Difference of Two Functions: $(f-g)(x) = f(x) - g(x)$

3. Product of Two Functions: $(f \cdot g)(x) = f(x) \cdot g(x)$

4. Quotient of Two Functions: $\left(\dfrac{f}{g}\right)(x) = \dfrac{f(x)}{g(x)}$ where $g(x) \neq 0$.

Graphing the Sum of Two Functions   pages 641-643

**Using a Graphing Calculator to Graph the Sum of
Two Functions**   pages 643-645

Section 8.2   Composition of Functions and Inverse Functions

Composite Functions   page 653

For two functions f and g, the **composite function** $f \circ g$ is
defined as follows: $(f \circ g)(x) = f\big(g(x)\big)$.
The domain of $f \circ g$ consists of those values of x in the
domain of g for which $g(x)$ is in the domain of f.

One-to-One Functions   page 655

A function is a **one-to-one function** (or **1–1 function**)
if for each value of y in the range there is only one
corresponding value of x in the domain.

Horizontal Line Test   page 655

A function is one-to-one if no **horizontal line** intersects
the graph of the function at more than one point.

Inverse Functions   page 658

If f is a 1–1 function with ordered pairs of the form (x, y),
then its **inverse function**, denoted as f^{-1}, is also a 1–1
function with ordered pairs of the form (y, x).

Section 8.2 Composition of Functions and Inverse Functions (cont.)

Inverse Functions page 659

If f and g are one-to-one functions and

$$f(g(x)) = x \qquad \text{for all } x \text{ in } D_g, \text{ and}$$
$$g(f(x)) = x \qquad \text{for all } x \text{ in } D_f$$

then f and g are **inverse functions**.
That is, $g = f^{-1}$ and $f = g^{-1}$.

To Find the Inverse of a 1–1 Function page 662

1. Let $y = f(x)$. (In effect, substitute y for $f(x)$.)
2. Interchange x and y.
3. In the new equation, solve for y in terms of x.
4. Substitute $f^{-1}(x)$ for y. (This new function is the inverse of f.)

Section 8.3 Exponential Functions

Exponential Functions page 669

An **exponential function** is a function of the form
$f(x) = b^x$ where $b > 0$, $b \neq 1$, and x is any real number.

Exponential Growth and Decay pages 670-673

1. If $b > 1$, then b^x increases to the right and is called an **exponential growth function**.
2. If $0 < b < 1$, then b^x decreases to the right and is called an **exponential decay function**.

Compound Interest page 674

Compound interest on a principal P invested at an annual interest rate r (in decimal form) for t years that is compounded n times per year can be calculated using the following formula:

$$A = P\left(1 + \frac{r}{n}\right)^{nt}$$

where A is the amount accumulated.

Section 8.3   Exponential Functions (cont.)

The Number e   page 677
The number e is defined to be
$$e = \lim_{n \to \infty} \left(1 + \frac{1}{n}\right)^n = 2.718281828459 \ldots$$

Continuously Compounded Interest   page 677
Continuously compounded interest on a principal P invested at an annual interest rate r for t years, can be calculated using the following formula:
$$A = Pe^{rt}$$
where A is the amount accumulated.

Section 8.4   Logarithmic Functions

Logarithm   page 684
For $b > 0$ and $b \neq 1$, $x = b^y$ is equivalent to $y = \log_b x$.
$y = \log_b x$ is read "y is the **logarithm** (base b) of x."

Basic Properties of Logarithms   page 685
For $b > 0$ and $b \neq 1$,
1. $\log_b 1 = 0$
2. $\log_b b = 1$
3. $x = b^{\log_b x}$, for $x > 0$
4. $\log_b b^x = x$

Graphs of Logarithmic Functions   page 687

Section 8.5 Properties of Logarithms

Properties of Logarithms   page 695

For $b > 0$, $b \neq 1$, $x, y > 0$, and any real number r:

1. $\log_b 1 = 0$
2. $\log_b b = 1$
3. $x = b^{\log_b x}$
4. $\log_b b^x = x$
5. $\log_b (xy) = \log_b x + \log_b y$ The Product Rule
6. $\log_b \dfrac{x}{y} = \log_b x - \log_b y$ The Quotient Rule
7. $\log_b x^r = r \cdot \log_b x$ The Power Rule

Section 8.6 Common Logarithms and Natural Logarithms

Common Logarithms (Base 10 Logarithms)   page 700
    Using a calculator to find common logarithms   page 701

Inverse Logarithms   page 702
    Using a calculator to find inverse logarithms   page 703

Natural Logarithms (Base e Logarithms)   page 704
    Using a calculator to find natural logarithms   page 705
    Using a calculator to find inverse natural logarithms   page 706

Section 8.7 Logarithmic and Exponential Equations and Change-of-Base

Properties of Equations with Exponents and Logarithms   page 709

For $b > 0$ and $b \neq 1$,

1. If $b^x = b^y$, then $x = y$.
2. If $x = y$, then $b^x = b^y$.
3. If $\log_b x = \log_b y$, then $x = y$ ($x > 0$ and $y > 0$).
4. If $x = y$, then $\log_b x = \log_b y$ ($x > 0$ and $y > 0$).

Change-of-Base   page 713

For $a, b, x > 0$ and $a, b \neq 1$, $\log_b x = \dfrac{\log_a x}{\log_a b}$.

Section 8.8   Applications

Applications

HAWKES LEARNING SYSTEMS: INTERMEDIATE ALGEBRA SOFTWARE

For a review of the topics and problems from Chapter 8, look at the following lessons from *Hawkes Learning Systems: Intermediate Algebra*

- Algebra of Functions
- Composition of Functions and Inverse Functions
- Exponential Functions and the Number *e*
- Logarithmic Functions
- Properties of Logarithmic Functions
- Common Logarithms and Natural Logarithms
- Exponential and Logarithmic Equations
- Applications: Exponential and Logarithmic Functions
- Chapter 8 Review and Test

Chapter 8 Review

8.1 Algebra of Functions

In Exercises 1 and 2, the graphs of two functions are given. Graph the sum of these two functions.

1. $f = \{(-2,3),(-1,4),(0,2),(1,0)\}$ 2. $f(x) = 2x - 1$

    $g = \{(-2,-1),(-1,2),(0,-3),(1,1)\}$ $g(x) = -x + 2$

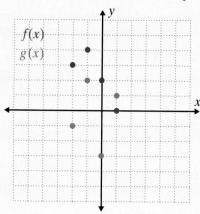   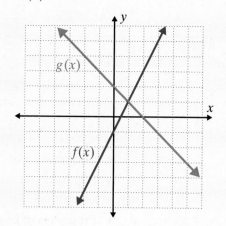

For Exercises 3 – 8, find ***a.*** $(f+g)(x)$ ***b.*** $(f-g)(x)$ ***c.*** $(f \cdot g)(x)$ ***d.*** $\left(\dfrac{f}{g}\right)(x)$.

3. $f(x) = 2x + 3,\ g(x) = x - 7$ 4. $f(x) = -5x,\ g(x) = x + 6$

5. $f(x) = 8x,\ g(x) = 3x + 5$ 6. $f(x) = x - 9,\ g(x) = -4x + 1$

7. $f(x) = x^2,\ g(x) = x^2 - 16$ 8. $f(x) = 2x^2 - x,\ g(x) = x^2 + 5$

In Exercises 9 – 12, let $f(x) = x^2 - 4$ *and* $g(x) = -x + 2$. *Find the value of the indicated expression.*

9. $f(3) + g(3)$ 10. $g(1) - f(1)$ 11. $(f \cdot g)(2)$ 12. $\left(\dfrac{f}{g}\right)(3)$

In Exercises 13 – 16, find the indicated function and state the domain, in interval notation, of each resulting function.

13. If $f(x) = \sqrt{x - 6}$ and $g(x) = x + 2$, find $(f+g)(x)$.

14. If $f(x) = x^2$ and $g(x) = x^2 - 3$, find $(f-g)(x)$.

15. For $f(x) = x - 3$ and $g(x) = \sqrt{x-3}$, find $\left(\dfrac{f}{g}\right)(x)$.

16. For $f(x) = \sqrt{x-1}$ and $g(x) = 3x + 2$, find $(f \cdot g)(x)$.

In Exercises 17 and 18, graph each pair of functions and the sum of these functions on the same set of axes.

17. $y = x + 3$ and $y = -2x + 3$

18. $y = x^2 - 4$ and $y = x + 4$

8.2   Composition of Functions and Inverse Functions

In Exercises 19 – 24, form the compositions $(f \circ g)(x)$ and $(g \circ f)(x)$.

19. $f(x) = 2x + 1$, $g(x) = 2x - 3$

20. $f(x) = x^2$, $g(x) = 3x - 1$

21. $f(x) = \sqrt{x}$, $g(x) = x^2 - 4$

22. $f(x) = \dfrac{1}{\sqrt{x}}$, $g(x) = x^2 - 9$

23. $f(x) = x^3$, $g(x) = \dfrac{1}{x}$

24. $f(x) = \dfrac{1}{x}$, $g(x) = \dfrac{1}{x^2}$

In Exercises 25 – 28, find the inverse of the given function. Graph both functions on the same set of axes and show the line $y = x$ as a dotted line on the graph.

25. $f(x) = 2x + 1$

26. $f(x) = -2x + 5$

27. $f(x) = \sqrt{x-6}, x \geq 6$

28. $f(x) = -x^2 + 2, x \geq 0$

Using the horizontal line test, determine which of the graphs in Exercises 29 – 32 are graphs of 1–1 functions. If the graph represents a 1–1 function, graph its inverse by reflecting the graph of the function across the line $y = x$.

29.

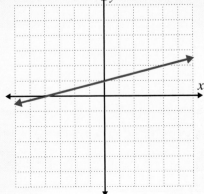

30.

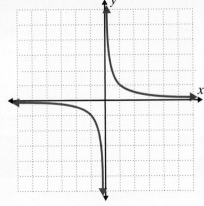

31.

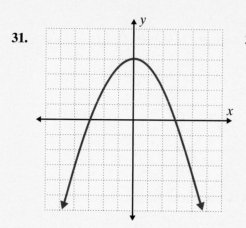

32.

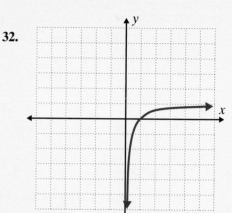

*For Exercises 33 – 36, find the inverse of the given function. Then use a graphing calculator to graph both the function and its inverse. Set the **WINDOW** so that it is square.*

33. $f(x) = x^3 - 1$

34. $f(x) = \dfrac{1}{x-2}$

35. $f(x) = \sqrt{x-5}, \, x \geq 5$

36. $f(x) = x^2, \, x \leq 0$

8.3 Exponential Functions

Sketch the graph of each of the exponential functions and label three points on each graph in Exercises 37 – 44.

37. $y = 3^x$

38. $y = \left(\dfrac{1}{4}\right)^x$

39. $y = 2^{-x}$

40. $y = -2^x$

41. $f(x) = 4 \cdot 2^x$

42. $g(x) = 2 \cdot 10^x$

43. $h(x) = \dfrac{1}{2} \cdot 3^x$

44. $y = \left(\dfrac{2}{5}\right)^x$

45. Use your calculator to find the value of $f(6)$ if $f(t) = 32.4 \cdot 2^{0.5t}$.

46. If $f(x) = \left(1 + \dfrac{1}{x}\right)^x$, use your calculator to find the value below rounded to five decimal places.

    a. $f(1000)$ **b.** $f(10,000)$ **c.** $f(1,000,000)$

47. **Savings accounts:** Find the amount, A, in a savings account of $5000 invested at 6% for 20 years if the interest is compounded:
    a. annually
    b. quarterly
    c. daily (use 360 days)
    d. continuously

48. **Sales revenue:** The revenue function is given by $R(x) = x \cdot p(x)$. If $p(x) = 25\left(2^{0.25x}\right)$ dollars, where x is the number of units sold, find the revenue if 50 units are sold.

49. **Automobile depreciation:** Cars depreciate in t years according to the formula $V = C(1-r)^t$ where V is the current value, C is the original cost, and r is the rate of depreciation. Find the current value of a car that depreciates at a rate of 15% over 5 years if its original cost was $40,000.

50. **Investing:** Suppose you have two accounts that are invested so that they compound continuously for 10 years. Each account started with $10,000. One was earning 4% interest and the other 8% interest. Find the value of each account after 10 years. Explain why the value of the account earning 4% interest is not simply one-half of the account earning 8% interest.

8.4   Logarithmic Functions

In Exercises 51 – 56, express each equation in logarithmic form.

51. $8^2 = 64$

52. $3^4 = 81$

53. $2^{-3} = \dfrac{1}{8}$

54. $3^{-2} = \dfrac{1}{9}$

55. $\left(\dfrac{1}{2}\right)^2 = \dfrac{1}{4}$

56. $\left(\dfrac{1}{5}\right)^3 = \dfrac{1}{125}$

In Exercises 57 – 62, express each equation in exponential form.

57. $\log_2 \dfrac{1}{2} = -1$

58. $\log_{10} \sqrt{10} = \dfrac{1}{2}$

59. $\log_5 625 = 4$

60. $\log_5 x = -2.5$

61. $\log_b x = y$

62. $\log_{\frac{1}{2}} 16 = -4$

Solve Exercises 63 – 68 by first changing each equation to exponential form.

63. $\log_8 x = \dfrac{1}{3}$

64. $\log_{25} 5 = x$

65. $\log_{10} 10^{2.35} = x$

66. $\log_2 2^{-5} = x$

67. $\log_4 4^{\log_2 64} = x$

68. $\log_3 3^{\log_3 27} = x$

In Exercises 69 – 72, graph each function and its inverse on the same set of axes. Label two points on each graph.

69. $f(x) = 5^x$

70. $f(x) = 2^{-x}$

71. $y = \log_6 x$

72. $g(x) = \log_{\frac{1}{4}} x$

8.5   Properties of Logarithms

In Exercises 73 – 82, use the properties of logarithms to expand the expressions as much as possible.

73. $\log_b \left(5x^4\right)$

74. $\log_b \left(xy\right)^{-2}$

75. $\log_2 \sqrt{7x}$

76. $\log_3 \left(xy^2\right)$

77. $\log_{10}\left(x^2 y^3\right)$ **78.** $\log_b\left(\dfrac{x}{y^4}\right)$ **79.** $\log_b\left(\dfrac{x^2}{y^3}\right)^5$ **80.** $\log_2\sqrt{\dfrac{5}{xy}}$

81. $\log_3\left(42 x^2 y^{-3}\right)$ **82.** $\log_5\sqrt[3]{\dfrac{x^2}{y^3}}$

In Exercises 83 – 90, use the properties of logarithms to write each expression as a single logarithm.

83. $2\log_2 3 + \log_2 x$ **84.** $5\log_3 x - \log_3 y$

85. $3\log_b x - 4\log_b y$ **86.** $\log_4\left(4x+1\right) + 2\log_4 x$

87. $\log_{10}\left(2x+1\right) + \log_{10}\left(x-5\right)$ **88.** $\log_{10}\left(x-2\right) + \log_{10}\left(x+2\right)$

89. $\log_b\left(x^2 + 2x - 3\right) - \log_b\left(x+5\right)$ **90.** $\log_2\left(x-3\right) - \log_2\left(2x^2 - 5x - 3\right) + \dfrac{1}{2}\log_2 x$

8.6   Common Logarithms and Natural Logarithms

Express each equation in logarithmic form in Exercises 91 – 96.

91. $10^0 = 1$ **92.** $10^{-5} = 0.00001$ **93.** $e^{1.2} = x$

94. $e^k = 10$ **95.** $e^x = 12$ **96.** $10^k = 8.4$

Express each equation in exponential form in Exercises 97 – 102.

97. $\ln 3 = x$ **98.** $\ln x = 5$ **99.** $\ln x = -1$

100. $\log 16 = x$ **101.** $\log x = 2.5$ **102.** $\log 100,000 = 5$

Use a calculator to evaluate the logarithms in Exercises 103 – 106 accurate to five decimal places.

103. $\log 275$ **104.** $\log 84$ **105.** $\ln(-3)$ **106.** $\ln 0.012$

Use a calculator to find the value of x in each equation in Exercises 107 – 110 accurate to five decimal places.

107. $\log x = 3.12$ **108.** $5\log x = -4$ **109.** $\ln x = -4.5$ **110.** $0.3\ln x = 66$

8.7   Logarithmic and Exponential Equations and Change-of-Base

Use the properties of exponents and logarithms to solve each of the equations in Exercises 111 – 124.

111. $\ln x^2 = \log_3 9$ **112.** $10^{2x-1} = 100^2$ **113.** $25^{x^2-3} = 5^x$ **114.** $2\log x - 6 = 0$

115. $\ln x + \ln x^2 = 3$ **116.** $e^{x+5} = 36$ **117.** $\ln x^2 + 5 = 0$

118. $3^{x+1} = 10$ **119.** $\log \sqrt{x^2 - 25} = 2$ **120.** $2\ln x + \ln 4 = \ln(4x - 1)$

121. $\log(x - 3) - 2 = \log(x + 2)$ **122.** $\log(x + 11) + \log x = \log 12$

123. $\log(x + 3) + \log(x - 3) = \log 27$ **124.** $\log(x^2 - 16) - \log(x - 4) = -1$

Use the change-of-base formula to evaluate each of the expressions or solve the equations in Exercises 125 – 128.

125. $\log_2 12$ **126.** $\log_4 100$ **127.** $2^x = 10$ **128.** $4^{x+2} = 400$

8.8   Applications

129. Investing: What will be the value of an investment of \$5000 after 20 years if it is invested at 5% and
  a. compounded quarterly (4 times a year)?
  b. compounded monthly (12 times a year)?
  c. compounded continuously?

130. Investing: How long does it take for \$10,000 to double when invested at 6% compounded continuously? What would be the doubling time of \$5000?

131. Half-life of strontium-90: Strontium-90 is an isotope of strontium present in radioactive fallout. Find the half-life of strontium-90 using the formula $y = y_0 e^{-0.0239t}$, where t is the time in years.

132. Computers: The value of a computer depreciates at 20% per year. After 3 years, what is the value of a computer that originally cost \$2000? (Use the formula $V = C(1 - r)^t$.)

133. Cancer cell growth: Certain cancer cells grow according to the formula $y = y_0 e^{0.5t}$ where y_0 is the initial number present and t is measured in days. If $y_0 = 200$, in how many days will the cells number 200,000?

134. Concentration of a drug: The concentration of a drug in the bloodstream is given by $C = 40e^{-0.6t}$ where 40 mg was the initial dose of the drug and t is the time in hours.
  a. How much of the drug will be left in the bloodstream after 5 hours?
  b. What is the half-life of this drug?

135. **Change in temperature:** Suppose that the room temperature is 72° and a cup of coffee with temperature 120° is placed in the room. In 6 minutes the temperature of the coffee is 100°. How long will it take for the coffee to cool to 90°? (Use Newton's law of cooling: $T = Ae^{-kt} + C$. C is the constant temperature of the surrounding medium. The values of A and k depend on the particular object that is cooling. See Section 8.8, Example 5.)

136. **Hydrogen ion concentration:** In chemistry, pH is a measure of the hydrogen ion concentration $[H^+]$ of a solution. The pH of a solution is given by the formula $pH = -\log[H^+]$. Find the hydrogen ion concentration of a solution with pH 6.3.

137. **Sound intensity:** The decibel level D of a sound of intensity I is measured by comparing it to a barely audible sound of intensity I_0 with the following formula:

$$D = 10\log\left(\frac{I}{I_0}\right).$$

   a. Find the decibel level of a lawn mower with intensity of $4.32 \times 10^9 I_0$.
   b. What is the intensity level of 140 dB (a loud gun)?

138. **The Richter scale:** The relationship between the magnitude R and intensity I on the Richter scale is given by the formula $R = \log I$. How much stronger was the 1992 earthquake in Kobe, Japan (7.2 on the Richter scale) than the 1971 earthquake in Los Angeles (6.6 on the Richter scale)?

Chapter 8 Test

*In Exercises 1 and 2, the graphs of two functions are given. Graph **a.** the sum (f + g) of these two functions and **b.** the difference (f − g) of these two functions on different graphs.*

1. $f = \{(-2,-2),(0,0),(1,1),(2,2)\}$

$g = \{(-2,-5),(0,-1),(1,-2),(2,-5)\}$

2. $f(x) = -2x$

$g(x) = (x+1)^2 - 3$

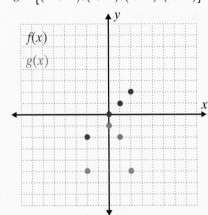

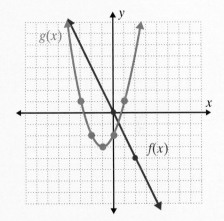

3. Given the two functions, $f(x) = \sqrt{x-3}$ and $g(x) = x^2 + 1$, find:

a. $(f+g)(x)$

b. $(f-g)(x)$

c. $(f \cdot g)(x)$

d. $\left(\dfrac{f}{g}\right)(x)$

4. If $f(x) = 2x - 5$ and $g(x) = 3 - 2x^2$, form the following compositions.

a. $f(g(x))$

b. $g(f(x))$

5. Use the horizontal line test to determine whether the graph on the right is a 1–1 function. If the graph represents a 1–1 function, graph its inverse by reflecting the graph of the function across the line $y = x$.

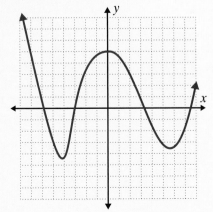

6. Determine, algebraically, whether or not each pair of functions are inverses of each other. Graph both functions on the same set of axes and show the line $y = x$ as a dotted line on each graph.

   a. $f(x) = x^2$ and $g(x) = -x^2$

   b. $f(x) = \dfrac{1}{x}$ and $g(x) = \dfrac{1}{x}$

7. Find $f^{-1}(x)$ if $f(x) = \dfrac{1}{x+1} - 3$. Graph both functions on the same set of axes and show the line $y = x$ as a dotted line on the graph.

8. Sketch the graphs of the exponential functions given in the following equations. Label three points on each graph.

   a. $y = -3^x$

   b. $y = 5^{x-2} - 1$

9. Express the following equations in logarithmic form.

   a. $10^5 = 100,000$

   b. $\left(\dfrac{1}{2}\right)^{-3} = 8$

   c. $e^x = 15$

10. Express the following equations in exponential form.

    a. $\log x = 4$

    b. $\log_3 \dfrac{1}{9} = -2$

    c. $\log_4 256 = x$

11. Solve the following equations by first changing them to exponential form.

    a. $\log_7 x = 3$

    b. $\log_9 27 = x$

12. Find the inverse of the function $y = \left(\dfrac{1}{2}\right)^x$. Graph both the function and its inverse on the same set of axes. Label two points on each graph.

13. Use the properties of logarithms to expand the following expressions as much as possible.

    a. $\ln(x^2 - 25)$

    b. $\log \sqrt[3]{\dfrac{x^2}{y}}$

14. Use the properties of logarithms to write each expression as a logarithm of a single expression.

    a. $\ln(x+5) + \ln(x-4)$

    b. $\log \sqrt{x} + \log x^2 - \log(5x)$

15. Use a calculator to find the value of x accurate to 5 decimal places.

    a. $x = \log 579$

    b. $5 \ln x = 9.35$

Use the properties of exponents and logarithms to solve each of the equations in Exercises 16 – 22 for x.

16. $7^3 \cdot 7^x = 7^{-1}$ **17.** $6^{x-1} = 36^{x+1}$

18. $10^{x+2} = 283$ **19.** $2e^{0.24x} = 26$ **20.** $4^x = 12$

21. $\log(2x + 3) - \log(x + 1) = 0$ **22.** $\ln(x^2 + 3x - 4) - \ln(x + 4) = 3$

23. Use the change-of-base formula to evaluate the following expression: $\log_6 25$.

Solve the following application problems. If necessary, round answers to two decimal places.

24. Bacteria growth: A scientist knows that a certain strain of bacteria grows according to the function $y = y_0 \cdot 3^{0.25t}$, where t is a measurement in hours. If she starts a culture with 5000 bacteria, how many will be present after 6 hours?

25. Investing: If $1000 is invested at 7% compounded continuously, when will the amount reach $3800?

26. Decomposition: A substance decomposes according to $A = A_0 e^{-0.0035t}$, where t is measured in years and A_0 is the initial amount.
 a. How long will it take for 800 grams to decompose to 500 grams?
 b. What is the half-life of this substance?

27. The Richter scale: The relationship between the magnitude R and intensity I on the Richter scale is given by the formula $R = \log I$. How much stronger was the 1985 earthquake in Mexico City (8.1 on the Richter scale) than the 2008 earthquake in the Sichuan province of China (7.8 on the Richter scale)?

28. Blacksmithing: The temperature of a piece of iron is 800°F when it is removed from a furnace. The temperature in the room is 80°F. It takes 20 minutes for the iron to cool to 600°F. How long will it take for the iron to cool to 200°F?

Cumulative Review: Chapters 1–8

Find the value of each expression in Exercises 1 and 2 by using the rules for order of operations.

1. $8 + \left[5 \cdot 6 - \left(9 \div 3 + 3 \right) \right]$

2. $2^3 \div 4 + 7 - 10 \div 5$

3. Find an equation in slope-intercept form for the line passing through the point $(-3, 4)$ with slope $m = -\dfrac{2}{3}$. Graph the line.

In Exercises 4 and 5, write the function as a set of ordered pairs for the given equation and domain.

4. $y = x^2 - 2x + 5$

$D = \{-2, -1, 0, 1, 2\}$

5. $y = x^3 - 5x^2$

$D = \{-2, -1, 0, 1, 2\}$

6. The graphs of three curves are shown. Use the vertical line test to determine whether or not each graph represents a function. Then state the domain and range of each graph.

a.
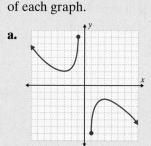

b.
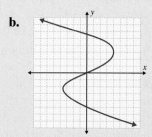

c.
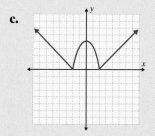

*For the systems of equations in Exercises 7 and 8: **a.** Determine whether each pair of lines is parallel, perpendicular, or neither. **b.** Solve the system of equations. **c.** Graph both lines.*

7. $\begin{cases} y = 5x - 6 \\ y = 5x + 1 \end{cases}$

8. $\begin{cases} y = -2x + 5 \\ y = \dfrac{1}{2}x \end{cases}$

9. Use Cramer's Rule to solve the system of linear equations: $\begin{cases} 2x - y = -1 \\ x + 2y = 12 \end{cases}$.

10. Use the Gaussian elimination method to solve the system of linear equations:
$$\begin{cases} x + y + z = 2 \\ 2x + y - z = -1 \\ x + 3y + 2z = 2 \end{cases}.$$

Perform the indicated operations in Exercises 11 – 14 and simplify the results.

11. $\dfrac{2x^2+7x+3}{x^2-3x-18} \cdot \dfrac{x^2-x-30}{2x^2+11x+5}$

12. $\dfrac{9-x^2}{x^2+7x+6} \div \dfrac{x-3}{x+6}$

13. $\dfrac{1}{x-1} + \dfrac{x-6}{x^2+3x-4}$

14. $\dfrac{2x}{2x+3} - \dfrac{7x+12}{2x^2+5x+3}$

Simplify Exercises 15 – 18. Assume all variables are positive.

15. $\sqrt{80x^2y^3}$

16. $\dfrac{\sqrt{8x^3y^2}}{\sqrt{5xy^3}}$

17. $\dfrac{x^{\frac{2}{3}}y^{\frac{1}{3}}}{x^{\frac{1}{2}}y^{\frac{2}{3}}}$

18. $\left(x^2y^{-1}\right)^{\frac{1}{2}}\left(4xy^3\right)^{-\frac{1}{2}}$

Change each expression in Exercises 19 and 20 to an equivalent exponential expression. Assume variables are positive.

19. $\sqrt{x^4y^3}$

20. $\sqrt[3]{x^2y^3}$

In Exercises 21 – 23, perform the indicated operations and simplify the results.

21. $\left(6-\sqrt{-4}\right)+\left(9+\sqrt{-64}\right)$ **22.** $\left(\sqrt{6}-\sqrt{2}i\right)\left(\sqrt{6}+\sqrt{2}i\right)$ **23.** i^{17}

24. Use the properties of logarithms to expand the expression $\log_b\left(x^3\sqrt{y}\right)$ as much as possible.

25. Use the properties of logarithms to write the expression $2\log_b x - \log_b y + \log_b 7$ as a single logarithm.

Solve each of the equations in Exercises 26 – 32.

26. $x^2+5x-2=0$ **27.** $4x^2-28x+49=0$ **28.** $x^4-13x^2+36=0$

29. $x-2=\sqrt{x+10}$ **30.** $\dfrac{2}{x-2}+\dfrac{3}{x-1}=1$

31. $6^{3x+5}=55$ **32.** $\ln\left(x^2-7x+10\right)-\ln\left(x-2\right)=2.5$

33. Solve the equation $\log_6 \dfrac{1}{216}=x$ by first changing it to exponential form.

In Exercises 34 – 37, solve the inequalities algebraically and graph each solution set on a real number line. Write the answers in interval notation.

34. $(x-3)(x+1) < 0$

35. $(x+2)(2x-5) > 0$

36. $2x^2 - x \geq 3$

37. $\left| \dfrac{1}{2}x + 3 \right| < \dfrac{3}{2}$

In Exercises 38 – 41, rewrite each of the quadratic functions in the form $y = a(x-h)^2 + k$ and find the vertex, range, and zeros of each function. Graph the function.

38. $y = x^2 - 6x - 2$

39. $y = 2x^2 + 8x + 3$

40. $y = x^2 - 8x + 16$

41. $y = x^2 - 2x - 6$

42. Given the two functions, $f(x) = x^2 + x$ and $g(x) = \sqrt{x+2}$, find:

    a. $(f+g)(x)$ **b.** $(f-g)(x)$ **c.** $(f \cdot g)(x)$ **d.** $\left(\dfrac{f}{g} \right)(x)$

43. Given the two functions, $f(x) = x^2 + 3$ for $x \geq 0$ and $g(x) = 2x + 1$ for $x \geq 0$, find:

    a. $f(g(x))$ **b.** $g(f(x))$

Which of the functions in Exercises 44 and 45 are 1–1 functions? If the graph represents a 1–1 function, graph its inverse by reflecting the graph of the function across the line $y = x$.

44.
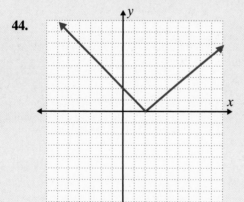

45.
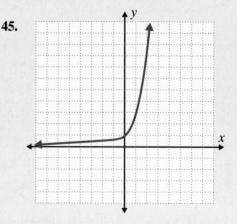

In Exercises 46 and 47, find the inverse of each function and graph both the function and its inverse on the same set of axes. Include the graph of the line $y = x$ as a dotted line in each graph.

46. $f(x) = x^2 - 4$ for $x \geq 0$

47. $g(x) = (x-2)^3$

48. The table below shows the number of millions of US households with cable television from 1985-2005.

Year	Households with Cable
1985	36.3
1990	51.9
1995	60.5
2000	68.6
2005	73.9

a. Plot these points on a graph.
b. Connect the points with line segments.
c. Find the slope of each line segment.
d. Interpret the slope as a rate of change.

Source: Nielsen

49. Publishing books: A book is available in both hardback and paperback. A bookstore sold a total of 43 books during the week. The total receipts were $297.50. If hardback books sell for $12.50 and paperbacks sell for $4.50, how many of each were sold?

50. Kayaking: Nadine kayaked 8 miles upstream and then returned. Her average speed on the return trip was 6 mph faster than her speed upstream. If her total travel time was 2.8 hours, find her rate each way.

51. Proofreading: Caleb can proofread a news article in 2 hours working alone. His coworker Tiffany can proofread the same article in $1\frac{1}{2}$ hours working alone. How long will it take if they work together?

52. Computer Screen: In order to display images properly, the width of a computer screen must be $\frac{4}{3}$ times the height of the screen. If a manufacturer wants the length of the diagonal of the screen to be 20 inches, find the width and height of the screen.

53. Light bulbs: Studies show that the fractional part, P, of light bulbs that have burned out after t hours of use is given by $P = 1 - 2^{-0.03t}$. What fractional part of the light bulbs have burned out after 100 hours of use?

54. Investing: If P dollars are invested at a rate, r (expressed as a decimal), and compounded k times a year, the amount, A, due at the end of t years is given by $A = P\left(1 + \frac{r}{k}\right)^{kt}$ dollars.

Find A if $500 is invested at 10% compounded quarterly for 2 years.

55. Decomposition of radium: Radium decomposes according to $A = A_0 e^{-0.04t}$, where t is measured in centuries and A_0 is the initial amount. Determine the half-life for radium.

56. Friction: When friction is used to stop the motion of a wheel, the velocity may be given by $V = V_0 e^{-0.35t}$, where V_0 is the initial velocity and t is the number of seconds the friction has been applied. How long will it take to slow a wheel from 75 ft/s to 15 ft/s?

57. Selling light fixtures: An electronics store sells a special type of light fixture. The owner estimates that by selling these fixtures for x dollars each, the sales will be $1200 - x$ each month.

   a. What price will yield the maximum monthly revenue?

   b. What will be the maximum monthly revenue?

58. Throwing a ball: A ball is projected vertically upward from the ground with an initial velocity of 144 ft/s. Use the formula $h = -16t^2 + v_0 t + h_0$ to answer the following questions.

   a. When will the ball reach its maximum height?

   b. What will be the maximum height?

   c. When will the ball be on the ground?

Use a calculator to evaluate the logarithms in Exercises 59 and 60 accurate to 5 decimal places.

59. $\log 54.6$

60. $\ln 10000$

In Exercises 61 – 63, use a graphing calculator to solve the following inequalities. Graph each solution set on a real number line and write the answers in interval notation. (Estimate endpoints when necessary.)

61. $x^2 > 2$

62. $x^2 + 4x > 0$

63. $x^2 + 5x - 6 \leq 0$

Conic Sections

Did You Know?

Euler

The mathematician who invented the notation for functions, $f(x)$, was Leonhard Euler (1707 – 1783) of Switzerland. Euler was one of the most prolific mathematical researchers of all time, and he lived during a period in which mathematics was making great progress. He studied mathematics, theology, medicine, astronomy, physics, and oriental languages before he began a career as a court philosopher-mathematician. His professional life was spent at St. Petersburg Academy by invitation of Catherine I of Russia, at the Berlin Academy under Frederick the Great of Prussia, and again at the St. Petersburg Academy under Catherine the Great. The collected works of Euler fill 80 volumes, and for almost 50 years after Euler's death the publications of the St. Petersburg Academy continued to include articles by him.

Euler was blind the last 17 years of his life, but he continued his mathematical research by writing on a large slate and dictating to a secretary. He was responsible for the conventionalization of many mathematical symbols such as $f(x)$ for function notation, i for $\sqrt{-1}$, e for the base of the natural logarithms, π for the ratio of circumference to diameter of a circle, and Σ for the summation symbol.

From the age of 20 to his death, Euler was busy adding to knowledge in every branch of mathematics. He wrote with modern symbolism, and his work in calculus was particularly outstanding. Euler had a rich family life, having fathered 13 children, and he not only contributed to mathematics, but reformed the Russian system of weights and measures, supervised the government pension system in Prussia and the government geographic office in Russia, designed canals, and worked in many areas of physics, including acoustics and optics. It was said of Euler, by the French academician François Arago, that he could calculate without any apparent effort "just as men breathe and eagles sustain themselves in the air."

An interesting story is told about Euler's meeting with the French philosopher Diderot at the Russian court. Diderot had angered the czarina by his antireligious views, and Euler was called to the court to debate Diderot. Diderot was told that the great mathematician Euler had an algebraic proof that God existed. Euler walked in towards Diderot and said, "Monsieur, $\dfrac{a+b^n}{n} = x$, therefore God exists, respond." Diderot, who had no understanding of algebra, was unable to respond.

"I have not hesitated in 1900, at the Congress of Mathematicians in Paris, to call the nineteenth century the century of the theory of functions."

Vito Volterra (1860 – 1940)

CHAPTER 9

A circular cone is a three-dimensional figure that can be generated by choosing one point on a line and rotating the line in a circular fashion about this point. (See Figure 9.7 on page 761.) [You can visualize a cone by holding a meter stick (or a yard stick) with two fingers and rotating the stick so that each end moves in a circle. Note that the cone has a top portion and a bottom portion.] If a plane intersects a cone, the intersections will be curves (or a single point) on the plane. These curves are called conic sections. The four conic sections we will discuss in this chapter are circles, ellipses, parabolas, and hyperbolas.

By setting a Cartesian coordinate system on the plane, the conic sections and their properties can be discussed algebraically. The study of geometry with algebraic equations is called analytic geometry. If you continue your studies in mathematics, you will find that analytic geometry can be applied in three dimensions as well as in two dimensions. Related figures would be spheres, ellipsoids (football shapes), and paraboloids (the reflective surfaces of telescopes).

9.1 Translations

- *Calculate and understand the difference quotient.*
- *Understand the concepts of horizontal and vertical translations.*
- *Graph translations and reflections of functions. That is, given the graph of a function y = f(x), graph translations and reflections of the form $y = \pm f(x - h) + k$.*

The Difference Quotient: $\dfrac{f(x + h) - f(x)}{h}$

We have already used function notation $f(x)$ (read "f of x") in Chapter 4 to represent and to evaluate polynomials, in Chapter 6 to represent and evaluate radical functions, and in Chapter 8 in dealing with inverse functions. In this section, we show how to use function notation to indicate a new value or a new function when the single variable x is replaced by an expression such as $(a + 1)$ or $(x + h)$. For example,

if $f(x) = 3x + 5$, then

$f(2) = 3(2) + 5 = 11$	x is replaced by 2.
$f(a) = 3(a) + 5$	x is replaced by a.
$f(a + 1) = 3(a + 1) + 5$	x is replaced by $a + 1$.
$f(x + h) = 3(x + h) + 5$	x is replaced by $x + h$.

We will see how these substitutions for x relate to formulas and to graphs of functions.

Example 1: Using $f(x)$ Notation

For the function $f(x) = 2x^2 - 4$, find:

a. $f(3)$

Solution: $f(3) = 2(3)^2 - 4 = 2 \cdot 9 - 4 = 14$

b. $f(a)$

Solution: $f(a) = 2(a)^2 - 4 = 2a^2 - 4$

c. $f(a + 1)$

Solution: Replace x with $(a + 1)$ and simplify.

$$f(a+1) = 2(a+1)^2 - 4$$
$$= 2(a^2 + 2a + 1) - 4$$
$$= 2a^2 + 4a + 2 - 4$$
$$= 2a^2 + 4a - 2$$

In higher level mathematics, we are interested in the formula

$$\frac{f(x+h) - f(x)}{h}$$

called the **difference quotient**. As shown in Figure 9.1, a geometric interpretation of the difference quotient is the **slope** of a line through two points on the graph of a function.

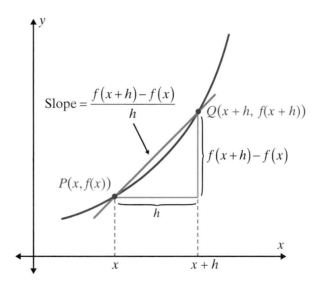

Figure 9.1

NOTES

The line illustrated in Figure 9.1 (through two points on the graph) is called a **secant line**. The secant line is used in calculus to help in understanding a new function called a **derivative**.

NOTES

Note carefully that $f(x + h)$ is not the same as $f(x) + h$. For example,

if

$$f(x) = x^2 + 2x + 3,$$

then

$$f(x+h) = (x+h)^2 + 2(x+h) + 3 = x^2 + 2xh + h^2 + 2x + 2h + 3$$

and

$$f(x) + h = x^2 + 2x + 3 + h.$$

Example 2: Finding the Difference Quotient $\dfrac{f(x+h) - f(x)}{h}$ for a Given Function

a.   Find the difference quotient for the function $f(x) = 2 - 6x$.

Solution: $f(x + h) = 2 - 6(x + h)$ and $f(x) = 2 - 6x$

Substituting gives:

$$\frac{f(x+h) - f(x)}{h} = \frac{[2 - 6(x+h)] - [2 - 6x]}{h}$$

$$= \frac{2 - 6x - 6h - 2 + 6x}{h}$$

$$= \frac{-6h}{h} = -6.$$

b.   Find the difference quotient for the function $f(x) = 2x^2 - 5x$.

Solution: $f(x+h) = 2(x+h)^2 - 5(x+h)$ and $f(x) = 2x^2 - 5x$

Substituting gives:

$$\frac{f(x+h) - f(x)}{h} = \frac{[2(x+h)^2 - 5(x+h)] - [2x^2 - 5x]}{h}$$

$$= \frac{2x^2 + 4xh + 2h^2 - 5x - 5h - 2x^2 + 5x}{h} \quad \text{Expand } (x+h)^2$$
$$\text{and multiply by 2.}$$

$$= \frac{4xh + 2h^2 - 5h}{h}$$

$$= \frac{h(4x + 2h - 5)}{h} \quad\quad\quad\quad\quad \text{Factor out } h.$$

$$= 4x + 2h - 5.$$

Horizontal and Vertical Translations

In Section 7.5, we discussed the graphs of parabolas of the form $y = ax^2$. We found that the graph of $y = a(x-h)^2$ is a **horizontal translation** (**horizontal shift**) of h units and the graph of $y = ax^2 + k$ is a **vertical translation** (**vertical shift**) of k units of the graph of $y = ax^2$.

In function notation,

if $f(x) = 2x^2$,

then $f(x-3) = 2(x-3)^2$ is a horizontal translation of 3 units to the right,

and $f(x) - 4 = 2x^2 - 4$ is a vertical translation of 4 units down.

Figure 9.2 shows the graphs of the three functions and their relationships. Note that the curves themselves are identical, only their positions are changed.

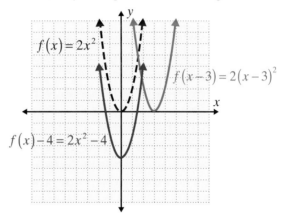

Figure 9.2

The following general approach can be used to help graph a horizontal and/or vertical translation of any function.

Horizontal and Vertical Translations

Given the graph of $y = f(x)$, the graph of $y = f(x-h) + k$ is

   1.   *a horizontal translation of h units and*

   2.   *a vertical translation of k units of the graph of $y = f(x)$.*

Think of the origin (0, 0) being moved to the point (h, k).

Then draw the graph of $y = f(x)$ in relation to (h, k) as if (h, k) were the origin, (0, 0).

This new graph will be the graph of $y = f(x-h) + k$.

A good example to use in illustrating translations is the function $f(x) = |x|$. First, we need to know what the graph of $f(x) = |x|$ or $y = |x|$ looks like. The definition of $|x|$ gives

$$f(x) = |x| = \begin{cases} x & \text{if } x \geq 0 \\ -x & \text{if } x < 0 \end{cases}.$$

The graph can be analyzed in two pieces.

First Piece:
The graph of $f(x) = x$ is a straight line, as shown in Figure 9.3a, but we want only the part where $x \geq 0$, as shown in Figure 9.3b.

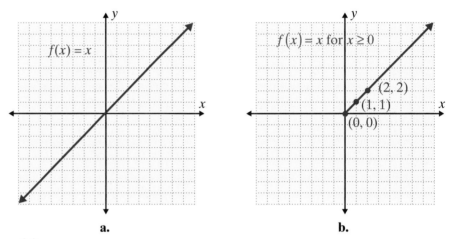

a.   **b.**

Figure 9.3

Second Piece:
The graph of $f(x) = -x$ is also a straight line, as shown in Figure 9.4a, but we want only the part where $x < 0$, as shown in Figure 9.4b.

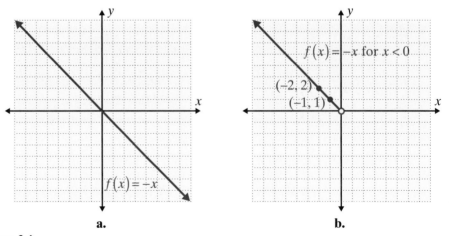

a.   **b.**

Figure 9.4

The two graphs in Figures 9.3b and 9.4b together give the graph of $f(x) = |x|$, as shown in Figure 9.5.

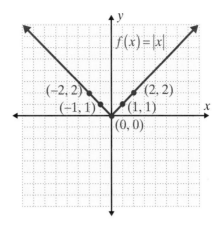

Figure 9.5

Now that we know what the graph of the function $f(x) = |x|$ or $y = |x|$ looks like, we can examine related horizontal and vertical shifts as in Example 3.

Example 3: Horizontal and Vertical Translations of the Function $y = |x|$

Graph each of the following functions. Use the graph in Figure 9.5 as a reference.

a. $y = |x - 3| + 2$

Solution: Here $(h, k) = (3, 2)$, so there is a horizontal translation of 3 units right and 2 units up. In effect, $(3, 2)$ is now the vertex of the new graph just as $(0, 0)$ is the vertex of the original graph. You should check that the points shown on the graph here do indeed satisfy the function.

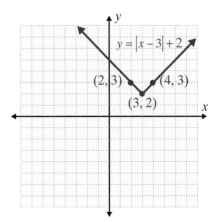

Continued on the next page...

b. $y = |x + 4| - 1$

Solution: Here $(h, k) = (-4, -1)$, so the horizontal translation is -4 (4 units left) and the vertical translation is -1 (1 unit down). The effect is that the vertex is now at the point $(-4, -1)$.

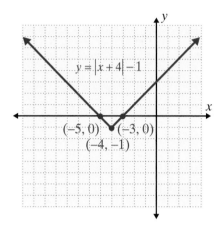

c. $y = |x + 2| + 7$

Solution: Here $(h, k) = (-2, 7)$, so the horizontal translation is -2 (2 units left) and the vertical translation is 7 (7 units up). The effect is that the vertex is now at the point $(-2, 7)$.

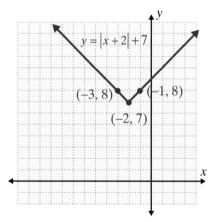

Reflections and Translations

In our discussion of the graphs of $y = ax^2$ and $y = |x|$, the leading coefficient, a, has not been changed. Changing the coefficient, a, can have an effect on the shape and direction of the basic function. For example, if a changes from positive to negative, the corresponding graph is reflected across the x-axis. In general, the graph of $y = -f(x)$ is a **reflection across the x-axis** of the graph of $y = f(x)$.

If we examine the function $y = |x|$ and the graph of the function $y = -|x|$, we see that the graph of each function is the mirror image of the other across the x-axis. The first "opens" upward and the second "opens" downward as illustrated in Figure 9.6. Both have the same vertex at $(0, 0)$.

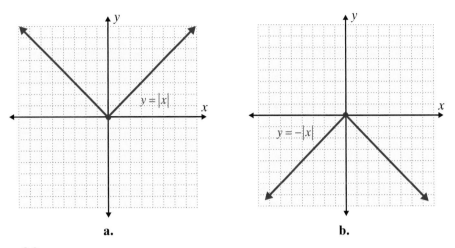

a.   **b.**

Figure 9.6

Example 4: Reflections and Translations of the Function $y = |x|$

Graph the function $y = -|x + 2| + 5$.

Solution: **The reflection is performed first, followed by the translations.**
Here $(h, k) = (-2, 5)$, and the graph is reflected across the x-axis. We show step-by-step how to "arrive" at the graph. (You should do these steps mentally and graph only the last step.)

Step 1:
Graph the reflection $y = -|x|$.

Step 2:
Translate the graph 2 units to the left.

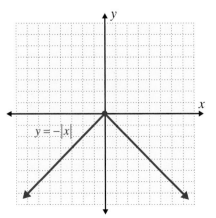

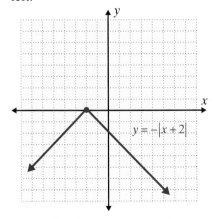

Continued on the next page...

Step 3:

Translate the graph 5 units up.

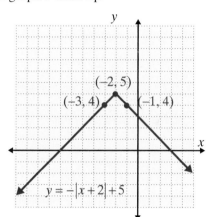

In a translation (horizontal or vertical) the shape of the graph of a function is unchanged. However, translations do change the position of the graph in the coordinate system. Example 5 illustrates two such cases.

Example 5: Translations of Functions with Graphs Given

a. The graph of $y = \sqrt{x}$ is given. Graph the function $y = \sqrt{x-2} + 1$.

Solution: If $y = \sqrt{x}$ is written $y = f(x)$, then $y = \sqrt{x-2} + 1$ is the same as $y = f(x-2) + 1$. So $(h, k) = (2, 1)$, and there is a horizontal translation of 2 units to the right and a vertical translation of 1 unit up.

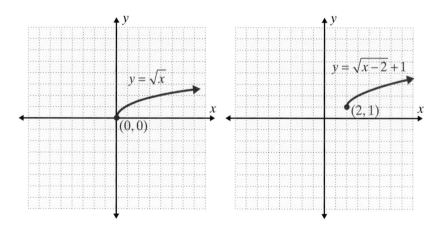

b.   The graph of $y = f(x)$ is given. Graph the function $y = f(x-3)-2$.

> **Solution:** Here $(h, k) = (3, -2)$, so translate the graph horizontally 3 units and vertically -2 units. (Add 3 to each x-value and -2 to each y-value.)

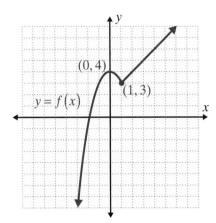
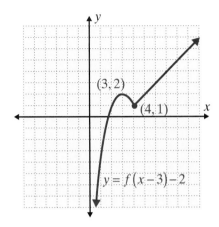

Practice Problems

1. For $f(x) = x^2 - 5$, find:

   a. $f(0)$ **b.** $f(a)$ **c.** $f(a + 2)$

2. If $g(x) = 3x + 7$, find:

   a. $g(0)$ **b.** $g(x + h)$ **c.** $\dfrac{g(x+h)-g(x)}{h}$

   (When evaluating the difference quotient, remember that $g(x+h) \neq g(x)+g(h)$.)

9.1 Exercises

1. For $f(x) = x^2 - 4$, find:

   a. $f(-2)$

   b. $f(a - 3)$

   c. $f(x + h)$

   d. $\dfrac{f(x+h)-f(x)}{h}$

2. For $g(x) = 2 - x^2$, find:

   a. $g(\sqrt{2})$

   b. $g(a - 1)$

   c. $g(x + h)$

   d. $\dfrac{g(x+h)-g(x)}{h}$

Answers to Practice Problems: **1. a.** $f(0) = -5$ **b.** $f(a) = a^2 - 5$ **c.** $f(a+2) = a^2 + 4a - 1$

                      2. a. $g(0) = 7$ **b.** $g(x + h) = 3x + 3h + 7$ **c.** $\dfrac{g(x+h)-g(x)}{h} = 3$

3. For $f(x) = 2x^2 - 3$, find:
   a. $f(0)$
   b. $f(a-2)$
   c. $f(x+h)$
   d. $\dfrac{f(x+h) - f(x)}{h}$

4. Let $f(x) = 3x^2 - x$, find:
   a. $f(4)$
   b. $f(a+2)$
   c. $f(x+h)$
   d. $\dfrac{f(x+h) - f(x)}{h}$

In Exercises 5 – 8, find and simplify the difference quotient, $\dfrac{f(x+h) - f(x)}{h}$, *for each function:*

5. $f(x) = x + 7$

6. $f(x) = 2x - 3$

7. $f(x) = 5 - 2x$

8. $f(x) = 4x - 3$

9. What particular information, if any, do you notice about the results in Exercises 5 – 8?

10. Analyze, in your own words, how the results in Exercise 9 relate to the graphs of the functions in relation to the secant line discussion in the text.

The graph of $y = |x|$ *is given along with a few points as aids. Graph the functions in Exercises 11 – 20 using your understanding of reflections and translations with no additional computations.*

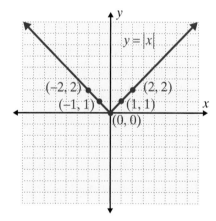

11. $y = |x - 1| - 2$

12. $y = |x - 2| + 6$

13. $y = -|x + 3|$

14. $y = -|x - 4|$

15. $y = -|x + 5| + 4$

16. $y = \left| x + \dfrac{3}{4} \right| - 3$

17. $y = |x - 3| + 5$

18. $y = |x + 2| - 3$

19. $y = \left| x + \dfrac{1}{2} \right| - \dfrac{3}{2}$

20. $y = \left| x - \dfrac{2}{3} \right|$

The graph of $y = \sqrt{x}$ is given along with a few points as aids. Graph the functions in Exercises 21 – 30 using your understanding of reflections and translations with no additional computations.

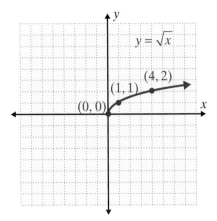

21. $y = \sqrt{x} - 2$ **22.** $y = \sqrt{x} + 1$ **23.** $y = -\sqrt{x+1}$ **24.** $y = -\sqrt{x-6}$

25. $y = \sqrt{x-4} - 3$ **26.** $y = \sqrt{x-2} - 4$ **27.** $y = \sqrt{x-3} + \dfrac{1}{2}$ **28.** $y = \sqrt{x + \dfrac{3}{2}} + 2$

29. $y = 5 + \sqrt{x+2}$ **30.** $y = \sqrt{x+4} - 3$

Using the graph of $y = \dfrac{1}{x}$, graph the functions in Exercises 31 – 40 using your understanding of reflections and translations and with no additional computations.

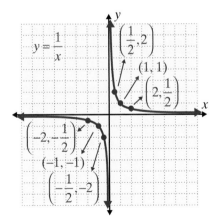

31. $y = \dfrac{1}{x} - 3$ **32.** $y = \dfrac{1}{x} + 5$ **33.** $y = \dfrac{1}{x-1}$ **34.** $y = \dfrac{1}{x+2}$

35. $y = \dfrac{1}{x-3} + 1$ **36.** $y = \dfrac{1}{x+5} - 2$ **37.** $y = \dfrac{1}{x+1} - 4$ **38.** $y = \dfrac{1}{x-2} + 3$

39. $y = \dfrac{1}{x+4} - 5$ **40.** $y = \dfrac{1}{x-5} + 2$

The graph of a function y = f(x) is given with the coordinates of four points. Graph the functions in Exercises 41 – 50 using your understanding of reflections and translations and with no additional computations. Label the new points that correspond to the four labeled points.

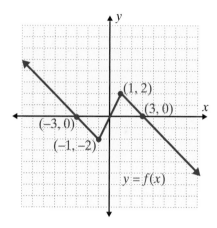

41. $y = f(x) - 1$ **42.** $y = f(x) + 2$ **43.** $y = f(x - 3)$

44. $y = f(x + 1)$ **45.** $y = -f(x)$ **46.** $y = -f(x - 4)$

47. $y = f(x + 5) + 3$ **48.** $y = f(x - 1) + 5$ **49.** $y = f(x + 2) - 4$

50. $y = f(x + 3) + 2$

In Exercises 51 – 54, match each equation with its corresponding graph.

51. $y = x^2 - 3$ **52.** $y = -x^2 + 5$ **53.** $y = (x - 1)^2 + 1$ **54.** $y = (x + 3)^2 - 2$

a.

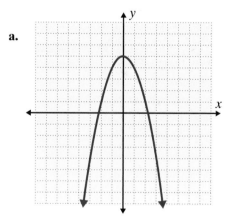

b.

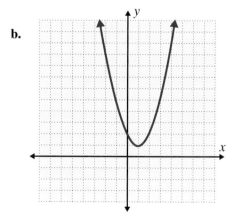

c. 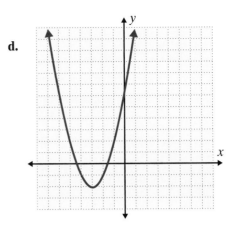   **d.**

The graph of $y = \log(x)$ is given. Graph the functions in Exercises 55 – 60 and state the domain and range of each function.

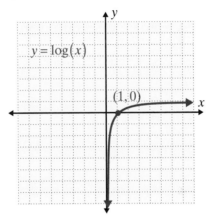

55. $y = \log(x+1)$ **56.** $y = 1 + \log x$ **57.** $y = -\log x$

58. $y = \log(-x)$ **59.** $y = \log(2-x)$ **60.** $y = -3 + \log x$

In Exercises 61 – 66, use a graphing calculator to graph each pair of functions on the same set of axes.

61. $y = 2x^2$ and $y = -3x^2$ **62.** $y = x^2 + 5$ and $y = (x-1)^2$

63. $y = (x+1)^2 - 4$ and $y = x^2 - 4$ **64.** $y = 2(x+3)^2 - 4$ and $y = 2x^2 + 3$

65. $y = -3(x-2)^2 + 1$ and $y = -x^2 + 1$ **66.** $y = 4x^2 + 4x - 4$ and $y = x^2 + x - 1$

Writing and Thinking About Mathematics

67. Explain, in your own words, how the graph of the function $y = f(x - h) + k$ represents a horizontal and a vertical shift of the graph of the function $y = f(x)$.

HAWKES LEARNING SYSTEMS: INTERMEDIATE ALGEBRA SOFTWARE

- Translations

9.2   Parabolas as Conic Sections

- *Graph parabolas that open left or right (horizontal parabolas).*
- *Find the vertices, y-intercepts, and lines of symmetry for horizontal parabolas.*

Conic sections are curves in a plane that are found when the plane intersects a cone. Four such intersections are the circle, ellipse, parabola, and hyperbola, as shown in Figure 9.7 respectively.

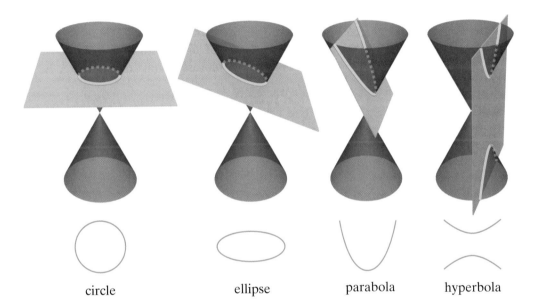

circle   ellipse   parabola   hyperbola

Figure 9.7

The corresponding equations for these conic sections are called quadratic equations because they are second-degree in x and/or y. After some practice you will be able to look at one of these equations and tell immediately what type of curve it represents and where the curve is located with respect to a Cartesian coordinate system. The technique is similar to that used in Chapter 2 in discussing straight lines.

By looking at a linear equation, you can identify
  a. the **slope** of the line,
  b. the **y-intercept** of the line, and
  c. **points** on the line.

By looking at a quadratic equation, you will be able to tell if the graph is
   a. a **circle** and identify the center and radius,
   b. an **ellipse** and identify the center and intercepts,
   c. a **parabola** and identify the vertex and line of symmetry, or
   d. a **hyperbola** and identify the vertices and asymptotes.

Parabolas

As discussed in Section 7.5, the equations of **quadratic functions** are of the basic form $y = ax^2$, and the corresponding graphs are parabolas. From the general view of **conic sections**, not all parabolas are functions. Parabolas that open upward or downward are functions, but those that open to the left or to the right are not functions.

The basic form for equations of parabolas that open left or right is $x = ay^2$, and several graphs of equations of this type are shown in Figure 9.8. As the vertical line test will confirm, these graphs do not represent functions.

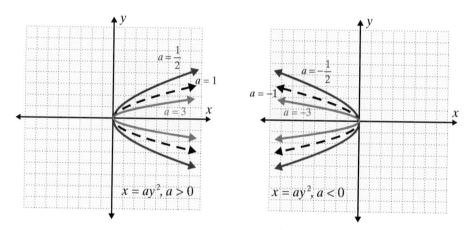

Figure 9.8

In general, the equations of **vertical parabolas** (parabolas that open upward or downward) are in the form

$$y = ax^2 + bx + c \quad \text{or} \quad y = a(x-h)^2 + k \quad (\text{where } a \neq 0)$$

where the parabolas open down if $a < 0$ or up if $a > 0$ and the vertex is at (h, k). The line $x = h$ is the line of symmetry.

By exchanging the roles of x and y, the equations of **horizontal parabolas** (parabolas that open to the left or right) can be written in the following form.

Equations of Horizontal Parabolas

Equations of horizontal parabolas (parabolas that open to the left or right) are of the form

$$x = ay^2 + by + c \quad or \quad x = a(y-k)^2 + h \quad (where \ a \neq 0).$$

The parabola opens left if $a < 0$ and right if $a > 0$.

*The vertex is at **(h, k)**.*

The line $y = k$ is the line of symmetry.

In a manner similar to the discussion in Section 9.1, adding h to the right hand side and replacing y with $(y-k)$ in the equation $x = ay^2$ gives an equation whose graph is a horizontal translation of h units and a vertical translation of k units of the graph of $x = ay^2$.

For example, the graph of $x = 2(y-3)^2 - 1$ is shown in Figure 9.9 with a table of y- and x-values. The vertex is at $(h,k) = (-1,3)$, and the line of symmetry is $y = 3$. The y-values in the table are chosen on each side of the line of symmetry.

y	$2(y-3)^2 - 1 = x$
3	$2(3-3)^2 - 1 = -1$
4	$2(4-3)^2 - 1 = 1$
2	$2(2-3)^2 - 1 = 1$
5	$2(5-3)^2 - 1 = 7$
1	$2(1-3)^2 - 1 = 7$

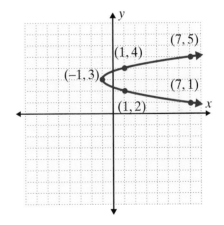

Figure 9.9

The graph of an equation of the form $x = ay^2 + by + c$ can be found by completing the square (as in Section 7.5) and writing the equation in the form

$$x = a(y-k)^2 + h.$$

Also, by setting $x = 0$ and solving the following quadratic equation

$$0 = ay^2 + by + c,$$

we can determine the **y-intercepts** (the points, if any, where the graph intersects the **y-axis**).

Example 1: Horizontal Parabolas

a. For $x = y^2 - 6y + 6$, find the vertex, the points where the graph intersects the y-axis, and the line of symmetry. Then sketch the graph.

Solution: To find the vertex, complete the square.

$$x = y^2 - 6y + 6$$

$$x = \left(y^2 - 6y + 9\right) - 9 + 6$$

$$x = \left(y - 3\right)^2 - 3$$

The vertex is at $(-3, 3)$.

To find the y-intercepts, let $x = 0$ and use the square root method as follows:

$$\left(y - 3\right)^2 - 3 = 0$$

$$\left(y - 3\right)^2 = 3$$

$$y - 3 = \pm\sqrt{3}$$

$$y = 3 \pm \sqrt{3}.$$

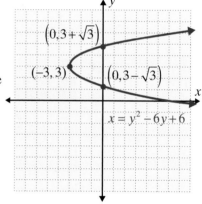

Since $a = 1$, the parabola has the same shape as $x = y^2$.

Vertex: $(-3, 3)$

y-intercepts: $\left(0, 3 + \sqrt{3}\right)$

and $\left(0, 3 - \sqrt{3}\right)$

Line of symmetry: $y = 3$

b. For $x = -2y^2 - 4y + 6$, find the vertex, the y-intercepts, and the line of symmetry. Then sketch the graph.

Solution: To find the vertex, complete the square.

$$x = -2y^2 - 4y + 6$$

$$x = -2\left(y^2 + 2y\right) + 6$$

$$x = -2\left(y^2 + 2y + 1 - 1\right) + 6$$

$$x = -2\left(y^2 + 2y + 1\right) + 2 + 6$$

$$x = -2\left(y^2 + 2y + 1\right) + 8$$

$$x = -2\left(y + 1\right)^2 + 8$$

The vertex is at $(8, -1)$.

To find the y-intercepts, let $x = 0$.

$$-2y^2 - 4y + 6 = 0$$

$$-2(y^2 + 2y - 3) = 0$$

$$-2(y + 3)(y - 1) = 0$$

$$y = -3 \quad \text{or} \quad y = 1$$

Since $a = -2$, the graph opens to the left and is slightly narrower than $x = y^2$.

Vertex: $(8, -1)$

y-intercepts: $(0, -3)$ and $(0, 1)$

Line of symmetry: $y = -1$

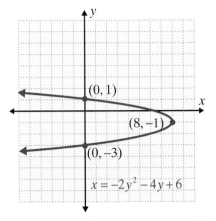

Using a TI-84 Plus Graphing Calculator to Graph Horizontal Parabolas

Horizontal parabolas are not functions and the graphing calculator is designed to graph only functions. Therefore, to graph a horizontal parabola, solve the given equation for y. For example, from the previous discussion we know that the graph of the equation $x = y^2 - 2$ is a horizontal parabola opening right with its vertex at $(-2, 0)$. To graph this equation using a graphing calculator, we must first solve for y since equations must be entered with y (to the first power) on the left hand side. By using the definition of square root, we can find two functions that we will designate as y_1 and y_2 as follows:

$$x = y^2 - 2$$

$$y^2 = x + 2 \qquad \text{First solve for } y^2.$$

$$\begin{cases} y_1 = \sqrt{x + 2} \\ y_2 = -\sqrt{x + 2} \end{cases} \qquad \begin{array}{l} \text{Solving for } y \text{ gives two equations} \\ \text{that represent two functions.} \end{array}$$

Graphing these equations individually gives the upper and lower halves of the parabola.

$y_1 = \sqrt{x + 2}$
(upper half)

$y_2 = -\sqrt{x + 2}$
(lower half)

Graphing both halves at the same time gives the entire parabola $x = y^2 - 2$.

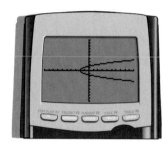

Example 2: Using a Calculator to Graph Horizontal Parabolas

Use a graphing calculator to graph the horizontal parabola $x = y^2 - 4y + 5$. Find the y-intercepts using the **CALC** features of the calculator.

Solution: To solve for y, complete the square and use the definition of square root as follows.

$$x = y^2 - 4y + 5$$

$$y^2 - 4y = x - 5$$

$$y^2 - 4y + 4 = x - 5 + 4 \qquad \text{Complete the square.}$$

$$(y - 2)^2 = x - 1$$

$$y - 2 = \pm\sqrt{x - 1} \qquad \text{Use the square root property.}$$

$$\begin{cases} y_1 = 2 + \sqrt{x - 1} \\ y_2 = 2 - \sqrt{x - 1} \end{cases} \qquad \text{Graph both of these equations.}$$

From this graph, we can determine that there are no y-intercepts.

Practice Problems

1. Write the equation $x = -y^2 - 10y - 24$ in the form $x = a(y-k)^2 + h$.

2. Find the vertex, y-intercepts, and line of symmetry for the curve $x = y^2 - 4$.

3. Find the y-intercepts for the curve $x = y^2 + 2y + 2$.

9.2 Exercises

*In Exercises 1 – 30, find **a.** the vertex, **b.** the y-intercepts, and **c.** the line of symmetry. Then draw the graph.*

1. $x = y^2 + 4$

2. $x = y^2 - 5$

3. $y + 3 = x^2$

4. $y - 2 = x^2$

5. $x = 2y^2 + 3$

6. $x = 3y^2 + 1$

7. $x = (y - 3)^2$

8. $x = (y - 2)^2$

9. $x - 4 = (y + 2)^2$

10. $x + 3 = (y - 5)^2$

11. $y + 1 = (x - 1)^2$

12. $y - 5 = (x - 3)^2$

13. $x = y^2 + 4y + 4$

14. $x = -y^2 + 10y - 25$

15. $x = y^2 - 8y + 16$

16. $x = y^2 + 6y + 1$

17. $y = -x^2 - 4x + 5$

18. $y = x^2 + 5x + 6$

19. $y = x^2 + 6x + 5$

20. $y = x^2 - 2x - 5$

21. $x = -y^2 + 4y - 3$

22. $x = y^2 + 8y + 10$

23. $y = 2x^2 + x - 1$

24. $y = -2x^2 + x + 3$

25. $x = 3y^2 + 6y - 5$

26. $x = 3y^2 + 5y + 2$

27. $x = -2y^2 + 5y - 2$

28. $x = 4y^2 - 4y - 15$

29. $y = 4x^2 - 12x + 9$

30. $y = -5x^2 + 10x + 2$

Answers to Practice Problems: **1.** $x = -(y + 5)^2 + 1$
2. Vertex: $(-4, 0)$, y-intercepts: $(0, 2)$ and $(0, -2)$, Line of symmetry: $y = 0$
3. There are no y-intercepts.

Use a graphing calculator to graph each of the parabolas in Exercises 31 – 40. Use the trace and zoom features of the calculator to estimate the y-intercepts of the parabolas.

31. $x = 2y^2 - 3$

32. $x = -3y^2 + 1$

33. $x = -y^2 + 2y$

34. $x = y^2 - 5y$

35. $x = 2y^2 + y + 1$

36. $x = -y^2 - 4y + 1$

37. $x = 4y^2 + 8y - 7$

38. $x = 3y^2 + 3y + 2$

39. $x = -2y^2 + 4y + 3$

40. $x = -5y^2 - 10y - 4$

In Exercises 41 – 44, use your knowledge of parabolas and equations to match the equation with the graph.

41. $x = 2(y - 3)^2 + 3$

42. $x = -(y + 1)^2 + 5$

43. $x = y^2 - 6$

44. $x = -y^2 - 1$

a.

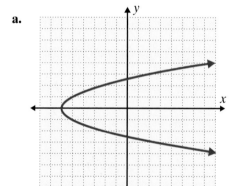

b.

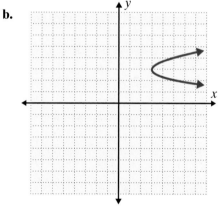

c.

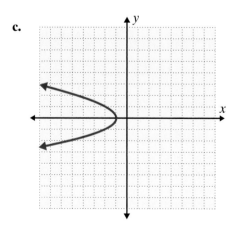

d.

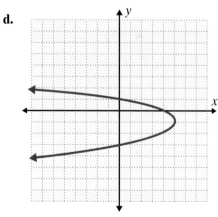

Writing and Thinking About Mathematics

45. For $x = ay^2 + by + c$ we know that the graph of the parabola opens to the right if $a > 0$ and to the left if $a < 0$. Discuss which values of a will cause the parabola to be wider and which will cause it to be narrower than the graph of $x = y^2$.

 HAWKES LEARNING SYSTEMS: INTERMEDIATE ALGEBRA SOFTWARE

- Parabolas as Conic Sections

<table>
<tr><td>**9.3**</td><td></td></tr>
</table>

Distance Formula and Circles

- *Find the distance between any two points in a plane.*
- *Write the equation of a circle given its center and radius.*
- *Graph circles centered at the point (h, k).*

Distance Between Two Points

The formula for the distance between two points in a plane is needed to develop the equations of circles. The **Pythagorean Theorem**, previously discussed in Sections 4.8 and 7.3, is the basis for the formula and is repeated here for easy reference. (Remember, the **hypotenuse** is the side opposite the right angle and is the longest side.)

The Pythagorean Theorem

In a right triangle, the square of the hypotenuse is equal to the sum of the squares of the legs.

$$c^2 = a^2 + b^2$$

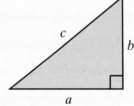

To find the distance between the two points $P(-1, 2)$ and $Q(5, 6)$, as shown in Figure 9.10a, form a right triangle, as shown in Figure 9.10b, and find the lengths of the sides a and b. Then, using a and b and the Pythagorean Theorem, we can find the length of the hypotenuse, which is the distance between the two points.

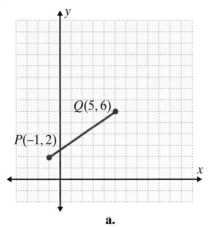

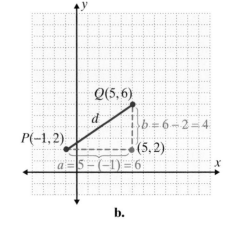

Figure 9.10

From Figure 9.10b,

$$d^2 = a^2 + b^2 = \left(5 - (-1)\right)^2 + (6-2)^2 = 6^2 + 4^2 = 36 + 16 = 52.$$

Taking the square root of both sides gives the distance d:

$$d = \sqrt{52} = 2\sqrt{13}. \quad (\, 2\sqrt{13} \approx 7.2111 \text{ estimated with a calculator})$$

In general, for points $P(x_1, y_1)$ and $Q(x_2, y_2)$ in a plane, with $a = |x_2 - x_1|$ and $b = |y_2 - y_1|$, the Pythagorean Theorem gives the following distance formula.

The Distance Formula

For two points $P(x_1, y_1)$ and $Q(x_2, y_2)$ in a plane, the distance between the points is

$$\boldsymbol{d = \sqrt{(x_2 - x_1)^2 + (y_2 - y_1)^2}}.$$

(See Figure 9.11)

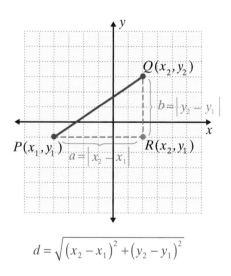

$$d = \sqrt{(x_2 - x_1)^2 + (y_2 - y_1)^2}$$

Figure 9.11

Note that in Figure 9.11, the calculations for a and b involve absolute values. These absolute values guarantee nonnegative values for a and b to represent the lengths of the legs. In the distance formula, the absolute values are disregarded because $x_2 - x_1$ and $y_2 - y_1$ are squared. **In the actual calculation of d, be sure to add the squares before taking the square root.**

Example 1: The Distance Formula

a. Find the distance between the two points $(3, 4)$ and $(-2, 7)$.

Solution: $d = \sqrt{\left[3-(-2)\right]^2 + (4-7)^2}$

$$= \sqrt{5^2 + (-3)^2} = \sqrt{25+9} = \sqrt{34}$$

b. Use the distance formula (3 times) and the Pythagorean Theorem to determine whether or not the triangle with vertices at $A(-5, -1)$, $B(2, 1)$, and $C(0, 7)$ is a right triangle.

Solution: Find the lengths of the three line segments $\overline{AB}$, $\overline{AC}$, and $\overline{BC}$, and decide whether or not the Pythagorean Theorem is satisfied.

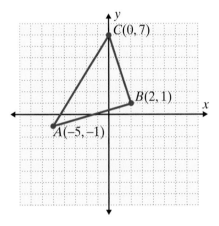

$$AB = \sqrt{(-5-2)^2 + (-1-1)^2} = \sqrt{(-7)^2 + (-2)^2}$$
$$= \sqrt{49+4} = \sqrt{53}$$

$$AC = \sqrt{(-5-0)^2 + (-1-7)^2} = \sqrt{(-5)^2 + (-8)^2}$$
$$= \sqrt{25+64} = \sqrt{89}$$

$$BC = \sqrt{(2-0)^2 + (1-7)^2} = \sqrt{(2)^2 + (-6)^2}$$
$$= \sqrt{4+36} = \sqrt{40}$$

The longest side is $AC = \sqrt{89}$.

The triangle is **not** a right triangle since $\left(\sqrt{89}\right)^2 \neq \left(\sqrt{53}\right)^2 + \left(\sqrt{40}\right)^2$ as $89 \neq 53+40$.

Equations of Circles

Circles and the terms related to circles (**center**, **radius**, and **diameter**) are defined as follows.

Circle, Center, Radius, and Diameter

*A **circle** is the set of all points in a plane that are a fixed distance from a fixed point.*

*The fixed point is called the **center** of the circle.*

*The distance from the center to any point on the circle is called the **radius** of the circle.*

*The distance from one point on the circle to another point on the circle measured through the center is called the **diameter** of the circle.*

Note: *The diameter is twice the length of the radius.*

The distance formula is used to find the equation of a circle. For example, to find the equation of the circle with its center at the origin $(0, 0)$ and radius 5, for any point on the circle (x, y) the distance from (x, y) to $(0, 0)$ must be 5. Therefore, using the distance formula,

$$\sqrt{\left(x_2 - x_1\right)^2 + \left(y_2 - y_1\right)^2} = d$$

$$\sqrt{\left(x - 0\right)^2 + \left(y - 0\right)^2} = 5$$

$$\sqrt{x^2 + y^2} = 5$$

$$x^2 + y^2 = 25. \qquad \text{Square both sides.}$$

Thus, as shown in Figure 9.12, all points on the circle satisfy the equation $x^2 + y^2 = 25$.

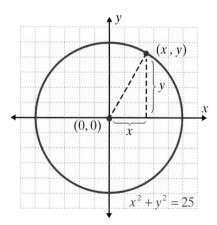

Figure 9.12

In general, any point (x, y) on a circle with center at (h, k) and radius $r > 0$ must satisfy the equation

$$\sqrt{(x-h)^2 + (y-k)^2} = r.$$

Squaring both sides of this equation gives the **standard form** for the equation of a circle:

$$(x-h)^2 + (y-k)^2 = r^2.$$

Equation of a Circle

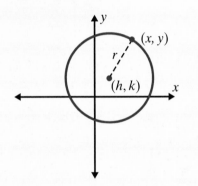

The equation of a circle with radius r and center at (h, k) is

$$(\boldsymbol{x-h})^2 + (\boldsymbol{y-k})^2 = \boldsymbol{r}^2.$$

If the center is at the origin, (0, 0), the equation simplifies to

$$\boldsymbol{x}^2 + \boldsymbol{y}^2 = \boldsymbol{r}^2.$$

Example 2: Equations of Circles

a. Find the equation of the circle with its center at the origin and radius $\sqrt{3}$. Are the points $\left(\sqrt{2}, 1\right)$ and $(1, 2)$ on the circle?

Solution: The equation of the circle is $x^2 + y^2 = 3$.

To determine whether or not the points $\left(\sqrt{2}, 1\right)$ and $(1, 2)$ are on the circle, substitute each of these points into the equation.

Substituting $\left(\sqrt{2}, 1\right)$ gives $\left(\sqrt{2}\right)^2 + (1)^2 = 2 + 1 = 3.$

Substituting $(1, 2)$ gives $(1)^2 + (2)^2 = 1 + 4 = 5 \neq 3.$

Therefore, $\left(\sqrt{2},1\right)$ is on the circle, but $(1,2)$ is not on the circle.

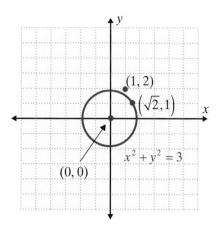

b. Find the equation of the circle with center at $(5,2)$ and radius 3. Is the point $(5,5)$ on the circle?

Solution: The equation of the circle is $(x-5)^2 +(y-2)^2 = 9$.

Substituting $(5,5)$ gives $(5-5)^2 +(5-2)^2 = 0^2 +3^2 = 9$.

Therefore, $(5,5)$ is on the circle.

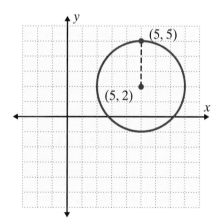

Continued on the next page...

c. Show that $x^2 + y^2 - 8x + 2y = 0$ represents a circle. Find its center and radius. Then graph the circle.

Solution: Rearrange the terms and complete the square for $x^2 - 8x$ and $y^2 + 2y$.

$$x^2 + y^2 - 8x + 2y = 0$$

$$x^2 - 8x + y^2 + 2y = 0$$

$$x^2 - 8x + 16 + y^2 + 2y + 1 = 16 + 1 \qquad \text{Add 16 and 1 to both sides.}$$

Completes the square

$$(x - 4)^2 + (y + 1)^2 = 17 \qquad \begin{array}{l}\text{Standard form for the} \\ \text{equation of a circle}\end{array}$$

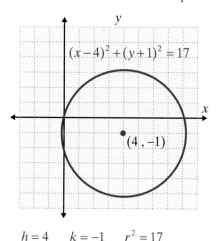

$$h = 4 \qquad k = -1 \qquad r^2 = 17$$

Center is at $(4, -1)$ and radius $= \sqrt{17}$.

Using a TI-84 Plus Graphing Calculator to Graph Circles

The equation of a circle does not represent a function. The upper half (upper semicircle) and the lower half (lower semicircle) do, however, represent separate functions. Therefore, to graph a circle, solve the equation for two values of y just as we did for horizontal parabolas in Section 9.2. For example, consider the circle with equation $x^2 + y^2 = 4$.

$$x^2 + y^2 = 4$$

$$y^2 = 4 - x^2 \qquad \text{First solve for } y^2.$$

$$\begin{cases} y_1 = \sqrt{4 - x^2} \\ y_2 = -\sqrt{4 - x^2} \end{cases} \qquad \begin{array}{l}\text{Solving for } y \text{ gives two equations that} \\ \text{represent two functions.}\end{array}$$

Graphing both of these functions gives the circle pictured in Figure 9.13.

Figure 9.13

The screen on a TI calculator is rectangular and not a square. The ranges for the standard viewing window are from −10 for both **Xmin** and **Ymin** to 10 for both **Xmax** and **Ymax**. That is, the horizontal scale (x values) and the vertical scale (y values) are the same. Since the rectangular screen is in the approximate ratio of 2 to 3, the graph of a circle using the standard **WINDOW** will appear flattened as the circle did in Figure 9.13.

To get a more realistic picture of a circle press the ⬭WINDOW key and set the **Xmin** and **Xmax** values to −6 and 6, respectively. Set the **Ymin** and **Ymax** values to −4 and 4, respectively. Since the numbers 4 and 6 are in the ratio of 2 to 3, the screen is said to show a "square window," and the graphs of y_1 and y_2 will now give a more realistic picture of a circle as shown in Figure 9.14. Alternatively, pressing the ⬭ZOOM key and choosing option **5:ZSquare** automatically "squares" the window.

Figure 9.14

Example 3: Using a Graphing Calculator

Use a graphing calculator with a "square window" to graph the circle $x^2 + y^2 = 9$.

Solution: Set the ⬭WINDOW scales to −6 and 6 for **Xmin** and **Xmax** and −4 and 4 for **Ymin** and **Ymax**, respectively.

Solving for y^2 gives: $y^2 = 9 - x^2$.

Continued on the next page...

Solving for y_1 and y_2 gives: $\begin{cases} y_1 = \sqrt{9 - x^2} \\ y_2 = -\sqrt{9 - x^2} \end{cases}$.

Graphing both y_1 and y_2 gives the following graph of the circle.

Practice Problems

1. *Find the equation of the circle with center at (−2, 3) and radius 6.*

2. *Write the equation in standard form and find the center and radius for the circle with equation* $x^2 + y^2 + 6y = 7$.

3. *Find the distance between the two points (5, 3) and (−1, −3).*

9.3 Exercises

In Exercises 1 – 12, find the distance between the two given points.

1. $(2,4),(6,7)$ **2.** $(1,0),(6,12)$ **3.** $(-3,2),(9,7)$

4. $(-6,3),(-2,0)$ **5.** $(1,7),(3,2)$ **6.** $(-2,1),(3,-4)$

7. $(4,-3),(7,-3)$ **8.** $(-2,6),(5,6)$ **9.** $(5,-2),(7,-5)$

10. $(6,4),(8,-5)$ **11.** $(-7,3),(1,-12)$ **12.** $(3,8),(-2,-4)$

Find equations for each of the circles in Exercises 13 – 32.

13. Center $(0,0)$; $r = 4$ **14.** Center $(0,0)$; $r = 6$ **15.** Center $(0,0)$; $r = \sqrt{3}$

16. Center $(0,0)$; $r = \sqrt{7}$ **17.** Center $(0,0)$; $r = \sqrt{11}$ **18.** Center $(0,0)$; $r = \sqrt{13}$

Answers to Practice Problems: **1.** $(x+2)^2 + (y-3)^2 = 36$ **2.** $x^2 + (y+3)^2 = 16$; center at $(0,-3)$ and radius 4   **3.** $\sqrt{72} = 6\sqrt{2}$

19. Center $(0,0)$; $r = \dfrac{2}{3}$ **20.** Center $(0,0)$; $r = \dfrac{7}{4}$ **21.** Center $(0,2)$; $r = 2$

22. Center $(0,5)$; $r = 5$ **23.** Center $(4,0)$; $r = 1$ **24.** Center $(-3,0)$; $r = 4$

25. Center $(-2,0)$; $r = \sqrt{8}$ **26.** Center $(5,0)$; $r = \sqrt{2}$ **27.** Center $(3,1)$; $r = 6$

28. Center $(-1,2)$; $r = 5$ **29.** Center $(3,5)$; $r = \sqrt{12}$

30. Center $(4,-2)$; $r = \sqrt{14}$ **31.** Center $(7,4)$; $r = \sqrt{10}$

32. Center $(-3,2)$; $r = \sqrt{7}$

Write each of the equations in Exercises 33 – 48 in standard form. Find the center and radius of the circle and then sketch the graph.

33. $x^2 + y^2 = 9$ **34.** $x^2 + y^2 = 16$ **35.** $x^2 = 49 - y^2$

36. $y^2 = 25 - x^2$ **37.** $x^2 + y^2 = 18$ **38.** $x^2 + y^2 = 12$

39. $x^2 + y^2 + 2x = 8$ **40.** $x^2 + y^2 - 4x = 12$ **41.** $x^2 + y^2 - 4y = 0$

42. $x^2 + y^2 + 6x = 0$ **43.** $x^2 + y^2 + 2x + 4y = 11$

44. $x^2 + y^2 - 4x + 10y + 20 = 0$ **45.** $x^2 + y^2 + 4x + 4y - 8 = 0$

46. $x^2 + y^2 - 6x - 8y + 9 = 0$ **47.** $x^2 + y^2 - 4x - 6y + 5 = 0$

48. $x^2 + y^2 + 10x - 2y + 14 = 0$

In Exercises 49 and 50, use the Pythagorean Theorem to decide if the triangle determined by the given points is a right triangle.

49. $A(1,-2)$, $B(7,1)$, $C(5,5)$ **50.** $A(-5,-1)$, $B(2,1)$, $C(-1,6)$

In Exercises 51 and 52, show that the triangle determined by the given points is an isosceles triangle (has two equal sides).

51. $A(1,1)$, $B(5,9)$, $C(9,5)$ **52.** $A(1,-4)$, $B(3,2)$, $C(9,4)$

In Exercises 53 and 54, show that the triangle determined by the given points is an equilateral triangle (all sides equal).

53. $A(1,0)$, $B\left(3,\sqrt{12}\right)$, $C(5,0)$ **54.** $A(0,5)$, $B(0,-3)$, $C\left(\sqrt{48},1\right)$

In Exercises 55 and 56, show that the diagonals $\left(\left|\overline{AC}\right| \text{ and } \left|\overline{BD}\right|\right)$ of the rectangle ABCD are equal.

55. $A(2,-2), B(2,3), C(8,3), D(8,-2)$ **56.** $A(-1,1), B(-1,4), C(4,4), D(4,1)$

In Exercises 57 and 58, find the perimeter of the triangle determined by the given points.

57. $A(-5,0), B(3,4), C(0,0)$ **58.** $A(-6,-1), B(-3,3), C(6,4)$

In Exercises 59 – 62, use a graphing calculator to graph the circles. Be sure to set a square window.

59. $x^2 + y^2 = 16$ **60.** $x^2 + y^2 = 25$

61. $(x+3)^2 + y^2 = 49$ **62.** $(x-2)^2 + (y-5)^2 = 100$

Writing and Thinking About Mathematics

63. For a given line and a point not on the line, a parabola is defined as the set of all points that are the same distance from the point and the line. The point is called the focus and the line is called the directrix. See the figure below.

a. Suppose that (x,y) is any point on a parabola and $(0,p)$ is the focus. Find the distance from (x,y) to the focus.

b. Suppose that (x,y) is any point on the same parabola in part **a.** and the line $y=-p$ is the directrix. Find the distance from (x,y) to the directrix.

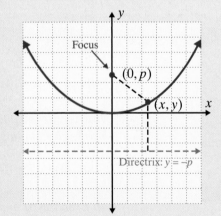

c. Show that the equation of the parabola is $x^2 = 4py$.

64. Using the equation developed in Exercise 63, find the equation of the parabola with focus at $(0,2)$ and line $y=-2$ as directrix. Draw the graph.

Writing and Thinking About Mathematics (cont.)

65. For a given line and a point not on the line, a parabola is defined as the set of all points that are the same distance from the point and the line. The point is called the focus and the line is called the directrix. See the figure below.

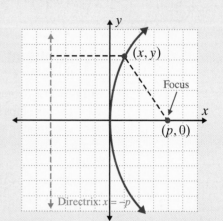

a. Suppose that (x, y) is any point on a parabola and $(p, 0)$ is the focus. Find the distance from (x, y) to the focus.

b. Suppose that (x, y) is any point on the same parabola in part **a.** and the line $x = -p$ is the directrix. Find the distance from (x, y) to the directrix.

c. Show that the equation of the parabola is $y^2 = 4px$.

66. Using the equation developed in Exercise 65, find the equation of the parabola with focus at $(-3, 0)$ and line $x = 3$ as directrix. Draw the graph.

HAWKES LEARNING SYSTEMS: INTERMEDIATE ALGEBRA SOFTWARE

▪ Distance Formula and Circles

<table>
<tr><td>**9.4**</td><td></td></tr>
</table>

Ellipses and Hyperbolas

- *Graph **ellipses** centered at the origin or at another point (h, k).*
- *Graph **hyperbolas** centered at the origin or at another point (h, k).*
- *Find the equations for the **asymptotes** of hyperbolas.*

Equations of Ellipses

Ellipses are curves in a plane that are oval in shape. To draw an ellipse, begin with two fixed points and a distance greater than the distance between the two points. Exercise 43 in the set of exercises is designed to help you understand how points in an elliptical path can be traced by using the following formal definition. You might want to go directly to that exercise now.

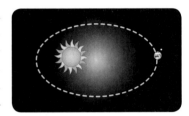

Ellipse

*An **ellipse** is the set of all points in a plane for which the sum of the distances from two fixed points is constant.*

*Each of the fixed points is called a **focus** (plural foci).*

*The **center** of an ellipse is the point midway between the foci.*

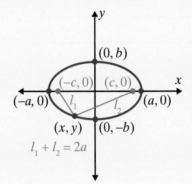

The graph of an ellipse with its center at the origin (0,0), foci at (−c,0) and (c,0), x-intercepts at (−a, 0) and (a, 0), and y-intercepts at (0, −b) and (0, b) (where $a^2 > b^2$)

Ellipses have many practical applications in the sciences, particularly in astronomy. For example, the planets in our solar system (including earth) travel in elliptical orbits (not circular orbits) and the sun is at one of the foci of each ellipse.

Initially, we will study ellipses with foci either on the x-axis or on the y-axis. Pay close attention to how the constants a, b, and c are used when defining ellipses. Note that the foci are not on the ellipse.

As an example, consider the equation

$$\frac{x^2}{25} + \frac{y^2}{9} = 1.$$

Several points that satisfy this equation are given in the following table and then are graphed in Figure 9.15.

x	y
5	0
−5	0
0	3
0	−3
3	$\frac{12}{5}$
3	$-\frac{12}{5}$
−3	$\frac{12}{5}$
−3	$-\frac{12}{5}$

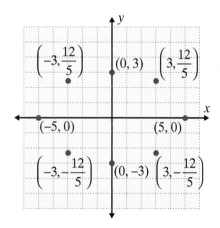

Figure 9.15

Joining the points in Figure 9.15 with a smooth curve, we get the graph of the ellipse shown in Figure 9.16. The points $(5, 0)$ and $(−5, 0)$ are the x-intercepts, and the points $(0, 3)$ and $(0, −3)$ are the y-intercepts.

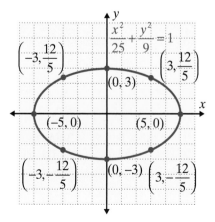

Figure 9.16

Equation of an Ellipse

The standard form for the equation of an ellipse with its center at the origin is

$$\frac{x^2}{a^2} + \frac{y^2}{b^2} = 1.$$

The points $(a, 0)$ and $(-a, 0)$ are the **x-intercepts** (called **vertices**).
The points $(0, b)$ and $(0, -b)$ are the **y-intercepts** (called **vertices**).

For $a^2 > b^2$:

1. The segment of length $2a$ joining the x-intercepts is called the **major axis**.
2. The segment of length $2b$ joining the y-intercepts is called the **minor axis**.

For $b^2 > a^2$:

1. The segment of length $2b$ joining the y-intercepts is called the **major axis**.
2. The segment of length $2a$ joining the x-intercepts is called the **minor axis**.

Note: In either case, the foci lie on the major axis.

Example 1: Equation of an Ellipse: Major Axis Horizontal

Graph the ellipse $4x^2 + 16y^2 = 64$.

Solution: First, divide both sides of the given equation by 64 to find the standard form.

$$4x^2 + 16y^2 = 64$$

$$\frac{4x^2}{64} + \frac{16y^2}{64} = \frac{64}{64}$$

$$\frac{x^2}{16} + \frac{y^2}{4} = 1$$

The curve is an ellipse. In this case $a = \sqrt{16} = 4$ and the major axis is $2a = 8$. Also, $b = \sqrt{4} = 2$ and the minor axis is $2b = 4$. The endpoints of the major axis are $(-4, 0)$ and $(4, 0)$. The endpoints of the minor axis are $(0, -2)$ and $(0, 2)$. The major and minor axes intersect at the center of the ellipse, $(0, 0)$.

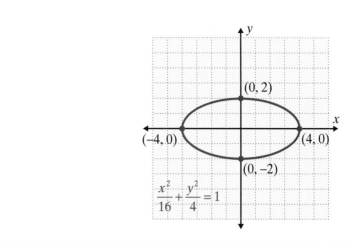

$$\frac{x^2}{16}+\frac{y^2}{4}=1$$

Example 2: Equation of an Ellipse: Major Axis Vertical

Graph the ellipse $\dfrac{x^2}{1}+\dfrac{y^2}{9}=1$.

Solution: The equation is in standard form with $b^2 = 9$ and $a^2 = 1$.

Because $b^2 > a^2$, we know the major axis is vertical. That is, the ellipse is elongated along the y-axis.

The points $(0,-3)$ and $(0,3)$ are the endpoints of the major axis.

The points $(-1,0)$ and $(1,0)$ are the endpoints of the minor axis.

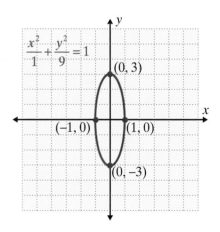

In the equation of an ellipse,

$$\frac{x^2}{a^2}+\frac{y^2}{b^2}=1,$$

the coefficients for x^2 and y^2 are both positive. If one of these coefficients is negative, then the equation represents a **hyperbola**.

Equations of Hyperbolas

Notice how the following definition of a hyperbola differs from the definition of an ellipse. Instead of the sum of the distances from two fixed points, we find the difference of the distances from two fixed points.

Hyperbola

A **hyperbola** is the set of all points in a plane such that the absolute value of the difference of the distances from two fixed points is constant.
Each of the fixed points is called a **focus** (plural foci).
The **center** of a hyperbola is the point midway between the foci.

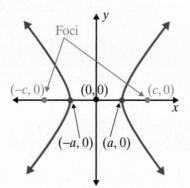

The graph of a hyperbola with its center at the origin $(0, 0)$, foci along the x-axis at $(-c, 0)$ and $(c, 0)$, and x-intercepts at $(-a, 0)$ and $(a, 0)$. There are no y-intercepts.

Several points that satisfy the equation

$$\frac{x^2}{25} - \frac{y^2}{9} = 1$$

and the curves joining these points (a hyperbola) are shown in Figure 9.17.

x	y
5	0
−5	0
7	$\dfrac{6\sqrt{6}}{5}$
7	$\dfrac{-6\sqrt{6}}{5}$
−7	$\dfrac{6\sqrt{6}}{5}$
−7	$\dfrac{-6\sqrt{6}}{5}$

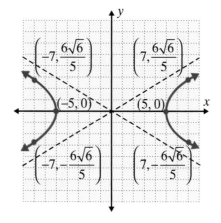

Figure 9.17

The two dotted lines shown in Figure 9.17 are called **asymptotes**. (Asymptotes were discussed in Chapter 8 with exponential and logarithmic functions.) These lines are not part of the hyperbola, but they serve as guidelines for sketching the graph. The curve gets closer and closer to these lines without ever touching them. In this figure the equations of the asymptotes are

$$y = \frac{3}{5}x \quad \text{and} \quad y = -\frac{3}{5}x.$$

With the equations of hyperbolas there is a positive term and a negative term. In the standard form for hyperbolas the roles of a and b are that a^2 is the denominator of the positive term and b^2 is in the denominator of the negative term. This relationship is very important in determining the equations of the asymptotes. The location of the negative sign determines whether the hyperbola "opens" left and right or "opens" up and down as illustrated in the following discussion.

Equations of Hyperbolas

In general, there are two standard forms for equations of hyperbolas with their centers at the origin.

1. $\dfrac{x^2}{a^2} - \dfrac{y^2}{b^2} = 1$

   x-intercepts (vertices) at (a, 0) and (−a, 0)

   No y-intercepts

   Asymptotes: $y = \dfrac{b}{a}x$ *and* $y = -\dfrac{b}{a}x$

   The curve "opens" left and right.

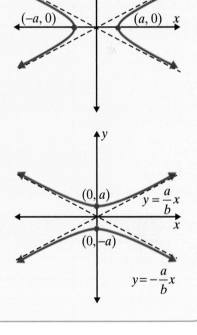

2. $\dfrac{y^2}{a^2} - \dfrac{x^2}{b^2} = 1$

   y-intercepts (vertices) at (0, a) and (0, −a)

   No x-intercepts

   Asymptotes: $y = \dfrac{a}{b}x$ *and* $y = -\dfrac{a}{b}x$

   The curve "opens" up and down.

A Geometric Aid for Sketching Asymptotes of Hyperbolas

As a geometric aid in getting the asymptotes in the correct positions, draw a rectangle with sides $2a$ and $2b$ centered at the origin. The diagonals of this fundamental rectangle are the asymptotes. Study these rectangles in Examples 3 and 4.

Example 3: Hyperbola Opening Left and Right

Graph the hyperbola $x^2 - 4y^2 = 16$.

Solution: Write the equation in standard form by dividing both sides by 16:

$$\frac{x^2}{16} - \frac{y^2}{4} = 1.$$

Here $a^2 = 16$ and $b^2 = 4$. So, using $a = 4$ and $b = 2$, the asymptotes are

$$y = \frac{2}{4}x = \frac{1}{2}x \ \text{ and } \ y = -\frac{2}{4}x = -\frac{1}{2}x.$$

The vertices are $(-4, 0)$ and $(4, 0)$ and the curve opens left and right. (Note that the fundamental rectangle has sides 8 and 4.)

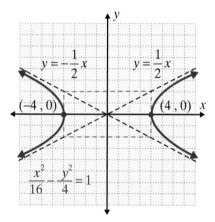

Example 4: Hyperbola Opening Up and Down

Graph the hyperbola $\dfrac{y^2}{1} - \dfrac{x^2}{9} = 1$.

Solution: Here $a^2 = 1$ and $b^2 = 9$. So, $a = 1$ and $b = 3$.

The asymptotes are $y = \dfrac{1}{3}x$ and $y = -\dfrac{1}{3}x$.

The vertices are $(0, -1)$ and $(0, 1)$ and the curve opens up and down. (Note that the fundamental rectangle has sides 6 and 2.)

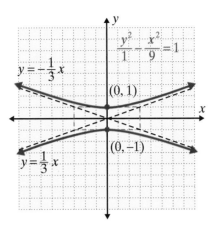

Ellipses and Hyperbolas with Centers at (*h*, *k*)

The discussion of translations in Section 9.1 indicates that replacing x with $x - h$ and y with $y - k$ in an equation gives the corresponding graph a horizontal shift of h units and a vertical shift of k units. These ideas were used in the discussion of circles in Section 9.3 for graphs of circles with center at (h, k). $\left[(x-h)^2 + (y-k)^2 = r^2 \right]$

The same procedure relates to the equations of ellipses and hyperbolas with centers at (h, k). That is, we can translate the graphs of ellipses and hyperbolas to center at (h, k) by replacing x with $x - h$ and y with $y - k$ in the standard forms. The roles of a^2 and b^2 are the same.

Ellipse with Center at (*h, k*)

The equation of an ellipse with its center at (h, k) is

$$\frac{(x-h)^2}{a^2} + \frac{(y-k)^2}{b^2} = 1.$$

Note: *a and b are distances from (h, k) to the vertices. (See Figure 9.18)*

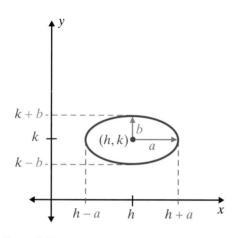 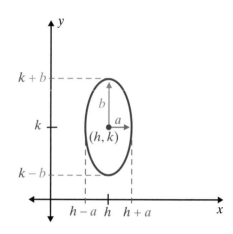

Figure 9.18

Example 5: Ellipse with Center at (h, k)

Graph the ellipse $\dfrac{(x+2)^2}{16}+\dfrac{(y-1)^2}{9}=1$.

Solution: The graph of $\dfrac{x^2}{16}+\dfrac{y^2}{9}=1$ is translated 2 units left and 1 unit up so that the center is at $(-2, 1)$ with $a = 4$ and $b = 3$. The graph is shown here with the center and vertices labeled.

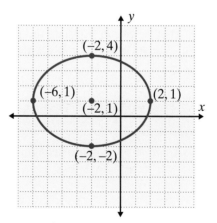

Hyperbola with Center at (h,k)

The equation of a hyperbola with its center at (h, k) is

$$\frac{(x-h)^2}{a^2} - \frac{(y-k)^2}{b^2} = 1 \; or \; \frac{(y-k)^2}{a^2} - \frac{(x-h)^2}{b^2} = 1.$$

Note: a and b are used as in the standard form but are measured from (h, k). (See Figure 9.19)

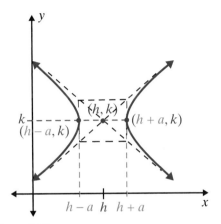

Figure 9.19

Example 6: Hyperbola with Center at (h,k)

Graph the hyperbola $\dfrac{(x-3)^2}{25} - \dfrac{(y+4)^2}{36} = 1$.

Solution: The graph of $\dfrac{x^2}{25} - \dfrac{y^2}{36} = 1$ is translated 3 units right and 4 units down so that the center is at $(3, -4)$ with $a = 5$ and $b = 6$. The graph is shown here with its asymptotes, the center, and the vertices labeled.

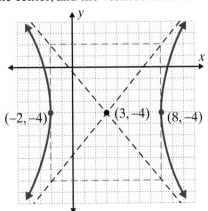

Practice Problems

1. Write the equation $2x^2 + 9y^2 = 18$ in standard form. State the length of the major axis and the length of the minor axis.

2. Write the equation $x^2 - 9y^2 = 9$ in standard form. Write the equations of the asymptotes.

3. Graph the ellipse $\dfrac{(x-2)^2}{4} + \dfrac{(y+1)^2}{1} = 1$.

4. Graph the hyperbola $\dfrac{x^2}{16} - \dfrac{(y-3)^2}{4} = 1$.

9.4 Exercises

Write each of the equations in Exercises 1 – 30 in standard form. Then sketch the graph. For hyperbolas, graph the asymptotes as well.

1. $x^2 + 9y^2 = 36$ 2. $x^2 + 4y^2 = 16$ 3. $4x^2 + 25y^2 = 100$

4. $4x^2 + 9y^2 = 36$ 5. $16x^2 + y^2 = 16$ 6. $25x^2 + 9y^2 = 36$

7. $x^2 - y^2 = 1$ 8. $x^2 - y^2 = 4$ 9. $9x^2 - y^2 = 9$

10. $4x^2 - y^2 = 4$ 11. $4x^2 - 9y^2 = 36$ 12. $9x^2 - 16y^2 = 144$

13. $2x^2 + y^2 = 8$ 14. $3x^2 + y^2 = 12$ 15. $x^2 + 5y^2 = 20$

16. $x^2 + 7y^2 = 28$ 17. $y^2 - x^2 = 9$ 18. $y^2 - x^2 = 16$

19. $y^2 - 2x^2 = 8$ 20. $y^2 - 3x^2 = 12$ 21. $y^2 - 2x^2 = 18$

22. $y^2 - 5x^2 = 20$ 23. $3x^2 + 2y^2 = 18$ 24. $4x^2 + 3y^2 = 12$

25. $4x^2 + 5y^2 = 20$ 26. $3x^2 + 8y^2 = 48$ 27. $3x^2 - 5y^2 = 75$

28. $4x^2 - 7y^2 = 28$ 29. $3y^2 - 4x^2 = 36$ 30. $9y^2 - 8x^2 = 72$

Answers to Practice Problems: 1. $\dfrac{x^2}{9} + \dfrac{y^2}{2} = 1$;

major axis: 6,

minor axis: $2\sqrt{2}$

2. $\dfrac{x^2}{9} - \dfrac{y^2}{1} = 1$;

asymptotes:

$y = \dfrac{1}{3}x$, $y = -\dfrac{1}{3}x$

3.

4.

In Exercises 31 – 36, match the equations with the given graphs.

31. $\dfrac{(x-1)^2}{4}+\dfrac{(y-3)^2}{25}=1$ **32.** $\dfrac{(x+1)^2}{4}+\dfrac{(y+3)^2}{25}=1$ **33.** $\dfrac{(x-1)^2}{25}+\dfrac{(y-3)^2}{4}=1$

34. $\dfrac{(x+1)^2}{25}+\dfrac{(y+3)^2}{4}=1$ **35.** $\dfrac{(x+1)^2}{25}-\dfrac{(y+3)^2}{4}=1$ **36.** $\dfrac{(y+3)^2}{4}-\dfrac{(x+1)^2}{25}=1$

a.
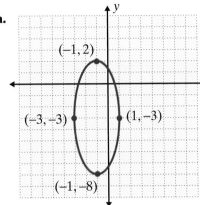

b.
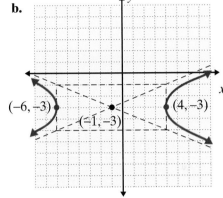

c.
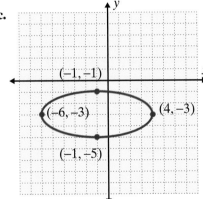

d.
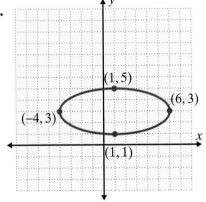

e.
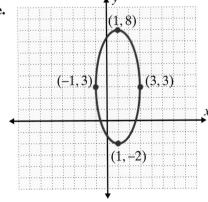

f.
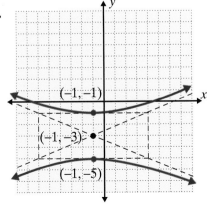

In Exercises 37 – 42, use your knowledge of translations to graph each of the following equations. These graphs are ellipses and hyperbolas with centers at points other than the origin.

37. $\dfrac{(x-2)^2}{25} + \dfrac{(y-1)^2}{9} = 1$ **38.** $\dfrac{(x+1)^2}{16} + \dfrac{(y-4)^2}{1} = 1$ **39.** $\dfrac{(x+5)^2}{1} - \dfrac{(y+2)^2}{16} = 1$

40. $\dfrac{(x-4)^2}{9} - \dfrac{(y-3)^2}{36} = 1$ **41.** $\dfrac{(x+1)^2}{49} + \dfrac{(y-6)^2}{100} = 1$ **42.** $\dfrac{(y-2)^2}{9} - \dfrac{(x+2)^2}{4} = 1$

Writing and Thinking About Mathematics

43. The definition of an ellipse is given in the text as follows:
An ellipse is the set of all points in a plane the sum of whose distances from two fixed points is constant.

  a. Draw an ellipse by proceeding as follows:
   Step 1: Place two thumb tacks in a piece of cardboard.
   Step 2: Select a piece of string slightly longer than the distance between the two tacks.
   Step 3: Tie the string to each thumb tack and stretch the string taut by using a pencil.
   Step 4: Use the pencil to trace the path of an ellipse on the cardboard by keeping the string taut. (The length of the string represents the fixed distance from points on the ellipse to the two foci.)

  b. Show that the equation of an ellipse with foci at $(-c, 0)$ and $(c, 0)$, center at the origin, and $2a$ as the constant sum of the lengths to the foci can be written in the form $\dfrac{x^2}{a^2} + \dfrac{y^2}{a^2 - c^2} = 1.$

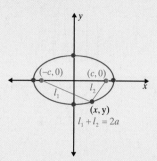

Writing and Thinking About Mathematics (cont.)

    c. In the equation in part **b.**, substitute $b^2 = a^2 - c^2$ to get the standard form for the equation of an ellipse. Show that the points $(0, -b)$ and $(0, b)$ are the y-intercepts and a is the distance from each y-intercept to a focus.

 HAWKES LEARNING SYSTEMS: INTERMEDIATE ALGEBRA SOFTWARE

    ▪ Ellipses and Hyperbolas

9.5

Nonlinear Systems of Equations

- *Solve systems of equations where one or both equations are nonlinear.*

The equations for the conic sections that we have discussed all have at least one term that is second-degree. These equations are called **quadratic equations**. (Only the equations for parabolas of the form $y = ax^2 + bx + c$ are **quadratic functions**.) A summary of the equations with their related graphs is shown in Figure 9.20.

Summary of Equations and Related Graphs

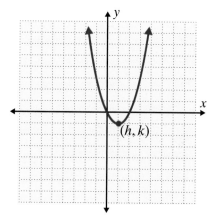

Parabola: $y = a(x-h)^2 + k$
$a > 0$, opens upward
$a < 0$, opens downward

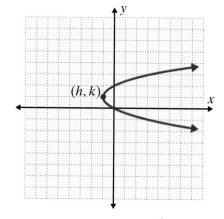

Parabola: $x = a(y-k)^2 + h$
$a > 0$, opens right
$a < 0$, opens left

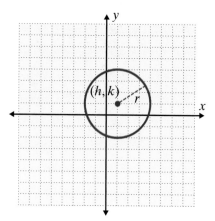

Circle: $(x-h)^2 + (y-k)^2 = r^2$

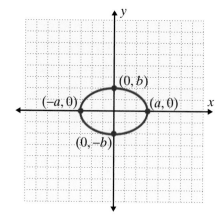

Ellipse: $\dfrac{x^2}{a^2} + \dfrac{y^2}{b^2} = 1$

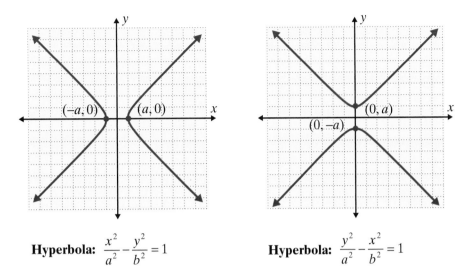

Hyperbola: $\dfrac{x^2}{a^2} - \dfrac{y^2}{b^2} = 1$ **Hyperbola:** $\dfrac{y^2}{a^2} - \dfrac{x^2}{b^2} = 1$

Figure 9.20

If a system of two equations has one quadratic equation and one linear equation, then the method of substitution should be used to solve the system. If the system involves two quadratic equations, then the method used depends on the form of the equations. The following examples show three possible situations. The graphs of the curves are particularly useful for approximating solutions and determining the exact number of solutions.

Examples 1 and 2 illustrate three possibilities and emphasize the value of graphing the curves to visualize the number of solutions and approximating these solutions.

Solving a System of One Quadratic Equation and One Linear Equation

If a system of two equations has **one quadratic equation and one linear equation**, then:

1. Solve the linear equation for one of the variables and substitute for this variable in the quadratic equation.

2. Solve the resulting second-degree equation and analyze the results.

3. Graph the curves to visualize the number of solutions and check that the solutions are reasonable and satisfy both equations.

Example 1: A System of One Quadratic and One Linear Equation

Solve the following systems of equations and graph both curves in each system.

a. A circle and a line: $\begin{cases} x^2 + y^2 = 25 & \text{Circle} \\ x + y = 5 & \text{Line} \end{cases}$

Solution: Solve $x + y = 5$ for y (or x). Then substitute into the other equation.

$$y = 5 - x$$

$$x^2 + (5 - x)^2 = 25$$

$$x^2 + 25 - 10x + x^2 = 25$$

$$2x^2 - 10x = 0$$

$$2x(x - 5) = 0 \qquad \text{Now solve for } x.$$

$$\begin{cases} x = 0 \\ y = 5 - 0 = 5 \end{cases} \quad \text{or} \quad \begin{cases} x = 5 \\ y = 5 - 5 = 0 \end{cases}$$

The solutions (points of intersection) are $(0, 5)$ and $(5, 0)$.

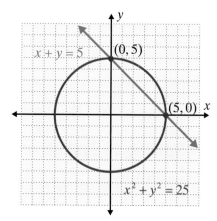

b. A line and a parabola: $\begin{cases} x + y = -7 & \text{Line} \\ y = x^2 - 4x - 5 & \text{Parabola} \end{cases}$

Solution: Solve the linear equation for y, then substitute into the quadratic equation. (In this case, the quadratic equation is already solved for y, so we could have chosen to make the substitution the other way.)

$$y = -x - 7$$

$$-x - 7 = x^2 - 4x - 5$$

$$0 = x^2 - 3x + 2$$

$$0 = (x - 2)(x - 1)$$

$$\begin{cases} x = 2 \\ y = -2 - 7 = -9 \end{cases} \qquad \text{or} \qquad \begin{cases} x = 1 \\ y = -1 - 7 = -8 \end{cases}$$

The solutions are $(2, -9)$ and $(1, -8)$.

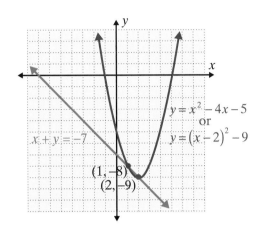

Solving a System of Two Quadratic Equations

If a system of two equations has **two quadratic equations**, then:

1. The method used depends on the form of the equations.

2. Substitution may work or addition may work.

3. Graph the curves to visualize the number of solutions and check that the solutions are reasonable and satisfy both equations.

Example 2: A System of Two Quadratic Equations

A hyperbola and a circle: $\begin{cases} x^2 - y^2 = 4 & \text{Hyperbola} \\ x^2 + y^2 = 36 & \text{Circle} \end{cases}$

Solution: Here addition will eliminate y^2.

$$\begin{array}{rcl} x^2 \quad - \quad y^2 &=& 4 \\ x^2 \quad + \quad y^2 &=& 36 \\ \hline 2x^2 \qquad\quad &=& 40 \\ x^2 \qquad\quad &=& 20 \end{array}$$

$$x = \pm\sqrt{20} = \pm 2\sqrt{5}$$

Continued on the next page...

If $x = 2\sqrt{5}$: $20 + y^2 = 36$ If $x = -2\sqrt{5}$: $20 + y^2 = 36$

$$y^2 = 16$$ $$y^2 = 16$$

$$y = \pm 4$$ $$y = \pm 4$$

There are four points of intersection:

$$\left(2\sqrt{5}, 4\right), \left(2\sqrt{5}, -4\right), \left(-2\sqrt{5}, 4\right), \text{ and } \left(-2\sqrt{5}, -4\right).$$

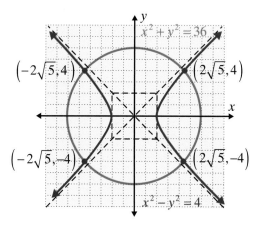

NOTES   In the examples and the exercises the curves do intersect. However, there are many situations where the curves do not intersect. This can be confirmed both algebraically and geometrically.

Practice Problems

Solve each of the following systems algebraically and geometrically.

1. $\begin{cases} y = x^2 - 4 \\ x - y = 2 \end{cases}$ **2.** $\begin{cases} x^2 + y^2 = 72 \\ x = y^2 \end{cases}$

Answers to Practice Problems: **1.** $(-1, -3)$ and $(2, 0)$ **2.** $\left(8, 2\sqrt{2}\right)$ and $\left(8, -2\sqrt{2}\right)$

9.5 Exercises

Solve each system of equations in Exercises 1 – 16. Sketch the graphs.

1. $\begin{cases} y = x^2 + 1 \\ 2x + y = 4 \end{cases}$

2. $\begin{cases} y = 3 - x^2 \\ x + y = -3 \end{cases}$

3. $\begin{cases} y = 2 - x \\ y = (x - 2)^2 \end{cases}$

4. $\begin{cases} x^2 + y^2 = 25 \\ y + x + 5 = 0 \end{cases}$

5. $\begin{cases} x^2 + y^2 = 20 \\ x - y = 2 \end{cases}$

6. $\begin{cases} x^2 - y^2 = 16 \\ 3x + 5y = 0 \end{cases}$

7. $\begin{cases} y = x - 2 \\ x^2 = y^2 + 16 \end{cases}$

8. $\begin{cases} x^2 + 3y^2 = 12 \\ x = 3y \end{cases}$

9. $\begin{cases} x^2 + y^2 = 9 \\ x^2 - y^2 = 9 \end{cases}$

10. $\begin{cases} x^2 + y^2 = 9 \\ x^2 - y + 3 = 0 \end{cases}$

11. $\begin{cases} 4x^2 + y^2 = 25 \\ 3x - y^2 + 3 = 0 \end{cases}$

12. $\begin{cases} x^2 - 4y^2 = 9 \\ x + 2y^2 = 3 \end{cases}$

13. $\begin{cases} x^2 + y^2 + 4x - 2y = 4 \\ x + y = 2 \end{cases}$

14. $\begin{cases} x^2 - y^2 = 9 \\ x^2 + y^2 - 2x - 3 = 0 \end{cases}$

15. $\begin{cases} x^2 - y^2 = 5 \\ x^2 + 4y^2 = 25 \end{cases}$

16. $\begin{cases} 2x^2 - 3y^2 = 6 \\ 2x^2 + y^2 = 22 \end{cases}$

Solve each system of equations in Exercises 17 – 30.

17. $\begin{cases} x^2 - y^2 = 20 \\ x^2 - 9y = 0 \end{cases}$

18. $\begin{cases} x^2 + 5y^2 = 16 \\ x^2 + y^2 = 4x \end{cases}$

19. $\begin{cases} x^2 + y^2 = 10 \\ x^2 + y^2 - 4y + 2 = 0 \end{cases}$

20. $\begin{cases} x^2 + y^2 = 20 \\ 4x + 8 = y^2 \end{cases}$

21. $\begin{cases} 2x^2 - y^2 = 7 \\ 2x^2 + y^2 = 29 \end{cases}$

22. $\begin{cases} y = x^2 + 2x + 2 \\ 2x + y = 2 \end{cases}$

23. $\begin{cases} 4y + 10x^2 + 7x - 8 = 0 \\ 6x - 8y + 1 = 0 \end{cases}$

24. $\begin{cases} x^2 + y^2 - 4x + 6y + 3 = 0 \\ 2x - y - 2 = 0 \end{cases}$

25. $\begin{cases} x^2 + y^2 - 4y = 16 \\ x - y = 0 \end{cases}$

26. $\begin{cases} 4x^2 + y^2 = 11 \\ y = 4x^2 - 9 \end{cases}$

27. $\begin{cases} x^2 - y^2 - 2y = 22 \\ 2x + 5y + 5 = 0 \end{cases}$

28. $\begin{cases} x^2 + y^2 - 6y = 0 \\ 2x^2 - y^2 + 15 = 0 \end{cases}$

29. $\begin{cases} y = x^2 - 2x + 3 \\ y = -x^2 + 2x + 3 \end{cases}$

30. $\begin{cases} y^2 = x^2 - 5 \\ 4x^2 - y^2 = 32 \end{cases}$

In Exercises 31 – 36, use a graphing calculator to graph and estimate the solution to each system of equations. If necessary, round values to two decimal places.

31. $\begin{cases} y = x^2 + 3 \\ x + y = 3 \end{cases}$

32. $\begin{cases} y = 1 - x^2 \\ x + y = -4 \end{cases}$

33. $\begin{cases} y = 3 - 2x \\ y = (x - 1)^2 \end{cases}$

34. $\begin{cases} x^2 + y^2 = 36 \\ y = x + 5 \end{cases}$

35. $\begin{cases} x^2 + y^2 = 10 \\ x - y = 1 \end{cases}$

36. $\begin{cases} x^2 + y^2 = 4 \\ x^2 - y^2 = 3 \end{cases}$

 HAWKES LEARNING SYSTEMS: INTERMEDIATE ALGEBRA SOFTWARE

- Nonlinear Systems of Equations

Chapter 9 Index of Key Ideas and Terms

Section 9.1   Translations

$f(x)$ Notation and Evaluating Functions   page 746

Difference Quotient   page 747

The formula $\dfrac{f(x+h)-f(x)}{h}$ is called the **difference quotient**. A geometric interpretation of the difference quotient is the **slope** of a line through two points on the graph of a function.

Horizontal and Vertical Translations   page 749

Given the graph of $y=f(x)$, the graph of $y=f(x-h)+k$ is
1. a **horizontal translation** of h units, and
2. a **vertical translation** of k units of the graph of $y=f(x)$.

Reflections   page 752

The graph of $y=-f(x)$ is a **reflection across the x-axis** of the graph of $y=f(x)$.

Section 9.2   Parabolas as Conic Sections

Conic Sections   page 761

Conic sections are curves in a plane that are found when the plane intersects a cone. Four such sections are the circle, ellipse, parabola, and hyperbola.

Horizontal Parabolas   page 763

Equations of **horizontal parabolas** (parabolas that open to the left or right) are of the form

$$x=ay^2+by+c \ \text{ or } \ x=a(y-k)^2+h \ \text{ where } \ a\neq 0.$$

The parabola opens left if $a<0$ and right if $a>0$.
The vertex is at (h,k), and $y=k$ is the line of symmetry.

Using a Graphing Calculator to Graph Horizontal Parabolas   pages 765-766

Section 9.3   Distance Formula and Circles

The Distance Formula   page 771
For two points $P(x_1, y_1)$ and $Q(x_2, y_2)$ in a plane, the
distance between the points is $d = \sqrt{(x_2 - x_1)^2 + (y_2 - y_1)^2}$.

Circles   page 773
A **circle** is the set of all points in a plane that are a fixed
distance from a fixed point. The fixed point is called the
center of the circle.

Radius   page 773
The distance from the center to any point on a circle is
called the **radius** of the circle.

Diameter   page 773
The distance from one point on a circle to another point
on the circle measured through the center is called the
diameter of the circle. The diameter is twice the length of
the radius.

Equation of a Circle   page 774
The equation of a **circle with radius r and center at (h, k)** is
$(x - h)^2 + (y - k)^2 = r^2$. If the center is at the origin, $(0, 0)$,
the equation simplifies to $x^2 + y^2 = r^2$.

Using a Graphing Calculator to Graph Circles   pages 776-777

Section 9.4   Ellipses and Hyperbolas

Ellipse   page 782
An **ellipse** is the set of all points in a plane for which the
sum of the distances from two fixed points is constant.
Each of the fixed points is called a **focus** (plural foci). The
center of an ellipse is the point midway between the foci.

Equation of an Ellipse   page 784

The standard form for the equation of an **ellipse with its center at the origin** is

$$\frac{x^2}{a^2} + \frac{y^2}{b^2} = 1.$$

The points $(a, 0)$ and $(-a, 0)$ are the **x-intercepts**.
The points $(0, b)$ and $(0, -b)$ are the **y-intercepts**.

For $a^2 > b^2$:
The segment of length $2a$ joining the x-intercepts is called the **major axis**. The segment of length $2b$ joining the y-intercepts is called the **minor axis**.

For $b^2 > a^2$:
The segment of length $2b$ joining the y-intercepts is called the **major axis**. The segment of length $2a$ joining the x-intercepts is called the **minor axis**.

Hyperbola   page 786

A **hyperbola** is the set of all points in a plane such that the absolute value of the difference of the distances from two fixed points is constant. Each of the fixed points is called a **focus** (plural foci). The **center** of a hyperbola is the point midway between the foci.

Equations of Hyperbolas   page 787

In general, there are two standard forms for equations of **hyperbolas with their centers at the origin**.

1. $\dfrac{x^2}{a^2} - \dfrac{y^2}{b^2} = 1$

x-intercepts (vertices) at $(a, 0)$ and $(-a, 0)$
No y-intercepts
Asymptotes: $y = \dfrac{b}{a}x$ and $y = -\dfrac{b}{a}x$
The curve "opens" left and right.

2. $\dfrac{y^2}{a^2} - \dfrac{x^2}{b^2} = 1$

y-intercepts (vertices) at $(0, a)$ and $(0, -a)$
No x-intercepts
Asymptotes: $y = \dfrac{a}{b}x$ and $y = -\dfrac{a}{b}x$
The curve "opens" up and down.

Section 9.4 Ellipses and Hyperbolas (cont.)

Ellipse with Center at (h, k)
page 789

The equation of an **ellipse with its center at (h, k)** is

$$\frac{(x-h)^2}{a^2} + \frac{(y-k)^2}{b^2} = 1.$$

Note: a and b are distances from (h, k) to the vertices.

Hyperbola with Center at (h, k)
page 791

The equation of a **hyperbola with its center at (h, k)** is

$$\frac{(x-h)^2}{a^2} - \frac{(y-k)^2}{b^2} = 1 \quad \text{or} \quad \frac{(y-k)^2}{a^2} - \frac{(x-h)^2}{b^2} = 1.$$

Note: a and b are used as in the standard form but are measured from (h, k).

Section 9.5 Nonlinear Systems of Equations

**Solving a System of One Quadratic Equation
and One Linear Equation**
page 797

If a system of two equations has **one quadratic equation
and one linear equation**, then:
1. Solve the linear equation for one of the variables and
   substitute for this variable in the quadratic equation.
2. Solve the resulting second-degree equation and analyze
   the results.
3. Graph the curves to visualize the number of solutions
   and check that the solutions are reasonable and satisfy
   both equations.

Solving a System of Two Quadratic Equations
page 799

If a system of two equations has **two quadratic equations**,
then:
1. The method used depends on the form of the equations.
2. Substitution may work or addition may work.
3. Graph the curves to visualize the number of solutions
   and check that the solutions are reasonable and satisfy
   both equations.

HAWKES LEARNING SYSTEMS: INTERMEDIATE ALGEBRA SOFTWARE

For a review of the topics and problems from Chapter 9, look at the following lessons from *Hawkes Learning Systems: Intermediate Algebra*

- Translations
- Parabolas as Conic Sections
- Distance Formula and Circles
- Ellipses and Hyperbolas
- Nonlinear Systems of Equations
- Chapter 9 Review and Test

Chapter 9 Review

9.1 Translations

*In Exercises 1 – 6, for each function, find **a.** f(6), **b.** f(a + 3) and **c.** the simplified difference quotient* $\dfrac{f(x+h)-f(x)}{h}$.

1. $f(x) = 3 - 5x$ **2.** $f(x) = 7x + 5$ **3.** $f(x) = 2x^2 - 6$

4. $f(x) = 8 - 3x^2$ **5.** $f(x) = x^2 - 3x + 1$ **6.** $f(x) = 2x^2 - x + 5$

7. Sketch the curve $y = x^2 + 2x - 3$.   Mark two points on the curve: $(-3, 0)$ and $(-1, -4)$. Show that the slope of the line through these two points (the secant line) is equal to the difference quotient where $x = -3$ and $h = 2$.

The graph of a function y = f(x) is given with the coordinates of four points. Graph the functions in Exercises 8 – 10 using your understanding of reflections and translations and with no additional computations. Label the new points that correspond to the four labeled points.

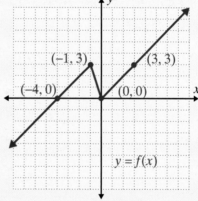

8. $y = f(x) + 3$ **9.** $y = f(x + 2)$ **10.** $y = f(x - 1) - 4$

In Exercises 11 and 12, use a graphing calculator to graph each pair of functions on the same set of axes.

11. $y = -2x^2$ and $y = 3x^2$ **12.** $y = 2(x - 1)^2 + 1$ and $y = 2x^2 + 1$

9.2 Parabolas as Conic Sections

*In Exercises 13 – 18, find **a.** the vertex, **b.** the y-intercepts, and **c.** the line of symmetry. Then draw the graph.*

13. $x = 2y^2$ **14.** $x = -\dfrac{1}{2}y^2$ **15.** $x = y^2 - 6y + 2$

16. $x = (y - 1)^2 + 2$ **17.** $x + 2 = (y - 2)^2$ **18.** $y = -2x^2 + 5x - 3$

Use a graphing calculator to graph each of the parabolas in Exercises 19 and 20. Use the trace and zoom features of the calculator to estimate the y-intercepts of the parabolas.

19. $x = -y^2 + 3$ **20.** $x = 2y^2 - 3y$

9.3   Distance Formula and Circles

In Exercises 21 – 28, find the distance between the two given points.

21. $(1, 3), (5, 6)$ **22.** $(-1, 2), (6, -10)$ **23.** $(-4, -2), (-4, 6)$

24. $(-5, 2), (6, 2)$ **25.** $(-3, -4), (1, 5)$ **26.** $(1, 7), (4, -2)$

27. $(-10, 20), (18, -21)$ **28.** $(3, 8), (-10, -16)$

Find equations for each of the circles in Exercises 29 – 32.

29. Center $(0, 0)$; $r = 2\sqrt{2}$ **30.** Center $(0, 0)$; $r = 3\sqrt{3}$

31. Center $(-2, 1)$; $r = 7$ **32.** Center $(5, 4)$; $r = 2\sqrt{5}$

Write each of the equations in Exercises 33 – 38 in standard form. Find the center and radius of the circle and then sketch the graph.

33. $x^2 + y^2 = 36$ **34.** $x^2 + y^2 = 18$ **35.** $x^2 + y^2 - 6y = 0$

36. $x^2 + y^2 + 4x + 2y = 4$ **37.** $x^2 + y^2 + 12x - 2y = -1$ **38.** $x^2 + y^2 - 10y + 5 = 0$

39. Use the Pythagorean Theorem to determine whether or not the triangle with the three given points is a right triangle: $A(-2, -4), B(1, -1), C(-3, 3)$.

40. Find the perimeter of the triangle determined by the given points: $P(-3, 5), Q(1, -5), R(3, -1)$. Is this triangle a right triangle? Why or why not?

9.4   Ellipses and Hyperbolas

In Exercises 41 – 50, write each equation in standard form, then sketch the graph. For hyperbolas, graph the asymptotes as well.

41. $x^2 + 4y^2 = 36$ **42.** $16x^2 + y^2 = 64$ **43.** $x^2 - y^2 = 81$

44. $x^2 - 9y^2 = 9$ **45.** $y^2 - 3x^2 = 9$ **46.** $4y^2 - x^2 = 36$

47. $4(x + 1)^2 + (y - 1)^2 = 100$ **48.** $(x - 3)^2 + y^2 = 25$

49. $(y + 2)^2 - 4x^2 = 16$ **50.** $9x^2 + (y - 4)^2 = 9$

9.5   Nonlinear Systems of Equations

Solve each system of equations in Exercises 51 – 60. Sketch the graphs.

51. $\begin{cases} y = x^2 - 1 \\ x - y = -5 \end{cases}$

52. $\begin{cases} y = -x^2 \\ x - 2y = 10 \end{cases}$

53. $\begin{cases} x^2 + y^2 = 4 \\ x + y = -2 \end{cases}$

54. $\begin{cases} x^2 - 4y^2 = 12 \\ x = 4y \end{cases}$

55. $\begin{cases} x^2 + y^2 = 16 \\ x^2 - y^2 = 16 \end{cases}$

56. $\begin{cases} y^2 - x^2 = 5 \\ 4x^2 + y^2 = 25 \end{cases}$

57. $\begin{cases} x^2 + y^2 - 6y = 18 \\ x - y = 0 \end{cases}$

58. $\begin{cases} y = x^2 - 4x + 5 \\ y = -x^2 + 4x + 5 \end{cases}$

59. $\begin{cases} x^2 - y^2 = 4 \\ x^2 + y^2 = -2y \end{cases}$

60. $\begin{cases} x^2 + y^2 = 10 \\ y^2 - x^2 = 8 \end{cases}$

In Exercises 61 – 64, use a graphing calculator to graph and estimate the solutions to each system of equations. If necessary, round values to two decimal places.

61. $\begin{cases} y = x^2 - 3 \\ x + y = -6 \end{cases}$

62. $\begin{cases} x^2 + y^2 = 9 \\ y = 2x + 5 \end{cases}$

63. $\begin{cases} y = 2x + 3 \\ y = -(x+1)^2 \end{cases}$

64. $\begin{cases} x^2 + 4y^2 = 5 \\ y = -x \end{cases}$

Chapter 9 Test

1. Let $f(x) = 5 - 7x$. Find:

    a. $f(2)$

    b. $f(a+2)$

    c. $\dfrac{f(x+h) - f(x)}{h}$

2. Let $f(x) = 2x^2 + 8x$. Find:

    a. $f(-3)$

    b. $f(a-1)$

    c. $\dfrac{f(x+h) - f(x)}{h}$

The graph of $y = |x|$ is given along with a few points as aids. Graph the functions in Exercises 3 – 5 using your understanding of reflections and translations with no additional computations.

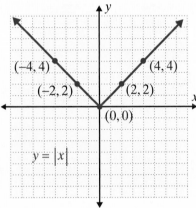

3. $y = -f(x)$

4. $y = f(x) + 3$

5. $y = f(x+1) - 1$

The graph of $y = g(x)$ is given along with a few points as aids. Graph the functions in Exercises 6 – 8 using your understanding of reflections and translations with no additional computations. Label the new points that correspond to the four labeled points.

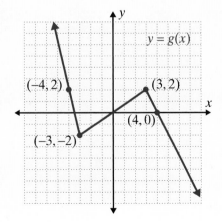

6. $y = -g(x)$

7. $y = g(x+2)$

8. $y = g(x-1) - 4$

*In Exercises 9 and 10, find **a.** the vertex, **b.** the y-intercepts, and **c.** the line of symmetry. Graph each curve.*

9. $x = y^2 - 5$

10. $x + 4 = y^2 + 3y$

11. Find the distance between the two points $(5, -2)$ and $(-4, 1)$.

12. The triangle determined by the three points $A(-4, -3)$, $B(0, 3)$, and $C(6, -1)$ is a right triangle. Use the distance formula and the Pythagorean Theorem to show that this is true.

13. Find the equation for the circle with radius 5 and center at (3, 4). Graph the circle.

14. Write the equation $x^2 + y^2 - 2y - 8 = 0$ in standard form. Then find the center and the radius of the circle, and sketch the graph.

In Exercises 15 and 16, write the equation for each ellipse in standard form and then sketch the graph.

15. $x^2 + 4y^2 = 9$

16. $25x^2 + 4y^2 = 100$

In Exercises 17 and 18, write the equation for each hyperbola in standard form and then sketch the graph and the asymptotes of the hyperbola.

17. $9x^2 - 4y^2 = 36$

18. $\dfrac{y^2}{9} - \dfrac{x^2}{16} = 1$

19. Graph the ellipse $\dfrac{(x+1)^2}{36} + \dfrac{(y-2)^2}{9} = 1$. Label the vertices and the center.

20. Graph given hyperbola and its asymptotes: $\dfrac{(x+1)^2}{16} - \dfrac{y^2}{9} = 1$.

21. State the definition of an ellipse.

In Exercises 22 – 24, solve each system of equations. Sketch the graphs.

22. $\begin{cases} x^2 + y^2 = 29 \\ x - y = 3 \end{cases}$

23. $\begin{cases} x^2 + 2y^2 = 4 \\ x = y^2 - 2 \end{cases}$

24. $\begin{cases} x^2 + y^2 = 25 \\ x^2 - y^2 = 7 \end{cases}$

Cumulative Review: Chapters 1–9

Simplify each of the expressions in Exercises 1 – 4. Assume all variables are positive.

1. $\dfrac{x^{-3} \cdot x}{x^2 \cdot x^{-4}}$

2. $\left(\dfrac{2x^{-1}y^2}{3x^3y^{-2}}\right)^{-2}$

3. $5x^{\frac{1}{2}} \cdot x^{\frac{1}{4}}$

4. $\left(4x^{-\frac{2}{3}}y^{\frac{2}{5}}\right)^{\frac{3}{2}}$

5. Write $\left(7x^3y\right)^{\frac{2}{3}}$ in radical notation.

6. Write $\sqrt[3]{32x^6y}$ in exponential notation.

Completely factor each expression in Exercises 7 – 9.

7. $2x^3 + 54$

8. $3x^2y - 48y$

9. $x^3 - 4x^2 + 3x - 12$

Perform the indicated operations and simplify in Exercises 10 – 12.

10. $2\sqrt{12} + 5\sqrt{108} - 7\sqrt{27}$

11. $3\sqrt{48x} - 2\sqrt{75x} + 5\sqrt{24}$

12. $2\sqrt{-28} + \sqrt{-175} - 3\sqrt{-112}$

13. Find an equation in standard form for the line parallel to $5x - 2y = 8$ and passing through $(-2, 3)$.

14. Find an equation in standard form for the line perpendicular to $4x + 3y = 8$ and passing through $(4, -1)$.

15. Divide by using the division algorithm. Write the answer in the form $Q + \dfrac{R}{D}$, where the degree of $R <$ the degree of D.

$$\frac{2x^3 - 8x^2 + 10x - 100}{x - 5}$$

Perform the indicated operations in Exercises 16 and 17.

16. $\dfrac{x}{2x^2 - 5x - 12} - \dfrac{x+1}{6x^2 + 5x - 6}$

17. $\dfrac{x^2 + 2x - 3}{x^2 + x - 2} \div \dfrac{9 - x^2}{x^2 - x - 6}$

18. Find the discriminant and determine the nature of the solution(s) of the quadratic equation $9x^2 + 6x + 1 = 0$.

Solve each of the equations in Exercises 19 – 23.

19. $6x^3 = 13x^2 + 5x$

20. $\left(x - 5\right)^2 = 16$

21. $3x^2 + 2x - 2 = 0$

22. $\dfrac{5}{x-3} - \dfrac{3}{x+2} = \dfrac{1}{x^2 - x - 6}$

23. $\sqrt{x + 14} - 2 = x$

24. Solve the following system of linear equations: $\begin{cases} 2x - 3y = 5 \\ -5x + y = 7 \end{cases}$.

25. Solve the following system by using Gaussian elimination: $\begin{cases} 2x - 3y + z = -4 \\ x + 2y - z = 5 \\ 3x + y + 2z = -5 \end{cases}$.

26. Evaluate the following logarithmic expressions.

   a. $\log_{10} 10,000$ **b.** $\log_2 64$ **c.** $\log_6 1$

27. Write each expression as a single logarithm.

   a. $3\log_b x - 2\log_b y$ **b.** $\dfrac{1}{3}\log 8 + \log 5 - \log y$

Solve the equations given in Exercises 28 – 31. Use a calculator to find the answers in decimal form (accurate to 4 decimal places).

28. $6^{4-3x} = 6^{x^2}$ **29.** $5^x = 20$

30. $\log(x-1) + \log(x-5) = 1$ **31.** $3\ln x - 10 = 0$

In Exercises 32 and 33, write the quadratic function in the form $y = a(x-h)^2 + k$. Find the line of symmetry, the vertex, the domain, the range, and the zeros. Graph the function.

32. $y = \dfrac{1}{2}x^2 - 3$ **33.** $y = -x^2 + 4x - 4$

34. For the equation $x = -y^2 - 3$, find **a.** the vertex, **b.** the y-intercepts, and **c.** the axis of symmetry. Then draw the graph.

Solve the inequalities in Exercises 35 – 37 algebraically. Write the answers in interval notation and then graph each solution set on a number line.

35. $\dfrac{x^2 - 4x - 5}{x} \geq 0$ **36.** $|2x - 5| \leq 1$ **37.** $2x^2 + 13x - 7 \leq 0$

38. Solve the following system of linear inequalities graphically: $\begin{cases} 3x + y < 4 \\ x - y \geq 2 \end{cases}$.

Graph each of the equations in Exercises 39 – 42.

39. $5x + 2y = 8$ **40.** $y = 2x^2 - 4x - 3$

41. $4x^2 - y^2 = 16$ **42.** $x^2 + 2x + y^2 - 2y = 4$

43. Given that $g(x) = \sqrt{x}$, graph both $g(x)$ and $h(x) = \sqrt{x-7} + 2$. State the domain and range for each function.

44. Given that $f(x) = 3x - 9$, find $f^{-1}(x)$ and graph both functions.

45. $f(x) = x^2 + 3$ and $g(x) = \sqrt{x-3}$ for $x \geq 0$.

    a. Find $f(g(x))$.

    b. Find $g(f(x))$.

    c. Are these two functions inverses of each other? Explain.

    d. Graph both functions.

46. For $f(x) = 4x - 7$, find:

    a. $f(3)$

    b. $f(x-1)$

    c. $f(x) + 3$

    d. $\dfrac{f(x+h) - f(x)}{h}$

47. For $g(x) = 2x^2 + 3x - 1$, find:

    a. $g(2)$

    b. $g(x) - 2$

    c. $g(x-2)$

    d. $\dfrac{g(x+h) - g(x)}{h}$

48. Graph the two functions $y = e^x$ and $y = \ln x$. Label two points on each graph. State which lines are asymptotes for each function.

49. The product of two consecutive positive odd integers is 323. Find the integers.

50. z varies directly as the cube of x and inversely as y. If $z = 10$ when $x = 3$ and $y = 5$, find z if $x = 6$ and $y = 4$.

51. The average of a number and its square root is 21. Find the number.

52. **Vote Counting:** Bethany and Walter need to count the votes for the school election. If Bethany can count all of the votes in 5 hours working alone, and Walter can count all of the votes in 4 hours working alone, how long will it take to count the votes working together?

53. **Investing:** Alisha has $9000 to invest in two different accounts. One account pays interest at the rate of 7%; the other pays at the rate of 8%. If she wants her annual interest to total $684, how much should she invest at each rate?

54. **Path of a ball:** The height of a ball projected vertically is given by the function $h = -16t^2 + 80t + 48$ where h is the height in feet and t is the time in seconds.

    a. When will the ball reach maximum height?

    b. What will be the maximum height?

55. **Sales revenue:** A store owner estimates that by charging x dollars for a certain shirt, he can sell $60 - 2x$ shirts each week. What price will give him maximum receipts?

56. **Growth of bacteria:** Determine the exponential function that fits the following information: $y_0 = 3000$ bacteria, and there are 243,000 bacteria after 3 days. Use the equation $y = y_0 b^t$ where t is measured in days.

57. **Sales revenue:** The revenue function is given by $R(x) = x \cdot p(x)$ dollars, where x is the number of units sold and $p(x)$ is the unit price. If $p(x) = 30(2)^{\frac{-x}{3}}$, find the revenue when 24 units are sold.

58. Find the perimeter of the triangle determined by the points $A(1, 3)$, $B(0, -5)$, and $C(4, 6)$. Explain why this triangle is or is not a right triangle.

Use a graphing calculator and the CALC and zero features to graph the functions in Exercises 59 – 62 and estimate the x-intercepts. Round zeros to two decimal places.

59. $y = 2x - 5$

60. $y = x^2 - 3$

61. $y = -x^2 + 5$

62. $y = (x + 3)^2 - 6$

In Exercises 63 and 64, use a graphing calculator to solve (or estimate the solutions of) the equations and inequalities. Write the solutions to the inequalities in interval notation and find any approximations accurate to two decimal places.

63. $f(x) = x^2 - 3x - 10$

    a. $x^2 - 3x - 10 = 0$

    b. $x^2 - 3x - 10 > 0$

    c. $x^2 - 3x - 10 < 0$

64. $f(x) = -2x^2 + 3x + 5$

    a. $-2x^2 + 3x + 5 = 0$

    b. $-2x^2 + 3x + 5 > 0$

    c. $-2x^2 + 3x + 5 < 0$

In Exercises 65 and 66, use a graphing calculator to graph both equations and estimate the solutions to the system. If necessary, round values to two decimal places.

65. $\begin{cases} x^2 + y^2 = 35 \\ x + y = 4 \end{cases}$

66. $\begin{cases} 3x^2 + 4y^2 = 12 \\ x = y^2 - 3 \end{cases}$

Sequences, Series, and the Binomial Theorem

Did You Know?

One of the outstanding mathematicians of the Middle Ages was Leonardo Fibonacci (Leonardo, son of Bonaccio), also known as Leonardo of Pisa (c. 1170 – 1250). Fibonacci's name is attached to an interesting sequence of numbers 1, 1, 2, 3, 5, 8, 13, . . . , x, y, $x + y$, . . . , the so-called Fibonacci sequence, where each term after the first two is obtained by adding the preceding two terms together. The sequence of numbers arises from a problem found in Fibonacci's writings. How many pairs of rabbits can be produced from a single pair in a year if every month each pair begets a new pair that from the second month on becomes productive? The answer to this odd problem is the sum of the first 12 terms of the Fibonacci sequence. Can you verify this?

The Fibonacci sequence itself has been found to have many beautiful and interesting properties. For example, of mathematical interest is the fact that any two successive terms in the sequence are

Fibonacci

relatively prime; that is, their greatest common divisor is one. In the world of nature, the terms of the Fibonacci sequence also appear. Spirals formed by natural objects such as centers of daisies, pine cone scales, pineapple scales, and leaves generally have two sets of spirals, one clockwise, one counterclockwise. Each set is made up of a specific number of spirals in each direction, the number of spirals being adjacent terms in the Fibonacci sequence. For example, in pine cone scales, 5 spiral one way and 8 spiral the other; on pineapples, 8 one way and 13 the other.

Leonardo Fibonacci was best known for the texts he wrote in which he introduced the Hindu-Arabic numeral system. He participated in the mathematical tournaments held at the court of the emperor Frederick I, and he used some of the challenge problems in the book he wrote. Fibonacci traveled widely and became acquainted with the different arithmetic systems in use around the Mediterranean. His most famous text, *Liber Abaci* (1202), combined arithmetic and elementary algebra with an emphasis on commercial applied problems. His attempt to reform and improve the study of mathematics in Europe was not too successful; he seemed to be ahead of his time. But his books did much to introduce Hindu-Arabic notation into Europe, and his sequence has provided interesting problems throughout the history of mathematics. In the United States, a Fibonacci Society exists to study the properties of this mysterious and intriguing sequence of numbers.

"There is no branch of mathematics, however abstract, which may not some day be applied to phenomena of the real world."

Nikolai Ivanovich Lobachevsky
(1792 – 1856)

Chapter 10 provides an introduction to a powerful notation using the Greek letter Σ, capital sigma. With this Σ notation and a few basic properties, we will develop some algebraic formulas related to sums of numbers and, in some cases, even infinite sums. The concept of having the sum of an infinite number of numbers equal to some finite number introduces the idea of limits. Consider adding fractions in the following manner:

$$\frac{1}{2} = \frac{1}{2}; \quad \frac{1}{2} + \frac{1}{4} = \frac{3}{4}; \quad \frac{1}{2} + \frac{1}{4} + \frac{1}{8} = \frac{7}{8}; \quad \frac{1}{2} + \frac{1}{4} + \frac{1}{8} + \frac{1}{16} = \frac{15}{16}; \quad \frac{1}{2} + \frac{1}{4} + \frac{1}{8} + \frac{1}{16} + \frac{1}{32} = \frac{31}{32}.$$

Continuing to add fractions in this manner, in which the denominators are successive powers of two, will give sums that get closer and closer to 1. We say that the sums "approach" 1, and that 1 is the **limit** of the sum. These fascinating ideas are discussed in Section 10.4 and are fundamental in the development of calculus and higher level mathematics.

Other topics in this chapter find applications in courses in probability and statistics, as well as in more advanced courses in mathematics and computer science.

10.1   Sequences

- *Write several terms of a **sequence** given the formula for its general term.*
- *Find the formula for the general term of a sequence given several terms.*
- *Determine whether a sequence is **increasing**, **decreasing**, or **neither**.*

In mathematics, a **sequence** is a list of numbers that occur in a certain order. Each number in the sequence is called a **term** of the sequence, and a sequence may have a finite number of terms or an infinite number of terms. For example,

$2, 4, 6, 8, 10, 12, 14, 16, 18$ is a **finite sequence** consisting of positive even integers less than 20.

$3, 6, 9, 12, 15, 18, \ldots$ is an **infinite sequence** consisting of the multiples of 3.

The infinite sequence of the multiples of 3 can be described in the following way.

For any positive integer n, the corresponding number in the list is $3n$. Thus we know that

$3 \cdot 6 = 18$ and 18 is the 6th number in the sequence,
$3 \cdot 7 = 21$ and 21 is the 7th number in the sequence,
$3 \cdot 8 = 24$ and 24 is the 8th number in the sequence, and so on.

Infinite Sequence

> *An **infinite sequence** (or a **sequence**) is a function that has the positive integers as its domain.*

NOTES   A finite sequence will be so indicated. The word sequence, used alone, indicates an infinite sequence.

Consider the function $f(n) = \dfrac{1}{2^n}$ where n is any positive integer.

For this function,

$$f(1) = \frac{1}{2^1} = \frac{1}{2}$$

$$f(2) = \frac{1}{2^2} = \frac{1}{4}$$

$$f(3) = \frac{1}{2^3} = \frac{1}{8}$$

$$f(4) = \frac{1}{2^4} = \frac{1}{16}$$

$$\vdots$$

$$f(n) = \frac{1}{2^n}$$

$$\vdots$$

Or, using ordered pair notation,

$$f = \left\{ \left(1, \frac{1}{2}\right), \left(2, \frac{1}{4}\right), \left(3, \frac{1}{8}\right), \left(4, \frac{1}{16}\right), ..., \left(n, \frac{1}{2^n}\right), ... \right\}.$$

The **terms** of the sequence are the numbers

$$\frac{1}{2}, \frac{1}{4}, \frac{1}{8}, \frac{1}{16}, ..., \frac{1}{2^n}, ...$$

Because the order of terms corresponds to the positive integers, it is customary to indicate a sequence by writing only its terms. In general discussions and formulas, a sequence may be indicated with subscript notation as

$$a_1, a_2, a_3, a_4, ..., a_n, ...$$

The general term a_n is called the nth **term** of the sequence. The entire sequence can be denoted by writing the nth term in braces as in $\{a_n\}$. Thus,

$$\{a_n\} \quad \text{and} \quad a_1, a_2, a_3, a_4, \ldots, a_n, \ldots$$

are both representations of the sequence with

a_1 as the first term,

a_2 as the second term,

a_3 as the third term,

$\vdots$

a_n as the nth term,

$\vdots$

Example 1: Writing Terms of a Sequence

a. Write the first three terms of the sequence $\left\{\dfrac{n}{n+1}\right\}$.

Solution: $a_1 = \dfrac{1}{1+1} = \dfrac{1}{2}$ $a_2 = \dfrac{2}{2+1} = \dfrac{2}{3}$ $a_3 = \dfrac{3}{3+1} = \dfrac{3}{4}$

b. If $\{b_n\} = \{2n-1\}$, find b_1, b_2, b_3, and b_{50}.

Solution: $b_1 = 2 \cdot 1 - 1 = 1$ $b_2 = 2 \cdot 2 - 1 = 3$

$b_3 = 2 \cdot 3 - 1 = 5$ $b_{50} = 2 \cdot 50 - 1 = 99$

Example 2: Finding the General Formula of a Sequence

a. Determine a_n if the first five terms of the sequence $\{a_n\}$ are $3, 5, 7, 9, 11$.

Solution: By studying the numbers carefully, we see that they are odd numbers. Odd numbers are of the form $2n+1$ or $2n-1$. Because the first term of the sequence is 3, we have $a_n = 2n+1$.

b. Determine a_n if the first five terms of the sequence $\{a_n\}$ are $0, 3, 8, 15, 24$.

Solution: In this case, study the numbers carefully and look for some pattern (or formula) that seems reasonable for one or two of the numbers. Then, after making this educated guess, check to see whether the remaining numbers fit your guess.

If not, guess again with some basic reasoning for at least one of the positions. In this case you might think as follows:

8 is the 3rd term, $3^2 = 9$ and $9 - 1 = 8$;

15 is the 4th term, $4^2 = 16$ and $16 - 1 = 15$.

So, a good guess seems to be $a_n = n^2 - 1$.

Checking: $a_1 = 1^2 - 1 = 0$

$a_2 = 2^2 - 1 = 3$ Although the formula for a_n may not be

$a_3 = 3^2 - 1 = 8$ obvious, with practice it becomes easier

$a_4 = 4^2 - 1 = 15$ to find.

$a_5 = 5^2 - 1 = 24$

We see that $a_n = n^2 - 1$ is indeed the correct formula.

Example 3: Application

A pick-up truck sells for $45,000 new. Each year its value depreciates by 15% of its value for the previous year. What will be its value at the end of each of the next three years?

Solution: Each year the value will be 85% of its value the previous year. (Because it depreciates by 15%, the new value will be $100\% - 15\% = 85\%$ of its previous value.) We can find the value at the end of each year with the following sequence where $v_0 = \$45,000$, the initial value:

Year 1: $v_1 = v_0 \cdot (0.85) = 45,000 \cdot (0.85) = \$38,250$
Year 2: $v_2 = v_1 \cdot (0.85) = 38,250 \cdot (0.85) = \$32,512.50$
Year 3: $v_3 = v_2 \cdot (0.85) = 32,512.50 \cdot (0.85) = \$27,635.63$

Alternating Sequence

*An **alternating sequence** is a sequence in which the terms alternate in sign.*

In an **alternating sequence**, if one term is positive, then the next term is negative. Example 4 illustrates an alternating sequence. Note that alternating sequences generally involve the expression $(-1)^n$ or $(-1)^{n+1}$ as a factor in the numerator or denominator.

Example 4: An Alternating Sequence

Write the first five terms of the sequence in which $a_n = \dfrac{(-1)^n}{n}$.

Solution: $a_1 = \dfrac{(-1)^1}{1} = -1$ $a_4 = \dfrac{(-1)^4}{4} = \dfrac{1}{4}$

$a_2 = \dfrac{(-1)^2}{2} = \dfrac{1}{2}$ $a_5 = \dfrac{(-1)^5}{5} = -\dfrac{1}{5}$

$a_3 = \dfrac{(-1)^3}{3} = -\dfrac{1}{3}$

Increasing and Decreasing Sequences

In subscript notation, the term a_{n+1} is the term following a_n, and this term is found by substituting $n + 1$ for n in the formula for the general term. For example,

$$\text{if } a_n = \frac{1}{3n}, \text{ then } a_{n+1} = \frac{1}{3(n+1)} = \frac{1}{3n+3}.$$

Similarly,

$$\text{if } b_n = n^2, \text{ then } b_{n+1} = (n+1)^2.$$

If the terms of a sequence grow successively smaller, then the sequence is said to be **decreasing**. If the terms grow successively larger, then the sequence is said to be **increasing**. The following definitions state these ideas algebraically. (Note that a sequence may be **neither** increasing nor decreasing.)

Decreasing Sequence

A sequence $\{a_n\}$ is

decreasing if $a_n > a_{n+1}$ for all n.

(Successive terms become smaller.)

Increasing Sequence

A sequence $\{a_n\}$ is

increasing if $a_n < a_{n+1}$ for all n.

(Successive terms become larger.)

Example 5: A Decreasing Sequence

Show that the sequence $\{a_n\} = \left\{\dfrac{1}{2^n}\right\}$ is decreasing.

Solution: Write the terms a_n and a_{n+1} in formula form and compare them algebraically:

$$a_n = \frac{1}{2^n} \quad \text{and} \quad a_{n+1} = \frac{1}{2^{n+1}} = \frac{1}{2 \cdot 2^n}.$$

Comparing denominators we see that $2^n < 2^{n+1}$.

Therefore,

$$\frac{1}{2^n} > \frac{1}{2^{n+1}} \quad \text{and} \quad a_n > a_{n+1}.$$

Note that the larger denominator gives a smaller fraction.

The sequence $\{a_n\} = \left\{\dfrac{1}{2^n}\right\}$ is decreasing.

Example 6: An Increasing Sequence

Show that the sequence $\{b_n\} = \{n+3\}$ is increasing.

Solution: In formula form: $b_n = n+3$ and $b_{n+1} = (n+1)+3 = n+4$.

Because $n+3 < n+4$, we have $b_n < b_{n+1}$ and the sequence is increasing.

Example 7: A Sequence that is Neither Increasing Nor Decreasing

Show that the sequence $\{c_n\} = \left\{2+(-1)^n\right\}$ is neither increasing nor decreasing.

Solution: The first four terms of the sequence are

$$c_1 = 2+(-1)^1 = 2-1 = 1$$

$$c_2 = 2+(-1)^2 = 2+1 = 3 \qquad \text{Here } c_1 < c_2.$$

$$c_3 = 2+(-1)^3 = 2-1 = 1 \qquad \text{Now } c_2 > c_3.$$

$$c_4 = 2+(-1)^4 = 2+1 = 3 \qquad \text{But } c_3 < c_4.$$

From this pattern we see that the sequence is $1, 3, 1, 3, \ldots$ which is neither increasing nor decreasing.

Practice Problems

Write the first three terms of each sequence.

1. $\{n^2\}$ **2.** $\{2n+1\}$ **3.** $\left\{\dfrac{1}{n+1}\right\}$

4. *Find a formula for the general term of sequence* $-1, 1, 3, 5, 7, \ldots$

5. *Determine whether the sequence* $\left\{(-1)^{n+1}\right\}$ *is increasing, decreasing, or neither. Is this sequence an alternating sequence?*

10.1 Exercises

Write the first four terms of each of the sequences in Exercises 1 – 18.

1. $\{2n-1\}$ **2.** $\{4n+1\}$ **3.** $\left\{1+\dfrac{1}{n}\right\}$

4. $\left\{\dfrac{n+3}{n+1}\right\}$ **5.** $\{n^2+n\}$ **6.** $\{n-n^2\}$

7. $\{2^n\}$ **8.** $\left\{\left(\dfrac{1}{2}\right)^n\right\}$ **9.** $\left\{(-1)^n\left(n^2+1\right)\right\}$

10. $\left\{(-1)^n\left(\dfrac{n}{n+1}\right)\right\}$ **11.** $\left\{(-1)^n\left(\dfrac{1}{2n+3}\right)\right\}$ **12.** $\left\{(-1)^{n-1}\left(3^n\right)\right\}$

13. $\{2^n-n^2\}$ **14.** $\left\{\dfrac{n(n-1)}{2}\right\}$ **15.** $\left\{\dfrac{1+(-1)^n}{2}\right\}$

16. $\left\{\left(\dfrac{2}{3}\right)^n\right\}$ **17.** $\left\{\left(-\dfrac{1}{2}\right)^{n+1}\right\}$ **18.** $\left\{\dfrac{2n}{n+1}\right\}$

Find a formula for the general term of each sequence in Exercises 19 – 28.

19. $2, 5, 8, 11, 14, \ldots$ **20.** $5, 9, 13, 17, 21, \ldots$ **21.** $6, 12, 18, 24, 30, \ldots$

22. $1, -3, 5, -7, 9, \ldots$ **23.** $-3, 7, -11, 15, -19, \ldots$ **24.** $1, 4, 9, 16, 25, \ldots$

25. $5, 10, 20, 40, 80, \ldots$ **26.** $\dfrac{1}{3}, \dfrac{1}{4}, \dfrac{1}{5}, \dfrac{1}{6}, \dfrac{1}{7}, \ldots$ **27.** $\dfrac{1}{2}, \dfrac{1}{4}, \dfrac{1}{8}, \dfrac{1}{16}, \dfrac{1}{32}, \ldots$

28. $2, 5, 10, 17, 26, \ldots$

For each of the sequences in Exercises 29 – 34, determine whether it is increasing or decreasing. Justify your answer by comparing a_n *with* a_{n+1}.

29. $\{n+4\}$ **30.** $\{1-2n\}$ **31.** $\left\{\dfrac{1}{n+3}\right\}$

Answers to Practice Problems: **1.** $1, 4, 9$ **2.** $3, 5, 7$ **3.** $\dfrac{1}{2}, \dfrac{1}{3}, \dfrac{1}{4}$ **4.** $a_n = 2n-3$ **5.** Neither, Yes

32. $\left\{ \dfrac{1}{3^n} \right\}$

33. $\left\{ \dfrac{2n+1}{n} \right\}$

34. $\left\{ \dfrac{n}{n+1} \right\}$

35. Show that the sequence of digits from the irrational number $\pi = 3.1415926535...$ is neither increasing nor decreasing. [**Note:** There is no formula for a_n.]

36. Show that the sequence $\left\{ \dfrac{(-1)^{n+1}}{3n} \right\}$ is neither increasing nor decreasing. Is this an alternating sequence?

Write the terms of the finite sequences described in Exercises 37 – 40. Then answer the stated question.

37. Buying a car: A certain automobile costs $40,000 new and depreciates at a rate of $\dfrac{3}{10}$ of its current value each year. What will be its value after 3 years?

38. Bacteria growth: A culture of bacteria triples everyday. If there were 100 bacteria in the original culture, how many would be present after 4 days?

39. Bouncing a ball: A ball is dropped from a height of 10 meters. Each time it bounces, it rises to $\dfrac{2}{5}$ of its previous height. How high will it bounce on its fourth bounce?

40. University enrollment: A local university is experiencing a declining enrollment of 6% per year. If the present enrollment is 20,000, what is the projected enrollment after 5 years?

Writing and Thinking About Mathematics

41. The famous Fibonacci sequence is an example of a recurrence sequence. That is, each term depends on previous terms.

   a. Write the first six terms of the Fibonacci sequence defined as
   $F_{n+2} = F_{n+1} + F_n$ where $F_1 = 1$ and $F_2 = 1$.

   b. Form the sequence of the differences of successive terms. What do you notice?

HAWKES LEARNING SYSTEMS: INTERMEDIATE ALGEBRA SOFTWARE

   ▪ Sequences

10.2 Sigma Notation

- *Write sums using Σ notation.*
- *Find the values of sums written in Σ notation.*

A sequence has an infinite number of terms. To find the sum of just a few terms of a sequence means to find the sum of a **finite sequence**. Such a sum is called a **partial sum** of the sequence and can be indicated by using **sigma notation** with the Greek letter capital sigma, Σ.

Partial Sums Using Sigma Notation

*The nth **partial sum**, S_n, of the first n terms of a sequence $\{a_n\}$ is*

$$S_n = \sum_{k=1}^{n} a_k = a_1 + a_2 + a_3 + ... + a_n.$$

*k is called the **index of summation**, and k takes the integer values 1, 2, 3, ..., n.*
*n is the **upper limit of summation**, and 1 is the **lower limit of summation**.*

NOTES
As we will see in Section 10.4, sigma notation can be used to indicate the sum of an entire sequence by using the symbol for infinity (∞), in place of n. Also, the upper and lower limits of summation can be adjusted to pick out a particular part of the sequence. For example, k can begin with 7 and stop with 10. There may be special times, because of the way formulas are written, when k would begin with 0.

To understand the concept of partial sums, consider the sequence $\left\{\dfrac{1}{n}\right\}$ and the following partial sums:

$$S_1 = a_1 = \frac{1}{1}$$

$$S_2 = a_1 + a_2 = \frac{1}{1} + \frac{1}{2}$$

$$S_3 = a_1 + a_2 + a_3 = \frac{1}{1} + \frac{1}{2} + \frac{1}{3}$$

$$\vdots$$

$$S_n = a_1 + a_2 + a_3 + ... + a_n = \frac{1}{1} + \frac{1}{2} + \frac{1}{3} + ... + \frac{1}{n}.$$

In some cases, the lower limit of summation in sigma notation may be an integer other than 1. Also, letters other than k may be used as the index of summation. The lower case letters $i, j, k, l, m,$ and n are commonly used.

For example, the sum of the second through sixth terms of the sequence $\{n^2\}$ can be written in sigma notation as

$$\sum_{i=2}^{6} i^2 = 2^2 + 3^2 + 4^2 + 5^2 + 6^2.$$

If the number of terms is large, then three dots are used to indicate missing terms after a pattern has been established with the first three or four terms. For example,

$$S_{100} = \sum_{k=1}^{100} (k-1) = 0 + 1 + 2 + 3 + \ldots + 99.$$

Example 1: Sigma Notation

Write the indicated sums of the terms and find the value of each sum.

a. $\displaystyle\sum_{k=1}^{4} k^3$

Solution: $\displaystyle\sum_{k=1}^{4} k^3 = 1^3 + 2^3 + 3^3 + 4^3 = 1 + 8 + 27 + 64 = 100$

b. $\displaystyle\sum_{k=5}^{9} (-1)^k k$

Solution: $\displaystyle\sum_{k=5}^{9} (-1)^k k = (-1)^5 \, 5 + (-1)^6 \, 6 + (-1)^7 \, 7 + (-1)^8 \, 8 + (-1)^9 \, 9$

$$= -5 + 6 - 7 + 8 - 9 = -7$$

Example 2: Writing Sums in Sigma Notation

Write each sum using sigma notation.

a. $4 + 9 + 16 + 25$

Solution: We note that each number is a square beginning with 2^2. Two possible forms are the following:

$$4 + 9 + 16 + 25 = \sum_{k=2}^{5} k^2 \quad \text{or} \quad 4 + 9 + 16 + 25 = \sum_{k=1}^{4} (k+1)^2.$$

b. $\dfrac{5}{2} + \dfrac{10}{3} + \dfrac{15}{4} + \dfrac{20}{5} + \dfrac{25}{6} + \dfrac{30}{7}$

Solution: We note that each numerator is a multiple of 5 and each denominator is 1 larger than its position. One possible form is as follows:

$$\frac{5}{2} + \frac{10}{3} + \frac{15}{4} + \frac{20}{5} + \frac{25}{6} + \frac{30}{7} = \sum_{k=1}^{6} \frac{5k}{k+1}.$$

Properties of Σ Notation

The following properties of Σ notation are useful in developing systematic methods for finding sums of certain types of finite and infinite sequences.

Properties of Σ Notation

For sequences $\{a_n\}$ and $\{b_n\}$ and any real number c:

I. $\displaystyle\sum_{k=1}^{n} a_k = \sum_{k=1}^{i} a_k + \sum_{k=i+1}^{n} a_k$ for any i, $1 \le i \le n-1$

II. $\displaystyle\sum_{k=1}^{n} \left(a_k + b_k\right) = \sum_{k=1}^{n} a_k + \sum_{k=1}^{n} b_k$

III. $\displaystyle\sum_{k=1}^{n} ca_k = c\sum_{k=1}^{n} a_k$

IV. $\displaystyle\sum_{k=1}^{n} c = nc$

These properties follow directly from the associative, commutative, and distributive properties for sums of real numbers.

I. $\displaystyle\sum_{k=1}^{n} a_k = a_1 + a_2 + \ldots + a_i + a_{i+1} + a_{i+2} + \ldots + a_n$

$$= \left(a_1 + a_2 + \ldots + a_i\right) + \left(a_{i+1} + a_{i+2} + \ldots + a_n\right)$$

$$= \sum_{k=1}^{i} a_k + \sum_{k=i+1}^{n} a_k$$

II. $\displaystyle\sum_{k=1}^{n} \left(a_k + b_k\right) = \left(a_1 + b_1\right) + \left(a_2 + b_2\right) + \ldots + \left(a_n + b_n\right)$

$$= \left(a_1 + a_2 + \ldots + a_n\right) + \left(b_1 + b_2 + \ldots + b_n\right)$$

$$= \sum_{k=1}^{n} a_k + \sum_{k=1}^{n} b_k$$

III. $\displaystyle\sum_{k=1}^{n} ca_k = ca_1 + ca_2 + \ldots + ca_n$

$$= c\left(a_1 + a_2 + \ldots + a_n\right)$$

$$= c\sum_{k=1}^{n} a_k$$

IV. $\displaystyle\sum_{k=1}^{n} c = \underbrace{c + c + c + \ldots + c}_{c \text{ appears } n \text{ times}} = nc$

Example 3: Properties of Σ Notation

a.   If $\sum_{k=1}^{7} a_k = 40$ and $\sum_{k=1}^{30} a_k = 75$, find $\sum_{k=8}^{30} a_k$.

   Solution: Since $\sum_{k=1}^{7} a_k + \sum_{k=8}^{30} a_k = \sum_{k=1}^{30} a_k$,

   then $40 + \sum_{k=8}^{30} a_k = 75$

   $$\sum_{k=8}^{30} a_k = 35.$$

b.   If $\sum_{k=1}^{50} 3a_k = 600$, find $\sum_{k=1}^{50} a_k$.

   Solution: Since $\sum_{k=1}^{50} 3a_k = 3\sum_{k=1}^{50} a_k$,

   then $3\sum_{k=1}^{50} a_k = 600$

   $$\sum_{k=1}^{50} a_k = 200.$$

Practice Problems

1.   Write the indicated sum of terms and find the value of the sum: $\sum_{k=1}^{4} \left(k^2 - 1\right)$.

2.   Write the sum $10 + 12 + 14 + 16 + 18$ in Σ notation.

3. $\sum_{k=1}^{5} a_k = 20$ and $\sum_{k=6}^{10} a_k = 30$. Find $\sum_{k=1}^{10} 2a_k$.

Answers to Practice Problems: **1.** $0 + 3 + 8 + 15 = 26$ **2.** $\sum_{k=5}^{9} 2k$ or $\sum_{k=1}^{5}(2k+8)$ **3.** 100

10.2 Exercises

For each of the sequences given in Exercises 1 – 10, write out the partial sums S_1, S_2, S_3, and S_4 and evaluate each partial sum.

1. $\{3k-1\}$ **2.** $\{2k+5\}$ **3.** $\left\{\dfrac{k}{k+1}\right\}$ **4.** $\left\{\dfrac{k+1}{k}\right\}$

5. $\{(-1)^{k-1}k^2\}$ **6.** $\{(-1)^k k^3\}$ **7.** $\left\{\dfrac{1}{2^k}\right\}$ **8.** $\left\{\left(\dfrac{2}{3}\right)^k\right\}$

9. $\left\{\left(-\dfrac{2}{3}\right)^k\right\}$ **10.** $\{k^2-k\}$

Write the indicated sums in Exercises 11 – 26 in expanded form and evaluate.

11. $\displaystyle\sum_{k=1}^{5} 2k$ **12.** $\displaystyle\sum_{k=1}^{11} k(k-1)$ **13.** $\displaystyle\sum_{k=2}^{6}(k+3)$ **14.** $\displaystyle\sum_{k=9}^{11}(2k+1)$

15. $\displaystyle\sum_{k=2}^{4}\dfrac{1}{k}$ **16.** $\displaystyle\sum_{k=1}^{3}\dfrac{1}{2k}$ **17.** $\displaystyle\sum_{k=1}^{3}2^k$ **18.** $\displaystyle\sum_{k=10}^{15}(-1)^k$

19. $\displaystyle\sum_{k=4}^{8}k^2$ **20.** $\displaystyle\sum_{k=1}^{4}k^3$ **21.** $\displaystyle\sum_{k=3}^{6}(9-2k)$ **22.** $\displaystyle\sum_{k=2}^{7}(4k-1)$

23. $\displaystyle\sum_{k=2}^{5}(-1)^k\left(k^2+k\right)$ **24.** $\displaystyle\sum_{k=1}^{6}(-1)^k\left(k^2-2\right)$ **25.** $\displaystyle\sum_{k=1}^{5}\dfrac{k}{k+1}$

26. $\displaystyle\sum_{k=3}^{5}(-1)^k\left(\dfrac{k+1}{k^2}\right)$

Write the sums in Exercises 27 – 35 in sigma notation. There may be more than one correct answer.

27. $1+3+5+7+9$

28. $16+25+36+49$

29. $-1+1+(-1)+1+(-1)$

30. $4+7+10+13+16$

31. $\dfrac{1}{8}-\dfrac{1}{27}+\dfrac{1}{64}-\dfrac{1}{125}+\dfrac{1}{216}$

32. $\dfrac{1}{8}+\dfrac{1}{16}+\dfrac{1}{32}+\dfrac{1}{64}+\dfrac{1}{128}$

33. $\dfrac{4}{5}+\dfrac{5}{6}+\dfrac{6}{7}+\ldots+\dfrac{15}{16}$

34. $8+15+24+35+48$

35. $\dfrac{6}{25}+\dfrac{7}{36}+\dfrac{8}{49}+\dfrac{9}{64}+\ldots+\dfrac{13}{144}$

Find the indicated sums in Exercises 36 – 45.

36. $\displaystyle\sum_{k=1}^{14} a_k = 18$ and $\displaystyle\sum_{k=1}^{14} b_k = 21$. Find $\displaystyle\sum_{k=1}^{14}\left(a_k + b_k\right)$.

37. $\displaystyle\sum_{k=1}^{19} a_k = 23$ and $\displaystyle\sum_{k=1}^{19} b_k = 16$. Find $\displaystyle\sum_{k=1}^{19}\left(a_k - b_k\right)$.

38. $\displaystyle\sum_{k=1}^{15} a_k = 19$. Find $\displaystyle\sum_{k=1}^{15} 3a_k$.

39. $\displaystyle\sum_{k=1}^{25} a_k = 63$ and $\displaystyle\sum_{k=1}^{11} a_k = 15$. Find $\displaystyle\sum_{k=12}^{25} a_k$.

40. $\displaystyle\sum_{k=1}^{18} a_k = 41$ and $\displaystyle\sum_{k=1}^{18} b_k = 62$. Find $\displaystyle\sum_{k=1}^{18}\left(3a_k - 2b_k\right)$.

41. $\displaystyle\sum_{k=1}^{21} a_k = -68$ and $\displaystyle\sum_{k=1}^{21} b_k = 39$. Find $\displaystyle\sum_{k=1}^{21}\left(a_k + 2b_k\right)$.

42. $\displaystyle\sum_{k=1}^{16} a_k = 56$ and $\displaystyle\sum_{k=17}^{40} a_k = 42$. Find $\displaystyle\sum_{k=1}^{40} a_k$.

43. $\displaystyle\sum_{k=13}^{29} a_k = 84$ and $\displaystyle\sum_{k=1}^{29} a_k = 143$. Find $\displaystyle\sum_{k=1}^{12} 5a_k$.

44. $\displaystyle\sum_{k=1}^{20} b_k = 34$ and $\displaystyle\sum_{k=1}^{20}\left(2a_k + b_k\right) = 144$. Find $\displaystyle\sum_{k=1}^{20} a_k$.

45. $\displaystyle\sum_{k=1}^{27} a_k = 46$ and $\displaystyle\sum_{k=1}^{10} a_k = 122$. Find $\displaystyle\sum_{k=11}^{27} 2a_k$.

Writing and Thinking About Mathematics

46. Use the sum of two expressions in sigma notation to represent the following sum: $-22 + 3 - 24 + 6 - 26 + 9 - 28 + 12 - 30 + 15$.

HAWKES LEARNING SYSTEMS: INTERMEDIATE ALGEBRA SOFTWARE

- Sigma Notation

10.3 Arithmetic Sequences

- *Determine whether or not a sequence is **arithmetic**.*
- *Find the general term for an arithmetic sequence.*
- *Find the specified terms of an arithmetic sequence.*
- *Find the sum of the first n terms of an arithmetic sequence.*

There are many types of sequences studied in higher levels of mathematics. In the next two sections, we will discuss two types of sequences: **arithmetic sequences** and **geometric sequences**. In this discussion, sigma notation is used and formulas for finding sums are developed. For arithmetic sequences, we can find sums of only a finite number of terms. For geometric sequences, we can find sums of a finite number of terms and, in some special cases, we define the sum of an infinite number of terms.

Arithmetic Sequences

The sequences

$$3, 5, 7, 9, 11, 13, \ldots$$

$$4, 5, 6, 7, 8, 9, \ldots$$

$$-2, -5, -8, -11, -14, -17, \ldots$$

all have a common characteristic. This characteristic is that any two consecutive terms in each sequence have the **same difference**.

$$3, \ 5, \ 7, \ 9, 11, 13, \ldots \qquad\qquad 5 - 3 = 2, 7 - 5 = 2, 9 - 7 = 2, \text{ and so on.}$$
$$2 \ \ 2 \ \ 2 \ \ 2 \ \ 2$$

$$4, \ 5, \ 6, \ 7, \ 8, \ 9, \ldots \qquad\qquad 5 - 4 = 1, 6 - 5 = 1, 7 - 6 = 1, \text{ and so on.}$$
$$1 \ \ 1 \ \ 1 \ \ 1 \ \ 1$$

$$-2, \ -5, \ -8, \ -11, \ -14, \ -17, \ldots \qquad -5 - (-2) = -3, -8 - (-5) = -3, \text{ and so on.}$$
$$-3 \ \ -3 \ \ -3 \ \ -3 \ \ -3$$

Such sequences are called **arithmetic sequences** or **arithmetic progressions**.

Arithmetic sequences are closely related to linear functions. To see this relationship we can plot the points of an arithmetic sequence and note that the rise from one point to the next is the constant difference, d. See Figure 10.1 as an illustration with a positive value for d. Note that the points do indeed lie on a straight line. The slope of the line is d.

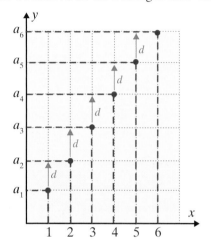

Figure 10.1

Arithmetic Sequences

A sequence $\{a_n\}$ is called an ***arithmetic sequence*** (or ***arithmetic progression***) if for any natural number k,

$$a_{k+1} - a_k = d \qquad \text{where } d \text{ is a constant.}$$

d is called the ***common difference***.

Example 1: An Arithmetic Sequence

Show that the sequence $\{2n - 3\}$ is arithmetic by finding d.

Solution: $a_k = 2k - 3$ and $a_{k+1} = 2(k+1) - 3 = 2k - 1$

Therefore, $a_{k+1} - a_k = (2k - 1) - (2k - 3) = 2k - 1 - 2k + 3 = 2$.

So $d = 2$, and the sequence $\{2n - 3\}$ **is arithmetic**.

Example 2: A Sequence that is Not Arithmetic

Show that the sequence $\{n^2\}$ is **not** arithmetic.

Solution: Consider $a_3, a_2,$ and a_1 as follows:

$$a_3 = 3^2 = 9, \ a_2 = 2^2 = 4, \ \text{and} \ a_1 = 1^2 = 1.$$

Therefore, $a_3 - a_2 = 9 - 4 = 5$ and $a_2 - a_1 = 4 - 1 = 3.$

So, there is no common difference between successive terms, and the sequence $\{n^2\}$ **is not arithmetic**.

NOTES

Note that in Example 2, we needed to show only one case in which the difference between two sets of consecutive terms was not the same. This is called finding a **counterexample**. However, to show that something is true in every case, we must use general formulas, as in Example 1.

The nth Term of an Arithmetic Sequence

To find a formula for the nth term of an arithmetic sequence, we can proceed as follows:

$$a_1 = a_1 \qquad\qquad \text{first term}$$
$$a_2 = a_1 + d \qquad\qquad \text{second term}$$
$$a_3 = a_2 + d = a_1 + 2d \qquad\qquad \text{third term}$$
$$a_4 = a_3 + d = a_1 + 3d \qquad\qquad \text{fourth term}$$
$$\vdots \qquad\qquad\qquad \vdots$$
$$a_n = a_{n-1} + d = a_1 + (n-1)d \qquad\qquad n^{th} \text{ term}$$

Formula for the nth Term of an Arithmetic Sequence

If $\{a_n\}$ is an arithmetic sequence, then the nth term has the form

$$a_n = a_1 + (n-1)d$$

where d is the common difference between consecutive terms.

Example 3: Finding a Specific Term in an Arithmetic Sequence

a. **Given a_1 and d:**
    If $a_1 = 5$ and $d = 3$, find a_{16}.

    Solution: To find a_{16}, let $n = 16$ in the formula for a_n:

    $$a_{16} = a_1 + (n-1)d$$
    $$= 5 + (16-1) \cdot 3$$
    $$= 5 + 15 \cdot 3$$
    $$= 50$$

b. **Given two consecutive terms:**
    If the first two terms are -2 and 8, find a_{20}.

    Solution: Knowing two consecutive terms, we can find d.
    Because the sequence is arithmetic, $d = a_2 - a_1 = 8 - (-2) = 10$.
    Now, using the formula for a_n, we have

    $$a_{20} = a_1 + (n-1)d$$
    $$= -2 + (20-1) \cdot 10$$
    $$= -2 + 19 \cdot 10$$
    $$= 188$$

c. **Using a system of equations given two terms:**
    If $a_3 = 6$ and $a_{21} = -48$, find a_1 and d for the arithmetic sequence.

    Solution: Using the formula $a_n = a_1 + (n-1)d$ and solving simultaneous equations, we have

    $$-48 = a_1 + 20d$$
    $$6 = a_1 + 2d$$

    $$\begin{array}{rcl} -48 &=& a_1 + 20d \\ \underline{-6} &=& \underline{-a_1 - 2d} \\ -54 &=& 18d \\ -3 &=& d. \end{array}$$

    Then,

    $$\begin{array}{rcl} 6 &=& a_1 + 2(-3) \\ 12 &=& a_1. \end{array}$$

    So, $a_1 = 12$ and $d = -3$.

Example 4: Finding *n* Given Certain Conditions

Given that $\{a_n\}$ is an arithmetic sequence, $a_{10} = -12$, $d = -3$, and $a_n = -72$, find *n*.

Solution: Because the sequence is arithmetic, we can use the conditions that $a_{10} = -12$ and $d = -3$ to find a_1 as follows:

$$a_n = a_1 + (n-1)d$$

$$-12 = a_1 + (10-1)(-3)$$

$$-12 = a_1 + (9)(-3)$$

$$-12 = a_1 - 27$$

$$15 = a_1$$

Now, using the formula again, we can solve for *n* as follows:

$$-72 = 15 + (n-1)(-3)$$

$$-72 = 15 - 3n + 3$$

$$-90 = -3n$$

$$30 = n$$

Thus, -72 is the 30th term in the sequence.

Partial Sums of Arithmetic Sequences

Consider the problem of finding the sum $S = \sum_{k=1}^{6}(4k-1)$. We can, of course, write all the terms and then add them:

$$S = \sum_{k=1}^{6}(4k-1) = 3 + 7 + 11 + 15 + 19 + 23 = 78.$$

However, to understand how the general formula is developed, we first write the sum and then write the sum again with the terms in reverse order. Adding vertically gives the same sum six times:

$$
\begin{array}{rcrcrcrcrcrcr}
S & = & 3 & + & 7 & + & 11 & + & 15 & + & 19 & + & 23 \\
S & = & 23 & + & 19 & + & 15 & + & 11 & + & 7 & + & 3 \\
\hline
2S & = & 26 & + & 26 & + & 26 & + & 26 & + & 26 & + & 26 \\
2S & = & 6 & \cdot & 26 \\
S & = & 78
\end{array}
$$

Using this same procedure with general terms in the subscript notation, we can develop the formula for the sum of any finite arithmetic sequence. Suppose that the *n* terms are

$$a_1, \quad a_2 = a_1 + d, \quad a_3 = a_1 + 2d, \quad ..., \quad a_{n-1} = a_n - d, a_n.$$

Thus, writing the terms in both ascending order and descending order and adding vertically, we have

$$S = a_1 + (a_1 + d) + (a_1 + 2d) + \ldots + (a_n - 2d) + (a_n - d) + a_n$$
$$S = a_n + (a_n - d) + (a_n - 2d) + \ldots + (a_1 + 2d) + (a_1 + d) + a_1$$
$$\overline{2S = (a_1 + a_n) + (a_1 + a_n) + (a_1 + a_n) + \ldots + (a_1 + a_n) + (a_1 + a_n) + (a_1 + a_n)}$$

$(a_1 + a_n)$ appears n times

$$2S = n(a_1 + a_n)$$
$$S = \frac{n}{2}(a_1 + a_n).$$

Partial Sums of Arithmetic Sequences

*The nth **partial sum**, S_n, of the first n terms of an arithmetic sequence $\{a_n\}$ is*

$$S_n = \sum_{k=1}^{n} a_k = \frac{n}{2}(a_1 + a_n).$$

A special case of an arithmetic sequence is $\{n\}$ and the corresponding sum of the first n terms:

$$\sum_{k=1}^{n} k = 1 + 2 + 3 + \ldots + n.$$

In this case, n = the number of terms, $a_1 = 1$, and $a_n = n$, so

$$\sum_{k=1}^{n} k = \frac{n}{2}(1 + n).$$

Gauss

German mathematician Carl Friedrich Gauss (1777 – 1855) understood and applied this sum at the age of 7 in order to solve an arithmetic problem given to him and his classmates as "busy" work. Gauss probably observed the following pattern when told to find the sum of the whole numbers from 1 to 100:

$$1 + 2 + 3 + \ldots + 98 + 99 + 100$$

101
101
101

He saw that 101 was a sum 50 times. Thus, to find the sum he simply multiplied $101 \cdot 50 = 5050$. Not bad for a 7-year old!

Example 5: Finding Partial Sums of Arithmetic Sequences

First show that the corresponding sequence is an arithmetic sequence by finding $a_{k+1} - a_k = d$. Then find the indicated sum using the formula.

a. $\displaystyle\sum_{k=1}^{75} k = 1 + 2 + 3 + \ldots + 75$

Solution: $a_{k+1} = k + 1$ and $a_k = k$

So, $d = a_{k+1} - a_k = k + 1 - k = 1$.

The upper limit of the summation is $n = 75$.

$$\sum_{k=1}^{75} k = \frac{75}{2}(1 + 75)$$

$$= \frac{75}{2}(76)$$

$$= 2850$$

b. $\displaystyle\sum_{k=1}^{50} 3k = 3 + 6 + 9 + \ldots + 150$

Solution: $a_k = 3k$ and $a_{k+1} = 3(k + 1) = 3k + 3$

$a_{k+1} - a_k = (3k + 3) - 3k = 3 = d$

So, $\{3k\}$ is an arithmetic sequence.

$$\sum_{k=1}^{50} 3k = \frac{50}{2}(3 + 150) \qquad \text{Here } n = 50,\ a_1 = 3,\ \text{and } a_{50} = 150.$$

$$= 25(153)$$

$$= 3825$$

Also, Property III of Section 10.2 can be used to find the sum.

$$\sum_{k=1}^{50} 3k = 3\sum_{k=1}^{50} k \qquad \text{Property III of Section 10.2}$$

$$= 3 \cdot \frac{50}{2}(1 + 50) \qquad \text{Here } n = 50,\ a_1 = 1,\ \text{and } a_{50} = 50.$$

$$= 3 \cdot 25 \cdot 51$$

$$= 3825$$

c. $\displaystyle\sum_{k=1}^{70} (-2k + 5) = 3 + 1 + (-1) + (-3) + \ldots + (-135)$

Solution: $a_k = -2k + 5$ and $a_{k+1} = -2(k + 1) + 5 = -2k + 3$,

$a_{k+1} - a_k = (-2k + 3) - (-2k + 5) = 3 - 5 = -2 = d$

So, $\{-2k + 5\}$ is an arithmetic sequence.

$$\sum_{k=1}^{70}(-2k+5) = \frac{70}{2}\big[3+(-135)\big] \qquad \text{Here } n = 70,\ a_1 = 3,\ a_{70} = -135.$$

$$= 35(-132)$$

$$= -4620$$

Also, Properties II, III, and IV of Section 10.2 and the sum of a finite arithmetic sequence from this section can be used to find the sum.

$$\sum_{k=1}^{70}(-2k+5) = \sum_{k=1}^{70}-2k + \sum_{k=1}^{70}5 \qquad \text{Property II}$$

$$= -2\sum_{k=1}^{70}k + \sum_{k=1}^{70}5 \qquad \text{Property III}$$

$$= -2\cdot\frac{70}{2}(1+70) + 70\cdot5 \qquad \begin{array}{l}\text{The sum of a finite arithmetic}\\ \text{sequence and Property IV}\end{array}$$

$$= -2\cdot35\cdot71 + 350$$

$$= -4970 + 350$$

$$= -4620$$

Example 6: An Application of an Arithmetic Sequence

Suppose that you are offered two jobs by the same company. The first job has a starting salary of $35,000, with a guaranteed raise of $2000 per year. The second job starts at $40,000 with a guaranteed raise of $1200 per year.

a. What would be your salary in the 10th year of each of these jobs?

b. If you were to stay 10 years with the company, which job would pay the most in total salary?

Solution: Since the salary would increase the same amount each year, the yearly salaries form arithmetic sequences and we can use the corresponding formulas for a_{10} and S_{10}.

a. **First job:** $a_{10} = a_1 + (10-1)d = 35{,}000 + 9(2000) = \$53{,}000$
  Second job: $a_{10} = a_1 + (10-1)d = 40{,}000 + 9(1200) = \$50{,}800$

In 10 years, you would be making a higher salary on the first job.

Continued on the next page...

b.   **First job:** $S_{10} = \dfrac{10}{2}(35,000 + 53,000) = \$440,000$

Second job: $S_{10} = \dfrac{10}{2}(40,000 + 50,800) = \$454,000$

At least for 10 years, the second job would pay more in total salary.

Practice Problems

1.   *Show that the sequence* $\{3n+5\}$ *is arithmetic by finding d.*
2.   *Find the 40th term of the arithmetic sequence with 1, 6, and 11 as its first three terms.*
3.   *Find* $\displaystyle\sum_{k=1}^{50}(3k+5).$

10.3 Exercises

Determine which of the sequences in Exercises 1 – 10 are arithmetic. Find the common difference and the nth term for each arithmetic sequence.

1. $2, 5, 8, 11, \ldots$ 2. $-3, 1, 5, 9, \ldots$ 3. $7, 5, 3, 1, \ldots$ 4. $5, 6, 7, 8, \ldots$

5. $1, 2, 3, 5, 8, \ldots$ 6. $2, 4, 8, 16, \ldots$ 7. $6, 2, -2, -6, \ldots$ 8. $4, -1, -6, -11, \ldots$

9. $0, \dfrac{1}{2}, 1, \dfrac{3}{2}, \ldots$ 10. $2, \dfrac{7}{3}, \dfrac{8}{3}, 3, \ldots$

In Exercise 11 – 20, write the first five terms of the sequence and determine which of the sequences are arithmetic.

11. $\{2n-1\}$ 12. $\{4-n\}$ 13. $\{(-1)^n(3n-2)\}$ 14. $\left\{n+\dfrac{n}{2}\right\}$

15. $\{5-6n\}$ 16. $\left\{\dfrac{1}{n+1}\right\}$ 17. $\left\{7-\dfrac{n}{3}\right\}$ 18. $\{(-1)^{n+1}(2n+1)\}$

19. $\left\{\dfrac{1}{2n}\right\}$ 20. $\left\{\dfrac{2}{3}n-\dfrac{7}{3}\right\}$

Answers to Practice Problems: **1.** $d = 3$ **2.** $a_{40} = 196$ **3.** 4075

Find the general term, a_n, for each of the arithmetic sequences in Exercises 21 – 30.

21. $a_1 = 1,\ d = \dfrac{2}{3}$ **22.** $a_1 = 9,\ d = -\dfrac{1}{3}$ **23.** $a_1 = 7,\ d = -2$

24. $a_1 = -3,\ d = \dfrac{4}{5}$ **25.** $a_1 = 10,\ a_3 = 13$ **26.** $a_1 = 6,\ a_5 = 4$

27. $a_{10} = 13,\ a_{12} = 3$ **28.** $a_5 = 7,\ a_9 = 19$ **29.** $a_{13} = 60,\ a_{23} = 75$

30. $a_{11} = 54,\ a_{29} = 180$

In Exercises 31 – 38, $\{a_n\}$ is an arithmetic sequence. Find the indicated quantity.

31. $a_1 = 8,\ a_{11} = 168.$ Find a_{15}.   **32.** $a_1 = 17,\ a_9 = -55.$ Find a_{20}.

33. $a_6 = 8,\ a_4 = 2.$ Find a_{18}.   **34.** $a_{16} = 12,\ a_7 = 30.$ Find a_9.

35. $a_{13} = 34,\ d = 2,\ a_n = 22.$ Find n.   **36.** $a_4 = 20,\ d = 3,\ a_n = 44.$ Find n.

37. $a_{10} = 41,\ d = 4,\ a_n = 77.$ Find n.   **38.** $a_3 = 15,\ d = -\dfrac{3}{2},\ a_n = 6.$ Find n.

Find the indicated sums in Exercises 39 – 50 by using the formula for partial sums of arithmetic sequences.

39. $-2 + 0 + 2 + 4 + \ldots + 24$ **40.** $3 + 6 + 9 + \ldots + 33$

41. $1 + 6 + 11 + 16 + \ldots + 46$ **42.** $5 + 9 + 13 + 17 + \ldots + 49$

43. $\displaystyle\sum_{k=1}^{9}(3k-1)$ **44.** $\displaystyle\sum_{k=1}^{12}(4-5k)$ **45.** $\displaystyle\sum_{k=1}^{11}(4k-3)$

46. $\displaystyle\sum_{k=1}^{10}(2k+7)$ **47.** $\displaystyle\sum_{k=1}^{13}\left(\dfrac{2k}{3}-1\right)$ **48.** $\displaystyle\sum_{k=1}^{28}(8k-5)$

49. $\displaystyle\sum_{k=1}^{9}\left(k+\dfrac{k}{3}\right)$ **50.** $\displaystyle\sum_{k=1}^{16}\left(9-\dfrac{k}{3}\right)$

Find the indicated sums in Exercises 51 – 54 by using the properties of sigma notation.

51. If $\displaystyle\sum_{k=1}^{33} a_k = -12,$ find $\displaystyle\sum_{k=1}^{33}(5a_k+7)$.   **52.** If $\displaystyle\sum_{k=1}^{15} a_k = 60,$ find $\displaystyle\sum_{k=1}^{15}(-2a_k-5)$.

53. If $\displaystyle\sum_{k=1}^{100}(-3a_k+4) = 700,$ find $\displaystyle\sum_{k=1}^{100} a_k$.   **54.** If $\displaystyle\sum_{k=1}^{50}(2b_k-5) = 32,$ find $\displaystyle\sum_{k=1}^{50} b_k$.

55. Construction: On a certain project, a construction company was penalized for taking more than the contractual time to finish the project. The company forfeited $750 the first day, $900 the second day, $1050 the third day, and so on. How many additional days were needed if the total penalty was $12,150?

56. Real estate: It is estimated that a certain piece of property, currently valued at $480,000, will appreciate as follows: $14,000 the first year, $14,500 the second year, $15,000 the third year, and so on. On this basis, what will be the value of the property after 10 years?

57. Rungs in a ladder: The rungs of a ladder decrease uniformly in length from 84 cm to 46 cm. What is the total length of the wood in the rungs if there are 25 of them?

58. Building blocks: How many blocks are there in a pile if there are 19 in the first layer, 17 in the second layer, 15 in the third layer, and so on, with only 1 block on the top layer?

59. Theater: Theater seats are arranged in semicircles so that there are 6 additional seats in each semicircular "row" moving away from the stage. The first row has 20 seats, and there are 20 rows. How many seats are in the last "row"? How many seats are in the theater?

60. Student loan repayment: Samantha accumulated $50,000 in student loans during her four years in college. She has agreed to pay back $1000 the first year and increase the payment by $500 each year thereafter. How much will she have paid back in 12 years?

Writing and Thinking About Mathematics

61. Explain why an alternating sequence (one in which the terms alternate being positive and negative) cannot be an arithmetic sequence.

 HAWKES LEARNING SYSTEMS: INTERMEDIATE ALGEBRA SOFTWARE

- Arithmetic Sequences

10.4

Geometric Sequences and Series

- *Determine whether or not a sequence is **geometric**.*
- *Find the general term of a geometric sequence.*
- *Find the specified terms of a geometric sequence.*
- *Find the sum of the first n terms of a geometric sequence.*
- *Find the sum of an infinite geometric **series**.*

Geometric Sequences

Arithmetic sequences are characterized by having the property that any two consecutive terms have the same difference. **Geometric sequences** are characterized by having the property that **any two consecutive terms are in the same ratio**. That is, if consecutive terms are divided, the ratio will be the same regardless of which two consecutive terms are divided. Consider the three sequences

$$\frac{1}{2}, \frac{1}{4}, \frac{1}{8}, \frac{1}{16}, \frac{1}{32}, \dots$$

$$3, 9, 27, 81, 243, \dots$$

$$-9, 3, -1, \frac{1}{3}, -\frac{1}{9}, \dots$$

As the following patterns show, each of these sequences has a **common ratio** when consecutive terms are divided.

$$\frac{1}{2}, \quad \frac{1}{4}, \quad \frac{1}{8}, \quad \frac{1}{16}, \quad \frac{1}{32}, \dots$$
$$\underbrace{\qquad}_{\frac{1}{2}} \underbrace{\qquad}_{\frac{1}{2}} \underbrace{\qquad}_{\frac{1}{2}} \underbrace{\qquad}_{\frac{1}{2}}$$

$$\frac{\frac{1}{4}}{\frac{1}{2}} = \frac{1}{2}, \frac{\frac{1}{8}}{\frac{1}{4}} = \frac{1}{2}, \frac{\frac{1}{16}}{\frac{1}{8}} = \frac{1}{2}, \text{ and so on.}$$

$$3, \quad 9, \quad 27, \quad 81, \quad 243, \dots$$
$$\underbrace{\qquad}_{3} \underbrace{\qquad}_{3} \underbrace{\qquad}_{3} \underbrace{\qquad}_{3}$$

$$\frac{9}{3} = 3, \frac{27}{9} = 3, \frac{81}{27} = 3, \text{ and so on.}$$

$$-9, \quad 3, \quad -1, \quad \frac{1}{3}, \quad -\frac{1}{9}, \dots$$
$$\underbrace{\qquad}_{-\frac{1}{3}} \underbrace{\qquad}_{-\frac{1}{3}} \underbrace{\qquad}_{-\frac{1}{3}} \underbrace{\qquad}_{-\frac{1}{3}}$$

$$\frac{3}{-9} = -\frac{1}{3}, \frac{-1}{3} = -\frac{1}{3}, \frac{\frac{1}{3}}{-1} = -\frac{1}{3}, \text{ and so on.}$$

Therefore these sequences are **geometric sequences** or **geometric progressions**.

Geometric Sequences

A sequence $\{a_n\}$ is called a **geometric sequence** (or **geometric progression**) if for any positive integer k,

$$\frac{a_{k+1}}{a_k} = r \qquad \text{where } r \text{ is constant and } r \neq 0.$$

r is called the **common ratio**.

Example 1: A Geometric Sequence

Show that the sequence $\left\{\dfrac{1}{3^n}\right\}$ is geometric by finding r.

Solution: $a_k = \dfrac{1}{3^k}$ and $a_{k+1} = \dfrac{1}{3^{k+1}}$

$$\frac{a_{k+1}}{a_k} = \frac{\dfrac{1}{3^{k+1}}}{\dfrac{1}{3^k}} = \frac{1}{3^{k+1}} \cdot \frac{3^k}{1} = \frac{3^k}{3 \cdot 3^k} = \frac{1}{3} = r$$

Example 2: A Sequence that is Not Geometric

Show that the sequence $\{n^2\}$ is not geometric.

Solution: To show that something is not true, we find a **counterexample**. In this case, we find two pairs of successive terms and show that they have different ratios.

Consider the ratio $\dfrac{a_3}{a_2} = \dfrac{3^2}{2^2} = \dfrac{9}{4}$ and the ratio $\dfrac{a_6}{a_5} = \dfrac{6^2}{5^2} = \dfrac{36}{25}$.

Because $\dfrac{9}{4} \neq \dfrac{36}{25}$ there is no common ratio for successive terms and the sequence $\{n^2\}$ is **not** geometric.

If the first term is a_1 and the common ratio is r, then the geometric sequence can be indicated as follows:

$$a_1 = a_1 \quad \rightarrow \quad a_1 \qquad\qquad \text{first term}$$

$$\frac{a_2}{a_1} = r \quad \rightarrow \quad a_2 = a_1 r \qquad\qquad \text{second term}$$

$$\frac{a_3}{a_2} = r \quad \rightarrow \quad a_3 = a_2 r = (a_1 r)r = a_1 r^2 \qquad\qquad \text{third term}$$

$$\frac{a_4}{a_3} = r \quad \rightarrow \quad a_4 = a_3 r = (a_1 r^2)r = a_1 r^3 \qquad\qquad \text{fourth term}$$

$$\vdots \qquad\qquad \vdots \qquad\qquad\qquad\qquad\qquad\qquad\qquad \vdots$$

$$\frac{a_n}{a_{n-1}} = r \quad \rightarrow \quad a_n = a_{n-1} \cdot r = (a_1 r^{n-2})r = a_1 r^{n-1} \quad n^{\text{th}} \text{ term}$$

$$\vdots \qquad\qquad \vdots \qquad\qquad\qquad\qquad\qquad\qquad\qquad \vdots$$

The Formula for the nth Term of a Geometric Sequence

If $\{a_n\}$ is a geometric sequence, then the nth term has the form

$$a_n = a_1 r^{n-1}$$

where r is the common ratio.

Example 3: Finding a Specified Term of a Geometric Sequence

a. If in a geometric sequence $a_1 = 4$ and $r = -\dfrac{1}{2}$, find a_8.

Solution: $a_8 = a_1 r^7 = 4\left(-\dfrac{1}{2}\right)^7 = 2^2\left(-\dfrac{1}{2^7}\right) = -\dfrac{1}{2^5} = -\dfrac{1}{32}$

b. Find the seventh term of the following geometric sequence: $3, \dfrac{3}{2}, \dfrac{3}{4}, \ldots$

Solution: Find r using the formula $r = \dfrac{a_{k+1}}{a_k}$ with $a_1 = 3$ and $a_2 = \dfrac{3}{2}$.

$$r = \frac{a_2}{a_1} = \frac{\frac{3}{2}}{3} = \frac{3}{2} \cdot \frac{1}{3} = \frac{1}{2}$$

Now, the seventh term is $a_7 = a_1 r^{7-1} = 3\left(\dfrac{1}{2}\right)^6 = \dfrac{3}{64}$.

Example 4: Finding a_1 and r for a Geometric Sequence

Find a_1 and r for the geometric sequence in which $a_5 = 2$ and $a_7 = 4$.

Solution: Using the formula $a_n = a_1 r^{n-1}$, we get

$$2 = a_1 r^4 \text{ and } 4 = a_1 r^6.$$

Now, dividing gives

$$\frac{\cancel{a_1} \overset{r^2}{\cancel{r^6}}}{\cancel{a_1} \cancel{r^4}} = \frac{4}{2}$$

$$r^2 = 2$$

$$r = \pm\sqrt{2}.$$

Using these values for r and the fact that $a_5 = 2$, we can find a_1.

For $r = \sqrt{2}$:   For $r = -\sqrt{2}$:

$$2 = a_1 \left(\sqrt{2}\right)^4 \qquad\qquad 2 = a_1 \left(-\sqrt{2}\right)^4$$

$$2 = a_1 \cdot 4 \qquad\qquad\qquad 2 = a_1 \cdot 4$$

$$\frac{1}{2} = a_1 \qquad\qquad\qquad\quad \frac{1}{2} = a_1$$

There are two geometric sequences with $a_5 = 2$ and $a_7 = 4$. In both cases, $a_1 = \frac{1}{2}$. The two possibilities are

$$a_1 = \frac{1}{2} \text{ and } r = \sqrt{2}$$

or $$a_1 = \frac{1}{2} \text{ and } r = -\sqrt{2}.$$

Partial Sum of a Geometric Sequence

The following discussion illustrates the method for finding the formula for the sum of the first n terms of a geometric sequence. Such a sum is called a **partial sum** of the sequence.

The sum of the first 6 terms of the geometric sequence $\left\{\frac{1}{3^n}\right\}$ can be indicated as $S = \sum\limits_{k=1}^{6} \frac{1}{3^k}$. To find the value of this partial sum, we can write the terms and simply add them:

$$S = \sum_{k=1}^{6} \frac{1}{3^k} = \frac{1}{3} + \frac{1}{3^2} + \frac{1}{3^3} + \frac{1}{3^4} + \frac{1}{3^5} + \frac{1}{3^6}$$

$$= \frac{3^5 + 3^4 + 3^3 + 3^2 + 3 + 1}{3^6} = \frac{364}{729}.$$

Adding a few terms is relatively easy. However, the following procedure will help in understanding the development of a general formula for partial sums. Write the terms in order and then multiply each term by $\frac{1}{3}$ (the common ratio, r). Arranging the terms in a vertical format and then subtracting gives the following results.

$$S = \frac{1}{3} + \frac{1}{3^2} + \frac{1}{3^3} + \frac{1}{3^4} + \frac{1}{3^5} + \frac{1}{3^6}$$

$$\frac{1}{3}S = \quad\quad \frac{1}{3^2} + \frac{1}{3^3} + \frac{1}{3^4} + \frac{1}{3^5} + \frac{1}{3^6} + \frac{1}{3^7}$$

$$S - \frac{1}{3}S = \frac{1}{3} - 0 - 0 - 0 - 0 - 0 - \frac{1}{3^7}$$

$$\left(1 - \frac{1}{3}\right)S = \frac{1}{3} - \frac{1}{3^7} \quad\quad\quad \text{Factor out the } S.$$

$$S = \frac{\frac{1}{3} - \frac{1}{3^7}}{1 - \frac{1}{3}} = \frac{\frac{1}{3} - \left(\frac{1}{3}\right)^7}{1 - \frac{1}{3}} \quad\quad \text{Divide by } \left(1 - \frac{1}{3}\right).$$

Using this technique with a general geometric sequence $\{a_1 r^{n-1}\}$ leads to the formula for partial sums of geometric sequences.

Consider the first n terms:

$$a_1, \quad a_2 = a_1 r, \quad a_3 = a_1 r^2, \quad a_4 = a_1 r^3, \quad \ldots, \quad a_{n-1} = a_1 r^{n-2}, \quad a_n = a_1 r^{n-1}.$$

Then, as before, the sum can be written

$$S = a_1 + a_1 r + a_1 r^2 + \ldots + a_1 r^{n-2} + a_1 r^{n-1}$$

$$rS = \quad a_1 r + a_1 r^2 + a_1 r^3 + \ldots + a_1 r^{n-1} + a_1 r^n \quad\quad \text{Multiply each term by } r.$$

$$S - rS = a_1 \quad -0 - 0 - 0 - \ldots - 0 - 0 - a_1 r^n \quad\quad \text{Subtract.}$$

$$(1 - r)S = a_1 \left(1 - r^n\right) \quad\quad\quad\quad\quad\quad\quad\quad \text{Factor.}$$

$$S = \frac{a_1 \left(1 - r^n\right)}{1 - r} \quad\quad\quad\quad\quad\quad\quad\quad\quad\quad \text{Simplify.}$$

Partial Sums of Geometric Sequences

The nth **partial sum**, S_n, of the first n terms of a geometric sequence $\{a_n\}$ is

$$S_n = \sum_{k=1}^{n} a_k = \frac{a_1\left(1-r^n\right)}{1-r} \qquad \text{where } r \text{ is the common ratio and } r \neq 1.$$

Example 5: Partial Sums of Geometric Sequences

First show that the corresponding sequence is a geometric sequence by finding $\dfrac{a_{k+1}}{a_k} = r$.

Then find the indicated sum by using the formula $\displaystyle\sum_{k=1}^{n} a_k = \frac{a_1\left(1-r^n\right)}{1-r}$.

a. $\displaystyle\sum_{k=1}^{10} \frac{1}{2^k}$

Solution: Represent both a_k and a_{k+1} and find the ratio of these two terms.

$$a_k = \frac{1}{2^k} \text{ and } a_{k+1} = \frac{1}{2^{k+1}}$$

$$\frac{a_{k+1}}{a_k} = \frac{\dfrac{1}{2^{k+1}}}{\dfrac{1}{2^k}} = \frac{1}{2^{k+1}} \cdot \frac{2^k}{1} = \frac{1}{2 \cdot 2^{\cancel{k}}} \cdot \frac{2^{\cancel{k}}}{1} = \frac{1}{2} = r$$

So, $\left\{\dfrac{1}{2^n}\right\}$ is a geometric sequence with $a_1 = \dfrac{1}{2}$ and $r = \dfrac{1}{2}$:

$$\sum_{k=1}^{10} \frac{1}{2^k} = \frac{\dfrac{1}{2}\left(1-\left(\dfrac{1}{2}\right)^{10}\right)}{1-\dfrac{1}{2}} = \frac{\dfrac{1}{2}\left(1-\dfrac{1}{1024}\right)}{\dfrac{1}{2}} = \frac{1023}{1024}.$$

b. $\displaystyle\sum_{k=1}^{5} (-1)^k \cdot 3^{\frac{k}{2}}$

Solution: Represent both a_k and a_{k+1} and find the ratio of these two terms.

$$\frac{a_{k+1}}{a_k} = \frac{(-1)^{k+1} \cdot 3^{(k+1)/2}}{(-1)^k \cdot 3^{k/2}} = \frac{\cancel{(-1)^k} (-1) \cdot \cancel{3^{k/2}} \cdot 3^{1/2}}{\cancel{(-1)^k} \cdot \cancel{3^{k/2}}}$$

$$= (-1) \cdot 3^{1/2} = -\sqrt{3} = r$$

So, $\left\{(-1^k) \cdot 3^{k/2}\right\}$ is a geometric sequence with $a_1 = -\sqrt{3}$ and $r = -\sqrt{3}$:

$$\sum_{k=1}^{5} (-1)^k \cdot 3^{k/2} = \frac{-\sqrt{3} \cdot \left(1-\left(-\sqrt{3}\right)^5\right)}{1-\left(-\sqrt{3}\right)} = \frac{-\sqrt{3}\left(1+9\sqrt{3}\right)}{1+\sqrt{3}}.$$

Example 6: Partial Sums of Geometric Sequences

The parents of a small child decide to deposit $1000 annually at the first of each year for 20 years for their child's education. If interest is compounded annually at 8%, what will be the value of the deposits after 20 years? (This type of investment is called an **annuity**.)

Solution: The formula for interest compounded annually is $A = P(1+r)^t$ where A is the amount in the account, r is the annual interest rate (in decimal form), and t is the time (in years).

The first deposit of $1000 will earn interest for 20 years:

$$A_{20} = 1000(1+0.08)^{20} = 1000(1.08)^{20}.$$

The second deposit will earn interest for 19 years:

$$A_{19} = 1000(1.08)^{19}.$$

$$\vdots$$

The last deposit will earn interest for only one year:

$$A_1 = 1000(1.08)^1.$$

The accumulated value of all deposits (plus interest) is the sum of the 20 terms of a geometric sequence.

Value at the end of twenty years:

$$= A_1 + A_2 + \cdots + A_{20} = \sum_{k=1}^{20} 1000(1.08)^k$$

$$= 1000(1.08)^1 + 1000(1.08)^2 + \cdots + 1000(1.08)^{20}$$

$$= \frac{1000(1.08)\left[1-(1.08)^{20}\right]}{1-1.08} \qquad \text{where } a_1 = 1000(1.08) \text{ and } r = 1.08$$

$$= \frac{1080\left[1-4.660957\right]}{-0.08}$$

$$= 49,423. \qquad\qquad \text{Rounded to the nearest dollar}$$

Thus the accumulated value of the annuity is $49,423.

Geometric Series

The indicated sum of all the terms (an infinite number of terms) of a sequence is called an **infinite series** (or simply a **series**). A thorough study of series is a part of calculus. In this text, we will be concerned only with special cases of **geometric series** where $|r| < 1$. We use the symbol for infinity (∞) to indicate that the number of terms is unbounded. The symbol ∞ does not represent a number.

Infinite Series

*The indicated sum of all terms of a sequence is called an **infinite series** (or a **series**). For a sequence $\{a_n\}$, the corresponding series can be written as follows:*

$$\sum_{k=1}^{\infty} (a_k) = a_1 + a_2 + a_3 + \ldots + a_n + \ldots$$

For geometric sequences in the case where $|r| < 1$, it can be shown, in higher level mathematics, that r^n approaches 0 as n approaches infinity. This does not mean that r^n is ever equal to 0, only that it gets closer and closer to 0 as n becomes larger and larger. In symbols, we write

$$r^n \rightarrow 0 \quad \text{as} \quad n \rightarrow \infty.$$

Thus we have the following result if $|r| < 1$:

$$S_n = \frac{a_1(1 - r^n)}{1 - r} \rightarrow \frac{a_1(1 - 0)}{1 - r} = \frac{a_1}{1 - r} \quad \text{as } n \rightarrow \infty.$$

Sum of an Infinite Geometric Series

If $\{a_n\}$ is a geometric sequence and $|r| < 1$, then the sum of the infinite geometric series is

$$S = \sum_{k=1}^{\infty} (a_k) = a_1 + a_1 r + a_1 r^2 + \ldots = \frac{a_1}{1 - r}.$$

Example 7: Geometric Series

Find the sum of each of the following geometric series.

a. $\displaystyle\sum_{k=1}^{\infty}\left(\frac{2}{3}\right)^{k-1}=\left(\frac{2}{3}\right)^{0}+\left(\frac{2}{3}\right)^{1}+\left(\frac{2}{3}\right)^{2}+\left(\frac{2}{3}\right)^{3}+\ldots$

$$=1+\frac{2}{3}+\frac{4}{9}+\frac{8}{27}+\ldots$$

Solution:　Here, $a_1=1$ and $r=\dfrac{2}{3}$. Substitution in the formula yields

$$S=\frac{1}{1-\dfrac{2}{3}}=\frac{1}{\dfrac{1}{3}}=3.$$

b.　$0.3333\ldots=0.\overline{3}$　　　　Recall that the bar over the 3 indicates a repeating pattern of digits in the decimal.

Solution:　$0.33333\ldots=0.3+0.03+0.003+0.0003+0.00003+\ldots$

This format shows that the decimal number can be interpreted as a geometric series with $a_1=0.3=\dfrac{3}{10}$ and $r=0.1=\dfrac{1}{10}$.

Applying the formula gives

$$S=\frac{\dfrac{3}{10}}{1-\dfrac{1}{10}}=\frac{\dfrac{3}{10}}{\dfrac{9}{10}}=\frac{3}{10}\cdot\frac{10}{9}=\frac{1}{3}.$$

In this way, an infinite repeating decimal can be converted to fraction form:

$$0.33333\ldots=\frac{1}{3}.$$

c.　$0.99999\ldots=0.\overline{9}$

Solution:　As shown in Example 7b, we can interpret the decimal number

$$0.99999\ldots=0.9+0.09+0.009+0.0009+\ldots$$

as a geometric series with $a_1=0.9=\dfrac{9}{10}$ and $r=0.1=\dfrac{1}{10}$.

Continued on the next page...

Applying the formula gives

$$S = \frac{\dfrac{9}{10}}{1 - \dfrac{1}{10}} = \frac{\dfrac{9}{10}}{\dfrac{9}{10}} = \frac{9}{10} \cdot \frac{10}{9} = 1.$$

This very interesting result shows that the infinite decimal notation 0.99999... is just another way of writing 1.

d. $5 - 1 + \dfrac{1}{5} - \dfrac{1}{25} + \dfrac{1}{125} - \dfrac{1}{625} + \ldots$

Solution: Here, $a_1 = 5$ and $r = -\dfrac{1}{5}$. A geometric series that alternates in sign will always have a negative value for r. Substitution in the formula gives

$$S = \frac{5}{1 - \left(-\dfrac{1}{5}\right)} = \frac{5}{1 + \dfrac{1}{5}} = \frac{5}{\dfrac{6}{5}} = \frac{5}{1} \cdot \frac{5}{6} = \frac{25}{6}.$$

Practice Problems

1. Show that the sequence $\left\{\dfrac{(-1)^n}{3^n}\right\}$ is geometric by finding r.

2. If in a geometric series $a_1 = 0.1$ and $r = 2$, find a_6.

3. Find the sum $\displaystyle\sum_{k=1}^{5} \dfrac{1}{2^k}$.

4. Represent the decimal $0.\overline{4}$ as a series using Σ notation.

5. Find the sum of the series in Problem 4.

Answers to Practice Problems: **1.** $r = -\dfrac{1}{3}$ **2.** $a_6 = 3.2$ **3.** $\dfrac{31}{32}$ **4.** $\displaystyle\sum_{k=1}^{\infty} \dfrac{4}{10^k}$ **5.** $\dfrac{4}{9}$

10.4 Exercises

In Exercises 1 – 10 determine which sequences are geometric. If the sequence is geometric, find its common ratio and write a formula for the nth term.

1. $2, 4, 6, 8, \ldots$

2. $\dfrac{1}{12}, \dfrac{1}{6}, \dfrac{1}{3}, \dfrac{2}{3}, \ldots$

3. $3, -\dfrac{3}{2}, \dfrac{3}{4}, -\dfrac{3}{8}, \ldots$

4. $5, 9, 13, 17, \ldots$

5. $\dfrac{32}{27}, \dfrac{4}{9}, \dfrac{1}{6}, \dfrac{1}{16}, \ldots$

6. $18, 12, 8, \dfrac{16}{3}, \ldots$

7. $\dfrac{14}{3}, \dfrac{2}{3}, \dfrac{2}{15}, \dfrac{2}{45}, \ldots$

8. $1, -\dfrac{2}{3}, \dfrac{4}{9}, -\dfrac{8}{27}, \ldots$

9. $48, -12, 3, -\dfrac{3}{4}, \ldots$

10. $4, -8, 12, -16, \ldots$

In Exercises 11 – 20 write the first four terms of the sequence and determine which of the sequences are geometric.

11. $\left\{ (-3)^{n+1} \right\}$

12. $\left\{ 3\left(\dfrac{2}{5}\right)^n \right\}$

13. $\left\{ \dfrac{2}{3}n \right\}$

14. $\left\{ (-1)^{n+1}\left(\dfrac{2}{7}\right)^n \right\}$

15. $\left\{ 2\left(-\dfrac{4}{5}\right)^n \right\}$

16. $\left\{ 1+\dfrac{1}{2^n} \right\}$

17. $\left\{ 3(2)^{n/2} \right\}$

18. $\left\{ \dfrac{n^2+1}{n} \right\}$

19. $\left\{ (-1)^{n-1}(0.3)^n \right\}$

20. $\left\{ 6(10)^{1-n} \right\}$

Find the general term a_n for each of the geometric sequences in Exercises 21 – 30.

21. $a_1 = 3, r = 2$

22. $a_1 = -2, r = \dfrac{1}{5}$

23. $a_1 = \dfrac{1}{3}, r = -\dfrac{1}{2}$

24. $a_1 = 5, r = \sqrt{2}$

25. $a_3 = 2, a_5 = 4, r > 0$

26. $a_4 = 19, a_5 = 57$

27. $a_2 = 1, a_4 = 9$

28. $a_2 = 5, a_5 = \dfrac{5}{8}$

29. $a_3 = -\dfrac{45}{16}, r = -\dfrac{3}{4}$

30. $a_4 = 54, r = 3$

In Exercises 31 – 38, $\{a_n\}$ is a geometric sequence. Find the indicated quantity.

31. $a_1 = -32, a_6 = 1$. Find a_8.

32. $a_1 = 20, a_6 = \dfrac{5}{8}$. Find a_7.

33. $a_1 = 18, a_7 = \dfrac{128}{81}$. Find a_5.

34. $a_1 = -3, a_5 = -48$. Find a_7.

35. $a_3 = \dfrac{1}{2}$, $a_7 = \dfrac{1}{32}$. Find a_4.

36. $a_5 = 48$, $a_8 = -384$. Find a_9.

37. $a_1 = -2$, $r = \dfrac{2}{3}$, $a_n = -\dfrac{16}{27}$. Find n.

38. $a_1 = \dfrac{1}{9}$, $r = \dfrac{3}{2}$, $a_n = \dfrac{27}{32}$. Find n.

In Exercises 39 – 56, find the indicated sums.

39. $3 + 9 + 27 + 81 + 243$

40. $-2 + 4 - 8 + 16$

41. $8 + 4 + 2 + \ldots + \dfrac{1}{64}$

42. $3 + 12 + 48 + \ldots + 3072$

43. $\displaystyle\sum_{k=1}^{3} -3\left(\dfrac{3}{4}\right)^{k}$

44. $\displaystyle\sum_{k=1}^{6} \left(\dfrac{-5}{3}\right)\left(\dfrac{1}{2}\right)^{k}$

45. $\displaystyle\sum_{k=1}^{5} \left(\dfrac{2}{3}\right)^{k}$

46. $\displaystyle\sum_{k=1}^{6} \left(\dfrac{1}{3}\right)^{k}$

47. $\displaystyle\sum_{k=4}^{7} 5\left(\dfrac{1}{2}\right)^{k}$

48. $\displaystyle\sum_{k=3}^{6} -7\left(\dfrac{3}{2}\right)^{k}$

49. $\displaystyle\sum_{k=1}^{\infty} \left(\dfrac{3}{4}\right)^{k-1}$

50. $\displaystyle\sum_{k=1}^{\infty} \left(\dfrac{5}{8}\right)^{k-1}$

51. $\displaystyle\sum_{k=1}^{\infty} \left(-\dfrac{1}{2}\right)^{k}$

52. $\displaystyle\sum_{k=1}^{\infty} \left(-\dfrac{2}{5}\right)^{k}$

53. $0.\overline{4}$

54. $0.\overline{6}$

55. $0.\overline{36}$

56. $0.\overline{81}$

57. Trust accounts: When Henry was born, his grandmother deposited $10,000 in a trust account bearing 5% interest compounded annually for him to use for college expenses when he became 21. How much money was in the account on his 21st birthday?

58. Automobile depreciation: An automobile that costs $18,500 new depreciates at a rate of 20% of its value each year. What is its value after 4 years?

59. Car repair: The radiator of a truck contains 20 liters of water. Four liters are drained off and replaced by antifreeze. Then 4 liters of the mixture are drained off and replaced by antifreeze, and so on. This process is continued until six drain-offs and replacements have been made. How much antifreeze is in the final mixture?

60. Investing: Suppose you deposit $1200 in a savings account each year for 8 years. If interest is compounded annually at 6%, what would be the value of the account at the end of 8 years?

61. Certificate of deposit: Kathleen buys a $1000 certificate of deposit each year for 10 years. If the annual interest rate on each CD is 4.5%, what will be the total value of these CDs after 10 years?

62. Decay rate: A substance decays at a rate of $\frac{2}{5}$ of its weight per day. How much of the substance will be present after 4 days if initially there are 500 grams?

63. Bouncing a ball: A ball rebounds to a height that is $\frac{3}{4}$ of its original height. How high will it rise after the fourth bounce if it is dropped from a height of 24 meters?

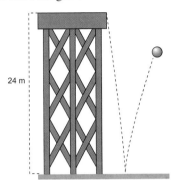

24 m

Writing and Thinking About Mathematics

64. Graph the first 8 partial sums of each geometric series as points to show how the sum of the series approaches a certain value. Show this value as a horizontal line in the graph.

    a. $\displaystyle\sum_{k=1}^{\infty}\left(\frac{1}{2}\right)^{k-1}$
    b. $\displaystyle\sum_{k=1}^{\infty}\frac{(-1)^{k+1}}{3^{k}}$

65. Consider the infinite series $4\cdot\displaystyle\sum_{k=1}^{\infty}\frac{(-1)^{k-1}}{2k-1}$. Write out several (at least 10 to 15) of the partial sums and their values until you can tell what number the partial sums "seem" to be approaching. What is this number?

66. Explain why there is no formula for finding the sum of an infinite geometric series when $r > 1$.

 HAWKES LEARNING SYSTEMS: INTERMEDIATE ALGEBRA SOFTWARE

    ▪ Geometric Sequences and Series

10.5	# The Binomial Theorem

- Calculate *factorials*.
- Expand binomials using the **Binomial Theorem**.
- Find specified terms in binomial expressions.

Factorials

The objective in this section is to develop a formula for writing powers of binomial expressions. This formula is called the **Binomial Theorem** (or **Binomial Expansion**). With the Binomial Theorem we can write products such as

$$(a+b)^3, \quad (x+y)^7, \quad \text{and} \quad (2x+5)^8$$

without having to actually multiply the binomial factors. We will find that the Binomial Theorem does this for us. For example, instead of multiplying three factors as follows,

$$(a+b)^3 = (a+b)(a+b)(a+b)$$
$$= (a^2 + 2ab + b^2)(a+b)$$
$$= a^3 + 2a^2b + ab^2 + a^2b + 2ab^2 + b^3$$
$$= a^3 + 3a^2b + 3ab^2 + b^3$$

knowledge of the Binomial Theorem will allow you to go directly to the final polynomial.

Before discussing the theorem itself, we need to understand the concept of **factorial**. For example, 6! (read "six factorial") represents the product of the positive integers from 6 to 1. Thus,

$$6! = 6 \cdot 5 \cdot 4 \cdot 3 \cdot 2 \cdot 1 = 720.$$

Similarly,

$$10! = 10 \cdot 9 \cdot 8 \cdot 7 \cdot 6 \cdot 5 \cdot 4 \cdot 3 \cdot 2 \cdot 1 = 3,628,800.$$

n Factorial (*n*!)

For any positive integer n,

$$n! = n(n-1)(n-2)\ldots 3 \cdot 2 \cdot 1.$$

n! is read as "n factorial."

To evaluate an expression such as

$$\frac{7!}{6!}$$

do **not** evaluate each factorial in the numerator and denominator. Instead, write the factorials as products and reduce the fraction.

$$\frac{7!}{6!} = \frac{7 \cdot \cancel{6} \cdot \cancel{5} \cdot \cancel{4} \cdot \cancel{3} \cdot \cancel{2} \cdot \cancel{1}}{\cancel{6} \cdot \cancel{5} \cdot \cancel{4} \cdot \cancel{3} \cdot \cancel{2} \cdot \cancel{1}} = 7$$

Note that $\quad n! = (n)(n-1)(n-2)...(3)(2)(1)$

and $\quad (n-1)! = (n-1)(n-2)(n-3)...(3)(2)(1).$

So $\quad\quad\quad n! = n(n-1)!$

In particular, $\quad \dfrac{7!}{6!} = \dfrac{7 \cdot (6!)}{6!} = 7.$

Also, for work with formulas involving factorials, zero factorial is defined to be 1.

0 Factorial (0!)

0! = 1

Using a Calculator to Calculate Factorials

Factorials can be calculated with the TI-84 Plus calculator by pressing the **MATH** *key and going to the menu under* **PRB**. *The fourth item in the list is the factorial symbol,* **!**. *For example, 6! can be calculated as follows:*

1. *Enter 6.*
2. *Press the* **MATH** *key.*
3. *Go to the* **PRB** *heading and press 4.* (**6 !** *will appear on the display.*)
4. *Press* **ENTER** *and* **720** *will appear on the display.*

Example 1: Factorials

Simplify the following expressions.

a. $\dfrac{11!}{8!}$

Solution: $\dfrac{11!}{8!} = \dfrac{11 \cdot 10 \cdot 9 \cdot (8 \cdot 7 \cdot 6 \cdot 5 \cdot 4 \cdot 3 \cdot 2 \cdot 1)}{(8 \cdot 7 \cdot 6 \cdot 5 \cdot 4 \cdot 3 \cdot 2 \cdot 1)} = 990$

or $\dfrac{11!}{8!} = \dfrac{11 \cdot 10 \cdot 9 \cdot 8!}{8!} = 990$

b. $\dfrac{n!}{(n-2)!}$

Solution: $\dfrac{n!}{(n-2)!} = \dfrac{n(n-1)(n-2)!}{(n-2)!} = n(n-1)$

c. $\dfrac{30!}{28!2!}$

Solution: $\dfrac{30!}{28!2!} = \dfrac{\overset{15}{\cancel{30}} \cdot 29 \cdot \cancel{28!}}{\cancel{28!} \cdot \cancel{2} \cdot 1} = 15 \cdot 29 = 435$

Binomial Coefficients

The expression in Example 1c can be written in the following notation.

$$\binom{30}{2} = \frac{30!}{2!28!} \quad \text{and} \quad \binom{30}{28} = \frac{30!}{28!2!}$$

Binomial Coefficient $\dbinom{n}{r}$

For non-negative integers n and r, with $0 \le r \le n$, we define

$$\binom{n}{r} = \frac{n!}{r!(n-r)!}.$$

Because this quantity appears repeatedly in the Binomial Theorem, $\dbinom{n}{r}$ *is often called a **binomial coefficient**.*

To get a formula for $\begin{pmatrix} n \\ n-r \end{pmatrix}$, we apply the formula for $\begin{pmatrix} n \\ r \end{pmatrix}$ and replace r with $n - r$. Thus,

$$\begin{pmatrix} n \\ n-r \end{pmatrix} = \frac{n!}{\left(n-(n-r)\right)!(n-r)!} = \frac{n!}{(n-n+r)!(n-r)!}$$

$$= \frac{n!}{r!(n-r)!} = \begin{pmatrix} n \\ r \end{pmatrix}.$$

Thus,

$$\begin{pmatrix} n \\ n-r \end{pmatrix} = \begin{pmatrix} n \\ r \end{pmatrix}.$$

Example 2: $\begin{pmatrix} n \\ r \end{pmatrix}$

Evaluate the following.

a. $\begin{pmatrix} 8 \\ 2 \end{pmatrix}$ and $\begin{pmatrix} 8 \\ 6 \end{pmatrix}$

Solution: $\begin{pmatrix} 8 \\ 2 \end{pmatrix} = \frac{8!}{2!6!} = \frac{\overset{4}{\cancel{8}} \cdot 7 \cdot \cancel{6!}}{\cancel{2} \cdot \cancel{6!}} = 28$ $\begin{pmatrix} 8 \\ 6 \end{pmatrix} = \frac{8!}{6!2!} = \frac{\overset{4}{\cancel{8}} \cdot 7 \cdot \cancel{6!}}{\cancel{6!} \cdot \cancel{2}} = 28$

b. $\begin{pmatrix} 17 \\ 0 \end{pmatrix}$

Solution: $\begin{pmatrix} 17 \\ 0 \end{pmatrix} = \frac{17!}{0!17!} = \frac{1}{1} = 1$

The Binomial Theorem

The expansions of the binomial $a + b$ from $(a+b)^0$ to $(a+b)^5$ are shown here.

$$(a+b)^0 = 1$$
$$(a+b)^1 = a+b$$
$$(a+b)^2 = a^2 + 2ab + b^2$$
$$(a+b)^3 = a^3 + 3a^2b + 3ab^2 + b^3$$
$$(a+b)^4 = a^4 + 4a^3b + 6a^2b^2 + 4ab^3 + b^4$$
$$(a+b)^5 = a^5 + 5a^4b + 10a^3b^2 + 10a^2b^3 + 5ab^4 + b^5$$

Three patterns are evident.

1. In each case, the **powers of *a* decrease by 1** in each term, and the **powers of *b* increase by 1** in each term.

2. In each term, the sum of the exponents is equal to the exponent on $(a + b)$.

3. A pattern called Pascal's Triangle is formed from the coefficients.

Pascal's Triangle

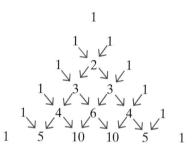

Pascal

In each case, the first and last coefficients are 1, and the other coefficients are the sum of the two numbers above to the left and above to the right of that coefficient. Thus, for $(a+b)^6$, we can construct another row of the triangle as follows:

$$
\begin{array}{ccccccccccccc}
& 1 & & 5 & & 10 & & 10 & & 5 & & 1 & \\
1 & & 6 & & 15 & & 20 & & 15 & & 6 & & 1
\end{array}
$$

and

$$(a+b)^6 = a^6 + 6a^5b + 15a^4b^2 + 20a^3b^3 + 15a^2b^4 + 6ab^5 + b^6.$$

Note that the coefficients can be written in factorial notation as follows:

$$\binom{6}{0} = \frac{6!}{0!6!} = 1 \qquad \binom{6}{1} = \frac{6!}{1!5!} = 6 \qquad \binom{6}{2} = \frac{6!}{2!4!} = 15 \qquad \binom{6}{3} = \frac{6!}{3!3!} = 20$$

$$\binom{6}{4} = \frac{6!}{4!2!} = 15 \qquad \binom{6}{5} = \frac{6!}{5!1!} = 6 \qquad \binom{6}{6} = \frac{6!}{6!0!} = 1$$

So, the expansion can be written in the following form:

$$(a+b)^6 = \binom{6}{0}a^6 + \binom{6}{1}a^5b + \binom{6}{2}a^4b^2 + \binom{6}{3}a^3b^3 + \binom{6}{4}a^2b^4 + \binom{6}{5}ab^5 + \binom{6}{6}b^6.$$

This last form is the form used in the statement of the Binomial Theorem, stated here without proof.

The Binomial Theorem

For real numbers a and b and a nonnegative integer n,

$$(a+b)^n = \binom{n}{0}a^n + \binom{n}{1}a^{n-1}b + \binom{n}{2}a^{n-2}b^2 + \ldots + \binom{n}{k}a^{n-k}b^k + \ldots + \binom{n}{n}b^n$$

In Σ notation, $(a+b)^n = \displaystyle\sum_{k=0}^{n}\binom{n}{k}a^{n-k}b^k$.

NOTES

1. There are $n+1$ terms in $(a+b)^n$.

2. In each term of $(a+b)^n$, the sum of the exponents of a and b is n.

Example 3: The Binomial Theorem

a. Expand $(x+3)^5$ by using the Binomial Theorem.

Solution: $(x+3)^5 = \displaystyle\sum_{k=0}^{5}\binom{5}{k}x^{5-k}3^k$

$$= \binom{5}{0}x^5 + \binom{5}{1}x^4 \cdot 3 + \binom{5}{2}x^3 \cdot 3^2 + \binom{5}{3}x^2 \cdot 3^3 + \binom{5}{4}x \cdot 3^4 + \binom{5}{5}3^5$$

$$= 1 \cdot x^5 + 5 \cdot x^4 \cdot 3 + 10 \cdot x^3 \cdot 9 + 10 \cdot x^2 \cdot 27 + 5 \cdot x \cdot 81 + 1 \cdot 243$$

$$= x^5 + 15x^4 + 90x^3 + 270x^2 + 405x + 243$$

b. Expand $(y^2-1)^6$ by using the Binomial Theorem.

Solution: $(y^2-1)^6 = \displaystyle\sum_{k=0}^{6}\binom{6}{k}(y^2)^{6-k}(-1)^k$

$$= \binom{6}{0}(y^2)^6 + \binom{6}{1}(y^2)^5(-1)^1 + \binom{6}{2}(y^2)^4(-1)^2$$

$$+ \binom{6}{3}(y^2)^3(-1)^3 + \binom{6}{4}(y^2)^2(-1)^4$$

$$+ \binom{6}{5}(y^2)^1(-1)^5 + \binom{6}{6}(-1)^6$$

$$= 1 \cdot y^{12} + 6 \cdot y^{10}(-1) + 15 \cdot y^8(+1) + 20 \cdot y^6(-1)$$

$$+ 15 \cdot y^4(+1) + 6 \cdot y^2(-1) + 1(+1)$$

$$= y^{12} - 6y^{10} + 15y^8 - 20y^6 + 15y^4 - 6y^2 + 1$$

Continued on the next page...

c.   Find the sixth term of the expansion of $\left(2x - \dfrac{1}{3}\right)^{10}$.

Solution: Since $\left(2x - \dfrac{1}{3}\right)^{10} = \displaystyle\sum_{k=0}^{10} \binom{10}{k}(2x)^{10-k}\left(-\dfrac{1}{3}\right)^{k}$, and the sum begins with

$k = 0$, the sixth term will occur when $k = 5$.

$$\binom{10}{5}(2x)^{10-5}\left(-\dfrac{1}{3}\right)^{5} = \dfrac{10!}{5!5!}(2x)^{5}\left(-\dfrac{1}{3}\right)^{5}$$

$$= \overset{28}{\cancel{252}}\cdot 32x^{5}\left(-\dfrac{1}{\underset{27}{\cancel{243}}}\right) = \dfrac{-896x^{5}}{27}$$

The sixth term is $\dfrac{-896x^{5}}{27}$.

d.   Find the fourth term of the expansion of $\left(x + \dfrac{1}{2}y\right)^{8}$.

Solution: $\left(x + \dfrac{1}{2}y\right)^{8} = \displaystyle\sum_{k=0}^{8}\binom{8}{k}x^{8-k}\left(\dfrac{1}{2}y\right)^{k}$

The fourth term occurs when $k = 3$.

$$\binom{8}{3}x^{8-3}\left(\dfrac{1}{2}y\right)^{3} = \dfrac{8!}{3!5!}\cdot x^{5}\cdot\dfrac{1}{8}y^{3} = 7x^{5}y^{3}$$

The fourth term is $7x^{5}y^{3}$.

e.   Using the binomial expansion, approximate $(0.99)^{4}$ to the nearest thousandth.

Solution: First rewrite 0.99 in the binomial form $(1 - 0.01)$ then proceed as follows:

$$(0.99)^{4} = (1 - 0.01)^{4} = \sum_{k=0}^{4}\binom{4}{k}(1)^{4-k}(-0.01)^{k}$$

$$= \binom{4}{0}\cdot 1^{4} + \binom{4}{1}\cdot 1^{3}\cdot(-0.01) + \binom{4}{2}\cdot 1^{2}\cdot(-0.01)^{2} + \binom{4}{3}\cdot 1\cdot(-0.01)^{3}$$

$$+ \binom{4}{4}\cdot(-0.01)^{4}$$

$$= 1 + 4(-0.01) + 6(-0.01)^{2} + 4(-0.01)^{3} + 1(-0.01)^{4}$$

$$= 1 - 0.04 + 0.0006 - 0.000004 + 0.00000001$$

$$= 0.96059601$$

$$\approx 0.961 \qquad\qquad \text{(to the nearest thousandth)}$$

Practice Problems

1. Simplify $\dfrac{10!}{7!}$.

2. Evaluate $\begin{pmatrix} 20 \\ 2 \end{pmatrix}$.

3. Expand $(x+2)^4$ by using the Binomial Theorem.

4. Find the third term of the expansion of $(2x-1)^7$.

10.5 Exercises

Simplify the expressions in Exercises 1 – 16.

1. $\dfrac{8!}{6!}$

2. $\dfrac{11!}{7!}$

3. $\dfrac{3!8!}{10!}$

4. $\dfrac{5!7!}{8!}$

5. $\dfrac{5!4!}{6!}$

6. $\dfrac{7!4!}{10!}$

7. $\dfrac{n!}{n}$

8. $\dfrac{n!}{(n-3)!}$

9. $\dfrac{(k+3)!}{k!}$

10. $\dfrac{n(n+1)!}{(n+2)!}$

11. $\begin{pmatrix} 6 \\ 3 \end{pmatrix}$

12. $\begin{pmatrix} 5 \\ 4 \end{pmatrix}$

13. $\begin{pmatrix} 7 \\ 3 \end{pmatrix}$

14. $\begin{pmatrix} 8 \\ 5 \end{pmatrix}$

15. $\begin{pmatrix} 10 \\ 0 \end{pmatrix}$

16. $\begin{pmatrix} 6 \\ 2 \end{pmatrix}$

Write the first four terms of the expansions in Exercises 17 – 28.

17. $(x+y)^7$

18. $(x+y)^{11}$

19. $(x+1)^9$

20. $(x+1)^{12}$

21. $(x+3)^5$

22. $(x-2)^6$

23. $(x+2y)^6$

24. $(x+3y)^5$

25. $(3x-y)^7$

26. $(2x-y)^{10}$

27. $(x^2-4y)^9$

28. $(x^2-2y)^7$

Using the Binomial Theorem, expand the expressions in Exercises 29 – 40.

29. $(x+y)^6$

30. $(x+y)^8$

31. $(x-1)^7$

32. $(x-1)^9$

33. $(3x+y)^5$

34. $(2x+y)^6$

35. $(x+2y)^4$

36. $(x+3y)^5$

37. $(3x-2y)^4$

38. $(5x+2y)^3$

39. $(3x^2-y)^5$

40. $(x^2+2y)^4$

Answers to Practice Problems: **1.** 720 **2.** 190 **3.** $x^4+8x^3+24x^2+32x+16$ **4.** $672x^5$

Find the specified term in each of the expressions in Exercises 41 – 46.

41. $(x - 2y)^{10}$, fifth term

42. $(x + 3y)^{12}$, third term

43. $(2x + 3)^{11}$, fourth term

44. $\left(x - \dfrac{y}{2}\right)^{9}$, seventh term

45. $(5x^2 - y^2)^{12}$, tenth term

46. $(2x^2 + y^2)^{15}$, eleventh term

Approximate the value of each expression in Exercises 47 – 53 correct to the nearest thousandth.

47. $(1.01)^6$　　　**48.** $(0.96)^8$　　　**49.** $(0.97)^7$　　　**50.** $(1.02)^{10}$

51. $(2.3)^5$　　　**52.** $(2.8)^6$　　　**53.** $(0.98)^8$

Writing and Thinking About Mathematics

54. Factor the polynomial: $x^4 + 8x^3 + 24x^2 + 32x + 16$.

HAWKES LEARNING SYSTEMS: INTERMEDIATE ALGEBRA SOFTWARE

- The Binomial Theorem

Chapter 10 Index of Key Ideas and Terms

Section 10.1 Sequences

Sequences   page 818
    A **sequence** is a list of numbers that occur in a certain
    order.

Terms   page 818
    Each number in the sequence is called a **term** of the
    sequence.

Infinite Sequences   page 819
    An **infinite sequence** (or **sequence**) is a function that has
    the positive integers as its domain.

Alternating Sequences   page 821
    An **alternating sequence** is a sequence in which the
    terms alternate in sign. Alternating sequences generally
    involve the expression $(-1)^n$ or $(-1)^{n+1}$.

Decreasing Sequences   page 822
    A sequence $\{a_n\}$ is **decreasing** if $a_n > a_{n+1}$ for all n.

Increasing Sequences   page 822
    A sequence $\{a_n\}$ is **increasing** if $a_n < a_{n+1}$ for all n.

Section 10.2 Sigma Notation

Partial Sum   page 826
    To find the sum of just a few terms of a sequence is
    called a **partial sum**.

Sigma Notation   page 826
    The Greek letter capital sigma, Σ, is used to indicate a
    partial sum.

Partial Sums Using Sigma Notation   page 826

The nth **partial sum**, S_n, of the first n terms of a sequence

$$\{a_n\} \text{ is } S_n = \sum_{k=1}^{n} a_k = a_1 + a_2 + a_3 + \ldots + a_n.$$

k is called the **index of summation**, and k takes the integer values $1, 2, 3, \ldots, n$.

n is the **upper limit of summation**, and 1 is the **lower limit of summation**.

Properties of Σ Notation   page 828

For sequences $\{a_n\}$ and $\{b_n\}$ and any real number c:

I. $\displaystyle\sum_{k=1}^{n} a_k = \sum_{k=1}^{i} a_k + \sum_{k=i+1}^{n} a_k$ for any i, $1 \le i \le n-1$

II. $\displaystyle\sum_{k=1}^{n} (a_k + b_k) = \sum_{k=1}^{n} a_k + \sum_{k=1}^{n} b_k$

III. $\displaystyle\sum_{k=1}^{n} ca_k = c\sum_{k=1}^{n} a_k$

IV. $\displaystyle\sum_{k=1}^{n} c = nc$

Section 10.3 Arithmetic Sequences

Arithmetic Sequences   page 833

A sequence $\{a_n\}$ is called an **arithmetic sequence** (or **arithmetic progression**) if for any natural number k, $a_{k+1} - a_k = d$ where d is a constant. d is called the **common difference**.

n^{th} **Term of an Arithmetic Sequence**   page 834

If $\{a_n\}$ is an arithmetic sequence, then the nth term has the form $a_n = a_1 + (n-1)d$ where d is the common difference between consecutive terms.

Partial Sums of Arithmetic Sequences   page 837

The nth **partial sum**, S_n, of the first n terms of an arithmetic sequence $\{a_n\}$ is $S_n = \displaystyle\sum_{k=1}^{n} a_k = \frac{n}{2}(a_1 + a_n)$.

Section 10.4 Geometric Sequences and Series

Geometric Sequences   page 844

A sequence $\{a_n\}$ is called a **geometric sequence** (or **geometric progression**) if for any positive integer k,

$\dfrac{a_{k+1}}{a_k} = r$ where r is constant and $r \neq 0$. r is called the **common ratio**.

n^{th} Term of a Geometric Sequence   page 845

If $\{a_n\}$ is a geometric sequence, then the nth term has the form $a_n = a_1 r^{n-1}$ where r is the common ratio.

Partial Sums of Geometric Sequences   page 848

The **n^{th} partial sum**, S_n, of the first n terms of a geometric sequence $\{a_n\}$ is $S_n = \displaystyle\sum_{k=1}^{n} a_k = \dfrac{a_1 \left(1 - r^n\right)}{1 - r}$

where r is the common ratio and $r \neq 1$.

Infinite Series (or Series)   page 850

The indicated sum of all the terms of a sequence is called an **infinite series** (or a **series**). For a sequence $\{a_n\}$, the corresponding series can be written as follows:

$$\sum_{k=1}^{\infty} \left(a_k\right) = a_1 + a_2 + a_3 + \ldots + a_n + \ldots$$

Sum of an Infinite Geometric Series   page 850

If $\{a_n\}$ is a geometric sequence and $|r| < 1$, then the sum of the infinite geometric series is

$$S = \sum_{k=1}^{\infty} \left(a_k\right) = a_1 + a_1 r + a_1 r^2 + \ldots = \dfrac{a_1}{1 - r}.$$

Section 10.5 The Binomial Theorem

n Factorial ($n!$)   page 856

For any positive integer n, $n! = n(n-1)(n-2) \ldots 3 \cdot 2 \cdot 1$. $n!$ is read as "n factorial."

0 Factorial ($0!$)   page 857

$0! = 1$

Using a Calculator to Calculate Factorials   page 857

Section 10.5 The Binomial Theorem (cont.)

Binomial Coefficient $\binom{n}{r}$ page 858

For non-negative integers n and r, with $0 \le r \le n$, we define

$$\binom{n}{r} = \frac{n!}{r!(n-r)!}.$$

Because this quantity appears repeatedly in the Binomial Theorem, $\binom{n}{r}$ is often called a **binomial coefficient**.

Pascal's Triangle   page 860

The Binomial Theorem   page 861

$$(a+b)^n = \binom{n}{0}a^n + \binom{n}{1}a^{n-1}b + \binom{n}{2}a^{n-2}b^2 + \ldots$$

$$+ \binom{n}{k}a^{n-k}b^k + \ldots + \binom{n}{n}b^n$$

In Σ notation, $(a+b)^n = \sum_{k=0}^{n}\binom{n}{k}a^{n-k}b^k$.

⟩ HAWKES LEARNING SYSTEMS: INTERMEDIATE ALGEBRA SOFTWARE

For a review of the topics and problems from Chapter 10, look at the following lessons from *Hawkes Learning Systems: Intermediate Algebra*.

- Sequences
- Sigma Notation
- Arithmetic Sequences
- Geometric Sequences and Series
- The Binomial Theorem
- Chapter 10 Review and Test

Chapter 10 Review

10.1 Sequences

Write the first four terms of each of the sequences in Exercises 1 – 6.

1. $\{2n-3\}$ **2.** $\left\{\dfrac{n+3}{n}\right\}$ **3.** $\left\{\dfrac{n^2}{n+1}\right\}$ **4.** $\left\{\dfrac{(-1)^n}{n^2}\right\}$

5. $\left\{6+(-1)^{n+1}\right\}$ **6.** $\left\{(-1)^{2n+1}\right\}$

Find a formula for the general term of each sequence in Exercises 7 – 10.

7. $10, 15, 20, 25, \ldots$

8. $3, -3, 3, -3, 3, \ldots$

9. $3, 6, 11, 18, 27, \ldots$

10. $3, 2, \dfrac{5}{3}, \dfrac{6}{4}, \dfrac{7}{5}, \ldots$

For each of the sequences in Exercises 11 – 16, determine whether it is increasing or decreasing. Justify your answer by comparing a_n with a_{n+1}.

11. $\{3^n\}$ **12.** $\{1-3n\}$ **13.** $\left\{\dfrac{1}{4n}\right\}$

14. $\left\{\dfrac{n^2+1}{2}\right\}$ **15.** $\left\{\dfrac{2n}{n+2}\right\}$ **16.** $\{-n^2\}$

17. Bouncing a ball: A ball is dropped from a height of 10 feet. It bounces to $\dfrac{2}{3}$ of its previous height on each subsequent bounce. How high will it bounce on the fourth bounce?

18. Value of a car: An automobile loses $\dfrac{1}{5}$ of its value each year. If its original cost was $30,000, what will be its value after four years? [**Hint:** If it loses $\dfrac{1}{5}$ of its value, it retains $\dfrac{4}{5}$ of its value.]

19. College tuition: You decide to save for college. Your plan is to save $100 the first month and then add an extra $5 each month thereafter for three years. (You save $100 the first month, $105 the second month, $110 the third month, and so on.) How much will you save the last month? What total amount will you save over the three years?

20. Allowances: Your father tells you he will:
   a. give you $10 each day for a month or
   b. give you $0.01 on the first day of the month and double it each day thereafter.
   Which should you choose for a 30 day month?

10.2 Sigma Notation

For each of the sequences given in Exercises 21 – 26, write out the partial sum S_4 and evaluate the partial sum.

21. $\{4n+1\}$

22. $\{2n+3\}$

23. $\left\{\dfrac{n+2}{n}\right\}$

24. $\{1+(-1)^n\}$

25. $\left\{\left(\dfrac{3}{4}\right)^n\right\}$

26. $\{n^2+n\}$

Write the indicated sums in Exercises 27 – 32 in expanded form and evaluate.

27. $\displaystyle\sum_{k=1}^{4} 5k$

28. $\displaystyle\sum_{k=3}^{7}(k+4)$

29. $\displaystyle\sum_{k=1}^{5}(-1)^{k+1}(2k+3)$

30. $\displaystyle\sum_{k=1}^{6}(-1)^k k^2$

31. $\displaystyle\sum_{k=8}^{10}(k^2+2k)$

32. $\displaystyle\sum_{k=1}^{4}\dfrac{1}{3k}$

Write the sums in Exercises 33 – 36 in sigma notation.

33. $3+5+7+9+11$

34. $\dfrac{1}{27}-\dfrac{1}{64}+\dfrac{1}{125}-\dfrac{1}{216}$

35. $0-3+8-15+24-35$

36. $2+\dfrac{3}{4}+\dfrac{4}{9}+\dfrac{5}{16}+\dfrac{6}{25}$

Find the indicated sums in Exercises 37 – 40.

37. $\displaystyle\sum_{k=1}^{4} a_k = 20$ and $\displaystyle\sum_{k=1}^{4} b_k = 30$. Find $\displaystyle\sum_{k=1}^{4}(a_k+b_k)$.

38. $\displaystyle\sum_{k=1}^{12} a_k = 78$ and $\displaystyle\sum_{k=1}^{12} b_k = 134$. Find $\displaystyle\sum_{k=1}^{12}(a_k-b_k)$.

39. $\displaystyle\sum_{k=1}^{10} a_k = 150$ and $\displaystyle\sum_{k=11}^{20} a_k = 230$. Find $\displaystyle\sum_{k=1}^{20} a_k$.

40. $\displaystyle\sum_{k=1}^{8} b_k = 64$ and $\displaystyle\sum_{k=1}^{8}(3a_k+b_k) = 94$. Find $\displaystyle\sum_{k=1}^{8} a_k$.

10.3 Arithmetic Sequences

Write the first five terms of each sequence in Exercises 41 – 46 and determine which of the sequences are arithmetic.

41. $\{1-2n\}$

42. $\left\{n+\dfrac{n}{3}\right\}$

43. $\{(n+1)^2\}$

44. $\left\{\dfrac{1}{n+2}\right\}$ **45.** $\left\{\dfrac{1}{2}n+\dfrac{3}{4}\right\}$ **46.** $\left\{6-\dfrac{n}{3}\right\}$

Find the general term, a_n, for each of the arithmetic sequences in Exercises 47 – 52.

47. $a_1 = 0$, $d = 2$ **48.** $a_1 = 10$, $d = \dfrac{1}{4}$ **49.** $a_1 = 10$, $a_5 = 6$

50. $a_1 = 20$, $a_3 = 6$ **51.** $a_5 = 2$, $a_{10} = 12$ **52.** $a_3 = -4$, $a_7 = -12$

Find the indicated sums in Exercises 53 – 58 by using the formula for partial sums of arithmetic sequences.

53. $\displaystyle\sum_{k=1}^{6}(2k+3)$ **54.** $\displaystyle\sum_{k=1}^{10}\left(\dfrac{k}{2}+4\right)$ **55.** $\displaystyle\sum_{k=1}^{100}(3k+10)$ **56.** $\displaystyle\sum_{k=8}^{20}(3k-7)$

57. If $\displaystyle\sum_{k=1}^{15}a_k = 120$, find $\displaystyle\sum_{k=1}^{15}(-2a_k+6)$.   **58.** If $\displaystyle\sum_{k=1}^{30}(2a_k+3) = 210$, find $\displaystyle\sum_{k=1}^{30}a_k$.

59. Building blocks: How many building blocks are needed to stack the blocks so that there are 29 blocks in the first layer, 27 blocks in the next layer, 25 blocks in the third layer, and so on with 1 block at the top?

60. Savings account: Jose has decided to deposit $200 in his savings each month and increase this amount by $5 each month. How much will he have deposited after 2 years?

10.4 Geometric Sequences and Series

In Exercises 61 – 66, determine which sequences are geometric. If the sequence is geometric, find its common ratio and write a formula for the nth term.

61. $12, 6, 3, \dfrac{3}{2}, \ldots$ **62.** $3, 6, 9, 12, \ldots$ **63.** $10, -2, \dfrac{2}{5}, -\dfrac{2}{25}, \ldots$

64. $\dfrac{1}{4}, \dfrac{1}{16}, \dfrac{1}{64}, \dfrac{1}{256}, \ldots$ **65.** $-2, 4, -8, 16, \ldots$ **66.** $5, -10, 15, -20, \ldots$

Write the first four terms of each of the following geometric sequences in Exercises 67 – 70.

67. $\left\{\dfrac{(-1)^n}{5^n}\right\}$ **68.** $\left\{2\left(\dfrac{2}{3}\right)^n\right\}$ **69.** $\left\{\left(\dfrac{3}{8}\right)^n\right\}$ **70.** $\left\{5(2)^{n+1}\right\}$

Find the general term, a_n, for each of the geometric sequences in Exercises 71 – 74.

71. $a_1 = 7$, $r = \dfrac{1}{2}$ **72.** $a_1 = 4$, $r = -3$ **73.** $a_3 = 4$, $a_4 = -6$ **74.** $a_2 = \dfrac{5}{8}$, $a_5 = 5$

In Exercises 75 – 78, $\{a_n\}$ is a geometric sequence. Find the indicated quantity.

75. $a_1 = -64$, $a_6 = 2$. Find a_8.

76. $a_1 = \dfrac{5}{8}$, $a_6 = 20$. Find a_7.

77. $a_5 = -3$, $a_7 = -12$. Find a_9.

78. $a_1 = \dfrac{1}{6}$, $r = \dfrac{3}{2}$, $a_n = \dfrac{81}{64}$. Find n.

In Exercises 79 – 82, find the indicated sums.

79. $\displaystyle\sum_{k=1}^{4} 4\left(\dfrac{1}{3}\right)^k$ **80.** $\displaystyle\sum_{k=1}^{5}\left(\dfrac{1}{6}\right)^k$ **81.** $\displaystyle\sum_{k=1}^{\infty}\left(-\dfrac{2}{3}\right)^k$ **82.** $\displaystyle\sum_{k=1}^{\infty}\left(\dfrac{3}{5}\right)^k$

83. Use a geometric series to show the following:

    a. $0.44444... = \dfrac{4}{9}$ **b.** $0.8888... = \dfrac{8}{9}$

84. **Annuities:** A professional athlete has a contract for 6 years and decides he might be wise to set up an annuity. (See Section 10.4, Example 6.) He decides to deposit $15,000 on the first of each year for 6 years. If interest is compounded annually at 6%, what will be the value of the deposits after 6 years?

85. **Automobiles:** An automobile that costs $45,000 new depreciates at a rate of 15% of its value each year. What will be its value after 5 years?

10.5   The Binomial Theorem

Simplify the expressions in Exercises 86 – 91.

86. $\dfrac{10!}{5!}$ **87.** $\dfrac{7!6!}{8!}$ **88.** $\dfrac{n!}{n(n-1)}$

89. $\dfrac{(n+3)!}{(n+1)!}$ **90.** $\dbinom{8}{5}$ **91.** $\dbinom{11}{0}$

Write the first four terms of the expansions in Exercises 92 – 95.

92. $(x+y)^8$ **93.** $(x+2)^9$ **94.** $(2x-y)^{10}$ **95.** $(x^2-4)^6$

Using the Binomial Theorem, expand the expressions in Exercises 96 – 99.

96. $(x+5)^4$ **97.** $(2x+3y)^3$ **98.** $(x^2-3y)^5$ **99.** $(x-1)^{10}$

Find the specified term in each of the expressions in Exercises 100 – 103.

100. $(x-2y)^9$, sixth term   **101.** $(3x+y)^7$, second term

102. $\left(\dfrac{x}{2}+y\right)^8$, third term   **103.** $(4x^2-y^2)^{12}$, eighth term

Chapter 10 Test

1. Write out and evaluate the partial sum S_5 for the sequence $\{n^2 - n\}$.

2. Write the first four terms of the sequence $\left\{\dfrac{1}{3n+1}\right\}$ and determine whether the sequence is arithmetic, geometric, or neither.

3. Find a formula for the general term of the sequence $\dfrac{1}{3}, \dfrac{2}{5}, \dfrac{3}{7}, \dfrac{4}{9}, \dots$

4. Write the following sum in Σ notation: $7 + 10 + 13 + 16$

5. Write the following sum in Σ notation: $1 + 8 + 27 + 64$

6. If $\displaystyle\sum_{k=1}^{50} a_k = 88$ and $\displaystyle\sum_{k=1}^{19} a_k = 14$, find $\displaystyle\sum_{k=20}^{50} a_k$.

Find each of the sums in Exercises 7 and 8.

7. $\displaystyle\sum_{k=1}^{8} (3k - 5)$

8. $\displaystyle\sum_{k=3}^{6} 2\left(-\dfrac{1}{3}\right)^k$

In Exercises 9 – 11, $\{a_k\}$ is an arithmetic sequence. Find the indicated quantity.

9. $a_1 = 5$, $d = 3$. Find a_8.

10. $a_2 = 4$, $a_7 = -6$. Find a_n.

11. $a_4 = 22$, $a_7 = 37$. Find $\displaystyle\sum_{k=1}^{9} a_k$.

In Exercises 12 – 14, $\{a_k\}$ is a geometric sequence. Find the indicated quantity.

12. $a_1 = 8$, $r = \dfrac{1}{2}$. Find a_7.

13. $a_4 = 3$, $a_6 = 9$. Find a_n.

14. $a_2 = \dfrac{1}{3}$, $a_5 = \dfrac{1}{24}$. Find $\displaystyle\sum_{k=1}^{6} a_k$.

15. Write the decimal number $0.\overline{15}$ in the form of an infinite series and find its sum in the form of a proper fraction.

16. Evaluate $\dbinom{11}{5}$.

17. Use the Binomial Theorem to expand $(2x - y)^5$.

18. Write the fifth term of the expansion of $(x + 3y)^8$.

19. Write the third term of the expansion of $(x - 5y)^{10}$.

20. **Buying a car:** A customer intends to buy a new car for $25,000 and anticipates that it will depreciate at a rate of 15% of its value each year. What will be the value of the car in 4 years when he wants to trade it in for another new car?

Cumulative Review: Chapters 1–10

Simplify each expression in Exercises 1 and 2.

1. $10x - \left[2x + (13 - 4x) - (11 - 3x)\right] + (2x + 5)$

2. $\left(x^2 + 5x - 2\right) - \left(3x^2 - 5x - 3\right) + \left(x^2 - 7x + 4\right)$

Factor completely in Exercises 3 – 5.

3. $64x^3 + 27$

4. $6x^2 + 17x - 45$

5. $5x^2(2x + 1) - 3x(2x + 1) - 14(2x + 1)$

Perform the indicated operations and simplify in Exercises 6 – 8.

6. $\dfrac{x+3}{3} + \dfrac{2x-1}{5}$

7. $\dfrac{x}{x^2 - 16} - \dfrac{x+1}{x^2 - 5x + 4}$

8. $\dfrac{x^2 - 9}{x^4 + 6x^3} \div \dfrac{x^3 - 2x^2 - 3x}{x^2 + 7x + 6} \cdot \dfrac{x^2}{x+3}$

Simplify each expression in Exercises 9 – 12. Assume that all variables are positive.

9. $\left(\dfrac{9x^{-1}y^{\frac{1}{3}}}{x^3 y^{-\frac{1}{3}}}\right)^{\frac{1}{2}}$

10. $\sqrt[3]{\dfrac{8x^4}{27y^3}}$

11. $\sqrt{12x} - \sqrt{75x} + 2\sqrt{27x}$

12. $\dfrac{1+3i}{2-5i}$

Solve the inequalities in Exercises 13 – 15. Graph the solution sets on real number lines.

13. $6(2x - 3) + (x - 5) > 4(x + 1)$

14. $|7 - 3x| - 2 \le 4$

15. $8x^2 + 2x - 45 < 0$

Solve the equations in Exercises 16 – 21.

16. $4(x - 7) + 2(3x + 2) = 3x + 2$

17. $2x^2 + 4x + 3 = 0$

18. $x - \sqrt{x} - 2 = 0$

19. $\dfrac{1}{2x} + \dfrac{5}{x+3} = \dfrac{8}{3x}$

20. $2\sqrt{6 - x} = x - 3$

21. $5 \ln x = 12$

22. Solve the formula for n: $P = \dfrac{A}{1 + ni}$

Solve the systems of equations in Exercises 23 and 24.

23. Solve the following system by using Cramer's rule.

$$\begin{cases} 9x + 2y = 8 \\ 4x + 3y = -7 \end{cases}$$

24. Solve the following system of equations by using the Gaussian elimination method.

$$\begin{cases} 3x + y - 2z = 4 \\ x - 4y - 3z = -5 \\ 2x + 2y + z = 3 \end{cases}$$

25. If $f(x) = 2x - 7$ and $g(x) = x^2 + 1$, find:

a. $f^{-1}(x)$

b. $f[g(x)]$

c. $g(x + 1) - g(x)$

26. The following is a graph of $y = f(x)$.   Sketch the graph of $y = f(x - 2) - 1$.

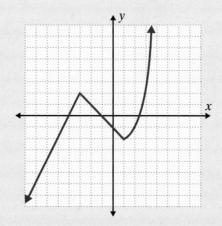

27. Solve the following system by graphing. $\begin{cases} 4x - 3y = 17 \\ 5x + 2y = 4 \end{cases}$

Graph each of the equations in Exercises 28 – 30.

28. $y = 4x^2 - 8x + 9$

29. $\dfrac{x^2}{4} + \dfrac{y^2}{16} = 1$

30. $x^2 - 4x + y^2 + 2y = 4$

31. If $\{a_n\}$ is an arithmetic sequence where $a_3 = 4$ and $a_8 = -6$,

a. find a_n.

b. find $\displaystyle\sum_{k=1}^{10} a_k$.

32. If $\{a_k\}$ is a geometric sequence where $a_1 = 16$ and $r = \dfrac{1}{2}$,

a. find a_n.

b. find $\displaystyle\sum_{k=1}^{6} a_k$.

33. Use the Binomial Theorem to expand $(x + 2y)^6$.

34. **Traveling by car:** Two cars start together and travel in the same direction, one traveling 3 times as fast as the other. At the end of 3.5 hours, they are 140 miles apart. How fast is each traveling?

35. **Grocery stores:** A grocer mixes two kinds of nuts. One costs $1.40 per pound and the other costs $2.60 per pound. If the mixture weighs 20 pounds and costs $1.64 per pound, how many pounds of each kind did he use?

36. **Swimming pools:** A rectangular yard is 20 ft by 30 ft. A rectangular swimming pool is to be built leaving a strip of grass of uniform width around the pool. If the area of the grass strip is 184 sq ft, find the dimensions of the pool.

37. **Isotope decomposition:** A radioactive isotope decomposes according to $A = A_0 e^{-0.0552t}$, where t is measured in hours. Determine the half-life of the isotope. Round the solution to two decimal places.

38. **Job salaries:** Joan started her job exactly 5 years ago. Her original salary was $12,000 per year. Each year she received a raise of 6% of her current salary. What is her salary after this year's raise?

39. **Liquid mixtures:** A tank holds 1000 liters of a liquid that readily mixes with water. After 150 liters are drained out, the tank is filled by adding water. Then 150 liters of the mixture are drained out and the tank is filled by adding water. If this process is continued 5 times, how much of the original liquid is left? Round the solution to one decimal place.

Use a graphing calculator to solve (or estimate the solutions of) the equations in Exercises 40 – 44. If necessary, round answers to four decimal places. Use either of the following two procedures:

  a. *Enter each of the functions indicated on each side of the equation and find the points of intersection.*

  b. *Get 0 on one side of the equation and find the zeros of the function on the other side.*

40. $x^4 = 2x + 1$

41. $e^x = -x^2 + 4$

42. $\ln x = x^2 - 2x - 1$

43. $x^3 = 3x^2 - 3x + 1$

44. $x^3 = 7x - 6$

A.1   Review of Fractions

Multiplication and Division with Fractions

Fractions are numbers written in the form $\dfrac{a}{b}$ where the **numerator** (top number) and the **denominator** (bottom number) are whole numbers and the denominator does not equal 0. It is important to note that the **denominator cannot be 0**. We say that the denominator b is nonzero or that $b \neq 0$ (b is not equal to 0).

To Multiply Fractions

1. Multiply the numerators.

2. Multiply the denominators.

$$\frac{a}{b} \cdot \frac{c}{d} = \frac{a \cdot c}{b \cdot d} \quad \textit{where } \mathbf{b, d \neq 0}.$$

Example 1: Multiplying Fractions

Find the following products:

a. $\dfrac{1}{3} \cdot \dfrac{5}{7}$

   Solution: $\dfrac{1}{3} \cdot \dfrac{5}{7} = \dfrac{1 \cdot 5}{3 \cdot 7} = \dfrac{5}{21}$

b. Find $\dfrac{2}{5}$ of $\dfrac{3}{5}$.

   Solution: Note that the word "of" indicates multiplication.

$$\frac{2}{5} \cdot \frac{3}{5} = \frac{2 \cdot 3}{5 \cdot 5} = \frac{6}{25}$$

Changing a fraction to lower terms is called **reducing** the fraction. **A fraction is reduced to lowest terms if the numerator and the denominator have no common factor other than 1.** To reduce a fraction, we use the fact that 1 is equal to any counting number divided by itself and "divide out" all common factors found in both the numerator and denominator.

Example 2: Multiplying and Reducing Fractions

Find the following products and reduce to lowest terms.

a. $\dfrac{16}{28} \cdot \dfrac{7}{10}$

Solution: $\dfrac{16}{28} \cdot \dfrac{7}{10} = \dfrac{2 \cdot 2 \cdot 2 \cdot 2 \cdot 7}{2 \cdot 2 \cdot 7 \cdot 2 \cdot 5} = \dfrac{2}{2} \cdot \dfrac{2}{2} \cdot \dfrac{2}{2} \cdot \dfrac{7}{7} \cdot \dfrac{2}{5} = 1 \cdot 1 \cdot 1 \cdot 1 \cdot \dfrac{2}{5} = \dfrac{2}{5}$

Note that by using prime factors for the numerators and denominators we are sure of finding all the common factors in the numerators and denominators.

b. $\dfrac{17}{25} \cdot \dfrac{50}{34} \cdot 4$

Solution: $\dfrac{17}{25} \cdot \dfrac{50}{34} \cdot 4 = \dfrac{\cancel{17} \cdot \cancel{2} \cdot \cancel{5} \cdot \cancel{5} \cdot 2 \cdot 2}{\cancel{5} \cdot \cancel{5} \cdot \cancel{2} \cdot \cancel{17}} = \dfrac{2 \cdot 2}{1} = \dfrac{4}{1} = 4$

In this case we have used prime factors.

We may also use common factors that are not prime numbers as follows:

$$\dfrac{17}{25} \cdot \dfrac{50}{34} \cdot 4 = \dfrac{\cancel{17} \cdot \cancel{2} \cdot \cancel{25} \cdot 4}{\cancel{25} \cdot \cancel{2} \cdot \cancel{17}} = \dfrac{4}{1} = 4.$$

The answer is the same.

If the product of two fractions is 1, then the fractions are called **reciprocals** of each other. For example,

$$\dfrac{3}{7} \text{ and } \dfrac{7}{3} \text{ are reciprocals because } \dfrac{3}{7} \cdot \dfrac{7}{3} = \dfrac{21}{21} = 1.$$

Division with Fractions

To divide by any nonzero number, multiply by its reciprocal.

In general,

$$\dfrac{a}{b} \div \dfrac{c}{d} = \dfrac{a}{b} \cdot \dfrac{d}{c} \quad \text{where} \quad b, c, d \neq 0.$$

Example 3: Dividing and Reducing Fractions

Divide and reduce to lowest terms:

a. $\dfrac{5}{6} \div 6$

Solution: Because the whole number $6 = \dfrac{6}{1}$, the reciprocal of 6 is $\dfrac{1}{6}$. To divide by 6, we multiply by $\dfrac{1}{6}$.

$$\frac{5}{6} \div 6 = \frac{5}{6} \cdot \frac{1}{6} = \frac{5}{36}$$

b. $\dfrac{8}{9} \div \dfrac{16}{27}$

Solution: The reciprocal of $\dfrac{16}{27}$ is $\dfrac{27}{16}$.

$$\frac{8}{9} \div \frac{16}{27} = \frac{8}{9} \cdot \frac{27}{16} = \frac{\cancel{8} \cdot 3 \cdot \cancel{9}}{\cancel{9} \cdot 2 \cdot \cancel{8}} = \frac{3}{2}$$

Example 4: Applications

A delivery truck is carrying 36 boxes of a certain size. This is $\dfrac{3}{4}$ of the truck's capacity for this size box.

a. Can the truck carry more than 36 of these boxes?

Solution: Yes, the truck can carry more than 36 boxes because $\dfrac{3}{4}$ is less than 1. (1 represents the entire capacity of the truck.)

b. If you were to multiply $\dfrac{3}{4}$ times 36, would the product be more or less than 36?

Solution: The product would be less than 36.

c. What is the carrying capacity of the truck for this size box?

Solution: To find the capacity of the truck divide:

$$36 \div \frac{3}{4} = \frac{36}{1} \cdot \frac{4}{3} = \frac{\cancel{3} \cdot 12 \cdot 4}{\cancel{3}} = \frac{48}{1} = 48.$$

The capacity of the truck is 48 boxes of this size.

(Checking we see that $\dfrac{3}{4}$ of 48 is 36: $\dfrac{3}{4} \cdot \dfrac{48}{1} = 36$.)

Addition and Subtraction with Fractions

To add (or subtract) fractions with the same denominator, we can proceed as follows.

To Add (or Subtract) Fractions with the Same Denominator

Addition	***Subtraction***
1. Add the numerators.	*1. Subtract the numerators.*
2. Keep the common denominator.	*2. Keep the common denominator.*
3. Reduce, if possible.	*3. Reduce, if possible.*

$$\frac{a}{b}+\frac{c}{b}=\frac{a+c}{b} \qquad\qquad \frac{a}{b}-\frac{c}{b}=\frac{a-c}{b}$$

Example 5: Adding (or Subtracting) Fractions with the Same Denominator

a. Find the sum and reduce if possible: $\dfrac{1}{15}+\dfrac{4}{15}$.

Solution: $\dfrac{1}{15}+\dfrac{4}{15}=\dfrac{1+4}{15}=\dfrac{5}{15}=\dfrac{\cancel{5}\cdot 1}{3\cdot\cancel{5}}=\dfrac{1}{3}$

b. Find the difference and reduce if possible: $\dfrac{15}{8}-\dfrac{9}{8}$.

Solution: $\dfrac{15}{8}-\dfrac{9}{8}=\dfrac{15-9}{8}=\dfrac{6}{8}=\dfrac{\cancel{2}\cdot 3}{\cancel{2}\cdot 4}=\dfrac{3}{4}$

To add (or subtract) fractions with different denominators, we find the least common denominator of the fractions, change each fraction to an equivalent fraction with this common denominator, and then add (or subtract). For example, if the denominators are 5 and 7, the least common denominator (LCD) is the least common multiple (LCM) which is $5\cdot 7=35$. Thus we proceed as follows to find the sum $\dfrac{1}{5}+\dfrac{2}{7}$:

$$\frac{1}{5}+\frac{2}{7}=\frac{1}{5}\cdot\frac{7}{7}+\frac{2}{7}\cdot\frac{5}{5}=\frac{7}{35}+\frac{10}{35}=\frac{17}{35}.$$

(Note that $\dfrac{7}{7}=\dfrac{5}{5}=1$ and multiplication by 1 gives an equivalent fraction.)

In general, **the least common denominator (LCD) is the least common multiple (LCM) of the denominators.**

To Add (or Subtract) Fractions with Different Denominators

1. *Find the least common denominator (LCD).*
2. *Change each fraction into an equivalent fraction with that denominator.*
3. *Add (or subtract) the new fractions.*
4. *Reduce, if possible.*

NOTES   Equal fractions are said to be **equivalent**.

Example 6: Adding (or Subtracting) Fractions with Different Denominators

a.   Find the sum and reduce if possible: $\dfrac{5}{8} + \dfrac{13}{12}$.

Solution: To find the LCD, find the LCM of the denominators.

$$\left.\begin{array}{l} 8 = 2 \cdot 2 \cdot 2 \\ 12 = 2 \cdot 2 \cdot 3 \end{array}\right\} \ \mathbf{LCD} = 2 \cdot 2 \cdot 2 \cdot 3 = 24$$

$$\frac{5}{8} + \frac{13}{12} = \frac{5}{8} \cdot \frac{3}{3} + \frac{13}{12} \cdot \frac{2}{2} = \frac{15}{24} + \frac{26}{24} = \frac{41}{24} \quad \text{Note that } 8 \cdot 3 = 24 \text{ and } 12 \cdot 2 = 24.$$

The fraction $\dfrac{41}{24}$ is in lowest terms because 41 and 24 have only 1 as a common factor.

b.   Find the difference and reduce if possible: $\dfrac{12}{55} - \dfrac{2}{33}$.

Solution: To find the LCD, find the LCM of the denominators.

$$\left.\begin{array}{l} 55 = 5 \cdot 11 \\ 33 = 3 \cdot 11 \end{array}\right\} \ \mathbf{LCD} = 3 \cdot 5 \cdot 11 = 165$$

$$\frac{12}{55} - \frac{2}{33} = \frac{12}{55} \cdot \frac{3}{3} - \frac{2}{33} \cdot \frac{5}{5} = \frac{36}{165} - \frac{10}{165} = \frac{26}{165} = \frac{2 \cdot 13}{3 \cdot 5 \cdot 11} = \frac{26}{165}$$

The answer $\dfrac{26}{165}$ does not reduce because 26 and 165 have only 1 as a common factor.

Perform the indicated operations and reduce, if possible.

1. $\dfrac{19}{8} \cdot \dfrac{14}{38} \cdot \dfrac{3}{4} \cdot \dfrac{5}{6}$ **2.** $\dfrac{63}{51} \div \dfrac{9}{7}$ **3.** $\dfrac{1}{10} + \dfrac{3}{5} + \dfrac{4}{25}$ **4.** $\dfrac{27}{48} - \dfrac{13}{36}$

Appendix 1 Exercises

In Exercises 1 – 20 find the indicated product or quotient. Reduce if possible.

1. $\dfrac{7}{8} \cdot \dfrac{4}{35}$ **2.** $\dfrac{23}{6} \cdot \dfrac{10}{46}$ **3.** $9 \cdot \dfrac{5}{24}$ **4.** $8 \cdot \dfrac{5}{36}$

5. $\dfrac{42}{22} \cdot \dfrac{33}{9} \cdot \dfrac{13}{52}$ **6.** $\dfrac{9}{40} \cdot \dfrac{7}{10} \cdot \dfrac{25}{36}$ **7.** $\dfrac{5}{8} \div \dfrac{15}{4}$ **8.** $\dfrac{2}{3} \div \dfrac{2}{5}$

9. $\dfrac{35}{21} \div 10$ **10.** $\dfrac{36}{70} \div 9$ **11.** $\dfrac{15}{20} \div \dfrac{1}{3}$ **12.** $\dfrac{16}{35} \div \dfrac{4}{7}$

13. $\dfrac{15}{18} \div \dfrac{25}{24}$ **14.** $\dfrac{25}{36} \div \dfrac{20}{24}$ **15.** $35 \div \dfrac{5}{8}$ **16.** $75 \div \dfrac{15}{2}$

17. $\dfrac{25}{8} \cdot \dfrac{16}{36} \cdot \dfrac{7}{15} \cdot 9$ **18.** $\dfrac{17}{4} \cdot \dfrac{5}{54} \cdot \dfrac{18}{51} \cdot 2$ **19.** $\dfrac{1}{4} \cdot \dfrac{3}{2} \cdot \dfrac{22}{63} \cdot \dfrac{9}{11}$ **20.** $\dfrac{69}{30} \cdot \dfrac{15}{14} \cdot \dfrac{28}{46} \cdot \dfrac{1}{3}$

In Exercises 21 – 40 find the indicated sum or difference. Reduce if possible.

21. $\dfrac{3}{16} + \dfrac{3}{16}$ **22.** $\dfrac{5}{6} + \dfrac{5}{6}$ **23.** $\dfrac{1}{10} + \dfrac{3}{10}$ **24.** $\dfrac{2}{15} + \dfrac{7}{15}$

25. $\dfrac{7}{12} - \dfrac{3}{12}$ **26.** $\dfrac{7}{32} - \dfrac{5}{32}$ **27.** $\dfrac{14}{25} - \dfrac{6}{25}$ **28.** $\dfrac{5}{6} - \dfrac{1}{6}$

29. $\dfrac{3}{5} + \dfrac{3}{10}$ **30.** $\dfrac{2}{7} + \dfrac{4}{21}$ **31.** $\dfrac{5}{8} - \dfrac{5}{16}$ **32.** $\dfrac{5}{7} - \dfrac{5}{14}$

33. $\dfrac{3}{7} + \dfrac{4}{21} + \dfrac{2}{3}$ **34.** $\dfrac{2}{3} + \dfrac{3}{4} + \dfrac{5}{6}$ **35.** $\dfrac{3}{9} + \dfrac{1}{5} - \dfrac{1}{3}$ **36.** $\dfrac{3}{8} + \dfrac{5}{12} - \dfrac{1}{4}$

Answers to Practice Problems: **1.** $\dfrac{35}{64}$ **2.** $\dfrac{49}{51}$ **3.** $\dfrac{43}{50}$ **4.** $\dfrac{29}{144}$

37. $1 - \dfrac{11}{16}$ **38.** $1 - \dfrac{5}{7}$ **39.** $\begin{array}{r} \dfrac{2}{27} \\[6pt] \dfrac{5}{18} \\[6pt] +\dfrac{1}{9} \\ \hline \end{array}$ **40.** $\begin{array}{r} \dfrac{5}{9} \\[6pt] \dfrac{2}{3} \\[6pt] +\dfrac{7}{15} \\ \hline \end{array}$

41. The product of $\dfrac{2}{3}$ with another number is $\dfrac{7}{9}$.

    a. In the given problem, which number is the product?

    b. Find the other number.

42. In a class of 40 students, 6 received a grade of A. What fraction of the class received a grade of A? What fraction of the class did not receive a grade of A?

43. A study showed that $\dfrac{3}{11}$ of the students in an elementary school are left-handed. If the school had an enrollment of 440 students, how many students are left-handed? How many are not left-handed?

44. There are 3000 students at Sierra High School and $\dfrac{1}{4}$ of these students are seniors. If $\dfrac{2}{5}$ of the seniors are in favor of the school forming a debate team and $\dfrac{1}{2}$ of the remaining students (not seniors) are also in favor of forming a debate team, how many of the students do not favor this idea?

45. If Tom's total income for the year was $36,000 and he spent $\dfrac{1}{3}$ of his income on rent and $\dfrac{1}{10}$ of his income on his car, what total amount did he spend on these two expenses?

Tom's Budget

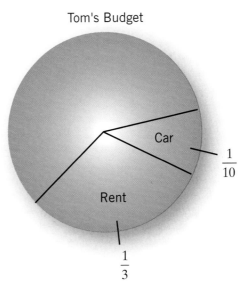

46. Of the personal computers (PCs) in use worldwide, the U.S. has $\frac{272}{1000}$, Japan $\frac{84}{1000}$, China $\frac{65}{1000}$, and Germany $\frac{56}{1000}$. What total fraction of the world's PCs are used in these four countries? What fraction is used in the rest of the world? (Source: *2006 World Almanac*)

PCs Worldwide

Other Countries

$\frac{272}{1000}$ — U.S

$\frac{84}{1000}$

$\frac{56}{1000}$ $\frac{65}{1000}$

Japan

Germany   China

47. Joseph's income is $5000 a month. He plans to budget $\frac{1}{5}$ of his income for rent and $\frac{1}{10}$ of his income for food.

a. What fraction of his monthly income does he plan to spend on these two expenses?

b. What amount of money does he plan to spend each month on these two expenses?

c. What is Joseph's annual income?

Joseph's Budget

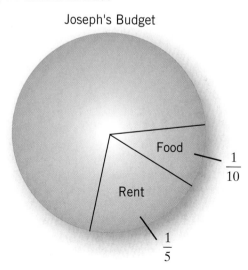

Food

$\frac{1}{10}$

Rent

$\frac{1}{5}$

48. The product of two numbers is 420.

   a. If one of the numbers is the fraction $\dfrac{3}{5}$, do you expect the other number to be larger or smaller than 420?

   b. What is the other number?

49. An airplane is carrying 144 passengers. This is $\dfrac{9}{10}$ of the capacity of the plane.

   a. Is the capacity of the plane more or less then 144?

   b. If you were to multiply 144 times $\dfrac{9}{10}$ would the product be more or less than 144?

   c. What is the capacity of the plane?

50. A student senate has 60 members, and $\dfrac{7}{15}$ of these are women. A change in the senate constitution is being considered, and at the present time a survey shows that $\dfrac{1}{4}$ of the women but only $\dfrac{3}{4}$ of the men are in favor of this change.

   a. If the change requires a $\dfrac{2}{3}$ majority vote in favor to pass, would the constitutional change pass if the vote were taken today?

   b. By how many votes would the change pass or fail?

 HAWKES LEARNING SYSTEMS: INTERMEDIATE ALGEBRA SOFTWARE

   ■ Review of Fractions

Synthetic Division and the Remainder Theorem

Synthetic Division

In Section 4.4, we divided polynomials using the division algorithm (long division). In the special case **when the divisor of a rational expression is a first-degree binomial with leading coefficient 1**, long division can be simplified by omitting the variables entirely and writing only certain coefficients. The procedure is called **synthetic division**. The following analysis describes how the procedure works for $\dfrac{5x^3 + 11x^2 - 3x + 1}{x + 3}$. (Note that $x + 3$ is first-degree with leading coefficient 1.)

a. With Variables

$$
\begin{array}{r}
5x^2 - 4x + 9 \\
x+3\overline{)5x^3 + 11x^2 - 3x + 1} \\
\underline{5x^3 + 15x^2} \\
-4x^2 - 3x \\
\underline{-4x^2 - 12x} \\
9x + 1 \\
\underline{9x + 27} \\
-26
\end{array}
$$

b. Without Variables

$$
\begin{array}{r}
5 \quad -4 \quad +9 \\
1+3\overline{)5 \quad +11 \quad -3 \quad +1} \\
\boxed{5} \;\; +15 \\
-4 \quad \boxed{-3} \\
\boxed{-4} \quad -12 \\
9 \quad \boxed{+1} \\
\boxed{9} \quad +27 \\
-26
\end{array}
$$

The boxed numbers in step **b.** can be omitted since they are repetitions of the numbers directly above them.

c. Boxed numbers omitted

$$
\begin{array}{r}
5 \quad -4 \quad +9 \\
1+3\overline{)5 \quad +11 \quad -3 \quad +1} \\
+15 \\
-4 \\
-12 \\
9 \\
+27 \\
-26
\end{array}
$$

d. Numbers moved up to fill in spaces

$$
\begin{array}{r}
5 \quad -4 \quad +9 \\
1+3\overline{)5 \quad +11 \quad -3 \quad +1} \\
+15 \quad -12 \quad +27 \\
\hline
-4 \quad +9 \quad -26
\end{array}
$$

Next, we omit the 1 in the divisor, change +3 to –3, and write the opposites of the boxed numbers (because the quotient coefficient will now be multiplied by –3 instead of +3), as shown in steps **e.** and **f.** This allows the numbers to be added instead of subtracted. The number 5 is written on the bottom line, and the top line is omitted. The quotient and remainder can now be read from the bottom line.

$$
\begin{array}{r}
5 \quad -4 \quad +9 \\
\textbf{e.} \quad 1+3\overline{)5 \quad +11 \quad -3 \quad +1} \\
\boxed{+15} \; \boxed{-12} \; \boxed{+27} \\
-4 \quad +9 \; -26
\end{array}
$$

$$
\begin{array}{r}
\textbf{f.} \quad -3\overline{)5 \; +11 \; -3 \; +1} \\
\downarrow -15 +12 -27 \\
\underline{5 \; - \; 4+ \; 9 \, -26}
\end{array}
$$

This represents

$$5x^2 - 4x + 9 + \frac{-26}{x+3}.$$

The numbers on the bottom now represent the coefficients of a polynomial of **one degree less than the dividend**, along with the remainder. The last number to the right is the remainder.

In summary, synthetic division can be accomplished as follows:

1. Write only the coefficients of the dividend and the opposite of the constant in the divisor.

$$\underline{-3|} \quad 5 \qquad 11 \qquad -3 \qquad 1$$

2. Rewrite the first coefficient as the first coefficient in the quotient.

$$
\begin{array}{c|cccc}
-3 & 5 & 11 & -3 & 1 \\
 & \downarrow & & & \\
\hline
 & 5 & & &
\end{array}
$$

3. Multiply the coefficient by the constant divisor and **add** this product to the second coefficient.

$$
\begin{array}{c|cccc}
-3 & 5 & 11 & -3 & 1 \\
 & \downarrow & -15 & & \\
\hline
 & 5 \nearrow & -4 & &
\end{array}
$$

4. Continue to multiply each new coefficient by the constant divisor and add this product to the next coefficient in the dividend.

$$
\begin{array}{c|cccc}
-3 & 5 & 11 & -3 & 1 \\
 & \downarrow & -15 & 12 & -27 \\
\hline
 & 5 \nearrow & -4 \nearrow & 9 \nearrow & -26
\end{array}
$$

5. The constants on the bottom line are the coefficients of the quotient and the remainder.

$$\frac{5x^3 + 11x^2 - 3x + 1}{x+3} = 5x^2 - 4x + 9 + \frac{-26}{x+3}$$

$$= 5x^2 - 4x + 9 - \frac{26}{x+3}$$

Example 1: Synthetic Division

Use synthetic division to write each expression in the form $Q + \dfrac{R}{D}$.

a. $\dfrac{4x^3 + 10x^2 + 11}{x + 5}$

Solution:

$$-5\overline{)\begin{array}{ccccc} 4 & 10 & 0 & 11 \\ \downarrow & -20 & 50 & -250 \\ \hline 4 & -10 & 50 & -239 \end{array}}$$

Since there is no x-term, 0 is the coefficient. The coefficient is 0 for any missing term.

$$\frac{4x^3 + 10x^2 + 11}{x + 5} = 4x^2 - 10x + 50 + \frac{-239}{x + 5}$$

$$= 4x^2 - 10x + 50 - \frac{239}{x + 5}$$

b. $\dfrac{2x^4 - x^3 - 5x^2 - 2x + 7}{x - 2}$

Solution:

$$2\overline{)\begin{array}{ccccc} 2 & -1 & -5 & -2 & 7 \\ \downarrow & 4 & 6 & 2 & 0 \\ \hline 2 & 3 & 1 & 0 & 7 \end{array}}$$

$$\frac{2x^4 - x^3 - 5x^2 - 2x + 7}{x - 2} = 2x^3 + 3x^2 + x + \frac{7}{x - 2}$$

NOTES

Remember that synthetic division is used only when the divisor is first-degree of the form $(x + c)$ or $(x - c)$.

The Remainder Theorem

Synthetic division can be used for several purposes, one of which is to find the value of a polynomial for a particular value of x. For example, we know (from Section 4.2) that if

$$P(x) = x^3 - 5x^2 + 7x - 10,$$

then

$$P(2) = 2^3 - 5 \cdot 2^2 + 7 \cdot 2 - 10 = -8.$$

With synthetic division of $x^3 - 5x^2 + 7x - 10$ by $x - 2$ we have

$$\begin{array}{r|rrrr} 2 & 1 & -5 & 7 & -10 \\ & & 2 & -6 & 2 \\ \hline & 1 & -3 & 1 & -8 \end{array} \quad \longleftarrow \text{Remainder}$$

The fact that the remainder is the same as $P(2)$ is not an accident. In fact, as the following theorem states, the remainder when a polynomial is divided by a first-degree factor of the form $(x - c)$ will always be $P(c)$.

The Remainder Theorem

If a polynomial, $P(x)$, is divided by $(x - c)$, then the remainder will be $P(c)$.

Proof:

By the division algorithm we know that $\dfrac{P(x)}{x - c} = Q(x) + \dfrac{R}{x - c}$ where R is a constant.

(Remember that the degree of the remainder must be less than the degree of the divisor.)

Now, multiplying through by $(x - c)$, we have

$$P(x) = (x - c) \cdot Q(x) + R$$

and substituting $x = c$ gives

$$\begin{aligned} P(c) &= (c - c) \cdot Q(c) + R \\ &= 0 \cdot Q(c) + R \\ &= 0 + R \\ &= R. \end{aligned}$$

The proof is complete.

Example 2: The Remainder Theorem and Synthetic Division

a. Use synthetic division to find $P(5)$ given $P(x) = -2x^2 + 15x - 50$.

Solution:
$$\begin{array}{r|rrr} 5 & -2 & 15 & -50 \\ & & -10 & 25 \\ \hline & -2 & 5 & -25 \end{array} \quad \longleftarrow \text{Remainder} = P(5)$$

Thus, $P(5) = -25$.

(Checking shows $P(5) = -2 \cdot 5^2 + 15 \cdot 5 - 50 = -50 + 75 - 50 = -25$.)

Continued on the next page...

b.   Use synthetic division to find $P(-3)$ given $P(x) = 3x^4 + 10x^3 - 5x^2 + 125$.

Note: To evaluate $P(-3)$ we must think of the divisor of the form $(x+3) = (x-(-3))$. That is, in the form $(x-c)$, $c = -3$.

Solution:

$$
\begin{array}{r|rrrrr}
-3 & 3 & 10 & -5 & 0 & 125 \\
   &   & -9 & -3 & 24 & -72 \\
\hline
   & 3 & 1 & -8 & 24 & 53
\end{array}
$$
 ← Remainder = $P(-3)$

Thus, $P(-3) = 53$.

c.   Use synthetic division to show that $(x-6)$ is a factor of $P(x) = x^3 - 14x^2 + 53x - 30$.

Solution:

$$
\begin{array}{r|rrrr}
6 & 1 & -14 & 53 & -30 \\
  &   & 6 & -48 & 30 \\
\hline
  & 1 & -8 & 5 & 0
\end{array}
$$
 ← Remainder = $P(6)$

Thus the remainder is $P(6) = 0$ and $x - 6$ **is a factor of** $P(x)$.

Note: The coefficients in the quotient tell us that $x^2 - 8x + 5$ is also a factor of $P(x)$.

Appendix 2 Exercises

In Exercises 1 – 20, divide by using synthetic division. ***a.*** *Write the answer in the form $Q + \dfrac{R}{D}$ where R is a constant.* ***b.*** *In each exercise, D = (x – c). State the value of c and the value of P(c).*

1. $\dfrac{x^2 - 12x + 27}{x - 3}$

2. $\dfrac{x^2 - 12x + 35}{x - 5}$

3. $\dfrac{x^3 + 4x^2 + x - 1}{x + 8}$

4. $\dfrac{x^3 - 6x^2 + 8x - 5}{x - 2}$

5. $\dfrac{4x^3 + 2x^2 - 3x + 1}{x + 2}$

6. $\dfrac{3x^3 + 6x^2 + 8x - 5}{x + 1}$

7. $\dfrac{x^3 + 6x + 3}{x - 7}$

8. $\dfrac{2x^3 - 7x + 2}{x + 4}$

9. $\dfrac{2x^3 + 4x^2 - 9}{x + 3}$

10. $\dfrac{4x^3 - x^2 + 13}{x - 1}$

11. $\dfrac{x^4 - 3x^3 + 2x^2 - x + 2}{x - 3}$

12. $\dfrac{x^4 + x^3 - 4x^2 + x - 3}{x + 6}$

13. $\dfrac{x^4 + 2x^2 - 3x + 5}{x - 2}$ **14.** $\dfrac{3x^4 + 2x^3 + 2x^2 + x - 1}{x + 1}$ **15.** $\dfrac{x^4 - x^2 + 3}{x - \dfrac{1}{2}}$

16. $\dfrac{x^3 + 2x^2 + 1}{x - \dfrac{2}{3}}$ **17.** $\dfrac{x^5 - 1}{x - 1}$ **18.** $\dfrac{x^5 - x^3 + x}{x + \dfrac{1}{2}}$

19. $\dfrac{x^4 - 2x^3 + 4}{x + \dfrac{4}{5}}$ **20.** $\dfrac{x^6 + 1}{x + 1}$

Collaborative Learning Exercise

21. The class should be divided into teams of 3 or 4 students. Each team should then develop answers to the following questions and be prepared to discuss these answers in class.

a. First use long division to divide the polynomial $P(x) = 2x^3 - 8x^2 + 10x + 15$ by $2x - 1$.

Then use synthetic division to divide the same polynomial by $x - \dfrac{1}{2}$.

Do the same process with two or three other polynomials and divisors. Next compare the corresponding long and synthetic division answers and explain how the answers are related.

b. Use the results from part a and explain algebraically the relationship of the answers when a polynomial is divided (using long division) by $ax - b$ and (using synthetic division) by $x - \dfrac{b}{a}$.

c. Show how the Remainder Theorem should be restated if $x - c$ is replaced by $ax - b$.

🦅 **HAWKES LEARNING SYSTEMS:** INTERMEDIATE ALGEBRA SOFTWARE

▪ Synthetic Division

A.3 Pi

As discussed in the text on page 11, π is an irrational number, and so the decimal form of π is an infinite nonrepeating decimal. Mathematicians even in ancient times realized that π is a constant value obtained from the ratio of a circle's circumference to its diameter, but they had no sense that it might be an irrational number. As early as about 1800 B.C., the Babylonians gave π a value of 3, and around 1600 B.C., the ancient Egyptians were using the approximation of 256/81, which has a decimal value of about 3.1605. In the third century B.C., the Greek mathematician Archimedes used polygons approximating a circle to determine that the value of π must lie between 223/71($\approx$3.1408) and 22/7($\approx$3.1429). He was thus accurate to two decimal places. About seven hundred years later, in the fourth century A.D., Chinese mathematician Tsu Chung-Chi refined Archimedes' method and expressed the constant as 355/113, which was correct to six decimal places. By 1610, Ludolph van Ceulen of Germany had also used a polygon method to find π accurate to 35 decimal places.

Knowing that the decimal expression of π would not terminate, mathematicians still sought a repeating pattern in its digits. Such a pattern would mean that π was a rational number and that there would be some ratio of two whole numbers that would produce the correct decimal representation. Finally, in 1767, Johann Heinrich Lambert provided a proof to show that π is indeed irrational and thus is nonrepeating as well as nonterminating.

Since Lambert's proof, mathematicians have still made an exercise of calculating π to more and more decimal places. The advent of the computer age in this century has made that work immeasurably easier, and on occasion you will still see newspaper articles pronouncing that mathematics researchers have reached a new high in the number of decimal places in their approximations. In 1988 that number was 201,326,000 decimal places. Within 1 year that record was more than doubled, and most recent approximations of π now reach beyond 1.24 trillion decimal places! For your understanding, appreciation and interest, the value of π is given in the table on the next page to a mere 3742 decimal places as calculated by a computer program. To show π calculated to one billion decimal places would take every page of nearly 400 copies of this text!

The Value of π

π =
3.14159265358979323846264338327950288419716939937510582097494459230781640628
62089986280348253421170679821480865132823066470938446095505822317253594081284
81117450284102701938521105559644622948954930381964428810975665933446128475648
23378678316527120190914564856692346034861045432664821339360726024914127372458
70066063155881748815209209628292540917153643678925903600113305305488204665213
84146951941511609433057270365759591953092186117381932611793105118548074462379
96274956735188575272489122793818301194912983367336244065664308602139494639522
47371907021798609437027705392171762931767523846748184676694051320005681271452
63560827785771342757789609173637178721468440901224953430146549585371050792279
68925892354201995611212902196086403441815981362977477130996051870721134999999
83729780499510597317328160963185950244594553469083026425223082533446850352619
31188171010003137838752886587533208381420617177669147303598253490428755468731
15956286388235378759375195778185778053217122680661300192787661119590921642019
89380952572010654858632788659361533818279682303019520353018529689957736225994
13891249721775283479131515574857242454150695950829533116861727855889075098381
75463746493931925506040092770167113900984882401285836160356370766010471018194
29555961989467678374494482553797747268471040475346462080466842590694912933136
77028989152104752162056966024058038150193511253382430035587640247496473263914
19927260426992279678235478163600934172164121992458631503028618297455570674983
85054945885869269956909272107975093029553211653449872027559602364806654991198
81834797753566369807426542527862551818417574672890977772793800081647060016145
24919217321721477235014144197356854816136115735255213347574184946843852332390
73941433345477624168625189835694855620992192221842725502542568876717904946016
53466804988627232791786085784383827967976681454100953883786360950680064425125
20511739298489608412848862694560424196528502221066118630674427862203919494504
71237137869609563643719172874677646575739624138908658326459958133904780275900
99465764078951269468398352595709825822620522489407726719478268482601476990902
64013636944374553050682034962524517493996514314298019065925093722169646151570
98583874105978859597729754989301617539284681382686838689427741559918559252459
53959431049972524680845987273644695848653836736222626099124608051243884390451
24413654976278079771569143599770012961608944169486855584840635342207222582848
86481584560285060168427394522674767889525213852254995466672782398645656961163
54886230577645498035593634568174324112515076069479451096596094025228879710893
14566913686722874894056010150330861792868092087476091782493858900971490967598
52613655497818931297848216829989487226558048575640142704775551323796414515237
46234364542858444795265867821051141354735739523113427166102135969536231442952
48493718711014576540359027993440374200731057853906219838744780847848968332144
57138687519435064302184531910484810053706146806749192781911979399520614196634
28754440643745123718192179998391015919561814675142691239784940907186494231961
56794520809514655022523160388193014209376213785959566389377870830390697920773
46722182562599661501421503068038447734549202605414665925201497442850732518666
00213243408819071048633173464965145390579626856100550810665879699816357473638
40525714591028971064140110971206280439039759515677157700420337869936007230558
76317635942187312514712053292819182618612586732157919841484882916447060957527
06957220917567116722910981690915280173506712748583222871835209353965725121083
57915136988209144421006751033467110314126711136990865851639831501970165151168
51714376576183515565088490998958598238734552833163550764791853589322618548963
21329330985706420467525907091548141654985946163718027098199430992448895757128
28905923233260972997120844335732654893823911932597...

A.4 Powers, Roots, and Prime Factorizations

No.	Square	Square Root	Cube	Cube Root	Prime Factorization
1	1	1.0000	1	1.0000	
2	4	1.4142	8	1.2599	prime
3	9	1.7321	27	1.4422	prime
4	16	2.0000	64	1.5874	2 · 2
5	25	2.2361	125	1.7100	prime
6	36	2.4495	216	1.8171	2 · 3
7	49	2.6458	343	1.9129	prime
8	64	2.8284	512	2.0000	2 · 2 · 2
9	81	3.0000	729	2.0801	3 · 3
10	100	3.1623	1000	2.1544	2 · 5
11	121	3.3166	1331	2.2240	prime
12	144	3.4641	1728	2.2894	2 · 2 · 3
13	169	3.6056	2197	2.3513	prime
14	196	3.7417	2744	2.4101	2 · 7
15	225	3.8730	3375	2.4662	3 · 5
16	256	4.0000	4096	2.5198	2 · 2 · 2 · 2
17	289	4.1231	4913	2.5713	prime
18	324	4.2426	5832	2.6207	2 · 3 · 3
19	361	4.3589	6859	2.6684	prime
20	400	4.4721	8000	2.7144	2 · 2 · 5
21	441	4.5826	9261	2.7589	3 · 7
22	484	4.6904	10,648	2.8020	2 · 11
23	529	4.7958	12,167	2.8439	prime
24	576	4.8990	13,824	2.8845	2 · 2 · 2 · 3
25	625	5.0000	15,625	2.9240	5 · 5
26	676	5.0990	17,576	2.9625	2 · 13
27	729	5.1962	19,683	3.0000	3 · 3 · 3
28	784	5.2915	21,952	3.0366	2 · 2 · 7
29	841	5.3852	24,389	3.0723	prime
30	900	5.4772	27,000	3.1072	2 · 3 · 5
31	961	5.5678	29,791	3.1414	prime
32	1024	5.6569	32,768	3.1748	2 · 2 · 2 · 2 · 2
33	1089	5.7446	35,937	3.2075	3 · 11
34	1156	5.8310	39,304	3.2396	2 · 17

No.	Square	Square Root	Cube	Cube Root	Prime Factorization
35	1225	5.9161	42,875	3.2711	5 · 7
36	1296	6.0000	46,656	3.3019	2 · 2 · 3 · 3
37	1369	6.0828	50,653	3.3322	prime
38	1444	6.1644	54,872	3.3620	2 · 19
39	1521	6.2450	59,319	3.3912	3 · 13
40	1600	6.3246	64,000	3.4200	2 · 2 · 2 · 5
41	1681	6.4031	68,921	3.4482	prime
42	1764	6.4807	74,088	3.4760	2 · 3 · 7
43	1849	6.5574	79,507	3.5034	prime
44	1936	6.6332	85,184	3.5303	2 · 2 · 11
45	2025	6.7082	91,125	3.5569	3 · 3 · 5
46	2116	6.7823	97,336	3.5830	2 · 23
47	2209	6.8557	103,823	3.6088	prime
48	2304	6.9282	110,592	3.6342	2 · 2 · 2 · 2 · 3
49	2401	7.0000	117,649	3.6593	7 · 7
50	2500	7.0711	125,000	3.6840	2 · 5 · 5
51	2601	7.1414	132,651	3.7084	3 · 17
52	2704	7.2111	140,608	3.7325	2 · 2 · 13
53	2809	7.2801	148,877	3.7563	prime
54	2916	7.3485	157,464	3.7798	2 · 3 · 3 · 3
55	3025	7.4162	166,375	3.8030	5 · 11
56	3136	7.4833	175,616	3.8259	2 · 2 · 2 · 7
57	3249	7.5498	185,193	3.8485	3 · 19
58	3364	7.6158	195,112	3.8709	2 · 29
59	3481	7.6811	205,379	3.8930	prime
60	3600	7.7460	216,000	3.9149	2 · 2 · 3 · 5
61	3721	7.8102	226,981	3.9365	prime
62	3844	7.8740	238,328	3.9579	2 · 31
63	3969	7.9373	250,047	3.9791	3 · 3 · 7
64	4096	8.0000	262,144	4.0000	2 · 2 · 2 · 2 · 2 · 2
65	4225	8.0623	274,625	4.0207	5 · 13
66	4356	8.1240	287,496	4.0412	2 · 3 · 11
67	4489	8.1854	300,763	4.0615	prime
68	4624	8.2462	314,432	4.0817	2 · 2 · 17
69	4761	8.3066	328,509	4.1016	3 · 23
70	4900	8.3666	343,000	4.1213	2 · 5 · 7
71	5041	8.4261	357,911	4.1408	prime

No.	Square	Square Root	Cube	Cube Root	Prime Factorization
72	5184	8.4853	373,248	4.1602	2 · 2 · 2 · 3 · 3
73	5329	8.5440	389,017	4.1793	prime
74	5476	8.6023	405,224	4.1983	2 · 37
75	5625	8.6603	421,875	4.2172	3 · 5 · 5
76	5776	8.7178	438,976	4.2358	2 · 2 · 19
77	5929	8.7750	456,533	4.2543	7 · 11
78	6084	8.8318	474,552	4.2727	2 · 3 · 13
79	6241	8.8882	493,039	4.2908	prime
80	6400	8.9443	512,000	4.3089	2 · 2 · 2 · 2 · 5
81	6561	9.0000	531,441	4.3267	3 · 3 · 3 · 3
82	6724	9.0554	551,368	4.3445	2 · 41
83	6889	9.1104	571,787	4.3621	prime
84	7056	9.1652	592,704	4.3795	2 · 2 · 3 · 7
85	7225	9.2195	614,125	4.3968	5 · 17
86	7396	9.2736	636,056	4.4140	2 · 43
87	7569	9.3274	658,503	4.4310	3 · 29
88	7744	9.3808	681,472	4.4480	2 · 2 · 2 · 11
89	7921	9.4340	704,969	4.4647	prime
90	8100	9.4868	729,000	4.4814	2 · 3 · 3 · 5
91	8281	9.5394	753,571	4.4979	7 · 13
92	8464	9.5917	778,688	4.5144	2 · 2 · 23
93	8649	9.6437	804,357	4.5307	3 · 31
94	8836	9.6954	830,584	4.5468	2 · 47
95	9025	9.7468	857,375	4.5629	5 · 19
96	9216	9.7980	884,736	4.5789	2 · 2 · 2 · 2 · 2 · 3
97	9409	9.8489	912,673	4.5947	prime
98	9604	9.8995	941,192	4.6104	2 · 7 · 7
99	9801	9.9499	970,299	4.6261	3 · 3 · 11
100	10,000	10.0000	1,000,000	4.6416	2 · 2 · 5 · 5

Answers

Chapter 1

Exercises 1.1, page 7

1. 2, 3, 5, 7, 11, 13, 17, 19, 23, 29, 31, 37, 41, 43, 47   **3.** A number which has more than two different factors.   **5.** $36 = 2^2 \cdot 3^2$
7. $48 = 2^4 \cdot 3$ **9.** $66 = 2 \cdot 3 \cdot 11$ **11.** $144 = 2^4 \cdot 3^2$ **13.** $270 = 2 \cdot 3^3 \cdot 5$ **15.** $336 = 2^4 \cdot 3 \cdot 7$ **17.** $550 = 2 \cdot 5^2 \cdot 11$
19. $675 = 3^3 \cdot 5^2$ **21.** 25   **23.** 225   **25.** 650   **27.** 28   **29.** 165   **31.** 70   **33.** 300   **35.** 576   **37.** 1210   **39.** 1440

Exercises 1.2, pages 21 - 24

1. $\left\{0, \sqrt{16}, 6\right\}$ **3.** $\left\{-8, -\sqrt{4}, 0, \sqrt{16}, 6\right\}$ **5.** $\left\{-8, -\sqrt{4}, -\frac{4}{3}, -1.2, 0, \frac{4}{5}, \sqrt{16}, 4.2, 6\right\}$ **7.** always   **9.** sometimes
11. sometimes   **13.** 0.625   **15.** $-2.\overline{3}$ **17.** 3.55   **19.** ⟵•—•—•—•—•—•—•⟶   **21.** ⟵—•—•—•—•—•—•—•••⟶
   0 1 2 3 4 5 6   −4 −3 −2 −1 0 1 2 3
23. ⟵—•—•—•—•—•⟶   **25.** ⟵•—•—•—•—•—•—•—•—•—•—•⟶   **27.** $\left\{x \mid 3 \le x < 5\right\}$ **29.** $\left\{x \mid x \ge -2.5\right\}$ **31.** ⟵——(——)——⟶
   1 2 3 4 5   1 2 3 4 5 6 7 8 9 10 11   2   8
33. ⟵——(——)——⟶   **35.** ⟵——[——]——⟶   **37.** ⟵——(——)——[——]——⟶   **39.** $\left\{x \mid -3 \le x \le 1\right\}$
       −√2 0   −1 0   −1.6 0   2 3.7
41. $\left\{x \mid -8 < x \le -2\right\}$ **43.** $\left\{x \mid x \le 1\right\}$ **45.** $\left\{x \mid -4 < x < 4\right\}$ **47.** Inverse property of addition   **49.** Identity property of addition
51. Distributive property   **53.** Inverse property of multiplication   **55.** Transitive property
57. Commutative property of addition   **59.** Distributive property   **61.** Identity property of multiplication
63. Transitive property   **65.** Associative property of addition   **67.** Inverse property of addition   **69.** $7 + x$
71. $x \cdot 6 + x \cdot y$ or $6x + xy$ **73.** $(3x) \cdot z$ **75.** $2 \cdot y + 2 \cdot 3$ or $2y + 6$ **77.** $\frac{1}{6}, 6 \cdot \frac{1}{6} = 1$ **79.** 7, $-7 + 7 = 0$ **81.** When graphing
integers, you graph single points on the line. When graphing real numbers, you include entire portions of the line.

Exercises 1.3, pages 35 - 38

1. 7   **3.** $\sqrt{5}$ **5.** −8   **7.** {7, −7}   **9.** {2, −2}   **11.** $\left\{\frac{4}{5}, -\frac{4}{5}\right\}$ **13.** $\varnothing$ **15.** $x \ge 0$ **17.** −11   **19.** −9   **21.** $\frac{1}{2}$ **23.** −3   **25.** 0   **27.** 13
29. $-\frac{6}{5}$ **31.** −6.9   **33.** 1   **35.** 0   **37.** 20   **39.** $-\frac{13}{6}$ **41.** $-\frac{23}{16}$ **43.** $-\frac{1}{30}$ **45.** 56   **47.** 480   **49.** $\frac{15}{16}$ **51.** 31.8   **53.** −3.92
55. 2   **57.** 13   **59.** Undefined   **61.** 0   **63.** −13.61   **65.** 0.676   **67.** 5   **69.** 6   **71.** 4   **73.** 70   **75.** −17   **77.** 3   **79.** −70   **81.** −18
83. −19   **85.** 1   **87.** 6   **89.** $705.68   **91.** Gained 51 yards   **93.** 484 fish   **95.** −226   **97.** 1   **99.** 6537.5   **101.** Answers may vary.

Exercises 1.4, pages 48 - 50

1. $4x + 3y$ **3.** $6x + y$ **5.** $3x - 4x^2$ **7.** $6x - 7$ **9.** $5x$ **11.** $6x$ **13.** $-x - 15$ **15.** $5x^2 - 11$ **17.** $x = 3$ **19.** $x = -5$ **21.** $x = 0.6$
23. $x = -0.7$ **25.** $x = 3$ **27.** $x = -25$ **29.** $x = 18$ **31.** All real numbers   **33.** $x = 7$ **35.** $x = 2$ **37.** No solution, $\varnothing$ **39.** $x = \frac{3}{4}$
41. $x = 10$ **43.** $x = 5$ **45.** $x = -12$ **47.** $x = 2$ **49.** $x = \frac{3}{2}$ **51.** $x = 4$ **53.** $x = -\frac{3}{2}$ **55.** $x = 9$ **57.** $x = 3$ **59.** $x = 3$
61. conditional   **63.** contradiction   **65.** conditional   **67.** conditional   **69.** identity   **71.** $x = 8$ or $x = -8$ **73.** No solution, $\varnothing$
75. $x = -1$ or $x = -5$ **77.** $x = \frac{9}{2}$ or $x = \frac{7}{2}$ **79.** $x = -\frac{1}{5}$ or $x = 1$ **81.** $x = -6$ or $x = 0$ **83.** $x = -\frac{1}{3}$ or $x = 3$
85. $x = -\frac{5}{2}$ or $x = \frac{3}{4}$ **87.** $x = -\frac{20}{7}$ or $x = \frac{4}{5}$ **89.** $x = -20$ or $x = \frac{40}{9}$

91. Any value substituted for the variable will make the two sides of the equation equal.

Exercises 1.5, pages 55 - 57

1. $P = \dfrac{I}{rt}$, $1650   **3.** $\alpha = 180 - \beta - \gamma$, $43°$ **5.** $w = \dfrac{P - 2l}{2}$, 13 ft   **7.** $a = P - b - c$, 61 in.   **9.** $A = \pi r^2$, 196π sq ft or 615.8 sq ft

11. $b = P - a - c$ **13.** $m = \dfrac{f}{a}$ **15.** $w = \dfrac{A}{l}$ **17.** $n = \dfrac{R}{p}$ **19.** $p = A - I$ **21.** $m = 2A - n$ **23.** $s = \dfrac{P}{4}$ **25.** $t = \dfrac{d}{r}$ **27.** $t = \dfrac{I}{Pr}$

29. $b = \dfrac{P - a}{2}$ **31.** $a = S - Sr$ **33.** $x = \dfrac{y - b}{m}$ **35.** $r^2 = \dfrac{A}{4\pi}$ **37.** $M = \dfrac{(IQ)C}{100}$ **39.** $h = \dfrac{3V}{\pi r^2}$ **41.** $I = \dfrac{E}{R}$ **43.** $L = \dfrac{R}{2A}$

45. $b = \dfrac{2A}{h} - a$ or $b = \dfrac{2A - ah}{h}$ **47.** $R = \dfrac{w_0 L}{Q}$ **49.** $r = \dfrac{S - a}{S}$ **51.** $P = \dfrac{WR - Wr}{2R}$ **53.** $R = \dfrac{nE - Inr}{I}$ **55.** $r = \dfrac{Sb - Sa + a}{L}$

Exercises 1.6, pages 66 - 71

1. 71   **3.** 9   **5.** −19   **7.** 7   **9.** 18   **11.** 8, 30   **13.** 46 ft by 84 ft   **15.** $1500   **17.** 78 minutes   **19.** 375 mph, 450 mph

21. 60 mph, 300 mi   **23.** $4\dfrac{4}{5}$ hr   **25.** 36 mph, 60 mph   **27.** $112.50   **29.** 90 half gallons   **31.** 62,500 pounds

33. 20 at $300; 24 at $250   **35.** $14,000 at 5%; $11,000 at 6%   **37.** 6.5% on $4000; 6% on $3000   **39.** $7000 at 6%; $9000 at 8%
41. a. 53.8 °F **b.** 44 °F **c.** 17 °F   **43. a.** $21.9 billion **b.** $15 billion **c.** Wal-Mart Stores and ING Group
45. See pages 58-59. Answers may vary.

Exercises 1.7, pages 84 - 87

1. Half-open Interval   **3.** Open Interval

5. Half-open Interval   **7.** Closed Interval

9. Half-open Interval   **11.** $x \leq -2$ Half-open Interval   **13.** $x \geq -4$ Half-open Interval

15. $-3 < x < 1$ Open Interval   **17.** $-2 \leq x \leq 2$ Closed Interval   **19.** $x < 3$ Open Interval   **21.** $(-\infty, 1)$

23. $(2, \infty)$ **25.** $\left(\dfrac{8}{3}, \infty\right)$ **27.** $[-0.4, \infty)$

29. $(-\infty, 2)$ **31.** $[3, \infty)$ **33.** $(-\infty, 2]$

35. $\left(-\dfrac{1}{3}, \infty\right)$ **37.** $\left(-\infty, -\dfrac{11}{2}\right)$ **39.** $(-\infty, 8]$

41. $\left(-\dfrac{9}{4}, \infty\right)$ **43.** $[-8, \infty)$ **45.** $(-\infty, 9)$

47. $[2, \infty)$ **49.** $\left(-\infty, -\dfrac{1}{28}\right)$ **51.** $[-4, \infty)$

53. $(-\infty, 1]$ **55.** $(-17, \infty)$ **57.** $(-\infty, -30)$

59. $\left(-\infty, -\dfrac{29}{2}\right]$ **61.** $(-9, 1)$ **63.** $\left[\dfrac{1}{2}, \dfrac{3}{2}\right]$

65. $[3, 15]$ **67.** $(-10, -5)$ **69.** $(-2.8, -0.3)$

71. a. The student would need a score higher than 102 points, which is not possible. Thus he cannot earn an A in the course.
b. The student must score at least 192 points. to earn an A in the course.
73. The second side must be more than 13 mm and less than 33 mm and the third side more than 5 mm and less than 15 mm.

75. $(-\infty, \infty)$ **77.** $\left[-\dfrac{4}{5}, \dfrac{4}{5}\right]$ **79.** $(-\infty, 1) \cup (5, \infty)$

81. No solution, $\varnothing$ **83.** $[-10, -2]$ **85.** No solution, $\varnothing$

87. $\left(-\infty, -\dfrac{2}{3}\right] \cup [6, \infty)$ **89.** $\left(-2, -\dfrac{4}{7}\right)$ **91.** $[-1, 10]$

93. $(-\infty, \infty)$ **95.** $\left\{\dfrac{7}{3}\right\}$ **97. a.**

b. $|x - 4| < 7$ **c.** $(-3, 11)$, Open Interval   **99. a.**   **b.** $|x + 1| \geq 3$ **c.** $(-\infty, -4] \cup [2, \infty)$

Chapter 1 Review, pages 96 - 99

1. $88 = 2^3 \cdot 11$ **2.** $95 = 5 \cdot 19$ **3.** $1000 = 2^3 \cdot 5^3$ **4.** $450 = 2 \cdot 3^2 \cdot 5^2$ **5.** $660 = 2^2 \cdot 3 \cdot 5 \cdot 11$ **6.** $300 = 2^2 \cdot 3 \cdot 5^2$ **7.** $150 = 2 \cdot 3 \cdot 5^2$
8. $195 = 3 \cdot 5 \cdot 13$ **9.** 210 **10.** 180 **11.** 350 **12.** 180 **13.** 204 **14.** 4620 **15.** 7920 **16.** $17,640$ **17. a.** $\{-1, 0, 5\}$ **b.** $\{-\sqrt{6}, \pi\}$
18. $0.\overline{428571}$ **19.**   **20.**   **21.** Zero Factor Law
22. Commutative Property of Addition   **23.** Additive Identity   **24.** Distributive Property   **25.** Multiplicative Inverse
26. Transitive Property   **27.** $\varnothing$ **28.** $\{-1.4, 1.4\}$ **29.** -72 **30.** Undefined   **31.** 0 **32.** $\dfrac{16}{3}$ **33.** -31.8 **34.** 0.5 **35.** 0 **36.** -5
37. $-\dfrac{25}{24}$ **38.** $\dfrac{15}{16}$ **39.** 21 **40.** -8 **41.** -4 **42.** 23 **43.** $\$840.13$ **44.** $16x^2 - 10x$ **45.** $-5y - y^2$ **46.** $23x$ **47.** $4x^2 - 2x - 2$
48. $x = 8$ **49.** $x = 15$ **50.** $x = 34$ **51.** $x = 37$ **52.** $x = -62$ **53.** $x = -6$ **54.** contradiction   **55.** identity   **56.** conditional
57. contradiction   **58.** identity   **59.** $x = -3, x = \dfrac{7}{3}$ **60.** $x = -10, x = 4$ **61.** $x = 0, x = 2$ **62.** No solution, $\varnothing$
63. $x = 3, x = 11$ **64.** $\alpha = 180 - \beta - \gamma, 72°$ **65.** $\dfrac{2A}{h} - b = c$, 7 meters   **66.** $P = \dfrac{k}{V}$ **67.** $h = \dfrac{2A}{b}$ **68.** $r = \dfrac{C}{2\pi}$
69. $a = P - 2b$ **70.** $y = 15 - 3x$ **71.** $y = 3x - 15$ **72.** $d = \dfrac{l}{N-1}$ **73.** $w = \dfrac{P - 2l}{2}$ **74.** $x = \dfrac{6 - 3y}{2}$
75. $a = \dfrac{2A}{h} - b$ **76.** 46 **77.** $x = -11$ **78.** 55 mph, 247.5 miles   **79.** $\$4000$ at 6%, $\$16,000$ at 8% **80. a.** 28.34 million
b. Everland   **c.** 2.8 million   **81.** $(-\infty, -4)$ **82.** $(-\infty, -6]$
83. $(1, \infty)$ **84.** $\left(-\infty, -\dfrac{8}{3}\right]$ **85.** $(-1, 2]$
86. $(-3.8, -0.3)$ **87.** No solution, $\varnothing$ **88.** $[-24, -10]$ **89.** $(0, 4)$
90. $(-\infty, \infty)$

Chapter 1 Test, pages 100 - 102

1. $1296 = 2^4 \cdot 3^4$ **2.** $575 = 5^2 \cdot 23$ **3.** 180 **4.** 240 **5. a.** $\{-2, 0, 2\}$ **b.** $\left\{-2, -\dfrac{5}{3}, 0, \dfrac{1}{2}, 2\right\}$
6. $0.41\overline{6}$ **7.**   **8.**   **9.**   **10.**
11. Distributive property   **12.** Associative property of addition   **13.** 0 **14.** $x < a$ **15. a.** $\{-5, 5\}$ **b.** $\varnothing$ **16.** 70 **17.** 8
18. $-\dfrac{15}{8}$ **19.** $-\dfrac{4}{15}$ **20.** -144 **21.** 38 **22.** 13 **23.** $6x + 16$ **24.** $7x - 12$ **25.** $x = -1$ **26.** No solution   **27.** $x = -6$
28. $x = -1.9$ or $x = 0.9$ **29.** $x = \dfrac{2}{3}$ or $x = \dfrac{8}{3}$ **30. a.** conditional   **b.** contradiction   **c.** identity   **31.** $A = \pi r^2$, 1134.1 ft^2
32. $c = \dfrac{2A}{h} - b$, 12m   **33.** $w = \dfrac{P - 2l}{2}$ **34.** $Q = \dfrac{2U}{V}$ **35.** 5 **36.** 4 hr   **37.** 16 packages at $\$1.25$; 26 packages at $\$1.75$
38. a. 175.05 million   **b.** 186.2 million   **c.** 128.5 million   **39.** Open Interval $x < 5$ **40.** Closed Interval $-1 \leq x \leq 4$

41. 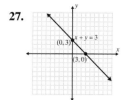 $(-5, \infty)$ **42.** $\left(-\infty, -\dfrac{7}{2}\right)$ **43.** $(-1, 5)$

44. $(2, 5)$ **45.** $(-\infty, -0.85) \cup (1.85, \infty)$

Chapter 2

Exercises 2.1, pages 112 - 117

1. $\{A(-4, 0), B(0, 0), C(5, -1), D(-3, -3), E(2, -4)\}$ **3.** $\{A(4, 0), B(-2, 3), C(-1, 0), D(4, 4), E(5, -3)\}$

5. $\{A(1, 1), B(0, -4), C(-2, -3), D\ (4, 3), E(3, -2)\}$ **7.**   **9.** 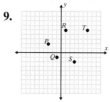   **11.**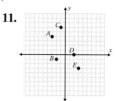

13. $(0, 0), (-2, 3), (2, -3)$; Answers may vary.

15. $(0, -2), (1, 1), (2, 4)$; Answers may vary.

17. $(-2, 0), (1, -2), (4, -4)$; Answers may vary.

19.

x	y
0	5
$\dfrac{5}{2}$	0
-2	9
1	3

21.

x	y
0	-4
$\dfrac{4}{3}$	0
2	2
3	5

23. a. $(0, 5)$ **b.** $\left(\dfrac{5}{2}, 0\right)$ **c.** $(2, 1)$ **d.** $(-1, 7)$

25. a. $(0, -3)$ **b.** $(2, 0)$ **c.** $(-2, -6)$ **d.** $(4, 3)$

27. 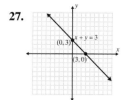 **29.** 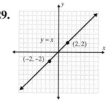 **31.** 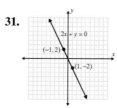 **33.** 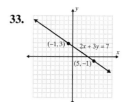 **35.**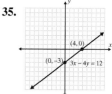

37. 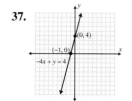 **39.** 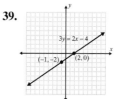 **41.** 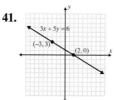 **43.** 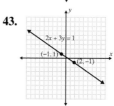 **45.**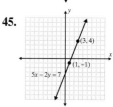

47. 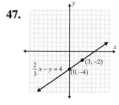 **49.** 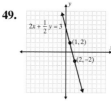 **51.** 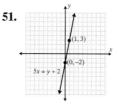 **53.** 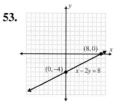 **55.**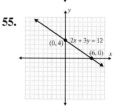

57. 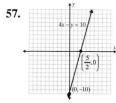 **59.** 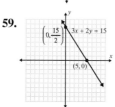 **61. a.** $60, 120, 240, 300, 420$ **b.**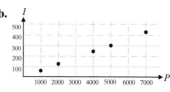

63. a. 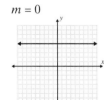 **b.** Yes, the more push-ups a person can do, it appears the more pull-ups he/she can do.

c. 8 pull-ups, 11 pull-ups, 12 pull-ups, 16 pull-ups; Answers may vary.

65. Plug the x and y values into the equation. Then evaluate both sides to see if the equation is true.

Exercises 2.2, pages 128 - 133

1. $m = 5$ **3.** $m = -\dfrac{8}{7}$ **5.** $m = 0$ **7.** $m = -\dfrac{3}{10}$ **9.** m is undefined. **11.** $m = \dfrac{1}{5}$

13. horizontal line; $m = 0$

15. vertical line; m is undefined.

17. horizontal line; $m = 0$

19. vertical line; m is undefined.

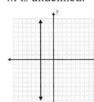
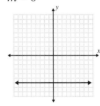
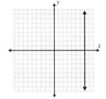

21. $y = 2x - 1; m = 2,$ y-int $= (0, -1)$

23. $y = -4x + 5; m = -4,$ y-int $= (0, 5)$

25. $y = \dfrac{2}{3}x - 3; m = \dfrac{2}{3},$ y-int $= (0, -3)$

27. $y = -x + 5; m = -1,$ y-int $= (0, 5)$

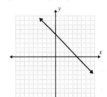

29. $y = -\dfrac{1}{5}x + 2; m = -\dfrac{1}{5},$ y-int $= (0, 2)$

31. $y = 4; m = 0,$ y-int $= (0, 4)$

33. $y = -4x; m = -4,$ y-int $= (0, 0)$

35. $y = \dfrac{2}{3}x - 2; m = \dfrac{2}{3},$ y-int $= (0, -2)$

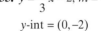

37. Cannot be written in slope-intercept form. $x = -3$; slope is undefined, no y-intercept

39. $y = \dfrac{5}{6}x - \dfrac{5}{3}; m = \dfrac{5}{6},$ y-int $= \left(0, -\dfrac{5}{3}\right)$

41. $y = -\dfrac{3}{4}x + \dfrac{5}{4}; m = -\dfrac{3}{4},$ y-int $= \left(0, \dfrac{5}{4}\right)$

43. $y = -\dfrac{3}{2}x - \dfrac{7}{4}; m = -\dfrac{3}{2},$ y-int $= \left(0, -\dfrac{7}{4}\right)$

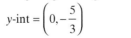

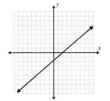

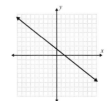
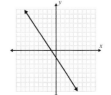

45. $y = \dfrac{1}{2}x + \dfrac{2}{3};$ **47.** $y = \dfrac{5}{2}x + \dfrac{5}{2};$ **49.** Answers may vary.

$m = \dfrac{1}{2},$ $m = \dfrac{5}{2},$

$y\text{-int} = \left(0, \dfrac{2}{3}\right)$ $y\text{-int} = \left(0, \dfrac{5}{2}\right)$

a. 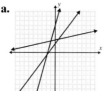 **b.** 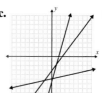 **c.** **d.**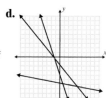

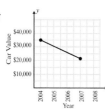   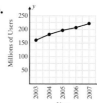

51. a. m is undefined **b.** no y-intercept **c.** $x = 3$

53. a. $m = -3$ **b.** $(0, -3)$ **c.** $y = -3x - 3$

55. a. $m = \dfrac{5}{4}$ **b.** $(0, 1)$ **c.** $y = \dfrac{5}{4}x + 1$

57. a. $m = 4$ **b.** $(0, -2)$ **c.** $y = 4x - 2$ **59.** yes **61.** yes

63. $4000/year

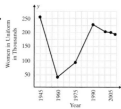

65. a. and b.

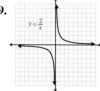

c. 23, 13, 12, 10

d. The number of internet users inc. 23 million people/year from '03-'04, 13 mppy from '04-'05, 12 mppy from '05-'06, and 10 mppy from '06-'07.

67. a. and b.

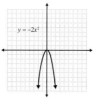

c. −15,647.07; 4354.53; 8676.67; −2441.7; −205.2; −1542.5

d. The number of female active duty military personnel dec. 15,647.07 women/year from 1945-1960; inc. 4354.53 wpy from '60-'75; inc. 8676.67 wpy from '75-'90; dec. 2441.7 wpy from '90-'00; dec. 205.2 wpy from '00-'05; and dec. 1542.5 wpy from '05-'07.

69.   **71.**   **73.**

Exercises 2.3, pages 141 - 143

1. a. $m = 2$ **b.** $(3, 1)$ **3. a.** $m = -5$ **b.** $(0, -2)$ **5. a.** $m = -1$ **b.** $(-8, 0)$ **7.** $2x + y = -3$

c. 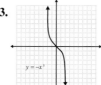   **c.**   **c.**

9. $y = -2$

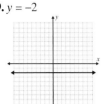

11. $x = -3$

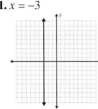

13. $x + 4y = -5$

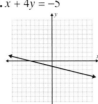

15. $2x - 3y = -5$

17. $y = \dfrac{1}{2}x + \dfrac{9}{2}$ **19.** $y = -\dfrac{1}{7}x + \dfrac{2}{7}$ **21.** $y = -\dfrac{5}{4}x + 2$ **23.** $y = -5$ **25.** $y = -x + 4$ **27.** $y = 6$ **29.** $y = 7$ **31.** $x = 2$

33. $y = 2x$ **35.** $y = 5x + 2$ **37.** $y = -\dfrac{3}{5}x + \dfrac{7}{5}$ **39.** $y = 7$ **41.** $y = -\dfrac{1}{2}x - 2$

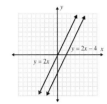

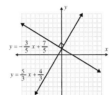

43.

45.

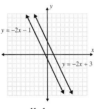

47.

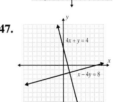

49.

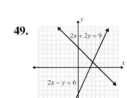

parallel   perpendicular   neither

51. The Americans with Disabilities Act recommends a slope of $\dfrac{1}{12}$ for business use and a slope of $\dfrac{1}{6}$ for residential use. Answers may vary.

Exercises 2.4, pages 158 - 164

1. $\{(-4,0),(-1,4),(1,2),(2,5),(6,-3)\}$; $D = \{-4,-1,1,2,6\}$; $R = \{-3,0,2,4,5,\}$; function

3. $\{(-5,-4),(-4,-2),(-2,-2),(1,-2),(2,1)\}$; $D = \{-5,-4,-2,1,2\}$; $R = \{-4,-2,1\}$; function

5. $\{(-4,-3),(-4,1),(-1,-1),(-1,3),(3,-4)\}$; $D = \{-4,-1,3\}$; $R = \{-4,-3,-1,1,3\}$; not a function

7. $\{(-5,-5),(-5,3),(0,5),(1,-2),(1,2)\}$; $D = \{-5,0,1\}$; $R = \{-5,-2,2,3,5\}$; not a function

9. $D = \{-3,0,1,2,4\}$

$R = \{-2,-1,0,5,6\}$

function

11. $D = \{-4,-3,1,2,3\}$

$R = \{4\}$

function

13. $D = \{-3,-1,0,2,3\}$

$R = \{1,2,4,5\}$

function

15. $D = \{-1\}$

$R = \{-2,0,2,4,6\}$

not a function

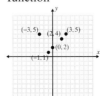

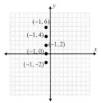

17. function; $D = (-\infty, \infty)$; $R = [0, \infty)$ **19.** function; $D = (-\infty, \infty)$; $R = (-\infty, \infty)$ **21.** not a function; $D = (-\infty, \infty)$; $R = (-\infty, \infty)$

23. not a function; $D = (-\infty, \infty)$; $R = (-\infty, \infty)$ **25.** function; $D = [-5, 5]$; $R = [-2, 2]$

27. not a function; $D = \{-3\}$; $R = (-\infty, \infty)$ **29.** $\left\{(-9, -26), \left(-\frac{1}{3}, 0\right), (0, 1), \left(\frac{4}{3}, 5\right), (2, 7)\right\}$

31. $\left\{(-2, -11), (-1, -2), (0, 1), (1, -2), (2, -11)\right\}$ **33. a.** -4 **b.** -16 **c.** -10 **35. a.** 0 **b.** 12 **c.** 56 **37. a.** -3 **b.** 0 **c.** 3

39. $D = (-\infty, \infty)$ **41.** $D = \left(-\infty, \frac{1}{2}\right) \cup \left(\frac{1}{2}, \infty\right)$ or $x \neq \frac{1}{2}$ **43.** $D = (-\infty, \infty)$

45. 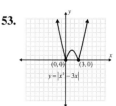 **47.** **49.** **51.** **53.**

55. **57.** **59.** 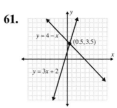 **61.** 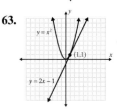 **63.**

65. y-intercept $= (0, -5)$ (should be $(0, 5)$) **67.** y-intercept $= (0, -9)$ (should be $(0, -2)$)

69. slope $= -3$ (should be the reciprocal, $-\frac{1}{3}$) and y-intercept $= (0, 2)$ (should be $(0, 0)$)

71. a. No restrictions, the domain of a linear function is $(-\infty, \infty)$.
b. Restrictions; the value under the radical cannot be negative. **c.** Restrictions; the denominator cannot equal zero.

Exercises 2.5, pages 171 - 172

1. 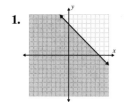 **3.** 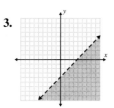 **5.** 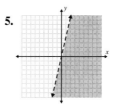 **7.** 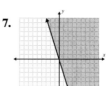 **9.**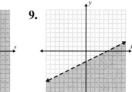

11. 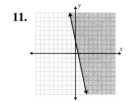 **13.** 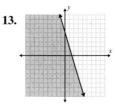 **15.** 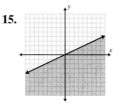 **17.** 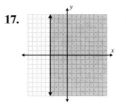 **19.**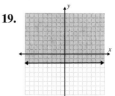

21. 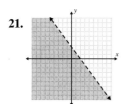 **23.** 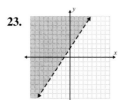 **25.** 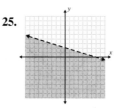 **27.** 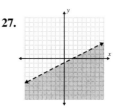 **29.**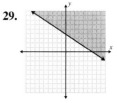

31. 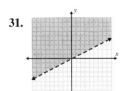 **33.** 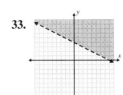 **35.** 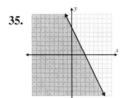 **37.** 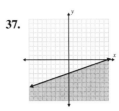 **39.**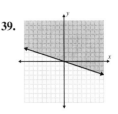

41. Test any point not on the line. If the test point satisfies the inequality, shade the half-plane on that side of the line. Otherwise, shade the other half-plane.

Chapter 2 Review, pages 177 - 183

1. $\{A(-1, 2), B(-5, -4), C(4, 4), D(3, 0), E(0, -5)\}$ **2.** $\{A(3, 1), B(0, 0), C(-1, -4), D(3, -2), E(-3, -1)\}$

3. $(-1, 3), (0, 2), (2, 0)$ Answers may vary. **4.** $(-1, -4), (0, -3), (3, 0)$ Answers may vary.

5. a. $(4, 0)$ **b.** $(1, 1)$ **c.** $\left(2, \dfrac{2}{3}\right)$ **d.** $(-2, 2)$ **6. a.** $(0, -7)$ **b.** $(2, -1)$ **c.** $(-2, -13)$ **d.** $\left(\dfrac{5}{3}, -2\right)$

7. a. $(4, 0)$ **b.** $(0, 8)$ **c.** $(3, 2)$ **d.** $(5, -2)$ **8. a.** $(0, 3)$ **b.** $(2, 0)$ **c.** $(-2, 6)$ **d.** $(6, -6)$

9. 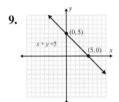 **10.** 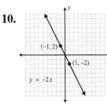 **11.** 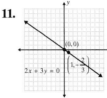 **12.** 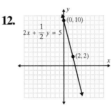 **13.**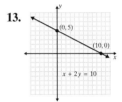

14. 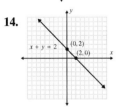 **15.** 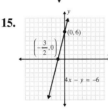 **16.** 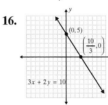 **17.** $m = -2$ **18.** $m = 3$ **19.** $m = -\dfrac{1}{2}$ **20.** $m = 4$

21. horizontal line; $m = 0$

22. vertical line; m is undefined.

23. vertical line; m is undefined.

24. horizontal line; $m = 0$

25. $y = -\dfrac{2}{5}x + 1;$ $m = -\dfrac{2}{5},$ y-int $= (0, 1)$

26. $y = \dfrac{4}{3}x - 2;$ $m = \dfrac{4}{3},$ y-int $= (0, -2)$

27. $y = -3x + 12;$ $m = -3,$ y-int $= (0, 12)$

28. $y = \dfrac{1}{2}x - 2;$ $m = \dfrac{1}{2},$ y-int $= (0, -2)$

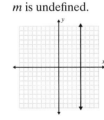 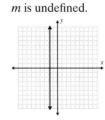 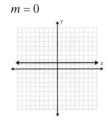

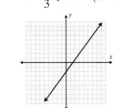 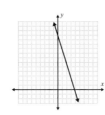

29. $y = x + 6$;
$m = 1$, y-int $= (0,6)$

30. $y = -2x - 8$;
$m = -2$, y-int $= (0,-8)$

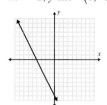

31. $y = 3$; $m = 0$,
y-int $= (0,3)$

32. $x = -3$; m is undefined,
no y-intercept

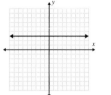

33. $y = \dfrac{2}{3}x + 2$; $m = \dfrac{2}{3}$,
y-int $= (0,2)$

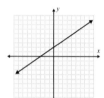

34. $y = -2x + \dfrac{7}{2}$; $m = -2$,
y-int $= \left(0, \dfrac{7}{2}\right)$

35. a. $m = -\dfrac{3}{2}$ **b.** $(0,5)$ **c.** $y = -\dfrac{3}{2}x + 5$

36. a. $m = -\dfrac{1}{4}$ **b.** $(0,-1)$ **c.** $y = -\dfrac{1}{4}x - 1$

37. a. m is undefined **b.** no y-intercept **c.** $x = -2$

38. a. $m = 0$ **b.** $(0,4)$ **c.** $y = 4$

39. a. $m = \dfrac{2}{3}$ **b.** $(0,-3)$ **c.** $y = \dfrac{2}{3}x - 3$

40. a. $m = -\dfrac{7}{2}$ **b.** $(0,-4)$ **c.** $y = -\dfrac{7}{2}x - 4$

41. a. $m = 3$ **b.** $(-1,7)$

c.

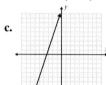

42. a. $m = -1$ **b.** $(1,-3)$

c.

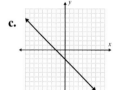

43. $3x + y = -11$

44. $5x - 2y = 19$

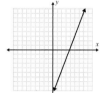

45. $x = -4$

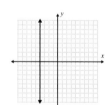

46. $y = -\dfrac{3}{2}$

47. $y = -2x - 1$ **48.** $y = -\dfrac{1}{2}x + 4$ **49.** $y = -\dfrac{1}{8}x + \dfrac{1}{12}$

50. $y = \dfrac{14}{15}x - \dfrac{1}{2}$ **51.** $y = 5$ **52.** $x = 5$

53. $y = \dfrac{1}{4}x - \dfrac{1}{2}$ **54.** $y = -x - 3$ **55.** $y = \dfrac{1}{2}x + \dfrac{1}{2}$

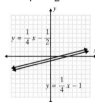

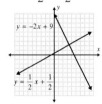

56. $y = 2x + 6$ **57.** $y = -\dfrac{2}{3}x - 4$ **58.** $y = -\dfrac{1}{3}x + 5$

59. parallel

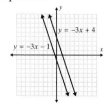

60. perpendicular

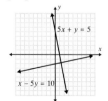

61. perpendicular

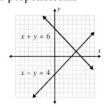

62. neither

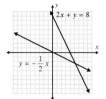

63. $D = \{0, -1, 4, 5, 6\}$
$R = \{3, 2\}$
function

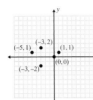

64. $D = \{-5, -3, 0, 1\}$
$R = \{-2, 0, 1, 2\}$
not a function

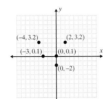

65. $D = \{-4, -3, 0, 2\}$
$R = \{-2, 0.1, 3.2\}$
not a function

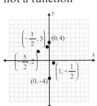

66. $D = \left\{-\dfrac{5}{2}, -\dfrac{1}{2}, 0, 1\right\}$
$R = \left\{-4, -\dfrac{1}{2}, 2, 3, 4\right\}$
not a function

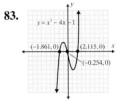

67. Function; $D:(-\infty,\infty)$, $R:(-\infty,\infty)$ **68.** Not a function; $D:(-\infty, -3] \cup [3, \infty)$, $R:(-\infty, \infty)$

69. Function; $D:(-\infty,\infty)$, $R:[-3,3]$ **70.** Function; $D:(-\infty,\infty)$, $R:(-\infty,4]$ **71.** Not a function; $D:(-\infty, 3]$, $R:(-\infty, \infty)$

72. Function; $D:(-\infty,-2] \cup [-1,\infty)$, $R:(-\infty,\infty)$ **73. a.** 5 **b.** 26 **c.** 3   **74. a.** 18 **b.** -17 **c.** -9   **75. a.** 6 **b.** 20 **c.** 6

76. a. 18 **b.** 108 **c.** 570   **77.** $D = (1,\infty)$ **78.** $D = (-\infty,\infty)$

79.
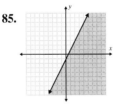

80.

81.

82.

83.

84.

85.
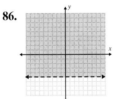

86.

87.

88.

89.
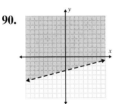

90.
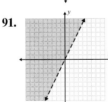

91.
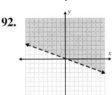

92.

93.
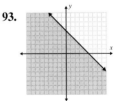

94.
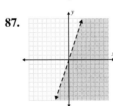
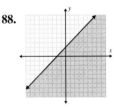

Chapter 2 Test, pages 184 - 185

1. a. $(0, 2)$ **b.** $\left(\dfrac{2}{3}, 0\right)$ **c.** $(-2, 8)$ **d.** $(3, -7)$ **2. a.** $\left(0, -\dfrac{6}{5}\right)$ **b.** $(6, 0)$ **c.** $(11, 1)$ **d.** $(-4, -2)$

3. 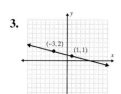 **4.** 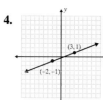 **5.** $m = \dfrac{9}{8}$ **6.** $m = -\dfrac{1}{5}$ **7.** $y = \dfrac{1}{3}x - \dfrac{4}{3}; m = \dfrac{1}{3},$

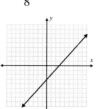 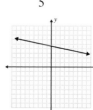

$y\text{-int} = \left(0, -\dfrac{4}{3}\right)$

8. $y = -\dfrac{4}{3}x + 1; m = -\dfrac{4}{3},$ **9.** $5x + 3y = 36$ **10.** $8x + 7y = 10$ **11.** $y = 6$ **12.** $y = -\dfrac{3}{2}x + 7$ **13.** $y = -2$

$y\text{-int} = (0,1)$ **14. a.**  **b.** $11, -14$ **c.** Sam's speed increased 11 mph the second hour, and decreased 14 mph the third hour.

15. See page 145. **16.** $f = \{A(-3,2), B(-2,-1), C(0,-3), D(1,2), E(3,4), F(5,0)\}$

$D = \{-3,-2,0,1,3,5\}$ $R = \{2,-1,-3,4,0\}$ function

17. not a function $D = (-\infty, \infty)$ $R = (-\infty, \infty)$ **18.** function $D = (-\infty, \infty)$ $R = [0, \infty)$ **19. a.** 13 **b.** 5 **c.** 4

20. $D = \left(-\infty, \dfrac{1}{2}\right) \cup \left(\dfrac{1}{2}, \infty\right)$

21. a. 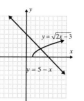 **22. a.** 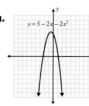 **23.** **24.**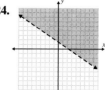

b. $(3.17, 1.83)$ **b.** $(-0.5, 5.5)$

Chapter 2 Cumulative Review, pages 186–190

1. $93 = 3 \cdot 31$ **2.** $300 = 2^2 \cdot 3 \cdot 5^2$ **3.** $188 = 2^2 \cdot 47$ **4.** $245 = 5 \cdot 7^2$ **5.** 180 **6.** 216 **7.** 2100 **8.** 840 **9. a.** $\{\sqrt{9}\}$ **b.** $\{0, \sqrt{9}\}$

c. $\left\{-10, -\sqrt{25}, -1.6, 0, \dfrac{1}{5}, \sqrt{9}\right\}$ **d.** $\{-10, -\sqrt{25}, 0, \sqrt{9}\}$ **e.** $\{-\sqrt{7}, \pi, \sqrt{12}\}$ **f.** $\left\{-10, -\sqrt{25}, -1.6, -\sqrt{7}, 0, \dfrac{1}{5}, \sqrt{9}, \pi, \sqrt{12}\right\}$

10. **11.** **12.** Associative Property of Addition **13.** Distributive Property

14. Inverse Property of Addition **15.** Identity Property of Multiplication **16.** Transitive Property of Order

17. Trichotomy Property of Order **18.** -20 **19.** 11 **20.** 22 **21.** -1 **22.** -0.8 **23.** $-\dfrac{1}{12}$ **24.** 84 **25.** -40 **26.** -2 **27.** $\dfrac{2}{3}$

28. Undefined **29.** 0 **30.** -2 **31.** -35 **32.** 26 **33.** -59 **34.** $-2x - 12$ **35.** 0 **36.** $3x^3 - x^2 + 4x - 1$ **37.** $-10x + 4$ **38.** $x = 2$

39. $x = -4$ **40.** $x = -4$ **41.** $x = -8$ **42.** $x = 2.3$ or $x = -3.3$ **43.** $x = \dfrac{10}{3}$ or $x = 2$ **44. a.** $n = 2A - m$ **b.** $f = \dfrac{\omega}{2\pi}$ **45.** $-\dfrac{1}{3}$

46. 7 hrs **47.** 10 miles **48.** 84 **49.** $(4, \infty)$ **50.** $[-12.8, \infty)$

51. $(-\infty, 2]$ **52.** $(-\infty, -5]$

53.  $\left(-\infty, -\dfrac{9}{5}\right)$ or $(1, \infty)$ **54.** $\left(-\dfrac{8}{3}, \dfrac{4}{3}\right)$

55. a. $(0, -4)$ **b.** $(2, 0)$ **c.** $(1, -2)$ **d.** $(3, 2)$ **56. a.** $(0, 2)$ **b.** $(6, 0)$ **c.** $\left(2, \dfrac{4}{3}\right)$ **d.** $(9, -1)$

57. 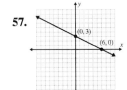   **58.** 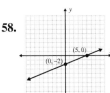   **59.**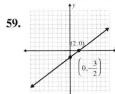

60. $y = -\dfrac{1}{5}x + 2$; $m = -\dfrac{1}{5}$, y-int $= (0, 2)$ **61.** $y = -3x + 1$; $m = -3$, y-int $= (0, 1)$ **62.** $y = \dfrac{3}{7}x - 1$; $m = \dfrac{3}{7}$; y-int $= (0, -1)$

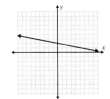   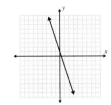   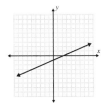

63. $2x - 5y = 17$ **64.** $4x - 3y = -10$ **65.** $2x - y = 0$ **66.** $x = 5$ **67.** $4x + 5y = 15$ **68.** $2x + y = 8$

69. $y = -\dfrac{3}{2}x + 6$ **70.** $x = 1$ **71.** $y = \dfrac{3}{4}x - 3$ **72.** $y = -\dfrac{5}{3}x + 8$

73. $\{A(-5, -2), B(-3, -2), C(-2, 4), D(1, -1), E(2, 4)\}$; $D = \{-5, -3, -2, 1, 2\}$; $R = \{-2, 4, -1\}$; function

74. not a function; $D = (-\infty, 0]$; $R = (-\infty, \infty)$ **75.** $D = [-2, \infty)$ **76.** $D = (-\infty, \infty)$ **77.** $D = (-\infty, 9) \cup (9, \infty)$ **78.** $f(5) = -1$

79. $g(3) = 4$ **80.** $F(0) = -10$ **81.** $G(-2) = -32$ **82.** 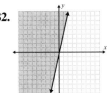   **83.**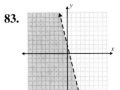

84. a. and **b.** 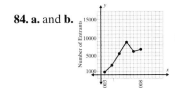   **c.** 1737; 3043; 3154; −2415; 486

d. The number of entrants inc. 1737 people/year from 2003-2004; rate of change inc. 3043 people/year from 2004-2005; rate of change inc. 3154 people/year from 2005-2006; rate of change dec. 2415 people/year from 2006-2007; rate of change inc. 486 people/year from 2007-2008.

85. a. 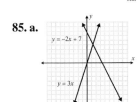   **b.** $(1.4, 4.2)$ **86. a.**   **b.** $(-1, -0.5)$ **87. a.** 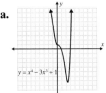   **b.** $(0.7648, 0)$, $(2.9615, 0)$

Chapter 3

Exercises 3.1, pages 201 - 203

1. c **3.** a **5.** $(1, 3)$ **7.** None **9.** $m_1 = -2$, $y\text{-int}_1 = (0, 4)$; $m_2 = -2$, $y\text{-int}_2 = \left(0, \dfrac{1}{2}\right)$

11. $m_1 = 4$, $y\text{-int}_1 = (0, -8)$; $m_2 = 4$, $y\text{-int}_2 = (0, -28)$

13. $(5, 0)$ **15.** No solution   **17.** $(3, 3)$ **19.** $(-1, 1)$ **21.** $(3, -4)$

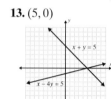 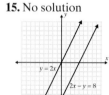 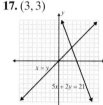 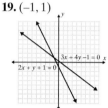 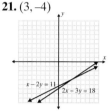

23. $(2, 1)$, consistent **25.** $(-1, -3)$, consistent **27.** No solution, inconsistent **29.** $(x, 8 - 3x)$, dependent **31.** $(1, 6)$, consistent

33. $\left(\dfrac{3}{2}, 1\right)$, consistent **35.** $\left(3, -\dfrac{3}{4}\right)$, consistent **37.** No solution, inconsistent **39.** $\left(\dfrac{1}{2}, \dfrac{1}{4}\right)$, consistent **41.** $(5, 2)$, consistent

43. $(x, 6x - 15)$, dependent **45.** $(7, 5)$, consistent **47.** $(4, 7)$, consistent **49.** $(2.6667, -2.3333)$ **51.** $(0.6667, -1.3333)$

53. $(-0.6, -2.2)$ **55.** $(13, -8)$ **57.** The solution to a consistent system of linear equations is a single point, which is easily written as an ordered pair.

Exercises 3.2, pages 207 - 211

1. $23, 79$ **3.** $80°, 100°$ **5.** 100 yards $\times$ 45 yards **7.** 40 liters of 12%, 50 liters of 30% **9.** 240 gal of 5%, 120 gal of 2%

11. \$87,000 in bonds, \$37,000 in certificates **13.** 325 at \$3.50/share, 175 at \$6.00/share **15.** 40 lb at \$3.90/lb, 30 lb at \$2.50/lb

17. 16 lb at \$0.70/lb, 4 lb at \$1.30/lb **19.** \$5.50 for paperback, \$9.00 for hardback

21. 205 legislators voted in favor of the bill **23.** 6:00 pm **25.** $a = 3$, $b = 1$; $3x + y = 7$ **27.** $a = -2$, $b = 3$; $-2x + 3y = 6$

29. $a = 10$, $b = -2$; $10x - 2y = -4$ **31.** \$11.20/hr labor, \$4.80/lb materials **33.** 9 lb of Ration I, 2 lb of Ration II

35. $65°, 65°, 50°$ **37.** Sue: 12 years old; Pat: 22 years old **39.** length: 15 m and width: 10 m

Exercises 3.3, pages 220 - 223

1. $x = 1$, $y = 0$, $z = 1$ **3.** $x = 1$, $y = 2$, $z = -1$ **5.** Infinite solutions **7.** $x = -2$, $y = 9$, $z = 1$ **9.** $x = 4$, $y = 1$, $z = 1$

11. $x = -2$, $y = 3$, $z = 1$ **13.** No solution **15.** Infinite solutions **17.** $x = \dfrac{1}{2}$, $y = \dfrac{1}{3}$, $z = -1$ **19.** $x = 2$, $y = 1$, $z = -3$

21. $34, 6, 27$ **23.** 18 ones, 16 fives, 12 tens **25.** $a = 1$, $b = 3$, $c = -2$; $y = x^2 + 3x - 2$ **27.** 19 cm, 24 cm, 30 cm

29. home: \$90,000, lot: \$22,000, improvements: \$11,000 **31.** savings: \$30,000, bonds: \$55,000, stocks: \$15,000

33. $100°, 30°, 50°$ **35.** 3 liters of 10%, 4.5 liters of 30%, 1.5 liters of 40% **37.** $A = 2$, $B = -1$, $C = 3$

Exercises 3.4, pages 236 - 238

1. $\begin{bmatrix} 2 & 2 \\ 5 & -1 \end{bmatrix}$, $\left[\begin{array}{cc|c} 2 & 2 & 13 \\ 5 & -1 & 10 \end{array}\right]$ **3.** $\begin{bmatrix} 7 & -2 & 7 \\ -5 & 3 & 0 \\ 0 & 4 & 11 \end{bmatrix}$, $\left[\begin{array}{ccc|c} 7 & -2 & 7 & 2 \\ -5 & 3 & 0 & 2 \\ 0 & 4 & 11 & 8 \end{array}\right]$ **5.** $\begin{bmatrix} 3 & 1 & -1 & 2 \\ 1 & -1 & 2 & -1 \\ 0 & 2 & 5 & 1 \\ 1 & 3 & 0 & 3 \end{bmatrix}$, $\left[\begin{array}{cccc|c} 3 & 1 & -1 & 2 & 6 \\ 1 & -1 & 2 & -1 & -8 \\ 0 & 2 & 5 & 1 & 2 \\ 1 & 3 & 0 & 3 & 14 \end{array}\right]$

7. $\begin{cases} -3x + 5y = 1 \\ -x + 3y = 2 \end{cases}$ **9.** $\begin{cases} x + 3y + 4z = 1 \\ 2x - 3y - 2z = 0 \\ x + y = -4 \end{cases}$ **11.** $x = -1, y = 2$ **13.** $x = -1, y = -1$ **15.** $x = -1, y = -2, z = 3$

17. $x = 1, y = 0, z = 1$ **19.** $x = 2, y = 1, z = -1$ **21.** $x = -2, y = 9, z = 1$ **23.** No solution   **25.** $x = 1, y = 2, z = 1$

27. $x = 1, y = -3, z = 2$ **29.** $52, 40, 77$ **31.** bacon: $3.09/lb, eggs: $4.03/doz, bread: $1.40/loaf   **33.** $x = 0, y = -4$

35. $x = 2, y = 1, z = 7$ **37.** $x = \dfrac{13}{12}, y = \dfrac{5}{4}, z = \dfrac{8}{3}$ **39.** $x = -\dfrac{1}{3}, y = 1, z = \dfrac{16}{3}$ **41.** Solving the second equation for z, we can back-substitute into the first equation, eliminating z. The result is a line, which contains an infinite number of solutions.

Exercises 3.5, pages 246 - 248

1. -22 **3.** -212 **5.** 11 **7.** 3 **9.** 47 **11.** 36 **13.** -3 **15.** -4 **17.** -25 **19.** 20 **21.** $x = 7$ **23.** $x = -7$ **25.** $x = -3$

27. $\begin{vmatrix} x & y & 1 \\ -2 & 1 & 1 \\ 5 & 3 & 1 \end{vmatrix} = 0$ which simplifies to $2x - 7y = -11$ **29.** $A = 1$ 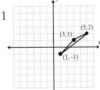   **31.** $A = \dfrac{31}{2}$

33. -25 **35.** -33.28

37. 0; Answers may vary. If you expand by minors using a row of all zeros, each minor will be multiplied by zero resulting in a sum of zero.

Exercises 3.6, pages 254 - 255

1. $x = 4, y = 3$ **3.** $x = \dfrac{2}{3}, y = -\dfrac{1}{4}$ **5.** No solution   **7.** $x = -\dfrac{1}{4}, y = \dfrac{3}{2}$ **9.** $x = \dfrac{31}{17}, y = \dfrac{2}{17}$ **11.** $x = \dfrac{39}{44}, y = \dfrac{41}{44}$

13. $x = \dfrac{18}{7}, y = -\dfrac{3}{7}$ **15.** $x = -\dfrac{7}{61}, y = \dfrac{266}{183}$ **17.** $x = \dfrac{210}{41}, y = -\dfrac{40}{123}$ **19.** $x = \dfrac{525}{124}, y = \dfrac{109}{93}$ **21.** $x = -2, y = 1, z = 3$

23. No solution   **25.** $x = 4, y = 2, z = 6$ **27.** Infinite solutions   **29.** $x = -\dfrac{2}{3}, y = \dfrac{11}{3}, z = 2$

31. $\begin{cases} x + y + z = 43 \\ 5 + 2x = y \\ x + y - 3 = z \end{cases}$; 6 feet, 17 feet, 20 feet   **33.** $\begin{cases} x + y = 6,000,000 \\ 0.02x + 0.04y = 170,000 \end{cases}$; $3,500,000 in mutual funds, $2,500,000 in stocks

Exercises 3.7, pages 261 - 262

1. 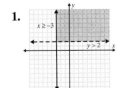   **3.** 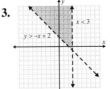   **5.** 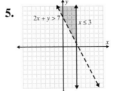   **7.** 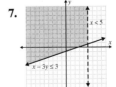   **9.**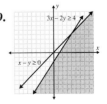

11. 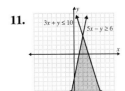   **13.** 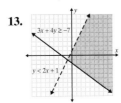   **15.** 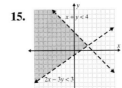   **17.** 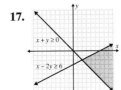   **19.**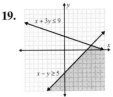

21. 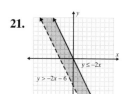   **23.** No solution, $\varnothing$ **25.**   **27.** 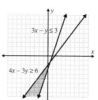   **29.**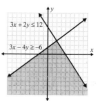

31. 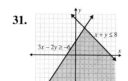   **33.**   **35.**

Chapter 3 Review, pages 269 - 273

1. $m_1 = 3$, $y\text{-int}_1 = (0,2)$ and $m_2 = 3$, $y\text{-int}_2 = (0,7)$. The two lines are parallel. The system is inconsistent. **2.** $(1,5)$

3. $(5,0)$ **4.** $(-2,-7)$ **5.** $(-1,-1)$ **6.** $(1,2)$

    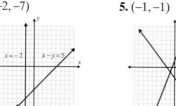

7. $(-2,1)$, consistent   **8.** $(6,-1)$, consistent   **9.** No solution, inconsistent   **10.** $\left(\dfrac{5}{2},\dfrac{1}{8}\right)$, consistent

11. No solution, inconsistent   **12.** $(x,-2x+3)$, dependent   **13.** $(1.5,2)$, consistent   **14.** $(1.2857,-4.6429)$, consistent

15. $(1,4)$, consistent   **16.** $(3,1)$, consistent   **17.** length = 31 meters; width = 19 meters   **18.** calculator = \$110; textbook = \$40

19. 8 gallons of 25% solution; 2 gallons of 50% solution   **20.** \$5000 at 8%; \$4000 at 6%   **21.** $a = 1$, $b = 7$; $x + 7y = 8$

22. 60 ounces of 40% acid solution; 40 ounces of 30% solution   **23.** 24 and 8

24. George drove 1.5 hrs for the first part and 2 hrs for the second part.   **25.** $x = 3$, $y = 1$, $z = 6$ **26.** $x = \dfrac{-1}{5}$, $y = 2$, $z = \dfrac{21}{5}$

27. $x = 0$, $y = 1$, $z = 5$ **28.** $x = 3$, $y = -4$, $z = \dfrac{1}{2}$ **29.** $x = 3$, $y = 5$, $z = -6$ **30.** No solution   **31.** 20, 21, and 24

32. \$10,000 in savings, \$15,000 in stocks, and \$25,000 in bonds   **33.** 8 nickels, 10 dimes, and 16 quarters

34. longest side = 20m, shortest side = 10m, other side = 12m   **35.** $\begin{bmatrix} 2 & -1 \\ 7 & 1 \end{bmatrix}$, $\left[\begin{array}{cc|c} 2 & -1 & 2 \\ 7 & 1 & 5 \end{array}\right]$ **36.** $\begin{bmatrix} 2 & -1 & 3 \\ 4 & 0 & -1 \\ 1 & -9 & 2 \end{bmatrix}$, $\left[\begin{array}{ccc|c} 2 & -1 & 3 & 5 \\ 4 & 0 & -1 & 1 \\ 1 & -9 & 2 & 3 \end{array}\right]$

37. $x = 4$, $y = -6$ **38.** $x = -4$, $y = 5$ **39.** $x = 11$, $y = 24$, $z = 22$

40. $x = 6$, $y = 1$, $z = -1$ **41.** $x = 1$, $y = -10$, $z = -6$ **42.** $x = 4$, $y = -1$, $z = -2$ **43.** $x = 5$, $y = 7$ **44.** $x = \dfrac{17}{11}$, $y = \dfrac{10}{11}$, $z = -\dfrac{16}{11}$

45. -1 **46.** -13 **47.** -49 **48.** -12 **49.** $\dfrac{-34}{3}$ **50.** 180   **51.** -11 **52.** $\dfrac{2}{15}$ **53.** $x = 0$ **54.** $x = -5$ **55.** $x = 2$; $y = 2$

56. $x = \dfrac{13}{7}$, $y = \dfrac{-10}{7}$ **57.** $x = -1$, $y = 8$, $z = 10$ **58.** $x = \dfrac{5}{28}$, $y = \dfrac{27}{28}$, $z = \dfrac{41}{28}$ **59.** $x = \dfrac{1}{2}$, $y = \dfrac{3}{2}$, $z = \dfrac{-1}{2}$

60. $x = -1$, $y = -1$, $z = -1$

61.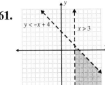

62.

63.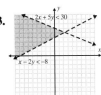

64.

65.

66.

67.

68.

69.

70.

Chapter 3 Test, pages 274 - 275

1. $(3, -1)$

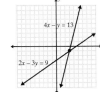

2. $(7, 2)$; consistent **3.** No solution; inconsistent **4.** $\left(\dfrac{1}{2}, \dfrac{1}{4}\right)$; consistent **5.** $(11, -7)$; consistent

6. $a = -3, b = 4; -3x + 4y = 17$ **7.** length = 23 ft; width = 8 ft **8.** $x = -1, y = 1, z = -2$

9. No solution **10. a.** $\begin{bmatrix} 1 & 2 & -3 \\ 1 & -1 & -1 \\ 1 & 3 & 2 \end{bmatrix}$ $\begin{bmatrix} 1 & 2 & -3 & | & -11 \\ 1 & -1 & -1 & | & 2 \\ 1 & 3 & 2 & | & -4 \end{bmatrix}$ **b.** $x = 1, y = -3, z = 2$

3×3 $\qquad$ 3×4

11. 41¢ stamps: 60; 58¢ stamps: 20; 75¢ stamps: 10 **12.** 42 **13.** 5 **14.** $x = 6$ **15.** $x = \dfrac{17}{3}$ **16.** $x = -\dfrac{78}{5}, y = \dfrac{38}{5}$

17. $x = -\dfrac{5}{26}, y = \dfrac{33}{26}, z = \dfrac{9}{26}$ **18.** $50°, 60°, 70°$ **19.** 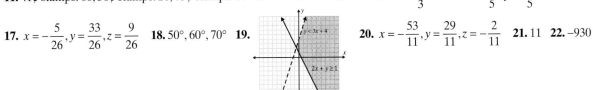 **20.** $x = -\dfrac{53}{11}, y = \dfrac{29}{11}, z = -\dfrac{2}{11}$ **21.** 11 **22.** -930

Chapter 3 Cumulative Review, pages 276 - 278

1. $82 = 2 \cdot 41$ **2.** $300 = 2^2 \cdot 3 \cdot 5^2$ **3.** $5000 = 2^3 \cdot 5^4$ **4.** 300 **5.** 616 **6.** 1440

7. a. $\{-3, 0, 1\}$ **b.** $\left\{-3, -\dfrac{1}{2}, 0, \dfrac{5}{8}, 1\right\}$ **c.** $\left\{-\sqrt{13}, \sqrt{2}, \pi\right\}$ **d.** $\left\{-\sqrt{13}, -3, -\dfrac{1}{2}, 0, \dfrac{5}{8}, 1, \sqrt{2}, \pi\right\}$ **8.** Inverse Property of Addition

9. Distributive Property **10.** 0 **11.** 320 **12.** $x = 2$ **13.** $x = -6$ **14.** $x = -\dfrac{29}{2}$ or $x = \dfrac{27}{2}$ **15.** $m = \dfrac{2gK}{v^2}$

16.  $(-\infty, 4)$ **17.** $[-144, \infty)$ **18.** $\left(-\dfrac{8}{3}, \dfrac{4}{3}\right)$

19. 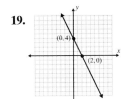 **20.** 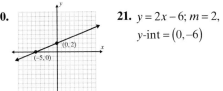 **21.** $y = 2x - 6; m = 2,$ y-int $= (0, -6)$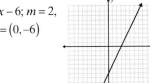

22. $x = \dfrac{2}{3}$; m is undefined, no y-intercept

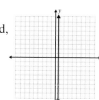

23. a. $m = -\dfrac{1}{2}$ **b.** $y = -\dfrac{1}{2}x - \dfrac{1}{2}$ **c.**

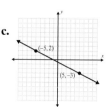

24. $y = \dfrac{2}{5}x$ **25.** $x = 3$

26. $D = \{-3, -1, 1, 4\}$ $R = \{-2, -1, 2\}$ not a function

27. a. 10 **b.** 22 **c.** -10 **28. a.** 0 **b.** 0 **c.** 20 **29.** $x = 1, y = 4$

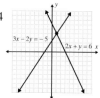

30. $x = 1, y = 3$ **31.** $x = 4, y = 3, z = 2$ **32.** 0 **33.** -11 **34.** $x = -3$ **35.** $x = 1, y = 1$ **36.** $x = 2, y = \dfrac{2}{5}, z = -\dfrac{1}{2}$

37.

38.

39. 3 batches of Choc-O-Nut; 2 batches of Chocolate Krunch
40. \$4200 at 7%; \$2800 at 8% **41.** $a = 2, b = 3, c = 4; y = 2x^2 + 3x + 4$
42. x-intercepts: $(-3.45, 0)$ and $(1.45, 0)$; maximum: $(-1, 6)$
43. $(1, -6)$ **44.** $x = \dfrac{8}{5}, y = \dfrac{27}{10}$ **45.** $x = -3, y = 0, z = 5$

Chapter 4

Exercises 4.1, pages 292 - 295

1. 49 **3.** 12 **5.** $\dfrac{1}{64}$ **7.** x **9.** $\dfrac{1}{y^5}$ **11.** x^3 **13.** x^4 **15.** $\dfrac{1}{x}$ **17.** 1 **19.** x^8 **21.** 1 **23.** $\dfrac{1}{y^5}$ **25.** y^5 **27.** x^3 **29.** 1 **31.** y^{11}
33. x^3 **35.** x^{10} **37.** x^{k+3} **39.** x^{3k+4} **41.** x^k **43.** x^{2k} **45.** x^{2k+1} **47.** x^{k-5} **49.** $-\dfrac{x^2 y^3}{3}$ **51.** $\dfrac{3y^4}{x^3}$ **53.** $x^{5n}y^{3-k}$
55. $\dfrac{b^7}{a^2}$ **57.** $\dfrac{x^{10}}{4y}$ **59.** $\dfrac{3y^7}{4x^2}$ **61.** $\dfrac{80x^{10}}{y^9}$ **63.** $x^4 y^5$ **65.** 472,000 **67.** 0.000 000 128 **69.** 0.00923 **71.** 4.79×10^5
73. 8.71×10^{-7} **75.** 4.29×10 **77.** 1.44×10^5 **79.** 6.0×10^{-4} **81.** 1.2×10^{-1} **83.** 5.0×10^3 **85.** 5.6×10^{-2}
87. 1.8×10^{12} cm/min; 1.08×10^{14} cm/hr **89.** 6.25×10^{-4} light year **91.** 1.9926×10^{-26} kg **93.** 1.0×10^2 **95.** 1.844×10^2
97. 1.2×10^6 **99.** See page 287.

Exercises 4.2, pages 302 - 304

1. Degree 3, binomial, leading coefficient 1 **3.** Not a polynomial **5.** Degree 3, trinomial, leading coefficient $\dfrac{5}{4}$
7. No degree, monomial, leading coefficient 0 **9.** Not a polynomial **11.** Not a polynomial **13.** $4x^2 - 3x - 6$
15. $-7x - 6$ **17.** $3x^2 - 4y^2$ **19.** $x^2 - x + 10$ **21.** $2x^4 - 5x^3 - x - 3$ **23.** $x^4 + 2x^2 + 8x - 1$ **25.** $6x^3 - 7x^2 + 8x + 3$
27. $9x^2 + 13xy - 6y^2$ **29.** $12x^3 - 21x^2 - 2x + 12$ **31.** $5x^3 - 2x^2$ **33.** $-2x^2 - 4xy + y^2$ **35.** $-3xy + 8x$ **37.** $2x^3 - 2x - 3$
39. $10x^4 + 2x^3 - 4x^2 + 2x + 1$ **41.** $5x^2 - 7x + 5$ **43.** $4x^3 + 4x^2 - 7x + 24$ **45.** $x^4 - 5x^3 - 8x^2 + 3x - 3$ **47.** 13 **49.** 1

51. -289 **53.** 7 **55.** -20 **57. a.** 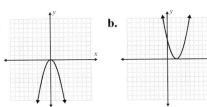 **b.** **59.** Any monomial or algebraic sum of monomials.

61. The largest of the degrees of its terms after like terms have been combined.

Exercises 4.3, pages 312 - 314

1. $5x^3 - 10x^2 + 15x$ **3.** $x^3y^2 + 4xy^3$ **5.** $x^2 - 3x - 18$ **7.** $x^2 - 9x + 8$ **9.** $2y^2 - 11y - 6$ **11.** $3x^2 - 19x + 20$

13. $6y^2 + 13y + 6$ **15.** $8x^2 - 37x - 15$ **17.** $27x^2 - 15x - 2$ **19.** $9x^2 + 6x + 1$ **21.** $25x^2 - 20xy + 4y^2$ **23.** $16x^2 - 49$

25. $4x^2 - 9y^2$ **27.** $9x^5 - 16x$ **29.** $x^3 - 1$ **31.** $x^3 + 9x^2 + 27x + 27$ **33.** $x^6 + 4x^3 + 4$ **35.** $4x^6 - 49$ **37.** $x^3 - 27y^3$

39. $x^4 - 36y^4$ **41.** $10x^4 - 13x^2y^2 - 3y^4$ **43.** $x^2 + 2xy + y^2 - 4$ **45.** $25x^2 - 10xy + y^2 + 30x - 6y + 9$

47. $x^2 - 4xy + 4y^2 + 8x - 16y + 16$ **49.** $x^{k+2} + 3x^2$ **51.** $x^{2k} - 2x^k - 15$ **53.** $x^{2k} + 5x^k + 4$ **55.** $3x^{2k} + 17x^k + 10$

57. $x^2 - \dfrac{25}{64}$ **59.** $y^2 - \dfrac{2}{5}y + \dfrac{1}{25}$ **61.** $x^2 - 6.25$ **63.** $6x^3 - 13x^2 - 23x - 6$ **65.** $2x^4 - 3x^3 - 6x^2 + 17x - 12$

67. $4x^4 - 6x^3 - 28x^2 + 26x + 24$ **69. a.** $10x^5 - 20x^4 + 10x^3$ **b.** $\dfrac{5}{16}$ **c.** 0.2048 **71. a.** $A(x) = 4x^2 + 150x + 1250$

b. $A(x) = 4x^2 + 150x$ **73.** As indicated in the diagram, $(x+5)^2 = x^2 + 2(5x) + 5^2$. Answers may vary.

Exercises 4.4, pages 320 - 322

1. $y^2 - 2y + 3$ **3.** $2x^2 - 3x + 1$ **5.** $10x^3 - 11x^2 + x$ **7.** $-4x^2 + 7x - \dfrac{5}{2}$ **9.** $y^3 - \dfrac{7}{2}y^2 - \dfrac{15}{2}y + 4$ **11.** $3x + 4 + \dfrac{1}{7x - 1}$

13. $x - 9$ **15.** $x^2 - 4x + 33 - \dfrac{265}{x + 8}$ **17.** $4x^2 - 6x + 9 - \dfrac{17}{x + 2}$ **19.** $x^2 + 7x + 55 + \dfrac{388}{x - 7}$ **21.** $2x^2 - 9x + 18 - \dfrac{30}{x + 2}$

23. $x^2 + 2x + 2 - \dfrac{12}{2x + 3}$ **25.** $7x^2 + 2x + 1$ **27.** $x^2 + 3x + 2 - \dfrac{2}{x - 4}$ **29.** $2x^2 + x - 3 + \dfrac{18}{5x + 3}$ **31.** $2x^2 - 8x + 25 - \dfrac{98}{x + 4}$

33. $3x^2 + 2x - 5 - \dfrac{1}{3x - 2}$ **35.** $3x^2 - 2x + 5$ **37.** $x^3 + 2x + 5 + \dfrac{17}{x - 3}$ **39.** $x^3 + 2x^2 + 6x + 9 + \dfrac{23}{x - 2}$

41. $x^3 + \dfrac{1}{2}x^2 - \dfrac{3}{4}x - \dfrac{3}{8} + \dfrac{45}{16\left(x - \dfrac{1}{2}\right)}$ **43.** $3x + 5 + \dfrac{x - 1}{x^2 + 2}$ **45.** $x^2 + x - 4 + \dfrac{-8x + 17}{x^2 + 4}$ **47.** $2x + 3 + \dfrac{6}{3x^2 - 2x - 1}$

49. $3x^2 - 10x + 12 + \dfrac{-x - 14}{x^2 + x + 1}$ **51.** $x^2 - 2x + 7 + \dfrac{-17x + 14}{x^2 + 2x - 3}$ **53.** $x^2 + 3x + 9$ **55.** $x^3 - x + \dfrac{x - 1}{x^2 + 1}$ **57.** $x^4 + x^3 + x^2 + x + 1$

59. $x^3 - \dfrac{14}{5}x^2 + \dfrac{56}{25}x - \dfrac{224}{125} + \dfrac{3396}{625\left(x + \dfrac{4}{5}\right)}$ **61.** $3x^3 + 4x^2 + 8x + 12$

63. a. $19;\ 2x^2 - 4x + 2 + \dfrac{19}{x - 2}$ **b.** $-5;\ 2x^2 - 10x + 20 - \dfrac{5}{x + 1}$ **c.** $55;\ 2x^2 + 10 + \dfrac{55}{x - 4}$

Exercises 4.5, pages 327 - 328

1. $3(x + 5)$ **3.** $-5x(x - 3)$ or $5x(-x + 3)$ **5.** $-9x(x - 3)$ or $9x(-x + 3)$ **7.** $7x^3(5x + 7)$ **9.** $xy(x - 2 + y)$ **11.** $4x^2y(2x - 1)$

13. $3xy(2x+1-3y)$ **15.** $-2x^2(x^2+15x+6)$ or $2x^2(-x^2-15x-6)$ **17.** $(a+4)(x+y)$ **19.** $(x-3)(y+4)$ **21.** $(x-3)(y+7)$
23. $(a^2-1)(b+c)$ **25.** $(3x+1)(x+y)$ **27.** Not factorable **29.** $(4x+3)(y-7)$ **31.** $(2x+3)(y+5)$ **33.** $(9x^2-5)(x-5)$
35. Not factorable **37.** $l=2x+30$ inches
39. $5xy+4uv-vy-20ux = 5xy-20ux+4uv-vy = 5x(y-4u)-v(y-4u) = (5x-v)(y-4u)$
   $5xy+4uv-vy-20ux = 5xy-vy-20ux+4uv = y(5x-v)-4u(5x-v) = (y-4u)(5x-v)$

Exercises 4.6, pages 337 - 339

1. $(x+3)(x+6)$ **3.** $(x-9)(x+3)$ **5.** $(x-25)(x-2)$ **7.** $(x+8)(2x-1)$ **9.** $(2x+3)(3x+2)$ **11.** Not factorable
13. $(x-4)(2x+1)$ **15.** $(2x-1)(9x+1)$ **17.** $2x(2x+5)(3x+2)$ **19.** Not factorable **21.** $(2x^3+y^2)(x^3+4y^2)$
23. $-2(9x^2-36x+4)$ or $2(-9x^2+36x-4)$ **25.** $-5(y-6)(y-2)$ **27.** $x^2(3x-4)(7x+8)$ **29.** $(5x^2+2y^2)(x^2+3y^2)$
31. $(x-5)(x-3)(x+7)$ **33.** $(x-6)(2x+1)(2x+3)$ **35.** $(2x+y-5)(2x+y-4)$ **37.** $(x+3y-16)(x+3y+2)$
39. $(6x-y+7)(6x-y+1)$ **41.** $4(2a+b+3)(2a+b-2)$ **43.** $l=2x+5$ inches **45. a.** $x(36-2x)(12-2x)$
b. $V(2)=512$ in.^{3} **c.** $V(4)=448$ in.^{3} **47.** $(5x+20)$ can have a five factored out.

Exercises 4.7, pages 345 - 347

1. $(x-5)(x+5)$ **3.** $2(x-8)(x+8)$ **5.** $4(x-2)(x+2)(x^2+4)$ **7.** Not factorable **9.** $-4(x^2+6x-2)$ **11.** $(3x-5)(3x+5)$
13. $(y-5)^2$ **15.** $(3y+2)^2$ **17.** $(4x-5)^2$ **19.** $4x(x-4)(x+4)$ **21.** $2xy(x+8)^2$ **23.** $(x+y+3)^2$ **25.** $(x-y-9)(x-y+9)$
27. $(x^k-10)^2$ **29.** $(x^2+5y)^2$ **31.** $(x-5)(x^2+5x+25)$ **33.** $(y+6)(y^2-6y+36)$ **35.** $(x+3y)(x^2-3xy+9y^2)$
37. $3(x+3)(x^2-3x+9)$ **39.** $(4x+3y)(16x^2-12xy+9y^2)$ **41.** $3x(x+5y)(x^2-5xy+25y^2)$
43. $x(xy-1)(x^2y^2+xy+1)$ **45.** $2x^2(1-2y)(1+2y+4y^2)$ **47.** $(x^2-4y)(x^4+4x^2y+16y^2)$
49. $(2x-y)(x^2-xy+y^2)$ **51.** $(x-3y-4z)(x^2-6xy+9y^2+4xz-12yz+16z^2)$
53. $(x+2y-4)(x^2+4x+xy+y^2-4y+16)$ **55.** $9(x+y-1)(3x^2+3x+3xy+y^2+3)$ **57.** $(x+5)(y+2)$
59. $(x-3)(y+7)$ **61.** $(2x+3)(y+5)$ **63.** $(3x+2)(2y-3)$ **65.** $(x+1-y)(x+1+y)$ **67.** $(x+2y-5)(x+2y+5)$
69. $(x+y+3)(x-y-3)$ **71.** $(2x^2-7)(2x-3)$ **73.** $(x-3)(x+3)^2$ **75.** $(x+7)(x+2)(x-2)$
77. $(x^k-2y)(x^k+2y)$ **79.** $(x^k+3y^k)(x^{2k}-3x^ky^k+9y^{2k})$ **81. a.** x^2-16 **b.** [ ] $x-4$
                                                                                                                 $x+4$
83. a. $xy+xy+x^2+y^2 = x^2+2xy+y^2 = (x+y)^2$ **b.**

$(x+y)(x+y) = (x+y)^2$

85. I. $abc = 100a+10b+c = (99+1)a+(9+1)b+c = 9(11a+b)+a+b+c$. So, if the sum $(a+b+c)$ is divisible by 3 (or 9),
then the number abc will be divisible by 3 (or 9).
II. $abcd = 1000a+100b+10c+d = (999+1)a+(99+1)b+(9+1)c+d = 9(111a+11b+c)+a+b+c+d$.
So, if the sum $(a+b+c+d)$ is divisible by 3 (or 9), then the number $abcd$ will be divisible by 3 (or 9).

Exercises 4.8, pages 357 - 360

1. $\left\{-\dfrac{1}{3}, 0, \dfrac{1}{2}\right\}$ **3.** $\{-1, 2, 6\}$ **5.** $\left\{-\dfrac{3}{2}, 0, \dfrac{1}{5}\right\}$ **7.** $x = -9$ or $x = -4$ **9.** $x = 6$ or $x = 8$ **11.** $x = -5$ or $x = 10$ **13.** $x = -7$ or $x = 7$

15. $x = \dfrac{2}{3}$ or $x = 5$ **17.** $x = \dfrac{4}{3}$ or $x = \dfrac{1}{2}$ **19.** $n = -6$ or $n = 6$ **21.** $z = -\dfrac{7}{2}$ or $z = \dfrac{7}{2}$ **23.** $z = \dfrac{1}{2}$ or $z = 3$

25. $x = -\dfrac{3}{4}$ or $x = \dfrac{2}{3}$ **27.** $n = -\dfrac{5}{2}$ or $n = \dfrac{7}{3}$ **29.** $y = -\dfrac{4}{5}$ or $y = \dfrac{4}{5}$ **31.** $x = -6$ or $x = 8$ **33.** $x = 2$ or $x = 6$

35. $x = -11$ or $x = 3$ **37.** $y = \dfrac{1}{6}$ or $y = \dfrac{2}{3}$ **39.** $x = -\dfrac{2}{9}$ or $x = \dfrac{6}{7}$ **41.** $x = -3$ or $x = -2$ or $x = 0$

43. $x = -\dfrac{5}{2}$ or $x = 0$ or $x = \dfrac{5}{2}$ **45.** $n = -\dfrac{5}{2}$ or $n = 0$ or $n = \dfrac{7}{3}$ **47.** $x = -\dfrac{3}{2}$ or $x = 0$ or $x = \dfrac{7}{5}$ **49.** $y^2 - y - 6 = 0$

51. $8x^2 - 10x + 3 = 0$ **53.** $x^3 - x^2 - 6x = 0$ **55.** $x^2 + 8x + 15 = 0$ **57.** $y^3 - 4y^2 - 3y + 18 = 0$ **59.** 7, 8 **61.** 21, 23

63. 10, 12, 14 or $-14, -12, -10$ **65.** 1, 2, 3, 4 or $-2, -1, 0, 1$ **67.** Length = 12 in. Width = 5 in. **69.** 20 miles, 21 miles

71. a. 4 s **b.** 2 s **c.** 2 s **73.** 128 in.^{2}; use the Pythagorean Theorem

Exercises 4.9, pages 366 - 367

1. $x = \pm 2$ **3.** $x \approx \pm 1.41$ **5.** $x = -2, 6$ **7.** $x \approx 2.46, -4.46$ **9.** $x \approx 4.70, -1.70$ **11.** no solution **13.** $x \approx 1.91, -1.57$ **15.** $x \approx -2.49,$
0.66, 1.83 **17.** $x = 0, 1, 3$ **19.** $x = -3, -1, 5$ **21.** $x \approx 3.40$ **23.** $x \approx -0.91, 1.71, 3.20$ **25.** $x \approx -1.73, 0, 1.73$ **27.** $x = -4, 7$

29. $x \approx -1.67, 3$ **31.** $x = -2, 0$ **33.** $x = -5, 1$ **35.** $x = 3.5$ **37.** $[-3, 3]$ **39.** $(-\infty, 3) \cup (7, \infty)$ **41.** $(-\infty, 1.33) \cup (4, \infty)$

43. $(-\infty, -8) \cup (16, \infty)$ **45.** $(-\infty, -16] \cup [-8, \infty)$

Chapter 4 Review, pages 374 - 378

1. $-6x^{10}$ **2.** x^2 **3.** x^{2k+1} **4.** $\dfrac{1}{y^8}$ **5.** $6y^6$ **6.** $\dfrac{1}{25x^8}$ **7.** $\dfrac{x^4}{y}$ **8.** $\dfrac{36b^2}{a^6}$ **9.** $\dfrac{a^5}{b^7}$ **10.** x^2y^{17} **11.** 4560 **12.** 0.000 059 1

13. 0.000 000 067 **14.** 8,235,000 **15.** 8.96×10^5 **16.** 9.92×10^{-7} **17.** 3.6×10^{-8} **18.** 6×10^{-1} **19.** 6×10^{-6}

20. 2.5×10^{-3} **21.** Not a polynomial **22.** Degree 3, trinomial, leading coefficient 5 **23.** $7x^2 + 8$ **24.** $x^3 + 4x^2 + 3x - 3$

25. $-40x - 28$ **26.** $-5x^2 + 8x + 18$ **27.** $-5xy + 3x^2$ **28.** $14x^2 + 3x - 7$ **29.** $2x^3 + x^2 + 2x - 9$ **30.** $3x^3 + 2x^2 - 33x + 24$

31. $-3x^2 + 10x - 9$ **32. a.** 19 **b.** 5 **33. a.** 2 **b.** -58 **34. a.** 7 **b.** 7 **35. a.** -9 **b.** 16 **36.** $-6x^5 + 10x^3$ **37.** $-35x^5 + 28x^4 + 7x^3$

38. $2y^3 - 7y^2 - 14y - 5$ **39.** $y^4 + 6y^3 - 2y^2 - 15y - 18$ **40.** $2x^2 + 7x + 5$ **41.** $6x^2 + 13x - 63$ **42.** $9x^2 + 12x + 4$

43. $4x^2 + 36x + 81$ **44.** $49y^2 - 112y + 64$ **45.** $x^2 - 196$ **46.** $9y^2 - 25$ **47.** $64y^2 - 1$ **48.** $3x^3 + 5x^2 - 19x - 21$

49. $2x^4 + 9x^3 + 4x^2 - 21x - 18$ **50.** $x^4 + 6x^3 + 11x^2 + 6x + 1$ **51.** $3y^2 + 2y + \dfrac{2}{3}$ **52.** $x^2 + 5x - 2$ **53.** $9x - 12 + \dfrac{4}{x}$

54. $4x + 8 + \dfrac{-5}{x}$ **55.** $2 + \dfrac{8}{y} + \dfrac{10}{y^2}$ **56.** $5x^2 + 4x + \dfrac{-6}{x}$ **57.** $\dfrac{1}{2}x^3 + 2x^2 - 3x + 1$ **58.** $4y^2 + 2y - 1 + \dfrac{11}{3y}$ **59.** $x + 3$

60. $x + 4$ **61.** $4x - 12 + \dfrac{-21}{x - 2}$ **62.** $8y - 14 + \dfrac{47}{y + 3}$ **63.** $16x^2 + 20x + 25$ **64.** $4x^2 - 6x + 9$ **65.** $3x + 4 + \dfrac{-13x - 7}{x^2 + 2}$

66. $x - 6 + \dfrac{5x - 32}{x^2 - 2x + 3}$ **67.** $6ax(x + 7)$ **68.** $4ax\left(x^2 + 2\right)$ **69.** $-6x^2\left(x^2 - 2x + 4\right)$ or $6x^2\left(-x^2 + 2x - 4\right)$

70. $-7x^2\left(x^2 - 2x - 4\right)$ or $7x^2\left(-x^2 + 2x + 4\right)$ **71.** $(x + 3)(y + 5)$ **72.** $(x - 6)(x - y)$ **73.** Not factorable **74.** Not factorable

75. $(x - 2)(y - 10)$ **76.** $(4x + 5)(y - 7)$ **77.** $(x - 2)(4x + 3)$ **78.** $(x + 7)(4x + 5)$ **79.** $5\left(x^2 + 4x + 1\right)$ **80.** $2x(2x - 3)(3x - 2)$

81. $(2x+5)(3x+1)$ **82.** $2(x+1)(x+5)$ **83.** $(x-5)(5x+2)$ **84.** $\left(x^3-4y^2\right)\left(2x^3-y^2\right)$ **85.** $\left(x^2-3y^2\right)\left(5x^2-2y^2\right)$

86. $-7by(y-1)(2y-1)$ **87.** $(x-6)(x-3)(2x+3)$ **88.** $(3x+y-7)(3x+y-1)$ **89.** $\left(x^3-10\right)\left(x^3+10\right)$

90. $(4x-5)(4x+5)$ **91.** Not factorable **92.** Not factorable **93.** $2(2x-11)(2x+11)$ **94.** $3x(x-4)(x+4)$

95. $-3(x+2)^2$ **96.** Not factorable **97.** $\left(x^k+5\right)^2$ **98.** $2\left(x^k-6\right)^2$ **99.** $(x-y+3)(x+y+3)$

100. $(3y-1-4x)(3y-1+4x)$ **101.** $\left(2x^3-y^2\right)\left(4x^6+2x^3y^2+y^4\right)$ **102.** $3(x+3)\left(x^2-3x+9\right)$

103. $4x(x+5)\left(x^2-5x+25\right)$ **104.** $\left(x^k-3y^{2k}\right)\left(x^{2k}+3x^ky^{2k}+9y^{4k}\right)$ **105.** $y\left(3x^2-3xy+y^2\right)$ **106.** $x=9$ or $x=8$

107. $x=-2$ or $x=-6$ **108.** $x=1$ (1 is a double root) **109.** $x=6$ or $x=-4$ **110.** $x=2$ or $x=-2$ **111.** $x=\dfrac{5}{3}$ or $x=-\dfrac{5}{3}$

112. $x=7$ or $x=1$ **113.** $x=-\dfrac{3}{2}$ or $x=-2$ **114.** $x=0$ or $x=-\dfrac{-1}{6}$ or $x=5$ **115.** $x=0$ or $x=\dfrac{2}{3}$ or $x=-4$

116. $x^2+x-20=0$ **117.** $x^2+10x+21=0$ **118.** $y^2+9y+8=0$ **119.** $9y^2-9y+2=0$ **120.** $32x^2+4x-15=0$

121. $6x^3+11x^2-2x=0$ **122.** $y^3+7y^2+8y-16=0$ **123.** $y^3-15y^2+75y-125=0$ **124.** 8, 9 **125.** 10, 12 or $-8, -6$

126. 5, 7, 9 **127.** 10, 11, 12, 13 or $-7, -6, -5, -4$ **128.** 20 meters **129.** 125 yards **130.** 3 sec **131.** $x=-0.5, 1$

132. $x\approx-3.87, 3.87$ **133.** $x\approx1.47, -1.14$ **134.** $x\approx-2.43$ **135.** $x=0, 4, -4$ **136.** $x=0, -2, 5$ **137.** $x\approx1, 0.33$ **138.** $x=1, -4$

139. $(-9, -5)$ **140.** $[-2, 6]$ **141.** $(-\infty, 25] \cup [35, \infty)$ **142.** $(-\infty, -1) \cup (2, \infty)$

Chapter 4 Test, pages 379 - 380

1. $\dfrac{-20x^3}{y^2}$ **2.** $\dfrac{x^3}{y^7}$ **3.** $\dfrac{27y^{13}}{2x^2}$ **4.** x **5. a.** -3.6×10^{-4} **b.** 9.1×10^9 **6.** Not a polynomial

7. Degree 3, binomial, leading coefficient 2 **8.** $5x^3-x^2-6$ **9.** $-x^3+4x^2+5x-3$ **10.** $3x^2+2x+13$

11. $3x^3+8x^2-5x-5$ **12. a.** 9 **b.** -12 **13.** $28x^2+47x+15$ **14.** $64x^2-9$ **15.** x^2+2x-y^2+1 **16.** $6x^3+5x^2-x-1$

17. $2x^2+8x+43+\dfrac{166}{x-4}$ **18.** $2x-3-\dfrac{x}{2x^2+1}$ **19.** $6xy(5x-3y+4)$ **20.** $(x+2y)(2x+1)$ **21.** $(x+7)(x-6)$

22. $(7x+2)(x-4)$ **23.** $(3x+5)(2x-1)$ **24.** $(x-12)(x+3)$ **25.** Not factorable **26.** $\left(x^k+y^k\right)\left(x^k-y^k\right)$ **27.** $3y(x-3y)^2$

28. $3(x+3y)\left(x^2-3xy+9y^2\right)$ **29.** $x=-\dfrac{5}{3}, 2$ **30.** $x=-1, 8$ **31.** $x=-3, \dfrac{5}{2}$ **32.** $x=-5, \dfrac{1}{3}$ **33.** $x^2+5x-24=0$

34. $4x^2-4x+1=0$ **35.** 11, 13, 15 or $-17, -15, -13$ **36.** hyp $= 25$, sides $= 24, 7$ **37.** $x=0, 1, 4$ **38.** $(-6, 12)$

Chapter 4 Cumulative Review, pages 381 - 384

1. A prime number is a positive integer with exactly two factors, 1 and itself. 2, 3, 5, 7, 11, 13, 17, 19, 23, 29, 31, 37, 41, 43, 47

2. 2 **3.** $2^3\cdot3^2\cdot7$ **4.** $2^5\cdot17$ **5.** $2^3\cdot5^4$ **6.** 120 **7.** 180 **8.** 300 **9.** $\dfrac{5}{6}$ **10.** $\dfrac{8}{5}$ **11.** 6 **12.** -138 **13.** $2x+17$ **14.** $12x-7$

15. $x=\dfrac{9}{10}$ **16.** 21 **17.** 21 **18.** $x=-5$ or $x=10$ **19.** 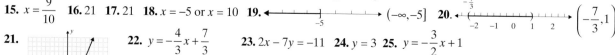 $(-\infty, -5]$ **20.** $\left(-\dfrac{7}{3}, 1\right)$

21. **22.** $y=-\dfrac{4}{3}x+\dfrac{7}{3}$ **23.** $2x-7y=-11$ **24.** $y=3$ **25.** $y=-\dfrac{3}{2}x+1$

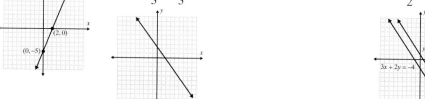

26. a. Function, $D=(-\infty, \infty) R=[0, \infty)$ **b.** Not a function, $D=(-\infty, \infty) R=(-\infty, \infty)$ **c.** Function $D=(-\infty, \infty) R=(-\infty, 0]$

27. a. 18 **b.** –10 **c.** –15 **28. a.** 0 **b.** –18 **c.** –38 **29.**

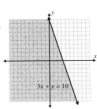

30. $x = 2, y = 4$ **31.** $x = 1, y = -3$

32. Infinite number of solutions **33.** $x = 4, y = -3$ **34.** 0 **35.**

36.

37. $64x^6 y^3$

38. $49x^{10} y^4$ **39.** $-\dfrac{1}{8x^9 y^6}$ **40.** $\dfrac{36x^4}{y^{10}}$ **41.** $\dfrac{-20.8x^3}{y^2}$ **42.** $\dfrac{y^2}{x^3}$ **43.** $\dfrac{9y^{14}}{x^4}$ **44.** x^5 **45. a.** 1.68×10^{-10} **b.** 8.37×10^{11}

46. Degree 4, monomial, leading coefficient 5 **47.** Not a polynomial **48.** $7x^2 + 7x - 2$ **49.** $5x^2 - 4x - 6$

50. $5x^2 - 37x - 24$ **51.** $4x^2 + 20x + 25$ **52.** $8x^2 + 14x + 3$ **53.** $4x - 5 + \dfrac{19x - 28}{x^2 - 3}$ **54.** $(3x + y)(y + 1)$ **55.** $(2x - 5)(2x + 3)$

56. $(3x - 2)(2x - 1)$ **57.** $2x(3x + 1)(x - 4)$ **58.** $\left(3x^3 + 2y\right)\left(3x^3 - 2y\right)$ **59.** $\left(2x + 5\right)\left(4x^2 - 10x + 25\right)$ **60.** $x = 7$ or $x = 3$

61. $x = -\dfrac{5}{2}$ **62.** $x = 4$ or $x = -2$ **63.** $x = -4, x = 0$ or $x = 5$ **64.** $x^2 + 15x + 50 = 0$ **65.** $x^3 - 17x^2 + 52x = 0$ **66.** \$4200 at 7%;

\$2800 at 8% **67.** $20, 24$ or $-24, -20$ **68.** $a = -1, b = 0, c = 1$ **69.** 10 liters of 6%, 20 liters of 15%

70. a.

b. $m_1 = 15; m_2 = -25; m_3 = 12; m_4 = -4$ **c.** Rate increased 15 mph, then decreased 25 mph, then increased 12 mph, then decreased 4 mph

71. $x \approx -1.77, 2.27$ **72.** $x \approx -0.24, 4.24$ **73.** $x \approx -1.19, 4.19$ **74.** $x \approx -1.25, 0.45, 1.80$

75. no solution **76.** $(-1, 11)$ **77.** $(-\infty, 0] \cup [4, \infty)$ **78.** $[-12, 8]$

Chapter 5

Exercises 5.1, pages 394 - 396

1. $\dfrac{3x}{4y}; x \neq 0; y \neq 0$ **3.** $\dfrac{2x^3}{3y^3}; x \neq 0; y \neq 0$ **5.** $\dfrac{1}{x - 3}; x \neq 0, 3$ **7.** $7; x \neq 2$ **9.** $-\dfrac{3}{4}; x \neq 3$ **11.** $-\dfrac{1}{2}; x \neq 4$ **13.** $\dfrac{x}{x - 1}; x \neq -6, 1$

15. $\dfrac{x - y}{3x}; x \neq 0, x \neq -y$ **17.** $\dfrac{x^2 - 4x + 16}{2x - 7}; x \neq -4, \dfrac{7}{2}$ **19.** $\dfrac{x^2 + 2x + 4}{y + 5}; x \neq 2, y \neq -5$ **21.** $\dfrac{ab}{y}$ **23.** $\dfrac{8x^2 y^3}{15}$ **25.** $\dfrac{x + 3}{x}$

27. $\dfrac{x - 1}{x + 1}$ **29.** $-\dfrac{1}{x - 8}$ **31.** $\dfrac{x - 2}{x}$ **33.** $\dfrac{4x + 20}{x(x + 1)}$ **35.** $\dfrac{x}{(x + 3)(x - 1)}$ **37.** $-\dfrac{x + 4}{x(x + 1)}$ **39.** $\dfrac{x + 2y}{(x - 3y)(x - 2y)}$ **41.** $\dfrac{x - 1}{x(2x - 1)}$

43. $\dfrac{1}{x + 1}$ **45.** $\dfrac{2x - 1}{2x + 1}$ **47.** $\dfrac{49y^3}{6x^5}$ **49.** $\dfrac{1}{12x}$ **51.** $\dfrac{7x}{x + 2}$ **53.** $\dfrac{x}{6}$ **55.** $\dfrac{x + 3}{x + 4}$ **57.** $\dfrac{x + 5}{x}$ **59.** $-\dfrac{x + 4}{x(2x - 1)}$ **61.** $\dfrac{x - 2}{x + 2}$

63. $\dfrac{3x^3 + 8x^2 - 9x - 18}{\left(6x^2 + x - 9\right)(x - 2)}$ **65.** $\dfrac{8x^2 - 14x + 5}{(x + 4)(x + 1)}$ **67.** $\dfrac{9x^3 - 6x^2 + 4x}{(3x - 2)^2 \left(9x^2 + 6x + 4\right)}$ **69.** $-\dfrac{3x - 2}{x^2}$ **71.** $\dfrac{x + 3}{2x - 1}$ **73.** $\dfrac{x - 1}{x - 8}$

75. a. A rational expression is an algebraic expression that can be written in the form $\dfrac{P}{Q}$ where P and Q are polynomials and $Q \neq 0$. **b.** $\dfrac{x - 1}{(x + 2)(x - 3)}$ Answers may vary. **c.** $\dfrac{1}{x + 5}$ Answers may vary. **77. a.** $x = 4$ **b.** $x = 10, x = -10$

Exercises 5.2, pages 405 - 407

1. 3 **3.** 2 **5.** 1 **7.** $\dfrac{2}{x-1}$ **9.** $\dfrac{14}{7-x}$ **11.** 4 **13.** $\dfrac{x^2-x+1}{(x+4)(x-3)}$ **15.** $\dfrac{x-2}{x+2}$ **17.** $\dfrac{4x+5}{2(7x-2)}$ **19.** $\dfrac{6x+15}{(x+3)(x-3)}$

21. $\dfrac{x^2-2x+4}{(x+2)(x-1)}$ **23.** $\dfrac{x^2+3x+6}{(x+3)(x-3)}$ **25.** $\dfrac{8x^2+13x-21}{6(x+3)(x-3)}$ **27.** $\dfrac{3x^2-20x}{(x+6)(x-6)}$ **29.** $\dfrac{-4x}{x-7}$ **31.** $\dfrac{4x^2-x-12}{(x+7)(x-4)(x-1)}$

33. $\dfrac{4}{(x+2)^2(x-2)}$ **35.** $\dfrac{-3x^2+17x+15}{(x-3)(x-4)(x+3)}$ **37.** $\dfrac{4x-19}{(7x+4)(x-1)(x+2)}$ **39.** $\dfrac{-7x-9}{(4x+3)(x-2)}$ **41.** $\dfrac{4x^2-41x+3}{(x+4)(x-4)}$

43. $\dfrac{-2x^2-4x+15}{(x-4)(x-2)}$ **45.** $\dfrac{x^2-4x-6}{(x+2)(x-2)(x-1)}$ **47.** $\dfrac{3x^2+26x-3}{(x+7)(x-3)(x+1)}$ **49.** $\dfrac{5x+12}{(x+3)(x+1)}$ **51.** $\dfrac{9x^3-19x^2+22x+12}{(3x+1)(x+3)(5x+2)(x-1)}$

53. $\dfrac{5x+1}{(y-3)(x+1)}$ **55.** $\dfrac{5xy-8y+2}{(y-4)(y+1)(x-2)}$ **57.** $\dfrac{6x^2-12x+5}{2(2x+1)(4x^2-2x+1)}$ **59.** $\dfrac{14x-43}{(x^2+6)(x-5)(x-2)}$

61. See page 398. Answers may vary.

Exercises 5.3, pages 411 - 412

1. $\dfrac{4}{5xy}$ **3.** $\dfrac{8}{7x^2y}$ **5.** $\dfrac{2x(x+3)}{2x-1}$ **7.** $\dfrac{7}{2(x+2)}$ **9.** $\dfrac{x}{x-1}$ **11.** $\dfrac{4x}{3(x+6)}$ **13.** $\dfrac{7x}{x+2}$ **15.** $\dfrac{x}{6}$ **17.** $\dfrac{24y+9x}{18y-20x}$ **19.** $\dfrac{xy}{x+y}$

21. $\dfrac{y+x}{y-x}$ **23.** $\dfrac{2(x+1)}{x+2}$ **25.** $\dfrac{x^2-3x}{x-4}$ **27.** $\dfrac{x+1}{2x-3}$ **29.** $-\dfrac{2x+h}{x^2(x+h)^2}$ **31.** $-(x-2y)(x-y)$ **33.** $\dfrac{2-x}{x+2}$ **35.** $\dfrac{-x^2y^2}{x+y}$

37. a. $\dfrac{8}{5}$ **b.** 1 **c.** $\dfrac{x^4+x^3+3x^2+2x+1}{x^3+x^2+2x+1}$

Exercises 5.4, pages 420 - 424

1. $x=7$ **3.** $x=4$ **5.** $x=-62$ **7.** $x=-23$ **9.** 360 defective computers **11.** 506 students **13.** $AC=2, ST=12$

15. $ST=8, TU=12, QR=24$ **17.** $QP=3, TU=12$ **19.** $AP=\dfrac{9}{2}$in., $PC=\dfrac{15}{2}$in. **21.** $x=\dfrac{1}{4}$ **23.** $x=\dfrac{3}{2}$ **25.** $x=4$

27. $x\neq0; x=\dfrac{10}{3}$ **29.** $x\neq0; x=\dfrac{1}{4}$ **31.** $x\neq0; x=-\dfrac{3}{16}$ **33.** $x\neq-\dfrac{2}{3},-\dfrac{1}{5}; x=-3$ **35.** $x\neq-\dfrac{7}{3},\dfrac{3}{5}; x=-39$ **37.** $x\neq-\dfrac{2}{3}; x=\dfrac{62}{7}$

39. $x\neq\dfrac{1}{2},4; x=-3$ **41.** $x\neq-4,-1; x=2$ **43.** $x\neq-1,\dfrac{1}{4}; x=\dfrac{2}{3}$ **45.** $x\neq-5,4; x=-2$ **47.** $x\neq-2,-\dfrac{2}{3}$; No solution

49. $x\neq-\dfrac{1}{2},\dfrac{1}{2},2; x=\dfrac{7}{11}$ **51.** $r=\dfrac{S-a}{S}$ **53.** $s=\dfrac{x-\bar{x}}{z}$ **55.** $R_{\text{total}}=\dfrac{R_1R_2}{R_1+R_2}$ **57.** $P=\dfrac{A}{1+r}$ **59.** $x=\dfrac{b-yd}{yc-a}$

61. a. $\dfrac{4x^2+41x-10}{x(x-1)}$ **b.** $x=\dfrac{1}{4}, 10$ and $x\neq0,1$ **63. a.** $\dfrac{10x-6}{(x-2)(x+2)}$ **b.** $x=\dfrac{7}{3}$ and $x\neq2,-2$

65. a. $\dfrac{x^2-153}{2(x-9)(x+9)}$ **b.** $x=3,-3$ and $x\neq9,-9$

Exercises 5.5, pages 432 - 435

1. 72, 45 **3.** 9 **5.** $\dfrac{6}{13}$ **7.** 36, 27 **9.** 7, 12 **11.** 45 shirts **13.** 12 defective **15.** 1875 miles **17.** $\dfrac{6}{5}$ or $1\dfrac{1}{5}$ hours **19.** 45 minutes

21. Beth: 52 mph, Anna: 48 mph **23.** Commercial airliner: 300 mph, Private plane: 120 mph

25. Jet: 500 mph, Propeller plane: 250 mph   **27.** 50 mph   **29.** $\dfrac{9}{4}$ or $2\dfrac{1}{4}$ hours   **31.** $\dfrac{9}{2}$ or $4\dfrac{1}{2}$ days, 9 days
33. Boat: 14 mph, Current: 2 mph   **35.** 14 kilometers per hour

Exercises 5.6, pages 442 - 447

1. $\dfrac{7}{3}$ **3.** 2   **5.** 10   **7.** 36   **9.** 24   **11.** $-\dfrac{27}{5}$ **13.** 40   **15.** 54   **17.** 32   **19.** 27   **21.** 232.3 pounds   **23.** 6 in.   **25.** 4.71 feet
27. 0.0073 cm   **29.** 6 m   **31.** 1890 lb   **33.** 3.6×10^{-10} N   **35.** 25,200 lb   **37.** 18,000 lb   **39.** 200 cm^3 **41.** 330 bar   **43.** 5.2 ohms
45. 10 ohms   **47.** $10\dfrac{2}{3}$ ft from the 120 lb weight (or $1\dfrac{1}{3}$ ft from the 960 lb weight)   **49.** 216 kg

Chapter 5 Review, pages 452 - 455

1. $\dfrac{3}{16x}$; $x \neq 0$; $y \neq 0$ **2.** $\dfrac{5}{3x(3x-1)}$; $x \neq 0, -2, \dfrac{1}{3}$ **3.** -1; $x \neq \dfrac{5}{2}$ **4.** $-\dfrac{4}{3}$; $x \neq 3$ **5.** $\dfrac{2x^2+4x+8}{x+2}$; $x \neq -2, 2$
6. $\dfrac{3x+15}{x-5}$; $x \neq 5$ **7.** $x+4$ **8.** $\dfrac{xy+2x-4y-8}{3(x+2)(x-2)}$; $x \neq -2, 2$ **9.** $\dfrac{2x-1}{x-3}$; $x \neq -\dfrac{5}{3}, 3$ **10.** $\dfrac{x-1}{x+5}$; $x \neq -5, -\dfrac{3}{2}$ **11.** $\dfrac{2x}{3(x-2)}$
12. $\dfrac{10x^2+6x}{5x-3}$ **13.** $\dfrac{2}{xy^3}$ **14.** $\dfrac{x}{21}$ **15.** $\dfrac{x^2-12x+35}{(x-2)(x+3)}$ **16.** $\dfrac{3x-9}{(x-4)(x+1)(x-2)}$ **17.** $\dfrac{5x+20}{4x(2x-1)}$ **18.** $\dfrac{x^2-4x}{(x+1)(x+6)}$
19. $\dfrac{2x^2-x}{2(2x+1)}$ **20.** $\dfrac{x^3+9x^2+17x+21}{2(x-3)(x+3)^2}$ **21.** 6   **22.** 8   **23.** $\dfrac{6x+10}{(x+2)(x-2)}$ **24.** $\dfrac{x+1}{x-11}$ **25.** $\dfrac{2x-8}{(x+5)(x-2)}$
26. $\dfrac{x^4-4x^3-2x^2+1}{(x-1)^2(x+1)^2}$ **27.** $\dfrac{7x^3+16x^2+12x+1}{(x+1)(3x+1)(x+2)(x+3)}$ **28.** $\dfrac{11}{2x-3}$ **29.** $\dfrac{9}{x+4}$ **30.** $\dfrac{20-x^2}{(x+4)(x+5)}$
31. $\dfrac{x^2}{(x-1)(x^2+x+1)}$ **32.** $\dfrac{8x^2+6x+2}{(x-1)(x+3)(3x+1)}$ **33.** $\dfrac{xy}{2}$ **34.** $\dfrac{x+6}{x}$ **35.** $\dfrac{-2}{(x+2)^2}$ **36.** $\dfrac{x-2}{x}$ **37.** $\dfrac{2(x-1)}{x^2}$ **38.** 1
39. $\dfrac{4x-12}{x^2-6x-9}$ **40.** $\dfrac{x}{x-5}$ **41.** $x = 40$ **42.** $x = 10$ **43.** $x = 5$ **44.** $x = 1, 12$ **45.** 60 gallons   **46.** 20 women
47. 600 times   **48.** 40 ft   **49.** $AB = 8, FE = 5$ **50.** $TU = 6, VW = 7$ **51.** $x \neq 0, 1$; $x = \dfrac{5}{2}$ **52.** $x \neq -3, 3$; $x = 12$
53. $x \neq -3, 4$; $x = 0, \dfrac{25}{6}$ **54.** $x \neq -6$; $x = -\dfrac{15}{2}, 4$ **55.** $x_1 = x - \dfrac{(y-y_1)}{m}$ **56.** $a_1 = \dfrac{a_n}{r^{n-1}}$ **57.** $\dfrac{6}{8}$ **58.** 21, 27
59. Alice takes 6 hrs, Judy takes 4 hrs   **60.** Karl takes 2 hrs, his son takes 6 hrs   **61.** 24 mph   **62.** Initial rate was 37.5 mph
and final rate was 62.5 mph   **63.** Initial rate was 400 mph and final rate was 430 mph
64. Raphael takes 1.5 hrs, Matilde takes 3 hrs   **65.** $\dfrac{490}{9}$ **66.** 31,250   **67.** 576 feet   **68.** 16.5 ohms   **69.** 37.68 cubic feet
70. 5 fc

Chapter 5 Test, pages 456 - 457

1. $\dfrac{x}{x+4}$; $x \neq -4, -3$ **2.** $-\dfrac{x^2+4x+16}{x+4}$; $x \neq -4, 4$ **3.** $\dfrac{1}{2x+5}$; $x \neq -\dfrac{5}{2}$ **4.** $\dfrac{x+3}{x+4}$ **5.** $\dfrac{3x-2}{3x+2}$ **6.** $\dfrac{-2x^2-13x}{(x+5)(x+2)(x-2)}$
7. $\dfrac{-2x^2-7x+18}{(3x+2)(x-4)(x+1)}$ **8.** $2x^2$ **9.** $\dfrac{x^2-7x+1}{(x+3)(x-3)}$ **10.** $\dfrac{-3x}{x-2}$ **11.** $\dfrac{12(x+2)}{(x-3)(x+3)^2}$ **12.** $x \neq -4$; $x = 21$ **13.** $x \neq 0$; $x = -\dfrac{7}{2}$
14. $x \neq -4, 1$; $x = -1$ **15.** $x \neq -2, -1$; $x = 1$ **16.** $n = \dfrac{2S}{a_1 + a_n}$ **17.** $x = \dfrac{y-b}{m}$ **18.** $\dfrac{2}{7}$ **19.** 4 hr

20. Carlos: 42 mph; Mario: 57 mph **21.** 11 mph **22.** $\dfrac{400}{9}$ **23.** $\dfrac{15}{2}$ cm **24.** $AC = 36, DC = 24$

Chapter 5 Cumulative Review, pages 458 - 460

1. a. $\left\{-9, 3, \sqrt{25}\right\}$ **b.** $\left\{-9, \dfrac{1}{4}, 3, \sqrt{25}\right\}$ **c.** $\left\{-\sqrt{7}, \dfrac{\pi}{2}\right\}$ **d.** $\left\{-9, -\sqrt{7}, \dfrac{1}{4}, \dfrac{\pi}{2}, 3, \sqrt{25}\right\}$ **2.** 840 **3.** $36x^2y^3$ **4.** $2(x+3)^2(x-3)$

5. $x = -\dfrac{9}{8}$ **6.** $x = \dfrac{29}{11}$ **7.** $x = -\dfrac{4}{3}, x = 4$ **8.** $\left(-\infty, -\dfrac{1}{4}\right]$ 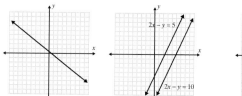 **9.** $\left(-\infty, -\dfrac{50}{3}\right)$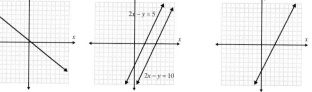

10. $\left[-\dfrac{8}{3}, 4\right]$ **11.** $y = -\dfrac{4}{5}x + \dfrac{2}{5}$ **12.** $2x - y = 10$ **13.** $y = 2x - 6$

14. a. Not a function **b.** Function; $D = \{-7, -3, 0, 5, 6\}$; $R = \{-2, 1, 4\}$ **c.** Function; $D = [-3, 3]$; $R = [0, 6]$ **15.** $x = 2, y = -1$

16. $x = 1, y = -5, z = 2$ **17.** $\det(A) = -102$ **18.** 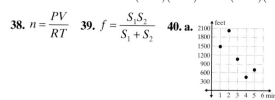 **19.** $\dfrac{y^6}{9x^4}$ **20.** $\dfrac{x^6}{y^6}$ **21.** $\dfrac{3}{2x^4y}$ **22.** 6.8 **23.** 1.05×10^6

24. $(3x+2)(5x+4)$ **25.** $(x+4)(2x^2+3)$ **26.** $2(2x-3)(4x^2+6x+9)$ **27.** $(2x+1)(x-3)$

28. a. -1 **b.** -12 **c.** 15 **29.** $x = -11$ **30.** $x = -7, 0, 2$ **31.** $x^2 + x - 20 = 0$

32. $6x^2 - 7x + 2 = 0$ **33.** $2x^2 + x - 3 + \dfrac{9}{x+2}$

34. $x - 4 + \dfrac{6x-15}{x^2-6}$ **35.** $\dfrac{2x-1}{(x+1)(x-1)}$ **36.** $\dfrac{x^2+15x-26}{(x-4)(x+2)(2x+3)}$ **37. a.** $x \neq -4, 4$; $x = -2, \dfrac{2}{3}$ **b.** $x \neq 2, 3$; $x = \dfrac{6}{5}$

38. $n = \dfrac{PV}{RT}$ **39.** $f = \dfrac{S_1 S_2}{S_1 + S_2}$ **40. a.**

[graph: feet vs min, points plotted from 300 to 2100]

b. $528, -968, -616, 264$ **c.** Juan is accelerating (his rate of change increased) at a rate of 528 feet per minute; decelerating (his rate of change decreased) 968 feet per minute; decelerating (his rate of change decreased) 616 feet per minute; accelerating (his rate of change increased) 264 feet per minute.

41. $a = 5, b = 1$ **42.** 256 feet **43.** 24 feet **44.** 3 and 11 **45.** 16, 18 **46.** $\dfrac{2}{3}$ hr or 40 minutes **47.** 120 yd^2

Chapter 6

Exercises 6.1, pages 474 - 476

1. 0.02 **3.** 32 **5.** 67.0820 **7.** 0.02 **9.** 2.3208 **11.** $2\sqrt{3}$ **13.** $7\sqrt{2}$ **15.** $-9\sqrt{2}$ **17.** $2\sqrt[3]{2}$ **19.** $3\sqrt[3]{4}$

21. Nonreal **23.** $2x^5 y\sqrt{6x}$ **25.** $ac\sqrt[3]{a^2b^2}$ **27.** Nonreal **29.** $5y$ **31.** $-8a^3$ **33.** $4x^2y^4\sqrt{2}$ **35.** $-2xy^2\sqrt[3]{3}$ **37.** $3xy^2\sqrt[3]{3x^2y}$

39. $2x^3y^4$ **41.** $9ab^2\sqrt[3]{ab^2}$ **43.** $5xy^3\sqrt{5x}$ **45.** $2bc\sqrt{3ac}$ **47.** $-5x^2y^3z^4\sqrt{3}$ **49.** $2xy^2z^3\sqrt[3]{3x^2y}$ **51.** $3 - \sqrt{3}$ **53.** $\dfrac{1+\sqrt{2}}{2}$

55. $\dfrac{1+\sqrt{6}}{2}$ **57.** $3 + x\sqrt{3}$ **59.** $\dfrac{4+a\sqrt{6a}}{4}$ **61.** 7 in. **63.** $\sqrt{6} \approx 2.45$ amperes **65.** $50\sqrt{5} \approx 111.80$ volts

67. A cube root has no restrictions.

Exercises 6.2, pages 486 - 488

1. 3 **3.** $\dfrac{1}{10}$ **5.** -512 **7.** Nonreal **9.** $\dfrac{3}{7}$ **11.** -4 **13.** -5 **15.** $\dfrac{1}{4}$ **17.** $-\dfrac{27}{8}$ **19.** $\dfrac{9}{16}$ **21.** $\dfrac{2}{5}$ **23.** $\dfrac{1}{81}$ **25.** $-\dfrac{1}{16,807}$

27. 2187 **29.** 4.6416 **31.** 0.0131 **33.** 158.7401 **35.** 7.7460 **37.** 1.0605 **39.** 0.2000 **41.** $8x$ **43.** $3a^2$ **45.** $8x^{\frac{5}{2}}$ **47.** $5a^{\frac{13}{6}}$

49. $a^{\frac{8}{5}}$ **51.** $x^{\frac{1}{2}}$ **53.** $a^{\frac{7}{6}}$ **55.** $a^{\frac{1}{4}}$ **57.** $\dfrac{y^{\frac{1}{2}}}{x^{\frac{4}{3}}}$ **59.** $\dfrac{1}{a^{\frac{3}{4}}b^{\frac{1}{2}}}$ **61.** $\dfrac{x^{\frac{3}{2}}}{16y^{\frac{2}{5}}}$ **63.** $\dfrac{x^2y^4}{z^4}$ **65.** $\dfrac{c^3}{3ab^2}$ **67.** $\dfrac{8b^{\frac{9}{4}}}{a^3c^3}$ **69.** $x^{\frac{1}{4}}y^{\frac{5}{4}}$ **71.** $\dfrac{y^{\frac{2}{3}}}{50x^{\frac{4}{3}}}$

73. $\dfrac{6a}{b^2}$ **75.** $\dfrac{x^{\frac{7}{4}}}{y^{\frac{3}{4}}}$ **77.** $\dfrac{3x^{\frac{1}{2}}}{20y^{\frac{2}{3}}}$ **79.** $x\sqrt[15]{x^4}$ **81.** $x\sqrt[12]{x}$ **83.** $\sqrt[12]{a^5}$ **85.** $\sqrt[10]{x}$ **87.** $\sqrt[4]{a}$ **89.** $a^9b^{18}c^3$ **91.** $a^{20}b^5c^{10}$

Exercises 6.3, pages 497 - 499

1. $-6\sqrt{2}$ **3.** $5\sqrt{x}$ **5.** $-3\sqrt[3]{7x^2}$ **7.** $10\sqrt{3}$ **9.** $-7\sqrt{2}$ **11.** $20\sqrt{3}$ **13.** $3\sqrt{6}+7\sqrt{3}$ **15.** $20x\sqrt{5x}$ **17.** $14\sqrt{5}-3\sqrt{7}$

19. $11\sqrt{2x}+14\sqrt{3x}$ **21.** $4\sqrt[3]{5}-13\sqrt[3]{2}$ **23.** $x^2\sqrt{2x}$ **25.** $4x^2y\sqrt{x}$ **27.** $3x-9\sqrt{3x}+8$ **29.** $24+10\sqrt{2x}+2x$ **31.** 2

33. $13+4\sqrt{10}$ **35.** -3 **37.** $6+\sqrt{30}-2\sqrt{3}-\sqrt{10}$ **39.** $5-\sqrt{33}$ **41.** $49x-2$ **43.** $9x+6\sqrt{xy}+y$ **45.** $12x+7\sqrt{2xy}+2y$

47. $\dfrac{2\sqrt{7}}{7}$ **49.** $\dfrac{\sqrt[3]{18}}{3}$ **51.** $\dfrac{-5\sqrt{x}}{x^2}$ **53.** $\dfrac{2\sqrt{3x}}{3y}$ **55.** $\dfrac{2a\sqrt{b}}{b}$ **57.** $\dfrac{\sqrt[3]{28xy^2}}{2y^2}$ **59.** $\dfrac{x\sqrt[3]{3xy^2}}{3y}$ **61.** $-\dfrac{3\sqrt{3}+15}{22}$

63. $-\dfrac{\sqrt{42}-3\sqrt{6}}{2}$ **65.** $4\sqrt{5}+4\sqrt{3}$ **67.** $-\dfrac{5\sqrt{2}+4\sqrt{5}}{3}$ **69.** $\dfrac{59+21\sqrt{5}}{44}$ **71.** $8+3\sqrt{6}$ **73.** $\dfrac{x^2+4x\sqrt{y}+4y}{x^2-4y}$ **75.** $\dfrac{2}{\sqrt{5}-1}$

77. $\dfrac{-1}{\sqrt{10}+\sqrt{2}}$ **79.** $\dfrac{1}{\sqrt{y}+\sqrt{2}}$ **81.** $\dfrac{9x-y^2}{9x\sqrt{x}+3xy}$ **83.** $\dfrac{1}{\sqrt{5+h}-\sqrt{5}}$ **85.** -3.1854 **87.** 38.7000 **89.** -2.9141 **91.** 0.6971

93. 0.3820 **95.** $5\sqrt{6}+\sqrt{170}\approx25.2859$ ft

Exercises 6.4, page 505

1. $x=3$ **3.** $x=13$ **5.** $x=3$ **7.** $x=2$ **9.** $x=-4, x=1$ **11.** $x=-5, x=\dfrac{5}{2}$ **13.** $x=-2$ **15.** $x=\pm5$ **17.** $x=2$

19. $x=4$ **21.** $x=0$ **23.** $x=5$ **25.** No solution **27.** $x=4$ **29.** $x=-1, x=3$ **31.** $x=5$ **33.** $x=2$ **35.** $x=6$

37. $x=-4$ **39.** $x=12$ **41.** No solution **43.** $x=40$ **45.** $(a+b)^2=(a+b)(a+b)=a^2+2ab+b^2\neq a^2+b^2$

Exercises 6.5, pages 513 - 516

1. a. $\sqrt{5}\approx2.2361$ **b.** 3 **c.** $5\sqrt{2}\approx7.0711$ **d.** 2 **3. a.** 3 **b.** -1 **c.** -2 **d.** $2\sqrt[3]{3}\approx2.8845$ **5.** $[-8,\infty)$ **7.** $\left(-\infty,\dfrac{1}{2}\right]$ **9.** $(-\infty,\infty)$

11. $[0,\infty)$ **13.** $(-\infty,\infty)$ **15.** E **17.** B **19.** A

21. 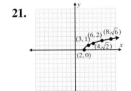 **23.** 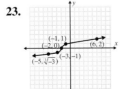 **25.** 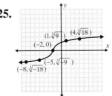 **27.** 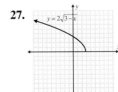 **29.**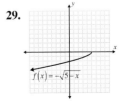

31.   **33.**   **35.**

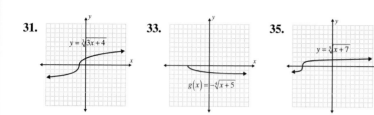

Exercises 6.6, pages 522 - 523

1. Real part is 4, imaginary part is –3   **3.** Real part is –11, imaginary part is $\sqrt{2}$ **5.** Real part is $\dfrac{2}{3}$, imaginary part is $\sqrt{17}$

7. Real part is $\dfrac{4}{5}$, imaginary part is $\dfrac{7}{5}$ **9.** Real part is $\dfrac{3}{8}$, imaginary part is 0   **11.** $7i$ **13.** $-8i$ **15.** $21\sqrt{3}$ **17.** $10i\sqrt{6}$

19. $-12i\sqrt{3}$ **21.** $11\sqrt{2}$ **23.** $10i\sqrt{10}$ **25.** $x=6, y=-3$ **27.** $x=-2, y=\sqrt{5}$ **29.** $x=\sqrt{2}-3, y=1$ **31.** $x=2, y=-8$

33. $x=2, y=-6$ **35.** $x=3, y=10$ **37.** $x=-\dfrac{4}{3}, y=-3$ **39.** $6+2i$ **41.** $1+7i$ **43.** $2-6i$ **45.** $14i$ **47.** $\left(3+\sqrt{5}\right)-6i$

49. $5+\left(\sqrt{6}+1\right)i$ **51.** $\sqrt{3}-5$ **53.** $-2-5i$ **55.** $11-16i$ **57.** 2   **59.** $3+4i$ **61. a.** Yes **b.** No

Exercises 6.7, pages 529 - 530

1. $16+24i$ **3.** $-7\sqrt{2}+7i$ **5.** $3+12i$ **7.** $1-i\sqrt{3}$ **9.** $5\sqrt{2}+10i$ **11.** $2+8i$ **13.** $-7-11i$ **15.** $13+0i$ **17.** $-3-7i$

19. $5+12i$ **21.** $5-i\sqrt{3}$ **23.** $21+0i$ **25.** $23-10i\sqrt{2}$ **27.** $\left(2+\sqrt{10}\right)+\left(2\sqrt{2}-\sqrt{5}\right)i$ **29.** $\left(9-\sqrt{30}\right)+\left(3\sqrt{5}+3\sqrt{6}\right)i$

31. $0+3i$ **33.** $0-\dfrac{5}{4}i$ **35.** $-\dfrac{1}{4}+\dfrac{1}{2}i$ **37.** $-\dfrac{4}{5}+\dfrac{8}{5}i$ **39.** $\dfrac{24}{25}+\dfrac{18}{25}i$ **41.** $-\dfrac{1}{13}+\dfrac{5}{13}i$ **43.** $-\dfrac{1}{29}-\dfrac{12}{29}i$ **45.** $-\dfrac{17}{26}-\dfrac{7}{26}i$

47. $\dfrac{4+\sqrt{3}}{4}+\left(\dfrac{4\sqrt{3}-1}{4}\right)i$ **49.** $-\dfrac{1}{7}+\dfrac{4\sqrt{3}}{7}i$ **51.** $0+i$ **53.** $-1+0i$ **55.** $0+i$ **57.** $1+0i$ **59.** $0-i$ **61.** x^2+9 **63.** x^2+2

65. $5y^2+4$ **67.** $x^2+4x+40$ **69.** $y^2-6y+13$ **71.** Given a complex number $(a+bi)$: $(a+bi)(a-bi)=a^2-abi+abi-b^2i^2$
$=a^2+b^2$ which is the sum of squares of real numbers. Thus, the product must be a positive real number.   **73.** $a^2+b^2=1$

Chapter 6 Review, pages 537 - 540

1. 5.7446   **2.** 8.4853   **3.** 22.3607   **4.** –13.4164   **5.** $2\sqrt{5}$ **6.** $6\sqrt{2}$ **7.** $-7\sqrt{2}$ **8.** $-9\sqrt{3}$ **9.** –3   **10.** $2\sqrt[3]{3}$ **11.** $-3\sqrt[3]{4}$

12. $2\sqrt[3]{5}$ **13.** Nonreal   **14.** Nonreal   **15.** $2ab\sqrt{3bc}$ **16.** $2xy^2\sqrt{2xy}$ **17.** $8x^2y^3\sqrt{2x}$ **18.** $6a^2b^4$ **19.** $4xy\sqrt[3]{y^2}$

20. $4x^4y^5\sqrt{5}$ **21.** 4   **22.** 9   **23.** $-\dfrac{1}{6}$ **24.** $\dfrac{1}{15}$ **25.** $\dfrac{2}{3}$ **26.** $\dfrac{9}{25}$ **27.** 64   **28.** $-\dfrac{1}{1000}$ **29.** 10.9027   **30.** 82.7037   **31.** 7.0711

32. 0.2408   **33.** $27x^2$ **34.** $8a^{\frac{3}{2}}$ **35.** $a^{\frac{11}{12}}$ **36.** $5y^{\frac{10}{3}}$ **37.** $\dfrac{9a^4b^2}{c^2}$ **38.** $\dfrac{1}{2y^{\frac{1}{2}}}$ **39.** $\sqrt[12]{x^5}$ **40.** $2\sqrt[3]{x}$ **41.** $-4\sqrt{11}$ **42.** $5\sqrt{x}$

43. $10\sqrt{7}+8$ **44.** $5\sqrt{2}-26\sqrt{3}$ **45.** 0   **46.** –1   **47.** $7+2\sqrt{10}$ **48.** $x-y$ **49.** $8\sqrt{2x}+15x+2$ **50.** $\dfrac{\sqrt{x}}{6x}$ **51.** $\dfrac{3\sqrt{y}}{5y}$

52. $\dfrac{\sqrt[3]{4x^2}}{4x}$ **53.** $\dfrac{\sqrt[3]{100xy^2}}{10y}$ **54.** $15-5\sqrt{5}$ **55.** $\dfrac{5\sqrt{7}-5\sqrt{3}}{4}$ **56.** $\dfrac{x^2-2x\sqrt{y}+y}{y-x^2}$ **57.** $\dfrac{13+3\sqrt{15}}{17}$ **58.** 25.4495

59. –1.6334   **60.** 0.2185   **61.** –28.0000   **62.** $x=7$ **63.** $x=14$ **64.** $x=2$ **65.** $x=3$ **66.** $x=\pm12$ **67.** $x=\pm9$

68. $x=9$ **69.** $x=4$ **70.** $x=9$ **71.** $x=2,4$ **72.** No solution   **73.** No solution   **74.** $x=27$ **75.** $x=11$ **76.** $x=20$

77. a. 4 **b.** 1   **78. a.** 0 **b.** 5   **79. a.** $\sqrt[3]{6}\approx1.8171$ **b.** $2\sqrt[3]{2}\approx2.5198$ **80. a.** $\sqrt{3}\approx1.7321$ **b.** $\sqrt{11}\approx3.3166$

81. $[-5,\infty)$ **82.** $\left[-\dfrac{1}{2},\infty\right)$ **83.** $(-\infty,\infty)$ **84.** $\left(-\infty,\dfrac{1}{5}\right]$ **85.** $\left(-\infty,\dfrac{3}{2}\right]$ **86.** $(-\infty,\infty)$

87. 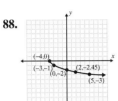 **88.** 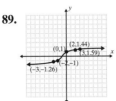 **89.** 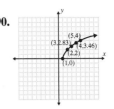 **90.** 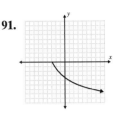 **91.**

92. 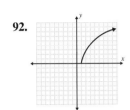 **93.** 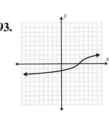 **94.** 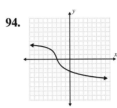 **95.** 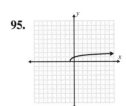 **96.**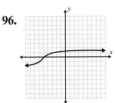

97. Real part is 5, imaginary part is -2 **98.** Real part is -10, imaginary part is $\sqrt{3}$ **99.** Real part is $\dfrac{3}{4}$, imaginary part is 0

100. Real part is $\dfrac{3}{2}$, imaginary part is $\dfrac{1}{2}$ **101.** $6i$ **102.** 12 **103.** $-6i\sqrt{2}$ **104.** $10\sqrt{29}$ **105.** $x=9, y=0$

106. $x=\sqrt{3}, y=\dfrac{4}{3}$ **107.** $x=-10, y=-7$ **108.** $x=5, y=8$ **109.** $5+8i$ **110.** $4+8i$ **111.** $-4+2i$ **112.** $4+\sqrt{2}-4i$

113. $2+\sqrt{5}+0i$ **114.** $15-25i$ **115.** $1+5i$ **116.** $5-i$ **117.** $4\sqrt{2}+12i$ **118.** $7-i$ **119.** $-5+12i$ **120.** $7+i\sqrt{3}$

121. $2+i$ **122.** $\dfrac{2}{25}-\dfrac{11}{25}i$ **123.** $\dfrac{2+7\sqrt{5}}{6}+\dfrac{\left(7-2\sqrt{5}\right)}{6}i$ **124.** $0-i$ **125.** $0-i$ **126.** $-1+0i$

Chapter 6 Test, pages 541 - 542

1. $4\sqrt{7}$ **2.** $2y^2\sqrt{30x}$ **3.** $2y\sqrt[3]{6x^2y^2}$ **4.** $3xyz\sqrt[3]{3x}$ **5.** $\sqrt[3]{4x^2}$ **6.** $2^{\frac{1}{2}}x^{\frac{1}{3}}y^{\frac{2}{3}}$ **7.** 4 **8.** $\dfrac{1}{27}$ **9.** $4x^{\frac{7}{6}}$ **10.** $\dfrac{7x^{\frac{1}{4}}}{y^{\frac{1}{3}}}$ **11.** $\dfrac{8y^{\frac{3}{2}}}{x^3}$

12. $\sqrt[12]{x^{11}}$ **13.** $17\sqrt{3}$ **14.** $(7x-6)\sqrt{x}$ **15.** $8\sqrt[3]{3}$ **16.** $5-2\sqrt{6}$ **17.** $-xy\sqrt{y}$ **18.** $30-7\sqrt{3x}-6x$ **19.** $7+\sqrt{7}$

20. $3-\sqrt{2}$ **21. a.** $\dfrac{y\sqrt{10x}}{4x^2}$ **b.** $1+\sqrt{x}$ **22.** $x=7$ **23.** $x=-4$ **24.** $x=1$ **25.** $x=0, 36$ **26. a.** $\left[-\dfrac{4}{3},\infty\right)$ **b.** $(-\infty,\infty)$

27. a. **b.** **28.** Real part is 6, imaginary part is $-\dfrac{1}{3}$ **29.** Real part is $-\dfrac{2}{5}$, imaginary part is $\dfrac{3}{5}$

30. $2i\sqrt{2}$ **31.** $39i$ **32.** $16+4i$ **33.** $-5+16i$ **34.** $23-14i$ **35.** $\dfrac{8}{13}-\dfrac{1}{13}i$ **36.** $x=\dfrac{11}{2}, y=3$ **37.** $0-i$ **38.** x^2+4

39. $x^2+6x+12$ **40.** 0.1250 **41.** 3.0711 **42.** 0.9758

Chapter 6 Cumulative Review, pages 543 - 546

1. Identity Property of Addition **2.** Distributive Property **3.** 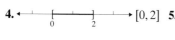 $(-\infty, 6)$

4. 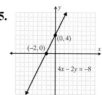 $[0, 2]$ **5.** 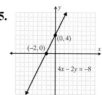 **6.** $y = -\dfrac{1}{6}x + 3$; $m = -\dfrac{1}{6}$, $y\text{-int} = (0,3)$ 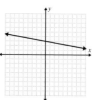 **7.** $x = 5$

8. a. Function $D = (-\infty, \infty), R = (-\infty, 0]$ **b.** Not a function $D = (-\infty, \infty), R = (-\infty, 5]$ **c.** Function $D = [-2\pi, 2\pi], R = [-2, 2]$

9. $(1, -2)$ **10.** $x = 1, y = 1, z = 1$ **11.** $x = \dfrac{85}{13}, y = \dfrac{60}{13}$ **12.** -333 **13.** **14.** 1×10^3

15. $2x^3 - 4x^2 + 8x - 4$ **16.** $6x^2 + 19x - 7$ **17.** $-5x^2 + 18x + 8$ **18.** $x - 9 + \dfrac{21x - 24}{x^2 + 2x - 1}$ **19.** $(4x + 3)(3x - 4)$

20. $(7 + 2x)(4 - x)$ **21.** $5(x - 4)(x^2 + 4x + 16)$ **22.** $(x + 4)(x + 1)(x - 1)$ **23.** $(x - 16)(x + 3) = 0, \ x = -3, 16$

24. $(2x + 3)(x - 2) = 0, \ x = -\dfrac{3}{2}, 2$ **25.** $(5x - 2)(3x - 1) = 0, \ x = \dfrac{1}{3}, \dfrac{2}{5}$ **26. a.** -12 **b.** 0 **c.** -168 **27. a.** $x^2 - 11x + 28 = 0$

b. $x^3 - 11x^2 - 14x + 24 = 0$ **28.** $\dfrac{4x^2 - 12x - 1}{(x + 3)(x - 2)(x - 5)}$ **29.** $\dfrac{4x^2 + 6x + 17}{(2x + 1)(2x - 1)(x + 3)}$ **30.** $\dfrac{x^2 - 1}{2x - 4}$ **31.** $\dfrac{x + 6}{x - 1}$

32. $x \neq -5, 0; \ x = -\dfrac{5}{7}$ **33.** $x \neq -7, 3; \ x = \dfrac{83}{4}$ **34.** $\dfrac{1}{16}$ **35.** $x^{\frac{7}{3}}$ **36.** $12\sqrt{2}$ **37.** $2x^2y^3\sqrt[3]{2y}$ **38.** $\dfrac{\sqrt{10y}}{2y}$ **39.** $2\sqrt{30} - 11$

40. $x = -1$ **41.** $x = 9$ **42.** $x = -1$ **43.** $14 - 2i$ **44.** $-\dfrac{8}{17} - \dfrac{19}{17}i$ **45.** $1 + 0i$ **46.** $n = \dfrac{s - a}{d} + 1$ or $\dfrac{s - a + d}{d}$ **47.** $p = \dfrac{A}{1 + rt}$

48. a. 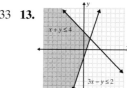 **b.** $m_1 = 8, m_2 = 24, m_3 = 0, m_4 = \dfrac{104}{5}$ **c.** Rate increased 8 ft/s, then increased 24 ft/s, then stayed constant, then increased $\dfrac{104}{5}$ ft/s.

49. $15, 17$ **50.** $\dfrac{50}{3}$ in.^{3} or $16.\overline{6}$ in.^{3} **51.** 8 ohms **52.** 10 mph **53.** \$35,000 in 6% and \$15,000 in 10%

54. 7.5 hours **55.** 60 pounds of first type (\$1.25 candy) and 40 pounds of second type (\$2.50 candy)

56. $x = 1$ **57.** $x = \pm 2\sqrt{3} \approx \pm 3.46, \pm\sqrt{6} \approx \pm 2.45$ **58.** $x \approx -1.372; \ x \approx 4.372$ **59.** $x = 1; \ x \approx -2.303; \ x \approx 1.303$

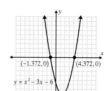

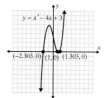

60. $x = -4; x = 1; x = 5$ **61.** 279,936.0000   **62.** 0.0400   **63.** 1.3027   **64.** 6.4552

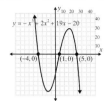

Chapter 7

Exercises 7.1, pages 558 - 560

1. $x = 3$ or $x = 15$ **3.** $y = -11$ or $y = 11$ **5.** $z = -13$ or $z = 7$ **7.** $x = -1$ or $x = 0$ or $x = 1$ **9.** $x = 0$ or $x = 2$ or $x = 3$

11. $x^2 - 12x + \underline{36} = \left(x - 6\right)^2$ **13.** $x^2 + 6x + \underline{9} = \left(x + 3\right)^2$ **15.** $x^2 - 5x + \underline{\dfrac{25}{4}} = \left(x - \dfrac{5}{2}\right)^2$ **17.** $y^2 + y + \underline{\dfrac{1}{4}} = \left(y + \dfrac{1}{2}\right)^2$

19. $x^2 + \dfrac{1}{3}x + \underline{\dfrac{1}{36}} = \left(x + \dfrac{1}{6}\right)^2$ **21.** $x = \pm 12$ **23.** $x = \pm 5i$ **25.** $x = \pm 3i\sqrt{2}$ **27.** $x = -1, x = 9$ **29.** $x = \pm 2\sqrt{3}$

31. $x = 1 \pm \sqrt{5}$ **33.** $x = -8 \pm 3i$ **35.** $x = 5 \pm i\sqrt{10}$ **37.** $x = -7, x = 1$ **39.** $x = 4 \pm \sqrt{13}$ **41.** $z = -2 \pm \sqrt{6}$

43. $x = -1 \pm i\sqrt{5}$ **45.** $y = 5 \pm \sqrt{21}$ **47.** $x = \dfrac{5 \pm \sqrt{5}}{2}$ **49.** $x = \dfrac{-1 \pm i\sqrt{7}}{2}$ **51.** $x = -2 \pm \sqrt{7}$ **53.** $x = -1 \pm i\sqrt{3}$

55. $y = -\dfrac{4}{3}, y = 1$ **57.** $x = \dfrac{-7 \pm \sqrt{17}}{8}$ **59.** $x = \dfrac{1 \pm i\sqrt{11}}{4}$ **61.** $y = \dfrac{-3 \pm i\sqrt{11}}{2}$ **63.** $x = 2 \pm \sqrt{2}$ **65.** $x = \dfrac{-5 \pm i\sqrt{7}}{2}$

67. $x^2 - 6 = 0$ **69.** $z^2 - 6z + 7 = 0$ **71.** $x^2 - 2x - 11 = 0$ **73.** $x^2 + 49 = 0$ **75.** $y^2 + 5 = 0$ **77.** $x^2 + 6x + 13 = 0$

79. $x^2 - 4x + 7 = 0$ **81.** See page 554.

Exercises 7.2, pages 567 - 569

1. 68; two real solutions   **3.** 0; one real solution   **5.** −44; two nonreal solutions   **7.** 4; two real solutions

9. 19,600; two real solutions   **11.** −11; two nonreal solutions   **13.** $x = -2 \pm 2\sqrt{2}$ **15.** $x = -\dfrac{2}{3}$ **17.** $x = 1 \pm i\sqrt{6}$

19. $x = -3, \dfrac{1}{2}$ **21.** $x = \dfrac{-3 \pm \sqrt{5}}{4}$ **23.** $x = \dfrac{-3 \pm i\sqrt{3}}{4}$ **25.** $x = \dfrac{-3 \pm \sqrt{29}}{2}$ **27.** $x = \dfrac{5 \pm \sqrt{17}}{2}$ **29.** $x = -\dfrac{1}{4}$ **31.** $x = -\dfrac{1}{6}, x = 1$

33. $x = \pm \dfrac{2\sqrt{3}}{3}$ **35.** $x = \dfrac{9 \pm \sqrt{65}}{2}, x = 0$ **37.** $x = \dfrac{-3 \pm \sqrt{5}}{2}, x = 0$ **39.** $x = \dfrac{2}{3}$ **41.** $x = \dfrac{-4 \pm i\sqrt{2}}{2}$ **43.** $x = \pm \sqrt{7}$

45. $x = -1, x = 0$ **47.** $x = \dfrac{7 \pm i\sqrt{51}}{10}$ **49.** $x = -2, x = \dfrac{5}{3}$ **51.** $x = \pm \dfrac{3}{2}i$ **53.** $x = \dfrac{2 \pm \sqrt{3}}{3}$ **55.** $x = \dfrac{7 \pm i\sqrt{287}}{12}$ **57.** $\dfrac{2 \pm \sqrt{2}}{2}$

59. $x = \dfrac{-7 \pm \sqrt{17}}{4}$ **61.** $k < 16$ **63.** $k = \dfrac{81}{4}$ **65.** $k > 3$ **67.** $k > -\dfrac{1}{36}$ **69.** $k = \dfrac{49}{48}$ **71.** $k > \dfrac{4}{3}$ **73.** $x \approx 2.5993, 60.4007$

75. $x \approx -0.7862, 2.0110$ **77.** $x \approx -4.1334, -0.5806$ **79.** $x \approx -2.6933, 2.6933$

81. $x^4 - 13x^2 + 36 = 0$; Multiplied $(x - 2)(x + 2)(x - 3)(x + 3)$; Answers may vary.

Exercises 7.3, pages 574 - 579

1. 6, 13 **3.** 4, 14 **5.** $-11, -6$ **7.** $-5 - 4\sqrt{3}$ **9.** $\dfrac{1 + \sqrt{33}}{4}$ **11.** $7\sqrt{2}$ cm, $7\sqrt{2}$ cm **13.** Mel: 30 mph, John: 40 mph

15. 4 feet $\times$ 10 feet **17.** 17 inches $\times$ 6 inches **19.** 70 trees **21.** 9 floors **23.** 3 inches **25.** 2 amperes or 8 amperes

27. $13 for shirts and $22 for pants **29. a.** $307.20 **b.** 45 cents or 35 cents **31.** 8 members **33.** 6 mph

35. 4 hours and 12 hours **37.** 10 lb of Grade A, 20 lb of Grade B **39. a.** 6.75 s **b.** 3 s, 3.75 s **41. a.** 7 s **b.** 2 s, 12 s

43. 14.14 cm **45. a.** $C = 94.20$ ft, $A = 706.50$ ft^2 **b.** $P = 84.85$ ft, $A = 450.00$ ft^2 **47. a.** No **b.** Home plate **c.** No **49.** 855.86 ft

51. See page 561.

Exercises 7.4, pages 584 - 586

1. $x = \pm 2$, $x = \pm 3$ **3.** $x = \pm 2, x = \pm\sqrt{5}$ **5.** $y = \pm\sqrt{7}, y = \pm 2i$ **7.** $y = \pm\sqrt{5}, y = \pm i\sqrt{5}$ **9.** $z = -\dfrac{1}{4}, z = \dfrac{1}{6}$

11. $x = 4, x = \dfrac{25}{4}$ **13.** $x = 1, x = 4$ **15.** $x = -8, x = \dfrac{1}{8}$ **17.** $x = \dfrac{1}{25}$ **19.** $x = -\dfrac{1}{3}, x = \dfrac{3}{8}$ **21.** $x = -27, x = -8, x = 0$

23. $x = 1, x = 2$ **25.** $x = -\dfrac{7}{2}, x = -3$ **27.** $x = 0, x = 8$ **29.** $x = -1, x = -\dfrac{1}{2}$ **31.** $x = \pm\sqrt{1 + i}, x = \pm\sqrt{1 - i}$

33. $x = \pm\sqrt{1 + 3i}, x = \pm\sqrt{1 - 3i}$ **35.** $x = \pm\sqrt{2 + i\sqrt{3}}, x = \pm\sqrt{2 - i\sqrt{3}}$ **37.** $x = \pm\dfrac{\sqrt{5}}{5}, x = \pm 1$ **39.** $x = \pm\dfrac{\sqrt{6}}{2}, x = \pm\dfrac{1}{3}i$

41. $x = -\dfrac{1}{4}$ **43.** $x = -4, x = 1$ **45.** $x = -\dfrac{1}{2}$ **47.** $x = -7, x = -\dfrac{3}{2}$ **49.** $x = \dfrac{26}{5}$ **51.** $x = 0, x = \pm 2\sqrt{2}, x = \pm 2i\sqrt{2}$

53. $x = 2, x = -1 \pm i\sqrt{3}$ **55.** $x = -10, x = 5 \pm 5i\sqrt{3}$ **57. a.** $l^2 - l - 1 = 0$, $l = \dfrac{1 + \sqrt{5}}{2}$ which is the golden ratio **b.** 97.08 feet

c. The yellow rectangle is "golden."

Exercises 7.5, pages 600 - 604

1. $x = 0; (0, -4); D = (-\infty, \infty); R = [-4, \infty)$ **3.** $x = 0; (0, 9); D = (-\infty, \infty); R = [9, \infty)$ **5.** $x = 0; (0, 1); D = (-\infty, \infty); R = (-\infty, 1]$

7. $x = 0; \left(0, \dfrac{1}{5}\right); D = (-\infty, \infty); R = \left(-\infty, \dfrac{1}{5}\right]$ **9.** $x = -1; (-1, 0); D = (-\infty, \infty); R = [0, \infty)$ **11.** $x = 4; (4, 0); D = (-\infty, \infty); R = (-\infty, 0]$

13. $x = -3; (-3, -2); D = (-\infty, \infty); R = [-2, \infty)$ **15.** $x = -2; (-2, -6); D = (-\infty, \infty); R = [-6, \infty)$

17. $x = \dfrac{3}{2}; \left(\dfrac{3}{2}, \dfrac{7}{2}\right); D = (-\infty, \infty); R = \left(-\infty, \dfrac{7}{2}\right]$ **19.** $x = \dfrac{4}{5}; \left(\dfrac{4}{5}, -\dfrac{11}{5}\right); D = (-\infty, \infty); R = \left[-\dfrac{11}{5}, \infty\right)$

21. a. **b.** **c.** **d.**

23. a. **b.** **c.** **d.**

25. $y = 2(x-1)^2; (1,0);$
$R = [0,\infty);$
Zeros: $x = 1$

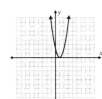

27. $y = (x-1)^2 - 4; (1,-4);$
$R = [-4,\infty);$
Zeros: $x = -1, x = 3$

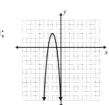

29. $y = (x+3)^2 - 4; (-3,-4);$
$R = [-4,\infty)$
Zeros: $x = -5, x = -1$

31. $y = 2(x-2)^2 - 3; (2,-3)$
$R = [-3,\infty);$
Zeros: $x = \dfrac{4 \pm \sqrt{6}}{2}$

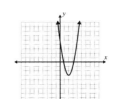

33. $y = -3(x+2)^2 + 3; (-2,3);$
$R = (-\infty,3];$
Zeros: $x = -1, x = -3$

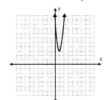

35. $y = 5(x-1)^2 + 3; (1,3);$
$R = [3,\infty);$
Zeros: none

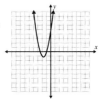

37. $y = -\left(x + \dfrac{5}{2}\right)^2 + \dfrac{17}{4}; \left(-\dfrac{5}{2}, \dfrac{17}{4}\right);$
$R = \left(-\infty, \dfrac{17}{4}\right];$ Zeros: $x = \dfrac{-5 \pm \sqrt{17}}{2}$

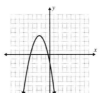

39. $y = 2\left(x + \dfrac{7}{4}\right)^2 - \dfrac{9}{8}; \left(-\dfrac{7}{4}, -\dfrac{9}{8}\right);$
$R = \left[-\dfrac{9}{8}, \infty\right);$
Zeros: $x = -\dfrac{5}{2}, x = -1$

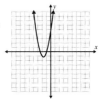

41. a. No **b.** Yes
c. Answers may vary.

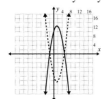

43. a. No **b.** Yes
c. Answers may vary.

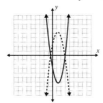

45. a. 3.5 s **b.** 196 ft **47. a.** 5 s **b.** 420 ft **49.** $20
51. a. $30 **b.** $1800 **53.** zeros: $x \approx -0.7321, x \approx 2.7321$
55. zeros: $x \approx -1.1583, x \approx 2.1583$ **57.** no real zeros

59. a. a parabola **b.** $x = -\dfrac{b}{2a}$ **c.** $x = -\dfrac{b}{2a}$
d. No. A graph can be entirely above or below the x-axis.

61. Domain: $(-\infty, \infty);$ For $a > 0$, Range: $[k, \infty);$ For $a < 0$, Range: $(-\infty, k]$.

Exercises 7.6, pages 618 - 621

1. $(-2,6)$

3. $\left(-\infty, \dfrac{2}{3}\right) \cup (5, \infty)$

5. $(-\infty, -7] \cup \left[\dfrac{5}{2}, \infty\right)$

7. $\left[-2, -\dfrac{1}{3}\right]$

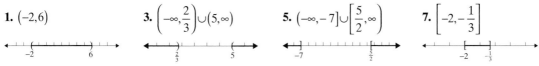

9. $\left(-\infty, -\dfrac{4}{3}\right) \cup (0,5)$

11. $\{-2\}$

13. $\left(-\infty, -\dfrac{5}{2}\right) \cup (3, \infty)$

15. $\left(-\dfrac{1}{4}, \dfrac{3}{2}\right)$

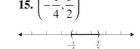

17. $\left(-\infty,\frac{1}{2}\right]\cup[2,\infty)$ **19.** $\left(-\frac{2}{3},-\frac{1}{2}\right)$ **21.** $\left(-\infty,\frac{5}{2}\right)\cup\left(\frac{5}{2},\infty\right)$ **23.** $\left[-\frac{5}{2},\frac{7}{4}\right]$

25. $(-1,0)\cup(3,\infty)$ **27.** $(0,1)\cup(4,\infty)$ **29.** $(-\infty,-1)\cup(4,\infty)$ **31.** $(-\infty,-2)\cup(-1,1)\cup(2,\infty)$

33. $[-3,-2]\cup[2,3]$ **35.** $(-\infty,-4]\cup[2,\infty)$ **37.** $\left(-\frac{2}{3},2\right)$ **39.** $\left(-\infty,-1-\sqrt{5}\right)\cup\left(-1+\sqrt{5},\infty\right)$

41. $\left(-\infty,-3-\sqrt{2}\right]\cup\left[-3+\sqrt{2},\infty\right)$ **43.** $\left(\frac{-5-\sqrt{13}}{6},\frac{-5+\sqrt{13}}{6}\right)$ **45.** $\left(-\infty,-\frac{1}{2}\right]\cup[0,4]$ **47.** $(-\infty,\infty)$

49. $\varnothing$ **51. a.** $x=\frac{7\pm\sqrt{89}}{2}$ **b.** $\left(-\infty,\frac{7-\sqrt{89}}{2}\right)\cup\left(\frac{7+\sqrt{89}}{2},\infty\right)$ **c.** $\left(\frac{7-\sqrt{89}}{2},\frac{7+\sqrt{89}}{2}\right)$

53. a. $x=-5,1$; **b.** $(-5,1)$; **c.** $(-\infty,-5)\cup(1,\infty)$ **55.** $(-\infty,-4]\cup(0,\infty)$

57. $(-\infty,-6)$ **59.** $(-\infty,-9)\cup(-3,\infty)$ **61.** $\left(2,\frac{5}{2}\right)$ **63.** $(7,\infty)$

65. $\left[\frac{7}{5},4\right)$ **67.** $\left(-9,-\frac{4}{3}\right)$ **69.** $(-\infty,-4]\cup[0,3)$ **71.** $(-2,0)\cup[5,\infty)$

73. $(-\infty,-3.1623)\cup(3.1623,\infty)$ **75.** $\varnothing$ **77.** $(-\infty,-3)\cup(0,3)$ **79.** $[-1.8868,\infty)$

81. $(-2.3344,-0.7420)\cup(0.7420,2.3344)$ **83. a.** $(-4,0)\cup(1,\infty)$ **b.** $(-\infty,-4)\cup(0,1)$ **c.** The function is undefined at $x=0$.

Chapter 7 Review, pages 627 - 630

1. $x=-9$ or $x=4$ **2.** $x=-2$ or $x=7$ **3.** $x=1$ or $x=11$ **4.** $x=-13$ or $x=0$ **5.** $x^2-6x+\underline{9}=\left(x-3\right)^2$

6. $x^2+10x+\underline{25}=\left(x+5\right)^2$ **7.** $y^2-\frac{1}{2}y+\underline{\frac{1}{16}}=\left(y-\frac{1}{4}\right)^2$ **8.** $x^2-x+\underline{\frac{1}{4}}=\left(x-\frac{1}{2}\right)^2$ **9.** $x=\pm 6i$ **10.** $x=\pm 10$ **11.** $x=\pm 2\sqrt{6}$

12. $x=-4\pm\sqrt{10}$ **13.** $x=3\pm\sqrt{5}$ **14.** $x=-1\pm 2\sqrt{2}$ **15.** $x=-1\pm i\sqrt{3}$ **16.** $x=\frac{1}{2},x=1$ **17.** $x=-1,x=\frac{1}{2}$

18. $x=\frac{3\pm i\sqrt{11}}{2}$ **19.** $x^2-3=0$ **20.** $x^2-2x-1=0$ **21.** $x^2+6=0$ **22.** $x^2+6x+14=0$ **23.** -23, two nonreal solutions

24. 25, two real solutions **25.** 9, two real solutions **26.** 0, one real solution **27.** $x=\frac{-1\pm\sqrt{11}}{5}$ **28.** $x=1\pm\sqrt{5}$

29. $x=\frac{2\pm i\sqrt{2}}{2}$ **30.** $x=\frac{\pm 13i}{2}$ **31.** $x=\frac{-1\pm i\sqrt{11}}{6}$ **32.** $x=-5$ **33.** $x=-\frac{1}{2},x=3$ **34.** $x=-1,x=7$ **35.** $x=-\frac{5}{3},x=1$

36. $x=4,x=7$ **37.** $x=-3\pm\sqrt{2}$ **38.** $x=\frac{-7\pm i\sqrt{23}}{4}$ **39.** $x=\frac{3\pm i\sqrt{3}}{4}$ **40.** $x=0,x=3\pm 2\sqrt{3}$ **41.** $x=0,x=\frac{5}{3}$

42. $x = \dfrac{\pm 2i}{3}$ **43.** $k > -9$ **44.** $k < -2$ or $k > 2$ **45.** 30 ft, 40 ft   **46.** 18   **47.** 8, 21   **48.** 120 seats   **49.** 9 m × 8 m   **50.** 1 m

51. 18 members, \$110   **52.** Martin: 15 mph, Milton: 20 mph   **53. a.** 40 s   **b.** 1.97 s, 38.03 s   **54. a.** No   **b.** 2.5 s

55. $x = \pm 2, x = \pm 3$ **56.** $x = \pm\sqrt{2}, x = \pm i\sqrt{6}$ **57.** $y = -\dfrac{1}{2}, y = 1$ **58.** $x = -\dfrac{1}{8}, x = 64$ **59.** $y = 1$ **60.** $x = -2$

61. $y = 7$ **62.** $x = \pm 3\sqrt{i}, \; x = \pm 3i\sqrt{i}$ **63.** $x = -8, x = -2$ **64.** $x = -10, x = -\dfrac{14}{3}$ **65.** $x = 0; (0, -3);$

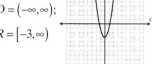

$D = (-\infty, \infty);$

$R = [-3, \infty)$

66. $x = 0; (0, 4);$ **67.** $x = -4; (-4, 0);$

$D = (-\infty, \infty);$ $D = (-\infty, \infty);$

$R = (-\infty, 4]$ $R = [0, \infty)$

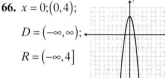
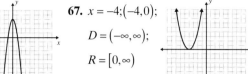

68. $x = 0; (0, 8);$ **69.** $x = -1; (-1, -4);$ **70.** $x = \dfrac{3}{2}; \left(\dfrac{3}{2}, -\dfrac{9}{2}\right);$

$D = (-\infty, \infty);$ $D = (-\infty, \infty);$ $D = (-\infty, \infty);$

$R = (-\infty, 8]$ $R = [-4, \infty)$ $R = \left(-\infty, -\dfrac{9}{2}\right]$

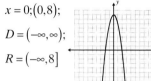

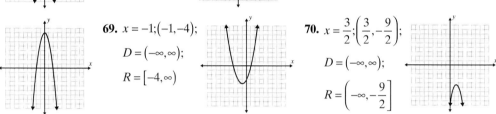

71. $y = 2(x - 3)^2; (3, 0);$ **72.** $y = (x - 1)^2 - 5; (1, -5);$ **73.** $y = (x - 2)^2 - 3; (2, -3);$ **74.** $y = -(x + 3)^2 + 7; (-3, 7);$

$R = [0, \infty);$ $R = [-5, \infty);$ $R = [-3, \infty);$ $R = (-\infty, 7];$

Zeros: $x = 3$ Zeros: $x = 1 \pm \sqrt{5}$ Zeros: $x = 2 \pm \sqrt{3}$ Zeros: $x = -3 \pm \sqrt{7}$

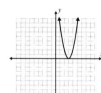

75. $y = 2\left(x + \dfrac{9}{4}\right)^2 - \dfrac{33}{8}; \left(-\dfrac{9}{4}, -\dfrac{33}{8}\right);$ **76.** $y = (x + 1)^2 + 3; (-1, 3);$

$R = \left[-\dfrac{33}{8}, \infty\right);$ $R = [3, \infty);$

Zeros: none

Zeros: $x = \dfrac{-9 \pm \sqrt{33}}{4}$

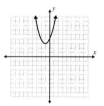

77. a. \$800   **b.** \$640,000   **78. a.** 300 copies   **b.** \$15,000   **79. a.** 6.5 s   **b.** 676 ft   **c.** 13 s   **80.** two sides 25 ft, third side 50 ft

81. $(-1, 5)$ **82.** $(-\infty, -3) \cup \left(\dfrac{1}{2}, \infty\right)$ **83.** $\left(-\infty, -\dfrac{3}{2}\right] \cup [1, \infty)$ **84.** $\{-3\}$

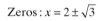
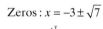

85. $\left(-\infty, -\dfrac{5}{3}\right) \cup (0, 4)$ **86.** $\left(-\infty, \dfrac{3}{2}\right) \cup \left(\dfrac{3}{2}, \infty\right)$ **87.** $\left(-\infty, \dfrac{1}{3}\right] \cup [1, \infty)$ **88.** $\left(-\infty, -1 - \sqrt{6}\right) \cup \left(-1 + \sqrt{6}, \infty\right)$

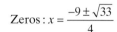
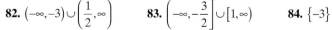

89. $\left(-\dfrac{1}{2},6\right]$

90. $(-\infty,1)\cup(4,\infty)$

91. $(-\infty,2)\cup(3,\infty)$

92. $(-\infty,-2)\cup\left[\dfrac{8}{3},\infty\right)$

93. $(-\infty,-2.8284)\cup(2.8284,\infty)$ **94.** $(-\infty,0)\cup(6,\infty)$ **95.** $[-1,6]$ **96.** $(-\infty,-1.1304)$

97. $(-\infty,-1.8478]\cup[-0.7654,0.7654]\cup[1.8478,\infty)$ **98.** $[-1.9422,0.5579]\cup[1.3844,\infty)$

Chapter 7 Test, pages 631 - 632

1. a. $x=\pm4$ **b.** $x=\pm4i$ **2.** $x=-\dfrac{1}{2},x=0$ **3. a.** $x^2-30x+\underline{225}=(x-15)^2$ **b.** $x^2+5x+\dfrac{25}{4}=\left(x+\dfrac{5}{2}\right)^2$ **4. a.** $x^2+8=0$

b. $x^2-2x-4=0$ **5.** $x=-4,x=2$ **6.** $x=-2\pm\sqrt{3}$ **7.** 73, two real solutions   **8.** $k=\pm2\sqrt{6}$ **9.** $x=\dfrac{-1\pm i\sqrt{7}}{4}$

10. $x=\dfrac{3\pm\sqrt{41}}{4}$ **11.** $x=\dfrac{2\pm i\sqrt{2}}{2}$ **12.** $x=\pm1,x=\pm3$ **13.** $x=-1,x=\dfrac{3}{2}$ **14.** $x=0,x=1$

15. $x=3;(3,-1)$
$D=(-\infty,\infty);$
$R=[-1,\infty)$

16. $x=\dfrac{3}{2};\left(\dfrac{3}{2},\dfrac{15}{2}\right)$
$D=(-\infty,\infty);$
$R=\left(-\infty,\dfrac{15}{2}\right]$

17. $x=3;(3,-9);$
$D=(-\infty,\infty);$
$R=[-9,\infty)$

18. $y=(x+2)^2-10;$
$(-2,-10);$
$R=[-10,\infty);$
zeros : $x=-2\pm\sqrt{10}$

19. $y=(x-3)^2-1;$
$(3,-1);$
$R=[-1,\infty);$
zeros : $x=2,4$

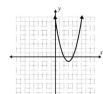
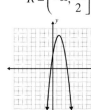
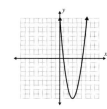
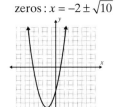
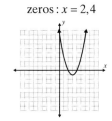

20. a. $(-\infty,-3)\cup(-2,\infty)$ **b.** $(-3,-2)$ **c.** $[-7,0]\cup[3,\infty)$

21. a. $\left(-\infty,-\dfrac{5}{2}\right]\cup(3,\infty)$ **b.** $\left(-\infty,-\dfrac{5}{3}\right)\cup\left(-\dfrac{1}{2},\infty\right)$

22. $(-5,0)\cup(2,\infty)$ **23.** 8 m, 15 m   **24. a.** $t=7$ seconds **b.** $t=6.46$ seconds

25. width $=12$ inches, length $=16$ inches   **26.** speed returning $=40$ mph, speed going $=50$ mph

27. $y=x(x-10);-5,5;$ minimum product $=-25$ **28.** $\dfrac{11}{2}$ in. $\times\dfrac{11}{2}$ in.

Chapter 7 Cumulative Review, pages 633 - 636

1. $\dfrac{1}{x^5}$ **2.** x^3y^6 **3.** $\dfrac{9\sqrt{y}}{x^2}$ **4.** $\dfrac{27x^2}{8y}$ **5.** $-3x^2y^2\sqrt[3]{y^2}$ **6.** $2x^2y^3\sqrt[4]{2xy^3}$ **7.** $\dfrac{9}{2}\sqrt{2}$ **8.** $\dfrac{4}{3}\sqrt{3}$ **9.** $(x-5)(y+2)$

10. $(8x+9)(8x-9)$ **11.** $2(x^2-6y)(x^4+6x^2y+36y^2)$ **12.** $5(\sqrt{3}-\sqrt{2})$ **13.** $(x+9)(\sqrt{x}-3)$ **14.** $4+2i$ **15.** $19+4i$

16. $\dfrac{-11+16i}{13}$ **17.** $\dfrac{x+3}{3(x-5)}$ **18.** $\dfrac{2x^2+3x+1}{2(2x-3)(x-1)}$ **19.** $x=6$ **20.** $x=\dfrac{7}{8}$ **21.** $x=-\dfrac{1}{2}, x=3$

22. $x=-\dfrac{3}{2}, x=\dfrac{2}{5}$ **23.** $x=\dfrac{-7\pm\sqrt{33}}{8}$ **24.** $x=-1$ **25.** $x=-2, x=\dfrac{4}{3}$ **26.** $x=-5,-3,3,5$

27. $x=-\dfrac{8}{27},-1$ **28.** $x=-\dfrac{38}{7}$ **29. a.** -2; **b.** 37   **30.** $4x^2+17x-15=0$ **31.** $x^2-2x-19=0$

32. $\left[-\dfrac{19}{2},\infty\right)$ 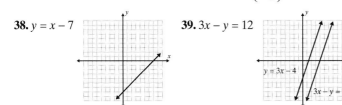   **33.** $\left(-\infty,\dfrac{7}{5}\right]$ **34.** $(-2,3)$

35. $\{4\}$ **36.** $(-\infty,-4)\cup\left(0,\dfrac{5}{3}\right)$ **37.** $(-2,6]$

38. $y=x-7$ **39.** $3x-y=12$

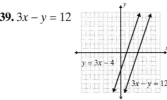

40. If any vertical line intersects the graph of a relation at more than one point, then the relation graphed is not a function. This works because it checks whether any first coordinate appears more than once.

41. $(-1,-2)$ 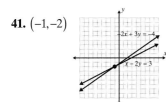   **42.** $x=5, y=4$ **43.** $x=1, y=-1, z=3$ **44.** 73   **45.**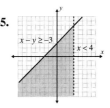

46. $x=-\dfrac{1}{2}, \left(-\dfrac{1}{2},-\dfrac{25}{4}\right); D=(-\infty,\infty);$ **47.** $x=\dfrac{1}{2}; \left(\dfrac{1}{2},\dfrac{49}{4}\right); D=(-\infty,\infty);$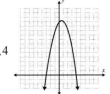

$R=\left[-\dfrac{25}{4},\infty\right);$ Zeros: $x=-3,2$ $R=\left(-\infty,\dfrac{49}{4}\right];$ Zeros: $x=-3,4$

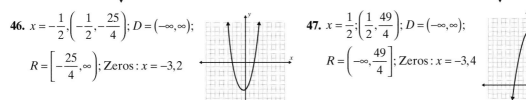

48. $x=\dfrac{9}{4}, \left(\dfrac{9}{4},-\dfrac{121}{8}\right); D=(-\infty,\infty);$ **49.** $y=-\dfrac{3}{2}x+3$ **50.** $y=-\dfrac{3}{2}x+10$ **51.** 100

$R=\left[-\dfrac{121}{8},\infty\right);$ Zeros: $x=-\dfrac{1}{2},5$ **52.** 15 mg of Drug A, 5 mg of Drug B   **53.** $35   **54.** 20 m × 26 m

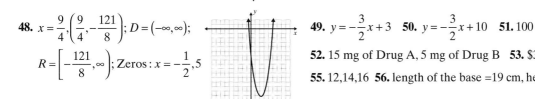   **55.** 12,14,16   **56.** length of the base =19 cm, height = 8 cm

57. a. 4.5 s   **b.** 324 ft   **c.** 9 s   **58. a.** 10,000 dozen   **b.** $995,000   **c.** 120,000 golf balls   **59. a.** 5.9136   **b.** 14.8306   **c.** −5.8284

60. x-int: $(-1.58,0)$ and $(1.58,0)$ 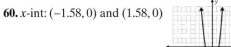   **61.** x-int: $(-0.62,0), (1,0),$ and $(1.62,0)$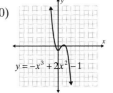

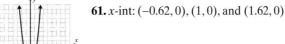

62. *x*-int: none   **63.** *x*-int: $(-0.60, 0)$ **64.** $D = [3, \infty)$; $R = [0, \infty)$ **65.** $D = (-\infty, 1]$; $R = [0, \infty)$ **66.** $D = [-1, \infty)$; $R = (-\infty, 0]$

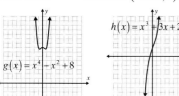 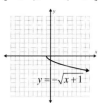

Chapter 8

Exercises 8.1, pages 646 - 651

1. a. 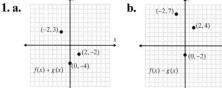   **3. a.** 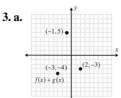   **5. a.**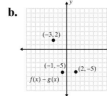

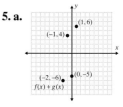

b. 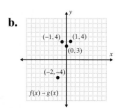   **7. a.** 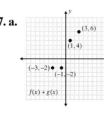   **b.** 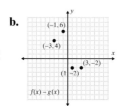   **9. a.** 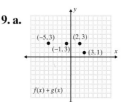   **b.**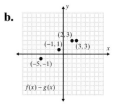

11. a. $2x - 3$ **b.** 7 **c.** $x^2 - 3x - 10$ **d.** $\dfrac{x+2}{x-5}$, $x \neq 5$ **13. a.** $x^2 + 3x - 4$ **b.** $x^2 - 3x + 4$ **c.** $3x^3 - 4x^2$ **d.** $\dfrac{x^2}{3x-4}$, $x \neq \dfrac{4}{3}$

15. a. $x^2 + x - 12$ **b.** $x^2 - x - 6$ **c.** $x^3 - 3x^2 - 9x + 27$ **d.** $x + 3, x \neq 3$ **17. a.** $3x^2 + x + 2$ **b.** $x^2 + x - 2$ **c.** $2x^4 + x^3 + 4x^2 + 2x$

d. $\dfrac{2x^2 + x}{x^2 + 2}$ **19. a.** $2x^2 + 2$ **b.** $8x$ **c.** $x^4 - 14x^2 + 1$ **d.** $\dfrac{x^2 + 4x + 1}{x^2 - 4x + 1}, x \neq 2 \pm \sqrt{3}$ **21.** 9 **23.** $-a^2 - a - 1$ **25.** 27 **27.** $\dfrac{8}{5}$ **29.** -31

31. $\sqrt{2x - 6} + x + 4$, Domain: $[3, \infty)$ **33.** $3x^2 - 19x - 14$, Domain: $(-\infty, \infty)$ **35.** $\dfrac{x - 5}{\sqrt{x + 3}}$, Domain: $(-3, \infty)$

37. $-3x\sqrt{x - 3}$, Domain: $[3, \infty)$ **39.** $\sqrt[3]{x + 3} + \sqrt{5 + x}$, Domain: $[-5, \infty)$ **41.** 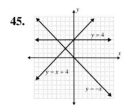   **43.**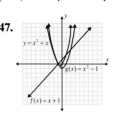

45. 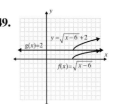   **47.** 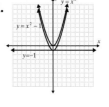   **49.**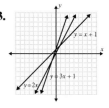

51. 4 **53.** -12 **55.** $-\dfrac{3}{7}$

57. 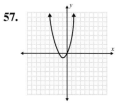   **59.** 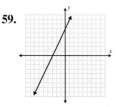   **61.** 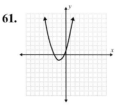   **63.** 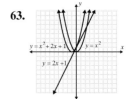   **65.**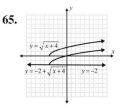

67.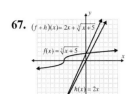

69.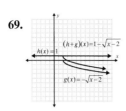

71. In general, subtraction is not commutative. Answers may vary.

73. a. 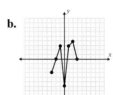   **b.**   **c.** $\dfrac{f}{g}$ is undefined at $x = -2$ as $g(-2) = 0$.

Exercises 8.2, pages 664 - 668

1. $f\big(g(2)\big) = 14; g\big(f(2)\big) = \dfrac{15}{2}$ **3.** $(f \circ g)(-5) = 49; (g \circ f)(-1) = 5$ **5.** $f\big(g(x)\big) = \dfrac{1}{5x - 8}; g\big(f(x)\big) = \dfrac{5}{x} - 8$

7. $f\big(g(x)\big) = \dfrac{1}{x^2} - 1; g\big(f(x)\big) = \dfrac{1}{(x-1)^2}$ **9.** $f\big(g(x)\big) = x^3 + 3x^2 + 4x + 3; g\big(f(x)\big) = x^3 + x + 2$

11. $f\big(g(x)\big) = \sqrt{x-2}; g\big(f(x)\big) = \sqrt{x} - 2$ **13.** $f\big(g(x)\big) = \sqrt{x^2} = |x|; g\big(f(x)\big) = \left(\sqrt{x}\right)^2 = x$

15. $f\big(g(x)\big) = \dfrac{1}{\sqrt{x^2 - 4}}; g\big(f(x)\big) = \dfrac{1}{x} - 4$ **17.** $f\big(g(x)\big) = x; g\big(f(x)\big) = x$ **19.** $f\big(g(x)\big) = (x-8)^{\frac{3}{2}}; g\big(f(x)\big) = \sqrt{x^3 - 8}$

21. a. 21 **b.** 2 **c.** No   **23. a.** 2 **b.** −5 **c.** No

25. 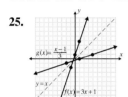   **27.** 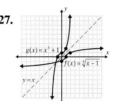   **29.** 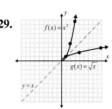   **31.** 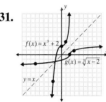   **33.**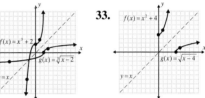

35. $f^{-1}(x) = \dfrac{x+3}{2}$ **37.** $g^{-1}(x) = x$ **39.** $f^{-1}(x) = \dfrac{x-1}{5}$ **41.** $g^{-1}(x) = \dfrac{1-x}{3}$ **43.** $f^{-1}(x) = x^2, x \le 0$

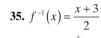   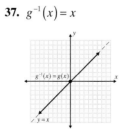      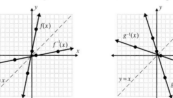   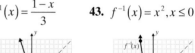   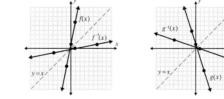   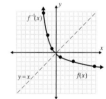

45. $f^{-1}(x) = \sqrt{x-1}$ **47.** $f^{-1}(x) = -x - 2$ **49.** 1–1   **51.** Not 1–1 **53.** Not 1–1

55. 1–1   **57.** 1–1

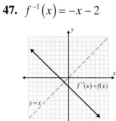   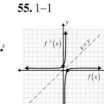   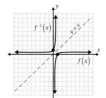   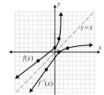

59. 1–1 **61.** Not 1–1 **63.** 1–1 **65.** 1–1 **67.** Not 1–1

69. $f^{-1}(x) = x^{\frac{1}{3}}$

71. $f^{-1}(x) = \dfrac{1}{x} + 3$

73. $g^{-1}(x) = (x-2)^{\frac{1}{3}}$

75. $f^{-1}(x) = \sqrt{x-2}$

77. $f^{-1}(x) = x^2 - 5, x \geq 0$

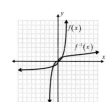

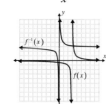

79. $f^{-1}(x) = \sqrt{1-x}$

81. The domain of $f(g(x))$ can only include values of x in the domain of g, while $g(f(x))$ can only include values of x in the domain of f. Ex: $f(x) = \sqrt{x}, g(x) = -x$ Answers may vary.

Exercises 8.3, pages 679 - 682

1. **3.** **5.** **7.** **9.**

11. **13.** **15.** **17.** **19.**

21. 48   **23.** 30   **25.** 6008.33; Answers may vary.   **27.** 20,000 bacteria   **29. a.** $2621.59 **b.** $2633.62 **c.** $2639.86 **d.** $2646.19

e. $2646.26 **31.** $3210.06 **33.** $53.33 **35.** $\dfrac{1}{64}$ **37.** $17,182.82 **39.** $1228.25 **41.** $y = 10{,}000 \cdot 2^t$

43. a. The graph is stretched vertically by a factor of a. **b.** The graph is tranlated horizontally h units to the right.
c. The graph is translated vertically k units up. **45.** The two functions are symmetric about the x-axis.

Exercises 8.4, pages 688 - 689

1. $\log_7 49 = 2$ **3.** $\log_5 \dfrac{1}{25} = -2$ **5.** $\log_2 \dfrac{1}{32} = -5$ **7.** $\log_{2/3} \dfrac{4}{9} = 2$ **9.** $\log_2 17 = x$ **11.** $\log_{10} 10 = 1$ **13.** $3^2 = 9$ **15.** $9^{\frac{1}{2}} = 3$

17. $7^{-1} = \dfrac{1}{7}$ **19.** $10^{1.74} = N$ **21.** $b^4 = 18$ **23.** $n^x = y^2$ **25.** $x = 16$ **27.** $x = 2$ **29.** $x = \dfrac{1}{6}$ **31.** $x = 11$ **33.** $x = 32$ **35.** $x = -2$

37. $x = 1.52$ **39.** $x = 3$ **41.** 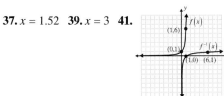   **43.** 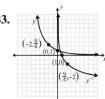   **45.** 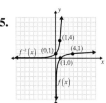   **47.**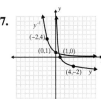

49.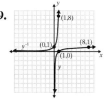

51. a. $D = (-\infty, \infty)$ **b.** $R = (0, \infty)$ **c.** $y = 0$ **d.** Answers may vary.

53. The two functions are symmetric about the line $y = x$.

Exercises 8.5, pages 698 - 699

1. a. 2   **b.** 3   **c.** 5   **d.** 6   **3.** 5   **5.** -2 **7.** $\dfrac{1}{2}$ **9.** 10   **11.** $\sqrt{3}$ **13.** $\log_b 5 + 4\log_b x$ **15.** $\log_b 2 - 3\log_b x + \log_b y$

17. $\log_6 2 + \log_6 x - 3\log_6 y$ **19.** $2\log_b x - \log_b y - \log_b z$ **21.** $-2\log_5 x - 2\log_5 y$ **23.** $\dfrac{1}{3}\log_6 x + \dfrac{2}{3}\log_6 y$

25. $\dfrac{1}{2}\log_3 x + \dfrac{1}{2}\log_3 y - \dfrac{1}{2}\log_3 z$ **27.** $\log_5 21 + 2\log_5 x + \dfrac{2}{3}\log_5 y$ **29.** $-\dfrac{1}{2}\log_6 x - \dfrac{5}{2}\log_6 y$

31. $-9\log_b x - 6\log_b y + 3\log_b z$ **33.** $\log_b\left(\dfrac{9x}{5}\right)$ **35.** $\log_2\left(\dfrac{7x^2}{9}\right)$ **37.** $\log_b\left(x^2 y\right)$ **39.** $\log_{10}\left(\dfrac{x^3}{y^2}\right)$ **41.** $\log_5\sqrt{\dfrac{x}{y}}$

43. $\log_2\left(\dfrac{xz}{y}\right)$ **45.** $\log_b\left(\dfrac{x}{y^2 z^2}\right)$ **47.** $\log_5\left(2x^2 + x\right)$ **49.** $\log_2\left(x^2 + 2x - 3\right)$ **51.** $\log_b\left(x + 1\right)$ **53.** $\log_{10}\left(\dfrac{1}{2x - 3}\right)$

55. Answers may vary.

Exercises 8.6, pages 707 - 708

1. $\log\dfrac{1}{1000} = -3$ **3.** $\log x = 1.5$ **5.** $\ln 27 = x$ **7.** $\ln 1 = 0$ **9.** $\log 3.2 = x$ **11.** $10^0 = 1$ **13.** $e^y = 5.4$ **15.** $e^{1.54} = x$

17. $10^x = 25.3$ **19.** $10^1 = 10$ **21.** 2.23805   **23.** 1.94645   **25.** -1.24185 **27.** 3.62434   **29.** Error (undefined)   **31.** -5.37953

33. 204.17379   **35.** 0.01995   **37.** 0.95719   **39.** 175.91484   **41.** 0.00025   **43.** 1.04029

45. $\log x$ is a base-10 logarithm. $\ln x$ is a base-e logarithm.   **47.** $D : \{x \mid x \neq 0\}$

Exercises 8.7, pages 715 - 717

1. $x = 11$ **3.** $x = 9$ **5.** $x = \dfrac{1}{6}$ **7.** $x = \dfrac{7}{12}$ **9.** $x = \dfrac{7}{2}$ **11.** $x = -2$ **13.** $x = -3$ **15.** $x = 4, x = -1$ **17.** $x = -1, x = \dfrac{3}{2}$

19. $x = -2, x = 3$ **21.** $x = 2$ **23.** $x = 0, x = -\dfrac{3}{2}$ **25.** $x = 3, x = -1$ **27.** $x \approx 0.7154$ **29.** $x \approx 7.5098$ **31.** $x \approx -3.6476$

33. $x \approx 24.7312$ **35.** $t \approx -653.0008$ **37.** $t \approx -1.5193$ **39.** $x \approx 3.3219$ **41.** $x \approx -1.4307$ **43.** $x = 1$ **45.** $x \approx 1.2203$
47. $x \approx -1.6467$ **49.** $x \approx 1.1292$ **51.** $x \approx 4.8540$ **53.** $x \approx 1.2520$ **55.** $x \approx 25.1189$ **57.** $x \approx 31.6228$ **59.** $x = 0.0001$
61. $x \approx 4.9530$ **63.** $x \approx \pm 0.3329$ **65.** $x = 3$ **67.** $x = 100$ **69.** $x = 6$ **71.** $x = 20$ **73.** $x = \dfrac{25}{2}$ **75.** $x = 1001$ **77.** No solution
79. $x \approx 1.1353$ **81.** $x \approx 22.0855$ **83.** $x = -10, x = 8$ **85.** 2.2619 **87.** 0.3223 **89.** -0.6279 **91.** 2.4391 **93.** -1.2222
95. $x = 2.3219$ **97.** $x = 3.1133$ **99.** $x = 0.8390$ **101.** Answers may vary. $x = \dfrac{1}{2}$ **103.** $7 \cdot 7^x = 7^{1+x}$ and $49^x = 7^{2x}$. Since
$1 + x \neq 2x$, in general, $7 \cdot 7^x \neq 49$. Answers may vary.

Exercises 8.8, pages 721 - 724

1. $4027.51   **3.** 11.55 years   **5.** $f \approx 0.3012$ **7.** 1.73 hours   **9.** 10.99 days   **11.** 7.44 hours   **13.** 2350.02 years   **15.** 2.31 days
17. 39.65 minutes   **19.** 7.00 years   **21.** 5600 years   **23. a.** 13.86 years **b.** 6.93 years   **25.** 100   **27.** 8.64 million
29. $2083, 2437, 2744$ **31. a.** 117.95 dB **b.** $3.16 \times 10^8 I_0$ **c.** $10^6 I_0$

Chapter 8 Review, pages 730 - 736

1. 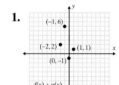   **2.** 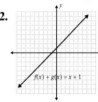   **3. a.** $3x - 4$ **b.** $x + 10$ **c.** $2x^2 - 11x - 21$ **d.** $\dfrac{2x + 3}{x - 7}, x \neq 7$

4. a. $-4x + 6$ **b.** $-6x - 6$ **c.** $-5x^2 - 30x$ **d.** $\dfrac{-5x}{x + 6}, x \neq -6$

5. a. $11x + 5$ **b.** $5x - 5$ **c.** $24x^2 + 40x$ **d.** $\dfrac{8x}{3x + 5}, x \neq -\dfrac{5}{3}$

6. a. $-3x - 8$ **b.** $5x - 10$ **c.** $-4x^2 + 37x - 9$ **d.** $\dfrac{x - 9}{-4x + 1}, x \neq \dfrac{1}{4}$ **7. a.** $2x^2 - 16$ **b.** 16 **c.** $x^4 - 16x^2$ **d.** $\dfrac{x^2}{x^2 - 16}, x \neq -4, 4$

8. a. $3x^2 - x + 5$ **b.** $x^2 - x - 5$ **c.** $2x^4 - x^3 + 10x^2 - 5x$ **d.** $\dfrac{2x^2 - x}{x^2 + 5}$ **9.** 4 **10.** 4 **11.** 0 **12.** -5

13. $\sqrt{x - 6} + x + 2$, Domain: $[6, \infty)$ **14.** 3, Domain: $(-\infty, \infty)$ **15.** $\sqrt{x - 3}$, Domain: $(3, \infty)$

16. $2\sqrt{x - 1} + 3x\sqrt{x - 1}$, Domain: $[1, \infty)$ **17.**   **18.**   **19.** $f\big(g(x)\big) = 4x - 5$
$g\big(f(x)\big) = 4x - 1$

20. $f\big(g(x)\big) = 9x^2 - 6x + 1$;
$g\big(f(x)\big) = 3x^2 - 1$

21. $f\big(g(x)\big) = \sqrt{x^2 - 4}$;   **22.** $f\big(g(x)\big) = \dfrac{1}{\sqrt{x^2 - 9}}$;   **23.** $f\big(g(x)\big) = \dfrac{1}{x^3}$;   **24.** $f\big(g(x)\big) = x^2$;
$g\big(f(x)\big) = x - 4$ $g\big(f(x)\big) = \dfrac{1}{x} - 9$ $g\big(f(x)\big) = \dfrac{1}{x^3}$ $g\big(f(x)\big) = x^2$

25. $f^{-1}(x) = \dfrac{x - 1}{2}$ **26.** $f^{-1}(x) = \dfrac{5 - x}{2}$ **27.** $f^{-1}(x) = x^2 + 6$ **28.** $f^{-1}(x) = \sqrt{2 - x}$

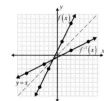   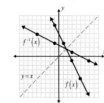   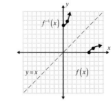   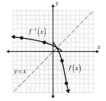

29. 1 – 1

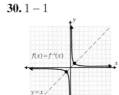

30. 1 – 1

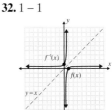

31. Not 1 – 1

32. 1 – 1

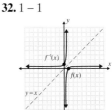

33. $f^{-1}(x) = (1+x)^{\frac{1}{3}}$

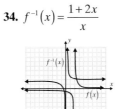

34. $f^{-1}(x) = \dfrac{1+2x}{x}$

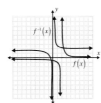

35. $f^{-1}(x) = x^2 + 5, x \geq 0$

36. $f^{-1}(x) = -\sqrt{x}$

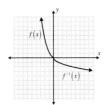

37.

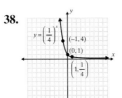

38.

39.

40.

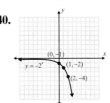

41.

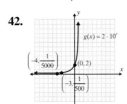

42.

43.

44.

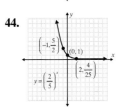

45. 259.2 **46. a.** 2.71692 **b.** 2.71815 **c.** 2.71828 **47. a.** \$16,035.68 **b.** \$16,453.31 **c.** \$16,598.92 **d.** \$16,600.58

48. \$7,240,773.44 **49.** \$17,748.21 **50.** \$14,918.25, \$22,255.41 **51.** $\log_8 64 = 2$ **52.** $\log_3 81 = 4$ **53.** $\log_2 \dfrac{1}{8} = -3$

54. $\log_3 \dfrac{1}{9} = -2$ **55.** $\log_{\frac{1}{2}} \dfrac{1}{4} = 2$ **56.** $\log_{\frac{1}{5}} \dfrac{1}{125} = 3$ **57.** $2^{-1} = \dfrac{1}{2}$ **58.** $10^{\frac{1}{2}} = \sqrt{10}$ **59.** $5^4 = 625$ **60.** $5^{-2.5} = x$

61. $b^y = x$ **62.** $\left(\dfrac{1}{2}\right)^{-4} = 16$ **63.** $x = 2$ **64.** $x = \dfrac{1}{2}$ **65.** $x = 2.35$ **66.** $x = -5$ **67.** $x = 6$ **68.** $x = 3$

69.

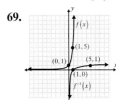

70.

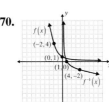

71.

72.

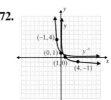

73. $\log_b 5 + 4\log_b x$ **74.** $-2(\log_b x + \log_b y)$ **75.** $\dfrac{1}{2}(\log_2 7 + \log_2 x)$ **76.** $\log_3 x + 2\log_3 y$ **77.** $2\log_{10} x + 3\log_{10} y$

78. $\log_b x - 4\log_b y$ **79.** $5(2\log_b x - 3\log_b y)$ **80.** $\dfrac{1}{2}(\log_2 5 - \log_2 x - \log_2 y)$ **81.** $\log_3 42 + 2\log_3 x - 3\log_3 y$

82. $\frac{1}{3}\left(2\log_5 x - 3\log_5 y\right)$ **83.** $\log_2\left(9x\right)$ **84.** $\log_3\left(\frac{x^5}{y}\right)$ **85.** $\log_b\left(\frac{x^3}{y^4}\right)$ **86.** $\log_4\left(4x^3 + x^2\right)$ **87.** $\log_{10}\left(2x^2 - 9x - 5\right)$

88. $\log_{10}\left(x^2 - 4\right)$ **89.** $\log_b\left(\frac{x^2 + 2x - 3}{x + 5}\right)$ **90.** $\log_2\left(\frac{\sqrt{x}}{2x + 1}\right)$ **91.** $\log 1 = 0$ **92.** $\log 0.00001 = -5$ **93.** $\ln x = 1.2$

94. $\ln 10 = k$ **95.** $\ln 12 = x$ **96.** $\log 8.4 = k$ **97.** $e^x = 3$ **98.** $x = e^5$ **99.** $x = e^{-1}$ **100.** $16 = 10^x$ **101.** $x = 10^{2.5}$

102. $100,000 = 10^5$ **103.** 2.43933   **104.** 1.92428   **105.** Error (undefined)   **106.** -4.42285 **107.** 1318.25674   **108.** 0.15849

109. 0.01111   **110.** 3.50579×10^{95} **111.** $x = \pm e$ **112.** $x = \frac{5}{2}$ **113.** $x = -\frac{3}{2}, 2$ **114.** $x = 1000$ **115.** $x = e$ **116.** $x \approx -1.41648$

117. $x \approx \pm 0.08208$ **118.** $x \approx 1.09590$ **119.** $x = \pm 5\sqrt{401} \approx \pm 100.12$ **120.** $x = \frac{1}{2}$ **121.** No solution   **122.** $x = 1$

123. $x = 6$ **124.** No solution   **125.** 3.58496   **126.** 3.32193   **127.** $x \approx 3.32193$ **128.** $x \approx 2.32193$ **129. a.** \$13,507.42

b. \$13,563.20   **c.** \$13,591.41   **130.** 11.55 years; 11.55 years   **131.** 29 years   **132.** \$1024   **133.** 13.82 days   **134. a.** 1.99 mg

b. 1.16 hours   **135.** 10.92 minutes   **136.** 5.01×10^{-7} **137. a.** 96.35 dB   **b.** $10^{14} I_0$ **138.** 3.98

Chapter 8 Test, pages 737 - 739

1. a. 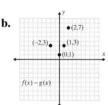   **b.**   **2. a.** 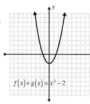   **b.**

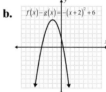

3. a. $\sqrt{x-3} + x^2 + 1$ **b.** $\sqrt{x-3} - x^2 - 1$ **c.** $\left(\sqrt{x-3}\right) \cdot \left(x^2 + 1\right)$ **d.** $\frac{\sqrt{x-3}}{x^2 + 1}$ **4. a.** $-4x^2 + 1$ **b.** $-8x^2 + 40x - 47$ **5.** Not $1 - 1$

6. a. Not inverse to each other   **b.** Inverse to each other   **7.** $f^{-1}(x) = \frac{1}{x+3} - 1$ **8. a.**

   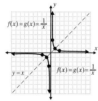   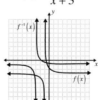   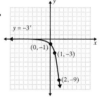   **b.**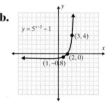

9. a. $\log 100,000 = 5$ **b.** $\log_{\frac{1}{2}} 8 = -3$ **c.** $x = \ln 15$ **10. a.** $10^4 = x$ **b.** $\frac{1}{9} = 3^{-2}$ **c.** $4^x = 256$ **11. a.** $x = 343$ **b.** $x = \frac{3}{2}$

12. $y^{-1} = \log_{\frac{1}{2}} x$

13. a. $\ln(x+5) + \ln(x-5)$ **b.** $\frac{2}{3}\log x - \frac{1}{3}\log y$ **14. a.** $\ln\left(x^2 + x - 20\right)$ **b.** $\log\left(\frac{x\sqrt{x}}{5}\right)$

15. a. 2.76268   **b.** 6.48830   **16.** $x = -4$ **17.** $x = -3$ **18.** $x \approx 0.4518$ **19.** $x \approx 10.6873$

20. $x \approx 1.7925$ **21.** No solution   **22.** $x \approx 21.0855$ **23.** 1.7965   **24.** 25,980.76 bacteria

25. 19.07 years   **26. a.** 134.29 years   **b.** 198.04 years   **27.** 2.00   **28.** 110.12 minutes

Chapter 8 Cumulative Review, pages 740 - 744

1. 32 **2.** 7 **3.** $y = -\frac{2}{3}x + 2$

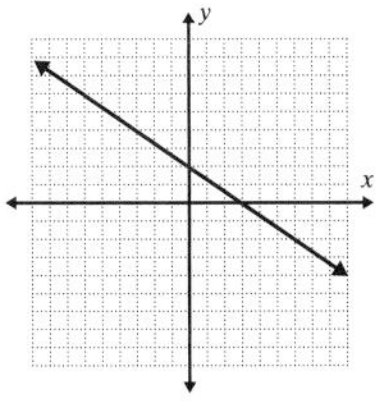

4. $\{(-2, 13), (-1, 8), (0, 5), (1, 4), (2, 5)\}$
5. $\{(-2, -28), (-1, -6), (0, 0), (1, -4), (2, -12)\}$
6. a. Function; $D = (-\infty, -1] \cup [1, \infty); R = (-\infty, -2] \cup [2, \infty)$
b. Not a function; $D = (-\infty, \infty); R = (-\infty, \infty)$
c. Function $D = (-\infty, \infty); R = [0, \infty)$

7. a. Parallel **b.** No solution **8. a.** Perpendicular **b.** $x = 2, y = 1$ **9.** $x = 2, y = 5$ **10.** $x = 1, y = -1, z = 2$

c.

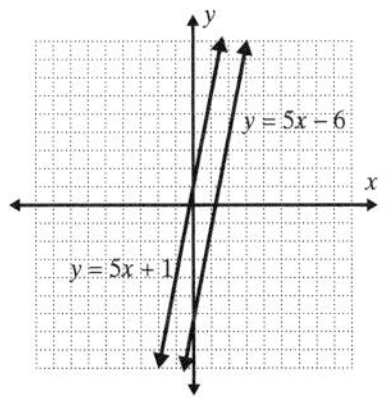

c.

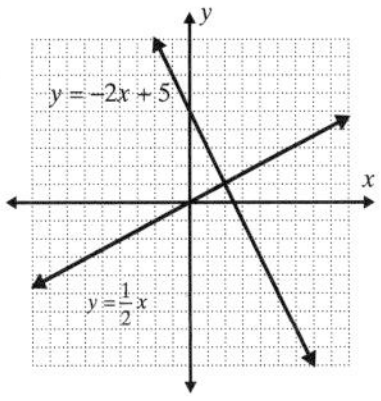

11. 1 **12.** $-\frac{x+3}{x+1}$ **13.** $\frac{2}{x+4}$ **14.** $\frac{x-4}{x+1}$ **15.** $4xy\sqrt{5y}$

16. $\frac{2x\sqrt{10y}}{5y}$ **17.** $\frac{x^{\frac{1}{6}}}{y^{\frac{1}{3}}}$ **18.** $\frac{x^{\frac{1}{2}}}{2y^2}$ **19.** $x^2y^{\frac{3}{2}}$ **20.** $x^{\frac{2}{3}}y$

21. $15 + 6i$ **22.** 8 **23.** i **24.** $3\log_b x + \frac{1}{2}\log_b y$ **25.** $\log_b\left(\frac{7x^2}{y}\right)$ **26.** $x = \frac{-5 \pm \sqrt{33}}{2}$ **27.** $x = \frac{7}{2}$ **28.** $x = \pm 2, x = \pm 3$

29. $x = 6$ **30.** $x = 4 \pm \sqrt{6}$ **31.** $x = -0.9212$ **32.** $x \approx 17.1825$ **33.** $x = -3$ **34.** $(-1, 3)$

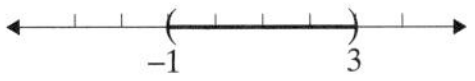

35. $(-\infty, -2) \cup \left(\frac{5}{2}, \infty\right)$

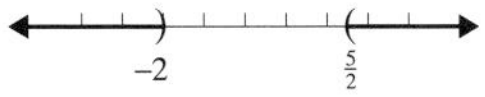

36. $(-\infty, -1] \cup \left[\frac{3}{2}, \infty\right)$

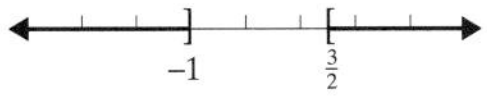

37. $(-9, -3)$

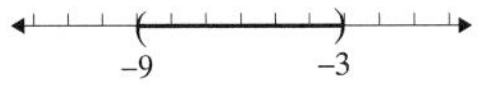

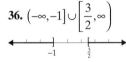

38. $y = (x-3)^2 - 11$
Vertex: $(3, -11)$
Range: $[-11, \infty)$
Zeros: $x = 3 \pm \sqrt{11}$

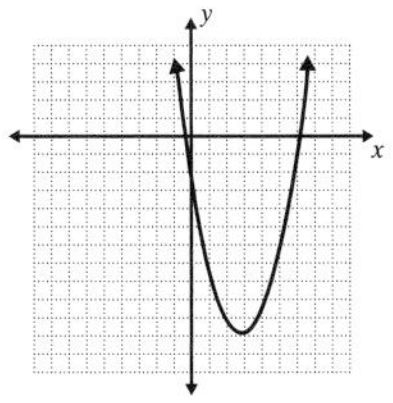

39. $y = 2(x+2)^2 - 5$
Vertex: $(-2, -5)$
Range: $[-5, \infty)$
Zeros: $x = -2 \pm \frac{\sqrt{10}}{2}$

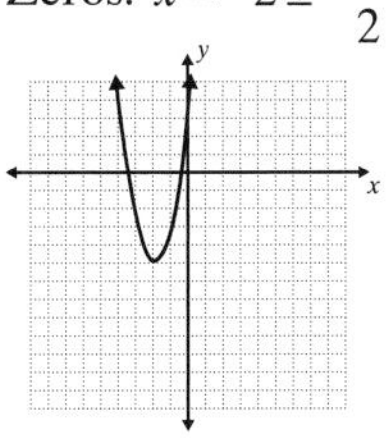

40. $y = (x-4)^2$
Vertex: $(4, 0)$
Range: $[0, \infty)$
Zeros: $x = 4$

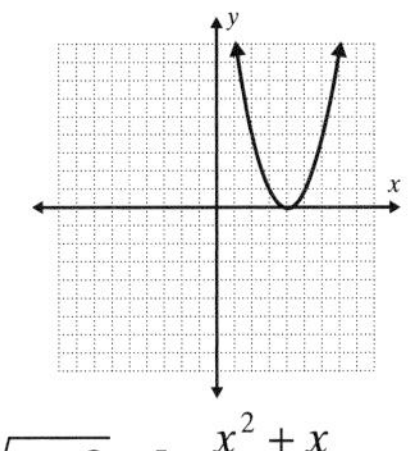

41. $y = (x-1)^2 - 7$
Vertex: $(1, -7)$
Range: $[-7, \infty)$
Zeros: $x = 1 \pm \sqrt{7}$

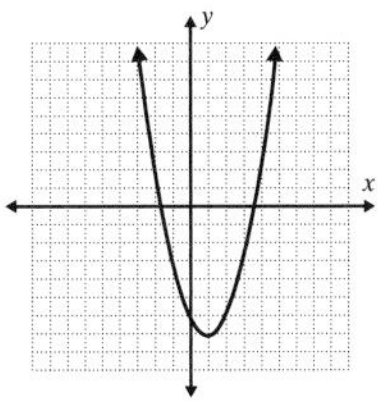

42. a. $x^2 + x + \sqrt{x+2}$ **b.** $x^2 + x - \sqrt{x+2}$ **c.** $(x^2 + x)\sqrt{x+2}$ **d.** $\frac{x^2 + x}{\sqrt{x+2}}, x > -2$ **43. a.** $4x^2 + 4x + 4$ **b.** $2x^2 + 7$

44. Is not a 1–1 function **45.** Is a 1–1 function **46.** $f^{-1}(x) = \sqrt{x+4}$ **47.** $g^{-1}(x) = \sqrt[3]{x} + 2$

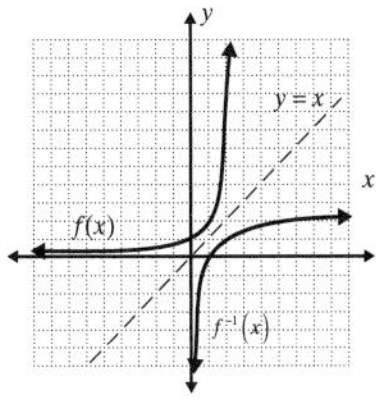

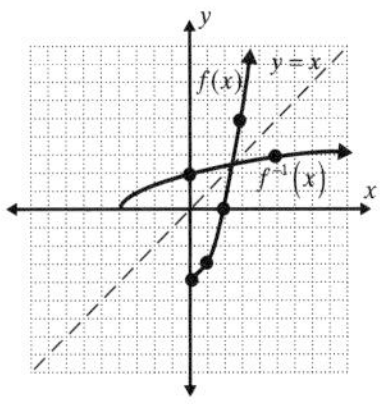

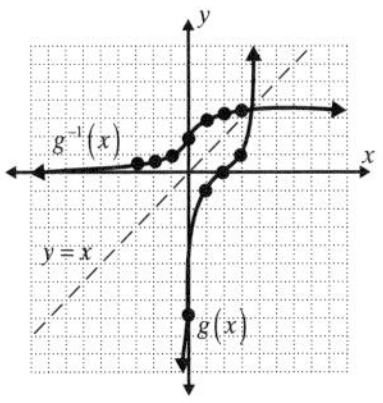

48. a. and **b.** 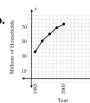 **c.** 3.12; 1.72; 1.62; 1.06 **d.** The number of users inc. 3.12 million from 1985-1990, 1.72 million from 1990-1995, 1.62 million from 1995-2000, and 1.06 million from 2000-2005. **49.** 13 hardback, 30 paperback **50.** 10 mph, 4 mph **51.** $\frac{6}{7}$ hours **52.** 12 inches, 16 inches **53.** $\frac{7}{8}$ **54.** $609.20 **55.** 17.33 centuries **56.** 4.6 s

57. a. $600 **b.** $360,000 **58. a.** 4.5 s **b.** 324 ft **c.** 9 s **59.** 1.73719 **60.** 9.21034

61. $\left(-\infty,-\sqrt{2}\right)\cup\left(\sqrt{2},\infty\right)$ **62.** $(-\infty,-4)\cup(0,\infty)$ **63.** $[-6,1]$

Chapter 9

Exercises 9.1, pages 755 - 760

1. a. 0 **b.** a^2-6a+5 **c.** $x^2+2xh+h^2-4$ **d.** $2x+h$ **3. a.** -3 **b.** $2a^2-8a+5$ **c.** $2x^2+4xh+2h^2-3$ **d.** $4x+2h$
5. 1 **7.** -2 **9.** The results equal the slopes of the lines.

11. **13.** 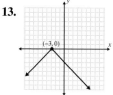 **15.** **17.** **19.**

21. 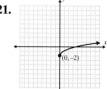 **23.** **25.** **27.** 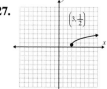 **29.**

31. 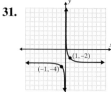 **33.** 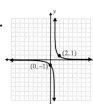 **35.** **37.** **39.**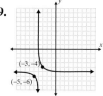

41. **43.** **45.** **47.** 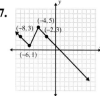 **49.**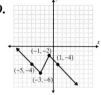

51. c **53.** b

55. 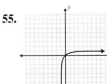 $D = (-1, \infty)$; **57.** 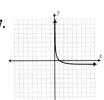 $D = (0, \infty)$; **59.** $D = (-\infty, 2)$;

$R = (-\infty, \infty)$ $R = (-\infty, \infty)$ $R = (-\infty, \infty)$

61.   **63.** 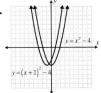   **65.**   **67.** The graph of the function $y = f(x - h) + k$ is the graph of the function $y = f(x)$ shifted h units to the right and k units up.

Exercises 9.2, pages 767 - 769

1. a. $(4, 0)$ **3. a.** $(0, -3)$ **5. a.** $(3, 0)$ **7. a.** $(0, 3)$ **9. a.** $(4, -2)$

  b. None   **b.** $(0, -3)$ **b.** None   **b.** $(0, 3)$ **b.** None

  c. $y = 0$ **c.** $x = 0$ **c.** $y = 0$ **c.** $y = 3$ **c.** $y = -2$

    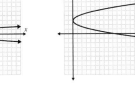

11. a. $(1, -1)$ **13. a.** $(0, -2)$ **15. a.** $(0, 4)$ **17. a.** $(-2, 9)$ **19. a.** $(-3, -4)$

  b. $(0, 0)$ **b.** $(0, -2)$ **b.** $(0, 4)$ **b.** $(0, 5)$ **b.** $(0, 5)$

  c. $x = 1$ **c.** $y = -2$ **c.** $y = 4$ **c.** $x = -2$ **c.** $x = -3$

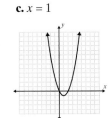 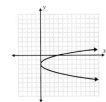 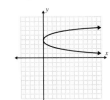 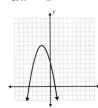 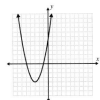

21. a. $(1, 2)$

  b. $(0, 1), (0, 3)$

  c. $y = 2$

23. a. $\left(-\dfrac{1}{4}, -\dfrac{9}{8} \right)$

  b. $(0, -1)$

  c. $x = -\dfrac{1}{4}$

25. a. $(-8, -1)$

  b. $\left(0, -1 + \dfrac{2\sqrt{6}}{3} \right)$, $\left(0, -1 - \dfrac{2\sqrt{6}}{3} \right)$

  c. $y = -1$

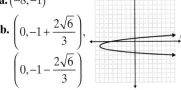

27. a. $\left(\dfrac{9}{8},\dfrac{5}{4}\right)$ **29. a.** $\left(\dfrac{3}{2},0\right)$ **31.** $(0,1.225),(0,-1.225)$ **33.** $(0,2),(0,0)$ **35.** No y-intercept

b. $\left(0,\dfrac{1}{2}\right),(0,2)$ **b.** $(0,9)$

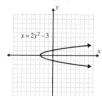

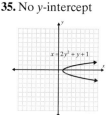

c. $y=\dfrac{5}{4}$ **c.** $x=\dfrac{3}{2}$

37. $(0,0.658),(0,-2.658)$ **39.** $(0,2.581),(0,-0.581)$ **41.** b   **43.** a   **45.** Values of a between -1 and 1 will cause the graph to be wider; values less than -1 or greater than 1 will cause the graph to be narrower.

Exercises 9.3, pages 778 - 781

1. 5   **3.** 13   **5.** $\sqrt{29}$ **7.** 3   **9.** $\sqrt{13}$ **11.** 17   **13.** $x^2+y^2=16$ **15.** $x^2+y^2=3$ **17.** $x^2+y^2=11$ **19.** $x^2+y^2=\dfrac{4}{9}$

21. $x^2+(y-2)^2=4$ **23.** $(x-4)^2+y^2=1$ **25.** $(x+2)^2+y^2=8$ **27.** $(x-3)^2+(y-1)^2=36$ **29.** $(x-3)^2+(y-5)^2=12$

31. $(x-7)^2+(y-4)^2=10$

33. $x^2+y^2=9$
   Center: $(0,0),r=3$

35. $x^2+y^2=49$
   Center: $(0,0),r=7$

37. $x^2+y^2=18$
   Center: $(0,0),r=3\sqrt{2}$

39. $(x+1)^2+y^2=9$
   Center: $(-1,0),r=3$

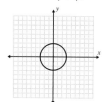

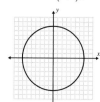

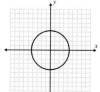

41. $x^2+(y-2)^2=4$
   Center: $(0,2),r=2$

43. $(x+1)^2+(y+2)^2=16$
   Center: $(-1,-2),r=4$

45. $(x+2)^2+(y+2)^2=16$
   Center: $(-2,-2),r=4$

47. $(x-2)^2+(y-3)^2=8$
   Center: $(2,3),r=2\sqrt{2}$

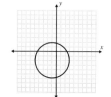

49. $AB=3\sqrt{5},AC=\sqrt{65},BC=2\sqrt{5}\,;\left(3\sqrt{5}\right)^2+\left(2\sqrt{5}\right)^2=\left(\sqrt{65}\right)^2$ Right Triangle   **51.** $AB=AC=4\sqrt{5}$

53. $AB = AC = BC = 4$ **55.** $AC = BD = \sqrt{61}$ **57.** $10 + 4\sqrt{5}$

59. 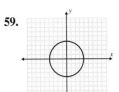   **61.**

63. a. $d = \sqrt{x^2 + (y - p)^2}$

b. $d = |y + p|$

c.
$$y + p = \sqrt{x^2 + (y - p)^2}$$
$$(y + p)^2 = x^2 + (y - p)^2$$
$$y^2 + 2py + p^2 = x^2 + y^2 - 2py + p^2$$
$$x^2 = 4py$$

65. a. $d = \sqrt{(x - p)^2 + y^2}$

b. $d = |x + p|$

c.
$$x + p = \sqrt{(x - p)^2 + y^2}$$
$$(x + p)^2 = (x - p)^2 + y^2$$
$$x^2 + 2px + p^2 = x^2 - 2px + p^2 + y^2$$
$$y^2 = 4px$$

Exercises 9.4, pages 792 - 795

1. $\dfrac{x^2}{36} + \dfrac{y^2}{4} = 1$ **3.** $\dfrac{x^2}{25} + \dfrac{y^2}{4} = 1$ **5.** $\dfrac{x^2}{1} + \dfrac{y^2}{16} = 1$ **7.** $\dfrac{x^2}{1} - \dfrac{y^2}{1} = 1$ **9.** $\dfrac{x^2}{1} - \dfrac{y^2}{9} = 1$

        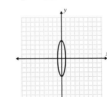   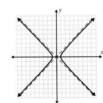   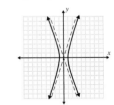

11. $\dfrac{x^2}{9} - \dfrac{y^2}{4} = 1$ **13.** $\dfrac{x^2}{4} + \dfrac{y^2}{8} = 1$ **15.** $\dfrac{x^2}{20} + \dfrac{y^2}{4} = 1$ **17.** $\dfrac{y^2}{9} - \dfrac{x^2}{9} = 1$ **19.** $\dfrac{y^2}{8} - \dfrac{x^2}{4} = 1$

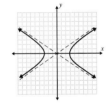   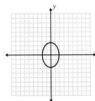   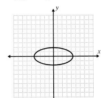   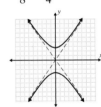

21. $\dfrac{y^2}{18} - \dfrac{x^2}{9} = 1$ **23.** $\dfrac{x^2}{6} + \dfrac{y^2}{9} = 1$ **25.** $\dfrac{x^2}{5} + \dfrac{y^2}{4} = 1$ **27.** $\dfrac{x^2}{25} - \dfrac{y^2}{15} = 1$ **29.** $\dfrac{y^2}{12} - \dfrac{x^2}{9} = 1$

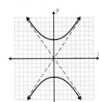   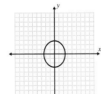   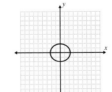   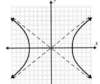   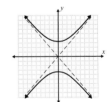

31. e   **33.** d   **35.** b   **37.** 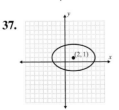   **39.**   **41.**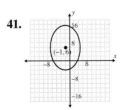

43. b. Pick an arbitrary point (x, y) located on the ellipse. The sum of the distances from (x, y) to the points $(-c, 0)$ and $(c, 0)$ (the foci) will equal $2a$. Using the distance formula we get the equation $\sqrt{(x+c)^2 + (y-0)^2} + \sqrt{(x-c)^2 + (y-0)^2} = 2a$. Now, to remove the radicals, simplify so that one side of the equation is a single radical and square both sides. Then, repeat the previous step and simplify. This gives $\dfrac{x^2}{a^2} + \dfrac{y^2}{a^2 - c^2} = 1$. **c.** At the y-intercept $(0, b)$ a right triangle is formed with hypotenuse a and sides b and c. The Pythagorean Theorem gives $a^2 = b^2 + c^2$.

Exercises 9.5, pages 801 - 802

1. $(-3, 10), (1, 2)$ **3.** $(1, 1), (2, 0)$ **5.** $(-2, -4), (4, 2)$ **7.** $(5, 3)$ **9.** $(-3, 0), (3, 0)$

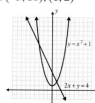 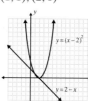 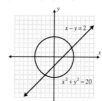 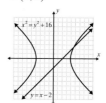 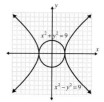

11. $(2, 3), (2, -3)$ **13.** $(-2, 4), (1, 1)$ **15.** $(-3, -2), (-3, 2), (3, -2), (3, 2)$ **17.** $\left(-3\sqrt{5}, 5\right), (-6, 4), \left(3\sqrt{5}, 5\right), (6, 4)$

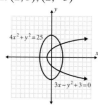 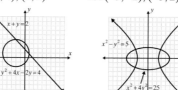

19. $(1, 3), (-1, 3)$

21. $\left(-3, \sqrt{11}\right), \left(-3, -\sqrt{11}\right), \left(3, \sqrt{11}\right), \left(3, -\sqrt{11}\right)$

23. $\left(-\dfrac{3}{2}, -1\right), \left(\dfrac{1}{2}, \dfrac{1}{2}\right)$ **25.** $(4, 4), (-2, -2)$ **27.** $(5, -3), (-5, 1)$ **29.** $(0, 3), (2, 3)$ **31.** $(-1, 4), (0, 3)$ **33.** $(1.41, 0.17), (-1.41, 5.83)$

35. $(2.68, 1.68), (-1.68, -2.68)$

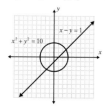 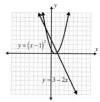

Chapter 9 Review, pages 808 - 810

1. a. -27 **b.** $-5a - 12$ **c.** -5 **2. a.** 47 **b.** $7a + 26$ **c.** 7 **3. a.** 66 **b.** $2a^2 + 12a + 12$ **c.** $4x + 2h$
4. a. -100 **b.** $-3a^2 - 18a - 19$ **c.** $-6x - 3h$ **5. a.** 19 **b.** $a^2 + 3a + 1$ **c.** $2x + h - 3$ **6. a.** 71 **b.** $2a^2 + 11a + 20$ **c.** $4x + 2h - 1$

7. 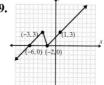   **8.**   **9.**   **10.**

11.

12.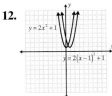

13. a. $(0,0)$ **b.** $(0,0)$
c. $y = 0$

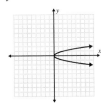

14. a. $(0,0)$ **b.** $(0,0)$
c. $y = 0$

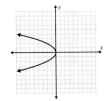

15. a. $(-7,3)$
b. $\left(0, 3 - \sqrt{7}\right), \left(0, 3 + \sqrt{7}\right)$ **c.** $y = 3$

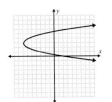

16. a. $(2,1)$
b. None **c.** $y = 1$

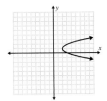

17. a. $(-2,2)$
b. $\left(0, 2 - \sqrt{2}\right), \left(0, 2 + \sqrt{2}\right)$ **c.** $y = 2$

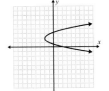

18. a. $\left(\dfrac{5}{4}, \dfrac{1}{8}\right)$
b. $(0,-3)$ **c.** $x = \dfrac{5}{4}$

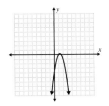

19. $(0, 1.7321), (0, -1.7321)$

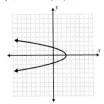

20. $(0,0), (0, 1.5)$

21. 5 **22.** $\sqrt{193}$ **23.** 8 **24.** 11 **25.** $\sqrt{97}$ **26.** $3\sqrt{10}$

27. $\sqrt{2465}$ **28.** $\sqrt{745}$ **29.** $x^2 + y^2 = 8$ **30.** $x^2 + y^2 = 27$

31. $(x+2)^2 + (y-1)^2 = 49$ **32.** $(x-5)^2 + (y-4)^2 = 20$

33. $x^2 + y^2 = 36$
Center : $(0,0), r = 6$

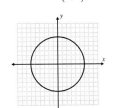

34. $x^2 + y^2 = 18$
Center : $(0,0), r = 3\sqrt{2}$

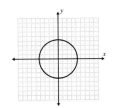

35. $x^2 + (y-3)^2 = 9$
Center : $(0,3), r = 3$

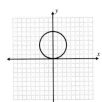

36. $(x+2)^2 + (y+1)^2 = 9$
Center : $(-2,-1), r = 3$

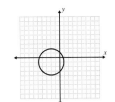

37. $(x+6)^2 + (y-1)^2 = 36$
Center : $(-6,1), r = 6$

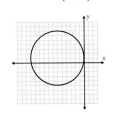

38. $x^2 + (y-5)^2 = 20$
Center : $(0,5), r = 2\sqrt{5}$

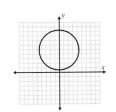

39. $AB = 3\sqrt{2}, AC = 5\sqrt{2}$
$BC = 4\sqrt{2}$
$\left(3\sqrt{2}\right)^2 + \left(4\sqrt{2}\right)^2 = \left(5\sqrt{2}\right)^2$
Right Triangle

40. $2\sqrt{29} + 2\sqrt{5} + 6\sqrt{2}$;
$PQ = 2\sqrt{29}$,
$QR = 2\sqrt{5}$,
$PR = 6\sqrt{2}$
$\left(2\sqrt{5}\right)^2 + \left(6\sqrt{2}\right)^2$
$\neq \left(2\sqrt{29}\right)^2$
Not a Right Triangle

41. $\dfrac{x^2}{36} + \dfrac{y^2}{9} = 1$

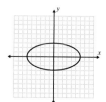

42. $\dfrac{x^2}{4} + \dfrac{y^2}{64} = 1$

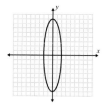

43. $\dfrac{x^2}{81} - \dfrac{y^2}{81} = 1$

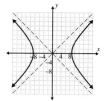

44. $\dfrac{x^2}{9} - \dfrac{y^2}{1} = 1$

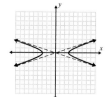

45. $\dfrac{y^2}{9} - \dfrac{x^2}{3} = 1$

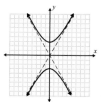

46. $\dfrac{y^2}{9} - \dfrac{x^2}{36} = 1$

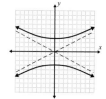

47. $\dfrac{(x+1)^2}{25} + \dfrac{(y-1)^2}{100} = 1$

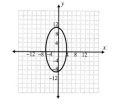

48. $\dfrac{(x-3)^2}{25} + \dfrac{y^2}{25} = 1$

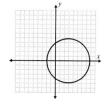

49. $\dfrac{(y+2)^2}{16} - \dfrac{x^2}{4} = 1$

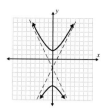

50. $\dfrac{x^2}{1} + \dfrac{(y-4)^2}{9} = 1$

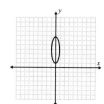

51. $(3, 8), (-2, 3)$

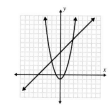

52. $(2, -4), \left(-\dfrac{5}{2}, -\dfrac{25}{4}\right)$

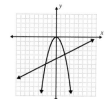

53. $(-2, 0), (0, -2)$

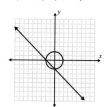

54. $(4, 1), (-4, -1)$

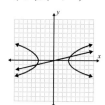

55. $(-4, 0), (4, 0)$

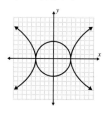

56. $(2, 3), (-2, 3),$
$(-2, -3), (2, -3)$

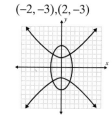

57. $\left(\dfrac{3+3\sqrt{5}}{2}, \dfrac{3+3\sqrt{5}}{2}\right),$
$\left(\dfrac{3-3\sqrt{5}}{2}, \dfrac{3-3\sqrt{5}}{2}\right)$

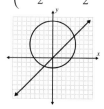

58. $(0, 5), (4, 5)$

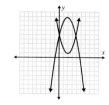

59. No solution

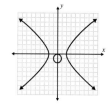

60. $(1, 3), (1, -3),$
$(-1, 3), (-1, -3)$

61. No solution

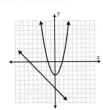

62. $(-1.11, 2.79)$,
$(-2.89, -0.79)$

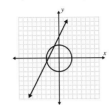

63. $(-2, -1)$

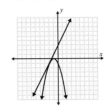

64. $(-1, 1), (1, -1)$

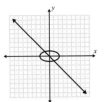

Chapter 9 Test, pages 811 - 812

1. a. -9 **b.** $-9 - 7a$ **c.** -7 **2. a.** -6 **b.** $2a^2 + 4a - 6$ **c.** $4x + 2h + 8$

3.

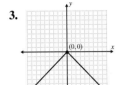

4.

5.

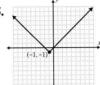

6.

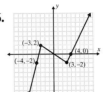

7.

8.

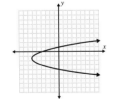

9. a. $(-5, 0)$ **b.** $\left(0, \sqrt{5}\right), \left(0, -\sqrt{5}\right)$ **c.** $y = 0$

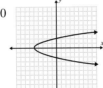

10. a. $\left(-\dfrac{25}{4}, -\dfrac{3}{2}\right)$ **b.** $(0, -4), (0, 1)$ **c.** $y = -\dfrac{3}{2}$ **11.** $3\sqrt{10}$

12. $AB = 2\sqrt{13}, AC = 2\sqrt{26},$
$BC = 2\sqrt{13}$
$\left(2\sqrt{13}\right)^2 + \left(2\sqrt{13}\right)^2 = \left(2\sqrt{26}\right)^2$

13. $(x - 3)^2 + (y - 4)^2 = 25$

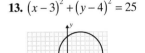

14. $x^2 + (y - 1)^2 = 9$
Center: $(0, 1), r = 3$

15. $\dfrac{x^2}{9} + \dfrac{y^2}{\left(\dfrac{9}{4}\right)} = 1$

16. $\dfrac{x^2}{4} + \dfrac{y^2}{25} = 1$

17. $\dfrac{x^2}{4} - \dfrac{y^2}{9} = 1$
$y = \dfrac{3}{2}x, y = -\dfrac{3}{2}x$

18. $\dfrac{y^2}{9} - \dfrac{x^2}{16} = 1$
$y = \dfrac{3}{4}x, y = -\dfrac{3}{4}x$

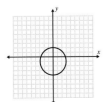

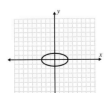

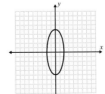

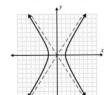

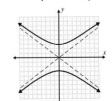

19.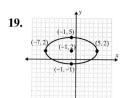

20.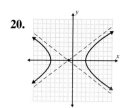

21. An ellipse is the set of all points in a plane for which the sum of the distances from two fixed points is constant.

22. $(-2, -5), (5, 2)$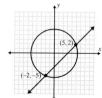

23. $\left(0, \sqrt{2}\right), \left(0, -\sqrt{2}\right), (-2, 0)$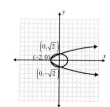

24. $(4, 3), (4, -3), (-4, 3), (-4, -3)$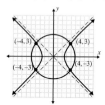

Chapter 9 Cumulative Review, pages 813 - 816

1. 1 **2.** $\dfrac{9x^8}{4y^8}$ **3.** $5x^{\frac{3}{4}}$ **4.** $\dfrac{8y^{\frac{3}{5}}}{x}$ **5.** $\sqrt[3]{\left(7x^3y\right)^2} = x^2\sqrt[3]{49y^2}$ **6.** $\left(32x^6y\right)^{\frac{1}{3}} = 2x^2\left(4y\right)^{\frac{1}{3}}$ **7.** $2(x+3)\left(x^2 - 3x + 9\right)$

8. $3y(x+4)(x-4)$ **9.** $(x-4)\left(x^2+3\right)$ **10.** $13\sqrt{3}$ **11.** $2\sqrt{3x} + 10\sqrt{6}$ **12.** $-3i\sqrt{7}$ **13.** $5x - 2y = -16$ **14.** $3x - 4y = 16$

15. $2x^2 + 2x + 20$ **16.** $\dfrac{2x^2 + x + 4}{(2x+3)(x-4)(3x-2)}$ **17.** -1 **18.** 0, one real solution **19.** $x = -\dfrac{1}{3}, 0, \dfrac{5}{2}$

20. $x = 1, 9$ **21.** $x = \dfrac{-1 \pm \sqrt{7}}{3}$ **22.** $x = -9$ **23.** $x = 2$ **24.** $x = -2, y = -3$ **25.** $x = \dfrac{1}{2}, y = \dfrac{1}{2}, z = -\dfrac{7}{2}$

26. a. 4 **b.** 6 **c.** 0 **27. a.** $\log_b \dfrac{x^3}{y^2}$ **b.** $\log \dfrac{10}{y}$ **28.** $x = -4$ or $x = 1$ **29.** 1.8614 **30.** 6.7417 **31.** 28.0316

32. $y = \dfrac{1}{2}x^2 - 3;$
$x = 0; \ (0, -3);$
$D = (-\infty, \infty); \ R = [-3, \infty);$
Zeros: $x = \pm\sqrt{6}$

33. $y = -(x - 2)^2$
$x = 2; (2, 0);$
$D = (-\infty, \infty); \ R = (-\infty, 0];$
Zeros: $x = 2$

34. a. $(-3, 0)$ **b.** None
c. $y = 0$
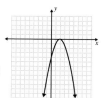

35. $[-1, 0) \cup [5, \infty)$

36. $[2, 3]$

37. $\left[-7, \dfrac{1}{2}\right]$

38.

39.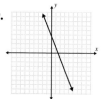

40.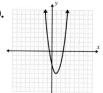

41.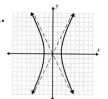

42.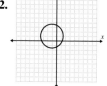

43. For $g(x)$: Domain: $[0, \infty)$, Range: $[0, \infty)$ **44.** $f^{-1}(x) = \dfrac{x+9}{3}$ **45. a.** x **b.** x **c.** yes **d.**

For $h(x)$: Domain: $[7, \infty)$, Range: $[2, \infty)$

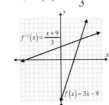

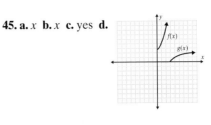

46. a. 5 **b.** $4x - 11$ **c.** $4x - 4$ **d.** 4   **47. a.** 13 **b.** $2x^2 + 3x - 3$ **c.** $2x^2 - 5x + 1$ **d.** $4x + 2h + 3$

48.

Asymptote for $y = \ln x$ is $x = 0$. Asymptote for $y = e^x$ is $y = 0$. **49.** 17, 19 **50.** 100 **51.** 36

52. $\dfrac{20}{9}$ or $2\dfrac{2}{9}$ hours   **53.** \$3600 at 7%; \$5400 at 8%   **54. a.** $t = 2\dfrac{1}{2}$ s **b.** 148 ft

55. \$15   **56.** $y = 3000\left(3\sqrt[3]{3}\right)^t$ **57.** \$2.81

58. $AB = \sqrt{65}, BC = \sqrt{137}, AC = 3\sqrt{2}; \left(\sqrt{65}\right)^2 + \left(3\sqrt{2}\right)^2 \neq \left(\sqrt{137}\right)^2$; Not a Right Triangle

59. 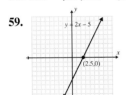 **60.** 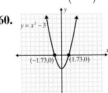 **61.** 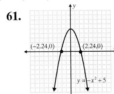 **62.**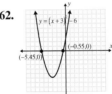

63. a. $x = -2, x = 5$ **b.** $(-\infty, -2) \cup (5, \infty)$ **c.** $(-2, 5)$ **64. a.** $x = -1, x = 2.5$ **b.** $(-1, 2.5)$ **c.** $(-\infty, -1) \cup (2.5, \infty)$

65. 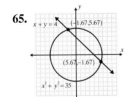 **66.**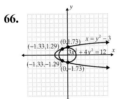

Chapter 10

Exercises 10.1, pages 824 - 825

1. 1, 3, 5, 7 **3.** $2, \dfrac{3}{2}, \dfrac{4}{3}, \dfrac{5}{4}$ **5.** 2, 6, 12, 20 **7.** 2, 4, 8, 16 **9.** $-2, 5, -10, 17$ **11.** $-\dfrac{1}{5}, \dfrac{1}{7}, -\dfrac{1}{9}, \dfrac{1}{11}$ **13.** 1, 0, -1, 0 **15.** 0, 1, 0, 1

17. $\dfrac{1}{4}, -\dfrac{1}{8}, \dfrac{1}{16}, -\dfrac{1}{32}$ **19.** $\{3n - 1\}$ **21.** $\{6n\}$ **23.** $\left\{(-1)^n (4n - 1)\right\}$ **25.** $\left\{5\left(2^{n-1}\right)\right\}$ **27.** $\left\{\dfrac{1}{2^n}\right\}$ **29.** Increasing **31.** Decreasing

33. Decreasing **35.** $a_1 = 3, a_2 = 1,$ and $a_3 = 4,$ so $a_1 > a_2,$ but $a_2 < a_3$ making the sequence neither increasing nor decreasing.

37. \$28,000 when $n = 1,$ \$19,600 when $n = 2,$ \$13,720 when $n = 3;$ It will be worth \$13,720 after 3 years.

39. 4 m when $n = 1,$ 1.6 m when $n = 2,$ 0.64 m when $n = 3,$ 0.256 m when $n = 4;$ It will bounce 0.256 m on the 4th bounce.

41. a. 1, 1, 2, 3, 5, 8 **b.** If we call the sequence of differences $D_n,$ then $D_n = F_{n-1}.$

Exercises 10.2, pages 830 - 831

1. $S_1 = 2, S_2 = 2 + 5 = 7, S_3 = 2 + 5 + 8 = 15, S_4 = 2 + 5 + 8 + 11 = 26$

3. $S_1 = \dfrac{1}{2}, S_2 = \dfrac{1}{2} + \dfrac{2}{3} = \dfrac{7}{6}, S_3 = \dfrac{1}{2} + \dfrac{2}{3} + \dfrac{3}{4} = \dfrac{23}{12}, S_4 = \dfrac{1}{2} + \dfrac{2}{3} + \dfrac{3}{4} + \dfrac{4}{5} = \dfrac{163}{60}$

5. $S_1 = 1, S_2 = 1 + (-4) = -3, S_3 = 1 + (-4) + 9 = 6, S_4 = 1 + (-4) + 9 + (-16) = -10$

7. $S_1 = \dfrac{1}{2}, S_2 = \dfrac{1}{2} + \dfrac{1}{4} = \dfrac{3}{4}, S_3 = \dfrac{1}{2} + \dfrac{1}{4} + \dfrac{1}{8} = \dfrac{7}{8}, S_4 = \dfrac{1}{2} + \dfrac{1}{4} + \dfrac{1}{8} + \dfrac{1}{16} = \dfrac{15}{16}$

9. $S_1 = -\dfrac{2}{3}, S_2 = -\dfrac{2}{3} + \dfrac{4}{9} = -\dfrac{2}{9}, S_3 = -\dfrac{2}{3} + \dfrac{4}{9} + \left(-\dfrac{8}{27}\right) = -\dfrac{14}{27}, S_4 = -\dfrac{2}{3} + \dfrac{4}{9} + \left(-\dfrac{8}{27}\right) + \dfrac{16}{81} = -\dfrac{26}{81}$ **11.** $2 + 4 + 6 + 8 + 10 = 30$

13. $5 + 6 + 7 + 8 + 9 = 35$ **15.** $\dfrac{1}{2} + \dfrac{1}{3} + \dfrac{1}{4} = \dfrac{13}{12}$ **17.** $2 + 4 + 8 = 14$ **19.** $16 + 25 + 36 + 49 + 64 = 190$

21. $3 + 1 + (-1) + (-3) = 0$ **23.** $6 + (-12) + 20 + (-30) = -16$ **25.** $\dfrac{1}{2} + \dfrac{2}{3} + \dfrac{3}{4} + \dfrac{4}{5} + \dfrac{5}{6} = \dfrac{71}{20}$ **27.** $\displaystyle\sum_{k=1}^{5}(2k-1)$ **29.** $\displaystyle\sum_{k=1}^{5}(-1)^k$

31. $\displaystyle\sum_{k=2}^{6}(-1)^k\left(\dfrac{1}{k^3}\right)$ **33.** $\displaystyle\sum_{k=4}^{15}\dfrac{k}{k+1}$ **35.** $\displaystyle\sum_{k=5}^{12}\dfrac{k+1}{k^2}$ **37.** 7 **39.** 48 **41.** 10 **43.** 295 **45.** −152

Exercises 10.3, pages 840 - 842

1. Arithmetic sequence; $d = 3, \{3n - 1\}$ **3.** Arithmetic sequence; $d = -2, \{9 - 2n\}$ **5.** Not an arithmetic sequence

7. Arithmetic sequence; $d = -4, \{10 - 4n\}$ **9.** Arithmetic sequence; $d = \dfrac{1}{2}, \left\{\dfrac{n-1}{2}\right\}$ **11.** 1, 3, 5, 7, 9; Arithmetic sequence

13. $-1, 4, -7, 10, -13$; Not an arithmetic sequence **15.** $-1, -7, -13, -19, -25$; Arithmetic sequence

17. $\dfrac{20}{3}, \dfrac{19}{3}, 6, \dfrac{17}{3}, \dfrac{16}{3}$; Arithmetic sequence **19.** $\dfrac{1}{2}, \dfrac{1}{4}, \dfrac{1}{6}, \dfrac{1}{8}, \dfrac{1}{10}$; Not an arithmetic sequence **21.** $\left\{\dfrac{2n+1}{3}\right\}$ **23.** $\{9 - 2n\}$

25. $\left\{\dfrac{17+3n}{2}\right\}$ **27.** $\{63 - 5n\}$ **29.** $\left\{\dfrac{81+3n}{2}\right\}$ **31.** 232 **33.** 44 **35.** 7 **37.** 19 **39.** 154 **41.** 235 **43.** 126 **45.** 231 **47.** $\dfrac{143}{3}$

49. 60 **51.** 171 **53.** −100 **55.** 9 days **57.** 1625 cm **59.** 134 seats, 1540 seats **61.** An arithmetic sequence has a constant difference d. If the terms alternated being positive and negative, then d would also have to alternate being positive and negative. This is a contradiction.

Exercises 10.4, pages 853 - 855

1. Not a geometric sequence **3.** Geometric sequence; $r = -\dfrac{1}{2}, \left\{3\left(-\dfrac{1}{2}\right)^{n-1}\right\}$ **5.** Geometric sequence; $r = \dfrac{3}{8}, \left\{\dfrac{32}{27}\left(\dfrac{3}{8}\right)^{n-1}\right\}$

7. Not a geometric sequence **9.** Geometric sequence; $r = -\dfrac{1}{4}, \left\{48\left(-\dfrac{1}{4}\right)^{n-1}\right\}$ **11.** $9, -27, 81, -243$; Geometric sequence

13. $\dfrac{2}{3}, \dfrac{4}{3}, 2, \dfrac{8}{3}$; Not a geometric sequence **15.** $-\dfrac{8}{5}, \dfrac{32}{25}, -\dfrac{128}{125}, \dfrac{512}{625}$; Geometric sequence

17. $3\sqrt{2}, 6, 6\sqrt{2}, 12$; Geometric sequence **19.** $0.3, -0.09, 0.027, -0.0081$; Geometric sequence **21.** $\left\{3(2)^{n-1}\right\}$

23. $\left\{\dfrac{1}{3}\left(-\dfrac{1}{2}\right)^{n-1}\right\}$ **25.** $\left\{(\sqrt{2})^{n-1}\right\}$ **27.** $\left\{\dfrac{1}{3}(3)^{n-1}\right\}$ **29.** $\left\{-5\left(-\dfrac{3}{4}\right)^{n-1}\right\}$ **31.** $\dfrac{1}{4}$ **33.** $\dfrac{32}{9}$ **35.** $\dfrac{1}{4}$ or $-\dfrac{1}{4}$ **37.** $n = 4$

39. 363 **41.** $\dfrac{1023}{64}$ **43.** $-\dfrac{333}{64}$ **45.** $\dfrac{422}{243}$ **47.** $\dfrac{75}{128}$ **49.** 4 **51.** $-\dfrac{1}{3}$ **53.** $\dfrac{4}{9}$ **55.** $\dfrac{4}{11}$ **57.** \$27,859.63 **59.** 14.76 liters

61. \$12,841.18 **63.** 7.59 meters **65.** π

Exercises 10.5, pages 863 - 864

1. 56 **3.** $\dfrac{1}{15}$ **5.** 4 **7.** $(n-1)!$ **9.** $(k+3)(k+2)(k+1)$ **11.** 20 **13.** 35 **15.** 1 **17.** $x^7 + 7x^6y + 21x^5y^2 + 35x^4y^3$

19. $x^9 + 9x^8 + 36x^7 + 84x^6$ **21.** $x^5 + 15x^4 + 90x^3 + 270x^2$ **23.** $x^6 + 12x^5y + 60x^4y^2 + 160x^3y^3$

25. $2187x^7 - 5103x^6y + 5103x^5y^2 - 2835x^4y^3$ **27.** $x^{18} - 36x^{16}y + 576x^{14}y^2 - 5376x^{12}y^3$

29. $x^6 + 6x^5y + 15x^4y^2 + 20x^3y^3 + 15x^2y^4 + 6xy^5 + y^6$ **31.** $x^7 - 7x^6 + 21x^5 - 35x^4 + 35x^3 - 21x^2 + 7x - 1$

33. $243x^5 + 405x^4y + 270x^3y^2 + 90x^2y^3 + 15xy^4 + y^5$ **35.** $x^4 + 8x^3y + 24x^2y^2 + 32xy^3 + 16y^4$

37. $81x^4 - 216x^3y + 216x^2y^2 - 96xy^3 + 16y^4$ **39.** $243x^{10} - 405x^8y + 270x^6y^2 - 90x^4y^3 + 15x^2y^4 - y^5$

41. $3360x^6y^4$ **43.** $1{,}140{,}480x^8$ **45.** $-27{,}500x^6y^{18}$ **47.** 1.062 **49.** 0.808 **51.** 64.363 **53.** 0.851

Chapter 10 Review, pages 869 - 872

1. $-1, 1, 3, 5$ **2.** $4, \dfrac{5}{2}, 2, \dfrac{7}{4}$ **3.** $\dfrac{1}{2}, \dfrac{4}{3}, \dfrac{9}{4}, \dfrac{16}{5}$ **4.** $-1, \dfrac{1}{4}, -\dfrac{1}{9}, \dfrac{1}{16}$ **5.** $7, 5, 7, 5$ **6.** $-1, -1, -1, -1$ **7.** $\{5n+5\}$ **8.** $\left\{3(-1)^{n+1}\right\}$

9. $\{n^2 + 2\}$ **10.** $\left\{\dfrac{n+2}{n}\right\}$ **11.** Increasing **12.** Decreasing **13.** Decreasing **14.** Increasing **15.** Increasing **16.** Decreasing

17. $\dfrac{160}{81}$ ft **18.** \$12,288 **19.** \$275, \$6750 **20.** b **21.** $5 + 9 + 13 + 17 = 44$ **22.** $5 + 7 + 9 + 11 = 32$ **23.** $3 + 2 + \dfrac{5}{3} + \dfrac{3}{2} = \dfrac{49}{6}$

24. $0 + 2 + 0 + 2 = 4$ **25.** $\dfrac{3}{4} + \dfrac{9}{16} + \dfrac{27}{64} + \dfrac{81}{256} = \dfrac{525}{256}$ **26.** $2 + 6 + 12 + 20 = 40$ **27.** $5 + 10 + 15 + 20 = 50$

28. $7 + 8 + 9 + 10 + 11 = 45$ **29.** $5 - 7 + 9 - 11 + 13 = 9$ **30.** $-1 + 4 - 9 + 16 - 25 + 36 = 21$ **31.** $80 + 99 + 120 = 299$

32. $\dfrac{1}{3} + \dfrac{1}{6} + \dfrac{1}{9} + \dfrac{1}{12} = \dfrac{25}{36}$ **33.** $\displaystyle\sum_{k=1}^{5}(2k+1)$ **34.** $\displaystyle\sum_{k=3}^{6}\dfrac{(-1)^{k+1}}{k^3}$ **35.** $\displaystyle\sum_{k=1}^{6}(-1)^{k+1}\left(k^2-1\right)$ **36.** $\displaystyle\sum_{k=1}^{5}\dfrac{k+1}{k^2}$ **37.** 50 **38.** -56 **39.** 380

40. 10 **41.** $-1, -3, -5, -7, -9$; Arithmetic sequence **42.** $\dfrac{4}{3}, \dfrac{8}{3}, 4, \dfrac{16}{3}, \dfrac{20}{3}$; Arithmetic sequence

43. $4, 9, 16, 25, 36$; Not an arithmetic sequence **44.** $\dfrac{1}{3}, \dfrac{1}{4}, \dfrac{1}{5}, \dfrac{1}{6}, \dfrac{1}{7}$; Not an arithmetic sequence

45. $\dfrac{5}{4}, \dfrac{7}{4}, \dfrac{9}{4}, \dfrac{11}{4}, \dfrac{13}{4}$; Arithmetic sequence **46.** $\dfrac{17}{3}, \dfrac{16}{3}, 5, \dfrac{14}{3}, \dfrac{13}{3}$; Arithmetic sequence **47.** $\{2n-2\}$ **48.** $\left\{\dfrac{n+39}{4}\right\}$

49. $\{11-n\}$ **50.** $\{27-7n\}$ **51.** $\{2n-8\}$ **52.** $\{2-2n\}$ **53.** 60 **54.** $\dfrac{135}{2}$ **55.** 16,150 **56.** 455 **57.** -150 **58.** 60

59. 225 blocks **60.** \$6180 **61.** Geometric sequence; $r = \dfrac{1}{2}, \left\{12\left(\dfrac{1}{2}\right)^{n-1}\right\}$ **62.** Not a geometric sequence

63. Geometric sequence; $r = -\dfrac{1}{5}, \left\{10\left(-\dfrac{1}{5}\right)^{n-1}\right\}$ **64.** Geometric sequence; $r = \dfrac{1}{4}, \left\{\left(\dfrac{1}{4}\right)^{n}\right\}$

65. Geometric sequence; $r = -2, \left\{(-2)^{n}\right\}$ **66.** Not a geometric sequence **67.** $-\dfrac{1}{5}, \dfrac{1}{25}, -\dfrac{1}{125}, \dfrac{1}{625}$ **68.** $\dfrac{4}{3}, \dfrac{8}{9}, \dfrac{16}{27}, \dfrac{32}{81}$

69. $\dfrac{3}{8}, \dfrac{9}{64}, \dfrac{27}{512}, \dfrac{81}{4096}$ **70.** $20, 40, 80, 160$ **71.** $\left\{7\left(\dfrac{1}{2}\right)^{n-1}\right\}$ **72.** $\left\{4(-3)^{n-1}\right\}$ **73.** $\left\{\dfrac{16}{9}\left(-\dfrac{3}{2}\right)^{n-1}\right\}$ **74.** $\left\{\dfrac{5}{16}(2)^{n-1}\right\}$ **75.** $\dfrac{1}{2}$

76. 40 **77.** -48 **78.** 6 **79.** $\dfrac{160}{81}$ **80.** $\dfrac{1555}{7776}$ **81.** $-\dfrac{2}{5}$ **82.** $\dfrac{3}{2}$ **83. a.** $\displaystyle\sum_{k=1}^{\infty}\dfrac{2}{5}\left(\dfrac{1}{10}\right)^{k-1}$ **b.** $\displaystyle\sum_{k=1}^{\infty}\dfrac{4}{5}\left(\dfrac{1}{10}\right)^{k-1}$ **84.** \$110,907.56

85. $\$19,966.74$ **86.** $30,240$ **87.** 90 **88.** $(n-2)!$ **89.** $(n+3)(n+2)$ **90.** 56 **91.** 1 **92.** $x^8 + 8x^7y + 28x^6y^2 + 56x^5y^3$

93. $x^9 + 18x^8 + 144x^7 + 672x^6$ **94.** $1024x^{10} - 5120x^9y + 11520x^8y^2 - 15360x^7y^3$ **95.** $x^{12} - 24x^{10} + 240x^8 - 1280x^6$

96. $x^4 + 20x^3 + 150x^2 + 500x + 625$ **97.** $8x^3 + 36x^2y + 54xy^2 + 27y^3$ **98.** $x^{10} - 15x^8y + 90x^6y^2 - 270x^4y^3 + 405x^2y^4 - 243y^5$

99. $x^{10} - 10x^9 + 45x^8 - 120x^7 + 210x^6 - 252x^5 + 210x^4 - 120x^3 + 45x^2 - 10x + 1$ **100.** $-4032x^4y^5$ **101.** $5103x^6y$

102. $\dfrac{7}{16}x^6y^2$ **103.** $-811,008x^{10}y^{14}$

Chapter 10 Test, pages 873 - 874

1. $0 + 2 + 6 + 12 + 20 = 40$ **2.** $\dfrac{1}{4}, \dfrac{1}{7}, \dfrac{1}{10}, \dfrac{1}{13}, ...;$ Neither **3.** $\left\{ \dfrac{n}{2n+1} \right\}$ **4.** $\displaystyle\sum_{k=1}^{4} 3k + 4$ **5.** $\displaystyle\sum_{k=1}^{4} k^3$ **6.** 74 **7.** 68 **8.** $-\dfrac{40}{729}$ **9.** 26

10. $a_n = 8 - 2n$ **11.** 243 **12.** $\dfrac{1}{8}$ **13.** $a_n = \left(\sqrt{3}\right)^{n-2}$ **14.** $\dfrac{21}{16}$ **15.** $\dfrac{15}{100} + \dfrac{15}{10,000} + ... = \dfrac{5}{33}$ **16.** 462

17. $32x^5 - 80x^4y + 80x^3y^2 - 40x^2y^3 + 10xy^4 - y^5$ **18.** $5670x^4y^4$ **19.** $1125x^8y^2$ **20.** $\$13,050.16$

Chapter 10 Cumulative Review, pages 875 - 877

1. $11x + 3$ **2.** $-x^2 + 3x + 5$ **3.** $(4x+3)(16x^2 - 12x + 9)$ **4.** $(3x-5)(2x+9)$ **5.** $(2x+1)(5x+7)(x-2)$ **6.** $\dfrac{11x+12}{15}$

7. $\dfrac{-6x-4}{(x+4)(x-4)(x-1)}$ **8.** $\dfrac{1}{x^2}$ **9.** $\dfrac{3y^{1/3}}{x^2}$ **10.** $\dfrac{2x^{\frac{4}{3}}}{3y}$ **11.** $3\sqrt{3x}$ **12.** $\dfrac{-13+11i}{29}$ **13.** $x > 3$

14. $\dfrac{1}{3} \le x \le \dfrac{13}{3}$ **15.** $-\dfrac{5}{2} < x < \dfrac{9}{4}$ **16.** $x = \dfrac{26}{7}$ **17.** $x = \dfrac{-2 \pm i\sqrt{2}}{2}$ **18.** $x = 4$ **19.** $x = \dfrac{39}{17}$

20. $x = 5$ **21.** $x = e^{2.4} \approx 11.02$ **22.** $n = \dfrac{A-P}{Pi}$ **23.** $x = 2, y = -5$ **24.** $x = 0, y = 2, z = -1$

25. a. $y = \dfrac{x+7}{2}$ **b.** $2x^2 - 5$ **c.** $2x + 1$ **26.** **27.** $(2, -3)$ **28.**

29. **30.** **31. a.** $a_n = 10 - 2n$ **32. a.** $a_n = 16\left(\dfrac{1}{2}\right)^{n-1}$

b. -10 **b.** $\dfrac{63}{2}$

33. $x^6 + 12x^5y + 60x^4y^2 + 160x^3y^3 + 240x^2y^4 + 192xy^5 + 64y^6$ **34.** 20 mph, 60 mph **35.** 16 lbs at $\$1.40$, 4 lbs at $\$2.60$

36. 16 ft by 26 ft **37.** 12.56 hrs **38.** $\$16,058.71$ **39.** 443.7 liters **40.** $x \approx -0.4746, 1.3953$ **41.** $x \approx -1.9646, 1.0580$

42. $x \approx 0.2408, 2.7337$ **43.** $x = 1$ **44.** $x = -3, 1, 2$

Appendix

Section A.1, pages 884 - 887

1. $\frac{1}{10}$ **3.** $\frac{15}{8}$ **5.** $\frac{7}{4}$ **7.** $\frac{1}{6}$ **9.** $\frac{1}{6}$ **11.** $\frac{9}{4}$ **13.** $\frac{4}{5}$ **15.** 56 **17.** $\frac{35}{6}$ **19.** $\frac{3}{28}$ **21.** $\frac{3}{8}$ **23.** $\frac{2}{5}$ **25.** $\frac{1}{3}$ **27.** $\frac{8}{25}$ **29.** $\frac{9}{10}$

31. $\frac{5}{16}$ **33.** $\frac{9}{7}$ **35.** $\frac{1}{5}$ **37.** $\frac{5}{16}$ **39.** $\frac{25}{54}$ **41. a.** $\frac{7}{9}$ **b.** $\frac{7}{6}$

43. 120 are left-handed students; 320 are not left-handed students **45.** \$15,600 **47. a.** $\frac{3}{10}$ **b.** \$1500 **c.** \$60,000

49. a. More than 144 **b.** Less than 144 **c.** 160 passengers

Section A.2, pages 892 - 893

1. a. $x - 9$ **b.** $c = 3, P(3) = 0$ **3. a.** $x^2 - 4x + 33 + \dfrac{-265}{x+8}$ **b.** $c = -8, P(-8) = -265$ **5. a.** $4x^2 - 6x + 9 + \dfrac{-17}{x+2}$ **b.** $c = -2, P(-2) = -17$

7. a. $x^2 + 7x + 55 + \dfrac{388}{x-7}$ **b.** $c = 7, P(7) = 388$ **9. a.** $2x^2 - 2x + 6 + \dfrac{-27}{x+3}$ **b.** $c = -3, P(-3) = -27$

11. a. $x^3 + 2x + 5 + \dfrac{17}{x-3}$ **b.** $c = 3, P(3) = 17$ **13. a.** $x^3 + 2x^2 + 6x + 9 + \dfrac{23}{x-2}$ **b.** $c = 2, P(2) = 23$

15. a. $x^3 + \dfrac{1}{2}x^2 - \dfrac{3}{4}x - \dfrac{3}{8} + \dfrac{45}{16\left(x - \dfrac{1}{2}\right)}$ **b.** $c = \dfrac{1}{2}, P\left(\dfrac{1}{2}\right) = \dfrac{45}{16}$ **17. a.** $x^4 + x^3 + x^2 + x + 1$ **b.** $c = 1, P(1) = 0$

19. a. $x^3 - \dfrac{14}{5}x^2 + \dfrac{56}{25}x - \dfrac{224}{125} + \dfrac{3396}{625\left(x + \dfrac{4}{5}\right)}$ **b.** $c = -\dfrac{4}{5}, P\left(-\dfrac{4}{5}\right) = \dfrac{3396}{625}$

21. a. $x^2 - \dfrac{7}{2}x + \dfrac{13}{4} + \dfrac{73}{4(2x-1)}; 2x^2 - 7x + \dfrac{13}{2} + \dfrac{73}{4\left(x - \dfrac{1}{2}\right)}$ Answers may vary. **b.** Answers may vary. **c.** Answers may vary.

Index

CHAPTER 6   Roots, Radicals, and Complex Numbers

Square Roots:

If $b^2 = a$, then b is called a **square root** of a $(a \geq 0)$.

If x is a real number, then $\sqrt{x^2} = |x|$. However, if $x \geq 0$, then we can write $\sqrt{x^2} = x$.

Properties of Square Roots:

If a and b are positive real numbers, then

1. $\sqrt{ab} = \sqrt{a}\sqrt{b}$

2. $\sqrt{\dfrac{a}{b}} = \dfrac{\sqrt{a}}{\sqrt{b}}$

For any nonnegative real number x and positive integer m,

3. $\sqrt{x^{2m}} = x^m$

4. $\sqrt{x^{2m+1}} = x^m\sqrt{x}$

Cube Roots:

If $b^3 = a$, then b is called the **cube root** of a. We write $\sqrt[3]{a} = b$.

Properties of Radicals:

1. If n is a positive integer and $b^n = a$, then $b = \sqrt[n]{a} = a^{\frac{1}{n}}$.

2. $a^{\frac{m}{n}} = \left(a^{\frac{1}{n}}\right)^m = \left(a^m\right)^{\frac{1}{n}}$ or, in radical notation, $a^{\frac{m}{n}} = \left(\sqrt[n]{a}\right)^m = \sqrt[n]{a^m}$

To Rationalize a Denominator Containing a Sum or Difference Involving Square Roots:

Rationalize the denominator by multiplying both the numerator and the denominator by the **conjugate of the denominator**.

1. If the denominator is of the form $a - b$, multiply both the numerator and denominator by $a + b$.

2. If the denominator is of the form $a + b$, multiply both the numerator and denominator by $a - b$.

The new denominator will be the difference of two squares and therefore not contain a radical term.

Definition of i:

$i = \sqrt{-1}$ and $i^2 = \left(\sqrt{-1}\right)^2 = -1$

If a is positive real number, $\sqrt{-a} = \sqrt{a} \cdot \sqrt{-1} = \sqrt{a}\,i = i\sqrt{a}$.

Complex Numbers:

The **standard form of a complex number** is $a + bi$ where a and b are real numbers. a is called the **real part** and b is called the **imaginary part**.

CHAPTER 7   Quadratic Equations and Quadratic Functions

Square Root Property:

If $x^2 = c$, then $x = \pm\sqrt{c}$.

If $(x - a)^2 = c$, then $x - a = \pm\sqrt{c}$ $\left(\text{or } x = a \pm \sqrt{c}\right)$.

Completing the Square:

To complete the square, find the third term of a perfect square trinomial when the first two terms are given. The trinomial should have the following characteristics:

1. The leading coefficient $\left(\text{the coefficient of } x^2\right)$ is 1.

2. The constant term is the square of $\dfrac{1}{2}$ of the coefficient of x.

Quadratic Formula:

For the quadratic equation $ax^2 + bx + c = 0$, where $a \neq 0$, the solutions are

$x = \dfrac{-b \pm \sqrt{b^2 - 4ac}}{2a}$.

Discriminant:

The expression $b^2 - 4ac$, the part of the quadratic formula that lies under the radical sign, is called the **discriminant**.

If $b^2 - 4ac > 0 \rightarrow$ There are two real solutions.

If $b^2 - 4ac = 0 \rightarrow$ There is one real solution, $x = -\dfrac{b}{2a}$.

If $b^2 - 4ac < 0 \rightarrow$ There are two nonreal solutions.

Projectiles:

$h = -16t^2 + v_0 t + h_0$, where h is the height of the object in feet, t is the time object is in the air in seconds, v_0 is the beginning velocity in feet per second, and h_0 is the beginning height in feet.

Parabolas:

Parabolas of the form $y = ax^2 + bx + c$:

  If $a > 0$, the parabola opens upward.

  If $a < 0$, the parabola opens downward.

  The bigger $|a|$ is, the narrower the opening.

  The smaller $|a|$ is, the wider the opening.

  Vertex: $\left(-\dfrac{b}{2a}, \dfrac{4ac - b^2}{4a}\right)$

  Line of Symmetry: $x = -\dfrac{b}{2a}$

Parabolas of the form $y = a(x - h)^2 + k$:

  Vertex: (h, k)

  Line of Symmetry: $x = h$

  The graph is a horizontal shift of h units and a vertical shift of k units of the graph of $y = ax^2$.

Minimum and Maximum Values:

For a parabola with its equation given in the form $y = a(x - h)^2 + k$:

1. If $a > 0$, then the parabola opens upward, (h, k) is the lowest point, and $y = k$ is called the **minimum value** of the function.

2. If $a < 0$, then the parabola opens downward, (h, k) is the highest point, and $y = k$ is called the **maximum value** of the function.

CHAPTER 8   Exponential and Logarithmic Functions

Algebraic Operations with Functions:

For functions $f(x)$ and $g(x)$ where x is in the domain of both functions,

$(f + g)(x) = f(x) + g(x)$

$(f - g)(x) = f(x) - g(x)$

$(f \cdot g)(x) = f(x) \cdot g(x)$

$\left(\dfrac{f}{g}\right)(x) = \dfrac{f(x)}{g(x)}, \quad g(x) \neq 0$

Composite Functions:

For functions $f(x)$ and $g(x)$, $(f \circ g)(x) = f(g(x))$.

Domain of $f \circ g$: The domain of $f \circ g$ consists of those values of x in the domain of g for which $g(x)$ is in the domain of f.

One-to-One Functions:

A function is a **one-to-one function** (or **1–1 function**) if for each value of y in the range there is only one corresponding value of x in the domain.

Horizontal Line Test:

A function is 1–1 if no horizontal line intersects the graph of the function at more than one point.

Inverse Functions:

If f is a 1–1 function with ordered pairs of the form (x, y), then its **inverse function**, denoted f^{-1}, is also a 1–1 function with ordered pairs of the form (y, x).

If f and g are 1–1 functions and $f(g(x)) = x$ for all x in D_g and $g(f(x)) = x$ for all x in D_f, then f and g are **inverse functions**.

Exponential Functions:

An **exponential function** is a function of the form $f(x) = b^x$ where $b > 0$, $b \neq 1$, and x is any real number.

Concepts of Exponential Functions:

For $b > 1$:

1. $b^x > 0$
2. b^x increases to the right and is called an **exponential growth function**
3. $b^0 = 1$, so $(0, 1)$ is on the graph
4. b^x approaches the x-axis for negative values of x
   (The x-axis is a horizontal asymptote.)

For $0 < b < 1$:

1. $b^x > 0$
2. b^x decreases to the right and is called an **exponential decay function**
3. $b^0 = 1$, so $(0, 1)$ is on the graph
4. b^x approaches the x-axis for positive values of x
   (The x-axis is a horizontal asymptote.)

Compound Interest:

Compound interest on a principal P invested at an annual interest rate r (in decimal form) for t years that is compounded n times per year can be calculated using the following formula:

$$A = P\left(1 + \frac{r}{n}\right)^{nt}$$

where A is the amount accumulated.

The Number e:

The number e is defined to be $e = 2.718281828459\ldots$

Continuously Compounded Interest:

Continuously compounded interest on a principal P invested at an annual interest rate r for t years, can be calculated using the following formula:

$$A = Pe^{rt}$$

where A is the amount accumulated.

Logarithms:

For $b > 0$ and $b \neq 1$, $x = b^y$ is equivalent to $y = \log_b x$.

Properties of Logarithms:

For $b, x, y > 0$, $b \neq 1$, and any real number r:

1. $\log_b 1 = 0$
2. $\log_b b = 1$
3. $x = b^{\log_b x}$
4. $\log_b b^x = x$
5. $\log_b (xy) = \log_b x + \log_b y$
6. $\log_b \dfrac{x}{y} = \log_b x - \log_b y$
7. $\log_b x^r = r \cdot \log_b x$

Properties of Equations with Exponents and Logarithms:

For $b > 0$ and $b \neq 1$:

1. If $b^x = b^y$, then $x = y$.
2. If $x = y$, then $b^x = b^y$.
3. If $\log_b x = \log_b y$, then $x = y$ ($x > 0$ and $y > 0$).
4. If $x = y$, then $\log_b x = \log_b y$ ($x > 0$ and $y > 0$).

Change of Base:

For $a, b, x > 0$ and $a, b \neq 1$, $\log_b x = \dfrac{\log_a x}{\log_a b}$.

CHAPTER 9   Conic Sections

Horizontal and Vertical Translations:

Given a graph $y = f(x)$, the graph of $y = f(x - h) + k$ is:

1. a horizontal translation of $f(x)$ by h units and
2. a vertical translation of $f(x)$ by k units.

Horizontal Parabolas:

Parabolas of the form $x = ay^2 + by + c$ or $x = a(y - k)^2 + h$

If $a > 0$, the parabola opens right.

If $a < 0$, the parabola opens left.

Vertex: (h, k)

Line of Symmetry: $y = k$

Distance Formula (distance between two points):

For two points, $P(x_1, y_1)$ and $Q(x_2, y_2)$, in a plane, the distance

between the points is $d = \sqrt{(x_2 - x_1)^2 + (y_2 - y_1)^2}$.

Circles:

A **circle** is the set of all points in a plane that are a fixed distance from a fixed point.

The fixed point is called the **center** of a circle.

The distance from the center to any point on the circle is called the **radius** of the circle.

The distance from one point on the circle to another point on the circle measured through the center is the **diameter** of the circle.

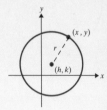

Standard form of a circle: $(x - h)^2 + (y - k)^2 = r^2$
The radius is r and the center is at (h, k).